Advances in Polymer Materials and Technology

Advances in Polymer Materials and Technology

Edited by
Anandhan Srinivasan • Sri Bandyopadhyay

CRC Press
Taylor & Francis Group
Boca Raton London New York

CRC Press is an imprint of the
Taylor & Francis Group, an **informa** business

CRC Press
Taylor & Francis Group
6000 Broken Sound Parkway NW, Suite 300
Boca Raton, FL 33487-2742

First issued in paperback 2020

© 2017 by Taylor & Francis Group, LLC
CRC Press is an imprint of Taylor & Francis Group, an Informa business

No claim to original U.S. Government works

ISBN 13: 978-0-367-57468-0 (pbk)
ISBN 13: 978-1-4987-1881-3 (hbk)

Library of Congress Cataloging-in-Publication Data

Names: Srinivasan, Anandhan, editor. | Bandyopadhyay, Sri, editor.
Title: Advances in polymer materials and technology / editors, Anandhan Srinivasan and Sri Bandyopadhyay.
Description: Boca Raton : Taylor & Francis, CRC Press, 2016. | Includes bibliographical references and index.
Identifiers: LCCN 2016002568 | ISBN 9781498718813 (alk. paper)
Subjects: LCSH: Polymers. | Polymeric composites. | Plastics.
Classification: LCC TA455.P58 A324 2016 | DDC 620.1/92--dc23
LC record available at https://lccn.loc.gov/2016002568

Visit the Taylor & Francis Web site at
http://www.taylorandfrancis.com

and the CRC Press Web site at
http://www.crcpress.com

Contents

Preface

POLYMERS HAVE OCCUPIED AN indispensable position in our everyday lives. They are one of the most important engineering materials used in a spectrum of applications, ranging from food to space. Although polymers were previously considered as low-strength materials with low stiffness/toughness, recent research has totally changed that notion. In fact, now polymers are the only materials that can act as matrices for incorporation of the widest range of ceramics, nanotubes, nanoparticles, and short as well as continuous fibers of various kinds to create new building and structural materials. At the same time, polymers are very important materials that can be used as biomaterials, sensors, and in electrical and electronic materials as insulators, conductors, or semiconductors.

This book highlights recent advancements in the field of polymeric materials and technology. Frontier areas such as polymers based on bio-sources, polymer-based ferroelectrics, polymer nanocomposites for capacitors, food packaging, and electronic packaging, organic field emission transistors, superhydrophobic materials, and electrospinning are discussed in this book. We do believe that this book will be suitable for a wide range of audiences working in various interdisciplinary fields of polymer technology.

The chapters are broadly organized into six sections of polymer materials and technology. Section I, "Novel Polymer Composites," discusses new developments in the field of functional nanofillers, layered double hydroxide-based nanocomposites, thermoset nanocomposites, hybrid composites, and fly ash-based composites. Section II, "Nanopolymer Technology," provides an overview of the fundamentals and applications of electrospinning, patterning of polymeric surfaces, electrospun polymer-based piezoelectric sensors, and superhydrophobic polymers. Section III, "Micro-, Macro-, Nanotesting and Characterization of Polymers," presents an analysis of the dynamic properties of polymers by split Hopkinson pressure bar apparatus, thermal conductivity and stability of exfoliated graphite nanoplatelets-based nanocomposites, effect of multiwall carbon nanotubes on miscibility and stability of specialty polymer blends, and effects of strain rate and temperature on mechanical properties of thermosets and their composites. Section IV, "Specialty Polymers," describes shape memory polymers, thermoplastic elastomers, multiferroic characteristics of poly(vinylidene fluoride), novel thiophene-based conducting polymers, and polymer-based dielectrics for capacitors. Section V, "Bio-Based and Biocompatible Polymer Materials," consists of chapters on biopolymers for endovascular applications, thermosets from natural resources, synthesis and properties of gum polysaccharides-based graft copolymers, rheological properties and self-assembly of cellulose nanowhiskers in biocomposites, and nanocellulose-based bio-nanocomposites. Finally, Section VI, "New Polymer Applications," discusses applications of polymer nanocomposites in biomedical applications, sensitive electronic device packaging, and food packaging. This section also features a novel topic on polymer quenchants for industrial heat treatment.

Anandhan Srinivasan
National Institute of Technology Karnataka, Mangalore, India

Sri Bandyopadhyay
University of New South Wales, Sydney, Australia

Acknowledgments

W E SINCERELY THANK DR. Gagandeep Singh, Taylor & Francis Group (CRC Press), for inviting us to edit this book and making its publication a reality. We also thank Marsha Pronin, Taylor & Francis Group (CRC Press), Boca Raton, Florida, for her constant support and assistance during various stages of the manuscript preparation.

In addition, we express our deep sense of gratitude to all contributors of the 26 chapters in the diverse but interconnected science and technology areas of polymers/plastics as materials and technology components. The contributors are renowned experts in their respective fields and hail from various academic institutes and industries around the world. We are grateful to them for their time and efforts to make this venture successful. We also appreciate their patience in waiting for the book to be published.

Professor Anandhan Srinivasan is grateful to his PhD students, Dr. Gibin and Dr. Sachin, for their assistance throughout the final stages of preparation of this book. Professor Sri Bandyopadhyay particularly thanks Professor Paul Munroe, his head of school; Professor Merlin Crossley, deputy vice-chancellor (Education); Professor Les Field AM, vice president/senior deputy vice-chancellor; and Professor Ian Jacobs, president/vice chancellor of UNSW Australia, for their continued encouragement. The editors thank various authors, editors, and journals for permitting them to reproduce copyright materials. The editors also thank their respective families, acknowledging their valuable support. Last but not the least, the editors thank their colleagues and research staff for their support.

Editors

Anandhan Srinivasan earned his PhD in polymer science and technology from the Indian Institute of Technology, Kharagpur, India (IIT KGP), in 2004. He was a postdoctoral fellow and lecturer in the Department of Applied Organic Materials Engineering, Inha University, Incheon, South Korea, during 2004–2005. He was an assistant professor in the Department of Materials Science at the Asian Institute of Medicine, Science & Technology (AIMST) University, Bedong, Malaysia, during 2005–2008. He has been serving at the National Institute of Technology Karnataka, Mangaluru, India, since February 2009. He researches in the areas of advanced ceramic and polymeric nanofibers, polymer blends, polymer composites, thermoplastic elastomers, and waste materials and fly ash utilization. Four students have completed their doctoral research under his supervision and three more students are continuing their PhD studies under his supervision. He was a gold medalist in both BSc and MSc. The Australia–India Council fellowship and Department of Science and Technology (DST, Government of India) fast-track award are noteworthy among the awards that he has received. He is on the editorial board of the *International Journal of Energy Engineering*, Hong Kong. He has been a referee for a number of reputed international journals. He is an author of 50 international journal articles, 2 patents, 7 book chapters, and 30 conference papers. Dr. Srinivasan has delivered invited talks in a number of international conferences, workshops, and seminars. His PhD students have won first prizes for their paper/poster in two international conferences. His biography has been featured in the 11th and 12th editions of *Marquis Who's Who in Science and Engineering*.

Sri Bandyopadhyay is now a senior visiting fellow in the School of Materials Science and Engineering, Faculty of Science, University of New South Wales (UNSW), Australia. He is a researcher in the fields of composites and nanocomposites. In August 2013, Australia's Campus Review Management selected him as "One of Top Five Australian Innovators" for his reinvention of coal power fly ash.

Sri Bandyopadhyay has about 140 refereed research publications plus 4 provisional patents, and 4 easy access intellectual properties on new composite materials, including metal matrix composites, polymer matrix composites, fly ash recycling, and carbon nanotube composites.

He is the chair of composites conferences known as ACUN (Australia, Canada, United States, and New Zealand), which happened on six occasions between 1999 and 2012 in UNSW and Monash University, Melbourne, Australia. Some of the ACUN conferences were ranked among top 5–10 in the world by attending delegates from more than 20 countries.

Sri Bandyopadhyay is the editor-in-chief of the *International Journal of Energy Engineering*, published by the World Academic Publishing Company. He served as an academic researcher at UNSW, Australia. Prior to that, he worked at the Defence Science and Technology Organisation (DSTO), Department of Defence, Materials Research Laboratories, Melbourne, Australia, where he was given the Best Scientist Award for his innovative research on "In Situ SEM Deformation and Fracture Studies of Polymers and Polymer–Matrix Composites." He also worked at the Australian Dental Standards Laboratory, Abbotsford, Victoria, Australia, and Indian Space Research Organisation, Bengaluru, India.

Sri Bandyopadhyay was invited as a visiting professor/academic at (1) École Polytechnique Fédérale de Lausanne (EPFL), Lausanne, Switzerland; Polymer Laboratory (Host: Professor H.H. Kausch); (2) Center for Composite Materials (Host: Professor R.P. Wool), University of Delaware, Newark, Delaware; (3) NanoScience Technology Center (Host: Professor Sudipta Seal), University of Central Florida, Orlando, Florida; (4) School of Physical Sciences (Host: Professor H.B. Bohidar), Jawaharlal Nehru University, New Delhi, India; (5) Department of Metallurgy and Materials Science (Hosts: Professors M.K. Banerjee and N.R. Bandyopadhyay), Indian Institute of Engineering Science and Technology (erstwhile Bengal Engineering and Science University) Shibpur, India; (6) Materials Science Center (Host: Professor Ajit K. Banthia), IIT, Kharagpur, India; and (7) School of Materials Science & Nanotechnology, Jadavpur University, Kolkata, India (UGC fellow with Professor Siddhartha Mukherjee).

Professor Sri Bandyopadhyay initiated the concept of today's Australia–India Science Research Funded (AISRF) scheme project by approaching India's 11th President Dr. A.P.J. Abdul Kalam in 2004, and Dr. Kalam kindly followed it through the Department of Science and Technology (Government of India), which was then taken over by the Department of Industry, Innovation, Science and Research (Australian Government).

Professor Bandyopadhyay was allotted a targeted allocation project on "Nanocomposite Materials in Clean Energy: Generation, Storage and Savings," involving six Australian and six Indian research academics, between 2008 and 2012. This project generated 115 publications, including 85 refereed journal paper publications/submissions and 4 provisional patent applications.

Contributors

Venimadhav Adyam
Cryogenic Engineering Centre
Indian Institute of Technology
Kharagpur, India

Yu Jin Ahn
Department of Advanced Materials
    Engineering for Information and
    Electronics
Kyung Hee University
Seoul, Republic of Korea

Hazizan Md Akil
School of Materials and Mineral Resources
    Engineering
Universiti Sains Malaysia
Nibong Tebal, Malaysia

Saeed M. Al-Zahrani
SABIC Polymer Research Center
Chemical Engineering Department
College of Engineering
King Saud University
Riyadh, Kingdom of Saudi Arabia

Arfat Anis
SABIC Polymer Research Center
Chemical Engineering Department
College of Engineering
King Saud University
Riyadh, Kingdom of Saudi Arabia

Gregório Guadalupe Carbajal Arizaga
Department of Chemistry
University of Guadalajara
Guadalajara, Mexico

Reza Arjmandi
Department of Polymer Engineering
Faculty of Chemical and Natural Resources
    Engineering
Universiti Teknologi Malaysia
Johor Bahru, Malaysia

Nicolai Aust
Department of Chemistry of Polymeric
    Materials
University of Leoben
Leoben, Austria

R. Rajesh Babu
Compound Development
Global R&D
Apollo Tyres Ltd
Chennai, India

Sri Bandyopadhyay
School of Materials Science and
    Engineering
University of New South Wales
Sydney, New South Wales, Australia

Arup R. Bhattacharyya
Department of Metallurgical Engineering
and Materials Science
Indian Institute of Technology Bombay
Mumbai, India

Suryasarathi Bose
Department of Metallurgical Engineering
and Materials Science
Indian Institute of Technology Bombay
Mumbai, India

and

Department of Materials Engineering
Indian Institute of Science
Bangalore, India

Santanu Chattopadhyay
Rubber Technology Centre
Indian Institute of Technology
Kharagpur, India

Cintil Jose Chirayil
School of Chemical Sciences
Mahatma Gandhi University
Kottayam, India

and

Department of Chemistry
Newman College
Thodupuzha, India

B. Deepa
Department of Chemistry
C.M.S. College
Kottayam, India

and

Department of Chemistry
Bishop Moore College
Mavelikara, India

Sheila Devasahayam
Faculty of Science and Technology
Federation University Australia
Ballarat, Victoria, Australia

D. Manjula Dhevi
Department of Chemistry
SRM University
Chennai, India

Shan Faiz
SABIC Polymer Research Center
Chemical Engineering Department
College of Engineering
King Saud University
Riyadh, Kingdom of Saudi Arabia

Gibin George
Department of Metallurgical and Materials
Engineering
National Institute of Technology
Karnataka
Mangalore, India

Patrick S. Grant
Department of Materials
University of Oxford
Oxford, United Kingdom

Satyajit Gupta
Department of Materials Engineering
and
Center for Nanoscience and Engineering
Indian Institute of Science
Bangalore, India

A. Hariharasubramanian
Organic Chemistry Division
School of Advanced Sciences
VIT University
Vellore, India

Azman Hassan
Department of Polymer Engineering
Faculty of Chemical and Natural Resources
Engineering
Universiti Teknologi Malaysia
Johor Bahru, Malaysia

I.M. Inuwa
Department of Polymer Engineering
Faculty of Chemical and Natural Resources
Engineering
Universiti Teknologi Malaysia
Johor Bahru, Malaysia

Sreeram K. Kalpathy
Department of Metallurgical and Materials
Engineering
Indian Institute of Technology Madras
Chennai, India

S. Kanagaraj
Department of Mechanical Engineering
Indian Institute of Technology
Guwahati, India

Devarshi Kashyap
Department of Mechanical Engineering
Indian Institute of Technology
Guwahati, India

Kap Jin Kim
Department of Advanced Materials
Engineering for Information and
Electronics
Kyung Hee University
Seoul, Republic of Korea

B. Sachin Kumar
Department of Metallurgical and Materials
Engineering
National Institute of Technology
Karnataka
Mangalore, India

Mohammad Luqman
Department of Basic Science
College of Applied Sciences
A'Sharqiyah University
Ibra, Sultanate of Oman

Giridhar Madras
Center for Nanoscience and Engineering
Indian Institute of Science
Bangalore, India

Amoghavarsha Mahadevegowda
Department of Materials
University of Oxford
Oxford, United Kingdom

Arunjunai Raj Mahendran
Team Green Composites
Kompetenzzentrum Holz GmbH
Linz, Austria

Arjun Maity
DST/CSIR National Centre for
Nanostructured Materials
Council for Scientific and Industrial
Research
Pretoria, South Africa

Hemant Mittal
Department of Applied Chemistry
University of Johannesburg
Johannesburg, South Africa

and

DST/CSIR National Centre for
Nanostructured Materials
Council for Scientific and Industrial
Research
Pretoria, South Africa

M.G. Murali
Department of Chemistry
National University of Singapore
Singapore

Golok B. Nando
Rubber Technology Centre
Indian Institute of Technology
Kharagpur, India

Vignesh Nayak
Department of Metallurgical and Materials
    Engineering
National Institute of Technology
    Karnataka
Mangalore, India

Mohd Firdaus Omar
CEGeoGTech
School of Materials Engineering
Universiti Malaysia Perlis
Arau, Malaysia

H. Pan
School of Materials Science and
    Engineering
University of New South Wales
Sydney, New South Wales, Australia

Akshata G. Patil
Department of Metallurgical and Materials
    Engineering
National Institute of Technology
    Karnataka
Mangalore, India

Laly A. Pothen
Department of Chemistry
C.M.S. College
Kottayam, India

and

Department of Chemistry
Bishop Moore College
Mavelikara, India

Petra Pötschke
Department of Functional
    Nanocomposites and Blends
Leibniz Institute of Polymer Research
Dresden, Germany

K. Narayan Prabhu
Department of Metallurgical and Materials
    Engineering
National Institute of Technology
Mangalore, India

A. Anand Prabu
Department of Chemistry
VIT University
Vellore, India

R. Rajesh
Organic Chemistry Division
School of Advanced Sciences
VIT University
Vellore, India

Praveen C. Ramamurthy
Department of Materials Engineering
and
Center for Nanoscience and Engineering
Indian Institute of Science
Bangalore, India

Pranesh Rao
Department of Metallurgical and Materials
    Engineering
National Institute of Technology
    Karnataka
Mangalore, India

Y. Dominic Ravichandran
Organic Chemistry Division
School of Advanced Sciences
VIT University
Vellore, India

Suprakas Sinha Ray
Department of Applied Chemistry
University of Johannesburg
Johannesburg, South Africa

and

DST/CSIR National Centre for
    Nanostructured Materials
Council for Scientific and Industrial
    Research
Pretoria, South Africa

Pieter Samyn
Faculty of Environment and Natural
    Resources
Freiburg Institute for Advanced Studies
and
Hermann Staudinger Graduate School
University of Freiburg
Freiburg, Germany

S. Saravanan
Department of Materials Engineering
and
Center for Nanoscience and Engineering
Indian Institute of Science
Bangalore, India

M. Selvakumar
Rubber Technology Centre
Indian Institute of Technology
Kharagpur, India

T. Senthil
Fujian Institute of Research on the
    Structure of Matter
Chinese Academy of Sciences
Key Laboratory of Design and Assembly of
    Functional Nanostructures
Fuzhou, Fujian, People Republic of China

Robert A. Shanks
School of Applied Sciences
RMIT University
Melbourne, Victoria, Australia

A.M. Shanmugharaj
Department of Chemical Engineering
Kyung Hee University
Seoul, Republic of Korea

S. Sindhu
Center for Nanoscience and Engineering
Indian Institute of Science
Bangalore, India

Anandhan Srinivasan
Department of Metallurgical and Materials
    Engineering
National Institute of Technology
    Karnataka
Mangalore, India

Hesam Taheri
Faculty of Environment and Natural
    Resources
Freiburg Institute for Advanced Studies
and
Hermann Staudinger Graduate School
University of Freiburg
Freiburg, Germany

Merin Sara Thomas
International and Interuniversity Centre
	for Nanoscience and Nanotechnology
Mahatma Gandhi University
and
Department of Chemistry
C.M.S. College
Kottayam, India

and

Department of Chemistry
MarThoma College
Thiruvalla, India

Sabu Thomas
School of Chemical Sciences
and
International and Interuniversity Centre
	for Nanoscience and Nanotechnology
Mahatma Gandhi University
Kottayam, India

D. Udayakumar
Department of Chemistry
National Institute of Technology
	Karnataka
Mangalore, India

Fernando Wypych
Department of Chemistry
Federal University of Paraná
Curitiba, Brazil

I

Novel Polymer Composites

Layered Hydroxide Salts as Alternative Functional Fillers for Polymer Nanocomposites

Fernando Wypych and Gregório Guadalupe Carbajal Arizaga

CONTENTS

1.1 LAYERED HYDROXIDE SALTS

Materials are defined as pure solid compounds or mixtures of them with the ability to satisfy human needs (Fahlman 2008). Polymers are a class of materials that have found applications in different areas such as the electronics, aerospace, biomedicine, agriculture, food, and home products industries. The wide range of uses for polymers is derived from the large number of properties associated with their composition, molecular weight, structural arrangement, and processing methods. These properties can be further modified by the addition of fillers.

The type of filler will determine the additional properties in the polymer. For instance, glass fiber or kaolinite increases mechanical resistance, and hydroxide compounds increase resistance to fire but do not improve mechanical properties. These types of fillers have been known for several years.

Recently, a family of materials, including layered double hydroxides (LDHs) and layered hydroxide salts (LHSs), has been proposed as polymer fillers to improve mechanical strength and resistance to fire as well as ultraviolet (UV) light absorption, to protect polymers against degradation, or to stabilize dyes for polymers (Cursino et al. 2011; Marangoni et al. 2008, 2009a,b). In other words, multiple properties or functions are added to the polymer by using LDHs or LHSs.

LDHs are synthetic inorganic structures in which layers are formed by a mixture of divalent and trivalent metal cations coordinated by hydroxyl groups. The general formula of an LDH is $[M^{2+}_{1-x}M^{3+}_{x}(OH)_2]^{x+}(A^{n-})_x \cdot y(H_2O)$, where $[M^{2+}_{1-x}M^{3+}_{x}(OH)_2]^{x+}$ represents the layer domain and $(A^{n-})_{x/n} \cdot y(H_2O)$ is the interlayer domain. When trivalent metals partially replace the divalent metals of the structure, a residual positive charge is generated in the layers, which are stabilized by anions located between them. These compounds have been widely studied, and every year hundreds of scientific articles related to these structures are published. As an example, a simple search in the ISI Web of Science database with the term "layered double hydroxide" on, March 21, 2016 showed 4487 results, 91 of them corresponding to "layered double hydroxide and polymer." This large number of annual publications reflects the ease of finding novel features by combining LDHs and polymers.

Regarding LHSs, they share two main features with LDHs, which are the layered structure and the anion-exchange property. These features allow employing LHSs to develop new materials for a several applications, among which their use as polymer fillers is included. Tracking the number of publications for applications of LHSs as fillers is not easy because the terms refer to these compounds are diverse, such as layered basic salts, layered hydroxy salts, or more commonly a specific name associated with the composition, such as zinc hydroxide nitrate, zinc hydroxide chloride (ZHC), zinc basic salt, copper hydroxide acetate, or copper basic salt. Despite the difficulty in finding reports of LHSs as polymer fillers, it can be estimated that there are not more than 10 publications per year. Therefore, LHSs represent an important opportunity to develop a new class of fillers or as a base to design multifunctional polymers.

1.1.1 Definition and History

Materials scientists and engineers study materials with diverse compositions and properties. Textbooks have proposed classifications of materials for better identification of their properties and applications. The main classification groups are polymers, metals, composites, and ceramics (Fahlman 2008), but additional or slightly different groups can be found.

First of all, LHSs are layered structures at the molecular level, like graphite or layered alumino silicates, single hydroxides (e.g., magnesium hydroxide or calcium hydroxide), or LDHs. These must not be confused with macro layered materials (layered composites) formed by heterogeneous mixtures of different compounds.

Most of the structures of LHSs reported are synthetic, but some naturally occurring materials can be found, such as felsobanyaite, hydrozincite, and gerhardite, with the compositions $Al_4(SO_4)(OH)_{10} \cdot 5H_2O$ or $Al(OH)_{2.5}(SO_4)_{0.25} \cdot 1.25H_2O$ (International Centre for Diffraction Data [ICDD] card 080068), $Zn_5(CO_3)_2(OH)_6$ or $Zn(OH)_{1.2}(CO_3)_{0.4}$ (ICDD card 191458), and $Cu_2(OH)_3NO_3$ or $Cu(OH)_{1.5}(NO_3)_{0.5}$ (ICDD card 140687), respectively.

General chemistry textbooks describe hydroxide salts as products of partial substitution of hydroxyls by protons in polyhydroxilated bases (Whitten et al. 2011). Although this statement is correct in terms of composition, not all hydroxide salts posses layered structures. An example of this found in nature is the mineral Scarbroite, represented by $Al_5(OH)_{13}(CO_3) \cdot 5H_2O$ (ICDD card 120727), in which there are no ion-exchange or intercalation properties as commonly occurred in layered structures. Some other natural minerals with composition of hydroxide salts are Georgeite, $Cu_2(OH)_2CO_3 \cdot 6H_2O$; Glaukosphaerite, $(Cu,Ni)_2(OH)_2CO_3$; Kolwezite, $(Cu,Co)_2(OH)_2CO_3$; Nullaginite, $Ni_2(OH)_2CO_3$; Rosasite, $(Cu,Zn)_2(OH)_2CO_3$; Claraite, $(Cu, Zn)_2(OH)_2CO_3 \cdot 4H_2O$; Kapellasite, $Cu_3Zn(OH)_6Cl_2$; and Haydeeite, $Cu_3Mg(OH)_6Cl_2$.

1.1.2 Layered Compounds

Layered compounds are formed by structures in which the covalent or ionic bonds are infinitely repeated along two axes of the crystal cell (O'Hare 1997), as represented in Figure 1.1, where the bonds occur along a and b axes.

The thickness of the layers in these compounds can be formed by only one atom, as in the case of graphene layers in graphite (Figure 1.1), by five atoms (Mg^{2+} between two hydroxyl groups in Brucite—magnesium hydroxide), or can even be more complex, as the structure of Pyrophyllite, whose layers are composed of two tetrahedrons of silicon

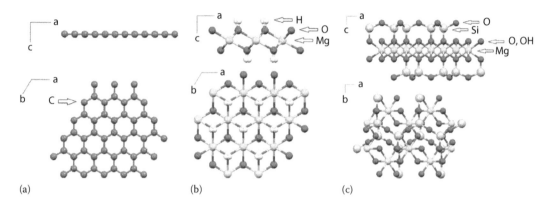

FIGURE 1.1 Lateral (top) and basal surface (bottom) views of natural layered materials' structures with different thickness: (a) graphite; (b) Brucite—magnesium hydroxide; and (c) pyrophyllite. In pyrophyllite's structure, hydrogen atoms have been omitted for better visualization (structures from CIF file numbers 1011060, 1000054, and 9000809 retrieved from the Crystallography Open Database).

and one octahedron of magnesium, coordinated by either oxygen or hydroxyl groups. The extension of the layers can reach the order of centimeters, so the thickness becomes negligible even in clays (it is said that the materials have a high aspect ratio).

However, the formation of particles in layered compounds results from the stacking of several layers interacting with weak forces; thus, the separation of layers with appropriate techniques, as well as intercalation of guest compounds between the layers, is possible.

To define the type of guest species that can be intercalated, it is necessary to know the electrostatic nature of the layered compounds. Figure 1.2 shows representative structures of compounds with neutral, positively, and negatively charged layers. Some layered chalcogenides are examples of compounds with neutral layers. In the structure of molybdenum disulfide ($2H\text{-}MoS_2$ polymorph) (depicted in Figure 1.2a), Mo^{4+} cations are neutralized by S^{2-} anions; thus, neutral molecules can be intercalated in the interlayer region. For instance, hydrazine molecules can be intercalated in layered chalcogenides (Oosawa et al. 1997). Also, aniline molecules can be intercalated in graphite, which is another electrostatically neutral compound (Amarnath et al. 2011).

Cationic intercalation is possible in structures with negative layers, like the titanate structure presented in Figure 1.2b, where the interlayer space is occupied by Mg^{2+} cations. Cationic exchange in layered titanates is possible, although in some cases a pre-exfoliation step is required (Kim et al. 2009).

Anionic layered structures are formed by layers with positive electrostatic charges that are stabilized by interlayer anions. Commonly, these anions are not directly bound to the cations within the layers, and therefore, they can be exchanged. One important family of anion exchangers is hydrotalcite-like compounds (anionic structure in Figure 1.2c), also named layered double hydroxides (LDHs). Hundreds of scientific investigations and patents related to new materials derived from LDH structures have been reported during the past two decades.

The anion-exchange capacity allows introducing organic or inorganic ions in the interlayer space, producing innumerous combinations of properties derived from these anions

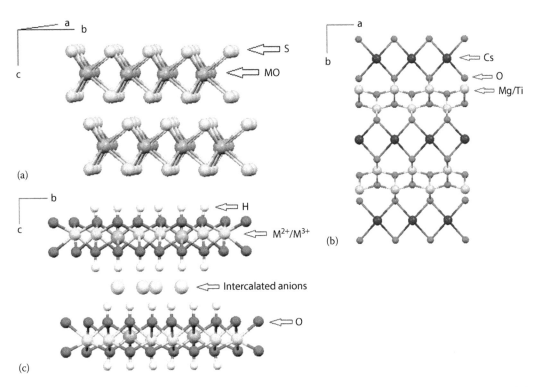

FIGURE 1.2 The electrostatic nature of layered structures: (a) neutral layered in dichalcogenides ($2H-MoS_2$), (b) positive ($Cs_{0.16}Mg_{0.09}Ti_{0.41}O$), and (c) negative (LDH) (structures from CIF file numbers 1010993, 7200323, and 9009272 retrieved from the Crystallography Open Database).

and the layers in only one material. The main condition to synthesize LDHs is the simultaneous presence of cations with two different valence states, which commonly are divalent and trivalent metal cations with ionic radii close to Mg^{2+} (Bravo-Suárez et al. 2004; Carlino 1997; Del Hoyo 2007; Rives and Ulibarri 1999), although in some cases cations with 1+ or 4+ valence state also allow preparation of stable LDH structures, such as Li, Ti, or Si (Hsieh et al. 2011; Jiang et al. 2013; Saber et al. 2005, 2011).

A second important family of layered structures with the anion-exchange capability is LHSs. These compounds share structural and physicochemical similarities with LDHs, except for the cation composition in the layers, where only one valence state is required. This chapter is focused on describing LHSs because these structures have been little investigated for applications compared to LDHs. Therefore, LHSs offer a good opportunity to develop new materials and academic studies, besides having advantages over LDHs for use in polymer composites because the morphology of LHS particles can be modified to influence their mechanical properties.

1.1.3 LHSs: Composition and Types of Structure

Hydroxides salts are metal hydroxides in which different anions substitute a fraction of the hydroxyl groups. Some compositions of hydroxide salts are capable of forming layered structures. The main condition to produce an LHS is the presence of one or more cations with the same valence state (Meyn et al. 1990). Layers formed by divalent cations are the most abundant group, although fewer examples are found in the literature with trivalent cations, especially lanthanides.

Layered hydroxide halide and hydroxide nitrates of divalent cations are the earliest compositions reported (Feitknecht 1945; Feitknecht and Bürki 1949; Louër et al. 1973; Stählin and Oswald 1970). Louër et al. (1973) identified structural arrangements and proposed theoretical arrangements to give a structural classification in the 1970s. The types of layered structures described in this section correspond to examples found in the literature, so these are the most likely structures to be used as polymer functional fillers.

The structural description of LHSs commonly starts with the Brucite-type structure present in hydroxides with compositions $Mg(OH)_2$, $Ca(OH)_2$, $Co(OH)_2$, $Zn(OH)_2$, $Ni(OH)_2$, or $Cu(OH)_2$, for example. These single metal hydroxides are formed by layered units like that depicted in Figure 1.1b for $Mg(OH)_2$. The same structure is represented with the polyhedral model in Figure 1.3, allowing visualization of two crystallographic sites: one cationic site containing the divalent metal cations (M^{2+}) in octahedral (Oh) coordination with six anionic positions with hydroxyl groups (OH^- site). The ordered sequence of octahedrons along two directions results in a layer where the cationic sites are located in the middle and hydroxyl groups remain exposed on the layer's surface.

The partial substitution of hydroxyl groups by different anions leads to the formation of hydroxide salts with three types of basic structures, represented in Figure 1.4. Structure I corresponds to the isomorphic substitution of OH^- groups by A^{n-} anions, in which the valence is $n = 1-$.

Structures II and III undergo an additional modification due to migration of M^{2+} cations from Oh sites to the surface of the layers, leading to the formation of tetrahedral (Td) sites with central cations coordinated by OH^- groups in the base, while the apex can be occupied by water molecules so that A^{n-} anions over the water molecules are needed to stabilize the charge of cations (structure II) or directly by A^{n-} anions (structure III).

1.1.3.1 Type I Structure

The composition of this type of LHS is represented by the general formula $M^{2+}(OH)_{2-x}(A^{n-})_{x/n}$. Copper hydroxide nitrate with composition $Cu_2(OH)_3NO_3$ is a hydroxide salt structure represented by the isomorphic substitution model (Figure 1.4I). The layers of this

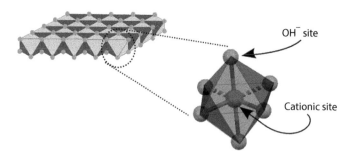

FIGURE 1.3 Polyhedral representations of a Brucite-like layer found in some hydroxides with composition $M(OH)_2$.

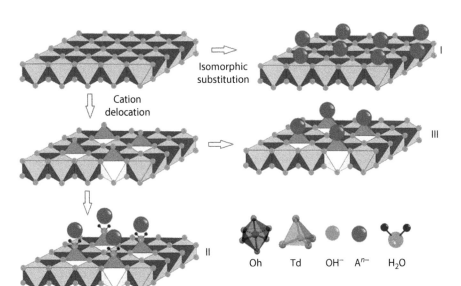

FIGURE 1.4   Structures of LHS derived from single-layered hydroxides obtained by isomorphic substitution of hydroxyl groups (OH⁻) or dislocation of M^{2+} cations from octahedral (Oh) sites to tetrahedral (Td) positions. In both cases, new anions (A^{n-}) are present.

LHS are composed by copper(II) cations in Oh coordination with hydroxyl groups, which are partially substituted by nitrate anions (Figure 1.5). This description has been called grafting by some authors. The increment of the interlayer space is one of the structural changes with the substitution of hydroxyl groups. This increase is proportional to the size of the new anion, for example, the basal spacing (see definition in Section 1.1.5) in $Zn(OH)_2$ of 4.42 Å increases to 6.93 Å after substitution with nitrate anions in $Zn(OH)_3(NO_3)_2$.

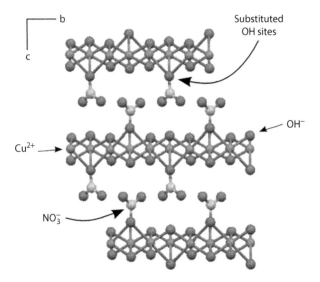

FIGURE 1.5   Structure of copper hydroxide nitrate with composition $Cu_2(OH)_3NO_3$ (CIF file 31353).

Other LHSs with type I structure are $Mg(OH)_{1.7}(CH_3COO)_{0.3}\cdot1.7H_2O$ (Nethravathi et al. 2011), $Zn_3(OH)_4(NO_3)_2$ (Biswick et al. 2005; Chouillet et al. 2004), $Zn(OH)(NO_3)\cdot1.7H_2O$ (Chouillet et al. 2004), and $Ni(OH)(NO_3)$ (Rajamathi et al. 2001), among others.

1.1.3.2 Type II Structure

The second type of structure depicted in Figure 1.4 is formed by a main layer with Oh units partially occupied by M^{2+} cations. The empty Oh sites are the base of two Td sites, where the apex corresponds to water molecules (Figure 1.6). The formula representing this structure is $M_{1-x}^{Oh}M_x^{Td}(OH)_{2-x}(A^-)_x\cdot y(H_2O)$. Here, the composition is restricted to divalent metal cations (M^{2+}) and monovalent anions (A^-). The most representative example of this structure is zinc hydroxide nitrate, with composition $Zn_3^{Oh}Zn_2^{Td}(OH)_8(NO_3)_2\cdot2H_2O$ (or $Zn_{1.25}(OH)_2(NO_3)_{0.5}\cdot0.5H_2O$), which spontaneously precipitates with the addition of a base to a solution of zinc nitrate.

The coordination of Td Zn^{2+} cations with water molecules avoids the need for direct grafting of nitrate ions to the layers. This type of structure can also be produced with ammonia instead of water molecules (Bénard et al. 2013). Both neutral molecules leave a residual electrostatic charge in the Td cations, which are stabilized by nitrate ions in the interlayer space.

1.1.3.3 Type III Structure

The third type of structure in LHSs also has empty Oh sites and metal cations in Td coordination as in type II structures, but no neutral molecules are involved in the coordination of cations. Thus, the A^- anion directly coordinates the Td cationic site; in other words, the anion is grafted to the layers, as chloride does in ZHC, represented in Figure 1.7. The composition of this type of structure is represented by $M_{1-x}^{Oh}M_x^{Td}(OH)_{2-y}(A^-)_y$. The reversible interconversion of structures II to III can be produced by heat treatment

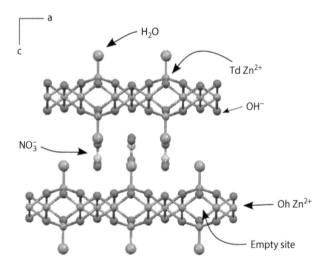

FIGURE 1.6   Section of a zinc hydroxide nitrate with structure type II (CIF file 16023).

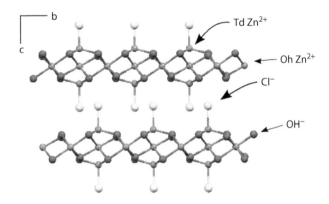

FIGURE 1.7   Structure of zinc hydroxide chloride (CIF file 34904).

to remove water or ammonia and graft anions to the layers, or by solvation to replace water in Td sites.

1.1.4 Methods of Synthesis

Some procedures are described in the literature to produce LHSs. These can be separated into synthesis to obtain hydroxide salts with oxoanions or small anions such as chloride and acetate, and synthesis to produce hybrid structures in which the interlayer anions are organic. This section describes the synthesis of LHS with oxoanions and small anions, which are used as the basis for further hybridization or reactions with polymers (described in Section 1.2).

1.1.4.1 Precipitation

Alkaline precipitation is one the most practical procedures to obtain LHSs, because it only requires an aqueous solution of the divalent metal salt and a Brönsted base to precipitate the LHS. Despite the simplicity of this procedure, our research group has detected that several parameters influence the composition, morphology, and size of the final particles, and these features are relevant for the use of LHSs as fillers in polymers. These parameters are the concentration of the starting reagent, stirring speed and duration during precipitation, temperature, concentration of the alkaline solution, and aging time, among others. Table 1.1 contains a list of relationships among some reaction conditions, the type of structure formed (according to the models in Figure 1.4), and the particle size.

This procedure is suitable for scaling and produces large amounts of the target LHS, because the reagents are mixed in only one reactor with water, and after the precipitation the LHS can be removed by decantation. Precipitation with alkaline solution spontaneously forms LSH structures at atmospheric pressure and room temperature when Zn(II), Cu(II), and Co(II) cations are used. When salts of magnesium, calcium, or nickel, for example, are subjected to alkaline compounds, the most probable result is the formation of single hydroxides (Wypych and Arizaga 2005), so alternative hydrolysis procedures should be followed, as described below.

TABLE 1.1 Relationship between Reaction Conditions and the Type of Structure Obtained by Precipitation

LHS	Type of Structure[a]	Chemicals and Concentration (mol L^{-1})	Base Solution	Temperature (°C)	Mean Particle Size (μm)[b]	Reference
$Zn_5(OH)_8(NO_3)_2 \cdot 2H_2O$	II	$Zn(NO_3)_2 \cdot 3H_2O$, 0.17	NH_4OH	25	0.2	Carbajal Arizaga (2008)
$Zn_5(OH)_8(NO_3)_2 \cdot 2H_2O$	II	$Zn(NO_3)_2 \cdot 3H_2O$, 0.17	NaOH	25	10	Elguera Hernández (2013)
$Zn_5(OH)_8(NO_3)_2 \cdot 2H_2O$, with excess of ZnO	II	$Zn(NO_3)_2$, 0.4	NaOH	4	6	Miao et al. (2006)
$Zn_5(OH)_8(Cl)_2$	III	$ZnCl_2$	NaOH	20	2[c]	Arízaga (2012)
$Cu_2(OH)_3(NO_3)$	I	$Cu(NO_3)_2 \cdot 3H_2O$, 0.17	$Mg(OH)_2$	24	1	Aguirre et al. (2011)
$Cu_2(OH)_3(Ac) \cdot H_2O$	I	$Cu(Ac)_2$, 0.1	NaOH	–	–	Fujita et al. (1999)

[a] Type of structure according to Figure 1.4.
[b] Size determined from the micrographs presented in the respective reference.
[c] Our samples analyzed by scanning electron microscopy, unpublished.

To avoid contamination, the solid should be centrifuged and the supernatant removed. The solid should be redispersed with a new portion of solvent using an ultrasound bath and the operation repeated for 4 or 5 times. A simple filtration and washing step is not efficient to remove all the soluble compounds embedded in the solid cake.

1.1.4.2 Hydrothermal Hydrolysis

Hydrolysis of metal salts can also produce LHSs. Because the temperature required to hydrolyze the salts is higher than the boiling point of water at atmospheric pressure, these reactions are conducted in hermetically sealed vessels (so this method is called hydrothermal). For example, layered hydroxide nitrates of magnesium, zinc, copper, and nickel have been produced by preparing a solution of metal nitrate salts and heating it between 150°C and 330°C for 12 h (Biswick et al. 2005). The peculiar result of this method is observed from the zinc hydroxide structure, which at room temperature spontaneously forms the type II structure (Figure 1.4), whereas the hydrothermal hydrolysis produce a type I LHS with composition $(Zn)_3(OH)_4(NO_3)_2$.

The hydrothermal route also helps to accelerate the synthesis, because the $(Zn)_3(OH)_4$ $(NO_3)_2$ compound can also be obtained by hydrolysis at 120°C and atmospheric pressure in a Petri dish, although this reaction takes 7 days (Chouillet et al. 2004). If this hydrolysis is performed under softer conditions (65°C for 14 days), the LHS will present a structure of the same type I but with composition $Zn(OH)(NO_3) \cdot H_2O$ (Chouillet et al. 2004).

Hydrothermal synthesis has been successful to produce acetate LHS of nickel and zinc (Choy et al. 1998; Tronto et al. 2006). For example, solutions between 0.015 and 0.030 mol L^{-1} of nickel or zinc acetate in hydrothermal treatments at 150°C for 24 h produce layered hydroxide acetates (Tronto et al. 2006). The acetate solution can also contain a mixture of zinc and nickel salt, and produce a zinc/nickel hydroxide acetate (Choy et al. 1998).

The disadvantage of this hydrothermal method is the higher cost for large-scale production in comparison with the precipitation method, due to the high thermal energy needed during long reaction times in addition to the need for special sealed and inert vessels to resist the high vapor pressure, which easily reaches 50 atm (Choy et al. 1998). However, the higher energy cost is compensated because this method produces LHSs that are not obtained by other techniques, as in the case of nickel or magnesium hydroxide salts, which by precipitation produce simple hydroxides (personal observation).

1.1.4.3 Polyol Hydrolysis

LHSs can also be prepared by hydrolysis of metal salts at high temperature in the presence of polyols. This requires the addition of a small amount of water to start the reaction. This process has been used to prepare layered magnesium hydroxide acetate. In this peculiar synthesis, a solution with 2 g of magnesium acetate in 100 mL of propylene glycol undergoes hydrolysis with the addition of 0.25 mL of water, followed by refluxing (temperature not stated) to produce magnesium hydroxide acetate (Nethravathi et al. 2011).

Layered hydroxide acetates of zinc, cobalt, and nickel by polyol hydrolysis have been prepared with acetate metal salts dissolved in diethyleneglycol 1,2-propanediol or ethanol at a concentration of around 0.1 mol L^{-1}. The temperature to dissolve the salts needs to be increased up to 160°C with diethyleneglycol and 1,2-propanediol, whereas the dissolution in ethanol is achieved close to 60°C. The hydrolysis is controlled with the addition of water. The water/cations molar ratio is one of the most important parameters to define the formation of LSHs and the quality of the crystals (Poul et al. 2000). Such ratios reported are >2 for zinc hydroxide acetate, >26 for cobalt hydroxide acetate, and >4 for nickel hydroxide acetate. The particles produced by this method are highly crystalline with size between 0.5 and 1.5 μm. The type of structures according to the models in Figure 1.4 are type I in nickel hydroxide acetate, type II in cobalt hydroxide acetate, and type III in zinc hydroxide acetate.

1.1.4.4 Urea Hydrolysis

The urea hydrolysis method is also named melt reaction, which is conducted with metal salts mixed with urea and water to form a slurry (Rajamathi and Vishnu Kamath 1998). Only a small quantity of water is required to homogenize the mixture and promote the decomposition of urea according to the reaction reported elsewhere (Henrist et al. 2003). The slurry is treated at high temperatures, as in the synthesis of nickel hydroxide nitrate, where 18 g of nickel nitrate is mixed with 2 g of urea and 2.4 mL of water. The heating of the slurry above 150°C for 2 h leads to the formation of LHS with composition $Ni(OH)_{1.4}(NO_3)_{0.6}$ with a type I structure (Rajamathi and Vishnu Kamath 1998). The same amount of urea and water mixed with 2 g of cobalt(II) nitrate produces selective layered materials with structures that depend on the temperature applied to the slurry. Because of the polyvalence of cobalt, redox reactions might occur, and if cobalt is oxidized to Co(III), the product would be a hydrotalcite-like compound, which occurs if the slurry is heated at 80°C. However, when the temperature is increased to 160°C, the cobalt cations retain the 2+ oxidation state and a type I LHS structure is formed with composition $Co_3(OH)_4(NO_3)_2$ (Rajamathi and Vishnu Kamath 2001).

The urea hydrolysis method has the main advantage of being selective for metals with different oxidation states. Other advantages are the smaller reaction volume needed and the ability to carry out the reaction without using hermetic vessels. Additionally, this hydrolysis method allows producing highly crystalline particles of $Cu(OH)NO_3$ with particles larger than 10 μm (Henrist et al. 2003).

1.1.5 Characterization

Two of the most important features of LHSs are (1) the layered structure with weak forces between layers, allowing expansion of the interlayer space and allocation of anions with different sizes and (2) the capability to exchange anions. Powder X-ray diffraction (XRD) is an essential tool to verify the crystalline structure after each synthesis or for quality control in the case the LHSs purchased from commercial suppliers. Due to the nature of this technique, the diffraction pattern is the result of the long-range ordered position of atoms regardless of the chemical bonds between atoms, that is, a layered sequence of atoms detected by XRD occurs if the layers are weakly joined by electrostatic or van der Waals forces like in LHS or if they are joined by covalent bonds as in the case of perovskites.

The identification of known structures by XRD is easily done by searching for the sample's diffraction pattern in the ICDD database. For new compositions, the layered structures require confirmation with an anion-exchange experiment to probe the ability to modify the interlayer space, which depends on the anion size.

1.1.5.1 Powder XRD

The XRD pattern of LHSs presents an intense reflection at low angles, related to the basal spacing (the interlayer spacing occupied by the intercalated anion, hydrated or not, plus the layer thickness). Due to the preferred orientation of the layered crystals in the sample holder, a series of basal peaks is common, which are related to the same basal spacing, using n times the wavelength in Bragg's equation. This spacing in LHSs is directly associated with the thickness of layers in single hydroxides such as $Ca(OH)_2$ and $Mg(OH)_2$ (Figure 1.8). In fact, the basal spacing of magnesium hydroxide (known as Brucite), where Mg^{2+} cations are octahedrally coordinated by hydroxyl groups, is taken as reference to

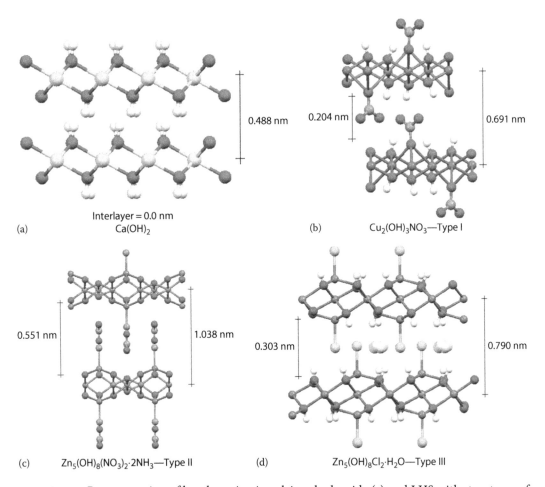

(a) Interlayer = 0.0 nm $Ca(OH)_2$ — 0.488 nm

(b) $Cu_2(OH)_3NO_3$—Type I — 0.204 nm, 0.691 nm

(c) $Zn_5(OH)_8(NO_3)_2 \cdot 2NH_3$—Type II — 0.551 nm, 1.038 nm

(d) $Zn_5(OH)_8Cl_2 \cdot H_2O$—Type III — 0.303 nm, 0.790 nm

FIGURE 1.8 Representation of basal spacing in calcium hydroxide (a) and LHS with structures of type I (b), II (c), and III (d).

determine the interlayer spacing in LHSs or related layered compounds (Carlino 1997). Because the structure of LHS is formed with divalent cations with cationic radii close to that of Mg^{2+}, the difference between the basal spacing of any LSH and the basal spacing (equivalent to the layer thickness) of magnesium hydroxide (0.487 nm) is equal to the interlayer spacing (Carlino 1997). The calcium hydroxide structure in Figure 1.8 presents a basal reflection at 18.15° (2θ) and represents a spacing of 0.488 nm according to the conversion with Bragg's equation (Figure 1.9).

This basal spacing is practically equal to that of Brucite, so the interlayer spacing is negligible. In the case of copper hydroxide nitrate with composition $Cu_2(OH)_3NO_3$, the basal spacing of 0.691 nm (Figure 1.9) includes the presence of nitrate ions between the two layers of copper cations octahedrally coordinated by hydroxyl groups (Figure 1.8), so when using nitrate this spacing is known as interlayer spacing and its dimension is calculated by subtracting the thickness of a Brucite layer from the basal spacing, giving a value of 0.204 nm. The interlayer spacing depends directly on the size of the anion (hydrated or not), which is clearly evidenced by comparing copper hydroxide nitrate ($d = 0.20$ nm) with copper hydroxide acetate ($d = 0.44$ nm) and copper hydroxide benzoate ($d = 1.07$ nm). These values were calculated from the basal spacing reported in the literature subtracting the thickness of the copper octahedron layer (Marangoni et al. 2001).

The basal spacing in structures classified as type II and III are larger than that of type I. The XRD patterns of $Zn_5(OH)_8Cl_2 \cdot H_2O$ present a basal reflection equivalent to a spacing between planes of 0.790 nm (Figure 1.9). The calculation of the interlayer spacing might need correction owing to the presence of cations in Td sites. However, the tetrahedrons

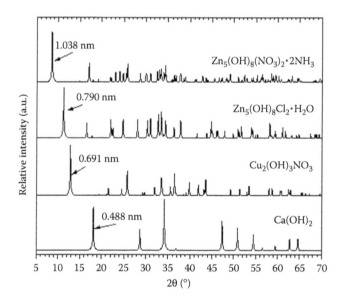

FIGURE 1.9 XRD patterns of $Ca(OH)_2$ (CIF file 1000045), $Cu_2(OH)_3NO_3$ (CIF file 31353), $Zn_5(OH)_8Cl_2 \cdot H_2O$ (CIF file 16973), and $Zn_5(OH)_8(NO_3)_2 \cdot 2NH_3$ (CIF file 77742). The values in the figure correspond to basal spacings calculated from Bragg's equation considering the X-ray radiation of $CuK\alpha = 0.1054$ nm.

do not cover the whole surface of the layer, so it is reasonable to consider that the spacing between the layers exclusively formed by cations in Oh coordination is the maximal restriction to allocating interlayer anions. This fact is relevant when the LHS is intercalated with organic anions. The assumption that the Oh layer is the limit to contain anions is based on experimental results demonstrating that anions can be linked to the octahedron layers of LHS (Arizaga et al. 2008a,b).

The interlayer space available to allocate anions in type III structures is thus determined by the difference between the interlayer spacing and the thickness of a Brucite layer, which in the case of $Zn_5(OH)_8Cl_2 \cdot H_2O$ is equal to 0.303 nm (Figure 1.8). Regarding LHS with type II structure and composition $Zn_5(OH)_8(NO_3)_2 \cdot 2NH_3$, the XRD pattern presents a basal reflection equivalent to a spacing of 1.038 nm (Figure 1.9). This LHS has the largest interlayer spacing (0.551 nm) among the structures in Figure 1.8, due to the coordination of water molecules with the cations in Td sites and the presence of nitrate ions over the water.

1.1.5.2 Infrared Spectroscopy

Spectroscopic vibrational analytic techniques give valuable information about the functional groups and the interaction or confinement effect of the intercalated species between the two-dimensional layers, to shed light on the composition of LHS and predict some properties. Although infrared (IR) spectroscopy is used to analyze interlayer anions, Raman spectroscopy provides more details on the vibrations of atoms forming the octahedrons of the rigid layers. The most common anions found in LHSs are nitrate, chloride, and carboxylates, according to the compositions described in the most representative articles (Bera et al. 2000; Louër et al. 1973; Meyn et al. 1993; Rajamathi and Vishnu Kamath 1998; Rajamathi et al. 2005).

Nitrate and chloride are recommended to formulate pristine compositions of LHSs because these are suitable anions to be displaced through an ionic-exchange reactions. They also facilitate the formation of hybrid derivatives (especially with carboxylic or sulfate organic acids) to produce polymer fillers with defined properties. However, carbonate is an anion produced during the synthesis of LHSs in aqueous solutions under air atmosphere, and due to the higher negative charge density in comparison with nitrate and chloride (Carlino 1997), carbonate reduces the anion-exchange capacity of LHS. Because of the relevance of these anions, the bands to identify them in IR spectra are shown in Figure 1.10.

Chloride in LHS is not detectable in IR spectra, not even in type III structures (Frost et al. 2002). However, the $Cl-M^{2+}$ stretching mode is weakly detected by Raman spectroscopy in type III structures between 410 and 350 cm^{-1} (Frost et al. 2002). The nitrate ions in IR spectra present two weak bands near to 830 and 1020 cm^{-1} (Figure 1.10), related to stretching modes (v_2 and v_1, respectively). These anions, as "free" solvated interlayer ions in LHSs with type II structure, belong to the D_{3h} symmetry group and present an intense signal near to 1380 cm^{-1} due to symmetric stretching (v_3) mode, whereas in type I and III structures, in which nitrate coordinates directly the metal cations, an additional signal of medium intensity appears close to 1440 cm^{-1}, related to the asymmetric stretching mode in

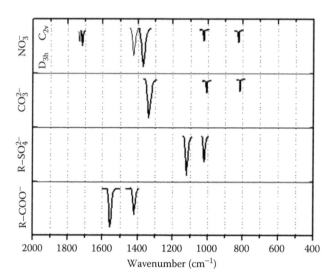

FIGURE 1.10 Representation of bands in IR spectra of the most significant interlayer anions in LHSs.

the C_{2v} symmetry (Arizaga et al. 2007). Finally, a signal near 1760 cm^{-1} appears sometimes due to the combination of v_1 and v_3 modes (Hindocha et al. 2009).

Regarding carbonate, the most intense band appears around 1365 cm^{-1}, related to the v_3 stretching mode. This band is close to the nitrate signal even in the v_1 and v_2 modes at 1040 and 830 cm^{-1}, respectively (Hindocha et al. 2009), but the bands produced by carbonate are shifted between 5 and 15 cm^{-1} to lower wavenumbers (da Silva et al. 2014).

Organic sulfates such as the dodecyl sulfate (DDS) intercalated in LHS present an antisymmetric stretching mode around 1210 cm^{-1} and a symmetric one near 1070 cm^{-1} (Otero et al. 2012), whereas the carboxylates are clearly identified due to the antisymmetric and symmetric modes at 1600–1560 and 1410–390 cm^{-1}, respectively (Hindocha et al. 2009).

1.1.6 Properties

1.1.6.1 Anion Exchange

The most significant property in LHS that enables the design of new multifunctional materials is the anion-exchange capacity, evidenced by a long list of articles following the systematic study of Meyn et al. (1993). This capacity is influenced by the charge density in A^{n-} anions observed in anionic layered matrices in the following order (Carlino 1997; Halajnia et al. 2013):

$$CO_3^{2-} > HPO_4^{2-} > HAsO_4^{2-} > CrO_4^{2-} > SO_4^{2-} > MoO_4^{2-} > OH^- > F^- > Cl^- > Br^- > NO_3^- > I^-$$

The anion-exchange phenomenon occurs regardless of whether the A^{n-} anion is directly bound to the layers, as in type I and III structures, or free within the interlayer space (Meyn et al. 1993). The exchange reaction is a key step to select the method to prepare LHSs as fillers in polymers, because this reaction will introduce monomers between the inorganic layers of LSH or introduce long-chain organic anions to improve compatibiliza-tion of the LHS in the polymer, as represented in Figures 1.11 through 1.13. An example

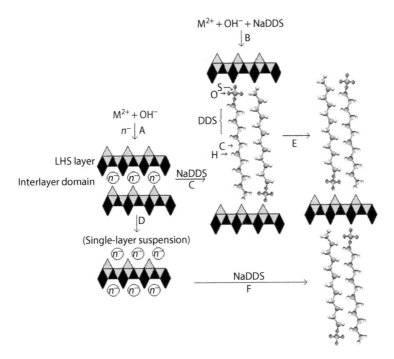

FIGURE 1.11 Experimental procedures to modify the surface of an LHS. A, B—synthesis by copre-cipitation; C, F—exchange reaction; D, E—exfoliation.

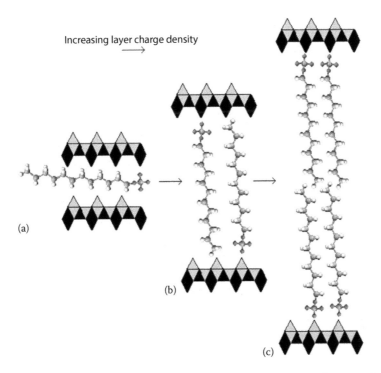

FIGURE 1.12 Arrangement of the DDS anion between the LHS layers. (a) flat single layer; (b) tilted monolayer; and (c) tilted double layer.

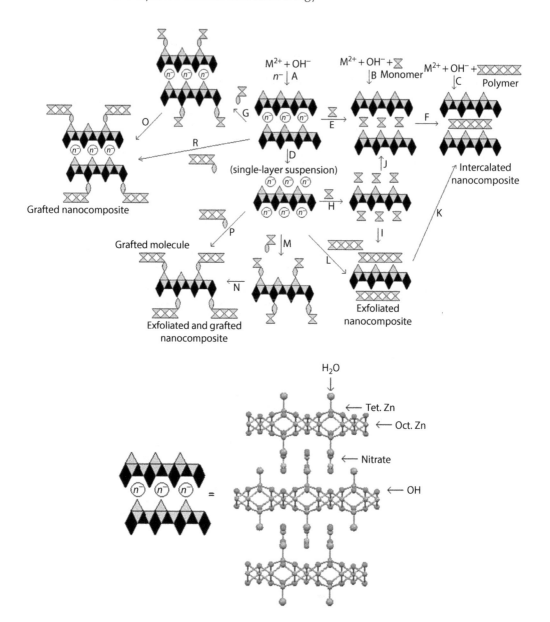

FIGURE 1.13 Different strategies to synthesize LHSs and disperse them in polymeric matrices.

of exchange reaction with organic anions is the experiment conducted with 1.5 g of zinc hydroxide nitrate with composition $Zn_5(OH)_8(NO_3)_2 \cdot 2H_2O$ dispersed in 50 mL of an aqueous solution with 0.59 g of succinic acid ($HOOC(CH_2)_2COOH$), 0.66 g of glutaric acid ($HOOC(CH_2)_3COOH$), or 0.73 g of adipic acid ($HOOC(CH_2)_4COOH$), which was adjusted to pH 7 with ammonia solution. The mass of the organic acids corresponded to an LHS/acid molar ratio of 1:2 (Wypych et al. 2005). After 72 h under stirring, the XRD patterns of the powders recovered by decantation presented increased basal reflection proportional to the chain length of the organic anions, thus confirming the exchange of nitrate with carboxylic anions.

1.1.6.2 Thermal Behavior

Metal hydroxides, LDHs, and LHSs are known for their thermal decomposition above 250°C, forming water as the first product and nonflammable gases derived from interlayer anions (Bera et al. 2000; Lopez et al. 1997). For years, magnesium hydroxide was used as a flame-retardant additive, although it is only effective at high loadings, thus reducing the mechanical properties of polymers (Kandare et al. 2006a). The ignition temperature of polymer composites with single hydroxides increases proportionally to the hydroxide loading in the polymer (Tang et al. 2013).

Such thermal stability can also be improved by using LHS at lower concentrations. A detailed study demonstrating the advantages of using LHS as flame retardant was conducted with poly(vinyl ester) (PVE) mixed with the layered copper hydroxide DDS with composition $Cu_2(OH)_3(CH_3(CH_2)_{11}OSO_3)$ (Kandare et al. 2006a). More details of this effect will be described in the last Section 1.2.3.4.

1.2 POLYMER/LHS NANOCOMPOSITES

Besides the possibility of promoting interlayer anion exchange with a broad range of organic or inorganic anions, another advantage of LHSs' layered surface is the reactivity. Because the surface is populated by hydroxyl groups, LHSs are also reactive to several inorganic or organic species, making the process of chemical modification very easy and broadening the industrial application.

1.2.1 Compatibilization of LHS Surface

In the case of LHS/polymer nanocomposites, one of the simplest ways to dramatically increase this surface and consequently improve the contact with the polymer is to delaminate the layered crystals (separate the layered crystals into a smaller number of stacked layers) or exfoliate them (separate the layered crystals into single layers) (Figure 1.11D and E). Several methods are used to produce single-layer suspensions, the most established being chemical and physical procedures, in either aqueous media or organic solvents.

As the homogeneous dispersion of the layered nanofiller and a suitable filler/polymer interface are the key factors to improve mechanical and physical properties of polymer matrix-based nanocomposites and because nanofillers are normally hydrophilic by nature, before adding them to a hydrophobic polymer a surface modification step is usually necessary (Figure 1.11C–F).

To improve the compatibility of the polymer with the layered crystals (ideally single layers), surface modification (organofunctionalization) is usually applied to the filler, but the pendant reactive groups of the polymer molecules can also be attached to the fillers' functional groups. Covalent, coordinative, electrostatic, or hydrogen bonds are involved in the interaction between the polymer and the filler. The functionalization of LHS can be by either physisorption or chemical reaction. The second is preferred because it is irreversible. Although preferred, it is only rarely expected in the usual preparation procedures.

The easiest procedure to promote chemical modification to turn the hydrophilic layered filler surface hydrophobic and produce a stable interfacial bond between the filler and the matrix is to replace the intercalated anion with an anionic surfactant such as DDS, for example. LHS surface functionalization with a surfactant can be achieved by reacting DDS with bulk crystals (Figure 1.11C) or single exfoliated inorganic layers (Figure 1.11F).

Another viable alternative is to synthesize the DDS-intercalated filler in a single step (Figure 1.11B) and, if needed, to exfoliate the hydrophobized LHS in a specific solvent (Figure 1.11E).

The intercalated organic species arrangement can be changed according to the charge density of the layers. In low-density charged layers, the surfactant chain anion can be arranged in a flat monolayer, and with increased layer charges, this conformation can change from a tilted monolayer, where the sulfate groups point to opposite positively charged layers, or even a tilted double-layer arrangement, where the sulfate groups point to the same layers. These changes in the surfactant positioning between the layers can be seen in Figure 1.12 (Naik et al. 2011).

The insertion of long-chain anions between LHS surfaces not only increases the interlayer space, allowing the polymer to migrate into the interlayer space, but also weakens the interlayer interaction, reducing the energy needed to cause the layered crystal to be delaminated/exfoliated into the polymeric matrix.

Organofunctionalization through covalent grafting is also used to introduce new functional groups onto surfaces. This procedure can be easily used to tailor the filler interface (or interphase), enhancing the compatibility between these two phases. Based on the number of possibilities and the type of polymer and filler, it is relatively easy to create the functional groups to optimize the filler/polymer interface and consequently optimize the desired polymer nanocomposite properties.

In LHSs such as LDHs, this procedure can be applied to modify the layered crystals or single-layer surfaces. The most frequently described procedure is to use silanes such as 3-aminopropyltriethoxysilane (APTES) or other alkoxysilanes. The process consists of hydrolysis of APTES to release ethanol and condensation with the LHS surface hydroxyl groups to release water molecules. Depending on the hydrolysis degree, the siloxane groups can be grafted by one, two, or three bonds to the layer surface. These or alternative functionalization processes can be applied to the previously prepared layered crystals or exfoliated single layers, or the grafted crystals can be prepared in a single step during the LHS synthesis. In the last case, the modified single layers can be restacked or not before their use as filler in hydrophobic polymers.

1.2.2 Procedures to Prepare Polymer/LHS Nanocomposites

To synthesize exfoliated layered fillers into a polymeric matrix, two different strategies can be adopted. The first is melt processing, which consists of hydrophobizing the filler and introducing this compound into the polymeric matrix. In this case, if shear forces are used during processing, the filler will be delaminated or exfoliated. The second is to obtain the hydrophobic filler in single layers and add this dispersion to the polymer dispersion in the same solvent. After mixing both dispersions, the solvent can be evaporated and the polymer nanocomposites can then be obtained by wet casting.

In spite of the several methods used to disperse the layered crystals in the polymeric matrix, melt processing is still the preferred method. This is explained by the possibility of using existing polymer processing facilities, the ease of the mixing process, the good results in terms of optimized nanocomposite properties, and the relatively low cost.

Figure 1.13 shows a compilation of different procedures that can be adopted to prepare LHS/polymer nanocomposites. Some of the proposed strategies presented in Figure 1.13 are also used for LDHs and very likely can be applied to LHS, representing an open field for investigation.

The steps involved in each route of Figure 1.13 are described as follows:

Procedure A: LHSs can be directly obtained by coprecipitation at constant pH when the appropriate metal salt is precipitated with alkaline solution in the presence of an anion to be intercalated or by following any method discussed in Section 1.4. Here, depending on the chosen metal, small variations in the final pH should be adopted and several different anions can be intercalated, ranging from inorganic to organic, including colored anionic azo dyes. In some cases, especially when the temperature is slightly increased, the intercalated species are not only intercalated but also grafted to the layered surfaces through covalent bonds (da Silva et al. 2012; Zimmermann et al. 2013).

Procedure B: LHSs intercalated with monomeric species can also be directly obtained when the desired anionic species is added to the reactor , which can be later polymerized *in situ* by thermal, chemical, or UV treatment to generate an intercalated LHS/polymer nanocomposite.

An example of this strategy is the preparation of styrene sulfonate-intercalated LDHs by coprecipitation followed by *in situ* polymerization of monomers in the interlayer spacing of LDH, generating poly(styrene sulfonate) intercalated between the inorganic layers (Si et al. 2004).

Another example reported in the literature is the intercalation of *m*-aminobenzoate between the layers of a Zn/Al LDH, after which the material is submitted to polymerization by heat treatment at 90°C for 2 days. According to the author, the orientation of the intercalated anions leads to a particular orientation of the resulting polymer, which was obtained by reacting the nitrogen atom of the amine group and the fourth carbon atom of the benzene ring of the neighboring molecule (Jamhour 2005).

Procedures C and F: Because the ideal polymer-based nanocomposites filled with layered compounds should be intercalated with a polymer or exfoliated into the polymer matrix, several strategies can be followed to achieve these conditions.

To obtain a layered filler intercalated with the polymer, two strategies can be adopted. The first is to perform the precipitation of the LHS in the presence of an anionic polymer (Figure 1.13C). Although this procedure has been only reported for LDH, this procedure can be easily adapted to LHSs. The reported procedure consisted of preparing M_2Al/lignosulfonate nanocomposites (M = Zn, Mg, and Co) by coprecipitation.

The second is to intercalate the monomer and perform the polymerization using chemical or thermal processes (Figure 1.13F) (Hennous et al. 2013).

Procedure D: To facilitate the dispersion of the inorganic species into the polymeric matrix, first the layered crystals can be exfoliated in an appropriate solvent. This procedure is widely applied to produce single-layer suspensions of different materials, including LDHs, but it is rarely applied to LHSs.

In the example found in the literature, *p*-aminobenzoate was intercalated into copper hydroxide salt by coprecipitation using NH_4OH solution with constant stirring, and the solid obtained was then exfoliated in water to form a translucent colloidal dispersion. The purpose of this study was to use the single-layer suspensions to obtain Cu_2O and CuO nanoparticles (Nethravathi et al. 2012).

The second example was applied to DDS-intercalated zinc hydroxide salt $Zn_5(OH)_8(DS)_2 \cdot mH_2O$. First, zinc hydroxide nitrate was prepared by mixing ZnO with $Zn(NO_3)_2$ solution at 60°C for 2 days. After drying, the resulting white solid was anion exchanged with sodium DDS. The resulting solid was exfoliated in 1-butanol and ethylene glycol. The single-layer suspension was used to prepare bimodal ZnO crystals having a nanotubular morphology decorated with nanolayers (Machado et al. 2010).

The third example involved the synthesis of zinc hydroxide nitrate, acetate, DDS, and benzoate using the coprecipitation technique. Zinc hydroxide DDS was also obtained by the ion-exchange procedure using zinc hydroxide nitrate as a solid precursor. After drying, the solid was dispersed in formamide or 1-butanol followed by shaking at room temperature, at 60°C, and under reflux for 24 h. Similarly, the suspensions obtained were used to prepare oriented films by casting, followed by solvent evaporation under reduced pressure at room temperature. The single-layer suspension was used to produce anisotropic ZnO nanoparticles, either needlelike particles prolonged in the [001] direction or disklike particles flattened along the (001) plane (Demel et al. 2011).

Procedures E and H: Instead of directly intercalating the anionic monomer for further polymerization, another precursor anion can be intercalated and later replaced by the anion of interest. This procedure was reported for the synthesis of layered (nickel, zinc) hydroxide acetates, and the intercalated acetate anions were exchanged by anions derived from different thiophene derivatives such as 2-thiophenecarboxylic acid, 3-thiophenecarboxylic acid, and 3-thiopheneacetic acid. All of the materials obtained were separated and washed with water by centrifugation, but the polymerization step was not reported (Tronto et al. 2006).

As it is not always easy to replace the intercalated anion through a polymer precursor anion, sometimes it is easier to exfoliate the precursor in a specific solvent to improve the exchange reaction. Normally, an ultrasonic bath can be used, where the ultrasonic waves generate cavitation bubbles that interact with the layered crystals, producing the exfoliation. The surface energy of the solvent and the layers of the crystal should be similar, but intercalated species also reduce the surface energy, an effect that still needs to be investigated. This exfoliation procedure will include an extra step, and the reaction can only be performed in a specific solvent, but many times the steric factor is the key limitation to the success of the reaction. Removing this limitation through exfoliation is sometimes the only way to obtain a reaction (Swanson et al. 2013; Zhao et al. 2011).

Procedures G and R: Grafting of layered material surfaces is a relatively common procedure and is widely reported in the literature, mainly related with the condensation reactions of LDHs with silanes and carboxylates. As far as we know, these procedures are rarely used to modify LHS or single layers of LDH and applied as polymer fillers. Examples of these are grafting of LDH with silanes (Lu 2010; Tao et al. 2009, 2014); grafting of LDH with other molecules (Guimarães et al. 2000; Kotal and Srivastava 2011; Liu et al. 2009; Wypych and Satyanarayana 2005); and covalent grafting in LHS (Arizaga et al. 2008a; Wypych et al. 2005).

Procedures I, J, K, and L: The single-layer suspension containing the exchangeable anion can be absorbed with the polymer (Figure 1.13L). The same layers, after incorporation of the monomer by exchange (Figure 1.13H), can be polymerized (Figure 1.13I) or restacked and then polymerized (Figure 1.13J and F). The single-layer suspension containing the adsorbed polymerized chain can also be restacked (Figure 1.13K). All these procedures need to be performed in solution.

Procedures M, P, N, O, and R: The grafting procedure can also be applied to single-layer suspensions to introduce new chemical functions to the single layers. These functions can carry reactive groups (Figure 1.13M) that can be further polymerized (Figure 1.13N) or the polymer can be directly grafted to the surface (Figure 1.13P), which can be restacked or not. The same procedures can be adopted for layered crystals, where only the surface of the crystals is expected to be modified (Figure 1.13O and R).

Most of the procedures described above are still open to investigation. The literature normally reports only the easiest ways, but more elaborate synthetic steps can be employed to produce more sophisticated compounds like those where the polymer is grafted to the surface of the single layers. This combination of layered compounds and polymer will certainly in the future allows researchers to tailor the properties of some compounds to meet consumer demands. Also, most of the LHSs' structures are still open to investigation for polymer composite applications (Arizaga et al. 2007; Kandare and Hossenlopp 2009; Moraes et al. 2014; Semsarilar and Perrier 2010; Tsujii et al. 2001).

1.2.3 LHS as Fillers in Thermoplastic/Thermoset Polymers

Only a small number of papers have reported the use of LHSs as fillers in thermoplastic/thermoset polymers. Some of these are summarized in Sections 1.2.3.1 through 1.2.3.5.

1.2.3.1 High-Density Polyethylene/LHS Nanocomposites

The procedure adopted to prepare ZHC and zinc hydroxide salts intercalated with azo dyes (methylorange—ZHMO and orange II—ZHOII) can be found in the literature (Zimmermann et al. 2013).

High-density polyethylene (HDPE) without any additives (reference HC 7260 LS, density of 0.959 g cm^{-3}) was supplied by Braskem (Brazil) and used without any further treatment. HDPE nanocomposites filled with ZHC, ZHMO, or ZHOII were prepared, and the filler content was varied (0.1, 0.2, 0.5, 1.0, or 2.0 wt.% in relation to the polymer mass).

The polymer and the filler were gently mixed and transferred to a HAAKE MiniLab II micro compounder (Thermo Scientific, USA) and submitted to processing using conical twin screws running in the same direction at 110 rpm. The temperature was fixed at 160°C and the residence time was 5 min.

After processing in the compounder, the samples were injected into a HAAKE MiniJet II injection molding machine. The cylinder and mold were fixed at temperatures of 160°C and 40°C, respectively, and the samples were injected at a pressure of 320 bar for 5 s, followed by compression for an additional 5 s. An ASTM D638 standardized mold was used and dog bone-shaped samples were obtained with dimensions of 100 mm long × 3.3 ± 0.1 mm thick × 3.18 mm wide. For statistical purposes, 10 specimens were prepared and evaluated for each composition, under identical operation conditions.

In the mechanical testing, 10 specimens prepared according to the ASTM D638 standard mold were tested at 20°C ± 3°C for each composition, and the results were reported as the average of the measurements along with the standard deviation.

Digital photographs of the HDPE/ZHMO and HDPE/ZHOII composites before the mechanical property assays (Figure 1.14) showed that the filler was homogeneously dispersed in the polymeric matrix in all the materials. The color intensity was proportional to the amount of added pigment.

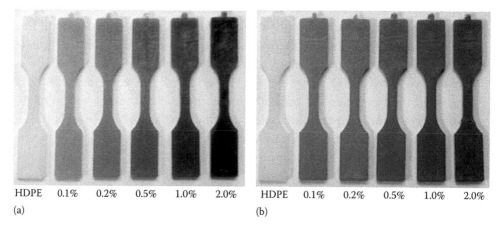

FIGURE 1.14 Digital photographs of the composite samples before the mechanical assays: (a) HDPE/ZHMO and (b) HDPE/ZHOII. (Reproduced from Zimmermann, A. et al., *J. Polym. Res.*, 20[9], 224, August 10, 2013. With permission.)

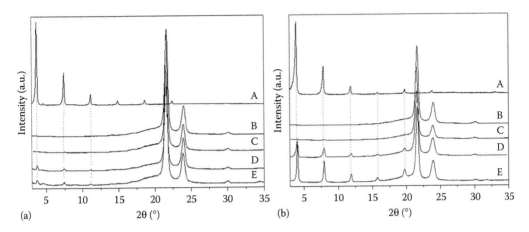

(a)

(b)

2θ (°)

FIGURE 1.15   XRD patterns of the filler (A), injected neat HDPE (B), and HDPE/filler composites with different filler contents (C: 0.1 wt.%, D: 1 wt.%, E: 2 wt.%). (a) ZHMO and (b) ZHOII. (Adapted from Zimmermann, A. et al., *J. Polym. Res.*, 20[9], 224, August 10, 2013.)

The XRD patterns (Figure 1.15a and b[A]) suggested that ZHMO and ZHOII are isostructural with $Zn_5(OH)_8(NO_3)_2 \cdot 2H_2O$ and presented basal spacings of 23.82 and 22.49 Å, respectively, which confirmed that the dyes MO and OII intercalated between the zinc LHS layers. Figure 1.16 shows the schematic representation of the dyes MO and OII intercalated between the zinc LHS layers, which are accommodated in a tilted interdigitated single-layer tilted arrangement.

By determining the content of anionic MO and OII intercalated into zinc LHS using UV-Vis spectroscopy and thermogravimetric analysis/differential thermal analysis (TGA/DTA) analysis (not shown) data, the formulas of ZHMO and ZHOII were estimated to be $Zn_5(OH)_8(MO)_{1.84}Cl_{0.16} \cdot 5.54H_2O$ and $Zn_5(OH)_8(OII)_{1.87}Cl_{0.13} \cdot 3.74H_2O$, respectively.

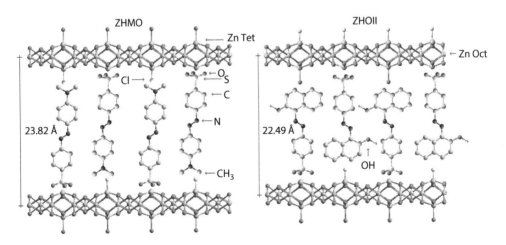

FIGURE 1.16   Schematic representation of the anionic dyes MO and OII intercalated between the zinc LHS layers. Hydrogen atoms have been removed to facilitate visualization.

The XRD pattern of neat HDPE (Figure 1.15a[B] and b[B]) displayed two intense peaks at 21.64° ($d = 4.11$Å) and 24.0° ($d = 3.71$Å), indexed as the (110) and (220) diffraction peaks of the orthorhombic unit cell. These peaks, apparently with the same intensity, appeared in the XRD pattern of all the samples containing HDPE (Figure 1.15a and b).

In the case of HDPE/ZHMO (Figure 1.15a), small broad peaks relative to the inorganic filler were detected at filler concentrations of 1% (Figure 1.15a[D]) and 2% (Figure 1.15a[E]), which showed that the filler was only highly delaminated, not exfoliated. The profile of the diffraction peaks recorded for HDPE/ZHMO confirmed the presence of crystals with reduced size, corroborating the scanning electron microscopy findings (not shown).

HDPE/ZHOII (Figure 1.15b) exhibited well-resolved and high-intensity peaks even at 1% of filler (Figure 1.15b[D]). Therefore, the more hydrophilic ZHOII interacted little with the polymer and the crystals' structure remained almost unaltered after their incorporation into HDPE.

As observed in Figure 1.17 and Table 1.2, in general the more hydrophilic filler ZHC reduced the ultimate tensile strength (UTS) and Young's modulus while improving strain at break, which is expected due to the weak interaction between this filler and the hydrophobic matrix HDPE. The more hydrophobic filler ZHMO elicited an opposite effect: it increased the UTS (+10.5% for 0.2 wt.% filler) and slightly increased Young's modulus (+10.7% for 1.0 wt.% filler), while reducing strain at break (−63.6% for 0.5 wt.% filler). Therefore, ZHMO was better dispersed in the HDPE matrix. This stiffer filler interacted with the polymer more strongly, overcoming the intermolecular hydrogen bonding between the HDPE molecules and the added filler increased the polymer crystals' sizes, as detected by differential scanning calorimetry (not shown).

ZHOII incorporation in HDPE reduced the UTS and Young's modulus. This filler also improved strain at break, especially at higher ZHOII loadings. The most pronounced negative impact on the UTS and Young's modulus was detected for a ZHOII concentration of 1 wt.% (−9.6% and −8.9%, respectively), whereas the most significant increase in strain at break was observed for a ZHOII loading of 2 wt.% (+192.9%).

Again, the poorer interaction between the filler ZHOII with the hydrophobic polymeric matrix HDPE accounts for this effect, generating polymer defects that propagate during the mechanical tests. The net result was a polymeric composite, HDPE/ZHMO, with improved mechanical properties.

1.2.3.2 Incorporation of Zinc-Layered Hydroxide Salts into Low-Density Polyethylene

The filler zinc hydroxide nitrate (called ZHN), $Zn_5(OH)_8(NO_3)_2 \cdot 2H_2O$, was obtained via a coprecipitation process in an alkaline medium at room temperature (Jaerger et al. 2014). The fillers intercalated with the DDS and the dodecylbenzene sulfonate (DBS) anions were also synthesized via direct coprecipitation with constant neutral pH at room temperature. The obtained white solids were washed/dried and used as fillers in low-density polyethylene (LDPE) powder (PB608 grade—melt flow rate = 30 g/10 min), donated by Braskem (Brazil). For comparison purposes, the sodium surfactant salts were also added to the polymers in the same proportions as the LHS fillers (NaDDS for sodium DDS and NaDBS for sodium DBS).

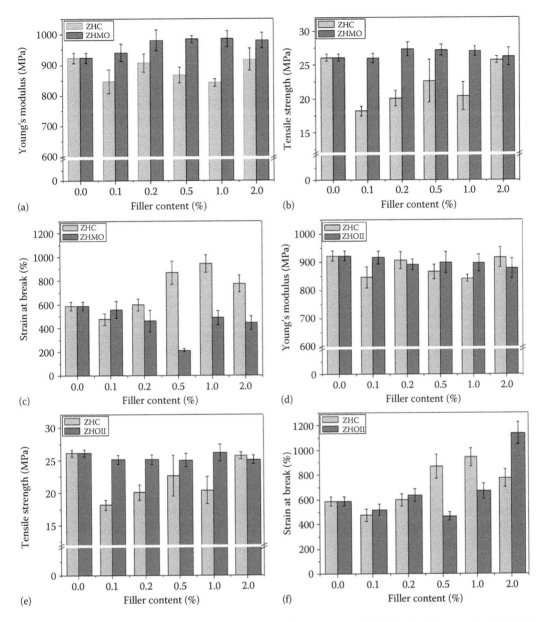

FIGURE 1.17 Mechanical properties of neat HDPE (percentage of filler) and of the samples HDPE/ZHMO (ZHMO), HDPE/ZHOII (ZHOII), and HDPE/ZHC (ZHC). Young's modulus of ZHC and ZHMO (a), ZHC and ZHOII (d); tensile strength of ZHC and ZHMO (b), ZHC and ZHOII (e) and strain at break of ZHC and ZHMO (c), ZHC and ZHOII (f). (Adapted from Zimmermann, A. et al., *J. Polym. Res.*, 20[9], 224, August 10, 2013.)

Composites were produced by melting the LDPE with the fillers using a HAAKE MiniLab II micro-extruder operating at 160 bar and 130°C. The material was then injected at 320 bar into a mold cavity (at 40°C) to produce ASTM D638 (type IV) samples using a HAAKE MiniJet II injection molding machine. The following LHS (and sodium surfactant salt) contents were used: 0.1%, 0.2%, 0.5%, 1.0%, and 2.0% in relation to the polymer weight.

TABLE 1.2   Young's Modulus (E), UTS (σu), and Strain at Break of the Investigated HDPE Composites

Filler (%)	HDPE/ZHMO	HDPE/ZHOII	HDPE/ZHC
		E (MPa)	
0	922.64 ± 17.27	922.64 ± 17.27	922.64 ± 17.27
0.1	940.06 ± 28.28	916.71 ± 22.24	846.15 ± 38.12
0.2	971.18 ± 31.41	891.65 ± 18.34	906.87 ± 29.86
0.5	983.94 ± 12.36	898.93 ± 37.15	866.50 ± 25.70
1	985.27 ± 26.27	897.13 ± 30.09	842.49 ± 12.47
2	979.43 ± 25.32	878.78 ± 34.30	926.31 ± 35.60
		σu (MPa)	
0	26.09 ± 0.51	26.09 ± 0.51	26.09 ± 0.51
0.1	26.01 ± 0.68	25.13 ± 0.68	18.16 ± 0.72
0.2	27.33 ± 1.03	25.10 ± 0.73	20.07 ± 1.14
0.5	27.18 ± 0.80	25.03 ± 1.03	22.64 ± 3.13
1	27.00 ± 0.74	26.16 ± 1.29	20.41 ± 2.07
2	26.21 ± 1.29	25.08 ± 0.69	25.73 ± 0.51
		Strain at Break (%)	
0	587.9 ± 36.5	587.9 ± 36.5	587.9 ± 36.5
0.1	524.1 ± 7.6	518.1 ± 45.5	494.9 ± 30.0
0.2	380.8 ± 3.2	635.9 ± 49.8	614.1 ± 38.3
0.5	213.4 ± 15.2	466.0 ± 35.0	830.6 ± 56.0
1	510.4 ± 39.0	669.5 ± 61.8	942.4 ± 73.0
2	420.9 ± 33.0	1135.1 ± 88.7	802.4 ± 37.9

Source: Adapted from Zimmermann, A. et al., *J. Polym. Res.*, 20(9), 224, August 10, 2013. With permission.

The synthesized LHS samples were characterized by XRD and showed the characteristic 9.66, 33.15, and 26.38 Å for the intercalated nitrate, DBS, and DDS, respectively. After the incorporation into LDPE, the samples were submitted to tensile testing in an Instron 5567 universal testing machine equipped with 1 kN load cell. For each family of samples, 10 specimens were tested at a speed of 10 mm min^{-1} and a minimum of five results was used to calculate the mean value.

Figure 1.18 shows the XRD patterns of the neat LPDE and the composites samples before the tensile testing.

The X-ray results of the ZHN/LDPE composites (Figure 1.18a) show diffraction peaks attributed to the planes (110) and (200) of the polyethylene orthorhombic structure and diffraction peaks attributed to ZHN. The ZHN structure was preserved, which was expected due to the weak interaction between the polymeric matrix and the hydrophilic filler. XRD patterns of the LDPE composites with LHS-DDS and LHS-DBS (Figure 1.18b and c) show absence of any diffraction peaks related to the layered materials, suggesting exfoliation of the layered crystals in the composites.

Homogeneous samples could be obtained by dispersing hydrophilic (ZHN) or hydrophobic (LHS-DDS and LHS-DBS) fillers by simple melt compounding in an extruder.

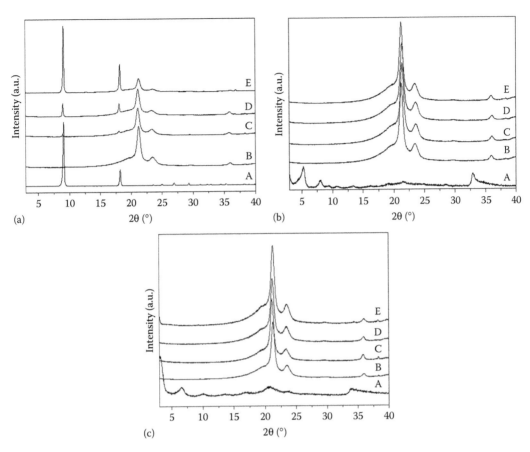

FIGURE 1.18 XRD patterns. (a) ZHN (A), LDPE (B), and after mixing 0.5% (C), 1% (D), and 2% (E) of ZHN with LDPE. (b) LHS-DBS (A), LDPE (B), and after mixing 0.5% (C), 1% (D), and 2% (E) of LHS-DBS with LDPE. (c) LHS-DDS (A), LDPE (B), and after mixing 0.5% (C), 1% (D), and 2% (E) of LHS-DDS with LDPE. (Adapted from Jaerger, S. et al., *Polímeros*, 24[6], 673–682, 2014.)

In general, Young's modulus decreased for the compounds with LHS-DDS and LHD-DBS anions intercalated with hydroxide salts regardless of the content, but small increments were observed when sodium salt surfactants were used, because they allow better interaction with the polymeric chains. Addition of the DDS and DBS sodium salts increased UTS in comparison with the pure LDPE, probably due to their hydrophobic character. However, addition of small amount of ZHN and LHS-DDS decreased ultimate strength values, although the original strength was recovered for larger filler (Figure 1.19 and Table 1.3). The lowest strength values were found for the LHS-DBS samples.

For the LHS-DDS samples, strain first increased then decreased for higher filler content, with a maximum at 0.1%–0.2%. However, addition of the LHS-DBS and the surfactant itself (NaDBS) increased strain for filler concentrations higher than 0.1%, and was even higher than that of the ZHN samples. In addition, the composites with sodium salt surfactants as fillers showed higher toughness.

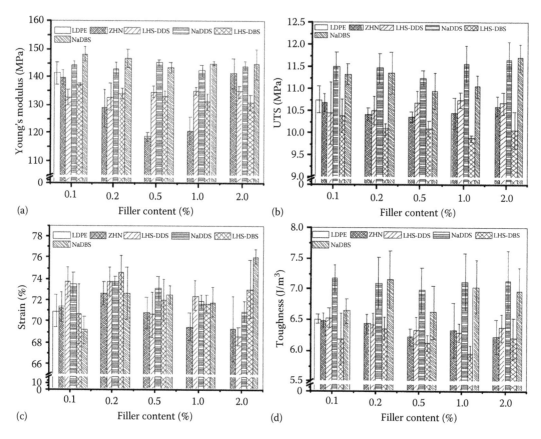

FIGURE 1.19 Mechanical properties of the LDPE composites using different fillers. Young's modulus (a); tensile strength (b); strain at break (c) and toughness (d). (Adapted from Jaerger, S. et al., *Polímeros*, 24[6], 673–682, 2014.)

However, addition of DDS- or DBS-intercalated hydroxide salts decreased toughness in relation to the pure polymer. The yield of the polymer in a high-deformation situation occurs as a result of shear, which is also related to the interaction between the polymer and the filler.

1.2.3.3 Poly(Methyl Methacrylate) Filled with Boron-Containing Layered Zinc Hydroxide Salt

Zinc hydroxide nitrate (ZHN—$Zn_5(OH)_8(NO_3)_2 \cdot 2H_2O$) was obtained by reacting ZnO with a zinc nitrate solution. After washing and characterizing procedures, nitrate anions intercalated into the obtained ZHN were exchanged with 4-(4,4,5,5-tetramethyl-1,3,2-dioxaborolan-2-yl) benzoate (TMDBB) anions to improve water dispersability. The TMDBB-exchanged LHS was mixed with poly(methyl methacrylate) (PMMA) in a Brabender Plasticorder chamber at a temperature of 185°C and the composites were prepared via melt blending. The proportions of the filler in relation to the polymer matrix were 3%, 5%, and 10%.

The conclusions of this study can be summarized as follows: TMDBB anions were successfully intercalated into the layered structure of ZHN by replacing the intercalated hydrated

TABLE 1.3   Mechanical Properties of the LDPE Composites

Filler Content (%)	Young's Modulus (MPa)				
	ZHN	LHS-DDS	LHS-DBS	NaDDS	NaDBS
0.0	141.39 ± 3.88	141.39 ± 3.88	141.39 ± 3.88	141.39 ± 3.88	141.39 ± 3.88
0.1	139.91 ± 2.69	132.79 ± 2.65	137.32 ± 0.28	144.39 ± 1.23	148.22 ± 2.63
0.2	128.82 ± 6.78	132.93 ± 4.93	134.10 ± 1.75	142.84 ± 2.41	146.58 ± 3.90
0.5	118.57 ± 1.82	134.78 ± 2.00	133.00 ± 2.20	145.07 ± 1.28	143.36 ± 1.94
1.0	120.84 ± 4.83	134.92 ± 1.44	131.10 ± 3.07	142.29 ± 2.21	144.99 ± 0.70
2.0	141.50 ± 5.34	134.87 ± 2.04	130.90 ± 2.70	144.07 ± 1.60	144.61 ± 5.25
	UTS (MPa)				
0.0	107.53 ± 0.32	107.53 ± 0.32	107.53 ± 0.32	107.53 ± 0.32	107.53 ± 0.32
0.1	106.78 ± 0.20	104.52 ± 0.72	103.84 ± 0.38	115.04 ± 3.23	113.37 ± 2.33
0.2	104.27 ± 0.14	105.04 ± 3.17	100.80 ± 0.13	114.73 ± 3.42	113.52 ± 4.74
0.5	103.63 ± 0.11	106.77 ± 2.58	101.00 ± 0.18	112.36 ± 1.74	109.35 ± 4.13
1.0	104.56 ± 0.36	107.48 ± 1.74	98.80 ± 0.07	115.71 ± 4.05	110.57 ± 2.58
2.0	106.02 ± 0.22	106.69 ± 2.55	100.61 ± 0.42	116.49 ± 4.08	117.02 ± 3.12
	Strain at Break (%)				
0.0	70.96 ± 1.56	70.96 ± 1.56	70.96 ± 1.56	70.96 ± 1.56	70.96 ± 1.56
0.1	71.34 ± 1.42	73.69 ± 1.38	70.65 ± 2.87	73.44 ± 1.14	69.13 ± 1.26
0.2	72.61 ± 1.15	73.74 ± 1.37	74.58 ± 1.63	73.75 ± 0.41	72.65 ± 2.42
0.5	70.82 ± 1.46	70.61 ± 2.19	72.00 ± 1.96	73.09 ± 1.16	72.52 ± 0.89
1.0	69.47 ± 1.29	72.38 ± 1.41	71.65 ± 0.90	71.84 ± 0.61	71.81 ± 1.41
2.0	69.27 ± 3.07	68.56 ± 0.89	73.03 ± 2.75	70.88 ± 0.96	76.10 ± 0.70
	Toughness (MPa)				
0.0	6.50 ± 0.07	6.50 ± 0.08	6.50 ± 0.07	6.50 ± 0.07	6.50 ± 0.07
0.1	6.48 ± 0.12	6.53 ± 0.15	6.18 ± 0.43	7.17 ± 0.22	6.64 ± 0.20
0.2	6.44 ± 0.16	6.43 ± 0.18	6.36 ± 0.18	7.10 ± 0.42	7.15 ± 0.46
0.5	6.22 ± 0.14	6.33 ± 0.22	6.12 ± 0.13	6.98 ± 0.37	6.62 ± 0.43
1.0	6.33 ± 0.44	6.28 ± 0.15	5.96 ± 0.11	7.12 ± 0.47	7.03 ± 0.43
2.0	6.23 ± 0.27	6.37 ± 0.22	6.20 ± 0.33	7.13 ± 0.49	6.96 ± 0.39

nitrate anions; the layered filler was poorly dispersed in the PMMA matrix; PMMA molecules were sometimes intercalated between the filler's layers; thermogravimetric results indicated that the thermal stability of the PMMA filled with the new layered compounds was improved in both N_2 and air atmosphere; and the degradation kinetics of PMMA and the composites showed that the filler increased the activation energies of the first step of the polymer degradation and reduced the peak release rate, despite pillaring of the char (Majoni et al. 2010).

1.2.3.4 Layered Copper Hydroxide DDS as Potential Fire Retardant for Poly(Vinyl Ester)

First, the layered copper hydroxide nitrate ($Cu_2(OH)_3NO_3$—CHN) was prepared by precipitating copper(II) nitrate with an aqueous ammonia solution. In the second step, interlayer nitrate anions were replaced by DDS by an anion-exchange reaction, obtaining the layered copper hydroxide DDS filler ($Cu_2(OH)_3(CH_3(CH_2)_{11}OSO_3$—CHDS). After washing and characterization, the solid was dispersed in the vinyl ester resin (bisphenol-A/novolac epoxy, mass fraction of 67% in styrene) and fire retardant using a mechanical stirrer for 3 h, followed by addition of the polymerization initiator (2-butanone peroxide—BuPo: 1.3%) and the mixture was stirred for a few minutes. The catalyst was added (cobalt naphthenate—CoNP: 0.3%) and the mixture was stirred until homogeneity. After mixing all the components, the samples were cured overnight at room temperature, and then postcured at 80°C for 12 h. The proportions of the filler in relation to the polymer matrix were 1%, 3%, 5%, 7%, and 10%.

The conclusions of this study can be summarized as follows: The layered copper hydroxide DDS filler was successfully synthesized and used as additive in PVE; addition of the filler in PVE resulted in a reduction of total hot release from cone calorimetry and increase of the amount of char, obtained in nitrogen atmosphere; addition of the filler in PVE resulted is no reduction in peak heat release rate; copper hydroxide DDS was then identified as a very promising filler for the improvement of thermal stability and fire retardancy of PVE (Kandare et al. 2006a).

1.2.3.5 Layered Copper Hydroxide Methacrylate as Filler into PMMA
In this chapter, the strategy was to prepare in a single step the layered copper hydroxide methacrylate, which was used as a filler in PMMA, prepared *in situ*.

First, layered copper hydroxide methacrylate was synthesized by adding NaOH 0.1 mol L^{-1} in a 0.1 mol L^{-1} solution of copper(II) methacrylate hydrate, adjusting the pH until 8.1 ± 0.1. After dispersing, filtering, and drying at room temperature, the solid was used to prepare PMMA nanocomposites using two different procedures. In the first, the solid was dispersed in an acetone PMMA solution and the samples were obtained by wet casting. In the second procedure, bulk polymerization was used, where the monomeric methyl methacrylate was mixed with the initiator (benzoyl peroxide—BPO) and the filler. The mixture was thermally treated until total polymerization.

The conclusions of this study can be summarized as follows: The copper hydroxide methacrylate was successfully synthesized and used as a filler in PMMA; the composites were obtained by two alternative procedures and the filler provided thermal stability to PMMA; the fillers prepared by both methods and used in the proportion of 3%–4% by mass resulted in a significant shift in the temperature for 50% mass loss in the TGA curves; in contrast to bulk polymerized samples, no significant difference was observed in the thermal heat release in cone calorimetry measurements for the composites prepared by wet casting (Kandare et al. 2006b).

REFERENCES

Aguirre, JM, A Gutiérrez, and O Giraldo. 2011. Simple route for the synthesis of copper hydroxy salts. *Journal of the Brazilian Chemical Society* 22 (3) (March): 546–551.
Amarnath, CA, CE Hong, NH Kim, B-C Ku, and JH Lee. 2011. Aniline- and *N*, *N*-dimethylformamide-assisted processing route for graphite nanoplates: Intercalation and exfoliation pathway. *Materials Letters* 65 (9) (May): 1371–1374.
Arizaga, GGC. 2008. Modificação química de superfícies de hidroxinitrato de zinco e hidróxidos duplos lamelares com ácidos mono e dicarboxílicos. Universidade Federal do Parana. Ph.D. thesis. Parana, Brasil.

Arízaga, GGC. 2012. Intercalation studies of zinc hydroxide chloride: Ammonia and amino acids. *Journal of Solid State Chemistry* 185 (January): 150–155.

Arizaga, GGC, AS Mangrich, JEF da Costa Gardolinski, and F Wypych. 2008a. Chemical modification of zinc hydroxide nitrate and Zn-Al-layered double hydroxide with dicarboxylic acids. *Journal of Colloid and Interface Science* 320 (1) (April 1): 168–176.

Arizaga, GGC, AS Mangrich, and F Wypych. 2008b. Cu^{2+} ions as a paramagnetic probe to study the surface chemical modification process of layered double hydroxides and hydroxide salts with nitrate and carboxylate anions. *Journal of Colloid and Interface Science* 320 (1) (April 1): 238–244.

Arizaga, GGC, K Satyanarayana, and F Wypych. 2007. Layered hydroxide salts: Synthesis, properties and potential applications. *Solid State Ionics* 178 (15–18) (June): 1143–1162.

Bénard, P, JP Auffrédic, and D Louër. 2013. An X-ray powder diffraction study of amine zinc hydroxide nitrate. *Powder Diffraction* 10 (01) (January 10): 20–24.

Bera, P, M Rajamathi, MS Hegde, and P Vishnu Kamath. 2000. Thermal behaviour of hydroxides, hydroxysalts and hydrotalcites. *Bulletin of Materials Science* 23 (2) (April): 141–145.

Biswick, T, W Jones, A Pacula, and E Serwicka. 2005. Synthesis, characterisation and anion exchange properties of copper, magnesium, zinc and nickel hydroxy nitrates. *Journal of Solid State Chemistry* 178 (3): 3810–3816.

Bravo-Suárez, JJ, EA Páez-mozo, and S Ted Oyama. 2004. Review of the synthesis of layered double hydroxides: A thermodynamic approach. *Quimica Nova* 27 (4): 601–614.

Carbajal, A and G Guadalupe. 2008. Modificação Química de Superfícies de Hidroxinitrato de Zinco E Hidróxidos Duplos Lamelares Com Ácidos Mono E Dicarboxílicos. Universidade Federal do Paraná, Brazil.

Carlino, S. 1997. The intercalation of carboxylic acids into layered double hydroxides: A critical evaluation and review of the different methods. *Solid State Ionics* 98: 73–84.

Chouillet, C, J-M Krafft, C Louis, and H Lauron-Pernot. 2004. Characterization of zinc hydroxynitrates by diffuse reflectance infrared spectroscopy—Structural modifications during thermal treatment. *Spectrochimica Acta Part A* 60 (3): 505–511.

Choy, J-H, Y-M Kwon, K-S Han, S-W Song, and SH Chang. 1998. Intra- and inter- layer structures of layered hydroxy double salts, $Ni_{1-x}Zn_{2x}(OH)_2(CH_3CO_2)_{2x} \cdot nH_2O$. *Materials Letters* 34: 356–363.

Cursino, ACT, AS Mangrich, JEF Da Costa Gardolinski, N Mattoso, and F Wypych. 2011. Effect of confinement of anionic organic ultraviolet ray absorbers into two-dimensional zinc hydroxide nitrate galleries. *Journal of the Brazilian Chemical Society* 22 (6) (June): 1183–1191.

da Silva, MLN, R Marangoni, ACT Cursino, WH Schreiner, and F Wypych. 2012. Colorful and transparent poly(vinyl alcohol) composite films filled with layered zinc hydroxide salts, intercalated with anionic orange azo dyes (methyl orange and orange II). *Materials Chemistry and Physics* 134 (1) (May): 392–398.

da Silva, V, MY Kamogawa, R Marangoni, AS Mangrich, and F Wypych. 2014. Hidróxidos duplos lamelares como matrizes para fertilizantes de liberação lenta de nitrato. *Revista Brasileira de Ciencias Do Solo* 38: 272–277.

Del Hoyo, C. 2007. Layered double hydroxides and human health: An overview. *Applied Clay Science* 36 (1–3) (April): 103–121.

Demel, J, J Pleštil, P Bezdička, P Janda, M Klementová, and K Lang. 2011. Layered zinc hydroxide salts: Delamination, preferred orientation of hydroxide lamellae, and formation of ZnO nanodiscs. *Journal of Colloid and Interface Science* 360 (2) (August 15): 532–539.

Elguera Hernández, MI. 2013. Síntesis Y Caracterización de Complejos de Zn Y Su Aplicación Como Catalizadores Heterogéneos En La Producción de Alquilésteres. Universidad Popular de la Chontalpa. Bachelor of Engineering Thesis. Tabasco, Mexico.

Fahlman, BD. 2008. *Materials Chemistry*. Dordrecht, the Netherlands: Springer.

Feitknecht, W. 1945. Die Struktur Der Cadmiumhydroxyhalogenide $CdCl_{0,67}(OH)_{1,33}$, $CdBr_{0,6}(OH)_{1,4}$, $CdJ_{0,5}(OH)_{1,5}$. *Experientia* 1 (7) (October): 230–231.

Feitknecht, W. and H. Bürki. 1949. Basische Salze organischer Säuren mit Schichtenstruktur. *Experientia* 5 (4) (April): 154–155.

Frost, RL, W Martens, JT Kloprogge, and PA Williams. 2002. Raman spectroscopy of the basic copper chloride minerals atacamite and paratacamite: Implications for the study of copper, brass and bronze objects of archaeological significance. *Journal of Raman Spectroscopy* 33 (10) (October): 801–806.

Fujita, W, K Awaga, and T Yokoyama. 1999. Controllable magnetic properties of layered copper hydroxides, $Cu_2(OH)_3X$ (X = carboxylates). *Applied Clay Science* 15: 281–303.

Guimarães, JL, R Marangoni, LP Ramos, and F Wypych. 2000. Covalent grafting of ethylene glycol into the Zn-Al-CO(3) layered double hydroxide. *Journal of Colloid and Interface Science* 227 (2) (July 15): 445–451.

Halajnia, A, S Oustan, N Naja, AR Khataee, and A Lakzian. 2013. Adsorption–desorption characteristics of nitrate, phosphate and sulfate on Mg–Al layered double hydroxide. *Applied Clay Science* 81: 305–312.

Hennous, M, Z Derriche, E Privas, P Navard, V Verney, and F Leroux. 2013. Lignosulfonate interleaved layered double hydroxide: A novel green organoclay for bio-related polymer. *Applied Clay Science* 71 (January): 42–48.

Henrist, C, K Traina, C Hubert, G Toussaint, A Rulmont, and R Cloots. 2003. Study of the morphology of copper hydroxynitrate nanoplatelets obtained by controlled double jet precipitation and urea hydrolysis. *Journal of Crystal Growth* 254 (1–2) (June): 176–187.

Hindocha, SA, LJ Mcintyre, and MF Andrew. 2009. Precipitation synthesis of lanthanide hydroxynitrate anion exchange materials, $Ln_2(OH)_5NO_3 \cdot H_2O$ (Ln = Y, Eu–Er). *Journal of Solid State Chemistry* 182 (3): 1070–1074.

Hsieh, Z-L, M-C Lin, and J-Y Uan. 2011. Rapid direct growth of Li–Al layered double hydroxide (LDH) film on glass, silicon wafer and carbon cloth and characterization of LDH film on substrates. *Journal of Materials Chemistry* 21 (6): 1880.

Jaerger, S, A Zimmermann, SF Zawadzki, SC Amico, and F Wypych. 2014. Zinc layered hydroxide salts: Intercalation and incorporation into low-density polyethylene. *Polímeros* 24 (6): 673–682.

Jamhour, RMAQ. 2005. Preparation and characterization of hybrid organic-inorganic composite material: Polymerization of M-aminobenzoic acid-intercalated into Zn/Al-layered double hydroxides. *American Journal of Applied Sciences* 2 (6): 1028–1031.

Jiang, J, Y Zhang, Y Zheng, and P Jiang. 2013. Transesterification of soybean oil with ethylene glycol, catalyzed by modified Li–Al layered double hydroxides. *Chemical Engineering & Technology* 36 (8) (August 24): 1371–1377.

Kandare, E, G Chigwada, D Wang, CA Wilkie, and JM Hossenlopp. 2006a. Nanostructured layered copper hydroxy dodecyl sulfate: A potential fire retardant for poly(vinyl ester) (PVE). *Polymer Degradation and Stability* 91 (8) (August): 1781–1790.

Kandare, E, H Deng, D Wang, and JM Hossenlopp. 2006b. Thermal stability and degradation kinetics of poly(methyl methacrylate)/layered copper hydroxy methacrylate composites. *Polymers for Advanced Technologies* 17 (4) (April): 312–319.

Kandare, E and JM Hossenlopp. 2009. Effects of hydroxy double salts and related nanodimensional-layered metal hydroxides on polymer thermal stability. In *Polymer Degradation and Performance*, MC Celina, JS Wiggins, and NC Billingham (eds.), Vol. 1004. ACS Symposium Series. Washington, DC: American Chemical Society.

Kim, TW, IY Kim, JH Im, H-W Ha, and S-J Hwang. 2009. Improved photocatalytic activity and adsorption ability of mesoporous potassium-intercalated layered titanate. *Journal of Photochemistry and Photobiology A: Chemistry* 205 (2–3) (June): 173–178.

Kotal, M and SK Srivastava. 2011. Synergistic effect of organomodification and isocyanate grafting of layered double hydroxide in reinforcing properties of polyurethane nanocomposites. *Journal of Materials Chemistry* 21 (46): 18540.

Liu, J, G Chen, J Yang, and L Ding. 2009. Improved thermal stability of poly(vinyl chloride) by nanoscale layered double hydroxide particles grafted with toluene-2,4-di-isocyanate. *Materials Chemistry and Physics* 118 (2–3) (December): 405–409.

Lopez, T, E Ramos, P Bosch, M Asomoza, and R Gomez. 1997. DTA and TGA characterization of sol-gel hydrotalcites. *Materials Letters* 30: 279–282.

Louër, M, D Louër, and D Grandjean. 1973. Etude structurale des hydroxynitrates de nickel et de zinc. I. Classification structurale. *Acta Crystallographica* 2 (1970): 1696–1703.

Lu, P. 2010. The polymerization of unsaturated polyester and silane-functionalized layered double hydroxides. *Polymer-Plastics Technology and Engineering* 49 (14) (November 23): 1450–1457.

Machado, J, N Ravishankar, and M Rajamathi. 2010. Delamination and solvothermal decomposition of layered zinc hydroxysalt: Formation of bimodal zinc oxide nanostructures. *Solid State Sciences* 12 (8) (August): 1399–1403.

Majoni, S, S Su, and JM Hossenlopp. 2010. The effect of boron-containing layered hydroxy salt (LHS) on the thermal stability and degradation kinetics of poly (methyl methacrylate). *Polymer Degradation and Stability* 95 (9) (September): 1593–1604.

Marangoni, R, M Bouhent, C Taviot-Guého, F Wypych, and F Leroux. 2009a. Zn_2Al layered double hydroxides intercalated and adsorbed with anionic blue dyes: A physico-chemical characterization. *Journal of Colloid and Interface Science* 333 (1) (May 1): 120–127.

Marangoni, R, GA Bubniak, MP Cantão, M Abbate, WH Schreiner, and F Wypych. 2001. Modification of the interlayer surface of layered copper(II) hydroxide acetate with benzoate groups: Submicrometer fiber generation. *Journal of Colloid and Interface Science* 240 (1) (August 1): 245–251.

Marangoni, R, LP Ramos, and F Wypych. 2009b. New multifunctional materials obtained by the intercalation of anionic dyes into layered zinc hydroxide nitrate followed by dispersion into poly(vinyl alcohol) (PVA). *Journal of Colloid and Interface Science* 330 (2) (February 15): 303–309.

Marangoni, R, C Taviot-Guého, A Illaik, F Wypych, and F Leroux. 2008. Organic inorganic dye filler for polymer: Blue-coloured layered double hydroxides into polystyrene. *Journal of Colloid and Interface Science* 326 (2) (October 15): 366–373.

Meyn, M, K Beneke, and G Lagaly. 1990. Anion-exchange reactions of layered double hydroxides. *Inorganic Chemistry* 29 (19): 5201–5207.

Meyn, M, K Beneke, and G Lagaly. 1993. Anion-exchange reactions of hydroxy double salts. *Inorganic Chemistry* 32: 1209–1215.

Miao, J, M Xue, H Itoh, and Q Feng. 2006. Hydrothermal synthesis of layered hydroxide zinc benzoate compounds and their exfoliation reactions. *Journal of Materials Chemistry* 16 (5): 474.

Moraes, SB, R Botan, and LMF Lona. 2014. Synthesis and characterization of polystyrene/layered hydroxide salt nanocomposites. *Quimica Nova* 37: 18–21.

Naik, VV, R Chalasani, and S Vasudevan. 2011. Composition driven monolayer to bilayer transformation in a surfactant intercalated Mg-Al layered double hydroxide. *Langmuir: The ACS Journal of Surfaces and Colloids* 27 (6) (March 15): 2308–2316.

Nethravathi, C, J Machado, UK Gautam, GS Avadhani, and M Rajamathi. 2012. Exfoliation of copper hydroxysalt in water and the conversion of the exfoliated layers to cupric and cuprous oxide nanoparticles. *Nanoscale* 4 (2) (January 21): 496–501.

Nethravathi, C, JT Rajamathi, P George, and M Rajamathi. 2011. Synthesis and anion-exchange reactions of a new anionic clay, α-magnesium hydroxide. *Journal of Colloid and Interface Science* 354 (2) (February 15): 793–797.

O'Hare, D. 1997. Inorganic intercalation compounds. In *Inorganic Materials*, DW Bruce and D O'Hare (eds.), Vol. II, pp. 171–254. England: John Wiley & Sons.

Oosawa Y, Y Gotoh, J Akimoto, M Sohma, T Tsunoda, H Hayakawa, and M Onoda. 1997. Preparation, characterization and intercalation of ternary chalcogenides with layered composite crystal structures formed in the Bi Ta S and Bi Ta Se systems. *Solid State Ionics* 101–103 (November): 9–16.

Otero, R, JM Fernández, MA Ulibarri, R Celis, and F Bruna. 2012. Adsorption of non-ionic pesticide S-Metolachlor on layered double hydroxides intercalated with dodecylsulfate and tetradecanedioate anions. *Applied Clay Science* 65–66 (September): 72–79.

Poul, L, N Jouini, and F Fievet. 2000. Layered hydroxide metal acetates (metal = zinc, cobalt, and nickel): Elaboration via hydrolysis in polyol medium and comparative study. *Chemisty of Materials* 12: 3123–3132.

Rajamathi, JT, N Ravishankar, and M Rajamathi. 2005. Delamination–restacking behaviour of surfactant intercalated layered hydroxy double salts, $M_3Zn_2(OH)_8(surf)_2\cdot2H_2O$ [M= Ni, Co and surf = dodecyl sulphate (DS), dodecyl benzene sulphonate (DBS)]. *Solid State Sciences* 7: 195–199.

Rajamathi, M, GS Thomas, and PV Kamath. 2001. Many ways of making anionic clays. *Proceedings of the Indian Academy of Sciences* 133 (5–6): 617–680.

Rajamathi, M and P Vishnu Kamath. 1998. On the relationship between α-nickel hydroxide and the basic salts of nickel. *Journal of Power Sources* 70 (1) (January): 118–121.

Rajamathi, M and P Vishnu Kamath. 2001. Urea hydrolysis of cobalt(II) nitrate melts: Synthesis of novel hydroxides and hydroxynitrates. *International Journal of Inorganic Materials* 3 (7) (November): 901–906.

Rives, V and MA Ulibarri. 1999. Layered double hydroxides (LDH) intercalated with metal coordination compounds and oxometalates. *Coordination Chemistry Reviews* 181: 61–120.

Saber, O, HM Gobara, and AA Al Jaafari. 2011. Catalytic activity and surface characteristics of layered Zn–Al–Si materials supported platinum. *Applied Clay Science* 53 (2) (August): 317–325.

Saber, O, B Hatano, and H Tagaya. 2005. Controlling of the morphology of Co–Ti LDH. *Materials Science and Engineering: C* 25 (4) (June): 462–471.

Semsarilar, M and S Perrier. 2010. "Green" reversible addition-fragmentation chain-transfer (RAFT) polymerization. *Nature Chemistry* 2 (10) (October): 811–820.

Si, L, G Wang, F Cai, Z Wang, and X Duan. 2004. Polymerization reaction in restricted space of layered double hydroxides (LDHs). *Chinese Science Bulletin* 49 (23) (December): 2459–2463.

Stählin, W and HR Oswald. 1970. The crystal structure of zinc hydroxide nitrate, $Zn_5(OH)_8(NO_3)_2\cdot2H_2O$. *Acta Crystallographica Section B Structural Crystallography and Crystal Chemistry* 26 (6) (June 15): 860–863.

Swanson, CH, T Stimpfling, A-L Troutier-Thulliez, H Hintze-Bruening, and F Leroux. 2013. Layered double hydroxide platelets exfoliation into a water-based polyester. *Journal of Applied Polymer Science* 128 (5) (June 5): 2954–2960.

Tang, H, X-B Zhou, and X-L Liu. 2013. Effect of magnesium hydroxide on the flame retardant properties of unsaturated polyester resin. *Procedia Engineering* 52: 336–341.

Tao, Q, H He, T Li, RL Frost, D Zhang, and Z He. 2014. Tailoring surface properties and structure of layered double hydroxides using silanes with different number of functional groups. *Journal of Solid State Chemistry* 213 (May): 176–181.

Tao, Q, J Yuan, RL Frost, H He, P Yuan, and J Zhu. 2009. Effect of surfactant concentration on the stacking modes of organo-silylated layered double hydroxides. *Applied Clay Science* 45 (4) (August): 262–269.

Tronto, J, F Leroux, M Dubois, C Taviot-Gueho, and JB Valim. 2006. Hybrid organic–inorganic materials: Layered hydroxy double salts intercalated with substituted thiophene monomers. *Journal of Physics and Chemistry of Solids* 67 (5–6) (May): 978–982.

Tsujii, Y, M Ejaz, K Sato, A Goto, and T Fukuda. 2001. Mechanism and kinetics of RAFT-mediated graft polymerization of styrene on a solid surface. 1. Experimental evidence of surface radical migration. *Macromolecules* 34 (26) (December): 8872–8878.

Whitten, KW, RE Davis, M Larry Peck, and GG Stanley. 2011. *Química*. Cengage Learning, Mexico City. http://books.google.com.mx/books/about/Química.html?id=Chw7tP3s7Z8C&redir_esc=y. Last accessed on April 13, 2016.

Wypych, F and GGC Arizaga. 2005. Intercalation and functionalization of brucite with carboxylic acids. *Quimica Nova* 28 (1) (March 13): 24–29.

Wypych, F, GGC Arízaga, and JEF da Costa Gardolinski. 2005. Intercalation and functionalization of zinc hydroxide nitrate with mono- and dicarboxylic acids. *Journal of Colloid and Interface Science* 283 (1) (March 1): 130–138.

Wypych, F and KG Satyanarayana. 2005. Functionalization of single layers and nanofibers: A new strategy to produce polymer nanocomposites with optimized properties. *Journal of Colloid and Interface Science* 285 (2) (May 15): 532–543.

Zhao, Y, W Yang, Y Xue, X Wang, and T Lin. 2011. Partial exfoliation of layered double hydroxides in DMSO: A route to transparent polymer nanocomposites. *Journal of Materials Chemistry* 21 (13): 4869.

Zimmermann, A, S Jaerger, SF Zawadzki, and F Wypych. 2013. Synthetic zinc layered hydroxide salts intercalated with anionic azo dyes as fillers into high-density polyethylene composites: First insights. *Journal of Polymer Research* 20 (9) (August 10): 224.

CHAPTER 2

Hybrid Filler Polymer Nanocomposites

Robert A. Shanks

CONTENTS

2.1 INTRODUCTION

Composites are materials of choice for most structural polymer compositions. In nature, composites abound in most structural biomaterials. A composite consists of at least two materials that retain their identity and properties, and are physically separable from the composite. A homogeneous mixture of two materials in contrast gives an averaging of morphologies and properties. Most composites have a disperse phase and a continuous or matrix phase, though cocontinuous composites can be formed. Typically, the disperse phase contributes the enhancement of properties required, which are often mechanical, including modulus, strength, and toughness. The matrix phase is the binder that converts the dispersed phase into a useful material through shape, space filling, and stress transfer [1]. The matrix phase of a polymer composite is either a thermoset or a thermoplastic with many variants and formulation enhancement available.

This chapter reviews composites in which there are a matrix polymer and two dispersed phases, typically with micrometer and nanometer dimensions, respectively. In this review, they are termed hybrid composites because they combine typical composites with nanocomposites in the same material. The review classifies the filler combinations according to their chemistry, shape, and morphology in the composites. The chapter is not concerned with fillers having broad particle size distribution extending over the micrometer–nanometer range. Hybrid systems will have two distinct filler size distributions. Figure 2.1 shows a generalized schematic of a hybrid polymer composite consisting of two fillers; it should be noted that the fillers differ in size by a factor of ~1000 so that each filler will contribute different characteristics to the composite.

The disperse phase has typically had one or more dimensions in the micrometer scale. Micrometer-sized particles and platelets of micrometer thickness are refined from minerals or synthetic sources. Fibers with micrometer diameter are most common because of their high aspect ratio that increased mechanical performance. After choice of the dispersed phase and matrix phase, the technology for composite preparation consists of dispersion of the disperse phase, enhancement of the adhesion between phases, and control of order of the disperse phase such as orientation. Design, manufacture, and fabrication of polymer composites has become a mature technology that is integrated into a vast range of materials [2]. Fibers in this category can be woven, felted, oriented uniaxially and biaxially, Z-pinned

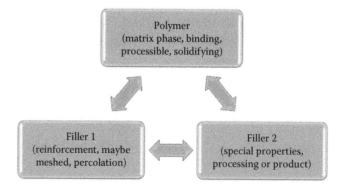

FIGURE 2.1   Schematic of a hybrid polymer composite consisting of two fillers.

or chopped, and randomized. A long fiber composite may have fibers extending throughout the composite product. Long fibers rely less on transfer of stress from matrix to fiber because the fiber continues throughout the stress field.

Composites with fillers in the nanometer range have been manufactured over a long time. A widespread example is carbon black-filled elastomer composition used in automotive tires. Nanofillers such as clays and silica are used as rheological agents in paint, adhesives, and colloidal products. The small nanometer dimension translates into a huge surface area, and the surface area increases the interface and hence binding between the dispersed and matrix phases [3]. Rheological agents often involve interparticle attractions resulting in reversible flocculation in which particles are separated during flow and then flocculated when in a quiescent state. The balance between separation and flocculation is critical as is the time dependence. Dispersion of nanoparticles is critical and difficult due to high surface area giving agglomerates that must be disrupted by shear for dispersion to occur. Some nanoparticles such as nanosilica are formed with particle aggregates that cannot be disrupted. The aggregates provide complex chain and branched particle shapes that contribute to rheology. Dispersions of these particles display shear thinning and thixotropic behaviors.

A hybrid is an entity composed of incongruous elements, such as the offspring of two plants or animals of different species or varieties, a person of mixed cultural origin, or a word with parts taken from different languages. In a chemical context, hybridization is the combination of two different atomic orbitals to give a molecular orbital, such as an s and three p orbitals to form an sp^3 orbital. In the context of this review, a hybrid polymer nanocomposite is a material that has a continuous polymer phase and a disperse phase consisting of two different materials, at least one of which has a dimension in the nanoscale. The two different dispersed phases may be both of nanoscale, a nano- and a microdimension material, an inorganic and an organic material or a nanoscale material, and a dispersed polymer blend phase.

Further distinctions between the nanodispersed phase and the other dispersed phase can be dimensionality, one-dimensional fiber or crystal, two-dimensional platelets or exfoliated layers, or a three-dimensional particulate material. The one- and two-dimensional materials will be further characterized by their aspect ratio. A hybrid composite with nanometer and micrometer filler components is a bimodal molar mass distribution polymer in which unique properties are contributed by the polymer. A single polymer with a broad molar mass distribution is not equivalent, though a mixture of low and high molar mass polymer is not the same as a reactor polymerized bimodal polymer. Perhaps bimodal fillers will be synthesized with dual peaks in a particle size distribution ranging from the nano- to micro-dimension range.

A composite is distinguished by consisting of two or more discrete phases in which each phase retains most of the properties of the isolated material. The composite properties will be a combination of the properties of its components, not an average of the components. A hybrid nanocomposite will consist of the combined individual properties of the continuous polymer, the nanoscale components, and at least one other component.

Composites are known to consist of a third phase called the interface. The interface is different from the continuous phase in that it consists of polymer molecules bound to a disperse phase and because of loss of molecular mobility exhibiting properties modified from the bulk continuous phase. In a nanocomposite, the huge surface area of the nano-dispersed phase will immobilize more of the continuous phase and therefore will have an intense contribution to properties at relatively low concentration. The interface in a hybrid nanocomposite will be more extensive and complex due to interactions between a nanoscale material and another different material. The nanoscale material and the other disperse phase may mutually interact.

Nanomaterials often consist of aggregates and agglomerates such that even when dispersed under high shear they exist in multiparticle aggregates. After dispersion, flocculation may reconstitute agglomerates and the agglomerates can exist reversibly under conditions of fluctuating strain, temperature, moisture absorption, and polymer matrix phase relaxation. Aggregates of nanosilica impart shear thinning and thixotropic behavior to a greater extent than if the nano-silica were dispersed as single particles. The aggregates contain about 20 primary particles in branched short chains. They flocculate readily giving rise to the rheological characteristics. The aggregates entrap more polymer than would be immobilized on the surface of individual particles. When agglomerates form, further polymer is immobilized within creating a pseudo increase in filler volume fraction. The shear and time dependence of agglomeration causes such nanocomposites to display nonlinear viscoelastic response with stress–strain hysteresis.

The aim of this chapter is to define, classify, and review hybrid polymer nanocomposite preparation, structure, morphology, properties, and consequential applications. An objective will be to consider the choice of materials, preliminary treatments, processing techniques, and conditions. A consequence of the complex material combinations is that rheology will become a significant determinant of suitability in choice and constraints of composition. Typical characterization techniques are utilized for analysis, and characterization and assessment of these will be a further review objective. Another objective is property determination and evaluation with emphasis on tangible or potential applications.

2.2 ORGANIC–INORGANIC FILLER REINFORCEMENT

Fillers depend on a group of physical characteristics for their properties and action in polymer composites as summarized in Figure 2.2. Dimension size is important for this review of hybrid composites. Sections 2.2.1 through 2.2.3 are classified into size and then shape and general function is described along with examples.

2.2.1 Micrometer-Dimensioned Fillers

2.2.1.1 Particulate Composites

These fillers have a three-dimensional shape according to the crystal morphology of the substance. The shape will mainly be isotropic, though some hexagonal crystalline forms are elongated, but not having an aspect ratio sufficient to be classified as a fiber. Interaction between the polymer and the filler, enhanced by surface area of the filler, produces a polymer bound to the filler layer that is distinct from the bulk polymer. This bound layer may have different glass transition temperature and is referred to as an interphase. In the case of amorphous polymers, the interphase is called a rigid amorphous phase to distinguish it from the bulk amorphous phase that exhibits properties typical of the polymer. A secondary

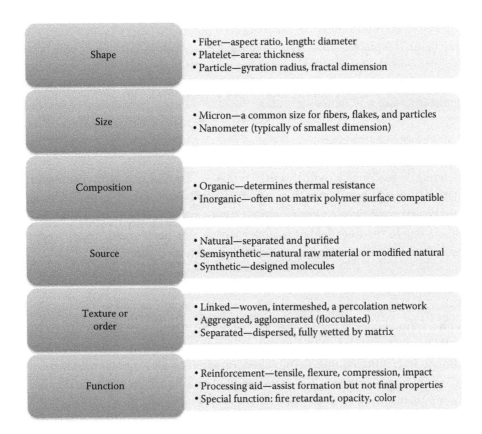

Shape	• Fiber—aspect ratio, length: diameter • Platelet—area: thickness • Particle—gyration radius, fractal dimension
Size	• Micron—a common size for fibers, flakes, and particles • Nanometer (typically of smallest dimension)
Composition	• Organic—determines thermal resistance • Inorganic—often not matrix polymer surface compatible
Source	• Natural—separated and purified • Semisynthetic—natural raw material or modified natural • Synthetic—designed molecules
Texture or order	• Linked—woven, intermeshed, a percolation network • Aggregated, agglomerated (flocculated) • Separated—dispersed, fully wetted by matrix
Function	• Reinforcement—tensile, flexure, compression, impact • Processing aid—assist formation but not final properties • Special function: fire retardant, opacity, color

FIGURE 2.2   Physical classification of fillers and their attributes and function.

effect of filler-bound polymer is nucleation of crystallization. As this complex polymer–filler morphology is portrayed, it becomes apparent that a two-component polymer–filler system is much more complex that a dispersed mixture of two components each contributes their characteristic properties to the composite.

Carbon in various shapes, such as particulate carbon blacks and fullerenes, platelet graphene, carbon fibers, nanofibers, and nanotubes, has been widely included in polymer composites. These carbon-based materials have been modified to increase adhesion to a polymer matrix and for ease of dispersion. Carbon materials enhance mechanical, viscoelastic, thermal, electrical, and barrier properties of the polymers in which they are dispersed [4].

2.2.1.2 Platelet Composites

The classification of platelet fillers is that they have a two-dimensional shape in which the thickness is on the order of 1 μm with the other two dimensions being larger. They will form layered composites with the filler oriented by processing shear fields. They will exhibit nonisotropic properties with lateral properties differing from those across the layer thickness. If the filler is isotropically oriented, then the volume fraction of filler and hence the bound polymer fraction will be limited by packing constraints. Isotropic platelet fillers adopt smetite or house-of-cards morphology due to their rigid layer structure. Such layers are better accommodated by stacking in a polymer matrix phase after orientation in a shear field. The platelets may agglomerate due to mutual attraction, which is greater than attraction to the polymer. Agglomeration is not desired, so surface treatments for compatabilization with the polymer are usually adopted to maintain dispersion and enhance the interface and hence stress transfer.

2.2.1.3 Short Fiber Composites

Fibers are of one-dimensional shape, with the cross section or diameter being in the micrometer scale for this classification. They have a high aspect ratio, even short fibers. A distinction of short fibers is that they have typically been cut from long fibers so they can be molded by extrusion or injection moulding. These molding techniques confer orientation on the fibers due to the flow field of polymer within the die or mold. Thus, properties are enhanced in the flow directions compared with the lateral directions.

2.2.1.4 Long Fiber Composites

Long fibers are used to construct composites with orientation in which the polymer is combined with a formed fiber array. In one dimension, pulltrusion is used to draw a bundle of fibers through a resin that is solidified by cooling or by curing. In two dimensions, the fibers can be simply intermingled, needle punched, or felted to give a nonwoven mat or woven in one of many weave patterns. The resin is added as a liquid prepolymer that is cured to form the composite. In three dimensions, there will be fiber stitching across a nonwoven or woven fiber mat. Usually, there will be multiple layers of mat that can be stitched together to increase lateral strength; the process is often called Z-stitching.

2.2.1.5 Laminated Textile Composites

Several layers of nonwoven or woven mat are assembled with alternate application of prepolymer resin to give a laminated composite. The resin is cured to give a multilayered structure. The interlayer bounding can be increased by Z-stitching or Z-pinning as described previously. Laminated structures are used to form thicker composites that cannot be readily formed in one step due to problems with resin impregnation and entrapped air. The layer-by-layer construction ensures that each layer can be fully impregnated with resin and that a high fiber-to-resin volume fraction can be obtained. Most composites are prepared as laminates because separate layers can be included to provide special properties. For instance, the surface layer or gel coat will contain a pigment for appearance or decoration, and often a different resin that is resistant to the environment. A sub-gelcoat and substances, such as moisture into the structural layers beneath that are designed to provide the strength and constitute the bulk of a laminate.

2.2.2 Nanometer-Dimensioned Fillers

2.2.2.1 Particulate Composites

These composites contain fillers with all dimensions in the nanometer range. Small volume fractions of fillers produce large changes in properties due to the high filler surface area to adsorb and immobilize much of the polymer. Nanometer fillers tend to agglomerate and aggregate because their high surface area can prefer self-interactions. High shear is required to disrupt agglomerates and prevent subsequent flocculation. Compatibilizers assist in dispersion by mediating between the surface properties of polymer and filler. Many nanofillers such as fumed silica contain aggregates that cannot be separated, so multiparticle clusters exist within most nanoparticle composites. Complete dispersion in an ideal situation would give a composite consisting of primary particles; however, this may not be a preferred situation because the aggregates contribute to both rheological and mechanical properties more effectively than separated or isolated primary particles within a continuous-phase polymer. Filler–filler interactions increase the apparent aspect ratio and entrap polymers within their clusters, giving unique viscoelastic properties. Fumed silica are widely used nanoparticles, although carbon black, alumina, and titanium dioxide are available. Other nanoparticles such as calcium carbonate and magnetite can be readily synthesized.

A unique nanoparticle filler is polyhedral oligomeric silsesquioxanes (POSSs) in which the filler particle is a discrete molecule. POSSs can be functionalized with groups to graft to a polymer or become copolymerized in a polymer chain. POSS composites with poly(styrene-*b*-butadiene-*b*-styrene) (SBS) have been formed by solution dispersion of dumbbell-shaped POSS formed by linking POSS via a difunctional carboxylic acid [5]. Reactive organic dyes were used to functionalize POSS, and the POSS–dye was used as a colored filler for SBS. The structure of the dye gave selectivity for either the polystyrene phase or the polybutadiene phase of the SBS [6].

2.2.2.2 Platelet Composites

These fillers have one dimension in the nanometer range, whereas the other dimensions can be on the order of micrometers. They have large surface area but also large lateral dimensions, so that small volume fractions (0%–5%·v/v) cause considerable change to properties. The first step in using platelet fillers is to separate the platelet layers because bonding is weak between layers they are strongly attracted. A chemical intercalation is usual to disrupt interlayer attraction. Then, exfoliation using combinations of shear, solvent attraction and envelopment by polymer results in single or few layers dispersed in solution or polymer. Much research and new products have utilized layered clays. Graphite has been used after oxidative intercalation followed by rapid heating to high temperature for exfoliation. The graphene layers formed in oxidized form with many edge attachments. High shear can create individual layer dispersions, while reduction can return graphene oxide to graphene. Graphene oxide has epoxy, hydroxyl, carbonyl, and carboxylic acid groups, some in the graphene plane with others such as carboxylic acid groups along the edges. These oxygen-containing groups provide sites for derivatization using many organic functional group reactions. Graphene derivatives have been used to prepare sensors in which selective attractions give specificity and high sensitivity.

A mixture of graphene and carbon nanotubes (CNTs) was combined with poly(vinylidene difluoride) to give enhanced conductivity compared with the individual components. Graphene has high conductivity, whereas the conductive CNTs provided a percolation network through which to conduct current [7]. Interaction between graphene and CNTs was found to enhance their dispersion in silicone rubber. The fillers have mutual interaction; however, they each have different interaction with the silicon matrix that facilitated dispersion [8].

Graphene layers have been separated by several methods to obtain individual or few layers where the ideal platelets are less than 1 nm thin, and they posses extraordinary modulus and strength. The platelets are characterized by wide-angle X-ray scattering to determine the proportion of multilayers and Raman spectroscopy to detect sp^2-bonded carbons of aromatic multiring layers and sp^3 carbons of nonaromatic or defective species. Graphene nanocomposites confer greatly enhanced properties on matrix polymers even at low fractions. A challenge is to uniformly disperse higher fractions of graphene without agglomeration [9].

2.2.2.3 Fiber Composites

Nanofibers are fillers with two dimensions in the nanometer range and a high aspect ratio. They have high surface area due to their nanodimensions, yet the fibers can be several micrometers or even millimeters long. Dispersion is difficult due to strong fiber–fiber interactions and entanglements. In common with other nanofillers, relatively low volume fractions produce large property changes. A currently much explored nanofiber is CNT in single or multiwalled variants. CNTs can be functionalized by reaction at cylinder edges or on surfaces. Similar to most aromatic forms of carbon, they are strongly adsorbing either matrix polymer or added functionalizing substances. Cellulose nanofibers and nanocrystals are derived from natural resources by selective partial hydrolysis. They are highly polar and strongly adsorbing polar substances, including water, although they can be functionalized. These cellulose materials can confer thixotropy and reinforcement when present in small volume fractions.

2.2.2.4 Long Fiber Composites

Electrospun intermingled fiber nonwoven mats are analogous to the micrometer needle-punched nonwoven mats. They can be produced with current production equipment and find application in mat form as filters or membranes, as well as reinforcement for laminated polymer composites. Typical long fibers used in composites are glass, carbon, cellulose, and synthetic fibers such as polyester or polyamide.

2.2.3 Hybrid Composite Products

There are many examples of composites with more than one filler in one material. Glass fiber composites often contain particulate fillers such as calcium carbonate, silica, or various pigments with titanium dioxide being the most common. This review is limited to composites with two fillers, one in the nanometer and the other in the micrometer dimension range. They may overlap in dimension scale because the nanofiller needs to only have one dimension in the nanometer range. Another distinction is surface area in which a small fumed nanosilica could have a surface area of 380 $m^2 \cdot g$. There will be an overlap with surface area because a micrometer-sized particle may have smaller surface area than a long nanofiber. However, in general, most nanofillers will have the highest surface areas for creating a large filler–polymer interface, which is the main functional distinction of nanofillers.

There are many hybrid composites that have been used or that are part traditional technologies. Nanofiller as a rheological agent for spray application in—unsaturated polyester, vinyl ester, epoxy, with glass or carbon fiber reinforcement. Any contribution of the nanofiller to mechanical properties is incidental to the intended rheological function.

Exfoliated layered clay is added to water-based polymer emulsions for rheological control, with micrometer-sized or pigment. Micrometer-sized particles give optimum light scattering for opacity and color intensity. Nanosized particles provide best rheological control. In the final product, both particle sizes are intermingled within the polymer to increase binding and abrasion resistance.

Each pair of nanometer and micrometer filler types as classified earlier can be combined to form hybrid nanocomposites; however, the advantage can best be obtained from certain combinations. In some cases, hybrid fillers can be prepared. An example of a hybrid filler is carbon fibers decorated with CNTs that protrude from the fiber surface, sometimes called CNT forests. The multitude of nanotubes create extremely high surface area for the carbon fibers to adsorb and bind with resin increasing the potential strength of carbon fiber–epoxy high–performance composites. The problem is wetting of densely packed CNTs with resin and expulsion of air from the system.

Sections 2.2.3.1 through 2.2.3.7 describe hybrid combinations where advantage can be obtained from disparate filler sizes. An omission from the combinations is platelet–platelet because nano- and microplatelets may have similar lateral dimensions although they differ in thickness and the nanoplatelets are flexible.

2.2.3.1 Nanoparticle–Microparticle Hybrid Composites

Particle-filled composites do not contribute greatly to tensile or flexure properties because these increase with the aspect ratio. Nanoparticles modify the flow and contribute thixotropy at small volume fractions. Microparticles contribute density, compressive strength, decreased coefficient of thermal expansion, decreased flammability, and related properties dependent on polymer dilution. Examples are nanosilica and calcium carbonate that have been used to form consolidated high filler volume fraction composites by centrifuging. Silica was dispersed first using sonication, then calcium carbonate was dispersed using a shear mixer into a bisphenol-A ethoxy diacrylate. Thermal initiator was added and initiation was not activated until after compaction by centrifuging. The compacted filler region was cut from the resin-rich region at the top of the centrifuge tube. A silica–talc composite was formed by the same process. An etched surface of the nanosilica–talc composite revealed a biomimetic composite that displayed an ordered brick-and-mortar nacre-like structure [10].

Nanosilica–epoxy dispersions were formed using sonication or a sol–gel approach assisted by sonication. Alumina or carbon nanofibers (CNFs) were then added to the silica–epoxy to form hybrid nanocomposites in which all fillers were in the nanoscale, though one or both were particles while another was fibers. The initial silica dispersion in the epoxy enhanced the dispersion of the second filler, and higher filler loading was able to be prepared [11].

Silica was modified by surface grafting of silsisquioxane originating from vinyltriethoxysilane. The modified silica with nanosurface grafted features was used to reinforce polydimethylsiloxane. The filler and PDMS were bonded together using a hydrosilation reaction for chain extension and cross-linking. The materials showed increased tensile modulus and strength [12].

2.2.3.2 Nanoparticle–Microfiber Hybrid Composites

Nanoparticles such as silica, POSS, carbon black, and other specially prepared minerals contribute to the binding of the polymer, thereby increasing viscosity and decreasing creep. Microfibers, such as glass, graphite, cellulose, and basalt, enhance tensile and flexural modulus and strength. The combination of particles and fibers is particularly suitable for thermosetting polymers that are molded as liquids and react to form a solid networked structure. The polymer and fibers exist as micrometer-dimensional phases, whereas the nanoparticles can reinforce the microregions of polymer giving a more rigid matrix phase.

Figure 2.3 summarizes the components of a thermoset resin, a nanoparticle, and a microfiber hybrid composite, and the main function and properties of the components. Typical resins are unsaturated polyesters, vinyl esters, epoxy resins, and bismaleimide resins. Other combinations of fillers, from silica particles and glass fibers, CNTs, nanofibers, or graphene layers combined with carbon fibers.

Cellulose fibers were combined with titania–poly(vinyl alcohol) (PVA) dispersion to form hierarchical porous titania–cellulose–PVA hybrids, which were constructed via a layer-by-layer assembly. Cellulose provided a template and was subsequently removed by dissolving in sodium hydroxide and urea. Alternatively, hierarchical PVA was prepared by dissolving titania in acid. These methods avoid often used calcination methods [13].

Polypropylene (PP)–glass fiber composites have been modified with nanosilica particles. PP was blended with poly(propylene-g-maleic anhydride) compatibilizer, and various types and fractions of nanosilica were used, including dimethylsilylated silica. Fiber–matrix interfacial shear strength was enhanced when the nanosilica fraction was 5%–7%·w.w. Nanosilica particles were proposed to increase the work of adhesion between PP and glass. A significant increase in elastic modulus and creep stability was found for the hybrid composites [14]. Figure 2.4 shows a schematic of glass fibers in a polymer composite

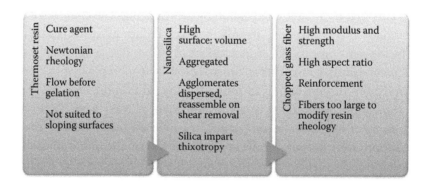

FIGURE 2.3   Components and their function for a thermoset resin nanoparticle–microfiber hybrid composite.

FIGURE 2.4   Schematic of a polymer microfiber composite with nanoparticles.

with the black dots representing a dispersed nanoparticle filler. The schematic should be compared with the actual scanning electron micrographs shown in subsequent figures.

Figure 2.5 shows microfibers in an all-poly(lactic acid) composite, where both the fibers and the matrix phase are poly(lactic acid) formed by hot compaction of a nonwoven mat substrate with added poly(lactic acid). Nanosilica dispersed in the added poly(lactic acid) is revealed on the fibers in Figure 2.5a and on a matrix-rich fracture surface in Figure 2.6.

2.2.3.3 Nanoplatelet–Microparticle Hybrid Composites

Nanoplatelets are much investigated as layered clays, such as montmorillonite (MMT) and related materials, and graphenes and their derivatives. They have dimensions of nanometer thickness, whereas the lateral dimensions may be on the order of micrometers. This means

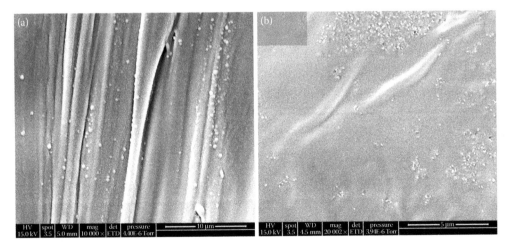

FIGURE 2.5   Poly(lactic acid) fibers in a composite with the same polymer as matrix with additional nanosilica particles: (a) delaminated fibers; (b) fibers embedded in the matrix surface.

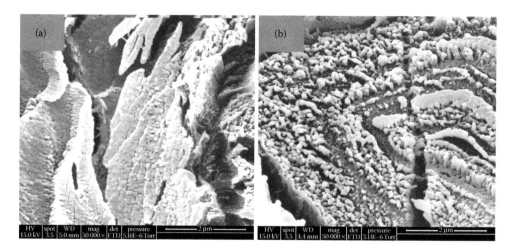

FIGURE 2.6 High-magnification micrograph of fractured poly(lactic acid) fibers showing (a) nanosilica on the fracture surfaces and (b) silica on the folds of fibers.

that the surface-to-volume ratio is extremely high. The large surface, when present in volume fractions as low as 0.01–0.05, is able to bind much polymer and create considerable increase in mechanical properties as well as decrease permeability to gases and liquids. In hybrids with microparticles, the latter mainly function as polymer diluents, providing changes in bulk properties as described previously. The challenge with layered nanofillers is dispersion because the layers hold strong attraction for each other relative to the polymer. Compatibilizing coatings on the layers assist with dispersion and prevention of agglomeration.

Nanoplatelets from layered clay have been found to be active fire retardants due to their limiting diffusion of flammable volatiles from a polymer exposed to combustion conditions. Microparticle fire-retardant chemicals such as zinc borate, magnesium hydroxide, or aluminum hydroxide, were added to provide fire retardants by complementary mechanisms such as water release or char formation in the case of zinc borate. These hybrid nano- and microfilled composites provided synergies to the current fire retardants by decreasing volatiles, decreasing gas migration/permeability, and assisting the formation of increased amounts of stable protective char [15].

Graphene with adsorbed magnetite particles has been used to form superparamagnetic composite sheets. The graphene was proposed to prevent agglomerates of the magnetite particles, thereby enhancing magnetism. The sheets were used to adsorb excess dyes from solution with ease of separation in an externally applied magnetic field [16].

2.2.3.4 Nanoplatelet–Microfiber Hybrid Composites

Microfiber composites with glass, carbon, or cellulose are prevalent and can be prepared with modulus and strength suitable for high-performance, critical applications such as construction and aerospace. Developers continue to seek further enhancement, so combining traditional composites with the extremely high mechanical properties of nanoplatelets such as layered clays and graphenes has potential to produce a new generation of advanced composite materials.

Cellulose fiber composites formed from nanofibrulated cellulose combined with MMT platelets were prepared using a paper-making process. The nanopaper was found to have increased toughness and self-extinguishing fire retardancy due to the presence of the clay. The morphology of the cellulose and clay exhibited layers that mimicked nacre, though with an inorganic content of 50%·w/w. Unusual ductility was exhibited, which allowed the inorganic content to be extended to 90%·w/w [17].

Aerogels were prepared from cellulose nanofibers prepared from tunicates, combined with MMT clay using a freeze–thaw method. A synergy between the two components enhanced the formation of a network structure, the morphology of which was determined during the freeze–thaw process. The clay–cellulose combination was compared with traditional materials, though in the nanoscale dimension [18].

PLA–wood flour composites have been modified by including MMT clay as a nanocomponent. Sodium MMT and organic treated MMT were used, with the latter reducing moisture absorption and swelling. The MMTs were shown to be adsorbed onto wood flour rather than diffused into the PLA matrix; hence, MMT contributed to increased interfacial strength between PLA and wood flour [19].

2.2.3.5 Nanofiber–Microparticle Hybrid Composites

Nanofibers of cellulose, CNTs, and mineral whiskers have huge specific surface area, high aspect ratio, and structural perfection that leads to exceptional mechanical strength. These fillers are difficult to disperse due to tendency to agglomerate and aggregate; however, they are effective reinforcements at relatively low volume fraction. Separately an inert particulate filler can be used to displace polymer, increase density, and increase compressive strength, while the nanofibers provide the mechanical enhancement.

CNFs and microsized silica particles were combined in epoxy resin to form a hybrid composite with enhanced mechanical damping, decreased coefficient of thermal expansion, and increased thermal stability. The synergy of combined filler system was accounted for with semiempirical models that predicted viscoelastic properties [20].

CNT together with rutile titanium dioxide or rutile nanotubes was combined to form hybrids because of their exceptional optical, mechanical, electrical, and thermal properties. Benzyl alcohol was used as a surfactant to enhance interaction between titanium dioxide and CNT. The hybrid fillers had potential application in photochemical, catalytic, and sensor devices [21].

2.2.3.6 Nanofiber–Micro-Platelet Hybrid Composites

Similar to the previous classification, nanofibers contribute to modulus and strength. Micro-platelets such as mica are better than particles for stiffening, energy damping, and decreasing permeability.

2.2.3.7 Nanofiber–Microfiber Hybrid Composites

Combinations of fiber with like chemical composition have the most potential. For instance, carbon fibers and CNFs can combine to further enhance carbon fiber composites. The nanofibers might attract to the micro-carbon fibers providing a roughened and enhanced area surface for polymer adsorption. Alternatively the nanofibers may reinforce polymer in micro-volumes between carbon fibers. The total volume fraction of nano- and micro-carbon fibers can be greater than that of only nanofibers because dispersion of nanofibers may be optimal at 0.05 v/v with a maximum of 0.10 v/v, whereas carbon microfibers are used at 0.50–0.6 v/v and the latter can be used as woven mats or pultruded strands.

Similar composite constructions can be prepared with glass fibers and glass nanofibers or whiskers. Cellulose fibers can be used in combination with microcrystalline cellulose, cellulose nanofibers, or cellulose nanocrystals. In each case, the nanomaterials have structural perfection leading to extraordinary modulus and strength.

Carbon fiber-reinforced polymer composites were prepared in an epoxy resin matrix with added multiwalled CNTs (MWCNTs). Damping of the hybrid composites was enhanced by the MWCNTs that adopted random orientations within the hybrid [22].

2.3 APPLICATIONS AND FUTURE DIRECTIONS

Carbon fiber–epoxy resin composites are high-performance materials used in the aerospace industry. There is potential for enhancement using nanoforms of carbon such as CNTs and CNFs. Carbon fibers with deposited CNFs or CNTs growing from the surface are being developed to create higher performance composites due to extra-high surface area and reinforcement at two levels of scale. Carbon fiber composites have high volume fractions of fibers that are difficult to increase; however, there is space between the fibers that can be occupied by nanofillers while still having all surfaces wetted by the resin.

Glass fiber composites can be prepared with nanosilica or sol–gel silica created *in situ*, to avoid high viscosity and dispersion problems. Such composites are prevalent already because nanosilica is added to the resin as a thixotropic agent in spray application so that the resin–glass fiber composite will not sag or drape when sprayed onto nonhorizontal surfaces. The reinforcement is assumed to derive from the glass fibers; however, with a suitable design, the combined glass fibers and nanosilica should both be contributors to the final mechanical properties. An alternative would be to combine typical glass fibers with nanoglass fibers. Coupling agents can be used to reactively link the surface of the hybrid filler combination to the resin network.

Many composites are relying on cellulose fibers for reinforcement due to availability of the fibers from renewable resources or waste materials and the biodegradability of cellulose. Cellulose nanofibers and nanocrystals have been added to cellulose fiber composites. The nanocellulose forms have much superior properties due to their crystal perfection and removal of impurities and noncrystalline materials. Nanocellulose fibers can enhance properties at low volume fraction, whereas microcellulose fibers can both enhance properties and provide a volume-filling component to reduce the volume of polymer in the composite. Interaction is likely between the fibers and the nanofibers giving an improvement in modulus and strength due to a network of fillers within the polymer matrix.

Synthesis of bimodal dimensioned fillers with size distribution maximum in the nano- and micro-dimension ranges. This approach would enable a single-step filler formation, though only certain filler types would be applicable to preparation of a bimodal filler. One situation is where a filler is formed as a precipitate in which nucleation will give nanoparticles and growth will give microparticles. Particle growth can be limited by enclosure in a surfactant layer. Another method is used to form a filler hybrid simultaneously via two different mechanisms, such as by precipitation and sol–gel reaction. Precipitation will give larger particles, whereas sol–gel reaction can be adjusted to give nanoparticles. A combined event can be where the precipitated particles attract the sol–gel reactants so that nanoparticles are formed on the surface of the microparticle. The resulting textured surface microparticles will have large surface area for interaction with polymer.

2.4 CONCLUSION

Composites must contain two phases in which each phase displays its individual properties and the phases are physically separable. Hybrid composites in the context of this review contain three phases with two being dispersed phases of differing dimensions. The dispersed phases will be chosen to exhibit complementary properties that should be synergistic. Nanocomposites have exceptional properties with even low filler volume fraction; however, together with traditional micrometer-scale fillers, enhanced properties can be obtained. Hybrid composites containing micro- and nanofillers are being investigated and developed for high-performance applications where a current composite can be enhanced by addition of a nanofiller. In some cases, the nanofillers are adsorbed onto the microfiller or grown from the micro-filler surface such as with CNT forests on carbon fibers. On a microscale, there is pure matrix polymer between the fillers, and this matrix can be further enhanced on a smaller scale with nanofiller to reinforce interstitial polymer. Opportunities are abundant for design of novel hybrid composites.

ACKNOWLEDGMENT

Izan Roshawaty provided scanning electron micrographs of nanosilica–hemp–poly(lactic acid) composites that illustrated the relative size and distribution of the two fillers in a hybrid composite.

ABBREVIATIONS

CNF	Carbon nanofiber
CNT	Carbon nanotube
MMT	Montmorillonite
MWCNT	Multiwalled carbon nanotube
POSS	Polyhedral oligomeric silsesquioxane
PP	Polypropylene
SBS	Poly(styrene-*b*-butadiene-*b*-styrene)

REFERENCES

1. Goh, K. L., R. M. Aspden and D. W. L. Hukins (2004). Review: Finite element analysis of stress transfer in short-fiber composite materials. *Composites Science and Technology*, 64: 1091–1100.
2. Cooper, G. A. (1971). The structure and mechanical properties of composite materials. *Review of Physics in Technology*, 2: 49–57.
3. Gan, Y. X. (2009). Effect of interface structure on mechanical properties of advanced composite materials. *International Journal of Molecular Sciences*, 10: 5115–5134.

4. Roy, N., R. Sengupta and A. K. Bhowmick (2012). Modifications of carbon for polymer composites and nanocomposites. *Progress in Polymer Science*, 37: 781–819.

5. Spoljaric, S., A. Genovese and R. A. Shanks (2012). Novel elastomer-dumbbell functionalized POSS composites: Thermomechanical and morphological properties. *Journal of Applied Polymer Science*, 123: 585–600.

6. Spoljaric, S. and R. A. Shanks (2012). Novel elastomer dye-functionalised POSS nanocomposites: Enhanced colourimetric, thermomechanical and thermal properties. *eXPRESS Polymer Letters*, 6: 354–372.

7. Shou, Q.-L., J.-P. Cheng, J.-H. Fang, F.-H. Lu, J.-J. Zhao, X.-Y. Tao, F. Liu and X.-B. Zhang (2013). Thermal conductivity of poly vinylidene fluoride composites filled with expanded graphite and carbon nanotubes. *Journal of Applied Polymer Science,* 127: 1697–1702.

8. Hu, H., L. Zhao, J. Liu, Y. Liu, J. Cheng, J. Luo, Y. Liang, Y. Tao, X. Wang and J. Zhao (2012). Enhanced dispersion of carbon nanotube in silicone rubber assisted by graphene. *Polymer*, 53: 3378–3385.

9. Young, R. J., I. A. Kinloch, L. Gong and K. S. Novoselov (2012). The mechanics of graphene nanocomposites: A review. *Composites Science and Technology*, 72: 1459–1476.

10. Daud, N., R. A. Shanks and I. Kong (2012). Compacted nanosilica-talc/calcium carbonate polyacrylate composites prepared using accelerated sedimentation. *World Journal of Engineering*, 9: 385–390.

11. Uddin, M. F. and C. T. Sun (2010). Improved dispersion and mechanical properties of hybrid nanocomposites. *Composites Science and Technology*, 70: 223–230.

12. Jia, L., Z. Du, C. Zhang, C. Li and H. Li (2008). Reinforcement of polydimethylsiloxane through formation of inorganic-organic hybrid network. *Polymer Engineering & Science*, 48: 74–79.

13. Gu, Y. and J. Huang (2009). Fabrication of natural cellulose substance derived hierarchical polymeric materials. *Journal of Materials Chemistry*, 19: 3764–3770.

14. Pedrazzoli, D. and A. Pegoretti (2013). Silica nanoparticles as coupling agents for polypropylene/glass composites. *Composites Science and Technology*, 76: 77–83. doi.org/10.1016/j.compscitech.2012.12.016.

15. Shanks, R. A. and A. Genovese (2009). Fire-retardant properties of polymer nanocomposites. In: *Recent Advances in Polymer Nano-Composites*. S. Thomas, G. E. Zaikov and S. V. Valsaraj (eds.), Koninklijke Brill NV, Leiden, the Netherlands, pp. 439–454.

16. Xie, G., P. Xi, H. Liu, F. Chen, L. Huang, Y. Shi, F. Hou, Z. Zeng, C. Shao and J. Wang (2012). A facile chemical method to produce superparamagnetic graphene oxide-Fe_3O_4 hybrid composite and its application in the removal of dyes from aqueous solution. *Journal of Materials Chemistry*, 22: 1033–1039.

17. Liu, A., A. Walther, O. Ikkala, L. Belova and L. A. Berglund (2011). Clay nanopaper with tough cellulose nanofiber matrix for fire retardancy and gas barrier functions. *Biomacromolecules*, 12: 633–641.

18. Gawryla, M. D., O. v. d. Berg, C. Weder and D. A. Schiraldi (2009). Clay aerogel/cellulose whisker nanocomposites: A nanoscale wattle and daub. *Journal of Materials Chemistry*, 19: 2118–2124.

19. Liu, R., J. Cao, S. Luo and X. Wang (2013). Effects of two types of clay on physical and mechanical properties of poly(lactic acid)/wood flour composites at various wood flour contents. *Journal of Applied Polymer Science*, 127: 2566–2573.

20. Khan, S. U., C. Y. Li, N. A. Siddiqui and J.-K. Kim (2011). Vibration damping characteristics of carbon fiber-reinforced composites containing multi-walled carbon nanotubes. *Composites Science and Technology*, 71: 1486–1494.

21. Eder, D. and A. H. Windle (2008). Morphology control of CNT-TiO_2 hybrid materials and rutile nanotubes. *Journal of Materials Chemistry*, 18: 2036–2043.

22. Jang, J.-S., J. Varischetti, G. W. Lee and J. Suhr (2011). Experimental and analytical investigation of mechanical damping and CTE of both SiO_2 particle and carbon nanofiber reinforced hybrid epoxy composites. *Composites Part A: Applied Science and Manufacturing*, 42: 98–103.

CHAPTER 3

Thermoset–Clay Nanocomposites

An Overview

M. Selvakumar, Golok B. Nando, and
Santanu Chattopadhyay

CONTENTS

3.1 INTRODUCTION

In the twentieth century, incorporation of clays into polymers for improving the technical properties of base polymer revolutionized both research and industrial sectors. The invention of nylon-6/silicate clay nanocomposites by the Toyota research group [1–3] is the inspiration for the development of utilizing mineral clays for advanced products, including aircrafts. Many works have been reported on the incorporation of clays into thermoplastics [4–7], thermosets, and elastomers [8–11]. Nanocomposites are a new class of materials, defined as those with dispersed phase having at least one of the dimensions ranging in 1–100 nm. Nanocomposites are not new to rubber technology, as carbon black (CB) and SiO_2 have been used as reinforcing fillers in rubbers over a few centuries now. Compared to conventional macro- or microcomposites, nanocomposites exhibit unique properties with low level of filler loading. At present, clays used are either derived from rocks or synthesized ones, for example, montmorillonite (MMT), kaolinite, sepiolite, and illite. MMT clays are layered aluminosilicates with a layer separation of few nanometers, and it is one of the naturally available nanostructured two-dimensional materials [12]. The compositions of the clays vary from one another. The basic procedure for producing polymer/clay nanocomposite is simply the insertion of polymers in-between clay layers or the polymerization of monomers in-between clay galleries. In general, polymer/clay nanocomposites can be classified into three different types: (1) intercalated nanocomposites, (2) flocculated nanocomposites, and (3) exfoliated nanocomposites. Typically, all these are schematically shown in Figure 3.1 In intercalated polymer composites, chains are inserted into galleries of silicate layers in a crystallographically regular fashion, with a few-nanometer repeat distance in the basal plane. In flocculated nanocomposites, stacked silicate layers are sometimes flocculated due to edge–edge interactions of the clay layers, especially–OH interaction. In exfoliation, complete delamination of individual layers in the polymer matrix, which results from extensive penetration of the polymer within the interlayer spacing, and also the average distance of delaminated layers, depends on the amount of clay loading, and there are no longer sufficient attractions between the silicate layers to maintain uniform layer spacing [13]. The properties of the exfoliated polymer/clay nanocomposites are better than the intercalated

FIGURE 3.1 Schematic diagram of three different types of the polymer/clay nanocomposite structure. (Reprinted from *Prog. Polym. Sci.*, 28, Sinha Ray, S. and Okamoto, M., Polymer/layered silicate nanocomposites: A review from preparation to processing, 1539–1641, 2003. Copyright 2012 with permission from Elsevier.)

ones [14,15]. The modification of clay is more important for improving compatibility between the polymer and the clay. This is because of the fact that generally clays are hydrophilic, but polymers are hydrophobic; due to this, we must modify clays with some organic surfactants. Modified clays exhibit good compatibility with the polymer matrix, leading to better properties compared to unmodified clays [16,17].

3.2 STRUCTURE OF CLAY

The most popular clay minerals used in polymer nanocomposites are of three types: kaolinite, MMT, and illite.

3.2.1 The Kaolinite Family

This family has three members (kaolinite, dickite, and nacrite). The chemical composition of kaolinite clay is as follows: kaolinite has a high content of alumina and a low content of silica, and a formula of $Al_2[Si_2O_5](OH)_4$. Their crystal structures consist of layers made up of tetrahedral sheets in which silicon is surrounded by four oxygen atoms and an octahedral aluminum sheet in which aluminum is surrounded by eight oxygen atoms in the ratio of 1:1 [18] as shown in Figure 3.2, where layers are packed up along the c-axis and extend to the plane in a and b axes. The closely stacked structure of kaolinite results in low cation-exchange capacity (CEC). Due to this, intercalation of any molecule or chain into the gallery is very difficult. This is a great disadvantage for many researchers because there is no scope for producing intercalated and exfoliated nanocomposites. It reveals that the kaolinite clay cannot be used for the preparation of nanocomposite [19], but it is mainly used as filler for paints. Its major use is in the paper industry to produce glossy paper.

3.2.2 The MMT Family

This family is composed of several minerals, including pyrophyllite, talc, vermiculite, sauconite, saponite, nontronite, and MMT. These are mostly differing in their chemical content. MMT has a low content of alumina and a high content of silica. Its crystal structure consists of layers made up of fused silica tetrahedral sheets sandwiching an edge-shared octahedral sheet of alumina in the ratio 2:1 as shown in Figure 3.3.

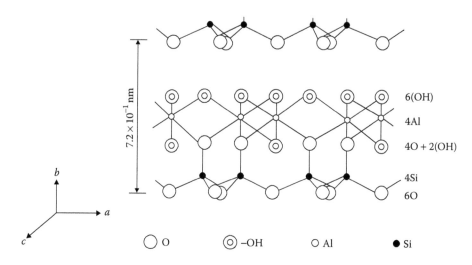

FIGURE 3.2 Diagrammatic sketch of kaolinite. (Reprinted from *Polymer-Layered Silicate and Silica Nanocomposites*, Ke, Y.C. and Stroeve, P. Copyright 2005, with permission from Elsevier.)

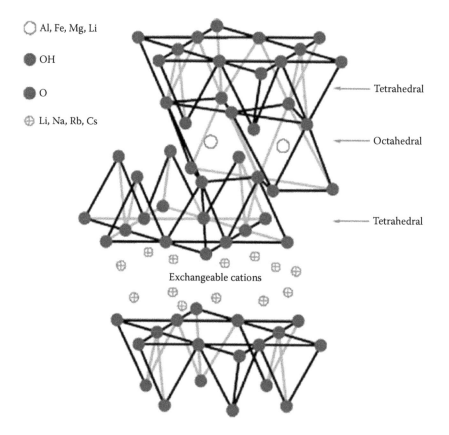

FIGURE 3.3 Diagrammatic sketch of MMT. (Reprinted from *Prog. Polym. Sci.*, 33, Pavlidou, S. and Papaspyrides, C.D., A review on polymer–layered silicate nanocomposites, 1119–1198. Copyright 2008, with permission from Elsevier.)

The layer thickness is around 1 nm, and the lateral dimensions may vary from 300 Å to several micrometers or larger than this, depending on the particular layered silicate [13]. The top and base layers of MMT are completely occupied by oxygen atoms, and relatively weak intermolecular forces hold these layers together. As a result, water molecules can easily intercalate and cause an expansion of lattice. Isomorphic substitution between the layers generates negative charges that are counterbalanced by alkali and alkaline earth cations situated inside the galleries. Thus, obviously the space between layers will increase, leading to good intercalation of polymer chains. MMT clay has been widely used by many researchers [7,16,17,19–23] compared to other clays such as illite and kaolin. Layered silicate is characterized by a moderate surface charge known as the CEC. In general, the properties of clay minerals are dependent on the crystal structure, chemical composition, and its gallery space. Some of the important characteristics of MMT are shown in Table 3.1 along with other types of clays.

3.2.3 The Illite Family

This family consists of host minerals of muscovite and biotite. It is also called as hydromica. It is one of the most abundant clay minerals in China. The general formula is $(K, H) Al_2(Si \cdot Al)_4 O_{10}(OH)_2 - xH_2O$, where x represents the variable amount of water that this group could contain [13]. The crystal structure of illite is similar to that of MMT. The major difference is that in illite, the isomorphic substitutions generally occur in the tetrahedral sheet with Al^{3+} for Si^{4+}. In most cases, up to one-fourth of the tetrahedral coordinated cations are Al^{3+}. Isomorphic substitution can also be found in the octahedral sheet, typically Mg^{2+} and Fe^{2+} for Al^{3+} [12]. When illite is suspended in an aqueous solution of surfactant, it will penetrate into the gallery of layered silicate causing expansion to improve the degree of intercalation or exfoliation. This kind of clay is widely used in latex blending method and *in situ* polymerization method for producing polymer/clay nanocomposites. Applications of this clay such in oil drilling, it is necessary to protect the surface of the wall well. Moreover, suspensions of illites are also used as a drilling fluid.

TABLE 3.1 Crystal Structure and Physical and Chemical Properties of Three Types of Clay Minerals

Clay Mineral	Layer Type	Distance between Adjacent Layers (10^{-1} nm)	Attraction Force between Layers	CEC (mmol/100 g Clay)	Comments
Kaolinite	1:1	7.2	Hydrogen bond, very strong	3–15	No chances for intercalation/exfoliation
MMT	2:1	9.6–40.0	Intermolecular force is weak	70–130	Degree of intercalation or exfoliation is high
Illite	2:1	10.0	Electrostatic interaction is strong	20–40	Degree of intercalation or exfoliation is moderate

Source: Reprinted from *Polymer-Layered Silicate and Silica Nanocomposites*, Ke, Y.C. and Stroeve, P. Copyright 2005, with permission from Elsevier.

3.2.4 Cation-Exchange Capacity

CEC plays a vital role in polymer/clay nanocomposites. Degree of intercalation or exfoliation of nanocomposites is determined by CEC. It refers to the total capacity of adsorbed cations at a pH value of 7, and its unit of measurement is mmol/100 g. If the negative charge is higher for layered silicates, it gives stronger capacity for hydration, swelling, and dispersion. It reveals that we can insert the polymer chain as much as in-between clay layers. CEC is generally measured by saturating the clay with NH_4^+ or Ba^{2+} and determining the amount held at pH ~ 7 by conductometric titration. Another method to find CEC of clay is saturating the clay with alkyl ammonium ions and evaluating the quantity of ions intercalated by igniting the sample using a simple thermogravimetric method [13].

3.2.5 Organic Clay Modifications

In most of the cases, layered silicates are only dispersible or compatible in hydrophilic polymers such as poly(vinyl alcohol), poly(methyl methacrylate), and poly(ethylene-*co*-vinyl acetate). For better filler dispersion and distribution of layered silicates with hydrophobic polymer must convert layered silicates into organophilic by suitable surface modification. Generally, it can be achieved by ion exchange with cationic surfactants (alkylammonium or alkylphosphonium cations). Alkylammonium or alkylphosphonium cations in the organosilicates lower the surface energy of the inorganic host and improve the wetting characteristics of the polymer matrix, and result in a larger interlayer spacing. The increase of interlayer spacing (d_{001}) can be confirmed by X-ray diffraction (XRD) at low angles as shown in Figure 3.4.

Figure 3.5 explains how the surface properties of the clay change from hydrophilic to hydrophobic. Chieng et al. [21] organically modified MMT by using octadecylammonium. They prepared it by adding a dispersed 20.00 g of sodium MMT in 800 mL of distilled water at 80°C into a solution of 13.476 g octadecylamine and 4.81 mL concentrated hydrochloric

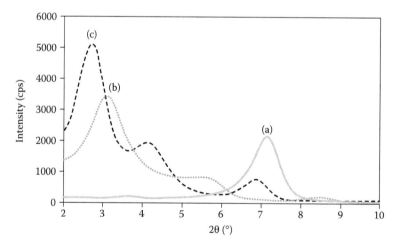

FIGURE 3.4 XRD of pristine clay (a) and organically modified clay (b–c). (Reprinted from Chieng, B.W., *eXPRESS Polym. Lett.*, 4, 404. Copyright 2010, with permission from Express Polymer Letter Publishing Group.)

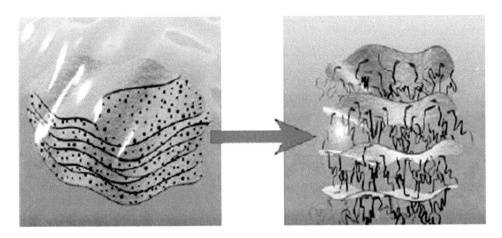

FIGURE 3.5   Schematic diagram of an ion-exchange reaction. (Reprinted from *Prog. Polym. Sci.*, 33, Pavlidou, S. and Papaspyrides, C.D., A review on polymer–layered silicate nanocomposites, 1119–1198. Copyright 2008, with permission from Elsevier.)

acid in 200 mL hot distilled water. The resulting suspension was vigorously stirred for 1 h. Then the precipitate was repeatedly filtered and washed with hot distilled water until no chloride ion was detected with 0.1 N $AgNO_3$ solutions. Subsequently, it was dried at 60°C for 24 h.

Additionally, the alkylammonium or alkylphosphonium cations can provide functional groups that can react with the polymer matrix, or in some cases initiate the polymerization of monomers to improve the strength of the interface between the inorganic and the polymer matrix. In practical preparation of organoclay, the added quantity of organic cation reagent should be less than the CEC value of the clay. An overload of surfactant will reduce the stability of organoclay.

3.3 ROUTES FOR SYNTHESIZING POLYMER/CLAY NANOCOMPOSITES

Several methods are used for the preparation of polymer/clay nanocomposites. So far, four methods are well exploited to develop a polymer/clay nanocomposite.

3.3.1 Solution Blending Method

It is a very simple and low-cost method. In this method, polymer is dissolved in a suitable solvent along with the clays or mixed together in suitable solvents. In solution, polymer chains well enter into the galleries of clay. After sometime, the clay gets dispersed and exfoliated; the solvent is to be evaporated at the boiling temperature of the solvent. Basically, the solvent is used to help in the mobility of the polymer chains, which in turn helps to intercalate the polymer chains in-between clay layers. Sometimes, ultrasonication is also being used to disperse the clay well. The schematic diagram for synthesizing polymer/clay nanocomposites is shown in Figure 3.6. Several works have been reported on solution blending [23,24].

3.3.2 Melt Intercalation Method

The melt intercalation technique for preparation of nanocomposites can be carried out in various kinds of internal mixer such as simple extruder, Brabender corotating twin-screw

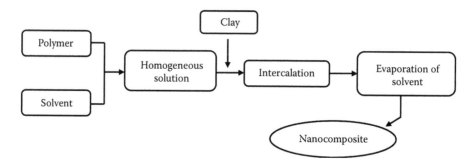

FIGURE 3.6   Schematic diagram showing synthesis of polymer/clay nanocomposites by the solution blending method.

mini extruder, and Haake Rheomix 600p, and also by using simple two-roll mill. In this method, polymer is melted with clay for optimizing the polymer/clay interaction. The mixture of polymer and clay is then annealed at a bit of above glass transition temperature of the polymer to form nanocomposite and to relieve thermal stress in nanocomposites. This method have more compensation like the degree of intercalation or exfoliation over *in situ* intercalative polymerization or solution intercalation. Moreover, it is more compatible with the current industrial process such as extrusion and injection molding, and it is also widely an accepted technique in industries, especially in tire industries [13]. The schematic diagram for synthesizing polymer/clay nanocomposites by this method is shown in Figure 3.7. Several works have been reported on elastomer-based clay nanocomposites prepared by melt mixing [3,4,8].

3.3.3 *In Situ* Intercalative Polymerization Method

This is the first method used for synthesizing polymer–clay nanocomposites based on nylon by Toyota research group. They have reported that modified Na^+-MMT to be swollen by the 1-caprolactam monomer at 100°C and subsequently initiate its ring-opening polymerization by initiator either by heat or radiation to produce nylon-6/MMT nanocomposites [13]. Usually, this method takes longer time compared to other techniques such as melt intercalation and solution blending. The actual time of preparation depends on the polarity of monomer, surface modification of monomer, and temperature. When the

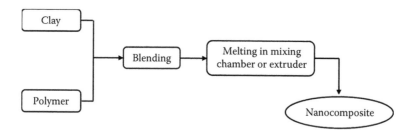

FIGURE 3.7   Schematic diagram showing synthesis of polymer/clay nanocomposites by the melt intercalation method.

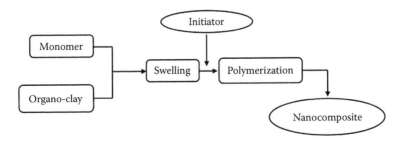

FIGURE 3.8   Schematic representation of *in situ* intercalative polymerization method.

reaction is initiated from the clay surface, high surface energy of the clay attracts polar monomer molecules so that it enters in-between the clay layers and is polymerized. This method is well adopted for both types of thermoplastic and thermoset nanocomposites [4,22,25]. The schematic representation of this method is shown in Figure 3.8.

3.3.4 Latex Intercalation Method

This method is another promising method for preparing polymer/clay nanocomposites [19,26,27]. Generally, latex is nothing but rubber particles, which are well dispersed in the aqueous medium. The clays are easily dispersed in water, and in that water they act as a swelling agent because of the hydration of the intergallery cations (usually sodium or potassium). The swelling capabilities of clays in water are not the same; they depend on the clay types and their CEC. Finally, the mixing of the latex with the clays followed by coprecipitation (coagulation) is a best route for producing polymer/clay nanocomposites, especially the elastomer-based nanocomposites because most of the rubbers are originating from latex form.

3.3.5 Advantages

It has gained many advantages such as good mechanical properties, improved heat stability, better optical properties without compromising its strength, and outstanding barrier properties without requiring multilayered design. Also it has an excellent thermal expanding coefficient, flow behavior at melt, adhesive strength, inflammability, good resistance to oil, chemical and weather resistance, improved biodegradability and biocompatibility, and very good electrical properties.

3.4 THERMOSET–CLAY NANO COMPOSITES

Various thermoset polymers have already been developed with various tuned properties. In this discussion, we almost covered all kinds of thermoset/clay nanocomposites, and some of the important results and preparations techniques are explained briefly with neat portfolios.

3.4.1 Natural Rubber

Natural rubber (NR) is a naturally existing rubber and so is the clay. Thus, we can consider NR-based clay nanocomposites as green composites. The literature reveals that a lot of research has been done on NR/clay nanocomposites [19,28–35]. Joly et al. [34] successfully synthesized and studied the properties of the two types of organically modified clay nanocomposites, namely, dimethyl dihydrogenated tallow ammonium MMT and dimethyl hydrogenated tallow (2-ethylhexyl) ammonium MMT with NR nanocomposites by the solution blending technique. In their study, they observed an increase in modulus even at rather low filler loadings. It is due to the organically modified galleries of MMTs were easily penetrated by NR chains, and leads to intercalated structures with partially exfoliation. Intercalated (increasing of basal plan spacing, d_{001}) and exfoliated structures were confirmed by XRD and transmission electron microscopy (TEM) micrographs. The intercalated structures confirmed by XRD and the tensile properties of nanocomposites are shown in Figures 3.9 and 3.10, respectively. However, Bhattacharya and Bhowmick [35,36] investigated the two types of ternary nanocomposites, namely, NR/nanoclay and NR/carbon nanofiber (CNF)/nanoclay nanocomposites combined with different loadings and different grades of CBs to obtain ternary nanocomposites. In their study, the mechanical properties such as tensile, tear, abrasion, wear, dynamic mechanical properties, viscoelasticity, and morphology were documented for tire applications.

They concluded that high-abrasion furnace (HAF) CB provides the best mechanical properties (modulus at 300% elongation, tensile strength) even in the presence of dual nanofillers such as clay and CNF. The reason behind the unique morphology of ternary nanocomposites are due to formation of a novel microstructure consists of "nanoblocks," small-size aggregates of CB, and "nanochannels" of free CB. This favorable electrostatic interaction cause enhanced filler dispersion and efficient stress transfer from the matrix, resulting in improved mechanical and dynamic mechanical properties. In addition, overbearing presence of carbon but also on significantly reducing the black loading and also the attachments of clay/nanofiber at the surface of CB weakens the

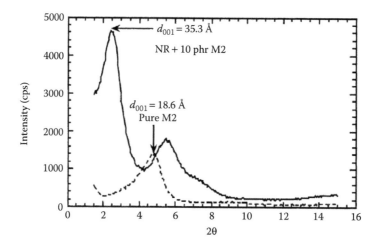

FIGURE 3.9 XRD spectra of pure OMMT and NR-based nanocomposites. (Reprinted with permission from Joly S, Garnaud G, Ollitrault R, Bokobza L, Mark JE. Organically modified layered silicates as reinforcing fillers for natural rubber. *Chemistry of Materials* 2002;14:4202. Copyright 2010 American Chemical Society.)

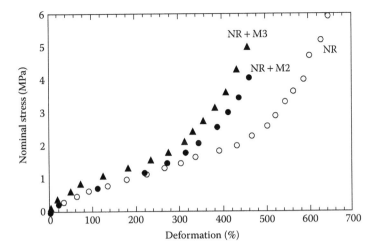

FIGURE 3.10 Tensile properties of different organoclays loaded (at 10 phr) in NR-based nanocomposites. (Reprinted with permission from Joly S, Garnaud G, Ollitrault R, Bokobza L, Mark JE. Organically modified layered silicates as reinforcing fillers for natural rubber. *Chemistry of Materials* 2002;14:4202. Copyright 2010 American Chemical Society.)

electrostatic interactions between CB particles due to a significant reduction in free surface area and surface energy, thereby preventing their reagglomeration of particles during curing. These kinds of ternary nanocomposites exhibit dramatically improved mechanical properties over the control microcomposites. The unique morphology of ternary nanocomposite is shown in Figure 3.11. NR-based nanocomposites containing the two fillers have been found to display good properties compared to single fillers. Moreover, the ternary nanocomposites exhibit less coefficient of friction and less temperature buildup.

Wear loss was reduced in the dual filler nanocomposites by 33% (over the CB microcomposite) under less stringent and 75% under severe wear conditions. The designed ternary nanocomposites show a strong wet grip and low rolling resistance, whereas abrasion resistance is improved stupendously. The possible potential applications of these nanocomposites were realized in vehicle tires.

3.4.2 Epoxy Resin and Epoxidized Rubber

Epoxy resin and epoxidized rubber-based clay nanocomposites made a great revolution in aircraft components manufacturing due to their unique properties, especially combined elastic and viscoelastic characteristics. Many researches were carried out in this area with different perspectives [25,37–42]. Abacha et al. [37] prepared epoxy resin with organically modified MMT nanocomposites; they thoroughly investigated the performance of epoxy/clay nanocomposites in a corrosive environment with different corrosive media such as H_2SO_4 and HCl, and also studied the mechanical properties of such nanocomposites under wet conditions in both water and corrosive media. They reported that epoxy/clay nanocomposites exhibit better mechanical properties under wet conditions in both water and

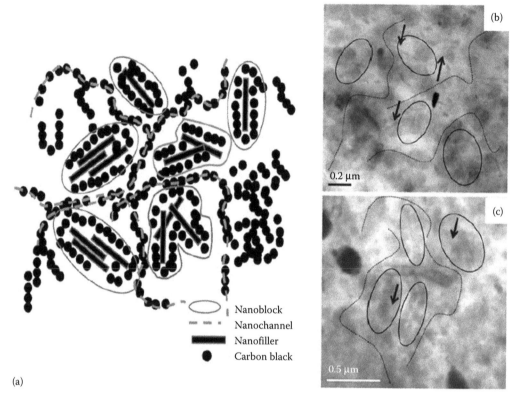

FIGURE 3.11 Microstructural development in nanoclay- and nanofiber-based nanocomposites in the presence of carbon black: (a) schematic representation, (b) corresponding morphology of nanoclay, and (c) nanofiber-based dual filler nanocomposites, as observed in transmission electron micrographs. (Reprinted with kind permission from Springer Science + Business Media: *J. Mater. Sci.*, Synergy in carbon black filled natural rubber nanocomposites. Part II: Abrasion and viscoelasticity in tire like applications, 45, 2010, 6139, Bhattacharya, M. and Bhowmick, A.K., 2010;45:6139–6150.)

acid as shown in Figure 3.12. For finding the mass uptake or solution content in the sample of H_2O or acid at a given time, they used the following equation:

$$M_t\,(\%)=\frac{W_t-W_0}{W_0}\times100 \tag{3.1}$$

where M_t, W_t, and W_0 are the solution content at a given time, the weight of the sample at the equilibrium time of the measurement, and the initial weight, respectively. Results were showing that equilibrium mass uptake decreases with increasing clay content in nanocomposites for both water and acid compared to neat epoxy resin. It was noticed that the incorporation of the organoclay increases the tortuosity of path due to high degree of intercalated morphology and thus decreases the equilibrium mass uptake. As a result, it significantly improved the barrier properties for nanocomposites against corrosive acid (H_2SO_4) as well as under H_2O.

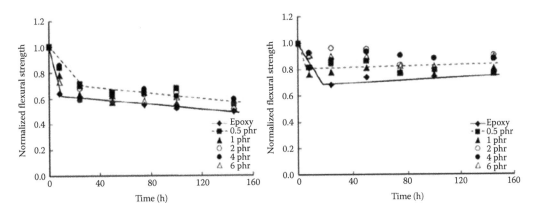

FIGURE 3.12   Flexural strength of epoxy–clay nanocomposites under wet conditions. (Reprinted from Abacha, N., *eXPRESS Polym. Lett.*, 1, 364. Copyright 2007, with permission from Express Polymer Letter Publishing Group.)

Turri et al. [38] investigated abrasion and scratch resistance of nanostructured epoxy coatings on the steel substrate by using top-down and bottom-up (sol–gel) methods. In this study, they thoroughly studied dynamic mechanical analysis, Buchholtz hardness indentation testing, taber abrasion testing as well as nanoscratch analysis of coating carried out by using an atomic force microscopy (AFM) tip followed by surface topographic analysis. Nanoscratch hardness was measured by the following equation [38]:

$$H_s = \frac{4F}{\pi\omega^2} \tag{3.2}$$

where:
   H_s represents the scratch hardness
   F is the normal load applied on the AFM tip
   ω is the width of the scratch

Figure 3.13 shows the surface topography of nanoscratch analysis of coating and its penetration depth by using AFM tool.

From the results, it has been inferred that more force is required to scratch the epoxy/clay coating compared to epoxy coating alone. In contrast, for clay/epoxy coating, the force was noticed as 112.3 MPa and the scratch depth was 14.4 nm, whereas for pristine epoxy coating, the force was noticed as 42.8 MPa and the scratch depth was 19.21 nm. The process of coating and curing reaction, mechanism are shown in Figure 3.14. Mohan and Velmurugan [39] investigated the effect of curing temperature on mechanical properties of the epoxy-alkyl ammonium-modified MMT clay nanocomposites prepared by the solution method. In this study, they approached different curing temperatures ranging from 60°C to 120°C to harden the nanocomposites. They claimed that composites' structure, morphology, and properties were largely governed by the curing temperature and the concentration of clay. Among the different curing conditions, samples cured at 120°C showed better improvement than that cured at other lower temperatures (<120°C).

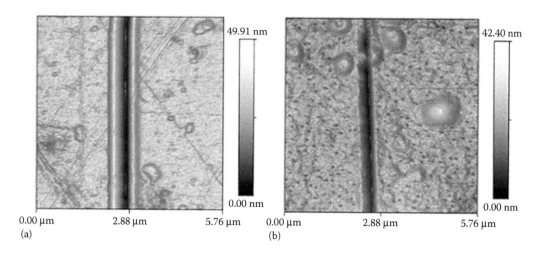

FIGURE 3.13 Topographic images of the nanoscratch analysis of coating: (a) epoxy coating alone; (b) epoxy/5% clay-loaded coating. (Turri, S., Torlaj, L., Piccinini, F., and Levi, M.: Abrasion and nanoscratch in nanostructured epoxy coatings. *J. Appl. Polym. Sci.* 2010. 118. 1720–1727. Copyright Wiley-VCH Verlag GmbH & Co. KGaA. Reprinted with permission.)

It has also been noted that the nanocomposites' properties depend on clay concentrations. Clays of more than 2% in epoxy matrix tends to decrease the tensile properties due to agglomerations. In 2% nanoclay-loaded epoxy nanocomposite, the crack surface has become rough, which suggests that the crack has taken a torturous path to failure. It shows that the crack has struggled more to propagate under the applied loading conditions, which in turn increases the strength to failure. In other weight percentage of clay loadings, the fracture surface was smooth, which is due to brittle failure and poor distribution of clay or because clay has got fully agglomerated. It makes the crack to more easily propagate and cause low strength values.

Chow [25] thoroughly investigated the optimization of process variables on the flexural properties of epoxy/organo-MMT (OMMT) nanocomposites by response surface methodology. In this study, they concluded that the degree of cross-linking will be higher for the longer postcuring time. In addition, flexural yield stress of epoxy/OMMT nanocomposites increases as the postcuring temperature increases due to the formation of higher degree of cross-linking. However, the increased speed of a mechanical stirrer during preparation of the composites is also responsible to decrease the flexural modulus. Because it improves the clay dispersion, the properties also improves significantly, as shown in Figure 3.15.

3.4.3 Ethylene Vinyl Acetate

Ethylene vinyl acetate (EVA) is the copolymer of ethylene and vinyl acetate. It approaches elastomeric materials in softness and flexibility, and it can be processed like other thermoplastics. Few works have been done on EVA/clay-based nanocomposites [11,43–46]. Mishra and Luyt [43] studied the effect of organic peroxides on the morphological, thermal, and tensile properties of EVA–organoclay nanocomposites. In this study, they compared the two types of peroxides, namely, dicumyl peroxide (DCP) and dibenzyl peroxide (DBP) as

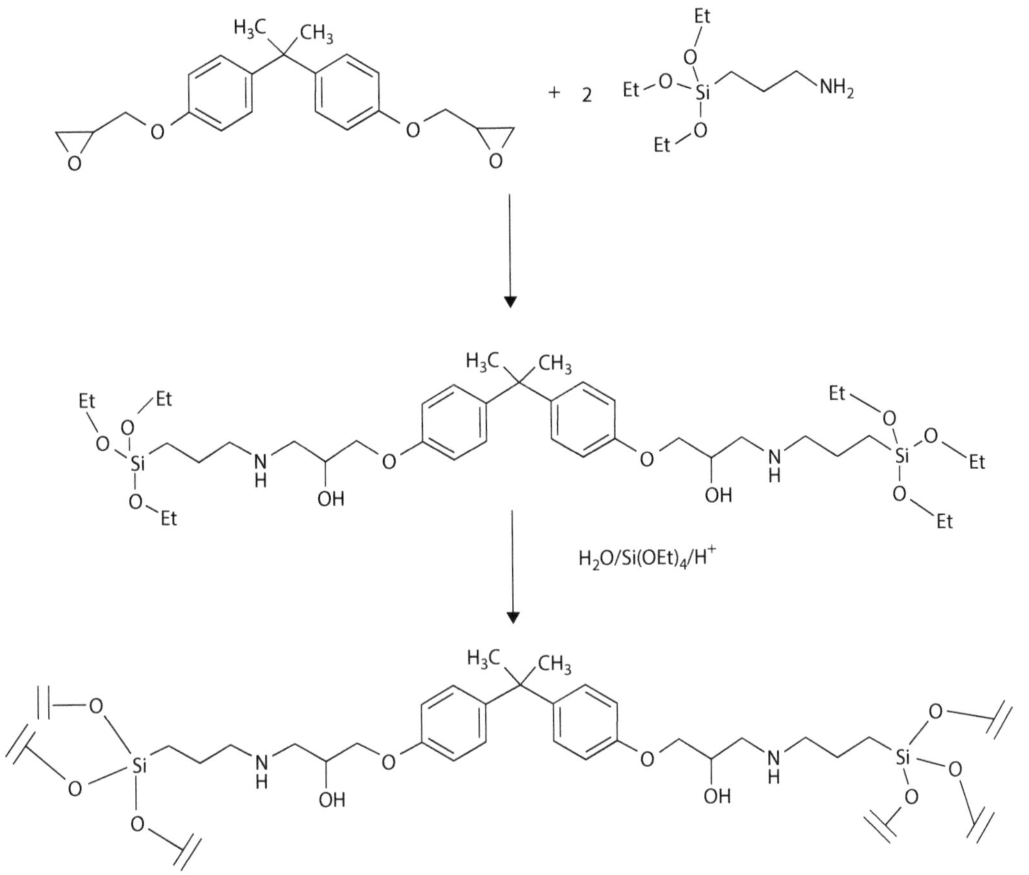

FIGURE 3.14 Scheme of the sol–gel reaction between diglycidyl ether of bisphenol A, 3-amino-propyltriethoxysilane, and tetraethylorthosilicate. (Turri, S., Torlaj, L., Piccinini, F., and Levi, M.: Abrasion and nanoscratch in nanostructured epoxy coatings. *J. Appl. Polym. Sci.* 2010. 118. 1720–1727. Copyright Wiley-VCH Verlag GmbH & Co. KGaA. Reprinted with permission.)

cross-linking agents. It seems that DCP initiated grafting between the polymer and the clay, which results in an exfoliated morphology, whereas DBP inhibited the initiation of EVA cross-linking by free radicals, because of the edge–edge interactions between the clay layers, which gives rise to a flocculated morphology and reduces the polymer–clay interaction. The better thermal and mechanical properties were obtained for DCP-cured clay nanocomposites, which is tabulated in Table 3.2.

Very few researches have successfully prepared nanocomposites with significantly improved dynamic mechanical properties, flames resistance, and thermal properties [11,44–46].

3.4.4 Polyurethane

Polyurethane (PU) is a synthetic polymer, which can be fit to meet the demands of modern technologies for a wide range of applications such as coatings, adhesives, reaction injection molding plastics, fibers, foams, rubbers, thermoplastic elastomers, and composites because

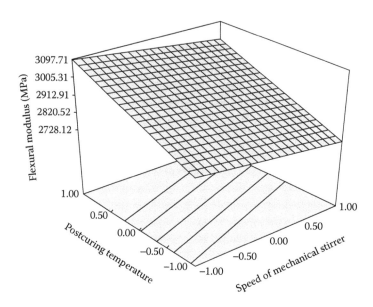

FIGURE 3.15 3D response surface plot of flexural modulus as a function of speed of mechanical stirrer and postcuring temperature in coded values. (Reprinted from Chow, W.S., *eXPRESS Polym. Lett.*, 2, 2, 2008. Copyright 2009, with permission from Express Polymer Letter Publishing Group.)

TABLE 3.2   Tensile Properties of the Different Nanocomposites

Sample	$\Sigma b \pm S\sigma b$ (MPa)	$E \pm SE$ (MPa)
Pure EVA	6.5 ± 1.5	24.8 ± 3.3
EVA/clay (1%)	6.9 ± 0.4	29.2 ± 1.9
EVA/clay (1%)/DCP	7.3 ± 1.6	29.0 ± 3.3
EVA/clay (1%)/DBP	6.0 ± 1.0	25.4 ± 2.8
EVA/clay (2%)	6.6 ± 0.2	28.0 ± 0.5
EVA/clay (2%)/DCP	7.1 ± 1.6	33.7 ± 4.7
EVA/clay (2%)/DBP	6.5 ± 0.4	26.8 ± 3.9
EVA/clay (3%)	6.0 ± 0.2	25.4 ± 2.3
EVA/clay (3%)/DCP	6.1 ± 0.5	25.9 ± 4.0
EVA/clay (3%)/DBP	6.4 ± 0.3	24.0 ± 0.7

Source: Reprinted from Mishra, S.B. and Luyt, A.S., *eXPRESS Polym. Lett.*, 2, 256–264. Copyright 2008, with permission from Express Polymer Letter Publishing Group.

of its versatility. The mechanical properties of PU can be modified by either varying the chemical structure of PU by tailoring their hard segments and soft segments or dispersing inorganic or organic filler into pristine PU. PU-based clay nanocomposites have been prepared, and the mechanical and barrier properties and flame retardantcy were studied by many researchers [47–59]. Kim et al. [55] prepared and characterized the nanoclay-reinforced rigid PU foams by XRD, scanning electron microscopy (SEM), and differential scanning calorimetry (DSC).

The thermal conductivity of the nanoclay-reinforced rigid PU closed cell foam can be approximated by a series model.

The conductive heat flux (q) through the composite wall is given by the following equation [55].

$$q = \frac{\Delta T}{R} \qquad (3.3)$$

where:
ΔT is the temperature drop across the foam
R is the conduction resistance

The conduction resistance can be calculated by the following equation [55]:

$$R = \sum_{i=1}^{n} \left(\frac{X_{W,i}}{k_{W,i}} + \frac{X_{G,i}}{k_{G,i}} \right) \qquad (3.4)$$

where:
$X_{W,i}$ and $X_{G,i}$ are the cell wall thickness and cell dimension, respectively
n is the number of polymer walls
k is the thermal conductivity

For uniform cells, the wall thickness and the cell dimension are constant, from the following equation [55]:

$$R = n \left(\frac{X_W}{k_W} + \frac{X_G}{k_G} \right) \qquad (3.5)$$

The conductivity of the polymer is much greater than that of the blowing gas. Therefore, the first term, polymer wall resistance, can be neglected to give the following equation [55]:

$$R = n \left(\frac{X_G}{k_G} \right) \qquad (3.6)$$

The analysis shows that the thermal insulation of closed cell foams increases linearly with the number of closed cells. Thermal conductivity (k) is one of the most important properties of foam for insulation applications. According to their reports, the thermal conductivity of foams decreases linearly with increasing clay content up to 1% clay due to the decreased cell size and increases with the number of cells (n). Overall, the effect of clay on thermal conductivity is based on the effect of clay on cell size.

3.4.5 Ethylene Propylene Diene Monomer

Ethylene propylene diene monomer (EPDM) is one of the most widely used synthetic rubbers. It offers an excellent resistance to heat, oxidation, ozone, and weathering. Because of these unique properties and structure, it finds its applications in automotive weather-stripping and seals, radiator, electrical insulation, roofing membrane, tubing, and belts.

Different processing conditions, morphology, and mechanical properties of EPDM/clay nanocomposites have been studied [60–64]. Ahmadi et al. [64] prepared nanocomposites, and the dispersion of layered silicate in the matrix was characterized by XRD and TEM. They also investigated the effect of gamma radiation on mechanical and dynamic mechanical thermal properties by using dynamic mechanical thermal analysis (DMTA) of nanocomposite and its conventional composite with pristine clay.

Figure 3.16 shows the effect of irradiation on the tensile strength values of nanocomposite (N), conventional composite(C), and unfilled EPDM (U).

The result seems that initially the tensile strength of nanocomposites (N) increases up to 200 kGy of radiation dose and then decreases, which leads to an increase in dose up to 1000 kGy. The tensile strength of N does not change even the irradiation dose is more than 1000 kGy. The tensile strength of conventional composite (C) and unfilled EPDM (U) also increase with increasing irradiation dose up to about 150 kGy and then decreases up to 300 kGy. The tensile strength values almost remain constant with an increase of irradiation dose up to 1000 kGy, and then a progressive reduction takes place at higher irradiation dose. Irradiation causes two types of reactions in nanocomposite: cross-link formation and scission of chains. The reduction in tensile strength for high irradiations due to chain scission lead to degradation of the polymer backbone. The results from Figure 3.16 suggest that the rate of reduction in tensile strength of C and U with an increase of irradiation dose is considerably high and is initiated in lower doses than that of N. It may be attributed to the effect of strong interaction between the modified nanoparticles and the EPDM matrix.

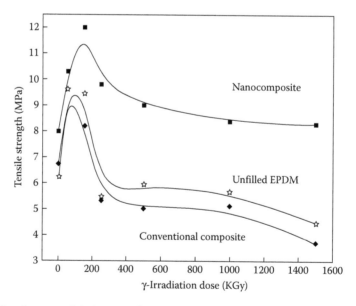

FIGURE 3.16   Tensile strength behavior of EPDM/organoclay nanocomposite, EPDM conventional composite, and unfilled EPDM vs. irradiation dose. (Reprinted from *Compos. Sci. Technol.*, 69, Ahmadi, S.J., Huang, Y.-D., Ren, N., Mohaddespour, A., and Ahmadi-Brooghani, S.Y., The comparison of EPDM/clay nanocomposites and conventional composites in exposure of gamma irradiation, 997–1003. Copyright 2009, with permission from Elsevier.)

3.4.6 Brominated Poly(Isobutylene-*co-para*-Methylstyrene)

Brominated poly(isobutylene-*co-para*-methylstyrene) (BIMS) is a relatively newly innovated one in the butyl rubber series. It is mainly used for making tire inner liners, tire-curing bags, and so on. Few researches have been conducted on BIMS/clay nanocomposites. Preparation, characterization, morphology, and improved mechanical and dynamic mechanical properties have been reported on BIMS/clay nanocomposites [65,66].

3.4.7 Acrylic Rubber

Acrylic-based polymers are usually referred to carboxylic acids polymer, and they are well known to react with multivalent metal cations and form a cross-link with metals, which are insoluble in water and most of the organic solvents. Many researches have been conducted on acrylic/clay-based nanocomposites [67–71]. Molu et al. [67] thoroughly studied the superabsorbent properties of poly(acrylic acid)/pillared clay nanocomposites prepared via graft copolymerization. In this study, the nanocomposites has been synthesized by graft copolymerization of two type's pillared clay by titrating aqueous 0.1 M NaOH with aqueous 0.1 M $AlCl_3 \cdot 6H_2O$ until the OH/Al ratio has equal to 2.0. The pillaring solution was aged for 2 h at 60°C and then kept overnight at 30°C. After aging, the resulting solution was reacted with a proper amount of aqueous suspension of KSF and K10 clay, keeping an Al/clay ratio of 10 mmol/g. The temperatures of the slurries were maintained at 60°C for 2 h followed by an aging period of 7 days at room temperature; then it was washed by centrifugation and dialysis until the absence of chloride and oven dried at 80°C for 18 h. It was calcined at 250°C for 2 h and then both pillared clay were taken for graft copolymerization reaction with acrylic acid with *N,N'*-methylene-bisacrylamide (MBA) as a cross-linker. The swelling behavior of Al-KSF- and Al-K10-based superabsorbent films seemed to be pH dependent as shown in Figure 3.17. The Fourier transform infrared spectroscopy (FTIR) results reveal that the clay layer structure remains unaffected by polymerization. It was confirmed by peak shifting of stretching vibrations of $-SiO_2$ tetrahedron and –OH bands supports the formation ester between acrylic network and pillared clay (Table 3.3).

Peila et al. [68] studied ultraviolet (UV)-cured acrylic nanocomposites based on modified organophilic MMT via photopolymerization. In this study, the photopolymerization reaction was monitored through real-time FTIR in order to check any influence of the nanofillers on the cure kinetics. The monomer was bisphenol-A ethoxylate.

(15EO/phenol) dimethacrylate (BEMA). Literature reveal that two types of commercially available MMT clays such as Cloisite Na⁺ (Clo-Na) and Cloisite 30B (Clo-30B) and 2-hydroxy-2-methyl-1-phenyl-1-propanone were used as a photoinitiator for photopolymerization. They strongly reported that the curing rate was not affected by the presence of nanoclay. The final conversion of the acrylic double bonds, even the maximum degree of conversion of the nanocomposites, was slightly higher with respect to that of the neat UV-cured resin by real-time IR as shown in Figure 3.18. TGA reveals that the initial degradation temperature of nanocomposites was slightly increased in the presence of nanoclay compared to the UV-cured neat acrylic resin as shown in Figure 3.19.

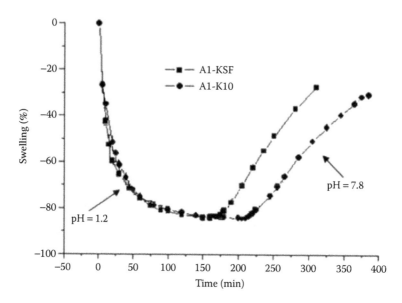

FIGURE 3.17 pH-dependent reversible swelling behavior of superabsorbents (equilibrated at pH = 7.8, then alternated between solutions at pH = 1.2 and 7.8). (Reprinted with kind permission from Springer Science + Business Media: *Polym. Bull.*, Preparation and characterization of poly(acrylic acid)/pillared clay superabsorbent composite, 64, 2009, 171, Molu, Z.B., Seki, Y., and Yurdakoç, K.)

Gao et al. [69] reported graft-acrylic acid superabsorbent nanocomposites by using glow-discharge electrolysis plasma. The glow-discharge electrolysis plasma polymerization procedure is as follows: A weighed amount of MMT clay along with a cross-linking agent MBA was dissolved in distilled water, and it was first placed in the reaction vessel and stirred for 40 min, followed by adding a desired amount of monomer into the above mixture solution and stirred again for 20 min to get mixed well. Then, the discharge was lasting for about 10 min. Discharge was stopped followed by reaction mixture and was warmed at 80°C in oil bath for 3 h after polymerization, then by cooling the resulting product at room temperature. The solid product has been cut into small pieces (diameter of 5 mm), and then treated with 1 mol/L sodium hydroxide (58.3–105 mL) to neutralize in part (50%–90%) the carboxylic groups of the grafted poly(acrylic acid) composite.

Finally, the product was washed many times with distilled water until it showed pH ~7. After removing water with methanol, the products were dried in the vacuum oven at 65°C to a constant weight, then milled and screened. The final products obtained are PAA/MMT superabsorbent nanocomposite. Literature reveals that two types of commercially available MMT clays such as Cloisite Na+ (Clo-Na) and Cloisite 30B (Clo-30B) and 2-hydroxy-2-methyl-1-phenyl-1-propanone were used as a photoinitiator for photopolymerization (Figures 3.20 and 3.21). However, prepared superabsorbent composite with excellent water retention, it would be expected to find applications like agricultural sectors. Sugimoto et al. [71] thoroughly studied UV-cured transparent acryl–clay nanohybrid films with a low thermal expansion coefficient. The coatings were done by doctor blade onto a polyethylene terephthalate (PET) film. In this study, they calculated the tensile strength,

TABLE 3.3   The Characteristic FTIR Data of the Samples

IR Bands	K10	Al-K10	Al-K10 Composite	KSF	Al-KSF	Al-KSF Composite
Al₂OH (octahedral layer) (cm⁻¹)	3,623	3,622–3,616	3,647	3,620	3,624	3,629
Stretching vibration of H₂O (cm⁻¹)	3,428	3,436	3,395–3,345	3,411–3,389	3,429	3,439–3,363
Stretching vibration of –CH₂ (cm⁻¹)			2,988–2,881			2,956–2,871
Stretching vibration of C=O (cm⁻¹)			1,750			1,722
Bending vibration of H₂O (cm⁻¹)	1,633	1,636	1,682	1,636	1,633	1,621
Asymmetric vibration of R-COOK (cm⁻¹)			1,539			1,554
Bending vibration of –CH₂ (cm⁻¹)			1,463			1,445
Symmetric vibration of R-COOK (cm⁻¹)			1,402			1,402
Asymmetric stretching vibration of SiO₂ tetrahedra (cm⁻¹)	1,046	1,041	1,076	1,046	1,041	1,075
Bending vibration of Al₂OH (cm⁻¹)	920	921	915	917	921	914
Stretching vibration of Al^IV tetrahedra (cm⁻¹)	798	796	800	794	794	806
Bending vibration of Si–O (cm⁻¹)	524,468	528,471	524,469	523,467	526,471	520,482

Source: Reprinted with kind permission from Springer Science + Business Media: *Polym. Bull.*, Preparation and characterization of poly(acrylic acid)/pillared clay superabsorbent composite, 64, 2009, 171, Molu, Z.B., Seki, Y., and Yurdakoç, K.

thermal properties, and morphologies of the fabricated films; orientation distribution (F(x)); and orientation coefficient (f) of clay platelets using the following equations [71]:

$$F(x) = \frac{\int_{x-\Pi/36}^{x+\Pi/36} I(\phi)d\phi}{\int_{\Pi/2}^{\Pi/2} I(\phi)d\phi} \quad (3.7)$$

$$\langle\cos^2\phi\rangle = \frac{\int_0^{\Pi/2} I(\phi)\cos^2\phi\sin\phi d\phi}{\int_0^{\Pi/2} I(\phi)\sin\phi d\phi} \quad (3.8)$$

$$f = \frac{3\langle\cos^2\phi\rangle - 1}{2} \quad (3.9)$$

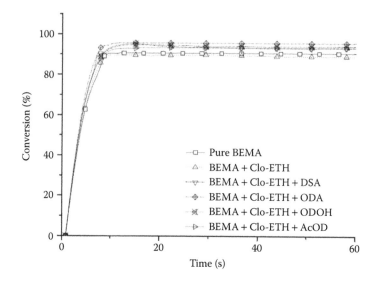

FIGURE 3.18   Real-time IR spectra of the UV-curable acrylic nanocomposite systems based on modified nanoclays. (Reprinted with kind permission from Springer Science + Business Media: *J. Therm. Anal. Calorim.*, Preparation and characterization of UV-cured acrylic nanocomposites based on modified organophilic montmorillonites, 97, 2009, 839, Peila, R., Malucelli, G., and Priola, A.)

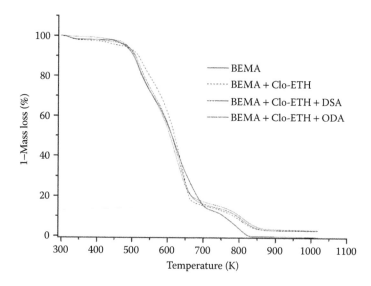

FIGURE 3.19   TGA curves in air of the BEMA nanocomposites based on the modified nanoclays. (Reprinted with kind permission from Springer Science + Business Media: *J. Therm. Anal. Calorim.*, Preparation and characterization of UV-cured acrylic nanocomposites based on modified organophilic montmorillonites, 97, 2009, 839, Peila, R., Malucelli, G., and Priola, A.)

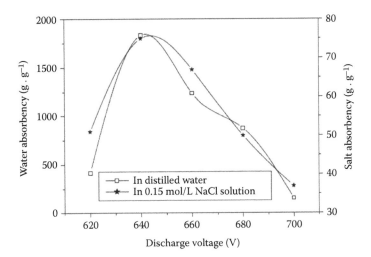

FIGURE 3.20 Effect of discharge voltage. (Reprinted with kind permission from Springer Science + Business Media: *Plasma Chem. Plasma Process.*, Synthesis and characterization of montmorillonite-graft-acrylic acid superabsorbent by using glow-discharge electrolysis plasma, 30, 2010, 873, Gao, J., Ma, D., Lu, Q., Li, Y., Li, X., and Yang, W.)

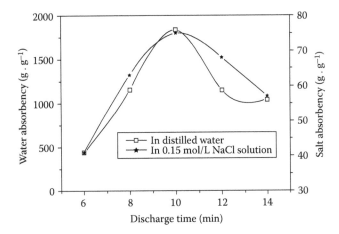

FIGURE 3.21 Effect of discharge time. (Reprinted with kind permission from Springer Science + Business Media: *Plasma Chem. Plasma Process.*, Synthesis and characterization of montmorillonite-graft-acrylic acid superabsorbent by using glow-discharge electrolysis plasma, 30, 2010, 873, Gao, J., Ma, D., Lu, Q., Li, Y., Li, X., and Yang, W.)

where ϕ is the tilt angle from the coating direction, which can be found using a wide-angle XRD measurement as illustrated in Figure 3.22. It was reported that the prepared nano-hybrid films show a low coefficient of thermal expansion, and the coefficient of thermal expansion is reduced to 10 ppm/K at 40 wt.% of clay.

3.4.8 Nitrile Rubber

The literature finds that very few types of clays were used for synthesizing the nitrile- and acrylonitrile-butadiene-based nanocomposites [72–75]. Xu and Karger-Kocsis [72]

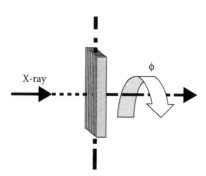

FIGURE 3.22 Wide-angle XRD measurement using a tilt angle (ϕ) from the coating direction. (Reprinted with kind permission from Springer Science + Business Media: *Colloid Polym. Sci.*, Transparent acryl–clay nanohybrid films with low thermal expansion coefficient, 288, 2010, 1131, Sugimoto, H., Nunome, K., Daimatsu, K., Nakanishi, E., and Inomata, K.)

prepared organophilic layered silicate/hydrogenated nitrile rubber nanocomposite. They studied the dry rolling and sliding friction and wear properties of nanocomposites. They reported that layered silicate/hydrogenated nitrile rubber nanocomposites exhibit better dry rolling and wear resistance compared to pristine nitrile rubber. However, network-related parameters such as apparent mean molecular mass between cross-links (M_c) and apparent network density (v_c) and stiffness (E') are shown in Figure 3.23 were deduced from DMTA. It shows that nanocomposites exhibit improved stiffness compared to pristine nitrile rubber. The obtained data are compiled in Table 3.4. The results are consistent with the intercalated and partially exfoliated morphology of nanocomposites.

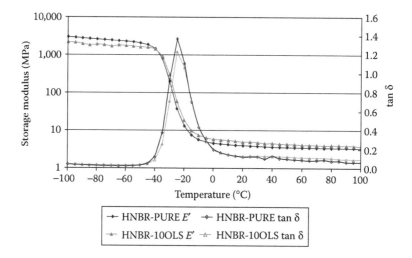

FIGURE 3.23 Storage modulus (E') and tan δ as a function of temperature for the layered silicate/hydrogenated nitrile rubber nanocomposites. (Reprinted with kind permission from Springer Science + Business Media: *J. Mater. Sci.*, Dry rolling and sliding friction and wear of organophilic layered silicate/hydrogenated nitrile rubber nanocomposite, 45, 2009, 1293, Xu, D. and Karger-Kocsis, J.)

TABLE 3.4 Structural and Wear Properties of the Pristine Nitrile Rubber and Layered Silicate/Hydrogenated Nitrile Rubber Nanocomposite

Properties	HNBR–PURE	HNBR–10OLS
Shore (Å)	42	52
Density (g/cm³)	1.057	1.050
M_c (g/mol)	2013	1564
υ_c (mol/dm³)	0.52	0.67
tan δ at T_g	1.37	1.23
Orbital–RBOP		
COF	4.23×10^{-2}	5.36×10^{-2}
W_s (mm³/N m)	2.95×10^{-4}	1.34×10^{-4}
POP		
COF	1.14	1.24
W_s (mm³/N m)	1.13×10^{-1}	3.68×10^{-2}
ROP		
COF	3.00	1.89
W_s (mm³/N m)	3.79×10^{-2}	7.83×10^{-3}
Fretting		
ROP	0.90	1.00
COF	1.50×10^{-3}	3.56×10^{-3}

Source: Reprinted with kind permission from Springer Science + Business Media: *J. Mater. Sci.*, Dry rolling and sliding friction and wear of organophilic layered silicate/hydrogenated nitrile rubber nanocomposite, 45, 2009, 1293, Xu, D. and Karger-Kocsis, J.

Wang et al. [74] attempted to explore sodium-MMT as a flame retardant in nitrile butadiene rubber nanocomposites. In this study, they reported that nanocomposites give higher flame retardancy compared to neat nitrile butadiene rubber.

Kader et al. [75] thoroughly studied the mechanical properties such as tensile, tear strength, and dynamic mechanical properties, and the thermal properties of nitrile rubber/MMT nanocomposites via latex blending. They strongly reported that significant improvements were obtained in the tensile properties such as the percentage of elongation, yield strength, and tear strength. They also claimed that decent improvement in the dynamic mechanical properties was obtained for the nanocomposites. Morphology also supported the property improvement. Intercalated and partially exfoliated morphology of nanocomposites has been revealed through TEM, which is illustrated in Figure 3.24.

3.4.9 Styrene-Butadiene-Based Rubber and Its Derivatives

Styrene-butadiene-based rubber (SBR) is another one type of synthetic rubber, which is commonly used for many commercial products, especially for tire industries. In general, the mechanical property of this rubber depends on the styrene content in SBR. With high styrene, the content leads to an increase of tensile strength along with modulus. Literature found that lots of reports are available on the styrene-butadiene-based rubber [10,26,76–80]. Ghasemi et al. [10] studied the effect of processing conditions and organoclay content on

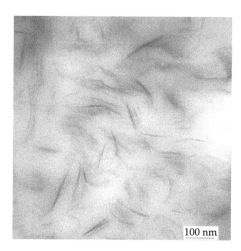

FIGURE 3.24   High-magnification transmission electron microscopic image of NBR/Na-MMT-intercalated/exfoliated nanocomposite (loading of Na-MMT = 5 phr). (Reprinted with kind permission from Springer Science + Business Media: *J. Mater. Sci.*, Preparation and properties of nitrile rubber/montmorillonite nanocomposites via latex blending, 41, 2006, 7341, Kader, M.A., Kim, K., Lee, Y.S., and Nah, C.)

the properties of SBR/organoclay nanocomposites by using response surface methodology. They reported that increasing temperature and mixing time in internal mixer contributed to better organoclay dispersion in matrix, which resulted in improved mechanical properties, and intercalation/exfoliation of the clay was observed for the compounds, which were mixed at higher temperature and longer mixing time. However, thermal degradation of the matrix occurred at high temperature around 140°C of mixing and mechanical properties were affected by this phenomenon. They also reported that increasing nanoclay proportion in nanocomposites decreases the scorch time (Figure 3.25). This could be one of the major disadvantages of the processing aspect.

irmohseni and Zavareh [76] conducted a research on epoxy/acrylonitrile-butadiene-styrene (ABS) copolymer/clay ternary nanocomposite as impact toughened epoxy. In this study, the main objective of their work is to incorporate both ABS copolymer and organically modified clay (Cloisite 30B) into the epoxy matrix with the aim of obtaining high impact strength than control epoxy. They reported that better impact resistance for nanocomposites due to intercalated and partially exfoliated morphology was revealed by using AFM, which is illustrated in Figure 3.26a and b.

Chakraborty et al. [26] studied *in situ* sodium-activated and organomodified bentonite clay/SBR nanocomposites by a latex blending technique. In this study, they strongly followed the different formulations for preparing nanocomposites. They claimed that clay nanocomposites exhibit better mechanical properties and also thoroughly studied the degradation kinetics of nanocomposite by using DSC followed by rheological behavior of rubber nanocomposites.

Sulfonated rubber-based clay nanocomposite is a unique kind of rubber. This is one of the less investigated research areas in the polymer/clay nanocomposites era. Swaminathan

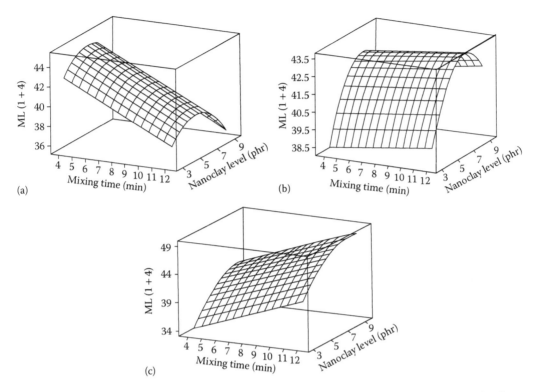

FIGURE 3.25 Mooney viscosity against mixing time and nanoclay content at three levels of mixing temperatures: (a) 80°C; (b) 110°C; (c) 140°C. (Reprinted from Ghasemi, I., *eXPRESS Polym. Lett.*, 4, 62. Copyright 2010, with permission from Express Polymer Letter Publishing Group.)

and Dharmalingam [79] studied the evaluation of sulfonated polystyrene ethylene butylene polystyrene (PSEBS)/MMT nanocomposites as a proton exchange membranes for fuel cell applications. In this study, sulfonated PSEBS (SPSEBS) were prepared by dissolving PSEBS in chloroform solvent with sulfonating agent chlorosulfonic acid was added drop wise simultaneously with constant stirring.

After 3 h, the sulfonation process was terminated by addition of methanol into the reaction vessel. The obtained SPSEBS with solvent was poured in a glass plate to allow the solvent to evaporate. The excess unreacted acid was removed by thoroughly washing off the dry ionomer with deionized water followed by drying. The whole reaction for preparing SPSEBS is illustrated in Figure 3.27. The weighed amount of dry ionomer was dissolved in tetrahydrofuran (THF) solvent followed by clay for preparing nanocomposites. Water absorption and methanol permeability were reduced by the interaction of the incorporated inorganic material and improved tensile strength compared to neat SPSEBS rubber.

3.4.10 Silicone Rubber (Q)

Silicone rubber (Q) has an excellent weather ability, good chemical stability, oxidation resistance, thermal stability, low-temperature toughness, electrical insulating properties, low surface energy, low toxicity, and high optical transparency. It is widely used for a variety of applications such as lubricants, sealants, adhesives, medical implants, and electrical insulating

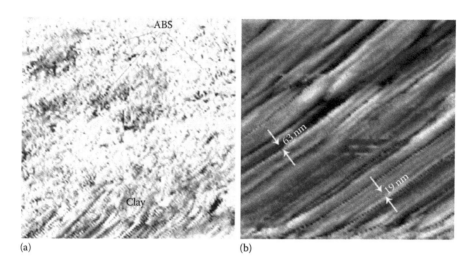

FIGURE 3.26 AFM-phase image of the ternary nanocomposite with optimum composition. (a) plane and (b) cross view. (Reprinted with kind permission from Springer Science + Business Media: *J. Polym. Res.*, Epoxy/acrylonitrile-butadiene-styrene copolymer/clay ternary nanocomposite as impact toughened epoxy, 17, 2009, 191, Mirmohseni, A. and Zavareh, S.)

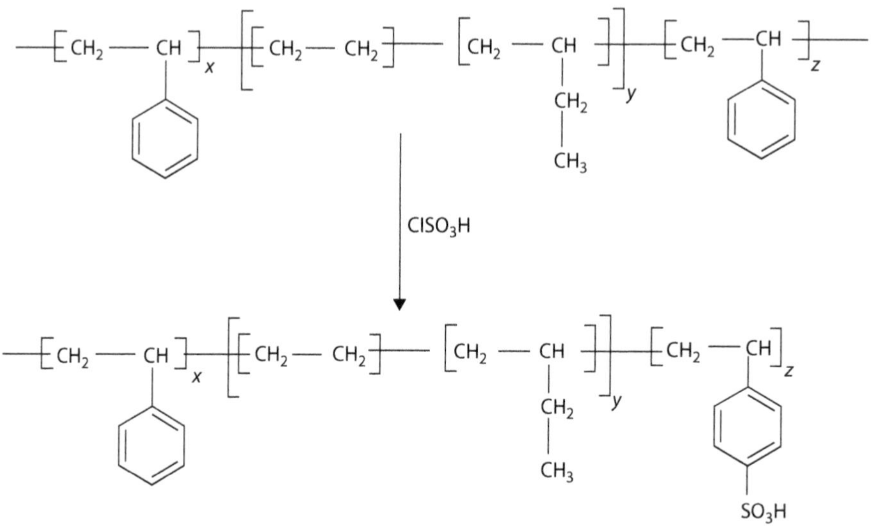

FIGURE 3.27 Sulfonation of PSEBS. (Reprinted with kind permission from Springer Science + Business Media: Int. J. Plast. Technol., Evaluation of sulfonated polystyrene ethylene butylene polystyrene-montmorillonite nanocomposites as proton exchange membranes, 13, 2010, 150, Swaminathan, E. and Dharmalingam, S.)

products. However, unfilled silicone elastomer usually has poor mechanical properties and low thermal/electrical conductivity. Nanoclay is one of the right candidates to overcome this problem. Many researches have been conducted on this rubber [2,81–87]. Wang and Chen [83] investigated high-temperature vulcanized silicone rubber (HTV-SR)/hyperbranched OMMT (HOMMT) nanocomposites. The HTV-SR is not beneficial to intercalate into

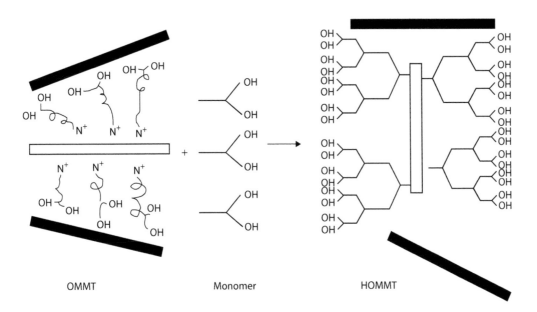

OMMT   Monomer   HOMMT

FIGURE 3.28 Scheme of hyperbranched reactions between OMMT and the monomer. (Reprinted with kind permission from Springer Science + Business Media: *J. Appl. Polym. Sci.*, Synthesis of hyperbranched organo-montmorillonite and its application into high-temperature vulcanized silicone rubber systems, 111, 2008, 648, Wang, J., Chen, Y., and Wang J.)

layered silicate galleries. To overcome this problem, they prepared HOMMT by condensation reaction between OMMT and *N,N*-dihydroxyl-3-aminmethyl propionate as a monomer. The reaction between those two nanocomposites is shown in Figure 3.28. The nanostructure formation occurs if the free energy change per interlayer volume (ΔF_V) is negative. ΔF_V can be expressed as follows [83]:

$$\Delta F_V = \Delta E_V - T\Delta S_V \tag{3.10}$$

where:
ΔE_V and ΔS_V are the enthalpy and entropy changes per interlayer volume
T is the temperature

Because there is no effect on the total entropy change when there is a small increase in the gallery spacing, intercalation will rather be driven by the changes in total enthalpy, which is expressed as follows [83]:

$$\Delta E_V = \varphi_1 \varphi_2 \frac{1}{Q} \left(\frac{2}{h_0} (\varepsilon_{sp} - \varepsilon_{sa}) + \frac{2}{r} \varepsilon_{ap} \right) \tag{3.11}$$

where:
φ_1 and φ_2 are the interlayer volume fractions of intercalated polymer and tethered surfactant chains, respectively
Q is a constant near unity

h_0 is the initial gallery height of organoclay

r is the radius of the intercalation surface of tethered surfactant chains

ε_{sp}, ε_{sa}, and ε_{ap} represent the intercalation energies per area between the layered silicate and the polymer, the layered silicate and the intercalation agent, and the intercalation agent and the polymer, respectively

When 1 phr of HOMMT was incorporated into the HTV-SR system, the interactions between the intercalation agent and the polymer, and the layered silicate and the polymer were also increased to the highest, leading to a most favorable ε_{ap} and ε_{sp}. Therefore, there is a favorable excess enthalpy to enhance the dispersion of the HOMMT, resulting in the formation of intercalated nanostructures. However, in the case of 1 phr OMMT, the interactions between the intercalation agent and the polymer, and the layered silicate and the polymer were decreased, resulting in a decreased ε_{ap} and ε_{sp}. Decrease of the enthalpy can decrease the dispersion degree of the OMMT.

3.4.11 Fluoroelastomers

Fluorocarbon rubbers are being widely used in critical applications because of their unbelievable heat and fluid resistance. The most important applications are in O-rings, gaskets, shaft seals, and fuel hoses. To further improve these performances, the fluoroelastomer nanocomposites are synthesized. The mechanical properties, morphology, and swelling behaviors were extensively studied by many researchers [88–93]. Maiti and Bhowmick [88] studied the morphology, mechanical, dynamic mechanical, and swelling properties of fluoroelastomer nanocomposites. Compared to the modified clay-filled system, the unmodified clay-filled system showed better mechanical properties. Because the better interaction was observed for, the unmodified clay as it can be seen in terms of surface energy. The values of work of adhesion, spreading coefficient, and interfacial tension is shown in Table 3.5, whereas the whole surface energy of the fluoroelastomer and the clay are reported in Table 3.6.

Generally, the surface energy of Cloisite Na+ is higher than that of fluorocarbon rubber. Assuming fluorocarbon rubber as a wetting polymer, its lower surface energy helps in wetting the solid clay. This is not in the case of the fluorocarbon rubber/modified clay system, where the surface energy of the modified clay was lower. These results are more obvious from the Δγ values. The Δγ value of Viton B-50 and the unmodified clay is much lower

TABLE 3.5 Different Surface Properties of Rubber and Clay

Sample Designation	Work of Adhesion (mJ/m²)	Spreading Coefficient (mJ/m²)	Interfacial Tension (mJ/m²)
Viton B-50 and Closite Na+	67.63	5.47	1.10
Viton B-50 and Closite 20A	51.42	−9.91	2.47

Source: Maiti, M. and Bhowmick, A.K.: Structure and properties of some novel fluoroelastomer/clay nanocomposites with special reference to their interaction. *J. Polym. Sci. Part B: Polym. Phys.* 2006. 44. 162–176. Copyright Wiley–VCH Verlag GmBH & CO. KGaA. Reprinted with permission.

TABLE 3.6 Surface Energy of Rubber and Clay

| Sample | Contact Angle (°) | | Surface Energy (mJ/m²) |
	Water	Formamide	
Viton B-50	71.8	66.8	31.51
Closite Na+	66.7	54.8	37.22
Closite 20A	84.9	87.8	22.38

Source: Maiti, M. and Bhowmick, A.K.: Structure and properties of some novel fluoroelastomer/clay nanocomposites with special reference to their interaction. *J. Polym. Sci. Part B: Polym. Phys.* 2006. 44. 162–176. Copyright Wiley–VCH Verlag GmBH & CO. KGaA. Reprinted with permission.

(5.71 mJ/m²) than that (9.13 mJ/m²) of Viton B-50 and the modified clay. A lower surface energy mismatch leads to better wetting, better interfacial adhesion, and increased diffusion of the polymers across the interface. The interfacial tension between Viton B-50 and Na is also much less. The positive spreading coefficient for this system definitely indicates better diffusion of the polymers into the clay interface. Hence, interaction was good in the case of the unmodified clay. In addition, the work of adhesion was also higher in the case of the rubber–unmodified clay system. Thus, the polymer chains can spread more easily on the surface of the unmodified clay than that of the modified clay.

The same research group again investigated the diffusion and sorption of methyl ethyl ketone (MEK) and THF through fluoroelastomer–clay nanocomposites [89] in the range of 30°C–60°C by using swelling experiment. The overall sorption value decreases with increasing the nanoclay content. However, it decreases significantly for the unmodified clay-filled (Table 3.7).

They also established a model to have more idea about how the aspect ratio of nanoclays affects swelling behavior. Addition of the layered nanoclay to a virgin polymer will restrict the permeability due to the following phenomena:

1. The available area for diffusion will decrease as a result of impermeable nanoclay replacing the permeable polymer.

2. When nanoclays are incorporated into the system, we may assume that the clay layers are randomly dispersed in the matrix. The diffusion of the solvent will detour around

TABLE 3.7 Equilibrium Sorption Values of Different Nanocomposite–Solvent Systems

| Sample (after Leaching out) | $M \infty$ (Mass Uptake), g | | | | | |
| | MEK | | | THF | | |
	30°C	45°C	60°C	30°C	45°C	60°C
F	0.260	0.271	0.281	0.274	0.284	0.294
FN4	0.087	0.104	0.115	0.139	0.153	0.168
F20A4	0.218	0.244	0.253	0.237	0.261	0.271

Source: Maiti, M. and Bhowmick, A.K.: Effect of polymer–clay interaction on solvent transport behavior of fluoroelastomer–clay nanocomposites and prediction of aspect ratio of nanoclay. *J. Appl. Polym. Sci.* 2007. 105. 435–445. Copyright Wiley-VCH Verlag GmbH & Co. KGaA. Reprinted with permission.

the impermeable clay layers. Diffusion will be diverted to pass a clay platelet in every layer, and thus, the solvent must have to travel a longer path (d_f) in the filled system compared with that (d_0) for the neat polymer.

Using the scaling concept, the permeability P can be written as follows:

$$P \sim \frac{A}{D} \tag{3.12}$$

where:
  A is the cross-sectional area available for diffusion
  D is the path length the solvent must travel to cross the sample

As a result, the permeability of nanocomposites (P_f) is reduced from that of the neat polymer (P_0) by the product of the decreased area and the increased path length as follows [89]:

$$\frac{P_0}{P_f} = \left(\frac{A_0}{A_f}\right)\left(\frac{d_f}{d_0}\right) \tag{3.13}$$

where:
  A_0 is the cross-sectional area available for diffusion in a neat polymer sample
  A_f is the cross-sectional area available for diffusion in a nanocomposite
  d_0 is the sample thickness (i.e., the distance a solvent molecule must travel to cross the neat polymer sample)
  d_f is the distance a solvent molecule must travel to cross the nanocomposite sample

Now, the above equation becomes [89]

$$\frac{P_0}{P_f} = \frac{V_0/d_0}{(V_0 - V_0)/d_f}\left(\frac{d_f}{d_0}\right) \tag{3.14}$$

where:
  V_0 is the total volume of the neat polymer sample
  V_f is the volume of nanoclays in the nanocomposite sample

The above equation is simplified to get the following equation:

$$\frac{P_0}{P_f} = \frac{V_0}{V_0 - V_f}\left(\frac{d_f}{d_0}\right)^2 = \frac{1}{1-\phi}\left(\frac{d_f}{d_0}\right)^2 \tag{3.15}$$

where ϕ is the volume fraction of filler.

When a solvent diffuses across a neat polymer, it must travel the thickness of the sample (d_0). When the same solvent diffuses through a nanocomposite film with nanoclays, its path length is increased by the distance it must travel around each clay layer it strikes. According to Lan et al. [94], the path length of a gas molecule diffusing through an exfoliated nanocomposite is

$$d_f = d_0 + \frac{d_0 L_\phi}{2 d_c} \tag{3.16}$$

where L and d_c are the length and thickness of a clay layer, respectively.

Substituting this value in Equation 3.15 [89], we get

$$\frac{P_0}{P_f} = \frac{1}{1-\phi}\left(1+\frac{L_\phi}{2 d_c}\right)^2 = \frac{1}{1-\phi}\left(1+\frac{\alpha\phi}{2}\right)^2 \tag{3.17}$$

where the aspect ratio $L/d_c = \alpha$.

The aspect ratio of the nanoclay in different samples has been calculated using Equation 3.17 and reported in Table 3.8.

3.4.12 Phenolic Resin

Phenolic and amino resins are synthetic types but exploited commercially, and they are produced from low-molecular-weight compounds. The novolac and resole belong to the phenolic resin family. Moreover, urea formaldehyde, melamine formaldehyde (MF), and furan-based resins are supplement as well as complement for phenolic resins. Since the curing is challenging, less work has been reported for phenolic resin/nanoclay-based nanocomposites. In general, phenolic resins are 3D structures; even prior to cure, clay intercalation is very much difficult to achieve. Moreover, most of the phenolic resins will have pores after curing since water as the by-product. For this reason, much attention is not paid to this field as we can see from the literature, because it is a vey less investigated area [94–100]. It finds its applications in various fields such as automobile, aerospace sector, aircraft, electrical insulation, and household appliance areas.

Kaynak and Tasan [95] studied the effect of production parameters on the structure–property relationship of resol-type phenolic resin/layered silicate nanocomposites. Production parameters such as two different types of phenolic resin (neat phenolic and additions of monoethyleneglycol [MEG] and diethyleneglycol [DEG]), two types of clay

TABLE 3.8 Average Aspect Ratio of Clay Layers Present in Different Nanocomposites

	Aspect Ratio	
Sample	Swelling	Morphology
FN4	146 ± 14	145 ± 6
F20A4	63 ± 5	53 ± 6
FN8	60 ± 4	60 ± 9
FN16	10 ± 1	6 ± 2

Source: Maiti, M. and Bhowmick, A.K.: Effect of polymer–clay interaction on solvent transport behavior of fluoroelastomer–clay nanocomposites and prediction of aspect ratio of nanoclay. *J. Appl. Polym. Sci.* 2007. 105. 435–445. Copyright Wiley-VCH Verlag GmbH & Co. KGaA. Reprinted with permission.

(organically modified MMT and pristine MMT) with different weight percentage load-ings, and also two types of curing system (acid and thermal or heat cure) were explored. As discussed earlier, while curing phenolic resin used to release water, it leads to the for-mation of micro voids in the final nanocomposites. The same problem was experienced in this work too, but they tuned it by heat curing route, by keeping the gel time long enough for the slow water vapor release. Hence, it completely prevents the formation of micro void formation. The microstructure of void formation and disappearance of voids are shown in Figure 3.29.

However, heat curing required very long heating schedules (as long as 3 days), lead-ing to the formation of separate clay-rich and resin-rich layers due to the density differ-ences. Moreover, in the acid curing route, cross-linking time was much quicker (4 h) for preventing the agglomeration or phase separation of the clay. Therefore, the structure and consequently mechanical properties of the acid-cured samples were better than those of the heat-cured ones. The effect of two different types of phenolic resins like ethylene glycol and modified resol type phenolic resins are much easier to cure. The required cure cycle time was much shorter for the modified phenolic resin. The SEM analysis revealed that macro and micro void formation was less likely for the pristine phenolic resin; and even when formed, these voids were smaller compared to those formed in modified resol along with that increase in density.

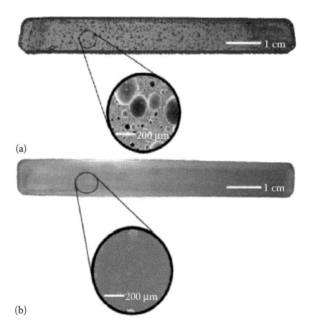

FIGURE 3.29   Specimen photo (a) and inset SEM image (b) showing macro voids formed due to by-product water molecules getting trapped during curing reaction in (a) and void-free phenolic resin structure obtained by the heat curing route in (b). (Reprinted from *Eur. Polym. J.*, 42, Kaynak, C. and Tasan, C.C., Effects of production parameters on the structure of resol type phenolic resin/layered silicate nanocomposites, 1908–1921. Copyright 2006, with permission from Elsevier.)

TABLE 3.9   Mechanical Properties of the Two Resol-Type Phenol Formaldehyde Neat Resin Specimens Produced by the Acid Curing Route

Resin Type	Flexural Strength (MPa)	Charpy Impact Strength (kJ/m²)	Fracture Toughness (MPa√m)
Pristine Phenol	101 ± 7	0.93 ± 0.10	0.87 ± 0.10
Additions of MEG and DEG	105 ± 7	1.08 ± 0.14	0.92 ± 0.02

Source: Reprinted from *Eur. Polym. J.*, 42, Kaynak, C. and Tasan, C.C., Effects of production parameters on the structure of resol type phenolic resin/layered silicate nanocomposites, 1908–1921. Copyright 2006, with permission from Elsevier.

TABLE 3.10   Mechanical Properties of the Modified Phenolic Resin with Different Modified Clay Loadings Produced by the Acid Curing Route

Resin Type	Flexural Strength (MPa)	Charpy Impact Strength (kJ/m²)	Fracture Toughness (MPa√m)
Modified phenolic resin without clay	105 ± 7	1.08 ± 0.14	0.92 ± 0.02
Resin + 0.5% clay	111 ± 2	1.08 ± 0.06	1.53 ± 0.29
Resin + 1% clay	108 ± 7	1.02 ± 0.16	0.73 ± 0.05
Resin + 1.5% clay	95 ± 4	1.12 ± 0.15	0.76 ± 0.19

Source: Reprinted from *Eur. Polym. J.*, 42, Kaynak, C. and Tasan, C.C., Effects of production parameters on the structure of resol type phenolic resin/layered silicate nanocomposites, 1908–1921. Copyright 2006, with permission from Elsevier.

Therefore, the structure and mechanical properties such as flexural strength, fracture toughness, and impact strength of the modified specimens were better than those of neat phenolic specimens as tabulated in Tables 3.9 and 3.10.

Generally, most of the thermoset resins are brittle in nature, so MEG and DEG act as a plasticizer for phenolic resin. In addition, organically modified clay nanocomposites show better mechanical properties compared unmodified ones because organically modified clays have increased interlayer distance platelet; it facilitates the entering of polymer chains into clay layers for intercalation or exfoliation.

A.R. Bahramian et al. [96] investigated high-temperature ablation of kaolinite layered silicate/phenolic resin/asbestos cloth nanocomposites. They compared the thermal degradation of nanocomposites theoretically as well as experimentally. In this work, the combination of solution and *in situ* intercalation technique was employed for nanocomposite preparation. In this study, ethyl alcohol was used to disperse the layered silicates, which were intercalated with dimethylsulfoxide, and simultaneously, they dissolved the phenolic resin too. To stack the layers, the crystallite was delaminated in ethyl alcohol. It is due to the weak van der Waals forces. Phenolic resin can then be adsorbed onto the delaminated individual layer. However, upon ethyl alcohol removal, the layers can reassemble to reform the stack with phenolic chains sandwiched in between, forming a well-ordered intercalated nanocomposite. Then immediately asbestos cloth was impregnated by phenolic resin/layered silicate intercalated. The final sample was precured at 120°C for 10 min, and then cured at 160°C and 3 bar for an hour in autoclave. After curing, the composite was postcured for 0.5 h at 150°C. Consequently, ethyl alcohol removed after phenolic resin polymerization. The final sample looked like a flat panel

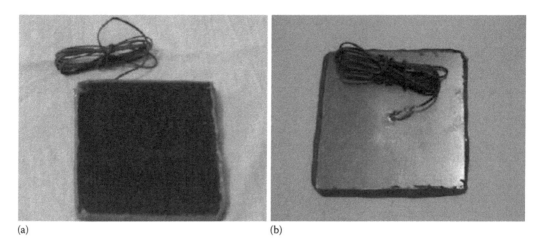

(a)   (b)

FIGURE 3.30 The asbestos cloth impregnated by phenolic resin/layered silicate intercalated nanocomposite sample on an aluminum substrate for oxyacetylene flame test: top (a) and back (b) surfaces of the sample. (Reprinted from *J. Hazard. Mater.*, 150, Bahramian, A.R., Kokabi, M., Famili, M.H., and Beheshty, M.H., High temperature ablation of kaolinite layered silicate/phenolic resin/asbestos cloth nanocomposite, 136–145. Copyright 2008, with permission from Elsevier.)

of dimensions of 6 × 100 × 100 mm with a sandwich structure, formed from 4 mm layer of nanocomposite and 2 mm layer of aluminum substrate. For oxyacetylene flame test, the nickel-chrome thermocouple was placed on the back of the sample as shown in Figure 3.30. Finally, the result shows improved thermal properties for modified kaolinite nanocomposites. Evidently, the thermal decomposition of the phenolic/asbestos-kaolinite nanocomposites shifts slightly toward the higher temperature range compared with that of pristine phenolic/asbestos composite, and also it confirms the enhancement of thermal stability of nanocomposite. After 600°C, mainly the inorganic residue was remained. It has been revealed by TGA as shown in Figure 3.31. The comparison between experimental and theoretical model values of kinetic parameters of thermal degradation such as activation energy and frequency factor values is almost similar as shown in Figure 3.32.

The ablation performance according to ASTM E285-80 and the flammability of the nanocomposite were studied by using oxyacetylene flame test and cone calorimetry according to ASTM E1354.

The ablation performance and flame resistance was improved significantly for the nanocomposites compared to neat polymer. The base of this ceramic layer is aluminum silicate that protected the ablation char layer against the thermal erosion effects as shown in Figures 3.33 and 3.34.

Park and Jeong [99] studied the cure kinetics of MF resin/clay/cellulose nanocomposites by using Ozawa and Kissinger models. The following equations explain the Ozawa and Kissinger models:

$$\log \beta = -2.135 - 0.456 \left(\frac{E}{RT_p} \right) + \log \left(\frac{ZE}{R} \right) - \log f(\alpha) \tag{3.18}$$

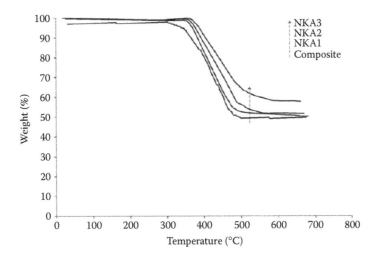

FIGURE 3.31 The TGA curves of composite and nanocomposite samples. (Reprinted from *J. Hazard. Mater.*, 150, Bahramian, A.R., Kokabi, M., Famili, M.H., and Beheshty, M.H., High temperature ablation of kaolinite layered silicate/phenolic resin/asbestos cloth nanocomposite, 136–145. Copyright 2008, with permission from Elsevier.)

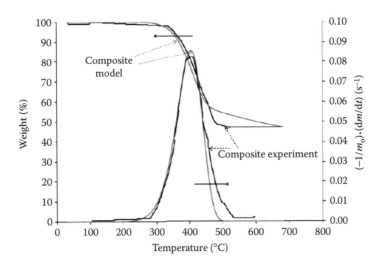

FIGURE 3.32 The thermogravimetric curves of asbestos/phenolic composite at 10°C/min in air, in comparison with the theoretical model. (Reprinted from *J. Hazard. Mater.*, 150, Bahramian, A.R., Kokabi, M., Famili, M.H., and Beheshty, M.H., High temperature ablation of kaolinite layered silicate/phenolic resin/asbestos cloth nanocomposite, 136–145. Copyright 2008, with permission from Elsevier.)

where:
β is the Arrhenius frequency factor (S^{-1}) or heating rate
E is the activation energy (J/mol)
R is the gas constant (8.314 J/mol K)
T is the absolute temperature (K)
α is the degree of conversion, which can be found by the following equation:

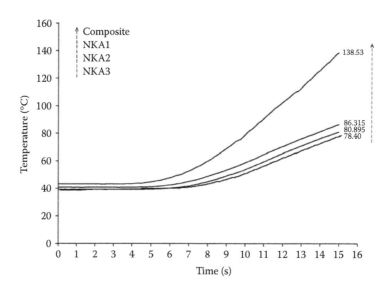

FIGURE 3.33 Temperature distribution of back surface of the asbestos/phenolic composite and NKA1, NKA2, and NKA3 nanocomposites; which measured in oxyacetylene flame test. (Reprinted from *J. Hazard. Mater.*, 150, Bahramian, A.R., Kokabi, M., Famili, M.H., and Beheshty, M.H., High temperature ablation of kaolinite layered silicate/phenolic resin/asbestos cloth nanocomposite, 136–145. Copyright 2008, with permission from Elsevier.)

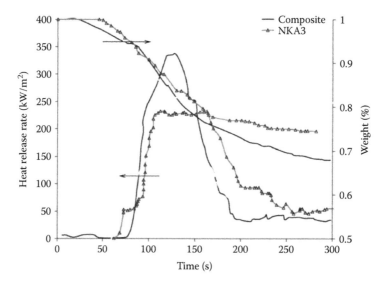

FIGURE 3.34 Comparison of the heat release rate and the mass loss plot for phenolic asbestos cloth composite and phenolic asbestos cloth kaolinite nanocomposite (NKA3) at 80 kW/m² heat flux. (Reprinted from *J. Hazard. Mater.*, 150, Bahramian, A.R., Kokabi, M., Famili, M.H., and Beheshty, M.H., High temperature ablation of kaolinite layered silicate/phenolic resin/asbestos cloth nano-composite, 136–145. Copyright 2008, with permission from Elsevier.)

$$\alpha = \frac{(\Delta H_p)_t}{\Delta H_0} \tag{3.19}$$

where:

$(\Delta H_p)_t$ is the heat released at time t

ΔH_0 is the total reaction heat of a reaction

Ozawa's method is based on a linear relationship between the logarithm of the heating rate and the inverse of the peak temperature of Equation 3.18, so the activation energy (E_a) can be calculated from a plot of heating rate log β versus $1/T_p$. Because the multiheating rate method uses the relationship between the peak exothermic temperature and its corresponding heating rate, this method is an appropriate choice for cure characterization of resins. It has been widely used by many researchers for different products such as polymer blends, thermoplastic elastomer, and thermoplastic vulcanizates.

$$-\ln\left(\frac{\beta}{T_p^2}\right) = \frac{E}{RT_p} - \ln\left(\frac{ZR}{E}\right) \tag{3.20}$$

This is the equation of a straight line between $\ln(\beta/T^2p)$ and $1/Tp$. The activation energy, E, can be calculated from the slope and the preexponential factor from the intercept. From these data with Arrhenius law, the rate constants have been derived. Moreover, their works diminishing the activation energies of both Ozawa and Kissinger methods showed that the E_a of the nanocomposite at the 0.5 wt.% nanoclay content reaches a maximum and then decreases thereafter. However, the Ozawa method provides greater E_a values than those of the Kissinger method. On the other hand, nanoclay loading follows similar trends for the overall E_a values from the both Ozawa and Kissinger methods as shown in Figure 3.35.

These results indicated that the exfoliation of layered nanoclay particles into MF resin leads to delay in the cure of MF resin/nanoclay/cellulose nanocomposites.

3.5 APPLICATIONS

The potential applications of thermoset/clay nanocomposites are plenty in various perspectives. The improvements in the mechanical properties have resulted in major interest in numerous automotive and general/industrial applications. It includes potential for utilization as mirror housings on various vehicle types, door handles, engine covers, and belt covers. More general applications such as improved gas barrier performance in packaging application, fuel cell, solar cell, fuel tank, and plastic container include usage of impellers and blades for vacuum cleaners, power tool housing, and cover for portable electronic equipment such as mobile phones and pagers.

3.5.1 Gas Barriers for Plastic Bottles, Packaging, and Sports Goods

At present, the role of clay nanocomposites in packaging and plastic bottle manufacturing industries is very important one. Applications of clay-based rubber nanocomposites

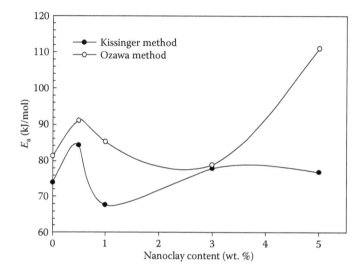

FIGURE 3.35 Changes of E_a values of curing reaction of the nanocomposites with different nanoclay contents determined by Ozawa and Kissinger models. (Reprinted from *J. Ind. Eng. Chem.*, 16, Park, B.-D. and Jeong, H.-W., Cure kinetics of melamine–formaldehyde resin/clay/cellulose nanocomposites, 375–379. Copyright 2010, with permission from Elsevier.)

existing in packaging field too. It reduces the packages weight and leads to cost reduction. Improved shelf life and lower package cost dominate the uses of nanotechnology in consumer packaging. Significant improvements in the gas barrier properties with the incorporation of a trace amount of nanoclay due to torturous zigzag diffusion path in an exfoliated polymer–clay nanocomposite as illustrated in Figure 3.36. Clay-based nanomaterial-reinforced rubber nanocomposites are playing a key role in packaging films for improving strength, good elongation, durability, barrier performance, long life, and quality. One innovation is clay-based polymer nanocomposite technology, which holds the key to future advances in flexible packaging film. According to Aaron Brody in a December 2003 Food Technology article,

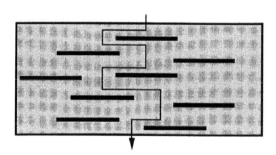

FIGURE 3.36 Torturous zigzag diffusion path in an exfoliated polymer/clay nanocomposite, when used as a gas barrier. (Reprinted from *Appl. Clay Sci.*, 15, LeBaron, P.C., Wang, Z., and Pinnavaia, T.J., Polymer-layered silicate nanocomposites: An overview, 11–29. Copyright 1999, with permission from Elsevier.)

clay based polymer nanocomposites appear capable of approaching the elusive goal of converting plastic into a super barrier the equivalent of glass or metal without upsetting regulators.

BRODY (2003)

The contribution of thermoset polymer nanocomposites in barrier and packaging applications is one of the greatest one. Many thermoset polymers have been used commercially for barrier applications such as butyl, SBR, EPDM, EVA, and ENGAGE (ethylene–octene copolymer). These polymers exhibit excellent barriers for many gases such as CO_2, O_2, and N_2. They also can be a barrier for many chemical solvents and acids such as toluene, HNO_3, H_2SO_4, and HCL. Due to excellent solvent barrier properties, they were utilized in chemical protective gloves in the medical field (which are thick with poor dexterity) in order to protect against chemical warfare agents and to avoid contamination from medicine. They have already a widely commercialized product.

Because of excellent gas and solvent barrier property, they have also been widely used in food packaging and plastic container products, both flexible and rigid. Specific examples include packaging of processed meat, cheese, cereals and dairy products, printer cartridge seals, medical container seals, medical container seals for blood collection tubes, stoppers for medical containers, stoppers for blood collection tubes, baby pacifiers, and drinking water bottles. The incorporation of layered silicate clay reduced the diffusion rate of small molecules, such as water and oxygen, through polymer films. It found exponential decay of permeability of oxygen with layered silicate content in the polymer up to a limit of 3% by volume, beyond which the permeability values did not change, apparently due to reduced particle–particle distance.

However, the clay-based polymer nanocomposites have been used in plastic bottle manufacturing industries for improving barrier and mechanical properties and shelf life of the product. To improve the bottle's performance and seek an alternative solution, plastic bottles have been tried in the beer package industry. The shelf life of a clay-based nanocomposite plastic beer bottle is more than 6 months. The first plastic beer bottle based on clay nanocomposites was introduced by Honeywell (2004) [12]. These hybrids can also be used for beer bottle manufacturing to solve many problems such as the beer colloid instability, including biological and nonbiological aspects, oxygen (O_2) permeation, and bad taste due to light exposure [12]. The relative vapor permeability of NBR/organoclay nanocomposites (in 10 phr organoclay) that was reduced up to 85% and 42% of water and methanol solvent compared to the neat polymer.

Recently, one of the commercialized sports goods is double-core Wilson tennis ball. The clay nanocomposite coating of the Wilson tennis balls maintained the internal pressure for an extended period of time. Figure 3.37 shows the Wilson tennis balls containing a double core. The core is coated with a butyl rubber-clay (vermiculite) nanocomposite that acts as a gas barrier and doubles its shelf life. Nanocomposite shows more coatings that are flexible with gas permeability of 30–300 times lower than butyl rubber. These coatings have been shown to be undamaged by strains up to 20%. The double-core new tennis balls using this

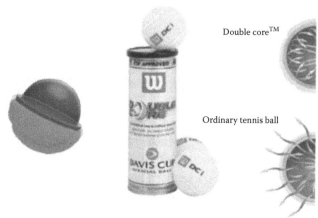

Double core™

Ordinary tennis ball

Butyl rubber nanocomposite core

FIGURE 3.37   The core of the Wilson tennis ball is covered by a polymer–clay nanocomposite coating that acts as a gas barrier, doubling its shelf life. (Reprinted from Joseph, H.K. (ed.), *Polymer Nanocomposites—Processing, Characterization and Applications.* Copyright 2006, with permission from McGraw-Hill Publishing Group.)

coating retain air longer and are able to bounce twice as long as ordinary balls (improvement in air retention). It is anticipated that this technology can be extended to the rubber industry and it can be incorporated into soccer balls and automobile/bicycle tires. The InMat Inc., New Jersey, first manufactured this wilson tennis ball [101]. Another important application in sports goods is basketball shoe pouch, which has been manufactured from EVA/clay nanocomposites. Figure 3.38 shows the clay nanocomposite pouch filled with helium inserts that fits into basketball shoe.

The ultimate property of the pouch is that it can give good resilience for a basket ball player while jumping; in the meantime, it exhibits excellent gas barrier properties. The converse system developed this process, but Triton Systems, Massachusetts, manufactured it first.

However, the tire industry is another very important application field for rubber/clay nanocomposites for improving barrier and mechanical properties of the tire and tube. The excellent air retention properties of butyl/halo butyl and chemically modified rubbers with incorporated clays are well known in the tire industries as these nanocomposites are widely used in the inner tube and the inner liner in automotive tires for improving many properties such as mechanical properties, namely, tensile, tear resistance, tire rolling loss, low-weight tires, dimensional stability, low colored or transparent tires, improved flame resistance, enhanced air retention, and tubeless tires without a conventional inner liner (in future).

3.5.2 Energy Storage Systems and Sensors

In general, polymer nanocomposite-based fuel cells act as an electrochemical device, which converts the chemical energy of carbon, hydrogen, and oxygen directly and efficiently into a useful electrical energy with heat and water as the only by-products. Due to the incorporation of nanomaterial, the efficiency is increasing many times. Sulfonated PSEBS/MMT

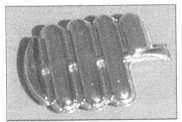

Nanocomposite pouch filled with helium

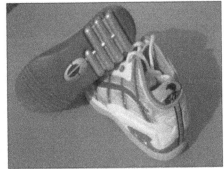

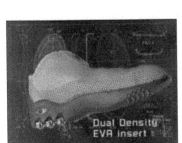

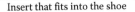

Insert that fits into the shoe

FIGURE 3.38 Converse basketball shoe pouch by Triton Systems. (Reprinted from Joseph, H.K. (ed.), *Polymer Nanocomposites—Processing, Characterization and Applications.* Copyright 2006, with permission from McGraw-Hill Publishing Group.)

nanocomposites are also used in proton exchange membranes for fuel cell applications due to its superior proton exchange capacity. Several polymers exist for serving as fuel cell applications such as hyperbranched polymer with a hydroxyl group at the periphery (HBP-OH), cross-linked sulfonated poly(ether ether ketone) (SPEEK), sulfonated polybenzimidazole copolymer, phosphoric acid-doped polybenzimidazole, sulfonated polyarylenethioethersulfone, and sulfonated polybenzimidazole. The clay-incorporated thermoset polymer fuel cells exhibit higher proton conductivity, better ion-exchange capacity, and higher rate of conductivity even at higher humidity without compromising their mechanical properties [102]. Another important storage application of clay nanocomposite is solar cells. It is a well-commercialized one now. Polymer/clay-based nanocomposite solar cells have the potential to become one of the leading technologies of the twenty-first century in conversion of sunlight into electrical energy. Because of their ease of processing from solution, fast and low-cost mass production of devices is possible in a roll-to-roll printing fashion. They exhibit high efficiency and enhanced light absorption, and possess an excellent barrier as well as mechanical strength [103]. Even though while keeping solar cells in outdoor application, we need not worry about the damage from any mechanical actions or UV rays of clay-based thermoset nanocomposites. Generally, it can be done by slot-die coating and plasma-enhanced chemical vapor deposition. Various polymers have been used for solar cell manufacturing such as semiconductor polymer, polyphenylene-vinylene, polythiopene, polyfluorene, polyaniline, polypyrrole for light absorption and charge transport, silicone, and PET. Sensor technology also made great impact in polymer-clay nanocomposites. The various polymers has involved such as poly(amidoamine) dendrimer, silicone, hyper branched polymers, poly(acrylamide)… and so on has been used for sensing gas, atmospheric moistures, detection of solvent leaking in pipe line application… and so on clay incorporated elastomers are being developed as sensors to detect fatigue, impact and large strain for aerospace applications.

3.5.3 Optical Glass and Membranes

The presence of filler incorporation at nanolevels has also been shown to have significant effects on the transparency and haze characteristics of films. In comparison with conventionally filled polymers, nanoclay incorporation has been shown to significantly enhance transparency and reduce haze. This effect is more prominent for the butyl rubber, polyamide, and acrylic rubber-based clay nanocomposites due to many factors such as crystallization behavior the formation of spherilitic domain microstructure. Similarly, clay-incorporated polymers have been shown, when employed to coat polymeric transparency materials, to enhance both toughness and hardness of these materials without sacrificing light transmission characteristics. An ability to resist high velocity impact combined with substantially improved abrasion resistance was demonstrated by Haghighat of Triton Systems. Because of the improved optical properties without scarifying the mechanical properties and light transmission, it was widely commercialized in contact lens and optical glass applications.

Polymer/clay nanocomposites can also be used to fabricate various types of membranes such as filtering of solvents, bacteria and virus, gas or solvent transportation, electrolytes for fuel cell, and membrane for gas separation. Organic membrane technology is used to design membranes for separating synthetic gas from natural gas. The transformation of natural gas into liquid materials requires design of inorganic membrane catalysts. A lot of research has been done on the membrane based on thermoset clay nanocomposites. Generally, polymers such as polyimide, polybenzimidazole, cross-linked SPEEK, and polyacrylate have been widely used in membrane applications. The creation of nanopores in nanocomposites using nanofillers is the main mechanism in gas or solvent separation and transportation. The nanopores (pore size) decide the sensitivity and quality of the membrane.

3.5.4 Flammability

The ability of nanoclay incorporation to reduce the flammability of polymeric materials was a major theme of the paper presented by Gilman of the National Institute of Standards and Technology in the United States. In his work, he demonstrated the extent to which flammability behavior could be restricted in polymers such as polypropylene with as little as 2% nanoclay loading. Clay-filled nanocomposites provide an attractive and alternative to conventional flame retardants. At present, the most common approach to improving flammability is the use of layered silicates such as clays. However, this improved fire performance is usually concerned with reduction only in heat release properties. The improvement in flammability resistance by incorporation of clay has been commercialized in various applications such as cable wire jacket, car seats (hard foams), packaging films, textile cloths, surface coatings for many steel product, and paints, and one of the higher

(a)   (b)

FIGURE 3.39 Comparison of the carbonaceous residues obtained after cone calorimeter tests: (a) compact char of epoxy–LDH (LDH nanocomposite) with in tumescent behavior (char thickness up to 50 mm); (b) fragmented char of Epoxy–MMT. (Adapted from Morgan, A.B. and Wilkie, C.A. eds., Flame retardant polymer nanocomposites, 2007, John Wiley & Sons.)

end applications was rocket ablative materials core manufacturing. Na-MMT in nitrile butadiene rubber nanocomposites shows higher flame retardancy with good oil resistance compared to control NBR sample [74]. Various instruments are available for calibrating flame-retardant capacity of the product such as limiting oxygen index (LOI), cone calorimeter, and gasification test. Figure 3.39 shows the comparison of the carbonaceous residues obtained after cone calorimeter tests between compact char of epoxy–layered double hydroxide (LDH) (LDH nanocomposite) and fragmented char of epoxy–MMT. The clay-reinforced epoxy was not fully degraded or increasing smoke density. The reduction in weight loss also for epoxy/clay nanocomposites was less compared to epoxy–LDH. In addition, PU/clay-based nanocomposites also exhibit superior flame retardancy, and they are already commercialized in the manufacturing of automobile seats (hard foam). Borden Chemical's SC-1008, a resole phenolic, was selected as the resin for manufacturing rocket ablative materials with MMT.

The nanoscale distribution of silicate layers leads to a uniform char layer that enhances the ablative performance. The formation of this char is mainly influenced by the type of organic modification on the silicate surface of specific interactions between the polymer and the aluminosilicate surface, such as end-tethering of a fraction of the polymer chains through ionic interaction with the layer surface. The same formulation has also been used for simulated solid rocket motor casing, and it exhibits excellent flame retardancy. The ablative test was performed at a temperature around 2200°C, and results showed that the motor has not been burnt completely. Close-up appearance of a posttest ablative specimen of simulated solid rocket motor casing is shown in Figure 3.40.

3.5.5 Electronics and Automobile Sectors

Applications of thermoset/clay nanocomposites in electronics and automobile sectors are another big milestone. The ability of the nanoclay incorporation is to reduce the solvent

FIGURE 3.40   Close-up of ablative material after SSRM testing. (Reprinted from Joseph, H.K. (ed.), *Polymer Nanocomposites—Processing, Characterization and Applications.* Copyright 2006, with permission from McGraw-Hill Publishing Group.)

transmission through many polymers such as specialty elastomers, polyimides, and PU. Industrialists found that significant reductions in fuel transmission with polyamide–6/66 and poly imide polymers based nanocomposites. The major driving force of a tire company is reduction of weight and processing costs. Clay is one of the naturally available materials with very low density; it facilitates the reduction in weight. The clay-incorporated tires exhibit excellent mechanical properties as well as improved gas barrier performance for tube applications compared to ordinary tires. Generally, styrene butadiene and NR-based clay nanocomposites are mostly preferred for automobile tire manufacturing followed by butyl rubber series for tubes. It is due to well-improved abrasive resistance and thermal properties of the tire with long life.

The addition of clay into conducting polymers such as polypyrrole, polythiophene, and polyaniline sensibly increases their conductivity nature with good mechanical properties such as elongation, impact, scratch resistance, impedance effect, and long life of the product. It finds many applications in solar cell, wind mill, electronic circuit board and battery manufacturing, microchips, and transistor. Nanocomposites consisting of spatially confined liquid crystals are also a great interest because of the prospects of their applications in optoelectronic devices, photonic crystals, depolarizers, scattering displays, information storage, and recording devices such as compact disk, universal serial bus storage, and windows with adjustable transparency [105]. In these systems, an applied external electric field caused switching between the scattering and transparent states and under some conditions; these states can be retained after switching off the field. In the case of spherical aerosol particles, it was found that the memory effect is achieved due to the formation of ordered branched network of the aerosol particles in the liquid crystal matrix. However, in the case of isometric particles such as clay mineral, an important contribution for the memory effect can be from the influence of the clay surface on the alignment of adjacent liquid crystal layers, which can be controlled by the application of hydrophobic organic modifier on the clay layers. The effect of modification of the MMT clay layers with different organic surfactant ions on the electrooptical properties of the MMT+5CB (4-pentyl-4′-cyanobiphenyl nematic liquid crystal) heterogeneous liquid crystal (LC)–clay nanosystems has been studied by Chashechnikova et al. [106]. Using IR and Raman spectroscopy techniques, it was shown that in the LC–clay nanocomposites consisting of organically modified MMT and 5CB, mutual influence of 5CB molecules and the clay particles occurred, which resulted in the ordering of the near-surface layers of both inorganic and organic components of the composites. Because of these van der Waals interactions, the system became transparent under the action of electrical field and maintained its original transparent state when the voltage was switched off, that is, the contrast and electrooptical memory effect was observed. They also reported that the use of polar additive (acetone) for the preparation of the nanocomposite increases the uniformity of the composites and considerably improves their electrooptical properties.

3.5.6 Coatings

The coating is the one of the most important processes, and it is associated with the surface phenomenon. Several strategies have been tried by researchers for improving the surface property of the product. One of the well-versed developments is nanoclay-mixed polymer coating. In general, clay-based thermoset polymer coating is defined as 2D coating developed by clay-incorporated thermoset polymers. It made a great attention to the industries too. Several works have been done in this area [38,57,105]. Nanoclay-incorporated thermoset polymer nano coating exhibits superior properties such as super hydrophobicity, improved wettability, excellent resistance for chemicals, corrosion resistance, improved weather resistance, better abrasion resistance, improved barrier properties and resist due to impact, and scratch. Coating thickness depends on process parameters such as dipping time, temperature, nature of surfactant, and purity of nanomaterial. Turri et al. [38] developed a nanostructured coating based on epoxy/clay on steel substrates. The scratch strength and depth of epoxy contained clay nanostructured coating shows 2 times greater than that of the epoxy coating alone. These kinds of coatings are best candidates for paint coating in various applications such as construction application, aerospace applications for thermal barrier, automobile, pipeline coating for marine applications, sometimes for decoration purpose, and textile. The clay and nanosilver-incorporated thermoset polymer nanocoatings sensibly improve antibacterial properties, and they have been widely applicable in medical sectors. At present, these kinds of hybrid coatings are one of the promising ones for improving shelf life and antibacterial properties in the medical field. Nanocrystals, for example, Sn_2Sb, and clay (dual filler)-incorporated polymer-based thin films are used as anode materials to improve the storage capacities in lithium batteries. These kinds of nanocomposites provide a greater flexibility in the design of new recording media compared to homogeneous materials. One of the important applications in electronics sector is that earlier the hard disk memory capacity was very less compared to now. Now we are using hard disk capacities such as more than 1 TB in a single chip. Moreover, the chip size is less than 1 cm × 1 cm. The reason behind that is additional filter mechanism and ability to better tune the photochemistry within composite materials, which makes nanocomposite polymeric systems highly attractive candidates for high-density optical storage media.

3.6 SUMMARY

Thermoset/clay nanocomposites are one of the most versatile classes of materials today, and their demand as a high-performance industrial material continues to grow. In this chapter, we have discussed various aspects of thermoset/clay nanocomposites such as different types of clay system with structure, different routes for synthesizing nanocomposites, various types of thermoset polymer followed by its characterization and examination of the properties of nanocomposites, and structure analysis, and finally we have dealt with some of the applications. We have also focused on discussing the corresponding thermal, dynamic mechanical, electrical, and surface properties of these materials. Now we realize that the thermoset/clay-based nanocomposite applications of this modern-day life are large. We hope that this chapter has given enough information to the readers for understanding the fundamental idea of thermoset/clay nanocomposites.

ABBREVIATIONS

ABS	Acrylonitrile–butadiene–styrene copolymer
AFM	Atomic force microscopy
α	Degree of conversion
BIMS	Brominated poly(isobutylene-*co-para*-methylstyrene)
β	Arrhenius frequency factor
CB	Carbon black
CEC	Cation-exchange capacity
DBP	Dibenzyl peroxide
DCP	Dicumyl peroxide
DEG	Diethyleneglycol
DMTA	Dynamic mechanical thermal analysis
DSC	Differential scanning calorimetry
E_a	Activation energy
EPDM	Ethylene–propylene–diene terpolymer
EVA	Ethylene vinyl acetate copolymer
FTIR	Fourier transform infrared
HAF	High abrasive furnace
HBP-OH	Hyperbranched polymer with a hydroxyl group
HOMMT	Hyperbranched organo-montmorillonite
HTV-SR	High-temperature vulcanized silicone rubber
IR	Infrared spectroscopy
LC	Liquid crystal
LDH	Layered double hydroxide
LOI	Limiting oxygen index
MEG	Monoethyleneglycol
MEK	Methyl ethyl ketone
MF	Melamine formaldehyde
MMT	Montmorillonite
NA	Natural

NBR	Nitrile butadiene rubber
NR	Natural rubber
PAA	Poly(acrylic acid)
PET	Polyethylene terephthalate
PSEBS	Polystyrene ethylene butylene polystyrene
PU	Polyurethane
R	Gas constant
SEM	Scanning electron microscopy
SPSEBS	Sulfonated polystyrene ethylene butylene polystyrene
SPEEK	Sulfonated poly (ether ether ketone)
TEM	Transmission electron microscopy
THF	Tetrahydrofuran
UV	Ultraviolet
XRD	X-ray diffraction
$\Delta\gamma$	Surface energy

REFERENCES

1. Abacha N, Kubouchi M, Sakai T, Tsuda K. Diffusion behavior of water and sulfuric acid in epoxy/organoclay nanocomposites. *Journal of Applied Polymer Science* 2009;112:1021–1029.
2. LeBaron PC, Wang Z, Pinnavaia TJ. Polymer-layered silicate nanocomposites: An overview. *Applied Clay Science* 1999;15:11–29.
3. Okada A, Usuki A. The chemistry of polymer-clay hybrids. *Materials Science and Engineering: C-Biomimetic* 1995;3:109–115.
4. Huang XY, Lewis S, Brittain WJ, Vaia RA. Synthesis of polycarbonate-layered silicate nanocomposites via cyclic oligomers. *Macromolecules* 2000;33:2000–2004.
5. Ito M, Nagai K. Thermal aging and oxygen permeation of nylon-6 and nylon-6/montmorillonite composites. *Journal of Applied Polymer Science* 2010;118:928–935.
6. Timochenco L. Swelling of organoclays in styrene. Effect on flammability in polystyrene nanocomposites. *eXPRESS Polymer Letters* 2010;4:500–508.
7. Strawhecker KE, Manias E. Structure and properties of poly(vinyl alcohol)/Na+ montmorillonite nanocomposites. *Chemistry of Materials* 2000;12:2943–2949.
8. Ahmadi SJ, Huang YD, Wei L. Synthesis of EPDM/organoclay nanocomposites: Effect of the clay exfoliation on structure and physical properties. *Iranian Polymer Journal* 2004;13:415–422.
9. Diez J, Bellas R, Ramírez C, Rodríguez A. Effect of organoclay reinforcement on the curing characteristics and technological properties of SBR sulphur vulcanizates. *Journal of Applied Polymer Science* 2010;118:566–573.
10. Ghasemi I. Evaluating the effect of processing conditions and organoclay content on the properties of styrene-butadiene rubber/organoclay nanocomposites by response surface methodology. *eXPRESS Polymer Letters* 2010;4:62–70.
11. Markanday SS, Stastna J, Polacco G, Filippi S, Kazatchkov I, Zanzotto L. Rheology of bitumen modified by EVA-organoclay nanocomposites. *Journal of Applied Polymer Science* 2010;118:557–565.
12. Ke YC, Stroeve P. *Polymer-Layered Silicate and Silica Nanocomposites.* Elsevier B.V., Amsterdam, the Netherlands, 2005.
13. Sinha Ray S, Okamoto M. Polymer/layered silicate nanocomposites: A review from preparation to processing. *Progress in Polymer Science* 2003;28:1539–1641.

14. Alam SMM. Preparation of polyimide-clay nanocomposites and their performance. *Journal of Scientific Research* 2009;1:326–333.

15. Barick AK, Tripathy DK. Preparation and characterization of thermoplastic polyurethane/organoclay nanocomposites by melt intercalation technique: Effect of nanoclay on morphology, mechanical, thermal, and rheological properties. *Journal of Applied Polymer Science* 2010;117:639–654.

16. Karger-Kocsis J. Melting and crystallization of in-situ polymerized cyclic butylene terephthalates with and without organoclay: A modulated DSC study. *eXPRESS Polymer Letters* 2007;1:60–68.

17. Kim W-S, Yi J, Lee D-H, Kim I-J, Son W-J, Bae J-W et al. Effect of 3-aminopropyltriethoxysilane and N,N-dimethyldodecylamine as modifiers of Na+-montmorillonite on SBR/organoclay nanocomposites. *Journal of Applied Polymer Science* 2010;116:3373–3387.

18. Pavlidou S, Papaspyrides CD. A review on polymer–layered silicate nanocomposites. *Progress in Polymer Science* 2008;33:1119–1198.

19. Bhowmick AK. *Current Topics in Elastomers Research*. CRC Press, Taylor & Francis Group, Boca Raton, FL; 2008. p. 1106.

20. Ray S, Bhowmick AK. Electron-beam-modified surface-coated clay: Influence on mechanical, dynamic mechanical and rheological properties of ethylene-octene copolymer. *Radiation Physics and Chemistry* 2002;65:259–267.

21. Chieng BW. Effect of organo-modified montmorillonite on poly(butylene succinate)/poly (butylene adipate-co-terephthalate) nanocomposites. *eXPRESS Polymer Letters* 2010;4:404–414.

22. Deng KL. Potassium diperiodatocuprate-mediated preparation of poly(methyl methacrylate)/organo-montmorillonite composites via in situ grafting copolymerization. *eXPRESS Polymer Letters* 2008;2:677–686.

23. Martucci JF, Ruseckaite RA. Biodegradable bovine gelatin/Na+-montmorillonite nanocomposite films. Structure, barrier and dynamic mechanical properties. *Polymer-Plastics Technology and Engineering* 2010;49:581–588.

24. Lau K, Lu M, Cheung H, Sheng F, Li H. Thermal and mechanical properties of single-walled carbon nanotube bundle-reinforced epoxy nanocomposites: The role of solvent for nanotube dispersion. *Composites Science and Technology* 2005;65:719–725.

25. Chow WS. Optimization of process variables on flexural properties of epoxy/organo-montmorillonite nanocomposite by response surface methodology. *eXPRESS Polymer Letters* 2008;2:2–11.

26. Chakraborty S, Sengupta R, Dasgupta S, Mukhopadhyay R, Bandyopadhyay S, Joshi M et al. Synthesis and characterization of in situ sodium-activated and organomodified bentonite clay/styrene-butadiene rubber nanocomposites by a latex blending technique. *Journal of Applied Polymer Science* 2009;113:1316–1329.

27. Wu Y-P, Wang Y-Q, Zhang H-F, Wang Y-Z, Yu D-S, Zhang L-Q et al. Rubber–pristine clay nanocomposites prepared by co-coagulating rubber latex and clay aqueous suspension. *Composites Science and Technology* 2005;65:1195–1202.

28. Karger-Kocsis J, Wu CM. Thermoset rubber/layered silicate nanocomposites. Status and future trends. *Polymer Engineering & Science* 2004;44:1083–1093.

29. Bala P, Samantaray BK, Srivastava SK, Nando GB. Effect of alkylammonium intercalated montmorillonite as filler on natural rubber. *Journal of Materials Science Letters* 2001;20:563–564.

30. Hrachová J, Chodák I, Komadel P. Modification and characterization of montmorillonite fillers used in composites with vulcanized natural rubber. *Chemical Papers* 2008;63:55–61.

31. Hernández M, Carretero-González J, Verdejo R, Ezquerra TA, López-Manchado MA. Molecular dynamics of natural rubber/layered silicate nanocomposites as studied by dielectric relaxation spectroscopy. *Macromolecules* 2010;43:643–651.

32. Ismail H, Munusamy Y, Mariatti M, Ratnam CT. The effect of trimethylol propane tetraacrylate (TMPTA) and organoclay loading on the properties of electron beam irradiated ethylene vinyl acetate (EVA)/natural rubber (SMR L)/organoclay nanocomposites. *Journal of Applied Polymer Science* 2010;117:865–874.

33. Liu Y, Li L, Wang Q, Zhang X. Fracture properties of natural rubber filled with hybrid carbon black/nanoclay. *Journal of Polymer Research* 2010;18:859–867.

34. Joly S, Garnaud G, Ollitrault R, Bokobza L, Mark JE. Organically modified layered silicates as reinforcing fillers for natural rubber. *Chemistry of Materials* 2002;14:4202–4208.

35. Bhattacharya M, Bhowmick AK. Synergy in carbon black-filled natural rubber nanocomposites. Part I: Mechanical, dynamic mechanical properties, and morphology. *Journal of Materials Science* 2010;45:6126–6138.

36. Bhattacharya M, Bhowmick AK. Synergy in carbon black filled natural rubber nanocomposites. Part II: Abrasion and viscoelasticity in tire like applications. *Journal of Materials Science* 2010;45:6139–6150.

37. Abacha N. Performance of epoxy-nanocomposite under corrosive environment. *eXPRESS Polymer Letters* 2007;1:364–369.

38. Turri S, Torlaj L, Piccinini F, Levi M. Abrasion and nanoscratch in nanostructured epoxy coatings. *Journal of Applied Polymer Science* 2010;118:1720–1727.

39. Mohan TP KK, Velmurugan R. Epoxy–clay nanocomposites—Effect of curing temperature in mechanical properties. *International Journal of Plastics Technology* 2009;13:123–132.

40. Chattopadhyay PK, Basuli U, Chattopadhyay S. Studies on novel dual filler based epoxidized natural rubber nanocomposite. *Polymer Composites* 2009;31:835–846.

41. Chattopadhyay PK, Das NC, Chattopadhyay S. Influence of interfacial roughness and the hybrid filler microstructures on the properties of ternary elastomeric composites. *Composites Part A: Applied Science and Manufacturing* 2011;42:1049–1059.

42. Chattopadhyay PK, Chattopadhyay S, Das NC, Bandyopadhyay PP. Impact of carbon black substitution with nanoclay on microstructure and tribological properties of ternary elastomeric composites. *Materials & Design* 2011;32:4696–4704.

43. Mishra SB, Luyt AS. Effect of organic peroxides on the morphological, thermal and tensile properties of EVA-organoclay nanocomposites. *eXPRESS Polymer Letters* 2008;2:256–264.

44. Peeterbroeck S, Alexandre M, Jérôme R, Dubois P. Poly(ethylene-co-vinyl acetate)/clay nanocomposites: Effect of clay nature and organic modifiers on morphology, mechanical and thermal properties. *Polymer Degradation and Stability* 2005;90:288–294.

45. Zhang Wa, Chen D, Zhao Q, Fang Y. Effects of different kinds of clay and different vinyl acetate content on the morphology and properties of EVA/clay nanocomposites. *Polymer* 2003;44:7953–7961.

46. Ahn Y, Kim H, Lee JW. Ultrasound assisted batch-processing of EVA-organoclay nanocomposites. *Korean Journal of Chemical Engineering* 2010;27:723–728.

47. Jisheng MA, Zhang S, Qi Z. Synthesis and characterization of elastomeric polyurethane—Clay nanocomposites. *Journal of Applied Polymer Science* 2001;82:1444–1448.

48. Zhang X, Xu R, Wu Z, Zhou C. The synthesis and characterization of polyurethane/clay nanocomposites. *Polymer International* 2003;52:790–794.

49. Song M, Hourston D, Yao KJ, Tay JKH, Ansarifar MA. High performance nanocomposites of polyurethane elastomer and organically modified layered silicate. *Journal of Applied Polymer Science* 2003;90:3239–3243.

50. Choi WJ, Kim SH, Jin Kim Y, Kim SC. Synthesis of chain-extended organifier and properties of polyurethane/clay nanocomposites. *Polymer* 2004;45:6045–6057.

51. Han B, Cheng A, Ji G, Wu S, Shen J. Effect of organophilic montmorillonite on polyurethane montmorillonite nanocomposites. *Journal of Applied Polymer Science* 2004;491:2536–2542.

52. Cao X, James Lee L, Widya T, Macosko C. Polyurethane/clay nanocomposites foams: Processing, structure and properties. *Polymer* 2005;46:775–783.
53. Song L, Hu Y, Tang Y, Zhang R, Chen Z, Fan W. Study on the properties of flame retardant polyurethane/organoclay nanocomposite. *Polymer Degradation and Stability* 2005;87:111–116.
54. Maji PK, Guchhait PK, Bhowmick AK. Effect of nanoclays on physico-mechanical properties and adhesion of polyester-based polyurethane nanocomposites: Structure–property correlations. *Journal of Materials Science* 2009;44:5861–5871.
55. Kim SH, Lee MC, Kim HD, Park HC, Jeong HM, Yoon KS et al. Nanoclay reinforced rigid polyurethane foams. *Journal of Applied Polymer Science* 2010;117:1992–1997.
56. Rama MS, Swaminathan S. Influence of structure of organic modifiers and polyurethane on the clay dispersion in nanocomposites via in situ polymerization. *Journal of Applied Polymer Science* 2010;118:1774–1786.
57. Jin H, Wie JJ, Kim SC. Effect of organoclays on the properties of polyurethane/clay nanocomposite coatings. *Journal of Applied Polymer Science* 2010;117:2090–2100.
58. Xu Z-b, Kong W-w, Zhou M-x, Peng M. Effect of surface modification of montmorillonite on the properties of rigid polyurethane foam composites. *Chinese Journal of Polymer Science* 2010;28:615–624.
59. Praveen S, Chattopadhyay PK, Jayendran S, Chakraborty BC, Chattopadhyay S. Effect of rubber matrix type on the morphology and reinforcement effects in carbon black-nanoclay hybrid composites—A comparative assessment. *Polymer Composites* 2009;31:97–104.
60. Acharya H, Pramanik M, Srivastava SK, Bhowmick AK. Synthesis and evaluation of high-performance ethylene-propylene-diene terpolymer/organoclay nanoscale composites. *Journal of Applied Polymer Science* 2004;93:2429–2436.
61. Zheng H, Zhang Y, Peng Z, Zhang Y. Influence of clay modification on the structure and mechanical properties of EPDM/montmorillonite nanocomposites. *Polymer Testing* 2004;23:217–223.
62. Mohammadpour Y, Katbab AA. Effects of the ethylene-propylene-diene monomer microstructural parameters and interfacial compatibilizer upon the EPDM/montmorillonite nanocomposites microstructure: Rheology/permeability correlation. *Journal of Applied Polymer Science* 2007;106:4209–4218.
63. Naderi G, Lafleur PG, Dubois C. Microstructure-properties correlations in dynamically vulcanized nanocomposite thermoplastic elastomers based on PP/EPDM. *Polymer Engineering & Science* 2007;47:207–217.
64. Ahmadi SJ, Huang Y-D, Ren N, Mohaddespour A, Ahmadi-Brooghani SY. The comparison of EPDM/clay nanocomposites and conventional composites in exposure of gamma irradiation. *Composites Science and Technology* 2009;69:997–1003.
65. Maiti M, Bandyopadhyay A, Bhowmick AK. Preparation and characterization of nanocomposites based on thermoplastic elastomers from rubber-plastic blends. *Journal of Applied Polymer Science* 2006;99:1645–1656.
66. Maiti M, Sadhu S, Bhowmick AK. Brominated poly(isobutylene-co-para-methylstyrene) (BIMS)-clay nanocomposites: Synthesis and characterization. *Journal of Polymer Science Part B: Polymer Physics* 2004;42:4489–4502.
67. Molu ZB, Seki Y, Yurdakoç K. Preparation and characterization of poly(acrylic acid)/pillared clay superabsorbent composite. *Polymer Bulletin* 2009;64:171–183.
68. Peila R, Malucelli G, Priola A. Preparation and characterization of UV-cured acrylic nanocomposites based on modified organophilic montmorillonites. *Journal of Thermal Analysis and Calorimetry* 2009;97:839–444.
69. Gao J, Ma D, Lu Q, Li Y, Li X, Yang W. Synthesis and characterization of montmorillonite-graft-acrylic acid superabsorbent by using glow-discharge electrolysis plasma. *Plasma Chemistry and Plasma Processing* 2010;30:873–883.
70. Zhang F, Guo Z, Gao H, Li Y, Ren L, Shi L et al. Synthesis and properties of sepiolite/poly(acrylic acid-co-acrylamide) nanocomposites. *Polymer Bulletin* 2005;55:419–428.

71. Sugimoto H, Nunome K, Daimatsu K, Nakanishi E, Inomata K. Transparent acryl–clay nanohybrid films with low thermal expansion coefficient. *Colloid & Polymer Science* 2010;288:1131–1138.
72. Xu D, Karger-Kocsis J. Dry rolling and sliding friction and wear of organophilic layered silicate/hydrogenated nitrile rubber nanocomposite. *Journal of Materials Science* 2009;45:1293–1298.
73. Zhan Y, Lei Y, Meng F, Zhong J, Zhao R, Liu X. Electrical, thermal, and mechanical properties of polyarylene ether nitriles/graphite nanosheets nanocomposites prepared by masterbatch route. *Journal of Materials Science* 2010;46:824–831.
74. Wang Q, Zhang X, Qiao J. Exfoliated sodium-montmorillonite in nitrile butadiene rubber nanocomposites with good properties. *Chinese Science Bulletin* 2009;54:877–879.
75. Kader MA, Kim K, Lee YS, Nah C. Preparation and properties of nitrile rubber/montmorillonite nanocomposites via latex blending. *Journal of Materials Science* 2006;41:7341–7352.
76. Mirmohseni A, Zavareh S. Epoxy/acrylonitrile-butadiene-styrene copolymer/clay ternary nanocomposite as impact toughened epoxy. *Journal of Polymer Research* 2009;17:191–201.
77. Ganguly A, Bhowmick AK. Effect of polar modification on morphology and properties of styrene-(ethylene-co-butylene)-styrene triblock copolymer and its montmorillonite clay-based nanocomposites. *Journal of Materials Science* 2008;44:903–918.
78. Li YM, Wei GX, Sue HJ. Morphology and toughening mechanisms in clay-modified styrene-butadiene-styrene rubber-toughened polypropylene. *Journal of Materials Science* 2002;37:2447–2459.
79. Swaminathan E, Dharmalingam S. Evaluation of sulphonated polystyrene ethylene butylene polystyrene-montmorillonite nanocomposites as proton exchange membranes. *International Journal of Plastics Technology* 2010;13:150–162.
80. Praveen S, Chattopadhyay PK, Jayendran S, Chakraborty BC, Chattopadhyay S. Effect of nanoclay on the mechanical and damping properties of aramid short fibre-filled styrene butadiene rubber composites. *Polymer International* 2009;59:187–197.
81. Peter C, LeBaron TJP. Clay nanolayer reinforcement of a silicone. *Chemistry of Materials* 2001;13:3760–3765.
82. Yang L, Hu Y, Lu H, Song L. Morphology, thermal, and mechanical properties of flame-retardant silicone rubber/montmorillonite nanocomposites. *Journal of Applied Polymer Science* 2006;99:3275–3280.
83. Wang J, Chen Y. Preparation of an organomontmorillonite master batch and its application to high-temperature-vulcanized silicone-rubber systems. *Journal of Applied Polymer Science* 2008;107:2059–2066.
84. Kaneko MLQA, Yoshida IVP. Effect of natural and organically modified montmorillonite clays on the properties of polydimethylsiloxane rubber. *Journal of Applied Polymer Science* 2008;108:2587–2596.
85. Wang J, Chen Y, Wang J. Synthesis of hyperbranched organo-montmorillonite and its application into high-temperature vulcanized silicone rubber systems. *Journal of Applied Polymer Science* 2008;111:658–667.
86. Voulomenou A, Tarantili PA. Preparation, characterization, and property testing of condensation-type silicone/montmorillonite nanocomposites. *Journal of Applied Polymer Science* 2010;118:2521–2529.
87. Vasilakos SP, Tarantili PA. The effect of pigments on the stability of silicone/montmorillonite prosthetic nanocomposites. *Journal of Applied Polymer Science* 2010;118:2659–2667.
88. Maiti M, Bhowmick AK. Structure and properties of some novel fluoroelastomer/clay nanocomposites with special reference to their interaction. *Journal of Polymer Science Part B: Polymer Physics* 2006;44:162–176.
89. Maiti M, Bhowmick AK. Effect of polymer–clay interaction on solvent transport behavior of fluoroelastomer–clay nanocomposites and prediction of aspect ratio of nanoclay. *Journal of Applied Polymer Science* 2007;105:435–445.

90. Maiti M, Bhowmick AK. New fluoroelastomer nanocomposites from synthetic montmorillonite. *Composites Science and Technology* 2008;68:1–9.
91. Maiti M, Bhowmick AK. Synthesis and properties of new fluoroelastomer nanocomposites from tailored anionic layered magnesium silicates (hectorite). *Journal of Applied Polymer Science* 2008;111:1094–1104.
92. Maiti M, Bhowmick AK. Dynamic viscoelastic properties of fluoroelastomer/clay nanocomposites. *Polymer Engineering & Science* 2007;47:1777–1787.
93. Valsecchi R, Viganò M, Levi M, Turri S. Dynamic mechanical and rheological behavior of fluoroelastomer-organoclay nanocomposites obtained from different preparation methods. *Journal of Applied Polymer Science* 2006;102:4484–4487.
94. Lan T, Kaviratna PD, Pinnavaia TJ. On the nature of polyimide clay hybrid composites. *Chemistry of Materials* 1994;6:573–575.
95. Kaynak C, Tasan CC. Effects of production parameters on the structure of resol type phenolic resin/layered silicate nanocomposites. *European Polymer Journal* 2006;42:1908–1921.
96. Bahramian AR, Kokabi M, Famili MH, Beheshty MH. High temperature ablation of kaolinite layered silicate/phenolic resin/asbestos cloth nanocomposite. *Journal of Hazardous Materials* 2008;150:136–145.
97. Wang D-C, Chang G-W, Chen Y. Preparation and thermal stability of boron-containing phenolic resin/clay nanocomposites. *Polymer Degradation and Stability* 2008;93:125–133.
98. Cai X, Riedl B, Wan H, Zhang SY, Wang X-M. A study on the curing and viscoelastic characteristics of melamine–urea–formaldehyde resin in the presence of aluminium silicate nanoclays. *Composites Part A: Applied Science and Manufacturing* 2010;41:604–611.
99. Park B-D, Jeong H-W. Cure kinetics of melamine–formaldehyde resin/clay/cellulose nanocomposites. *Journal of Industrial and Engineering Chemistry* 2010;16:375–379.
100. Hu J, Situ Y, Xu L, Huang H, Fu H, Zeng H et al. Synthesis and properties of novel epoxidized soybean oilmodified phenolic resin/montmorillonite nanocomposites. *Journal of Wuhan University of Technology—Materials Science Edition* 2008;23:431–435.
101. Joseph HK (ed.). *Polymer Nanocomposites—Processing, Characterization and Applications.* The McGraw-Hill, New York, 2006.
102. Bai H, Ho WSW. New sulfonated polybenzimidazole (SPBI) copolymer-based proton-exchange membranes for fuel cells. *Journal of the Taiwan Institute of Chemical Engineers* 2009;40:260–267.
103. Loos J. Volume morphology of printable solar cells. *Materials Today* 2010;13:14–20.
104. Alexander B, Morgan CAW (eds.). Flame retardant polymer nanocomposites, John Wiley & Sons, Hoboken, NJ, 2007.
105. Xing W, Song L, Lv X, Wang X, Hu Y. Preparation, combustion and thermal behaviors of UV-cured coatings containing organically modified α-ZrP. *Journal of Polymer Research* 2010;18:179–185.
106. Chashechnikova I, Dolgov L, Gavrilko T, Puchkovska G, Shaydyuk Y, Lebovka N et al. Optical properties of heterogeneous nanosystems based on montmorillonite clay mineral and 5CB nematic liquid crystal. *Journal of Molecular Structure* 2005;744–747:563–571.

CHAPTER 4

Fly Ash-Based Polymer Matrix Composites

Akshata G. Patil, Anandhan Srinivasan, and Sri Bandyopadhyay

CONTENTS

4.1 INTRODUCTION

Since the early 1960s, there has been a great demand for stiffer, stronger, and lightweight materials used in transportation, aerospace, and construction industries. Composite materials have attracted a lot of interest for more than a decade because of their outstanding properties, such as high elastic stiffness, high strength, and low density, compared with traditional engineering materials (Chawla 2012). There is an extensive research and development in these high-performance engineering materials.

One can consider plastics as nonrenewable materials, as many of the widely used ones are derived from petroleum and hold up to 95% of the market share. The efficacy of plastic consumption can be reduced in many cases by adding inexpensive fillers, which contribute to the volume of the final product, at the same time providing the adequate strength to the product for intended use. Another way of reducing the usage of plastics to a large extent is by adding nano/microsized reinforcements (sometimes referred to as fillers) to the polymer matrix through which the several properties of the matrix can be controlled meticulously. By doing so, these materials can outperform the conventional pristine plastics with a lower consumption rate. For instance, addition of layered fillers to polyethylene terephthalate (PET) improves the gas barrier properties and strength; therefore, the thickness of the PET bottles can be reduced to meet the barrier properties of the current requirements compared with the pure PET.

Nano/microcomposites of polymers have been extensively studied using a large variety of nanofillers with different sizes, shapes, aspect ratios, and origins. Many nano/microfillers exhibit multifunctionality in polymer matrices, as their addition causes improvement in crystallinity, mechanical properties, electrical properties, wear resistance, flammability properties, and so on. Fillers originating from nature are considered to be safer than the synthetic ones, because the risk assessment of many types of fillers has not been done till date. Polymer matrix composites (PMCs) exhibit a combination of properties of the fillers and matrices. In addition to that, the large surface area of the nano/microparticles enables them to uniformly distribute in the polymer matrix, at the same time the properties of a rigid material in a flexible matrix. In most of the PMCs, the fillers are rigid than the matrix, and also nano/micromaterials exhibit properties in between that of atoms or molecules and that of bulk materials. The synergic interaction between the filler and the matrix results in an improvement in the properties of composites.

The basic difference between natural and synthetic polymers is that natural polymers occur in nature, are biodegradable, and are often water based. Examples for natural polymers are silk, wool, DNA, cellulose, and proteins. Synthetic polymers are produced by modern chemical technology to meet the needs of industrial applications such as high strength and toughness. Examples of synthetic polymers include nylon, polyethylene (PE), polyester, poly(tetrafluoroethylene) (Teflon), and epoxy.

Fly ash (FA) is an industrial by-product generated during the combustion of pulverized coal in coal-fired power stations. It is recognized as an environmental pollutant, and disposal of FA has become a major concern due to the higher demand of energy requirement in the world (Blissett and Rowson 2012). It consists of inorganic, incombustible matter, which is a mixture of nonmetal and metal oxides present in coal that fuses during combustion into an amorphous structure.

The physicochemical properties of FA are functions of the source, the type and mineral matter in the coal, the combustion conditions, and postcombustion cooling. During the combustion process, the heat causes the inorganic mineral to become fluid or volatile or to react with oxygen. During cooling, it may form spherical amorphous particles, crystalline solids, or cluster of agglomerates, or condense as coatings on particles.

FAs are heterogeneous, which include large amounts of silicon dioxide (SiO_2), alumina (Al_2O_3), calcium oxide (CaO), and iron oxide (Fe_2O_3). It is important to ensure its safe disposal and reduce its accumulation, as it contains traces of toxic elements, such as lead, arsenic, beryllium, selenium, strontium, cadmium, mercury, chromium, thallium, cobalt, manganese, molybdenum, boron, and vanadium, which are hazardous to people living in the vicinity of disposal sites (Sushil and Batra 2006). According to ASTM C 618-03, FA is classified into two classes: class F and class C. The FA that contains more than 70% oxides of SiO_2, Al_2O_3, and Fe_2O_3 of the total composition, with Fe_2O_3 content being higher than that of CaO belong to class F. Utilization of FA has received a great deal of attention over the past two decades. The majority of FA is utilized in cement and concrete industries. There are several other applications that mainly include structural fill and construction materials, roadway and pavement, lightweight aggregate, environmental engineering and reclamation of damaged areas, ceramic industry, metallurgy and valuable metal extraction, and agriculture (Iyer and Scott 2001).

In order to provide stronger interfacial interaction between the filler and the polymer, functional group-based polymers may be selected (Nath et al. 2010d). Poly(vinyl alcohol) (PVA), poly(butylene succinate), and polylactide are some of the biodegradable and water-soluble biopolymers used in the fabrication of environmentally compatible composites.

The performance of a composite is determined by the properties of the individual constituents in the composites. Various characterization techniques, such X-ray diffraction (XRD), scanning electron microscopy (SEM), transmission electron microscopy (TEM), field emission SEM (FESEM), Fourier transform infrared (FTIR) spectroscopy, and X-ray photoelectron spectroscopy (XPS), have extensively been utilized to characterize macro- and microstructural features of FA-based polymer matrix composites.

4.2 PHYSICOCHEMICAL CHARACTERISTICS OF FA

FA is a mineral residue, consisting of an inorganic, incombustible matter present in coal that has been fused during combustion into a glassy, amorphous structure. The physical properties such as color, density, particle size, and morphology of the FA, and the chemical property such as the composition of FA are the key factors, which determine the characteristics and performance of the FA-based polymer composites.

FA contains unburnt carbon and iron particles, which impart gray-black color to FA, and the composites made out of this FA will have less brighter and reflecting surfaces. One group of researchers suggested that the color of FA is principally controlled by the iron content and the content of unburnt carbon. By contrast, another group indicated that only particle size and its shape determine the color of FA. The quantitative and qualitative effects of chemical composition, unburnt carbon content, and particle size/distribution are the criteria that determine the color of FA (Zaeni et al. 2010).

To understand the basic characteristics of FA, many researchers have characterized it by various techniques. Paul et al. (2007), Wang et al. (2012), Sharma (2012), Mollah et al. (1994), Swami et al. (2009), and Patil and Anandhan (2012) analyzed the crystallite size of FA by XRD, the morphology by SEM micrographs, chemical functional groups by FTIR spectroscopy, chemical composition by XRF, and chemical states of several elements by XPS. The FA samples in the aforementioned works were obtained from various thermal power plants, such as Raichur Thermal Power Station, Karnataka, India; Tuticorin Thermal Power Station, Tamil Nadu, India (Patil et al. 2015); Swanbank Coal Fire Plant, Cement Australia, Queensland (Nath et al. 2010b); Kolaghat Thermal Power Station, West Bengal, India (Sengupta et al. 2011); and Saudi Electricity Company, Riyadh, Saudi Arabia (Khan et al. 2011).

4.2.1 X-Ray Diffraction

The XRD pattern of FA obtained to study the composition and crystallite size of major intense peaks (quartz peak) is shown in Figure 4.1. The average crystallite size was calculated from Scherrer's formula using full width at half maximum (FWHM) of the XRD peak:

$$D = \frac{k\lambda}{\beta\cos\theta} \tag{4.1}$$

where:

D is the crystallite size
β is the FWHM of the diffraction peak
θ is the diffraction angle
λ is the wavelength of the X-rays
k is Scherrer's constant of the order of unity for usual crystals (Cullity 2001)

FA exhibited a lower degree of crystallinity and few numbers of crystalline peaks in the diffraction pattern, which were analyzed by X'Pert HighScore software. Mullite (aluminosilicate) peaks were observed at 25.82°, 27.20° at 2θ values (*d* spacing of 3.45 and 3.4 Å).

The quartz peaks at 2θ values of 20.73°, 26.52°, 26.66°, 40.66°, 49.96° (*d* spacing of 4.28, 3.36, 3.34, 2.21, 1.82 Å) and iron oxide and calcium oxide at 33.08° and 60.42° at 2θ values (*d* spacing of 2.70 and 1.53 Å) were observed. An amorphous hump was observed in

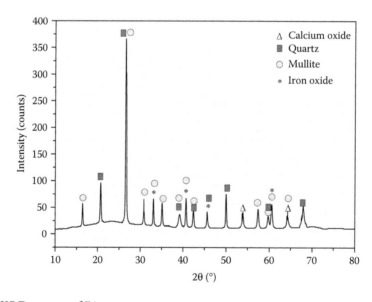

FIGURE 4.1 XRD pattern of FA.

the diffraction pattern between 2θ values of approximately 14°–35°, which is due to the presence of amorphous materials. The crystallite sizes of quartz phase were 37.58 nm (Patil and Anandhan 2012) and 33 nm (Sharma 2012).

4.2.2 Fourier Transform Infrared Spectroscopy

The chemical functional groups of FA were characterized by FTIR as shown in Figure 4.2. The FTIR spectrum of FA showed characteristic stretching vibration bands at 1092, 797, 570, and 465 cm^{-1}. The strong broad band at 1092 cm^{-1} was related to Si–O–Si asymmetric stretching vibration, whereas 797 cm^{-1} band was related to Si–O–Si symmetric stretching vibration as well as AlO$_4$ vibrations. The Si–O–Si bending was observed at 465 cm^{-1} (Smith 1998).

4.2.3 Morphological Studies

Micro-morphology observations revealed that the FA particles are predominantly spherical in shape and consist of solid spheres, cenospheres (hollow spheres), irregular-shaped debris, and porous unburnt carbon as shown in Figure 4.3.

In fluidized bed combustion ash, spherical particles are rarely observed and most of the particles exhibit irregular shapes, primarily because most minerals in the coal do not undergo melting, but only soften under the relatively low boiler temperature of 850°C–900°C (Yao et al. 2015). The irregular fragments consist mainly of unburnt carbon, anhydrate, and calcite. The morphology of FA particles is controlled by the combustion temperature and the cooling rate. The furnace operating temperature of pulverized coal-fired boiler is usually higher than 1400°C. At these high temperatures, the inorganic material in coal becomes fluidlike and solidifies, leading to different morphologies of generated FA. Due to rapid cooling, interparticle fusion occurs and FA particles agglomerate, which results in an irregular shape. The size of the particles ranges from less than 1 μm to greater than 200 μm (Kutchko and Kim 2006; Xue and Lu 2008).

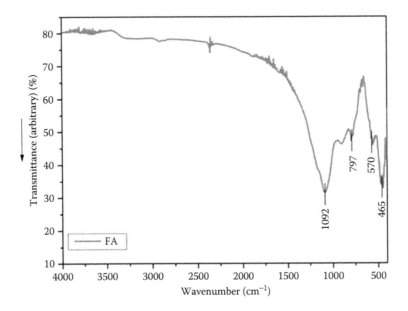

FIGURE 4.2   FTIR spectrum of FA.

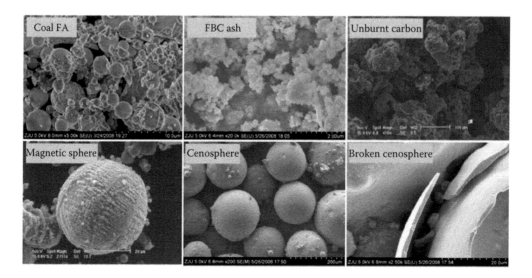

FIGURE 4.3 SEM images of coal FA. FBC, fluidized bed combustion. (From Yao, Z.T. et al., *Earth Sci. Rev.*, 141, 105–121, 2015. With permission.)

The elemental composition of FA was obtained from energy-dispersive X-ray spectroscopy (EDS) as shown in Figure 4.4. It indicates that 50%–70% of the FA consists of amorphous aluminosilicate and the rest consists of oxides of other elements such as iron, calcium, magnesium, titanium, and potassium (Patil et al. 2016).

Similarly, XRF results showed that the oxide contents of silica, alumina, and iron oxide were 62.68%, 27.05% and 3.73%, respectively, whereas Na_2O, K_2O, P_2O_5, CaO, MgO, and TiO_2 account for minor contribution to the composition of FA (Patil et al. 2015). The results of the EDS analysis are complementary to those of the XRF analysis.

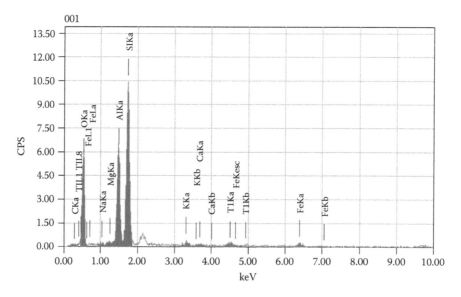

FIGURE 4.4 EDS spectrum of fresh FA. (From Patil, A.G. et al., *Silicon*, 8, 165, 2016. With permission.)

4.2.4 X-Ray Photoelectron Spectroscopy

XPS was used to understand the changes in the chemical states of several typical elements of the FA particles. The wide-scan XPS spectrum of FA is shown in Figure 4.5. The primary components in the spectrum were oxygen, carbon, silicon, iron, calcium, and aluminum (Mollah et al. 1994). Other elements such as sodium and potassium were not significantly observed because of their weak binding energy signals. The high-resolution spectra of the elements at the core levels of C 1s, O 1s, Si 2p, and Al 2p are shown in Figure 4.6. The XPS spectrum was fixed using Gaussian cross-product function after subtracting the background by an inelastic Shirley method. Upon deconvolution, the binding energy of each peak varied in the range of ±0.5 eV to obtain the optimum curve resolution (Patil et al. 2015).

The relative concentrations of the constituents of the FA were determined by quantitative analysis (in atomic percentage, at.%) by XPS. This was done by measuring the area under the peak using the following equation (Shanmugharaj et al. 2002):

$$C_j = \frac{A_i S_i}{\sum_j^m A_j S_j} \qquad (4.2)$$

where:
A_i is the area under the peak of element i
S_i is the sensitivity factor of element i
m is the number of elements in the sample
i and j represent the elements

The atomic percentages of C, O, Si, and Al are given in Table 4.1.

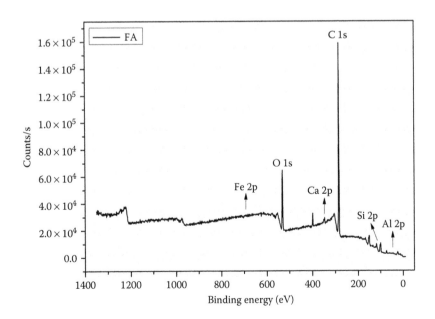

FIGURE 4.5   Wide-scan XPS spectrum of FA.

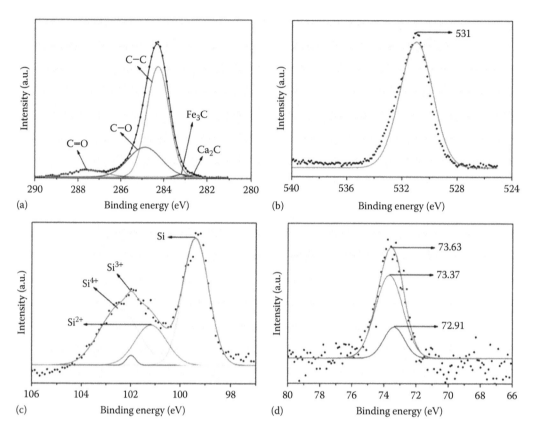

FIGURE 4.6 Deconvoluted high-resolution spectra of FA: (a) C 1s; (b) O 1s; (c) Si 2p; and (d) Al 2p. (From Patil, A.G. et al., *Powder Technol.*, 272, 246–249, 2015. With permission.)

TABLE 4.1 XPS Results

Elements	Type of Photoelectron	Binding Energy (eV)	ASF	Atomic Percentage
		FA		FA
C	1s	284.50	0.250	75.58
O	1s	531.01	0.660	13.36
Si	2p	99.31	0.270	7.70
Al	2p	73.63	0.185	2.89

ASF, atomic sensitivity factor.

4.3 SURFACE MODIFICATION OF FA

Incorporation of FA as a filler into a polymer matrix has improved the properties of PMCs. Compared to other particulate fillers, FA has the advantages of being easily available, low cost, and density (density of FA = 2.2 g/cc) (Patil et al. 2014). However, some properties of the composites may deteriorate depending on the characteristics of the filler, such as particle size, shape, distribution and dispersion, volume fraction, aspect ratio, and the intrinsic adhesion between the surfaces of filler and polymer. Generally, the spherical particles of FA reduce the mechanical strength of PMCs due to easy slippage of polymer matrix over the smooth surface of FA. Thus, reducing filler dimension and increasing filler content will significantly improve the specific surface area of the filler and the aspect ratio, which in turn could improve the yield strength. The interface between the filler and the matrix is also an important characteristic of the composite, which results in increased rigidity. This is because the filler should be capable of supporting high local stress transferred from the polymer matrix.

In order to provide stronger interfacial bonding between the filler and the polymer, the surface reactivity of the FA particles has been enhanced to improve the overall performance of the composites. Many researchers have modified the surface of FA by chemical activation, mechanical activation, or mechanochemical activation. In chemical activation, the surfaces of FA particles are modified in the presence of a surfactant or coupling agent (CA); mechanical activation includes the wet or dry ball milling process by fracturing the FA particles to lower the size; and mechanochemical activation is a combined process of chemical and mechanical activation.

4.3.1 Chemical Activation

Lot of research has been carried out on the activation of FA using chemical activators. The chemical activation is done by treating the FA particles with a surfactant, coupling agent (CA), and/or alkali activation or sulfate activation. Alkali activation involves the surface modification of FA particles, which improves the reactivity without any change in the inherent property of FA. Sulfate activation is based on the ability of sulfates to react with aluminum oxide in the glassy phase of FA to form sulfate active sites at the surface. Alkaline compounds, such as sodium hydroxide, sodium carbonate, potassium hydroxide, and calcium hydroxide, are frequently used. The widely used surfactants and CAs include sodium lauryl sulfate (SLS) and triethoxyvinylsilane (TEVS). Such activations are used to improve the interfacial interaction and the compatibility of FA with different types of polymer matrix.

FA particles were chemically modified with calcium hydroxide by Parvaiz et al. (2011) to improve the interfacial interaction of FA with poly(ether ether ketone) (PEEK) matrix. The modification was carried out at various concentrations of 10, 20, and 30 wt.% solution of calcium hydroxide for every 100 g of FA. The interaction of FA surface with calcium hydroxide is shown in Figure 4.7. The calcium hydroxide-modified FA showed better interaction with PEEK matrix, as the SEM micrographs showed more uniform dispersion of filler throughout the matrix. This resulted in improved mechanical strength; however, the similar effect was not observed for unmodified FA/PEEK composites.

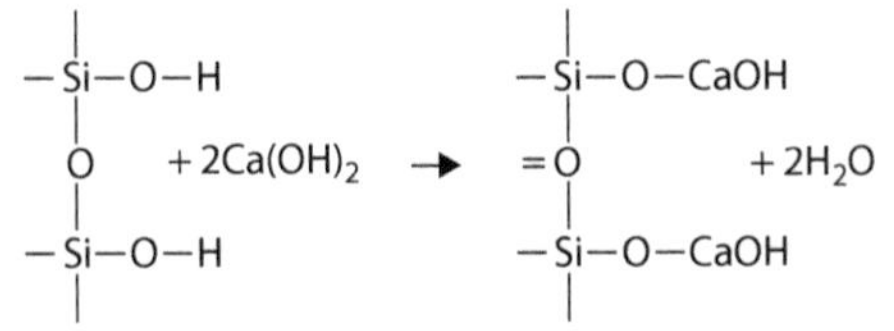

FIGURE 4.7   Interaction of FA surface with calcium hydroxide.

Sengupta et al. (2011) used furfuryl palmitate (FP) CA as a surface modifier for FA particles. FP is an acid ester of furfuryl alcohol and a renewable chemical, which has flame-retardant properties. The concentration of the CA was varied from 1 to 5 wt.% with respect to the weight of FA. It was observed that the true porosity (%) of the particles increased and the surface roughness of the FA particles decreased by treating the FA surface with FP. Low-cost chemicals, such as FP, can be used as an effective CA for FA in place of expensive synthetic CAs.

The effects of surface treatment on FA particles by silane CA *N*-2(aminoethyl)-3-aminopropyltrimethoxysilane (KBM-603) is shown through a schematic in Figure 4.8 (Chaowasakoo and Sombatsompop 2007). The mechanical properties improved at a critical content of KBM-603 (0.5 wt.%), because of the relatively high interfacial bonding between the FA particles and the epoxy resin (ER). This is attributed to the chemical reactions in which ether linkages are formed by the reaction between the CA and FA particles, and also the N–C linkages of the amino groups in the CA and the ER. The decrease in the tensile properties at high KBM-603 contents (above 0.5 wt.%) was related to a self-condensation reaction of the hydrolyzed or partially hydrolyzed CAs, resulting in a formation of flexible polysilanol molecules on the FA surfaces. These flexible polysilanol molecules probably reduced the reactivity of the ER and the CA, thus resulting in a decreased diffusion of the ER into the polysiloxane network.

Surface modifications of FA by NaOH, NaOH/NH$_4$HCO$_3$, ethylenediaminetetraacetic acid (EDTA), and HCl were reported by Sarbak and Kramer-Wachowiak (2002). After modification, there was an increase in the surface area and the pore volume of FA particles by HCl and

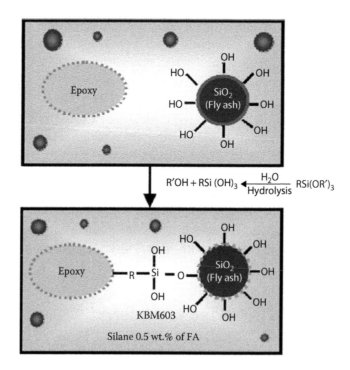

FIGURE 4.8　A model of the treated FA/epoxy composites using KBM-603 concentration of 0.5 wt.% of the FA.

EDTA, respectively. The SEM micrographs revealed that modification with EDTA resulted in a transformation of ball-shaped FA particles into a rodlike structure. The treatment with HCl resulted in the appearance of agglomerations of undefined shape. The above-described modifications can find the appropriate application of the modified FA particles as adsorbents and catalysts in chemical industries. Additionally, these surface modifiers cannot improve the color, restricting the wide use of FA as a filler. In order to meet the demands of the plastics industry, the color of FA and the specific surface area are of an important research keystone. The surface modification technology for color change and increased specific surface area was studied by Yang (Yang et al. 2006). By the addition of calcium hydroxide solution with high basicity, the outer layer of FA was corroded and the active silica and alumina reacted with hydrated lime. These form a variety of silicate and aluminate reactants, which deposit, nucleate, and grow on the FA particle surface, resulting in rough surfaces and large specific surface areas of FA. This improves the whiteness and appearance of FA particles shown in Figure 4.9.

A study by Shawabkeh et al. (2011) showed enhancement in surface properties of oil FA (OFA) by chemical treatment. A mixture of sulfuric and nitric acids was used to modify the surface of OFA with carboxylic acid groups. The goal of surface modification of OFA is to make its surface more compatible with polar polymers in order to produce OFA/polymer composite materials with improved dispersion of OFA and to increase the OFA surface area in order to support its use as adsorbent materials in adsorptive separation and purification applications. Evaluation of different structural changes during the surface modification was investigated by XRD analysis; the XRD plot is shown in Figure 4.10. The XRD analysis showed an increase in the carbon crystalline structure by surface modification of FA. The increase was even better when air was introduced and the concentration of HNO_3 increased up to 15%.

FIGURE 4.9 Comparison of the appearance of the different samples. (a) FA with less than 45 μm grain size, (b) FA after removing magnetic pearls, (c) FA after removing magnetic pearls, and unburned carbon, (d) unburned carbon, (e) magnetic pearls, and (f) composite fly ash. (From Yang, Y.-F. et al., *J. Hazard. Mater.*, 133, 276–282, 2006. With permission.)

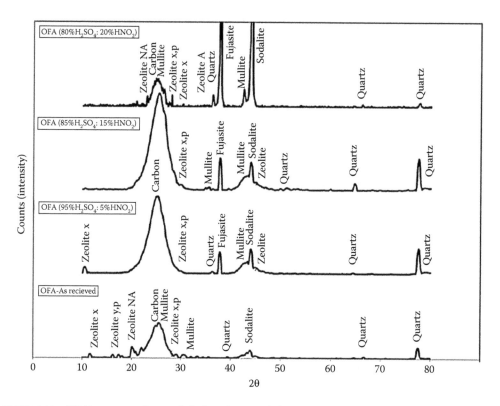

FIGURE 4.10   XRD patterns for modified and unmodified OFAs.

Anandhan et al. (2012) used TEVS as a FA surface modifier and reinforced ethylene–octene copolymer (EOC) matrix with the surface-modified FA. The modification was confirmed by FTIR analysis in which two new peaks were observed for modified FA at 2997 and 1656 cm^{-1}, which correspond to the C–H stretching vibration and C–O–C unsaturation of the vinyl group from the modifier TEVS shown in Figure 4.11. This confirms that the OH groups of the FA surface have successfully reacted with TEVS modifier.

Three different types of treatments were attempted on the surface of the FA with the aim of modifying its characteristics of adhesion to the epoxy matrix system by Kishore et al. (2002): in the first batch, a simple procedure of cleaning the filler surface with acetone; in the second batch, the FA particles were coated with a silane-bearing system; and in the third batch, the FA particles were exposed to paraffin oil. Among the various treatments, the silane treatment was found to be effective. Acetone-cleaned FA particulate also yielded comparable results. Treatment with paraffin oil reduced the adhesion, and thus also the energy and load-bearing capacities during impact.

Nath et al. (2010d) modified the surface of FA by SLS. This modified FA was used as a filler in biodegradable PVA matrix, which improved the properties of the polymer composites. The enhancement of the properties was due to the elimination of particle–particle interaction and a better distribution of modified FA within the polymer matrix, which was confirmed from the SEM micrographs. The SEM images of FA particles before and after coating with SLS are shown in Figure 4.12, in which the unmodified FA particles are

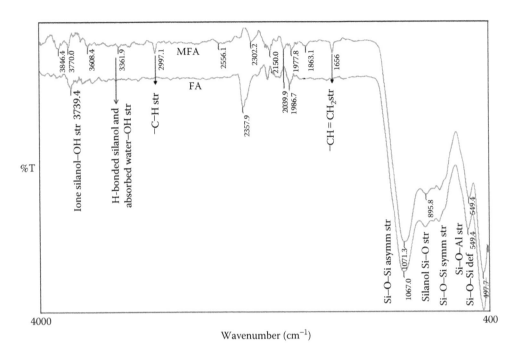

FIGURE 4.11 FTIR spectra of FA and modified FA.

FIGURE 4.12 SEM morphology images of FAs: (a) FA and (b) SLS-FA. (From Nath, D.C.D. et al., *Appl. Surf. Sci.*, 256, 2761, 2010d. With permission.)

of scattered spherical and nonspherical shapes, whereas the modified ones are of multiple stacked layers with mostly irregular shape. The aspect ratio was high for multiple stacked layers of FA particles. The surfactant forms a coating on the surface of FA particles and contributes to the superior light-reflecting quality of the particles. The thin layers of SLS on FA particles reduce the attractive forces between the particles and aid to disperse in the aqueous/or polymer matrix uniformly. The orderly distributed filler in the polymer matrix enhances the mechanical properties of the composites.

Similar features were observed in the TEM images of modified and unmodified FA particles as shown in Figure 4.13. The TEM micrographs of FA particles showed some

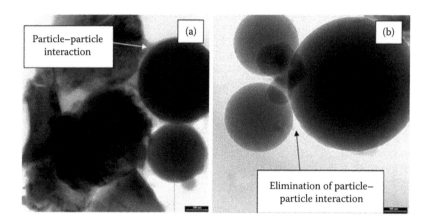

FIGURE 4.13   TEM images of FAs: (a) FA and (b) SLS-FA.

empirical results because they were not exactly the same particles before and after modification. The existing interfacial interactions between the unmodified FA particles were disappeared in the modified FA particles.

4.3.2   Mechanical Activation

Mechanical activation is a solid-state powder processing technique that involves repeated fracturing, welding, and rewelding of powder particles in a high-energy ball mill. It is a top-down process of producing nanoparticles by high-energy milling as a versatile alternative to other processing routes. It occurs due to compression, shear (attrition), impact (stroke), and impact (collision) forces acting between two grinding media beads or between grinding media beads and the inner wall of the vial. The high-energy milling process is greatly affected by a number of factors, such as the type of mill, the milling temperature, the extent of vial filling, the type and size of balls, the ball-to-powder weight ratio, and the time of milling (Baláž 2008). The properties of the milled powders, such as the particle size distribution, the final stoichiometry, and the degree of disorder or amorphization, depend on the abovementioned factors. The smooth, glassy, and inert surface of the FA can be altered to a rough and more reactive state by this technique. Swami et al. (2009) milled the FA for 30 h to obtain a nanostructured FA. The sample was taken out after every 5 h of milling, and the nanostructured FA (NFA) was characterized for its crystallite size, lattice strain, and percentage of crystallinity using an X-ray diffractometer. It was found that after 30 h of milling, the crystallite size was reduced from 92 to 29 nm, and the percentage of crystallinity was reduced from 63% to 38%. The size, shape, and texture of the FA as well as the NFA were studied using SEM micrographs. Initially, FA particles were of spherical shape with smooth surface, whereas the 30 h milled particles were of irregular shape with a rough surface morphology.

4.3.3   Mechanochemical Activation

Intensive (dry/wet) milling of powders in the presence of a surfactant and a medium is considered to be a way of applying "mechanochemical activation," which involves dispersion of solids, generation and migration of defects in the bulk, and plastic deformation of particles. Mechanochemical activation refers to enhancing the FA reactivity through physicochemical changes occurring due to the combined effects of decreased crystallite size and average particle size, and increased specific surface area. The process of activation starts by loading the sample along with a surfactant and a medium into the ball mill. The sample is then milled for the desired length of time based on the requirements. Paul et al. (2007) subjected a class F FA to high-energy ball milling and converted it into a nanostructured material. Ball milling was carried out in the presence of a surfactant, namely, SLS, and toluene medium for 60 h. The samples were taken out at every 10 h of milling and analyzed for their particle size and specific surface area. Figure 4.14a and b shows that the particle size gradually decreased and the specific surface increased for milling the FA particles during 10–60 h. The particle size got reduced from 60 μm to 148 nm, and the surface area increased from 0.249 to 25.53 m^2/g for 60 h milled FA. The crystallite size was reduced from 36.22 to 23.01 nm for quartz and from 33.72 to 16.38 nm for mullite during ball milling for 60 h. During mechanochemical activation, high stresses were exerted on FA particles by the milling media that damage the crystalline structure, causing the formation of amorphous segments as observed by Patil and Anandhan (2012). The FA obtained from Tuticorin Thermal Power Station was subjected to high-energy milling for 60 h. The average crystallite size was determined by Scherrer's equation, and the reduction in the crystallite size of the quartz phase was from 37.58 to 9.25 nm with a increase in the milling duration. Morphological studies revealed that the surface of the NFA was irregular in shape with a rough and uneven surface compared with FA particles, which often have a spherical shape (Patil and Anandhan 2012).

Similarly, a class F FA was subjected to high-energy ball milling-induced mechanochemical activation aided by a surfactant for 48 h (Patil et al. 2015). Ethyl acetate as the milling medium, a ball-to-powder ratio of 12:1, and 2 wt.% of surfactant reduced the average particle size of FA to 329 nm and led to a specific surface area of 8.73 m^2/g. The decrease in the crystallite size of the mechanochemically activated FA (MCA-FA) was confirmed

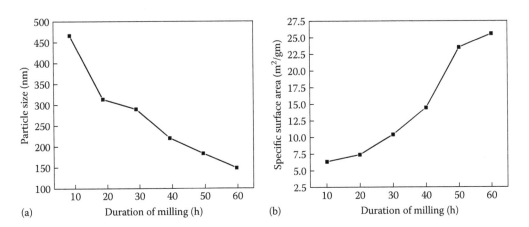

FIGURE 4.14 Variation in (a) particle size and (b) specific surface area of FA with milling time (in hours).

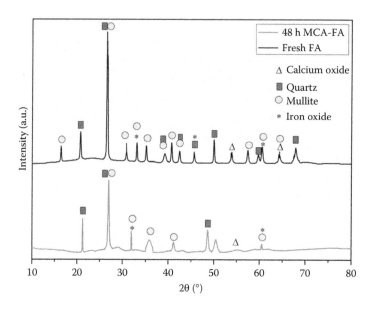

FIGURE 4.15   XRD patterns of FA and MCA-FA.

from a reduction in the peak intensity with a broadened amorphous phase by XRD studies. It can be clearly observed from Figure 4.15 that the peak intensities of quartz phase are reduced considerably and the peaks broaden after 48 h of milling. The crystallite size of the quartz phase present in FA was 28 nm, which was reduced to 7.7 nm after 48 h of milling. The chemical state of several typical elements of the FA particles after mechanochemical activation was characterized by XPS analysis.

This characterization illustrated that the peak area of major elements (O, Si, and Al) increased after milling. Figure 4.16a–d shows the assignment of deconvoluted high-resolution XPS peaks of C, O, Si, and Al in MCA-FA. Deconvolution of the carbon peak reveals the presence of C=O groups (287.5 eV), C–O groups (284.8 eV), C–C groups (284 eV), Fe_3C groups (283 eV), and Ca_2C (282.5 eV) in FA (Figure 4.15a) (Xu et al. 2013). There was a decrease in the atomic percentage of carbon groups from 75.58% to 38.16% for MCA-FA (Figure 4.16a). This was attributed to the pretreatment of FA, which was washed with distilled water to remove the carbon that creamed up on the surface. However, the presence of carbon was still due to the addition of the surfactant during the milling process. Oxygen (O 1s) peaks of MCA-FA are shown in Figure 4.16b. Upon deconvolution, the peak maxima occur at around 531.01 and 531.27 eV for MCA FA. The presence of more –OH groups leads to a higher peak intensity, which attributes to the high surface reactivity of MCA-FA. Figure 4.16c shows the deconvoluted high-resolution XPS peaks of silicon, corresponding to four types of oxidation states of Si: Si^0, Si^{2+} (SiO), Si^{3+} (Si_2O_3), and Si^{4+} (SiO_2) (Logofatu et al. 2011). There was an increase in the atomic percentage of silicon from 7.7% to 13.74% for MCA-FA. In the case of MCA-FA, the Si peak at 101.9 eV corresponds to either Si–OH or Si–O–Si bond. Also, a similar observation was made for Al, as the content of aluminol (Al–OH) at 73.5 eV and the (Al–O) bond was quite low in Al atoms of MCA-FA (Figure 4.16d).

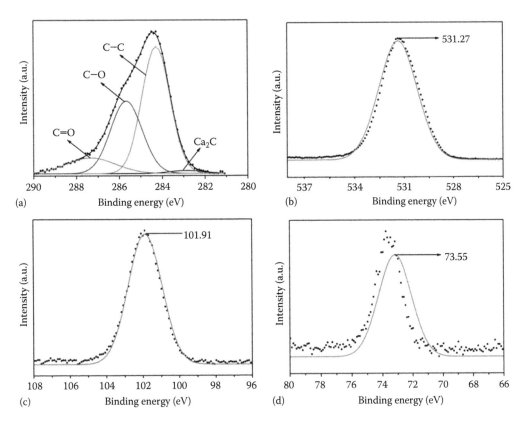

FIGURE 4.16 Deconvoluted high-resolution spectra of the core levels for MCA-FA: (a) C 1s; (b) O 1s; (c) Si 2p; and (d) Al 2p.

Dispersion and aggregation are regarded to be the two main processes taking place during grinding of solids. The reaggregation of particles is a major concern with the ball milling process, which significantly increases the particle size of MCA-FA. From Figure 4.17, the fractal dimension of MCA-FA was studied by Patil et al. (2015). The spherical particles of FA with smooth glassy surfaces are shown in Figure 4.17a. After mechanochemical activation, the surface of FA has become rough, and particles are of irregular shape (Figure 4.17d). Because agglomerates are considered as fractal-like structures, it is acceptable to quantify their irregularity with the fractal dimension, D_f (Figure 4.17c) as originally proposed by Mandelbrot (1982). The number of primary particles in the agglomerate (n) can be theoretically estimated from the fractal geometry given in the following equation:

$$n \approx \left(\frac{R_a}{a} \right)^{D_f} \tag{4.3}$$

where:
  R_a is the radius of the agglomerate
  a is the radius of the particle
  D_f is the fractal dimension

FIGURE 4.17 SEM micrographs of (a) FA, (b) particle agglomerate of MCA-FA, (c) diffusion-limited particle–cluster agglomerate ($D_f = 2.5$), and (d) MCA-FA at 8000×.

This study focused on the agglomerates of nanoparticles having a fractal structure with a fractal dimension D_f close to 2.5, in agreement with the diffusion-limited agglomeration model. The $D_f = 2.5$ in the diffusion-limited particle–cluster agglomerate (Lapuerta et al. 2010), when the values of $a = 1–0.1$ µm and $R_a = 25$ µm were used in Equation 4.3. An estimate of 3,000 to more than 10,00,000 primary particles was observed in the cluster assembly even though the mechanical energy involved in the ball milling process is apparently critical to the breakage of large-sized FA agglomerates. Again, the milled fine agglomerates with a reduced cluster size are suspected to reagglomerate favorably during the multiple collisions of particles in the milling process.

Similarly, the adsorption characteristics of coal FA were enhanced by mechanochemical activation with a high-energy monoplanetary ball mill by Stellacci et al. (2009). The FA was mechanochemically activated for 4 h in N_2 atmosphere containing a higher carbon percentage. The adsorption capacity of phenol from an aqueous solution was compared between the MCA-FA and the powdered activated carbon. It was observed that MCA-FA had encouraging results, such as favorable adsorption isotherms, improved specific adsorption capacity, and very fast adsorption rate. This provides new opportunities for utilizing FA in environmental protection applications, such as the stabilization/solidification treatment of hazardous waste and contaminated soil.

4.4 FA–POLYMER COMPOSITES

Polymeric materials are the most useful damping materials, but they cannot be used alone for high-damping structural fields due to their limited properties, such as poor mechanical properties, high shrinkage ratio, and limited acclimatization. Therefore, an inorganic/polymer composite is a possible way of overcoming some of these problems. Inorganic FA particles have been widely used as fillers in the polymer matrix to produce particulate-reinforced polymer composites.

PMCs consist of either a thermoset or a thermoplastic matrix, which binds the reinforcement or filler together, and transfer the applied stresses from the matrix to the reinforcement. Thermosets are plastics that cannot be melted once cured, which include resins such as polyesters, epoxies, and phenolics. Thermoplastics are melt-processable plastics that can be repeatedly melted, thus enabling them to be recycled. Commonly used thermoplastics include poly(vinyl chloride), PE, and polypropylene (PP).

FA has been widely used as a filler in various polymer matrixes, such as epoxy (Kulkarni and Kishore 2002; Chaowasakoo and Sombatsompop 2007; Gu et al. 2007), polyester (Devi et al. 1998; Guhanathan et al. 2001; Rohatgi et al. 2009), EOC (Anandhan et al. 2011; Patil et al. 2016), PVA (Yunsheng et al. 2006; Nath et al. 2010c; Patil et al. 2014), PP (Satapathy et al. 2011; Sengupta et al. 2011), PEEK (Parvaiz et al. 2011), polyurea (PU) (Qiao and Wu 2011), and acrylonitrile–butadiene–styrene (ABS) terpolymer (Kulkarni et al. 2014).

4.4.1 Epoxy Polymer

Epoxies are the most widely used family of thermosets. They are widely used in various engineering and structural applications, such as industrial tooling and composites, electrical systems and electronics, and marine and aerospace industry. ER has been extensively employed worldwide because of its better heat resistance, moisture resistance, chemical resistance, high strength, low volatility, and good adhesion to different substrates. Therefore, ER filled with FA can be used to obtain advanced low-density composites.

The hydrophobic thermosetting epoxy polymer and hydrophilic FA particles were incompatible and led to poor interfacial bonding, resulting in inferior mechanical properties. Hydroxyl groups on the FA surfaces cause clustering or agglomeration among themselves and lead to a strong filler–filler interaction in polymer matrices. Therefore, many silane CAs are applied to promote adhesion between the thermosets and FA. The mechanism of the polymer–FA interaction involves the formation of siloxane bonding (Kulkarni and Kishore 2002; Chaowasakoo and Sombatsompop 2007).

PMCs containing silane-treated FA displayed improvement in strength and modulus for all volume fractions of the filler with varying average particle sizes. The better compatibility with the matrix was responsible for the improvement in properties, observed through microscopic examination of the compression-failed sample (Figure 4.18). The composites reinforced with filler of lower average particle size showed lesser debonding and displayed a better adhesion to matrix. Thus, from SEM micrograph (Figure 4.18a), the smaller particles marked "1" are displaying less debonding compared with the bigger ones (marked "2"). The tensile strength of the composites with larger particles was almost the same as that with

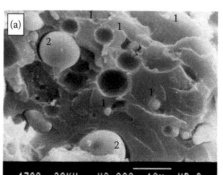

FIGURE 4.18 (a) SEM micrograph showing less of debonds at smaller particles in lower average particle size sample and (b) SEM picture illustrating a cluster containing large particles with debonds. (From Kulkarni, S.M. and Kishore, *J. Mater. Sci.*, 3, 4321–4326, 2002. With permission.)

lower volume fractions of smaller particles, but the strength values of the former diminished at higher volume fraction of the filler. This change is attributed to the clustering of particles and debonding around such clusters (marked by arrows), which were observed only for larger particles (Figure 4.18b).

Conventional thermal and microwave curing methods were utilized to cure FA/epoxy composites, and the mechanical and morphological properties of the composites were evaluated by Chaowasakoo and Sombatsompop (2007). The results suggested that the tensile and flexural moduli of the composites increased with increasing FA content, whereas the effect became opposite for tensile, flexural, and impact strengths, and tensile strain at break. An improved mechanical property of the composite was obtained by addition of the CA N-2(aminoethyl)-3-aminopropyltrimethoxysilane at an optimum value of 0.5 wt.%. Beyond 0.5 wt.%, the mechanical properties greatly reduced, except for the flexural modulus. The comparative results indicated that the composites prepared by the microwave cure consumed shorter cure time and had higher ultimate strengths (especially impact strength), and strain at break than those obtained by the conventional thermal cure. The composites with higher tensile and flexural moduli could be obtained by the conventional thermal cure.

The impact strength of carboxyl-terminated butadiene acrylonitrile copolymer (CTBN)-modified ER was superior to that of the pure ER. The addition of FA to the ER composites showed a negative effect on the impact property. It was observed that the impact strength of the ER with FA decreased as the FA content increased up to 8 phr (Figure 4.19a). The increase in CTBN concentration improved the impact strength for the FA fixed up to 8 phr, and also the hybrid composite containing CTBN and FA exhibited a positive effect on the impact strength (Figure 4.19b and c). However, at a higher CTBN concentration of 15 phr and FA, there was a decline in the impact strength (Figure 4.19d). For the hybrid composites with FA, an improvement in the impact properties was noticed only in the presence of CTBN (Ramos et al. 2005).

Gu et al. (2007) characterized the damping properties of low-density epoxy/FA composites; a series of epoxy composites filled with different volume fractions of FA were prepared in their work. Damping tests of cured epoxy composites were performed in the temperature range from −40°C to 150°C and in the frequency range from 10 to 800 Hz

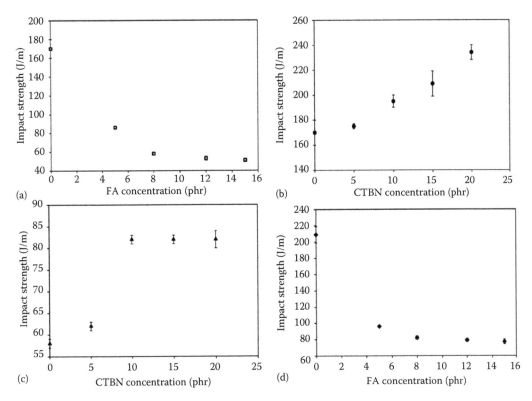

FIGURE 4.19 Impact strength of ER with different concentrations of CTBN and FA: (a) ER with different FA concentration; (b) ER with different CTBN concentration; (c) ER with different CTBN concentration and FA content fixed in 8 phr; and (d) ER with different FA concentration and CTBN fixed in 15 phr.

using a tension–compression mode. The results showed that the values of tangent delta (tan δ) reach their peak values at the glass transition temperatures for the composites with 30–50 vol.% of FA. The tan δ values attenuated slowly with the increased frequency, which indicated that the damping property of such composites was better than those of other composites. The degree of dispersion of FA particulates was uniform due to surface treatment and enhanced interfacial adhesion with the matrix. The decomposition temperatures ascended with an increase in the volume fraction of FA, which indicated that the heat resistance of the matrix could be optimized by controlling the FA content.

The impact behavior of epoxy specimens reinforced with treated and untreated FA was compared with the neat epoxy. The FA was treated with three different chemicals: acetone, combination of both acetone and silane, and paraffin oil. Among the various treatments, the silane treatment was found to be quite effective. Acetone-cleaned FA particulate also yielded comparable results, but treatment with paraffin oil reduced the adhesion and thus also the energy and load-bearing capacities during impact (Kishore et al. 2002).

4.4.2 Polyester

Polyester is a long-chain polymer, which is a transesterification product of an ester and a dihydric alcohol and dimethyl terephthalate. Polyesters are hydrophobic in nature, widely used for insulation by manufacturing hollow fibers, because, they are very durable and resistant to most chemicals. They also exhibit low stretching and shrinkage, wrinkle resistance and abrasion resistance.

Devi et al. (1998) used FA and CaCO₃ as fillers in polyester resin (PR) and compared the mechanical properties of those composites. It was found that FA-filled unsaturated PR (FPR) was inferior to calcium carbonate-filled PR (CPR) and neat PR with respect to tensile and flexural strengths. However, FPR was found to have a higher flexural modulus than CPR and PR. The decrease in mechanical properties was due to poor adhesion at the interface between the organic polymer matrix and inorganic FA particles. They suggested that the compatibility between the polymer matrix and FA can be improved by treating the surface of FA with silane or chrome CA.

Therefore, the surface of FA was treated with two different silane CAs, such as adenosine monophosphate (AMP) and vinyltriethoxysilane (VES). The mechanical properties of FA/general-purpose unsaturated PR (GPR) particulate composites were studied by Guhanathan et al. (2001). The properties of FA–CA–GPR were also compared with those of GPR and CaCO₃–GPR. The increase in the CA content from 0.5% to 2.0% enhanced the tensile, flexural, compressive, and impact strengths, and the hardness of the composites. However, the tensile and flexural moduli decreased for FA–CA–GPR, when FA was surface treated with silane CAs. In the presence of CAs, the bonding (adhesive strength) was higher than the elasticity between the FA and the polyester matrix, resulting in increased elongation and decreased modulus. The FA treated with 2% of CA showed comparable properties with CaCO₃-filled GPR composites. Thus, it was concluded that the CA AMP was better than VES in modifying the FA–GPR composite.

Rohatgi et al. (2009) studied the compressive properties of polyester/FA cenosphere composites. The composite material was fabricated by releasing FA cenospheres in a tube filled with PR. Such fabrication method resulted in functionally graded structure, where the volume fraction of FA cenosphere was varied from 0 to 29.5 vol.% in the composite. Results showed that the compressive modulus increased and the compressive strength decreased as the volume fraction of FA cenosphere increased in the composite. However, numerous defects such as irregular shape, rough surface, and porosity in the wall of FA cenospheres contribute to their early fracture and reduce the strength of the composite materials. As such, their studies reveal that FA cenosphere can be used as a filler to enhance the compressive modulus and to reduce the density of polyester matrix composite.

4.4.3 Ethylene–Octene Random Copolymer

EOC is a new random copolymer of ethylene and 1-octene that is commercially available as a polyolefin plastomer. It possesses the high flexibility and mechanical properties of a synthetic rubber and the melt processability of thermoplastics. Because of its high filler loading capability, it is widely used in making soft polymer composites and thermoplastic vulcanizates, wires and cables, and automotives.

Anandhan et al. (2012) reinforced FA particles modified by TEVS into EOC. In EOC–FA composites, the tensile strength decreased at a filler loading of 5% and increased to the maximum value at 15% of filler loading, beyond which it decreased. There was no major improvement in the tensile strength of EOC on addition of FA, which attributes to the interaction between dissimilar polarities of the EOC matrix and FA particles. The EOC/modified FA composites exhibited increased tensile strength in proportional to the volume fraction of filler. The improvement was nearly 150% in tensile strength at a filler loading of 20%. This indicates the effectiveness of modified FA as reinforcement in the EOC matrix.

In a work by Patil et al. (2016), an attempt was made to improve the compatibility between the nonpolar EOC matrix and the polar filler by modifying FA by mechanochemical activation. A good interfacial adhesion was observed in the case of MCA-FA/EOC composites than fresh FA/EOC composite, which in turn was reflected as the improvement in the tensile properties of the composites. The property of EOC composite was improved even at a low filler loading. The irregular and rough surface of MCA-FA improved the mechanical interaction with the matrix, and the surfactant wrapping on the particles increased the compatibility of the matrix and the MCA-FA.

4.4.4 Poly(Vinyl Alcohol)

PVA is one of the biodegradable plastics and is a water-soluble semicrystalline polymer produced by the alcoholysis of poly(vinyl acetate). Modification of PVA with inorganic fillers is common to improve its properties, such as mechanical strength, thermal resistance, and permeability properties when used in a film form. The composites of FA/PVA exhibit three kinds of interaction: the interaction between FA particles and PVA, the interaction between the inorganics and water, and the interaction among the inorganics (Liu et al. 2007). These interactions have significant effects on the performance of the final composite films.

Composite films of PVA and chemically modified FA (MFA) by sodium hydroxide were prepared with 5, 10, 15, 20, and 25 wt.% of MFA treated with 1 wt.% cross-linking agent (glutaraldehyde [GLA]). The effect of FA modification on PVA–MFA composite films was investigated by the study on tensile properties, and the results were compared with those of PVA-FA composites. The composite films with MFA showed higher tensile strength than those with FA. The addition of MFA to neat PVA proportionally improved the tensile strength up to 20 wt.% MFA content. Further addition of filler (25 wt.%) led to a significant decrease in the tensile strength. The decrease was assumed from the supersaturation of particles in composite films, which led to an enhancement of particle–particle interaction rather than particle–PVA interaction. The increase was up to 289% at 20 wt.% MFA in comparison with FA as shown in Figure 4.20a. The formation of interfacial interactions played a significant role in the improvement of mechanical strength compared with neat PVA (Nath et al. 2010a).

Similarly, composite films of PVA reinforced with 5, 10, 15, 20, and 25 wt.% of SLS-modified FA (SLS-FA), along with 1 wt.% cross-linking agent, GLA, were prepared by the aqueous casting method. The tensile strengths of the composite films increased proportionally with the addition of SLS-FA. The composite films with SLS-FA exhibited higher tensile strengths than those with FA (Figure 4.20b).

The –OH groups on the surface of FA particles form hydrogen bonds with PVA chains. The higher strength of the films with SLS-FA was due to the formation of stronger interfacial

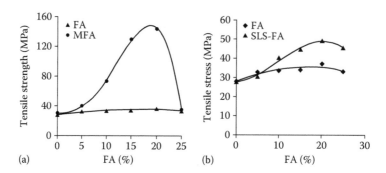

FIGURE 4.20 (a) Effect of chemical modification on the relationship of tensile stress and FA in composite films and (b) relationship plots of tensile stress and FA in the composite films.

interactions by the chemical reaction of GLA with PVA and SLS-FA. This leads to greater load transfer between the polymer and the filler, and enables greater stress generation in the material before failure (Nath et al. 2010c).

Patil et al. (2014) prepared PVA/MCA-FA composites using MCA-FA milled for 30 and 60 h. The good interfacial interaction between the filler and the matrix at low filler loading improved the mechanical strength remarkably compared with the composites loaded with fresh FA, as the filler loading increased to a certain critical loading. The properties were reduced above the critical filler loading, because, the filler–filler interactions were more than the filler–polymer interactions, and also the stress concentration at agglomerations. In the case of composite with fresh FA, the polymer chains slip over the smooth surface of FA, ultimately resulting in an early failure of the corresponding composite. This observation was further corroborated by Nielsen's equation, which relates the strain at break with the filler volume fraction of the composites:

$$\varepsilon_c = \varepsilon_m \left(1 - V_f^{1/3}\right) \tag{4.4}$$

where:
  ε_c is the strain at break of the composite
  ε_m is the strain at break of the polymer matrix
  V_f is the volume fraction of filler

The theoretical and experimental values of elongations at break with respect to the volume fraction of the filler were compared as shown in Figure 4.21a–c. It was observed that on reinforcing PVA with fresh FA and 30 h MCA-FA, elongation at break decreased with an increase in the volume fraction of the filler (V_f) (Figure 4.21a and b). However, the theoretical and experimental values of elongations at break of these composites had a close match between them. The abrupt decrease in elongation at break of the composite with 1 wt.% of MCA-FA milled for 60 h (Figure 4.21c) signifies interruption of crack propagation through the polymer matrix by the filler particles. However, at a filler loading of 5 wt.% of MCA-FA, filler particles might have promoted craze formation in the PVA matrix, which resulted in a sudden increase in elongation at break.

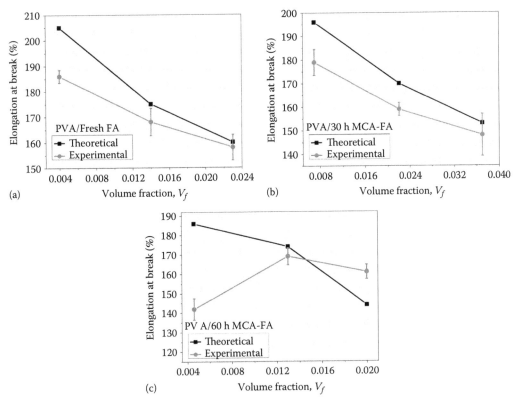

FIGURE 4.21 Elongation at break vs. volume fraction of filler of (a) PVA/fresh FA, (b) PVA/30 h MCA-FA, and (c) PVA/60 h MCA-FA composites.

Yunsheng et al. (2006) studied the impact behavior and microstructural characteristics of PVA fiber-reinforced FA-based geopolymer boards prepared by the extrusion technique. It was observed that the spherical shape of FA greatly improved the extrudability of the geopolymer mixture, which resulted in a denser microstructure of geopolymer boards at a lower percentage of FA. At higher percentage of filler loading, it was not the same. When the composite was incorporated with high volume fraction of PVA, there was a great increase in impact toughness and the mode of failure transformed from a brittle to a ductile pattern. The combination of short PVA fibers along with low percentage of FA possessed very high impact strength and stiffness of FA-geopolymer boards. At higher loading of FA, the impact resistance decreased.

4.4.5 Polyethylene

It is a family of polyolefin resins obtained from the polymerization of ethylene. It is the most common thermoplastic material having long hydrocarbon chains, which is widely used as packing materials. Some of the widely used grades of PE are low-density PE, medium-density PE, and high-density PE (HDPE).

The reports of Satapathy et al. (2013) discussed the results of various modifications onto the HDPE-FA/NFA composites. It was proved that FA is a valuable reinforcing filler for

HDPE, and its size reduction to nanolevel is a more effective criterion for its future use. The three modifications were maleic anhydride (MA) grafting of the matrix, electron beam irradiation of the composite, and irradiation of the FA/NFA. The electron beam irradiation of HDPE-FA/NFA composite yielded excellent physicomechanical, thermal, and dynamic mechanical properties.

Similarly, the three modifications mentioned previously were carried out for FA and NFA-filled waste PE (WPE) composites. It was observed that the FA/NFA at 5 wt.% imparted enhanced mechanical properties. Of the three methods of modification, the composite irradiated by electron beam gave the best balance in the physicomechanical properties. The tensile and flexural strengths of WPE increased from 21.2 and 25.4 MPa to 33.0 MPa (57.8%) and 45.8 MPa (72%), respectively, for 5 wt.% of FA-filled WPE composites, which further increased to 34.5 MPa (65%) and 47.7 MPa (87.8%), respectively, for 5 wt.% of NFA-filled WPE composites, after electron beam irradiation. The fractured surfaces were examined under SEM and were correlated with the mechanical properties. The results indicated that FA and NFA in these composites acted as excellent stress reducers, preventing crack propagation and enhancing their performance on electron beam irradiation (Satapathy et al. 2012).

Deepthi et al. (2010) also prepared HDPE/silanated FA cenosphere lightweight composites by using maleated HDPE as an interfacial modifier. The tensile strength values of the composites exceeded that of neat HDPE due to the addition of surface-treated cenospheres and the compatibilizer. This signifies that the interfacial modifier has effectively anchored the two immiscible phases by undergoing reactive blending between the ester group of the compatibilizer and the amine groups of silane-treated cenospheres.

Sharma and Mahanwar (2010) studied the effect of particle size of FA on recycled PET (RPET)/FA composites. The addition of FA to RPET matrix resulted in an improvement in the mechanical properties at a critical concentration of FA. The variations in the tensile strength and the elongation at break of RPET/FA composite are shown in Figure 4.22a and b.

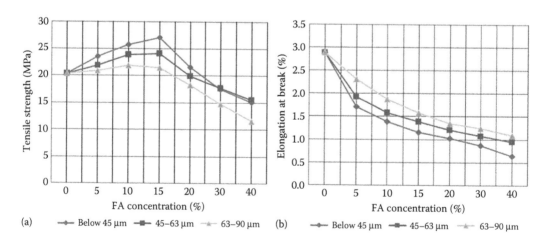

FIGURE 4.22   (a) Tensile strength and (b) elongation at break of RPET/FA composites.

The tensile strength gradually increased as the concentration of the filler to a maximum value at 15 wt.% loading and decreased beyond that loading (Figure 4.22a). A similar trend was observed for smaller particle sizes (below 43 μm) than for larger particle sizes (43–63 and 43–90 μm). This signifies that as the particle size decreases, the interfacial area per unit volume of filler increases, so the probability of finding large flaws existing within this volume decreases, resulting in higher tensile strength. From Figure 4.22b, it can be seen that the elongation at break drastically decreases with filler concentration. The rate of decrease of elongation was more for smaller particles (below 43 μm) than for larger particles (43–63 and 63–90 μm). On increasing the filler loading, a saturation level was attained, which is influenced by the aggregation of filler particles in the polymer matrix. Thus, the mechanical properties of these composites were a function of particle size, dispersion, and interaction between the filler particles and the polymer matrix.

Similarly, OFA was used as a filler for the preparation of low-density PE composite by Khan et al. (2011). The surface of FA was functionalized with the COOH group, and PE-grafted MA (PE-g-MA) was used as a compatibilizer in these composites. The composite samples prepared with four different filler loadings of 1, 2, 5, and 10 wt.% exhibited improvement in Young's modulus and yield strength. The elongation at break and toughness decreased with an increase in filler concentration. This shows that the polymer becomes stiffer by addition of filler; meanwhile, its plasticity decreases. The modification of FA and the addition of compatibilizer in these composites led to an improvement in the mechanical properties at lower filler loading compared with that of unmodified FA. This is attributed to the uniform dispersion of filler and better interlink between the filler and the polymer matrix.

FA has been widely used as a filler material, but it also served as a heat conductor, a decomposition inhibitor, and a lubricating agent. The FA not only improved the mechanical properties of the PET composite but also improved the melting and mixing processes (Li et al. 1998).

4.4.6 Polypropylene

Sengupta et al. (2011) modified the surface of FA with FP at various concentrations of 1, 2, 3, and 5 wt.%, and subsequently used it as a filler in recycled PP matrix composites. The highest enhancement in properties was observed in 2 wt.% FP (FP2)-coated FA-filled composites. This improved mechanical strength was correlated with the interaction between the filler and the matrix using SEM micrographs. In RFAFP0 (R—recycled PP, FA—fly ash, FP—wt.% of FP coated) (Figure 4.23a), the adhesion of the FA particles with the matrix was poor compared with the rest of the composite samples. In RFAFP1 (Figure 4.23b) and RFAFP2 (Figure 4.23c), much better interfacial bonding was evident. In RFAFP3 (Figure 4.23d), a ductile failure of the matrix was observed, which supports the highest strain of these composites. In RFAFP5, it is apparent from Figure 4.23e that the entanglement of the polymer chains around the FA particles is more pronounced that leads to larger aggregates, but interfacial bonding is poor.

The mechanical properties of chemically activated FA (CFA)/PP composite were superior to those of FA/PP composites (Yang et al. 2006). This is attributed to the surface roughness

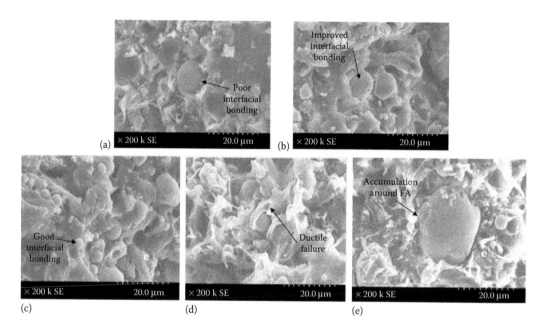

FIGURE 4.23   SEM micrographs of composite samples: (a) RFAFP0; (b) RFAFP1; (c) RFAFP2; (d) RFAFP3; and (e) RFAFP5.

of CFA due to surface modification explained in Section 4.3.1. Bandyopadhyay et al. (2010) also reinforced PP with near whitened FA particles to enhance its mechanical properties and esthetic look of the composites.

Satapathy et al. (2011) observed a ductile-to-brittle transition in cenosphere-filled PP composites. The incompatibility between the hydrophobic polymer and the hydrophilic FA cenospheres led to inferior interfacial adhesion, resulting in low performance of the composites. In contrast, the hydroxyl groups on FA cenosphere surface cause cluster or agglomeration due to filler–filler interaction in the matrices. This attributes to non-uniform macroscale distributions, consequentially giving rise to mechanical anisotropy. Thus, they studied unmodified hollow and rigid microsphere-filled PP systems as examples for crack toughness behavior of multicomponent composites with a higher extent of modulus mismatch between the matrix and the reinforcing spherical filler, and established the mechanistic correlations of a systematic semiductile-to-ductile-to-brittle transition in polymer matrix composites. The SEM micrographs of PP composites with 0, 5, 10, 15, and 20 wt.% cenospheres are shown in Figure 4.24a–e. The unfilled PP and the composites with 5–10 wt.% cenosphere content underwent a low fracture-strain semiductile failure accompanied by plastic and shear deformation/flow-induced fibril formation (Figure 4.24a–c). The fracture surface micrographs of the composites with cenosphere content below 10 wt.% revealed the presence of primary cenosphere particles, indicating the absence of filler clustering. However, the composites with cenosphere content greater than 10 wt.% showed the presence of secondary structures formed by clustering of cenosphere particles (Figure 4.24c–e). The formation of microvoids around the

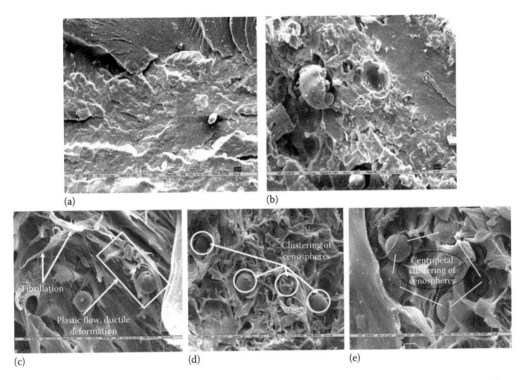

FIGURE 4.24   (a) Fractured surface of PP and PP composite with (b) 5 wt.%, (c) 10 wt.%, (d) 15 wt.%, and (e) 20 wt.% of filler loadings.

particle clusters facilitated interparticle matrix yielding, thereby increasing the toughness of polymers at 15 and 20 wt.% of filler loading.

4.4.7   Metakaolin-Based Geopolymeric and ABS Terpolymer

There exists a novel class of materials called geopolymers, which are the inorganic, aluminosilicate-based ceramic materials. The different types of geopolymers are metakaolin (MK)-based, silica-based, and sol–gel-based (synthetic) geopolymers. Geopolymers are obtained by the reaction between the alkaline liquid and an aluminosilicate, such as MK or FA with subsequent curing at a moderate temperature. The geopolymerization mechanism involves dissolution, migration, gelation, reorganization, and polymerization of Al and Si precursor species. The latest applications of geopolymer are in waste management, including radioactive waste management and immobilization of toxic metals.

ABS is the most common thermoplastic terpolymer, obtained by polymerizing styrene and acrylonitrile in the presence of polybutadiene. The most important mechanical properties of ABS are impact resistance and toughness together with chemical resistance, surface appearance, and processability.

Cenospheres are hollow spherical particles of FA, which have wide applications in various industries due to their low densities (0.2–0.8 g/cm^3), low thermal conductivities

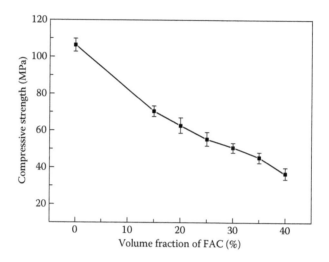

FIGURE 4.25   Compressive strength of the composites vs. volume fraction of FAC.

(about 0.065 W/m/K), and excellent stability in alkali solution at high temperatures. Cenospheres were used as fillers in MK-based geopolymer (Wang et al. 2011) and ABS (Kulkarni et al. 2014).

The compressive strength of MK-based geopolymers as a function of FA content is shown in Figure 4.25. The compressive strength of geopolymeric matrix was 106.2 MPa. At 40 vol.% of cenosphere loading, the compressive strength of the cenosphere/MK-based geo-polymeric composite was 36.5 MPa. The compressive strength decreased with an increase in the cenosphere content due to the highly porous nature of censopheres. The mechanical properties of cenosphere/MK-based geopolymeric composites decreased monotonically with increasing filler content and exhibited the minimum value at 40 vol.% of cenosphere.

Similarly, Kulkarni et al. (2014) studied the effect of particle size (100, 150, and 300 mesh) variations of cenospheres and their different concentrations (0–40 wt.%) on the mechanical properties of ABS. The composites filled with smaller size particles exhibited better properties in comparison with those containing larger size particles.

The yield stress data were compared using the following equations:

$$\frac{\sigma_c}{\sigma_p} = 1 - k\phi_F^{2/3} \tag{4.5}$$

where σ_c/σ_p is the ratio of the yield stress of the composite (c) to that of the unfilled polymer (p).

The dependence of the relative yield stress (σ_c/σ_p) on the volume fraction of cenospheres ϕ_F is shown in Figure 4.26.

For the spherical-shaped particles, $k = 0$ for perfect adhesion and $k = 1.21$ for no adhesion. The data for cenosphere (150 and 300 mesh) filled ABS composites lie in between the curves with $k = 0$–0.6, whereas the data for cenosphere (100 mesh) filled ABS composites lie closer to the curve with $k = 0.6$. As the filler loading was increased, the mechanical properties decreased due to agglomeration of filler particles in the polymer matrix.

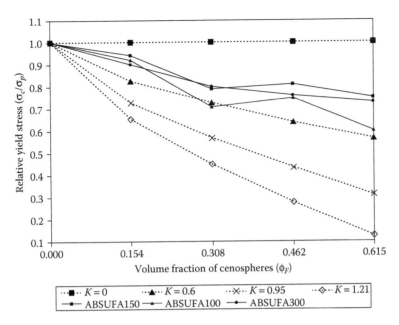

FIGURE 4.26   Variation in relative yield stress (σ_c/σ_p) of an ABS/cenosphere for 150, 100, and 300 mesh composites against volume fraction of cenosphere (ϕ_F). Dotted curves represent the predicted behavior according to Equation 4.5.

This signifies that the mechanical properties of the polymer filled with cenospheres is a function of the particle size, filler dispersion, and interfacial interaction between the filler particles and the polymer matrix.

4.4.8   Poly(Ether Ether Ketone)

Ketone-based resins, such as PEEK, are semicrystalline engineering thermoplastics. PEEK exhibits high strength and toughness, and excellent chemical and wear resistance. Typical applications include nuclear, automotive, marine, electronics, medical, and aerospace industries. The easy processability of PEEK along with its Young's modulus of 3.6 GPa and its tensile strength of 90–100 MPa has created a great interest in manufacturing composites.

The PEEK/FA composites reinforced with MFA with calcium hydroxide have a good interfacial interaction of FA with the PEEK matrix (Parvaiz et al. 2011). The concentration of the chemical modifier was optimized in this work to improve the surface properties of FA discussed in Section 4.3.1. It was observed that the tensile strength and tensile moduli were increased at 20 wt.% for calcium hydroxide-modified FA. This signifies the role of calcium hydroxide, which resulted in uniform distribution of filler and increased the interaction of FA particles with the PEEK matrix. This phenomenon is due to the formation of reactive radicals on the surface of FA upon calcium hydroxide treatment, which formed the chemical linkage at the interface. Even the flexural strength and moduli were improved at 20 wt.% of calcium hydroxide-modified FA. However, the unmodified FA-filled PEEK composites did not exhibit this effect.

4.4.9 Polyurea

PU is a product from the chemical reaction between an isocyanate and an amine. The elastomeric PU exhibits excellent properties, including high thermal stability, abrasion and corrosion resistance, and superior mechanical properties. They are insensitive to humidity and low temperatures. Some PUs are able to reach tensile strengths of 6000 psi and strains of over 500%. The properties of PU can be enhanced by addition of FA particles to obtain low-density and high-strength composites.

Figure 4.27 shows the stress–strain curves of unfilled PU and FA/PU composites at different volume fractions of FA. Both the unfilled PU and the FA/PU composites showed typical elastomeric tensile behavior. All the stress–strain curves of the composite were divided into three regions: an initial linear elastic region, a plateau region, and a terminal nonlinear increasing region before rupture. The initial regions of the FA/PU composites and the unfilled PU are almost superposed, which indicates that Young's modulus does not change after introducing FA particles into the PU matrix. In the second region of the stress–strain curves, the strain has increased from 35% to 250%, whereas the stress remains nearly constant. In this region, necking starts at a localized point in the specimen, and then increases in length as stretching continues. As a result of this process, molecules are highly orientated. Upon further stretching of the necked specimen, the stress rapidly increases and failure occurs (Qiao and Wu 2011).

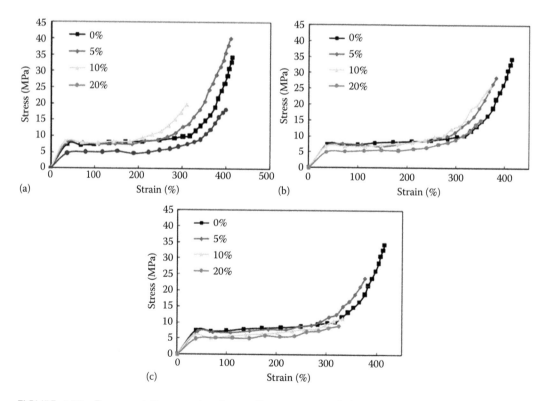

FIGURE 4.27 Representative quasistatic tensile constitutive behavior of FA/PU composites as a function of volume fraction of FA: (a) FA/PU composites filled with small FA particles; (b) FA/PU composites filled with medium FA particles; and (c) FA/PU composites filled with large FA particles.

4.5 CONCLUSIONS

There is a growing trend toward the development of composites with low environmental impact and good commercial viability. One of the principal problems that occurred whenever fly ash has been used as filler in polymers is the interface formation between the polymer and fly ash. The chemical, mechanical, and mechanochemical treatments to the fly ash are proven to improve its surface properties, which in turn enhanced the filler dispersion and interfacial interaction between fly ash and polymer matrices. Surface-modified fly ash could be a value-added filler for the polymer industries as reinforcement in composites.

REFERENCES

Anandhan, S., S. Madhava Sundar, T. Senthil, A. R. Mahendran, and G. S. Shibulal. 2012. Extruded poly(ethylene-co-octene)/fly ash composites—Value added products from an environmental pollutant. *J Polym Res* 19: 1–11.

Anandhan, S., H.G. Patil, and R.R. Babu. 2011. Characterization of Poly(ethylene-Co-Vinyl Acetate-Co-Carbon Monoxide)/layered Silicate Clay Hybrids Obtained by Melt Mixing. *J Mater Sci* 46(23): 7423–7430.

ASTM C 618-03. 2003. Standard specification for coal fly ash and raw or calcined natural pozzolan for use in concrete. West Conshohocken, PA: ASTM International.

Baláž, P. (ed.) 2008. High-energy milling. In: *Mechanochemistry in Nanoscience and Minerals Engineering*, pp. 103–132. Springer, Berlin, Germany, 2013.

Bandyopadhyay, S., A. Zaeni, D. Nath, A. Yu, Q. Zeng, D. Blackburn, and C. White. 2010. Advanced utilization of as received and near whitened fly ash in polypropylene polymer to improve mechanical, notched impact and whiteness colour properties. *Int J Plast Technol* 14: 51–56.

Blissett, R. S. and N. A. Rowson. 2012. A review of the multi-component utilisation of coal fly ash. *Fuel* 97: 1–23.

Chaowasakoo, T. and N. Sombatsompop. 2007. Mechanical and morphological properties of fly ash/epoxy composites using conventional thermal and microwave curing methods. *Compos Sci Technol* 67: 2282–2291.

Chawla, K. K. 2012. *Composite Materials: Science and Engineering*. Springer Science & Business Media, New York.

Cullity, B. D. 2001. *Elements of X-Ray Diffraction*, 3 edn. Upper Saddle River, NJ: Prentice Hall.

Deepthi, M. V., M. Sharma, R. R. N. Sailaja, P. Anantha, P. Sampathkumaran, and S. Seetharamu. 2010. Mechanical and thermal characteristics of high density polyethylene–fly ash cenospheres composites. *Mater Design* 31: 2051–2060.

Devi, M. S., V. Murugesan, K. Rengaraj, and P. Anand. 1998. Utilization of fly ash as filler for unsaturated polyester resin. *J Appl Polym Sci* 69: 1385–1391.

Gu, J., G. Wu, and Q. Zhang. 2007. Preparation and damping properties of fly ash filled epoxy composites. *Mater Sci Eng A* 452–453: 614–618.

Guhanathan, S., M. Saroja Devi, and V. Murugesan. 2001. Effect of coupling agents on the mechanical properties of fly ash/polyester particulate composites. *Appl Polym Sci* 82: 1755–1760.

Iyer, R.S. and J. A. Scott. 2001. Power station fly ash—A review of value-added utilization outside of the construction industry. *Resour Conserv Recycl* 31: 217–228.

Khan, M. J., A. A. Al-Juhani, R. Shawabkeh, A. Ul-Hamid and I. A. Hussein. 2011. Chemical modification of waste oil fly ash for improved mechanical and thermal properties of low density polyethylene composites. *J Polym Res* 18(6): 2275–2284.

Kishore, S. M. Kulkarni, D. Sunil, and S. Sharathchandra. 2002. Effect of surface treatment on the impact behaviour of fly-ash filled polymer composites. *Polym Int* 51: 1378–1384.

Kulkarni, M. B., V. A. Bambole, and P. A. Mahanwar. 2014. Effect of particle size of fly ash ceno-spheres on the properties of acrylonitrile butadiene styrene-filled composites. *J Thermoplast Compos Mater* 27(2): 251–267.

Kulkarni, S. M. and Kishore. 2002. Effects of surface treatments and size of fly ash particles on the compressive properties of epoxy based particulate composites. *J Mater Sci* 3: 4321–4326.

Kutchko, B. G. and A. G. Kim. 2006. Fly ash characterization by SEM–EDS. *Fuel* 85(17–18): 2537–2544.

Lapuerta, M., F. J. Martos, and G. Martín-González. 2010. Geometrical determination of the lacu-narity of agglomerates with integer fractal dimension. *J Colloid Interface Sci* 346: 23–31.

Li, Y., D. J. White, and R. L. Peyton. 1998. Composite material from fly ash and post-consumer PET. *Resour Conserv Recycl* 24(2): 87–93.

Liu, M., B. Guo, M. Du, and D. Jia. 2007. Drying induced aggregation of halloysite nanotubes in polyvinyl alcohol/halloysite nanotubes solution and its effect on properties of composite film. *Appl Phys A* 88(2): 391–395.

Logofatu, C., -C. Negrila, R. Ghita, F. Ungureanu, C. Cotirlan, A.-S. Manea, M.-F. and Lazarescu. 2011. Study of SiO_2/Si interface by surface techniques. In: *Crystalline Silicon—Properties and Uses* B. Sukumar (ed.), 23–42 (Chapter 2), InTech, India.

Mandelbrot, B. B. 1982. *The Fractal Geometry of Nature*, 1 edn. San Francisco, CA: W. H. Freeman & Company.

Mollah, M., A. Yousuf, T. R. Hess, and D. L. Cocke. 1994. Surface and bulk studies of leached and unleached fly ash using XPS, SEM, EDS and FTIR techniques. *Cement Concrete Res* 24: 109–118.

Nath, D. C. D., S. Bandyopadhyay, P. Boughton, A. Yu, D. Blackburn, and C. White. 2010a. Chemically modified fly ash for fabricating super-strong biodegradable poly(vinyl alcohol) composite films. *J Mater Sci* 45: 2625–2632.

Nath, D. C. D., S. Bandyopadhyay, P. Boughton, A. Yu, D. Blackburn, and C. White. 2010b. High-strength biodegradable poly(vinyl alcohol)/fly ash composite films. *J Appl Polym Sci* 117: 114–121.

Nath, D. C. D., S. Bandyopadhyay, J. Campbell, A. Yu, D. Blackburn, and C. White. 2010c. Surface-coated fly ash reinforced biodegradable poly(vinyl alcohol) composite films: Part 2—Analysis and characterization. *Appl Surf Sci* 257: 1216–1221.

Nath, D. C. D., S. Bandyopadhyay, S. Gupta, A. Yu, D. Blackburn, and C. White. 2010d. Surface-coated fly ash used as filler in biodegradable poly(vinyl alcohol) composite films: Part 1—The modification process. *Appl Surf Sci* 256: 2759–2763.

Parvaiz, M. R., S. Mohanty, S. K. Nayak, and P. A. Mahanwar. 2011. Effect of surface modifica-tion of fly ash on the mechanical, thermal, electrical and morphological properties of poly-etheretherketone composites. *Mater Sci Eng, A* 528: 4277–4286.

Patil, A. G. and S. Anandhan. 2012. Ball milling of class-F Indian fly ash obtained from a thermal power station. *Int J Energy Eng* 2: 57–62.

Patil, A. G., A. Mahendran, and S. Anandhan. 2016. Nanostructured fly ash as reinforcement in a plastomer-based composite: A new strategy in value addition to thermal power station fly ash. *Silicon* 8:159–173.

Patil, A. G., M. Selvakumar, and S. Anandhan. 2014. Characterization of composites based on bio-degradable poly(vinyl alcohol) and nanostructured fly ash with an emphasis on polymer-filler interaction. *J Thermoplast Compos Mater*, DOI: 10.1177/0892705714563130.

Patil, A. G., A. M. Shanmugharaj, and S. Anandhan. 2015. Interparticle interactions and lacunarity of mechano-chemically activated fly ash. *Powder Technol* 272: 241–249.

Paul, K. T., S. K. Satpathy, I. Manna, K. K. Chakraborty, and G. B. Nando. 2007. Preparation and characterization of nano structured materials from fly ash: A waste from thermal power sta-tions, by high energy ball milling. *Nanoscale Res Lett* 2: 397.

Qiao, J. and G. Wu. 2011. Tensile properties of fly ash/polyurea composites. *J Mater Sci* 46: 3935–3941.

Ramos, V. D., H. M. da Costa, V. L. P. Soares, and R. S. V. Nascimento. 2005. Hybrid composites of epoxy resin modified with carboxyl terminated butadiene acrylonitrile copolymer and fly ash microspheres. *Polym Test* 24: 219–226.

Rohatgi, P. K., T. Matsunaga, and N. Gupta. 2009. Compressive and ultrasonic properties of polyester/fly ash composites. *J Mater Sci* 44: 1485–1493.

Sarbak, Z. and M. Kramer-Wachowiak. 2002. Porous structure of waste fly ashes and their chemical modifications. *Powder Technol* 123: 53–58.

Satapathy, B. K., A. Das, and A. Patnaik. 2011. Ductile-to-brittle transition in cenosphere-filled polypropylene composites. *J Mater Sci* 46: 1963–1974.

Satapathy, S., A. Nag, and G. B. Nando. 2012. Effect of electron beam irradiation on the mechanical, thermal, and dynamic mechanical properties of flyash and nanostructured fly ash waste polyethylene hybrid composites. *Polymr Compos* 33(1): 109–119.

Satapathy, S., G. B. Nando, A. Nag, and K. V. S. N. Raju. 2013. HDPE-fly ash/nano fly ash composites. *J Appl Polym Sci* 130(6): 4558–4567.

Sengupta, S., K. Pal, D. Ray, and A. Mukhopadhyay. 2011. Furfuryl palmitate coated fly ash used as filler in recycled polypropylene matrix composites. *Compos Part B Eng* 42: 1834–1839.

Shanmugharaj, A. M., S. Sabharwal, A. B. Majali, V. K. Tikku, and A. K. Bhowmick. 2002. Surface characterization of electron beam modified dual phase filler by ESCA, FT-IR and surface energy. *J Mater Sci* 37: 2781–2793.

Sharma, A. 2012. Modification in properties of fly ash through mechanical and chemical activation. *Am Chem Sci J* 2(4): 177–187.

Sharma, A. K. and P. A. Mahanwar. 2010. Effect of particle size of fly ash on recycled poly (ethylene terephthalate)/fly ash composites. *Int J Plast Technol* 14: 53–64.

Shawabkeh, R., M. J. Khan, A. A. Al-Juhani, H. I. Al-Abdul Wahhab, and I. A. Hussein. 2011. Enhancement of surface properties of oil fly ash by chemical treatment. *Appl Surf Sci.* 258: 1643–1650.

Smith, B. C. 1998. *Infrared Spectral Interpretation: A Systematic Approach*. CRC Press, USA.

Stellacci, P., L. Liberti, M. Notarnicola, and P. L. Bishop. 2009. Valorization of coal fly ash by mechano-chemical activation. Part I. Enhancing adsorption capacity. *Chem Eng Sci* 149: 11–18.

Sushil, S. and V. S. Batra. 2006. Analysis of fly ash heavy metal content and disposal in three thermal power plants in India. *Fuel* 85(17–18): 2676–2679.

Swami, P. N., B. Nooka Raju, D. Venkata Rao, and J. Babu Rao. 2009. Synthesis and characterization of nano-structured fly ash: A waste from thermal power plant. *J Nanoeng Nanosys* 223: 35–44.

Wang, C., J. Liu, H. Du, and A. Guo. 2012. Effect of fly ash cenospheres on the microstructure and properties of silica-based composites. *Ceram Int* 38: 4395–4400.

Wang, M.-R., D.-C. Jia, P.-G. He, and Y. Zhou. 2011. Microstructural and mechanical characterization of fly ash cenosphere/metakaolin-based geopolymeric composites. *Ceram Int* 37: 1661–1666.

Xu, G., L. Wang, J. Liu, and J. Wu. 2013. FTIR and XPS analysis of the changes in bamboo chemical structure decayed by white-rot and brown-rot fungi. *Appl Surf Sci* 280: 799–805.

Xue, Q.-F. and S.-G. Lu. 2008. Microstructure of ferrospheres in fly ashes: SEM, EDX and ESEM analysis. *J Zhejiang Univ Sci A* 9: 1595–1600.

Yang, Y.-F., G.-S. Gai, Z.-F. Cai, and Q.-R. Chen. 2006. Surface modification of purified fly ash and application in polymer. *J Hazard Mater* 133: 276–282.

Yao, Z. T., X. S. Ji, P. K. Sarker, J. H. Tang, L. Q. Ge, M. S. Xia, and Y. Q. Xi. 2015. A comprehensive review on the applications of coal fly ash. *Earth Sci Rev* 141: 105–121.

Yunsheng, Z., S. Wei, and L. Zongjin. 2006. Impact behavior and microstructural characteristics of PVA fiber reinforced fly ash-geopolymer boards prepared by extrusion technique. *J Mater Sci* 41: 2787–2794.

Zaeni, A., S. Bandyopadhyay, A. Yu, J. Rider, C. S. Sorrell, S. Dain, D. Blackburn, and C. White. 2010. Colour control in fly ash as a combined function of particle size and chemical composition. *Fuel* 89: 399–404.

II

Nanopolymer Technology

CHAPTER 5

Electrospinning

From Fundamentals to Applications

T. Senthil, Gibin George, and Anandhan Srinivasan

CONTENTS

5.1 INTRODUCTION

Over the past several years, nanotechnology based on polymers has attracted a lot of interest because objects of size in the range of 1–100 nm often exhibit unique physical and chemical properties (Gao and Xu 2012). Nanotechnology has been regarded as the next great new frontier of materials science and is one of the enabling technologies of the twenty-first century. The commercial importance of polymers has been increasing in various fields, such as aerospace, automotive, marine, infrastructure, military, consumer commodities, electronics, and information. Nanostructuring is regarded as a new method to mend many properties, which include improved strength, enhanced modulus, decreased thermal expansion coefficient, increased thermal stability, outstanding barrier properties, improved solvent and heat resistance, decreased flammability, and reduced gas permeability compared with the conventional microfabrication methods. Electrospinning is one of the nanofabrication processes that have found a plethora of applications in various fields.

5.2 ELECTROSPINNING

Electrospinning significantly improves the properties of polymeric fibers owing to their size reduction to the nanometer level (Feng et al. 2008). High surface area, superior mechanical properties, and remarkable surface properties are the attractive features of electrospun nanofibers (Figure 5.1). Because of the several amazing characteristics of electrospun fibers, they have been used in tissue engineering, drug delivery, fiber-based sensors, medicine, photovoltaic, filtration membranes, advanced photonic applications, wound healing, and composite materials (Jian et al. 2012). This chapter describes the theory behind the electrospinning technique, the effect of various process parameters, and the applications of electrospun fibers.

5.2.1 Theory of Electrospinning

Electrospinning is not only a unique and fascinating process for the production of polymeric nanofibers with diameters ranging from 3 nm to 20 μm, but also the simplest and inexpensive technique to fabricate ultrafine continuous polymeric fibers. Electrospinning is also known as electrostatic spinning and electrospraying. In the last few years, electrospinning has become popular among academic researchers and industries as it overcomes the various processing difficulties in the other nanofiber-forming techniques. Some of the other techniques for the production of polymer nanofibers are drawing, template synthesis, phase separation, and self-assembly (Table 5.1). While electrospinning can produce nanofibers, its conventional counterparts, such as melt spinning, dry spinning, gel spinning, and wet spinning, can only produce microfibers (Figure 5.2) (Gibson et al. 2001). Almost all the straight chain homopolymeric materials have been actively investigated for developing nanofibers by electrospinning. Meanwhile, various

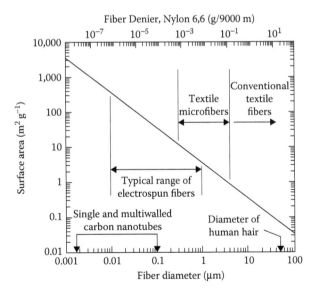

FIGURE 5.1 Electrospun nanofibers have high surface area. (Reprinted from *Colloids Surf., A*, 187–188, Gibson, P., Gibson, H.S., and Rivin, D., Transport Properties of Porous Membranes Based on Electrospun Nanofibers, 469–481. Copyright 2001, with permission from Elsevier.)

TABLE 5.1   Comparison of Different Nanofiber Fabrication Techniques: Advantages and Disadvantages

Fabrication Technique	Advantages	Disadvantages
Drawing	Single nanofibers Simple instrumentation	Time consuming Non scalable Fiber dimensions cannot be controlled Operator dependent
Template synthesis	Uniform fibers can be obtained Fiber dimensions can be varied Different shapes can be achieved	Not suitable for mass production Replacement of templates are necessary
Temperature-induced phase separation	Simple procedure Minimal equipment complexity	Time consuming Not suitable for all polymeric materials Nonscalable Difficult to control fibers dimension
Molecular self-assembly	Smaller nanofibers can be fabricated	Time consuming in processing Continuous nanofibers Only suitable for short fibers

Electrospinning	Simple equipment and easy alignment process
	Low cost compared to the bottom-up method
	Uniform and long-length fibers
	Do not require expensive purification
	High surface area per unit mass and
	  complex porous structure fibers
	Mass production of one-by-one continuous
	  nanofibers from several polymers
		Mechanisms of jet thinning
		Dependent on experimental
		  parameters
		Skilled operator is needed
		High power consumption

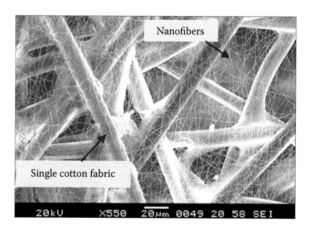

FIGURE 5.2   Scanning electron microscopy (SEM) image of PVA nanofibers spun across surface of surgical tape (cotton fabric).

semicrystalline and amorphous polymers, natural and synthetic biopolymers, polymer blends, and block copolymers can also be converted into nanofibers via the electrospinning technique.

Electrospinning has a number of advantages over other conventional nanofiber fabrication methods listed in Table 5.1.

5.3 THE HISTORY OF ELECTROSPINNING

In the late sixteenth century, William Gilbert observed that when a suitably charged piece of amber was held near a droplet of water, it would form a cone shape and small droplets would be ejected from the tip of the cone. This is the first ever recorded observation on electrospinning. In 1745, Bose set out to describe the process of electrohydrodynamic spraying of fluids (Malakhov et al. 2009); later in 1882, Rayleigh (Salata 2005) expanded the domain by investigating the behavior of thin liquid jets when placed in an electric field and their stability criterion. In February and July 1902, Cooley (1902) and Morton (1902) separately developed the first device to spray fluids with the assistance of electrical charges and those were patented for the process of electrospinning. In 1914, John Zeleny developed a mathematical model on the behavior of liquids under electrostatic forces. The first patent was issued to Formhals in the year 1934 on electrospinning (Formhals 1934) for his invention on creating synthetic fibers using electrostatic forces and his patent states that he had made a device that successfully electrospun cellulose acetate (CA) nanofibrous structures.

In 1936, Norton found that fibers can be spun from a melt rather than a solution with the help of an air blast. Further, Rozenblum and Petryanov-Sokolov, in 1938 (Malakhov et al. 2009), produced electrospun fibers for filter materials that were known as "Petryanov filters," by adjusting processing variables, which is regarded as a major footstep in the field of electrospinning. In 1939, Formhals developed a new process in which a number of fibers were spun from a single polymeric solution using multiple needles concurrently and in his second patent, problems to adjust the distance between the collector plate and spinning site were eliminated. Later, Formhals (1940) generated composite fibers by spinning a polymer solution directly onto a moving object. In 1952, Vonnegut and Neubauer produced uniform droplets of about 0.1 mm diameter by using few tenths of millimeter-sized glass tube filled with water, which was connected to a high-voltage (5–10 kV) electric field. In 1955, Drozin captured aerosols from dispersion of a series of liquids under an applied high electric field by following the electrospinning technique used by Vonnegut and Neubauer (Huang et al. 2003).

In 1968, Simons produced nonwoven fabric with the help of electrical spinning method under a high voltage. His patent states that fibers produced from a highly viscous solution were continuous ones and those obtained from a low viscous solution were short. In 1969, Taylor (Yarin et al. 2001) developed a mathematical model to study the fundamentals of fiber-forming process from fluid droplet under an applied electric field, later acknowledged as the "Taylor cone." This shape is specifically a cone with a semivertical angle of 49.3° or an apex angle of 98.6°. Taylor studied the role of surface tension and electrostatic forces on electrospinning by considering successful jet formation from the fiber initiating surface into account. His report states that when a solution is charged, it induces a charge on the surface of the solution. At a certain critical applied voltage, the surface charges overcome the surface tension of the solution and result in a change in the shape of the solution; this change in shape is in the form of a cone known as the Taylor cone. Doshi and Reneker studied different process parameters in electrospinning. In 1971, Baumgarten found that an increase in solution viscosity leads to an increase in fiber diameter and also determined that the solution and processing parameters have their own role in fiber diameter. He has devised machinery to spin acrylic fibers in the range of 500–1100 nm by controlling process parameters. In his experiments, solutions with viscosities ranging from 1.7 to 215 Poise were used (Huang et al. 2003). The list of U.S. patents filed before 2000 is given in Table 5.2.

Later on, in 1981, Larrondo and Mandley (Doshi and Reneker 1995) studied the association between the melt temperature and fiber diameter of polyolefin fibers. They

TABLE 5.2   The List of U.S. Patents Issued for Electrospinning

Patent Issued Date	Patent Number	Inventor	Title of the Patent
February 1902	US 692631	J.F. Cooley	Apparatus for electrically dispersing fluids
July 1902	US 705691	W.J. Morton	Method of dispersing fluids
November 1903	US 745276	J.F. Cooley	Electrical method of dispersing fluids
January 1929	US 1699615	K. Hagiwara	Process for manufacturing artificial silk and other filaments by applying electric current
October 1934	US 1975504	A. Formhals	Process and apparatus for preparing artificial threads
July 1936	US 2048651	C.L. Norton	Method and apparatus for producing fibrous or filamentary material
April 1937	US 2077373	A. Formhals	Production of artificial fibers
February 1938	US 2109333	A. Formhals	Artificial fiber construction
May 1938	US 2116942	A. Formhals	Method and apparatus for the production of fibers
July 1938	US 2123992	A. Formhals	Method and apparatus for the production of fibers
May 1939	US 2158415	A. Formhals	Method of producing artificial fibers
May 1939	US 2158416	A. Formhals	Method and apparatus fob the production of artificial fibers
June 1939	US 2160962	A. Formhals	Method and apparatus for spinning
August 1939	US 2168027	E.K. Gladding	Apparatus for the production of filaments, threads, and the like
January 1940	US 2187306	A. Formhals	Artificial thread and method of producing same
January 1940	US 2185417	C.L. Norton	Method and apparatus for forming fibrous material
June 1943	US 2323025	A. Formhals	Production of artificial fibers form fiber forming liquids
December 1943	US 2336745	F.W. Manning	Method and apparatus fob making unwoven and composite fabrics
January 1944	US 2338570	H.R. Childs	Process of electrostatic spinning
May 1944	US 2349950	A. Formhals	Method and apparatus for spinning (Formhals 1944)
April 1953	US 2636216	W.C. Huebner	Method and means of producing threads or filaments electrically
October 1959	US 2908545	J.D. Teja	Spinning nonfused glass fibers from an aqueous dispersion
October 1966	US 3280229	H.L. Simons	Process and apparatus for producing patterned non-woven fabrics
October 1969	US 3475198	E.W. Drum	Method and apparatus for applying a binder material to a prearranged web of unbound, non-woven fibers by electrostatic attraction
January 1970	US 3490115	J.E. Owens and S.P. cheinberg	Apparatus for collecting charged fibrous material in sheet form

(Continued)

TABLE 5.2 (*Continued*)   The List of U.S. Patents Issued for Electrospinning

Patent Issued Date	Patent Number	Inventor	Title of the Patent
June 1972	US 3670486	G.L. Murray	Electrostatic spinning head funnel
September 1972	US 3689608	H.L Hollberg and J.E. Owens	Process for forming a nonwoven web
August 1975	US 3901012	V. Safar	Method and device for processing fibrous material
November 1976	US 3994258	W. Simm	Apparatus for the production of filters by electrostatic fiber spinning
August 1977	US 4044404	G.E. Martin, I.D. Cockshott, and K.T. Fildes	Fibrillar lining for prosthetic device
November 1978	US 4127706	G.E. Martin, I.D. Cockshott, and K.T. McAloon	Porous fluoropolymeric fibrous sheet and method of manufacture
October 1980	US 4230650	C. Guignard	Process for the manufacture of a plurality of filaments
April 1982	US 4323525	A. Bornat	Electrostatic spinning for tubular products
August 1982	US 4345414	A. Bornat and R.M. Clarke	Shaping process
September 1984	US 4468922	P.E. Mecrady and R.B. Reif	Apparatus for spinning textile fibers
December 1984	US 4486365	B. Kliemann and M. Stoll	Process and apparatus for the preparation of electrets filaments, textile fibers and similar articles
November 1985	US 4552707	T.V. How	Synthetic vascular grafts, and methods of manufacturing such grafts
October 1986	US 4618524	D. Groitzsch and E. Fahrbach	Microporous multilayer nonwoven material for medical applications
August 1987	US 4689186	A. Bornat	Production of electrostatically spun products
November 1989	US 4878908	G.E. Martin, I.D. Cockshott, and F.J.T. Fildes	Fibrillar product
October 1990	US 4965110	J.P. Berry	Electrostatically produced structures and methods of manufacturing
June 1991	US 5024789	J.P. Berry	Method and apparatus for manufacturing electrostatically spun structure
February 1992	US 5088807	C.M Waters, T.J. Noales, I. Pavey, and C. Hitomi	Liquid crystal devices

concluded that a larger fiber diameter results from melt spinning rather than solution spinning, due to the high temperature in melt spinning. Also, the fibers produced would be extremely crystalline as a result of the very high jet velocities, 275–380 m/s, as reported by Baumgarten. In 1987, Hayati et al. studied the relationship between the solution conductivity and the applied voltage; it has been pointed that the larger fiber diameter distribution occurs from unstable jets of highly conducting liquids with an increase in applied voltage and stable jets are obtained from insulating and semiconducting fluids. The potential applications of electrospun fibers, such as membranes and protective clothing, filter media, electrode materials, electronic and optical devices, biomedical usage, sensors, and

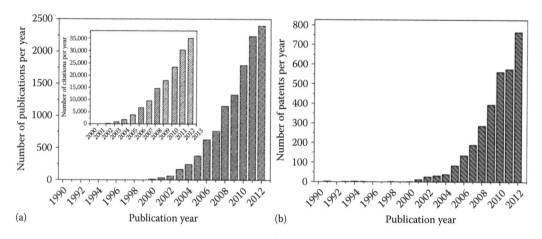

FIGURE 5.3 Number of (a) scientific articles (inset shows the number of citations) and (b) patents on nanofibers from 1990 to 2012. (Reprinted from Doyle, J.J. et al., *Conf. Pap. Sci.*, 2013, e269313. Copyright 2013 under Creative Commons Attribution License.)

sacrificial templates were identified in the recent past. Since then, the number of publications, scientific articles, and patents on electrospinning has been increasing exponentially every year. Figure 5.3 explains that the current trend and growth in this area by plotting the number of patents and scientific papers in the last 10 years.

In the early 1995s, many research groups studied different process parameters in electrospinning and established that many organic polymers could be electrospun in the nanometer dimensions. In 1990, Ijima studied the incorporation of carbon nanotubes (CNTs) into electrospun polymer nanofibers to get better mechanical strength and electrical conductivity of nanometer fibers. In 1993, Kaiser explained that when a solution is being electrospun using either negative or positive polarity, dielectrophoretic forces are present in it as a result of a high nonuniform field acting on an uncharged fluid. In 1996, Reneker and Chun reported the possibility of the production of nanometer-sized polymer fibers. In 1999, Fong studied the formation of beads on the electrospun nanofibers with the variation in the applied voltage. Figure 5.4 shows the countries that are most active in electrospinning research.

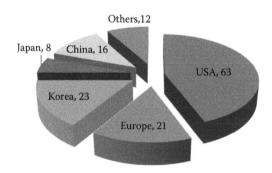

FIGURE 5.4 Country of origin for papers containing the keyword "Electrospinning" in percentage by 2013.

In 2001, Deitzel et al. (2001) studied the effects of accelerating voltage and solution concentration on the structure and morphology of electrospun poly(ethylene oxide) (PEO) fibers systematically. They found that significant changes in fiber diameter, size distribution, and morphology accompanied the changes in these variables, and such analogies have been drawn to similar experiments that are found in the electrospinning literature. In 2002, Loscertales proved the possibility of obtaining uniform fibers instead of particles using an electrospinning technique with adequate selection of process parameters. The systemic and process parameters vary with different polymeric systems and in most cases lend themselves to modification, thereby enabling tailoring of nanofibers for specific end uses (Shin et al. 2001, Fridrikh et al. 2003). Sundarary (2004) developed controlled and aligned poly(styrene) (PS) and acrylic nanofibers with the help of moving and conducting needle and an insulating rotating drum. This technique also offers the opportunity to have control over thickness and composition of nanofibers along with porosity of the nanofiber meshes using a relatively simpler experimental setup (Doshi and Reneker 1993, Reneker and Chun 1996, Dzenis 2004).

Through the electrospinning process, fibers ranging from 10 to 1000 nm or greater (Reneker and Chun 1996, Shin et al. 2001, Fridrikh et al. 2003) can be produced by applying a high voltage to a polymeric solution (Hohman et al. 2001). In 2011, Milleret et al. (2011) reported that 3D-fiber fleeces can be produced with controlled morphology, which satisfies the requirements for successful scaffolds in tissue engineering applications. In his work, he showed that fiber diameter increased either with an increase in flow rate or a decrease in working distance and collector velocity, whereas fiber alignment is increased with the working distance and the collector velocity but decreased with an increase in flow rate.

5.4 ELECTROSPINNING FUNDAMENTALS AND PROCESS

Electrospinning provides a simple and convenient method for generating polymer fibers, inorganic fibers, and ceramic fibers with diameters ranging from tens of nanometers to several micrometers (Figure 5.5). The process exists since the last 80 years and has undergone a very turbulent history until now. In 1960s, the fundamental studies on the

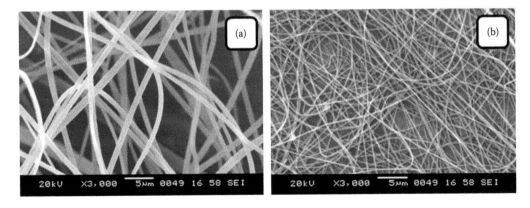

FIGURE 5.5   SEM photographs of electrospun fibers. (a) Micrometer fibers and (b) nanometer fibers.

jet-forming process were initiated by Taylor. Natural and synthetic polymers in addition to composites and blends have been effectively electrospun in the diameter range of ~3 nm to 6 μm. Table 5.3 shows the different polymers, which can be spun into nanofibers by electrospinning with controlled parameters. Current research focuses on the processing of synthetic and biological polymeric nanofibers with tailored properties and functions for various applications.

TABLE 5.3   List of Polymers and the Solvents for Electrospinning

Polymers	Solvents	References
PVA	Distilled water	Adomavičiūtė and Milašius (2007), Ding et al. (2002)
PEO	Distilled water, ethanol	Wan et al. (2007)
PAAc	Distilled water, ethanol	Gestos et al. (2010)
PVP	Distilled water, ethanol	Yang et al. (2004)
PEI	Distilled water	Khanam et al. (2007)
PAM	Distilled water	Li et al. (2010), Liu et al. (2009)
PEG	Distilled water	Han et al. (2011)
PLLA/PDLA	DCM, DMF	Schofer et al. (2011)
PCL	Chloroform	Kanani and Bahrami (2011)
PGA	HFIP	Wulkersdorfer et al. (2010)
Poly-β-hydroxyalkanoates (PHA)	DMF, chloroform	Zahedi et al. (2010)
CA	Acetone, DMF, DMAc, formic acid, acetic acid	Han et al. (2008a)
EC	TFE, THF	Jeun et al. (2007)
Chitin	HFIP	Min et al. (2004b)
Chitosan	HFIP, formic acid, ethanoic acid, trifluoroacetic acid (TFA)	Homayoni et al. (2009)
Dextran	DMF, deionized water	Shawki et al. (2010)
Collagen	HFIP	Matthews et al. (2002)
Gelatine	TFE, HFIP, formic acid	Huang et al. (2004)
Silk and silk-like polymer with Fibronectin functionality (SLPF)	Formic acid/HFIP	Min et al. (2004a)
Zein (CASP)	Distilled water, ethanol	Yao et al. (2007)
PC	THF and DMF	Kattamuri and Sung (2004)
PET	TFA and DCM	Veleirinho and Lopes-da-Silva (2009)
PBT	HFIP, TFA	Mathew et al. (2005)
PSU	DMAc DMF	Yao et al. (2006)
PAN	DMF	Heikkilä and Harlin (2009)
PS	DMF/THF, DMAc	An et al. (2006)
PMMA	Acetone, chloroform, DCM, THF, toluene, HFIP and TFE	Qian et al. (2010)
PVC	DMF/THF	Chiscan et al. (2012)
PVDF	Acetone and DMAc	Huang et al. (2010)

(*Continued*)

TABLE 5.3 (*Continued*)   List of Polymers and the Solvents for Electrospinning

Polymers	Solvents	References
PU	DMF	Zhuo et al. (2008)
PVAc	Ethanol	Park et al. (2008)
PI	DMAc	Zhang et al. (2006)
Nylon 6 (PA6)	Formic acid	Ojha et al. (2008)
PBI	DMAc	Kim and Reneker (1999)

Spinning solution may be prepared from a wide range of polymers including poly(vinyl alcohol) (PVA), PEO, poly(vinyl chloride) (PVC), PS, poly(acrylic acid) (PA), poly(urethane) (PU), and fluorocarbon polymers. A typical schematic diagram of electrospinning setup is shown in Figure 5.6. The electrospinning unit incorporates three basic components, namely a high-voltage DC electric supply unit, a syringe with a metal spinneret of small diameter, and a grounded rotating drum/static metallic plate as collector. The distance between the collector and the spinneret is adjustable, because it can influence the fiber diameter and morphology. Also, the electrospun fibers drawn through the spinneret stick together if sufficient time is not given for the evaporation of the solvent before the solution jets reach the collector. To maintain a constant flow rate of the polymeric solution, a syringe pump is used. A high-voltage DC power source is used to draw an electrically charged jet of polymeric solution through the spinneret at a constant feed rate under high electric potential in the range of 0–50 kV. The spinneret is usually attached to the positive terminal of the high-voltage electric supply and the collector is simply grounded as shown in Figure 5.6. Initially, at the tip of the spinneret, a droplet of the fluid is held by means of surface tension, and then the electric field creates charges on the surface of the polymeric droplet. Once a threshold voltage is applied on the polymer solution, electrostatic force overcomes the surface tension

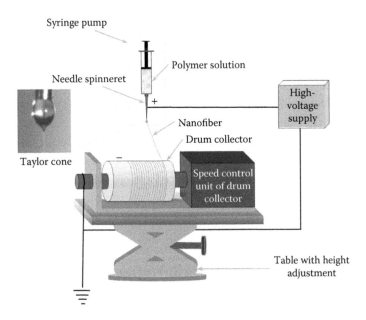

FIGURE 5.6   Schematic of an electrospinning unit.

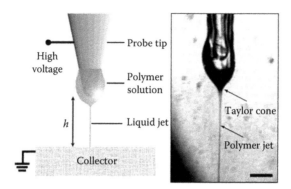

FIGURE 5.7 Schematic representation of a Taylor cone and a photograph of Taylor cone formation under applied voltages. (Reprinted with permission from Sun, D., C. Chang, S. Li et al., Near-field electrospinning, *Nano Lett.,* 6, 839. Copyright 2006, American Chemical Society.)

of the pendant droplet and thereby initiates the formation of nanofibers. While raising the electric field the semicircular surface of the solution at the tip of the needle stretches to form a conical shape known as the Taylor cone (cone angle ~ 49.3°), shown in Figure 5.7.

The charged jet becomes unstable due to electrostatic forces, and this in turn induces a whipping effect, which reduces the diameters of fibers to as low as 30 nm. Under ambient conditions, almost all the solvent is evaporated from the fibers before they reach the collector.

5.5 PARAMETERS OF ELECTROSPINNING PROCESS

The production of nanofibers is not a trivial task. In 1987, Hayati et al. studied the effects of electric field, experimental conditions, and the factors affecting the fiber stability and atomization (Subbiah et al. 2005). With optimized control variables, spinning of a large variety of polymeric fibers with nanoscale dimension has been carried out successfully. Several factors have been widely considered and investigated so as to obtain better control over the electrospinning process and to manipulate electrospun fiber diameter. These factors are generally classified into two main categories: the properties of the electrospinning polymeric solution and the process variables. The most important factors are surface tension (γ), conductivity (κ), viscosity (η_o), electric voltage (V), tip-to-collector distance (H), and solution flow rate (Q). Table 5.4 details the parameters affecting the electrospinning process.

Several studies have been conducted on the interconnectivity of various parameters and morphology. Deitzel and coworkers evaluated the effects of two important processing parameters, spinning voltage and solution concentration, on the morphology of the fibers formed (Deitzel et al. 2001). They observed that the spinning voltage and solution concentration have been strongly correlated with the formation of bead defects in the fibers while increasing the solution concentration. The diameters of fibers increased based on a power law relationship and it produced a bimodal and reminiscent distribution of fiber sizes. Among the various factors, two main factors driving the electrospinning mechanisms are the length of the straight part of the jet before bending instability occurs and the initial angle of the looping envelope. These most important factors affecting the electrospinning process are the solution properties and the electric potential.

TABLE 5.4 The Parameters Affecting the Electrospinning Process

Independent Parameters	
Material parameters	Type of polymer, molecular weight of the polymer, type of solvent
	Concentration of polymer solution, solvent volatility of polymer solution
	Rheological characteristics of polymer solution
	Temperature of polymer solution, surface tension of polymer solution
	Electric conductivity of polymer solution
	Dielectric constant of polymer solution
Process parameters	Electrostatic potential and electric field strength
	Feed rate, orifice diameter, applied voltage, velocity of supporting material
	Type and geometry of spinning and collector electrode
	Distance between electrodes, velocity (rpm) of spinning electrode
	Type of supporting material and its electric properties
Ambient parameters	Relative humidity, temperature
	Velocity of air, local atmosphere flow
	Pressure, atmospheric composition
Dependent Parameters	
Process parameters	Number of Taylor cones per area, average life of jets
	Electric current (per jet and total), length of jets
Material parameters	Average nanofibers diameter, fiber diameter distribution
	Amount and character of nonfibrous formations
	Area weight of nanofibers layer, amount of additives

5.5.1 Solution Properties

Solution properties include the viscosity/concentration, surface tension, solution conductivity, average molecular weight of polymer, boiling point of solvent, dipole moment, and dielectric constant. The effects of the solution properties cannot be easily isolated because varying any one of these parameters can generally affect the other solution properties.

5.5.1.1 Viscosity/Concentration

Viscosity is one of the main parameters influencing the diameter of electrospun fibers. One of the conditions for the electrospinning process is that the polymer should have sufficient viscosity to prevent the breakage of the fibers formed. Deitzel et al. (2001) showed that the average fiber diameter increases with increasing polymer concentration and that happens according to a power law relationship. A high concentration of the polymeric solution represents a large number of polymer chains and hence more entanglements are present, which in turn prevents the breakage of the fibers.

Low concentrations/viscosities yielded defects in the form of beads and junctions, although increasing concentration/viscosity yields uniform fibers but with a small number of beads (Figure 5.8). In some cases, the surface tension of the polymeric solution can also be affected by an increase in the concentration of polymer in the solution (Martinez et al. 2004). A very high concentration of the polymer in the respective solvent leads to failure of the fibers, the average distance between beads on the fibers will be less and the beads turn to bigger sizes, when the viscosity is too high.

FIGURE 5.8 (a) At high viscosity, the solvent molecules are distributed over the entangled polymer molecules and (b) with a lower viscosity, the solvent molecules tend to congregate under the action of surface tension.

5.5.1.2 Surface Tension

The surface tension (γ) of a polymer solution has a significant effect on the morphology of the electrospun fibers. To initiate the electrospinning process, charges in the polymer solution/melt must be high enough to overcome the surface tension of the solution and this will lead to the stretching of the fluid. Surface tension reduces the surface area per unit mass of a fluid. The beaded morphology has been well recognized for electrospun fibers and is affected by the processing conditions; hence, the reduced surface tension favors the formation of nanofibers without mushroom-like structures (Figure 5.9). Beads can be considered as capillary break-up of the jets by surface tension (Pham et al. 2006). The low surface tension correlates with the formation of beaded fibers in a way that is related to raising the solution viscosity, as shown in Figure 5.9. At higher concentration, there is a superior interaction between polymer molecules and solvent.

5.5.1.3 Solution Conductivity/Charge Density

Electrospinning involves stretching of the polymer solution droplet by the induced charges at the surface; the higher the net charge density thinner the fibers without

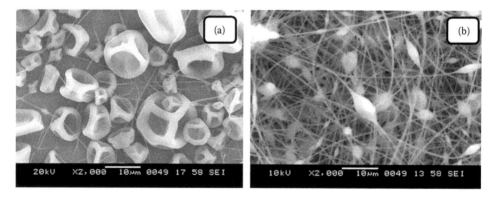

FIGURE 5.9 (a) The morphology of a mushroom-like structure and (b) beaded fibers versus higher surface tension.

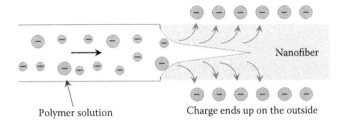

Polymer solution

Nanofiber

Charge ends up on the outside

FIGURE 5.10   Zone of solidification.

defects (Teo et al. 2011). This means that if the conductivity of the solution is high, more charges will develop on the surface of the polymer droplet, and hence it will become easier to stretch the polymer droplet (Figure 5.10). Beads are formed because of the lack of surface charges, and the polymer droplet is not stretched sufficiently. Many researchers have found that increasing the solution conductivity can be used to produce more uniform fibers with fewer beads. For example, the addition of zinc acetate to a PVA/water solution was found to increase the net charge density carried by the spinning jet (Yang et al. 2004). The conductivity can be increased by the addition of any chemical species, which produces ions, and hence voltage required to produce smooth fibers reduces (Pham et al. 2006) (Figure 5.11). The ion size will also affect the spinning process. For instance, salts with smaller ionic radii produce smaller fibers (~200 nm), while salts with larger ionic radii yield larger fibers (~1000 nm). Usually, smaller ions (e.g., Na^+) move very fast compared to bigger ions (e.g., Cl^-). Thus, charge densities carried by the jet, surface tension, and viscoelastic properties of the solution are significant parameters. The addition of a salt (i.e., NaCl) to enhance whipping is a solution conductivity effect. Finer fibers are formed due to the enhanced whipping effect. The effects that lead to increased conductivity enhance the formation of bead-free fibers and also the fibers with thinner diameters. The addition of cationic surfactants has produced smaller

FIGURE 5.11   SEM photographs for a variation of fibers as net charge density changes due to the addition of metal acetate (a) PVA nanofibers and (b) PVA/zinc acetate composite fibers.

diameter fibers with no beads (Beachley and Wen 2009). The net charge density can be determined as follows:

$$\text{Net charge density} = (\text{solution concentration}) \times (\text{solution density})$$
$$\times (\text{collecting time}) \times (\text{jet current})/(\text{mass of dry polymer}) \quad (5.1)$$

5.5.1.4 Polymer Molecular Weight

The molecular weight has an important role in determining the structure of electrospun nanofibers. At a fixed concentration, as the molecular weight increases the morphology of the electrospun structure changes from beads to beaded fibers and then to fibers. However, above a certain critical concentration, uniform fibers are obtained irrespective of the molecular weight. Figure 5.12 shows the dependence of molecular weight on morphology at a constant concentration.

5.5.1.5 Effect of Solvent

Solvents are chosen based on the solubility parameter. If a solvent is good, there is a good interaction between the solvent molecules and polymer chains, which leads to the dissolution of the polymer. In electrospinning, selection of solvents is also important, because wise selection of a solvent is necessary to decide the morphology of the electrospun fibers. Figure 5.13 depicts the change in morphology of the electrospun fibers with the solvents.

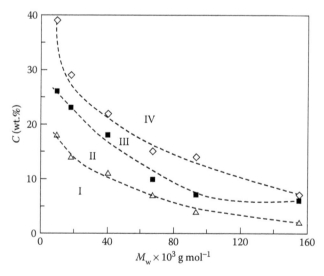

FIGURE 5.12 Regimes for various morphologies observed in the electrospun PVA polymer. I: Beads, II: beaded fibers, III: complete fibers, and IV: flat fibers. The symbols and the accompanying lines correspond to the transition point from one structure to the other. (Reprinted from *Mater. Lett.*, 61, Tao, J. and Shivkumar, S., Molecular weight dependent structural regimes during the electrospinning of PVA, 2325–2328. Copyright 2007, with permission from Elsevier.)

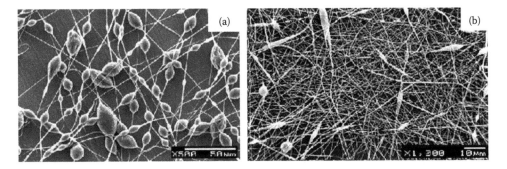

FIGURE 5.13   PBS electrospun fibers (a) CF and (b) chloroform (CF)/2-chloroethanol (CE) (7/3 w/w) as solvents keeping all other conditions same. (Liu, Y., He, J.-H., Yu, J. et al., Controlling numbers and sizes of beads in electrospun nanofibers. *Polym. Int.* 2008. 57. 632–636. Copyright Wiley-VCH Verlag GmBH & CO. KGaA. Reprinted with permission.)

5.5.1.6 Amount of Additives

The presence of additives in the precursor solution can alter the nanofiber morphology. The additives can be nanoparticles, salts, or any other polymer itself. For example, the addition of poly(ethylene glycol) (PEG) in poly(ε-caprolactone) (PCL) dissolved in chloroform changed the porosity of the fibers (Pant et al. 2011b), shown in Figure 5.14.

Similarly, the addition of salts such as NaCl, KBr, and $CaCl_2$ can also change the morphology of the fibers irrespective of the polymer used. Barakat et al. (2009) reported the formation of spider net within the N6, PVA, and PU electrospun nanofibers by the addition of salts (Figure 5.15).

As the concentration is increased beyond a certain limit, morphology is further changed and microsized balls are formed in the fiber network as shown in Figure 5.16.

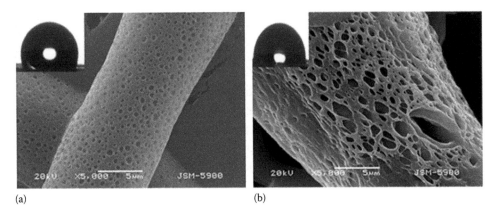

FIGURE 5.14   (a) Pristine PCL and (b) MPEG/PCL (2 g MPEG) electrospun mats. (Reprinted from *Colloids Surf., B*, 88, Pant, H.R., Neupane, M.P., Pant, B. et al., Fabrication of highly porous poly (ε-caprolactone) fibers for novel tissue scaffold via water-bath electrospinning, 587–592. Copyright 2011b, with permission from Elsevier.)

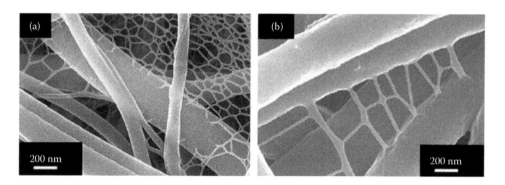

FIGURE 5.15 Electrospun nanofiber mat from the solution containing 1.5 wt.% NaCl (a) nylon 6 and (b) PU. (Reprinted from *Polymer*, 50, Barakat, N.A.M., Kanjwal, M.A., Sheikh, F.A. et al., Spider-net within the N6, PVA and PU electrospun nanofiber mats using salt addition: Novel strategy in the electrospinning process, 4389–4396. Copyright 2009, with permission from Elsevier.)

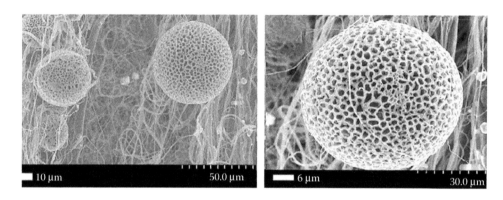

FIGURE 5.16 Nylon 6 fiber mat containing micro balls at 2.5 wt.% of NaCl at different scales. (Reprinted from *Polymer*, 50, Barakat, N.A.M., Kanjwal, M.A., Sheikh, F.A. et al., Spider-net within the N6, PVA and PU electrospun nanofiber mats using salt addition: Novel strategy in the electro-spinning process, 4389–4396. Copyright 2009, with permission from Elsevier.)

5.5.2 Process Parameters

5.5.2.1 Flow Rate

Feed/flow rate will determine the amount of solution available for the electrospinning process. For a given voltage, a higher feed rate will lead to a higher fiber diameter and the solvent takes more time to dry. An increased feed rate leads to larger fiber diameters and the formation of beads, whereas a lower feed rate leads to the formation of thinner fibers without defects (Soliman et al. 2010). The authors also observed a similar effect (Figure 5.17).

5.5.2.2 Field Strength/Voltage

The crucial point during the electrospinning process is the time of application of the voltage. The electric current due to the ionic conduction of charges in the polymer solution is usually considered small enough to be negligible (Kim and Reneker 1999). A very high

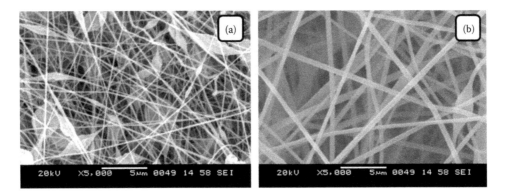

FIGURE 5.17 SEM micrographs of nanofibers electrospun from (a) higher flow rate and (b) lower flow rate.

voltage is applied during the electrospinning process; as a result, the polymer solution gets charged and the Taylor cone is formed. As the voltage is increased, the volume of the droplet decreases. In the same way, increasing the applied voltage is found to reduce the fiber diameter (Deitzel et al. 2001); it also reduces flight time between the collector and tip, proper flight time increases the crystallinity too.

The fibers produced under low-voltage conditions have a high density of bead defects. In fact, the fiber bead defect density decreases very sharply with voltage and can be further minimized through control of solution flow rates (Figure 5.18).

5.5.2.3 Distance between the Tip and Collector

In the case of electrospinning, the lower the distance between the tip and collector lower will be the flight time and higher will be the electric field. The lower flight time means that there is insufficient time for stretching and efficient evaporation of the solvent, which may lead to the formation of beads. As the tip-to-collector distance is increased, the jet acceleration will also increase, which favors more uniform bead-free fibers. Hence, the distance between the tip and collector should be optimum.

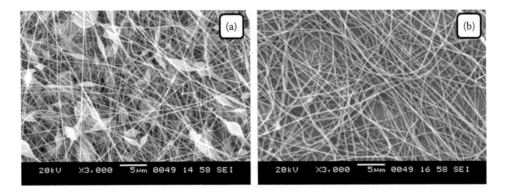

FIGURE 5.18 Scanning electron micrographs of nanofibers electrospun from (a) lower voltage and (b) higher voltage PVA in water.

5.5.2.4 Needle Tip Design and Placement

The smaller diameter of the orifice reduces the formation of beads as less volume of the solution is collected at the tip, which in turn reduces fiber diameter. The smaller the droplet higher is the surface tension. Also, the average fiber diameter decreases with decreasing orifice size. For a fixed voltage, the time for stretching increases and fiber elongates before it reaches the collector. If the diameter of the orifice is too small, extrusion is not possible. A large inner diameter of the needle can lead to the clogging at the tip due to the significant exposure of the solution to the air (Tong and Wang 2010).

5.5.2.5 Collector Composition and Geometry

To initiate the electrospinning process, there must be a proper electric field between the collector and tip. The collector used is conductive in nature such as aluminum foil/carbon paper that creates a greater electrostatic field effect, forming a thick membrane structure. The charges collected will be dissipated at the collector and allow fibers to collect. A non-conducting plate may lead to lower packing of fibers compared to a conductive plate, which also favors more uniform fibers with fewer defects. The accumulation of charges at the collector may lead to the formation of a honeycomb structure (Pham et al. 2006).

5.5.3 Ambient Parameters

5.5.3.1 Temperature

The solution viscosity and vapor diffusivity can be affected by a change in temperature, which in turn affects the fiber diameter and morphology. An increase in the temperature leads to a decrease in the fiber diameter, and it is ascribed to the reduction in the viscosity of the polymer solution at higher temperatures (Hardick et al. 2011).

5.5.3.2 Humidity

Electrospun nanofiber diameters are determined by the interaction between solvent evaporation and humidity level, because it affects the distribution of electric charges on the surface of the electrospinning jet. At high humidity, water will condense on the fibers that in turn will affect the fiber morphology. A relative humidity of less than 25% will yield smooth fibers; but as the humidity increases to 30%–45%, small circular pores appear on the surface of the fibers and the further increase in humidity leads to the coalescence of the small circular pores leading to irregular large-sized pores. The formation of pores at high humidity is explained by the evaporative cooling effect. A high humidity can lead to the entanglement of nanofibers on the collector; also, under this condition, the evaporation of the solvents from the fibers is not good enough, which can lead to the fusing of fibers to form bigger fibers (Hardick et al. 2011). At lower humidity thinner fibers are produced, because the electric charge densities on the polymer jets are high, and in addition to that the volatile solvents evaporate very quickly.

5.6 CHARACTERIZATION OF NANOFIBERS

The characterization of the different properties of electrospun nanofibers is carried out by different analytical techniques. The properties include the geometrical properties, surface chemical properties, and molecular structure. The geometrical properties are fiber morphology, size and distribution, and alignment. SEM, TEM, and AFM have been widely used to characterize the geometrical properties. The surface chemical properties have been identified by water contact angle measurement and X-ray photoelectron spectroscopy. The molecular structure has been identified using Fourier transform infrared spectroscopy, nuclear magnetic resonance spectroscopy, X-ray diffraction, and differential scanning calorimetry (Kulkarni et al. 2010). Moreover, the special features of the electrospun fibrous mats are identified by the direct implementation of the mats into the subsequent applications. The typical applications include tissue engineering, fuel cells, batteries, solar cells, filtration, and composites.

5.7 TYPES OF ELECTROSPINNING UNITS

A variety of electrospun nanofibers with different shapes and morphologies have been fabricated by researchers all around the world. These morphologies have been achieved either by the modification of conventional fiber-forming techniques or by the optimal control of the process parameters that were already discussed in Section 5.5. In this section, a few modified electrospinning units that are often employed to develop special featured nanofibers, nanofibrous mats, yarns, and so on, have been discussed. The modifications in electrospinning units are sated either by modifying the spinneret or by modifying the substrate on which the fibers are collected.

5.7.1 Melt Electrospinning

The conventional electrospinning method demands immense measure of solvents. The reuse of solvents is essential to make a process cost-effective and eco-friendly. However, it is not so easy to recycle the solvents used during electrospinning. Melt electrospinning is also an electrospinning technique in which the polymers or precursors in their molten state are ejected from the spinneret (Lee and Obendorf 2006). The fibers are cooled and solidified as they reach the collector. In a common melt electrospinning unit, there are two heating input divisions: one is to melt the polymer at the spinneret or syringe and the other is to control the temperature at the guiding chamber (Figure 5.19), which ensures that the electrospun fibers are not deviating from the center line.

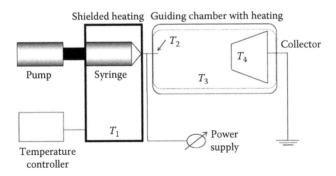

FIGURE 5.19   Melt electrospinning unit. (Lee, S. and Obendorf, S.K.: Developing protective textile materials as barriers to liquid penetration using melt-electrospinning. *J. Appl. Polym. Sci.* 2006. 102. 3430–3437. Copyright Wiley-VCH Verlag GmbH & Co. KGaA. Reprinted with permission.)

Melt electrospinning can be used to electrospin polymers that do not have any optimal solvents at room temperature. It is a cost-effective electrospinning technique and more suitable from the industrial production outlook. Even though it overcomes the difficulties associated with the solvents, it has its own drawbacks. In melt electrospinning, the accurate control of temperatures at both the heating divisions, the high viscosity of the polymer melts, chances of degradation of polymers, and the wide melting ranges of semicrystalline and amorphous polymers are the major constraints.

5.7.2 Coaxial/Core–Shell Electrospinning

In coaxial electrospinning unit, the spinneret with a single capillary is replaced with a coaxial spinneret as shown in Figure 5.20a. Both the inner and the outer tube jointly act as anode. Coaxial electrospinning has been used to fabricate core-sheath nanofibers as well as hollow nanotubes. The core portion of the spinneret, that is, mineral oil in Figure 5.20a, can be replaced with different precursors to get core–shell nanotubes with cores of different functionalities. The removal of mineral oil after spinning will end up with a nanotube or a hollow nanofiber as shown in Figure 5.20b. The flow of fluid in a coaxial electrospinning is shown in Figure 5.21a, and the detailed construction of the spinneret used in coaxial spinning is shown in Figure 5.21b.

5.7.3 Multichannel Electrospinning

Multichannel electrospinning can be considered as a modification of the spinneret in the coaxial electrospinning. In a multichannel electrospinning unit, the spinneret holds multiple metallic tubes inside a metallic outer tube (Figure 5.22a). The metallic inner tubes are arranged symmetrically around the axis of the outer tube and the whole unit acts as an anode during electrospinning. The inner and outer tubes are filled with immiscible fluids (mineral oil). The successful removal of the inner tube fluids will provide multichanneled fibers (Figure 5.22b).

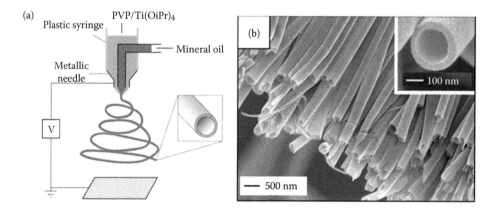

FIGURE 5.20 (a) Coaxial electrospinning unit and (b) TEM images of hollow nanotubes. (McCann, J.T., D. Li, and Y. Xia. 2005. Electrospinning of nanofibers with core-sheath, hollow, or porous structures. *J. Mater. Chem.* 15:735–738. Reprinted by permission of The Royal Society of Chemistry.)

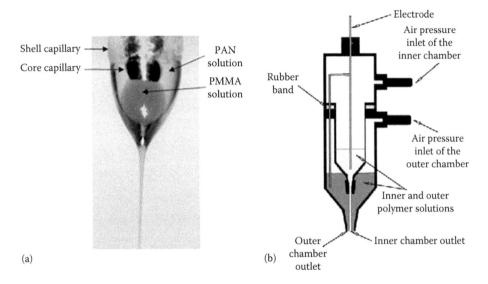

FIGURE 5.21 (a) Flow pattern in the compound droplet attached to the core–shell spinneret with PAN solution (shell) and PMMA solution dyed with malachite green (core). (b) A double-compartment plastic syringe used as the spinneret core–shell electrospinning. (Yarin, A.L., E. Zussman, J.H. Wendorff et al. 2007. Material encapsulation and transport in core–shell micro/nanofibers, polymer and carbon nanotubes and micro/nanochannels. *J. Mater. Chem.* 17:2585–2599. Reprinted by permission of The Royal Society of Chemistry.)

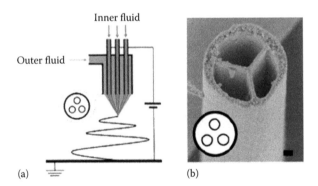

FIGURE 5.22 (a) Multichannel electrospinning unit. (b) Multichannel nanotube. (Reprinted with permission from Zhao, Y., X. Cao, and L. Jiang. Bio-mimic multichannel microtubes by a facile method. *J. Am. Chem. Soc.* 129:764–765. Copyright 2007 American Chemical Society.)

5.7.4 Drum Collector

The drum collector comprises a rotating drum accompanying a speed control unit (Figure 5.23). The rotating drum plays the role of the substrate on which the fibers are deposited. A high negative potential is applied on the drum, thus making it a cathode. The drum collector is employed to generate aligned fibers (Sonehara et al. 2008). The continuous rotation of the drum makes the nanofibers to be uniaxially aligned. It also helps the polymer molecules to align themselves along the axis of the fiber thereby enhancing

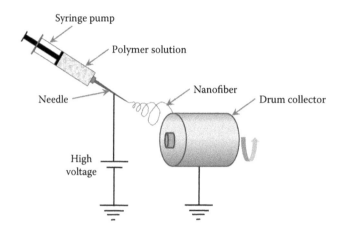

FIGURE 5.23   Electrospinning unit with a drum collector.

the axial strength of the fiber. The speed of the drum must be always greater than the speed with which the fibers are attracted to the drum.

5.7.5 Near-Field Electrospinning

In near-field electrospinning, a single nanofiber is directly written on a movable substrate. The motion of the substrate in the X- and Y-directions is accomplished by an X–Y motion stage (Figure 5.24). The anode of a high-voltage DC source is connected to the spinneret, and the cathode is connected to the substrate, fastened on the motion stage (Huang et al. 2011c). The idea of near-field electrospinning is that the substrate is retained at the close premises of the spinneret, within a distance of 0.5–3 mm, so that straight fibers are deposited on the substrate before whipping and splitting occurs. The straightness of the fiber is

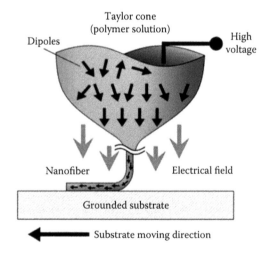

FIGURE 5.24   Schematic diagram showing a near-field electrospinning process. (Qi, Y. and M.C. McAlpine. 2010. Nanotechnology-enabled flexible and biocompatible energy harvesting. *Energy Environ. Sci.* 3:1275–1285. Reprinted by permission of The Royal Society of Chemistry.)

entirely dependent on the substrate motion speed. The substrate motion speed must be close to or higher than the speed of electrospinning to get straight fibers.

5.7.6 Yarn Spinning

Yarn spinning is exploited to procure either a single continuous nanofiber or a uniaxial nanofiber bundle, which is vital in plenty of applications. There are two methods to electrospin nanofiber yarns. In the first method, the substrate that collects the electrospun fibers is a liquid reservoir (Smit et al. 2005). The metal plate placed at the bottom of the reservoir acts as the cathode. At first, the yarn is drawn by hand slowly; once the fiber is drawn on the liquid reservoir surface, it is loaded to a motorized take-up roller running at a slow speed of about 20 rpm. A detailed schematic of yarn spinning unit with water bath is shown in Figure 5.25.

In the second method, the traditional metal plate is replaced with twisting tubes and a winding tube as shown in Figure 5.26. All the tubes are connected to motors and grounded; the motors connected to the twisting tubes run in the opposite directions and are exactly similar to the rotating drum, as discussed in the previous section. The drawn nanofibers are twisted with the help of a rotating twisting tube as the spinning progresses; at the same time, the winding tube is inserted at the middle of the twisting tubes and the rotation of the winding tube collects the twisted yarn from the twisting tubes in tandem with different twist angles α (Figure 5.27) (Yan et al. 2011).

The twist angle can be adjusted by controlling the speed of the twisting drives. The vertices of the cone must be controlled near the winding tube to avoid the untwisted nanofibers, usually known as hairs in textile industries.

5.7.7 Centrifugal Electrospinning

The centrifugal electrospinning system is a hybrid system in which aligned fibers are produced on a large scale. In this process, electrospinning is assisted by centrifugal forces.

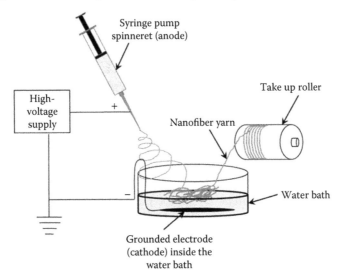

FIGURE 5.25   Yarn spinning using water bath.

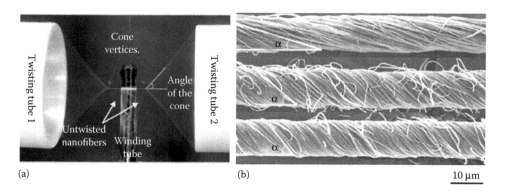

FIGURE 5.26 (a) Twisting of aligned nanofibers as yarns and (b) SEM image of the nanofiber yarns. (Reprinted from *Mater. Lett.*, 65, Yan, H., Liu, L., and Zhang, Z., Continually fabricating staple yarns with aligned electrospun polyacrylonitrile nanofibers, 2419–2421. Copyright 2011, with permission from Elsevier.)

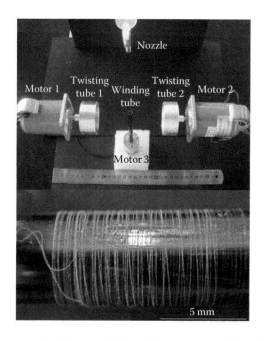

FIGURE 5.27 Yarn spinning unit. (Reprinted from *Mater. Lett.*, 65, Yan, H., Liu, L., and Zhang, Z., Continually fabricating staple yarns with aligned electrospun polyacrylonitrile nanofibers, 2419–2421. Copyright 2011, with permission from Elsevier.)

A centrifugal electrospinning unit is shown in Figure 5.28. The rotating spinneret is placed at the center of the nonconducting cylindrical housing and connected to a variable speed motor. The aluminum plates, which are placed equidistantly on the rim of the cylindrical housing on which the fibers will be collected, are grounded. As the spinning commences, the centrifugal force causes elongation and thinning of the solution jet and aligns them on the static cylindrical collector (Edmondson et al. 2012).

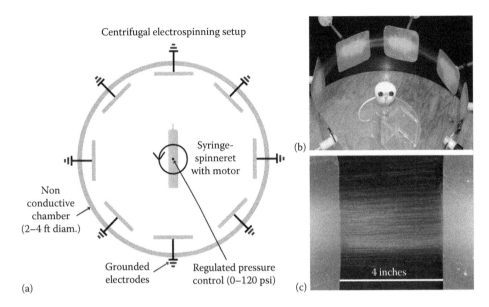

FIGURE 5.28 (a) Schematic of the centrifugal electrospinning setup (b) photograph of a centrifugal electrospinning chamber and (c) photograph of collected nanofibers. (Edmondson, D., A. Cooper, S. Jana et al. 2012. Centrifugal electrospinning of highly aligned polymer nanofibers over a large area. *J. Mater. Chem.* 22:18646–18652. Reprinted by permission of The Royal Society of Chemistry.)

5.8 APPLICATIONS OF ELECTROSPUN NANOFIBERS

The various plausible applications of electrospun nanofibers are listed in Table 5.5.

5.8.1 Biomedical Applications

5.8.1.1 Tissue Engineering

The primary focus of tissue engineering is to enable the natural healing of the damaged tissues (see Table 5.5). The fundamental feature of a synthetic scaffold for tissue engineering application is its ability to mimic the extracellular matrix (ECM) architecture in terms of physical, chemical, and biological characteristics. In another way, the synthetic scaffolds must have good porosity with even pore size distribution, high surface area, nontoxicity, biocompatibility, biodegradability, structural integrity, and wettability among cells.

The ECM has a complex role in regulating the behavior of the cells that are in contact with it, influencing their survival, development, migration, proliferation, shape, and functions. The ECM has a complex molecular composition. The ECM has a system of nonliving tissues made of cellulose in plants and proteins, minerals, and certain carbohydrates in animals. The macromolecules that constitute the ECM are mainly produced locally by cells in the matrix, for example, the fibroblast cells (Figure 5.29). Synthetic scaffolds are made of biodegradable and biocompatible materials. The advantage of using a biodegradable material is that a second surgery is not involved in the removal of scaffold; simultaneously, biocompatibility omits the chances of rejection of it from the body by the immune system. Electrospun ECM performs as a temporary scaffold in tissue engineering applications. Electrospinning has widely been used in the fabrication of nanofibers, with features

TABLE 5.5   Various Applications of Electrospun Fibers

Ground of Application	Use	Uniqueness of Electrospun Nanofibers
Automobile and aerospace	Air filters Lightweight wings for micro-air vehicles	Meek and low-cost technique for nanofibers production. Precise position and control of the nanofiber geometry Industrial scalability and mass production Functionalization of organic molecules into inorganic fibers Increase in electron conduction Mechanical strength Electrochemical stability Sensing and actuating capabilities Reliability and selectivity Selective permeability in membranes High surface area huge active sites Even pore size distribution in electrospun nanofiber mats
Biomedical	Biosensors Tissue engineering Artificial heart valves Medical devices Neural prostheses Wound healing Controlled drug delivery Cables for implantable Catalysts and enzyme carriers	
Environmental	Water filters Affinity membranes Separation membranes Air filters Personal protection masks	
Energy and Electronics	Batteries Photovoltaic cells Hydrogen storage Fuel cell membranes Supercapacitors Printable electronics	
Others	Catalysis Military garments Composites Sound proofing	

that are crucial for the synthetic scaffolds satisfying the aforesaid criterion. Electrospinning gives room for the synthesis of biocompatible nanofibers from biodegradable base materials without biocompatibility by adding biological materials into it. In other words, electrospinning can be used to fabricate polymer fibers blended with biological materials, such as albumin or protein. It is also possible to produce fibers of biological materials alone in a simple and a reasonably low-cost route, which will create a favorable environment for the cells. Electrospun nanofibers exhibit morphological similarities to the naturally occurring ECM.

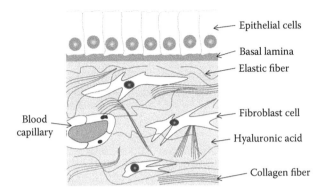

FIGURE 5.29   Connective tissue underlying an epithelium contains the ECM.

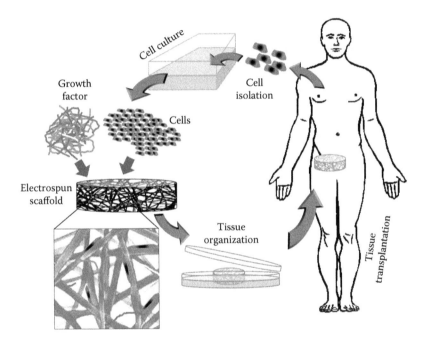

FIGURE 5.30   Schematic representation of steps involved in tissue engineering.

Electrospun composite nanofibrous scaffolds can support cell adhesion, proliferation, and differentiation, and it is important that the degradation time of the scaffolds must be comparable with the new tissue formation.

Ceramic and polymeric nanofibers have been generally used in tissue engineering applications (Figure 5.30), because of their biocompatibility, light weight, noncorrosiveness, and bonding properties. Naturally occurring polymers such as silk have also been used for the electrospinning of tissue scaffolds (Meechaisue et al. 2007). Numerous studies have been performed on the potential of electrospun fibers as tissue scaffolds. Electrospun fibers have been used as synthetic scaffolds for bones, muscle, nerve, cartilages, blood vessels, skin, and so on, and a few electrospun polymers used in tissue engineering applications are listed in Table 5.6.

TABLE 5.6   Electrospun Fibers Used in Tissue Engineering Applications

Nanofiber	Solvents	Applications	References
Collagen–Chitosan–TPU (tecoflex EG-80A)	HFIP/TFA	Vascular grafts, muscle cells	Huang et al. (2011a) and Chen et al. (2009)
PCL-collagen	HFIP/chloroform	Vascular grafts, tendon/ligament, nerve cells, bone	Zamarripa et al. (2009), Schnell et al. (2007), Corey et al. (2007), and Yoshimoto et al. (2003)
PLGA-co-PCL/phospholipid polymer/rapamycin	Acetone/methanol mix	Vascular grafts	Kim et al. (2009)
PLGA	DMAc	Vascular grafts	Park et al. (2010)
PLA/PCL	Chloroform	Vascular grafts	Vaz et al. (2005)
	HFIP	Skin tissue, skeletal muscle	Choi et al. (2008a) and Jin et al. (2011)
	DMAc	Bone	Guarino et al. (2008)
PLLA-co-PCL/collagen	HFIP	Vascular grafts	He et al. (2005)
PVP	Ethanol/bovine serum albumin (BSA)	Skin patch	Karatas et al. (2010)
Human tropoelastin	HFIP	Skin	Kovacina et al. (2011)
PEG-PDLA	Chloroform	Skin	Yang et al. (2011)
PLGA	Chloroform/DMF	Tendon	Chen et al. (2009)
	THF/DMF	Skin	Kumbar et al. (2008)
	HFIP	Neural tissue	Lee et al. (2009b)
PLLA/collagen	HFIP	Tendon	Theisen et al. (2010)
PLLA	Chloroform/ethanol	Tendon	Yin et al. (2010)
PVA/PEG	Distilled water	Cartilage tissue	Chen et al. (2009)
Gelatin modified PLLA	DCM/DMF	Cartilage tissue	Chen and Su (2011)
PLLA-co-PCL	Acetone	Muscle/endothelial cell	Mo et al. (2004)
PMGI	Cyclopentanone/ tetrahydrofurfuryl alcohol	Cardiac tissue	Orlova et al. (2011)
PEG–PDLLA diblock polymer	HFIP	Heart tissue	Zong et al. (2005)
Chitosan-based nanofiber	TFA:DCM	Cardiac tissue	Hussain et al. (2010)
PCL	Chloroform and methanol	Cardiac grafts, brain implants	Shin et al. (2004) and Nisbet et al. (2009)
Poly(1,4-dioxan-2-one) (PDO)	—	Right ventricular outflow tract in heart	Kalfa et al. (2010)
PU	DCM	Cardiac tissue	Rockwood et al. (2008)
PEG-PLA diblock copolymer	Chloroform	Cardiac muscle cell	Hai-ling et al. (2011)
Surface modified PLLA	HFIP	Nerve regeneration	Kakinoki et al. (2011)
Tussah silk fibroin (TSF)	Lithium thiocyanate	Nerve regeneration	Zhang et al. (2010)
PLLA blended with PANI	HFIP	Nerve regeneration	Prabhakaran et al. (2011)
PLLA	Chloroform	Nerve regeneration	Wang et al. (2009)
PAN-MA	DMF	Neural tissue	Mukhatyar et al. (2011)
PVDF–TrFE	1:1 DMAc/acetone	Neural tissue	Lee et al. (2011)

(Continued)

TABLE 5.6 *(Continued)* Electrospun Fibers Used in Tissue Engineering Applications

Nanofiber	Solvents	Applications	References
PES	DMSO	Neural tissue	Christopherson et al. (2009)
PVDF–TrFE	Methyl-ethyl-ketone	Spinal chord	Lee et al. (2009c)
PCL/nano hydroxyapatite and nano beta tricalcium phosphate composite	DCM and DMF	Bone	Patlolla et al. (2009)
PVA–collagen-hydroxyapatite biocomposite	Deionized water	Bone tissue	Asran et al. (2010)
PLGA/gelatin/hydroxyapatite composite	HFIP	Bone	Lee et al. (2010a)
Silk fibroin	Formic acid, HFIP	Bone scaffold, cartilage	Meechaisue et al. (2007) and Wang et al. (2006)
PBS extended with 1,6-diisocyanatohexane	DCM/TFA	Bone scaffold	Sutthiphong et al. (2009)
Lecithin blended polyamide-6	Formic acid	Osteoblast cell culture	Nirmala et al. (2011)
Hydroxyapatite/chitosan nanofibers	TFA	Bone	Frohbergh et al. (2012)
PU/gelatin blend	HFIP	Blood vessels, muscle cells	Vatankhah et al. (2014a,b)
Chitosan/gelatin/PVA/gum arabic nanofibers	Aqueous acetic acid	Mesenchymal stem cell proliferation	Tsai et al. (2015)
CNT/CA assembly	Acetone/DMF	Fibroblast proliferation	Luo et al. (2013)
Bioactive glass and hydroxyapatite in PCL/chitosan nanofibers	—	Fibroblast cells	Shalumon et al. (2013)
Hemoglobin/gelatin/fibrinogen	TFE and HFIP	Cardiac tissue	Ravichandran et al. (2013)
Poly(L-lactic)-*co*-PCL/laminin	HFIP	Nerve tissue	Kijeńska et al. (2014)
PHBV/collagen	HFIP	PHBV/collagen	Prabhakaran et al. (2013)
Chitosan/PCL core shell fibers	TFA/DCM for chitosan, chloroform/methanol for PCL	Vascular tissue	Du et al. (2012)
PANI/PCL	HFIP	Skeletal muscle	Chen et al. (2013)
Sodium alginate/PEO core–shell	Water	Fibroblast cell proliferation	Ma et al. (2012)

5.8.1.2 Drug Delivery

Controlled release of drugs has always been a challenge in medical treatment. The concentration of drugs in the blood must be always higher than the minimum effective level to achieve the curative effect. Each drug has its own degradation time in a biological system leading to a decline of its concentration in blood after a few hours of intake, which will be less than the minimum effective level. To surmount the problem due to degradation, drugs are consumed at higher doses at regular intervals. The blood transports these drugs, which are at a high level, to all parts of the body that can impart toxicity to the healthy regions of the body, typically known as side effects. Controlled release of drugs can retain the desired drug level in blood for a longer duration of time without reaching a toxic level or dropping below the minimum effective level.

Biodegradable polymers have been used to cover drugs as we can see in capsules; when the capsule reaches certain abdominal regions, the polymer coatings are degraded and the drug is released suddenly; thus it can cause side effects. Biodegradable electrospun nanofibers with a measured degradability can be used for the controlled release of drugs, because electrospun fiber mats have a high porosity and surface area that help enhance the effectiveness of drugs (Jiang et al. 2014). Not only biodegradable polymers, but nonbiodegradable polymers have also been used in controlled drug delivery. In the case of biodegradable polymers, the release of drugs takes place by means of diffusion and degradation; on the other hand, in nonbiodegradable polymers, it is through diffusion alone. The fibrous nature will enhance the solubility of drugs while protecting them from gastric acids and enzymes. It is possible to load the drugs into electrospun nanofibers during the electrospinning process itself, or it can be used as a template for generating drug-loaded nanotubes (Figure 5.31), or it can be used in the form of core-sheath fibers. Drugs loaded to the fibers at the time of processing are released from the fibers as the polymer fiber degrades (Meng et al. 2011). Because of the controlled degradation of the polymer fibers, the drugs are released in proportion with the polymer degradation. The sensitivity of the polymers to the nature of the gastrointestinal fluids (acidic or basic) aids the targeted delivery of drugs.

The suitability of a number of electrospun polymer nanofibers has been studied for drug delivery applications, and a few of them are listed in Table 5.7.

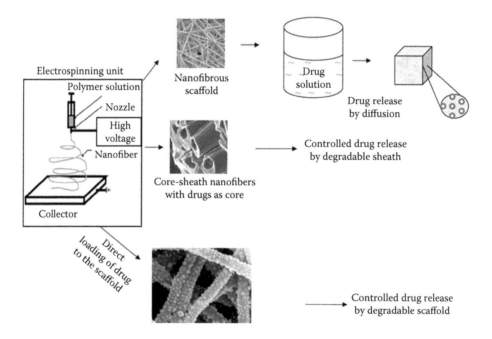

FIGURE 5.31 Schematic representation of the three familiar strategies of electrospun fibers in drug delivery applications.

TABLE 5.7   Electrospun Polymeric Fibers in Drug Delivery

Polymer/Composite	Solvent	Loaded Drug	Remark	Reference
PLGA/gelatin	TFE	Fenbufen (FBF), anti-inflammatory drug	Increased hydrophilicity and drug release	Meng et al. (2011)
PEO	Acetonitrile and water	Nabumetone (Relafen®), anti-inflammatory drug	Superior drug release comparable with nanoparticulate drug delivery, but not suitable for long-term storage	Ignatious et al. (2010)
PVA	Distilled water	Sodium salicylate, diclofenac sodium, naproxen, iindomethacin (nonsteroidal, anti-inflammatory drug)	The drug release rate is a decreasing function of molecular weight	Taepaiboon et al. (2006)
PVA	Water/ methanol	Ketoprofen (nonsteroidal, anti-inflammatory drug)	High release as temperature increases	Kenawy et al. (2007)
PVA	Distilled water	Caffeine and riboflavin	Fast-dissolving delivery system	Li et al. (2013b)
PLGA	Chloroform/ DMF/water	Rhodamine B	Unique morphology and sustained release rate	Liao et al. (2008)
PCL	Chloroform/ DMF	Biteral (abdominal adhesion)	Improved healing with a slow biodegradability	Bölgen et al. (2007)
EVA/PLA	Chloroform	Tetracycline hydrochloride (a model drug)	Smooth release of drugs in 5 days	Kenawy et al. (2002)
CA	Acetone/N,N-dimethyl-acetamide	Naproxen, indomethacin, ibuprofen, sulindac	The drug release rate from the fiber mats is higher than the corresponding casted films	Tungprapa et al. (2007)
Poly(L-Lactide) acid/chitosan core shell fibers	DCM and aqueous acetic acid	—	—	Ji et al. (2013)
Surface functionalized PCL fibers	DCM/DMF	Rhodamine 6G hydrochloride (R6G) and doxorubicin hydrochloride (DOX)	pH-responsive drug delivery	Jiang et al. (2014)
PMMA/ Indomethacin	Ethanol	Indomethacin	Colonic drug delivery	Akhgari et al. (2013)
MPEG-b-PLA micelles/chitosan/ PEO nanofibers	Aqueous acetic acid	$CaCl_2$ and cefradine	Controlled release of both hydrophobic and hydrophilic drugs	Hu et al. (2014)

5.8.1.3 Wound Dressing

There is always a continual demand for skin regeneration products (Figure 5.32). Wounds that happen on the periphery of the dermis are capable of self-regeneration, but sadly the body cannot heal deep injuries extended to the connective tissues. In such cases, where there are no remaining sources of cells for regeneration except at the wound periphery, complete rehabilitation therefore requires extended time and ample care. Also, the possibility of scar marks cannot be omitted. In the case of severe wound or burning, survival of the patient or complete recovery is possible only by the quick closure of the wounds.

FIGURE 5.32 Nanofibers in wound dressing. (Reprinted from *Compos. Sci. Technol.*, 63, Huang, Z.-M., Zhang, Y.-Z., Kotaki, M. et al., A review on polymer nanofibers by electrospinning and their applications in nanocomposites, 2223–2253. Copyright 2003, with permission from Elsevier.)

The introduction of bioengineered skin changed the scenario to a large extent. However, the production of the bioengineered skin involves tedious processes, and also it depends on the availability of the donor site. The production of bioengineered skin requires experienced personnel and involves large-scale investment.

Electrospinning can play a key role to conquer the complexities involved in the bioengineered skin. The porous nature of the nonwoven electrospun fibrous mats can allow cells, drugs, and genes, making them ideal for drug delivery. Electrospun fibers have architecture and morphologies exactly similar to the ECM of the skin. Electrospinning can be used for the fabrication of nanofibers from materials that are biodegradable, biocompatible, and commercially available at a low cost (Choi et al. 2008b). All these properties enhance the proliferation of cells all the way through it. Incorporation of nanoparticles, such as silver, can keep the microorganisms away; at the same time, the loading of nanosized drugs accelerates the healing rate. In many cases, the artificial dermal analogs can fail because of their lack of flexibility, permeability, high thickness, and so on. Nowadays, electrospun nanofibrous mats have started taking over the places of other artificial dermal-forming techniques (Table 5.8).

5.8.1.4 Coatings

There are a wide range of coatings we come across in our day-to-day life. Coatings for the industrial applications are also equally important as those in household applications. The electrospun nanofibrous composite coatings with selective barrier properties can be used in chemical industries, war field, mines, and so on; these coatings have a remarkable breathability while preserving the maximum protection against chemicals. Quite a few coatings have also been used in the biomedical applications. These are silver-loaded antibacterial

TABLE 5.8   Electrospun Fibers Used in Wound Dressing

Fiber Material	Solvent/Precursor	Application	Reference
Polyglycerol	Methanol and DMF	Wound dressing	Vargas et al. (2010)
PCL–PEG block copolymer	Methanol and chloroform	Wound healing	Choi et al. (2008b)
Silk/PEO blend	Water	Wound dressing	Schneider et al. (2009)
PCL/gelatin	TFE	Wound healing and dermal reconstitution	Chong et al. (2007)
PVA/silver nanoparticles	PVA/silver nitrate (AgNO₃) aqueous solution followed by heat treatment	Antimicrobial wound dressing	Hong et al. (2006)
EVOH Ag nanoparticles	EVOH/AgNO₃ follwed by heat treatment	Skin wound healing	Xu et al. (2011)
PLGA/collagen	HFIP	Wound dressing	Liu et al. (2010)
Collagen	HFIP	Wound healing	Miao et al. (2011)
Chitosan/sericin composite nanofibers	TFA	Antibacterial wound dressings	Zhao et al. (2014)
Chitosan/silver-NPs/PVA	Aqueous acetic acid	Antimicrobial wound dressing	Abdelgawad et al. (2014)

coatings, coating on implants to make them biocompatible, wound dressing, and so on. Bioglass-ceramic materials are primarily used in bone replacement and bone engineering. These materials are brittle and have low fracture strength. Because the processing temperatures of such materials are high, it is difficult to encapsulate drugs or growth factors for cell adhesion and proliferation. It is proven that electrospun biodegradable polymer coating on such glass will overcome the above difficulties. Poly(3-hydroxybutyrate) and poly(3-hydroxybutyrate-co-3-hydroxyvalerate) (PHBV), PVA, and PCL are generally used as the biodegradable electrospun polymeric materials for this application (Bretcanu et al. 2009).

Electrospun nanofibers of PEO/PEG containing silver nanoparticles can be used as protective coating against *Escherichia coli* bacteria. These coatings have potential applications in wound dressing and coating for implants (Rujitanaroj et al. 2007).

5.8.2 Filters

Filters are used in appliances, such as water purifiers, protective masks, air conditioners, and automobiles. Filters are also important in many of the engineering and chemical industries for absorption, separation, waste water treatment, and so on. For all these applications, the performance of the filters can be improved tremendously by increasing the fineness of the fiber structure used in the filter. Figure 5.33 depicts the large improvement in the effective area available for the filtration as the size of the fibers constituting the filter reduces. The same filter efficiency is achieved by a small area as the fineness of the fiber increases.

A filter is said to be efficient when it can remove the unfavorable particles with sizes in all the ranges including submicron ranges. Submicron particles can be removed by the filter only when it has the structure as well as the channel size in that range (Huang et al. 2003). Electrospun nanofibers can be used for the development of effective filters. The electrospun nanofibrous-structured filters have a high surface-to-volume ratio, thereby a

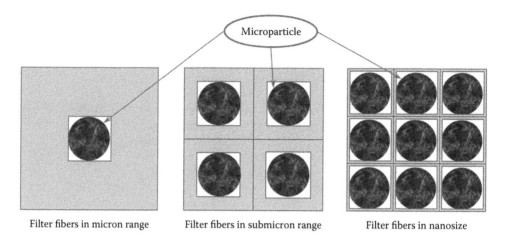

Filter fibers in micron range   Filter fibers in submicron range   Filter fibers in nanosize

FIGURE 5.33   The efficiency of a filter increases with decrease in fiber diameter.

higher surface cohesion, which entrap tiny particles in the submicron range. The strength of the nanofiber web alone is often too low to be used as a filter; so spun-bonded nonwoven or melt-blown nonwoven fabrics are always used as sublayers to support the nanofibrous web (Qin and Wang 2006). Electrostatically charged electrospun fibers enable the electrostatic attraction of the particles, improving the filtration efficiency. The selective filtration is also possible by incorporating specificity agents during the production of the nanofibers using electrospinning. These filters can be used as molecular filters or as a counter measure to chemical and biological weapons and their detection.

There are many electospun nanofibrous filters, which are introduced for the removal of contaminants from water and Table 5.9 lists some of them.

5.8.3 Energy Applications

5.8.3.1 Batteries

Batteries are the center of attraction in this century as an energy storage medium. The low energy density and charging is always a concern when we are dealing with batteries. The performance of batteries can be significantly improved by increasing the contact area of the electrode and the electrolyte especially in lithium-ion batteries, which have a high energy density and low self-discharge as compared with the other electrochemical storage devices. One method to increase the contact area between the electrode and electrolyte is to integrate the fibrous structure to the electrodes or to modify the electrolyte or separator. Electrospinning can be used to fabricate such an electrode and also for the electrolyte modification. The electrodes fabricated by electrospinning have a web-like structure, which has a significant role in improving the number of contact points on the electrodes, energy density, transparency for the ions, and the internal conductivity of the electrode in an easy and at a low-cost line (Rubacek and Duchoslav 2008).

5.8.3.1.1 Electrodes   When polymers in bulk form are used as electrodes in electrochemical systems, such as capacitors and batteries, a filler material is required to improve the electrical conductivity of the polymeric material and a binder to fix them on polymer,

TABLE 5.9   Electrospun Fibers in Filtration

Fiber	Precursor/Solvent	Application	Reference
PAA, PSU, PEI, and polyamide (PA) assemblage	—	Water filtration. Clean water permeability and strength comparable with commercially available membrane	Bjorge et al. (2009)
MWCNT incorporated PAN	DMF	Removal of fisinfection byproducts (DBPs) from water	Singh et al. (2010)
Beta-cyclodextrin (β-CD) incorporated PS	DMF	Filtration/purification/separation purposes with high efficiency	Uyar et al. (2009)
Chloridized-PVC	—	Filters heavy metals by adsorption from ground water	Sanga et al. (2008)
Chitosan nanofibers	Chitosan, deactylation (DDA), PEO. and acetic acid were used as the precursors	Water purification and air filtration of aerosol. Antimicrobial	Desai et al. (2009)
Cyclodextrin functionalized PMMA	PMMA and β-CD in DMF	Removal of organic vapors in waste management systems	Uyar et al. (2010)
PEO/PVA	Water and ethanol	Microfiltration membranes to remove air particle	Wang et al. (2007)
PAN	PAN is dissolved into DMF	Filtration of nanoparticles	Yun et al. (2007)
PAA/PVA/MWCNTs	Aqueous medium, Fe(III) ions used to modify the fibers	Removal of heavy metal ions, from water	Xiao et al. (2011)
PES	DMF and NMP	Water purification	Yoon et al. (2009)
SAN	DMF	Chemical filtration	Senthil et al. (2013)
PVA/TiO$_2$	Distilled water	Particulates and volatile organic compounds	Chuang et al. (2014)

making the fabrication of polymer electrodes tedious. Electrospinning can be used for the production of electrodes in a single step with fillers as additives that can improve the conductivity of the polymer thereby reducing the complication involved in the production process.

5.8.3.1.2 Electrolytes   Polymer electrolytes hold good attention due to their high electrochemical stability and affinity to the electrolytes than their competing counter parts (Figure 5.34). Porous polymer electrolytes are established in light-weight high-performance batteries. Electrospun nanofibers can have the porosities above 90% making them the ideal material for the electrolytic application in batteries. Electrospun polymer electrolytes have an improved ionic conductivity, electrochemical stability, lower interfacial resistance, and improved cycle performance than the conventional ones (Kim et al. 2004). All these improvements are due to the high surface area of the electrolyte and the wettability by the electrolyte solution.

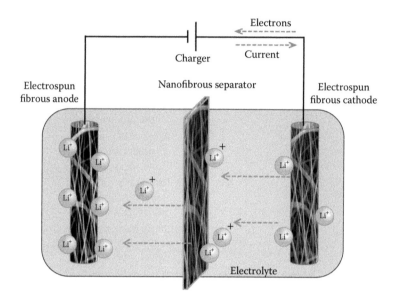

FIGURE 5.34 Application of electrospun fibers in various components of a lithium-ion battery.

5.8.3.1.3 Separators Separators are porous sheets used in batteries to prevent the physical contact between the anode and cathode. In the meantime, separators act as the electrolyte reservoir to enable free ionic transport. The structure of the separator can be either a membrane or a fibrous mat. The main features that a separator must have are good porosity and wettability to electrolyte. The separator does not take part in electrochemical reactions; still its properties affect energy density, power density, cycle life, and safety of the battery. Separators are generally made from polymers. To achieve highly porous fiber mats, electrospinning is the most effective method. Table 5.10 lists the electrospun fibers used as electrodes and electrolytes in batteries.

5.8.3.2 Photovoltaic Cells

Photovoltaic cells are solid-state devices, which convert light energy to electrical energy. The conventional solar cells are p–n junction devices made of silicon-based materials. The costs of these solar cells are quite high, limiting them to specific applications. Dye-sensitized solar cells (DSSCs) are credible alternatives for p–n photovoltaic cells, both economically and technically. The heart of a DSSC is a nanocrystalline metal oxide mesoporous layer on which the dye molecules are sintered. As the light falls on the dye, the photoexcitation of the dye will take place leading to the transfer of electrons to the nanostructured metal oxide, which in turn is connected to the electrodes of the cell. The restoration of dye to its original state takes place by the electron transfer from the electrolyte (Grätzel 2003). Recent studies on DSSCs show that nanorods or wires of metal oxides can be used in place of their other nanostructured counterparts, because the latter confines the high efficiency of DSSCs due to their disordered geometrical structure and interfacial effects in electron transport (Figure 5.35). Electrospinning is a versatile technique for the fabrication of metal oxide mats of TiO_2, ZnO, and so on in an ordered array enabling fast and effective transport of electrons (Kim et al. 2007).

TABLE 5.10   Electrospun Fibers Used as Electrodes and Electrolytes in Batteries

Polymer/Composite	Solvent/Precursor	Remarks	Application	Reference
Porous carbon nanofibers as anode	Carbonization of electrospun PAN/SiO_2 composite nanofiber, followed by removing SiO_2 nanoparticles with HF acid	Large capacities and good cyclability	Electrode	Ji et al. (2009)
MnOx/carbon nanofiber composites as anode	PAN, DMF, manganese acetate $[Mn(CH_3COO)_2]$, and sodium sulfate (Na_2SO_4)	High reversible capacity with good versatility, good capacity maintenance after cycling	Electrode	Lin et al. (2010b)
Carbon–cobalt composite nanofibers for anode	PAN and $Co(CH3COO)_2$ in DMF	High conductivity, a large reversible capacity. and good cycling performance	Electrode	Wang et al. (2008a)
C/Fe3O4 composite nanofibers	PAN dissolved in DMF/ferric acetylacetonate as precursor	High reversible capacity, good cycling performance. and excellent rate capability	Electrode	Wang et al. (2008b)
PVDF	DMAc	Improved physical properties and porosity	Electrolyte	Choi et al. (2004)
PVdF	Acetone/DMAc	Good electrochemical stability and longer storage time	Electrolyte	Kim et al. (2004)
PVdF-HFP/PMMA	(DMF)/acetone	The suppressed crystallinity of PVDF with high electrolyte intake and ionic conductivity	Electrolyte	Ding et al. (2009)
PVdF–PAN–ESFM	(DMF)/acetone	Good electrochemical stability, electrolyte uptake, ionic transparency and interfacial characteristics	Electrolyte	Gopalan et al. (2008)
PVDF-HFP	Acetone/DMAc	Excellent cycle-life	Electrolyte/ seperator	Lee et al. (2006)
PVDF–PVC	DMF/THF	High electrolyte uptake and ionic conductivity	Electrolytes	Zhong et al. (2012)
PAN	—	Higher ionic conductivities and better cycle lives along with a high rate capability	Seperator	Cho et al. (2008)

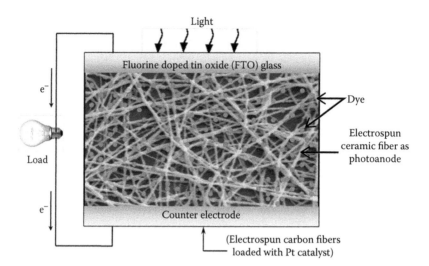

FIGURE 5.35 Schematic of a typical DSSC in which electrospun fibers are used.

The volatility of the organic solvents carrying the electrolyte in DSSCs is high, which is the second limitation in DSSC technology. It is overcome by the introduction of gel polymer electrolytes, but the mechanical strength of such electrolytes is generally poor and their preparation is also complex. To overcome the difficulties, electrospun polymer fibers can be used in electrolytic applications in DSSCs (Park et al. 2011). Flexible solar cells attached to cloths to power low-energy devices are sold in the market by several companies. Silicon and conjugated polymers have been used in such devices. The use of electrospinning to fabricate nonwoven solar cloth is an innovative approach in this area.

The fabrication of the organic solar cloth was done by electrospinning of poly(3-hexylthiophene-2,5-diyl) (P3HT)/phenyl-C61-butyric acid methyl ester (PCBM) as core and poly(vinyl pyrrolidone) (PVP) as shell in a core–shell setup with a subsequent etching of PVP shell with anhydrous ethanol. The efficiency of the solar cloth is very less, but it can be improved by the careful control of the processing parameters (Sundarrajan et al. 2010). Semiconductor $Cu(In_xGa_{1-x})Se_2$ nanofibers produced by electrospinning can be used in the making of thin film solar cells (Chen et al. 2011b). The application of different polymers in this field still needs to be uncovered. The application of electrospun polymeric electrolytes such as PEO, ethylene oxide/epichlorohydrin copolymer, poly(aniline) (PANI), and so on still needs to be analyzed (Table 5.11).

5.8.3.3 Supercapacitors

Supercapacitors are energy storage systems with high power density, rapid charging and discharging abilities, and a long cycle life. They have a power density more or less comparable to batteries. The capacitance values of supercapacitors can be improved by increasing either the energy density or the operating voltages. By increasing the surface area of the capacitor electrodes, an increase in capacitance can be achieved in electrical double-layer capacitors. Polymers loaded with activated carbon materials are excellent materials for the supercapacitor electrodes (Figure 5.36). Also, by blending other polymers and additives to

TABLE 5.11 Electrospun Fibers in Solar Cells

Application	Electrospun Fiber	Precursor/Solvent	Advantages	Reference
Electrode	Cu nanofibers	A mixture of PVAc and copper acetate with a suitable solvent is used as precursor. The fibers are calcined to obtain CuO fibers.	The excellent flexibilities and stretch abilities. Highly transparent solar cell electrode with high power efficiency.	Wu et al. (2010)
	TiO$_2$ fibers	Titanium n-propoxide and PVAc in DMF.	Long electron lifetime in nanorod electrode contributes to the enhanced effective photocarrier collection as well as the conversion efficiency.	Lee et al. (2009a)
	Silver nanoparticle doped TiO$_2$ nanofiber	The precursor consisted of titanium isoproxide, silver nitrate, ethanol, acetic acid and PVP.	The conversion efficiency is improved by 25%.	Li et al. (2011)
	ZnO fiber mat	Calcination of fibers obtained from the precursor consists of DMF as solvent, PVAc, zinc acetate. and acetic acid as a catalyst.	High absorption.	Kim et al. (2007)
Electrolyte	PVDF–HFP)/PS blend nanofibers	Acetone/DMAc	High ionic conductivity, electrolyte uptake. and porosity.	Park et al. (2011)
	PAN	DMF	Comparable efficiencies to those with volatile solution electrolyte-based DSSCs.	Dissanayake et al. (2014)

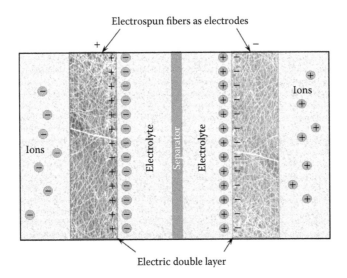

FIGURE 5.36 Schematic of a supercapacitor.

TABLE 5.12   Electrospun Fibers in Supercapacitor

Electrospun Fibers	Preparation	Use/Advantages	Reference
Activated carbon nanofibers	PAAc is electospun using THF/methanol as solvent. The fibers are carbonized at 1000°C and activated by steam at 800°C.	Electrode for electrical double layer capacitors (EDLCs). A high specific capacitance even at high current density.	Kim et al. (2004)
	Carbonization of PBI-based nanofibers in nitrogen. Activation is done in the presence of steam as N_2 as the carrier gas at 850°C.	Electrode in EDLCs. High specific surface area and a high capacitance.	Kim (2005)
	Fibers of PAI are used as the precursor for carbonization. Activated by CO_2 at 900°C at atmospheric pressure.	Electrode in EDLCs. ACNFs afforded good electronic conductivity, higher specific surface, suitable pore size, and higher content of surface oxygen functional groups, thereby a higher capacitance.	Seo and Park (2009)
PEDOT	PVP/pyridine electrospun oxidant nanofibers act as the precursor for the PEDOT fibers from EDOT monomers.	Active material for the textile supercapacitors. Suited for the fabrication of textile supercapacitors by staking layers of different function. PEDOT serves the role of electrodes.	Laforgue (2011)
Nickel/carbon nanofibers composite electrode.	Electrospun the mixture of PAN and nickel acetate in DMF, as solvent. Carbonization at 1000°C in the presence of nitrogen leads to the formation of the nanofibers composite.	Electrode of supercapacitors. Ni-loaded carbon nanofibers electrode have a capacitance three times than that of a normal carbon nanofiber electrode.	Li et al. (2009)

the base polymer can improve the capacitance. Composite electrospun fibers are also used in such applications (Miao et al. 2010). Electrospun nanofibers used in supercapacitors are listed in Table 5.12.

5.8.3.4 Piezoelectric Materials

Piezoelectric materials convert electrical energy to mechanical energy and vice versa. The common forms of piezoelectric materials are film and bulk. Piezoelectric materials are also used in sensors, actuators, MEMS systems, muscle and neural scaffolds, and electronics applications. Piezoelectric electrospun fibers can be used in potentially wearable "smart clothes" to produce electrical energy from the body movements, which is enough to run low-power electronic devices with high efficiency. Electrospun poly(vinylidene fluoride) (PVDF) piezoelectric fibers are used as nanogenerators with high energy conversion efficiency. The piezoelectric coefficient of a single PVDF nanofiber is double that of a thin film (Pu et al. 2010).

The applications of piezoelectric materials are not limited to the generation of electricity alone. Piezoelectric fibers can help in the nerve regeneration because these can act as simulators for neurite extension. Electrospun composite nanofibers of poly(vinyliene fluoride-*co*-trifluroethylene) (PVDF-TrFE), a piezoelectric material, have found potential application in neural tissue engineering (Lee et al. 2011). The electrospun nanofibers of the above-mentioned material are also used in sensing and actuating. A single PVDF fiber produced by near-field electrospinning can be used as a generator with a consistent production 7.2 pW with the external strain (Chang et al. 2009). Needle-less electrospun nanofiber webs are useful in mechanical-to-electrical energy harvest devices. The β-crystal phase in PVDF nanofibers leads to an enhanced mechanical-to-electrical energy conversion (Fang et al. 2013). The addition of silver nanoparticles (Li et al. 2013a) and CNTs (Liu et al. 2013) to PVDF nanofibers can enhance the piezoelectric properties of the nanofibers. Poly(γ-benzyl, L-glutamate) (West et al. 2014) and PVDF-TrFE (Persano et al. 2013) are two other promising polymers that exhibit piezoelectric behaviors in the form of electrospun nanofibers.

5.8.3.5 Thermoelectric Materials

Thermoelectric materials produce electricity using the temperature difference between a heat source and a heat sink. Thermoelectric power generators provide another environment-friendly solution for power production. Thermoelectric power generators are used in devices with low energy consumption, such as wrist watches and hearing aids, using the body temperature. Thermoelectric materials are suitable for waste heat recovery in many applications. The thermoelectric properties of the materials can be improved substantially by forming it into nanofibers using electrospinning.

A number of thermoelectric materials are electrospun for the energy conversion and electronic applications. Sodium cobalt oxide ($NaCo_2O_4$), a widely used thermoelectric material, is spun into nanofibrous form. The fibers are obtained by the calcination of the electrospun fibers from the homogeneous solution of poly(acrylonitrile) (PAN), *N,N*-dimethylformamide (DMF), sodium acetate trihydrate, and cobalt acetate tetrahydrate (Maensiri and Nuansing 2006). Thermoelectric materials, such as Ag_2S, are also electrospun as composite fibers with polymer materials. (PVP)/Ag_2S composite fibers are fabricated using electrospinning the mixture of $AgNO_3$, PVP, and CS_2 in ethanol as solvent (Lu et al. 2005). The electrospun Janus-type Co_3O_4/TiO_2 nanofibers, the combination of p-type Co_3O_4 on one side and the n-type TiO_2 on the other forming a p–n junction at the interface, are potential materials for the p–n junction in nanoscale thermoelectric devices (Zeng et al. 2014). The well-defined orientation of PANI chains in PANI/CNT electrospun composite fibers improves the carrier mobility in the composite fibers; thereby, an enhancement in anisotropic thermoelectric properties is observed in the direction of orientation (Wang et al. 2012). Because of the rapid thermal annealing of ZnO/PVP composite nanofibers, the obtained ZnO nanofibers have mesoscale grains. The presence of mesoscale grains in these fibers results in a higher Seebeck coefficient than those obtained by conventional annealing, which can ultimately contribute to an enhancement of the thermoelectric properties (Lee et al. 2015).

5.8.3.6 Hydrogen Storage

Hydrogen is considered to be the most dominant alternative fuel, which has the potential to replace fossil fuels. The advantages of hydrogen as fuel are as follows: it provides green source of energy, it is renewable, it has high energy density, it is pollution free, and it is also abundant in nature. Hydrogen is also used as a catalyst, reagent, and so on. In most of the cases, the application is not fully achieved because of the lack of an efficient storage medium. There are many materials in use to store hydrogen, such as metals, intermetallic compounds, and ceramics; however, the storage capacity and the reversibility of stored hydrogen at the required environmental condition are limited (Im et al. 2008). Highly porous ceramics and carbon-based materials such as CNTs, fullerenes, and activated carbon are efficient hydrogen storage media, but the cost-effective production of such materials has not been achieved till date. Carbon nanomaterials have an extremely high surface area, which helps them adsorb hydrogen on their surfaces.

Electrospinning is a cost-effective mass production technique for the fabrication of porous, carbon-based nanofibers as well as ceramic nanofibers. Electrospinning the nanofibers with optimized structure and high porosity will enhance the adsorption rate of hydrogen (Table 5.13). Chemical activation of the electrospun nanofiber surface will control the pore size and specific surface area, which contributes to the excellent hydrogen adsorption on the nanofibers (Im et al. 2009a). It is also possible to incorporate materials, which have a high hydrogen adsorption capability with the electrospun fibers. Porous ceramic fibers for hydrogen storage can also be prepared using electrospinning, but to release the stored hydrogen for the application needs a high temperature.

5.8.3.7 Fuel Cells

Fuel cells are devices that convert chemical energy to electrical energy. Fuel cell offers a clean source of energy; it is broadly considered for transportation and portable power applications. In a fuel cell, there are three chief components, namely a cathode, an anode, and an electrolyte. Usually, the cathode and anode are made with porous carbon comprising platinum catalyst. There are different types of fuel cells named after the electrolytes used in it. Some of them are a phosphoric acid fuel cell, molten carbonate fuel cell, solid oxide fuel cell, direct methanol fuel cell, regenerative fuel cell, microbial fuel cell, alkaline fuel cell, and proton exchange membrane (PEM) fuel cell. Among these fuel cells, the proton exchange fuel cell drags more attraction because it has a high power density and low operating temperature range.

Nowadays, polymers, in particular, with acid side groups that can pass protons from anode to cathode, are used as electrolyte in many fuel cells, known as polymer electrolyte fuel cells. The key benefits of polymer electrolyte fuel cells are low operating temperature, high conversion efficiency, and high power density. In a polymer electrolyte fuel cell, there is an assembly of five membrane layers, namely a PEM, two catalyst layers, and two gas diffusion layers. Undoubtedly, the catalyst layers, their supporting substrate, and the PEM have the major role in a fuel cell.

Electrospun membranes can be used as catalysts, catalyst supporting material, and electrode and PEMs in fuel cells. Catalyst aids the reaction between fuels. Platinum nanoparticles used in commercial fuel cells have less durability. To improve the performance of

TABLE 5.13 Electrospun Fibers in Hydrogen Storage

Material	Preparation	Advantage	Reference
Activated electrospun carbon nanofibers (ACNF)	PAN dissolved into DMF is electrospun to nanosized fibers. The fibers are carbonized at a temperature of 1323 K to obtain carbon fiber. Chemical activation is done by dipping the carbon fiber to NaOH solution.	High specific surface area and pore volume increase the hydrogen storage significantly. A better storage capacity is observed than CNTs and fullerenes.	Im et al. (2009b)
	Phenol-formaldehyde polymer (PF)/PE blend is used as the precursor. Carbonization is done at 1023 K in nitrogen atmosphere. Activation is done by KOH and NaOH.	Activated CNF has high surface area than the CNF produced by CVD. High hydrogen storage is achieved by their proper packing density. Increase in storage capacity with pressure has been reported.	Suarez-Garcia et al. (2009)
PANI	Aniline monomer is mixed with camphosulfonic acid, a surfactant, and ammonium persulfate, an oxidant. The polymer solution is directly used in electrospinning to get the PANI nanofibers.	PANI fibers show a swelling effect after the absorption of hydrogen. This can lead to high hydrogen storage.	Srinivasan et al. (2010)
Palladium carbon nanofibers	PAN dissolved in DMF solution is electrospun to nanofibers. The carbonization of fiber is done at 800°C in argon atmosphere. Palladium chloride is used for coating palladium on the fiber.	The carbon fibers have a subnanoporous structure with a high surface area in the order of 815.6–1120.8 m^2/g. The hydrogen adsorption is 2.8% by weight at 77 K	Kim et al. (2011)

the commercial fuel cells, nanosized platinum fibers or nanosized catalyst particles loaded on a highly porous carbon-based substrate can be used. In both the cases, electrospinning can be applicable. Nafion membranes are used as PEMs in polymer electrolyte fuel cells because they can provide an excellent path for ion flow (Figure 5.37). The major drawback of the nafion membranes is the crossover of fuel through it, leading to the loss of fuel. The properties of the nafion membranes can be improved by incorporating electrospun nanofibers within it. Because nafion does not exhibit good electrospinning properties, it is used as a blend with other polymers by providing good mechanical strength and checks the crossover of fuels from anode to cathode, which is necessary in PEMs (Dong et al. 2011). Table 5.14 presents a few examples of electrospun materials in fuel cells.

5.8.4 Electronic Components

Nanosized fibers are the forthcoming foundation materials for the new generation electronic devices, because they can achieve a wide range of properties that are required by the electronic devices. Electrospinning can be employed in the fabrication of nanofiber-based electronic components due to its light weight, controllable sizes and morphologies, high

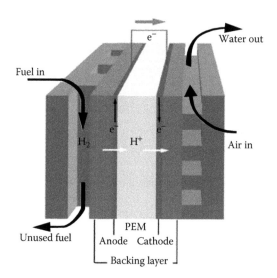

FIGURE 5.37 Diagram of PEM fuel cell. (Reprinted from *J. Power Sources,* 196, Dong, Z., Kennedy, S.J., and Wu, Y., Electrospinning materials for energy-related applications and devices, 4886–4904. Copyright 2011, with permission from Elsevier.)

specific area, and so on. Porous electrospun fibrous mats are used in electronic devices, thermal management system of electronic devices, electronic packaging, and so on. Continuity and smoothness of electrospun nanofibers along with porosity, high surface area, and mechanical properties facilitate their use in electronic applications.

The discrete passive components on printed circuit boards (PCBs) are always troublesome in microelectronics. The 90% of the production time is taken to fit the discrete passive components in microelectronics and they also occupy 40% of the PCB surface. The solder joints used to fix the discrete elements are the major sources of errors in microelectronics. The further development of microelectronics is only possible by replacing the discrete components by the embedded ones. The requirement of embedded systems in microelectronics demands the modification in such a way that the materials in use must possess manufacturability, electrical performance, reliability, and low cost. Electrospinning is a reliable method for the manufacturing of the embedded microelectronic components such as capacitors and FETs. Ferroelectric and paraelectric materials are commonly used as dielectric materials in capacitors (Carlberg et al. 2007). Light-emitting electrospun nanofibers can be fabricated using polymers such as PMMA, PS, PEO, and PVP as matrices and embedding them with light-emitting quantum dots. Another approach is blending of light-emitting polymers such as poly[2-methoxy-5-(2-ethylhexyloxy)-1,4-phenylenevinylene] (MEH-PPV), poly-3-dodecylthiophene, and poly(2,5-dialkoxy-pphenylene ethynylene) with PMMA, PS, and so on (Camposeo et al. 2013).

Miniaturization and the high power levels generate enormous heat in electronic components. Proper thermal management is always necessary to cool down such components to protect themselves from failure. The conventional cooling method includes a heat sink and a fan, but as the size of the components reduces, this method is no more effective. At this situation, nanostructured thermal interface materials come into picture as a heat

TABLE 5.14 Electrospun Materials in Fuel Cells

Electrospun Fiber	Precursor/Solvent	Remarks	Application	Reference
PtRh	$H_2PtCl_6 \cdot 6H_2O$, $RuCl_3 \cdot xH_2O$, in PVP	High mass activity	Catalyst	Dong et al. (2011)
Pt	$H_2PtCl_6 \cdot H_2O$, in PVP	High power density	Catalyst	Dong et al. (2011)
Nitrogen-doped ultrathin carbon nanofibers	Carbonizing electrospun PAN nanofibers in NH_3	High electrocatalytic activity, simple production, low cost, absence of metal catalysts, and free standing	Electrocatalysts for oxygen reduction in fuel cells	Qiu et al. (2011)
Graphene-modified carbon fiber mats	Electrospun PAN fibers dipped ia a solution of GO followed by a heat treatment	Improved catalytic activity and long-term stability of the Pt catalyst	Catalyst support membrane	Chang et al. (2011)
Carbonized nanofibers with embedded CNTs	PAN in DMF	Good fuel cell performance.	Anode in glucose fuel cells	Prilutsky et al. (2010)
Electrospun carbon fiber mat	PAN in DMF	High anodic current density	Anode in microbial fuel cells	Chen et al. (2011d)
Nafion-impregnated electrospun PVDF composite membranes	PVDF is electrospun with acetone/dimethylacetamide as solvent. Nafion is dispersed into isopropanol and added over the electrospun membrane of PVDF	PVDF prevents without compromising the ionic conductivity	Proton exchange membrane	Choi et al. (2008c)
PS	THF used as solvent	High ion exchange capacity	Ion exchanger in fuel cells	An et al. (2006)
Nafion/PVA fiber composite membrane	Aqueous PVA	Less cost and better performance than nafion	Proton exchange membrane in direct methanol fuel cells	Lin et al. (2010a)
PES-based membranes	Electrospun PES using DMF as solvent impregnated with nafion	Reduced methanol permeability	Proton exchange membranes in DMFC	Sadrabadi et al. (2011)
Sulfonated PI	DMF	High proton conductivity, good membrane stability, and low cross-over	Fuel cell electrolytes	Takemori and Kawakami (2010)

removal tool. Nanostrucured thermal interface materials have a high surface area, which can improve the contact area between the electronic device and themselves, leading to larger heat dissipation effects (Xu et al. 2008). Carbon-loaded electrospun polymer nano-fibers have been gaining attraction due to their high thermal conductivity, mechanical strength, and light weight. Nanofibers of CNT-loaded CA have a significant improvement in thermal conductivity. The extent to which CNTs are aligned in a polymer decides its thermal conductivity. The addition of CNTs increases the thermal conductivity of polymers by a factor of 30 (Datsyuk et al. 2011). Table 5.15 lists the major applications of electrospun fibers in electronics.

5.8.5 Textile Fabrics

Textile fabrics are used everywhere, starting from clothing to electronic applications. There are different possibilities of modifying conventional textiles in one way or the other. The next generation smart fabrics exhibit unique characteristics, such as hydrophobicity, selective transparency, sensing, antimicrobial properties, energy generation, and good mechanical strength. The major focus of many research organizations is on smart fabrics, and there are many smart fabrics already available in the market. The production costs of such materials are very high and the affordability of such clothes is an important concern. Electrospinning can play a major role in developing smart fabrics in an inexpensive way.

TABLE 5.15   Electrospun Fibers in Electronics

Material	Preparation	Application	Reference
PEO	PEO is electrospun using water/ethanol as solvent	Flexible electronic manufacturing	Zheng et al. (2007)
ZnO	PVAc, zinc acetate and water solution is electrospun and calcined	Field effect transistor (FET)	Wu et al. (2008a)
P3HT	P3HT is dissolved in chloroform	Field effect transistor	González and Pinto (2005)
P3HT-PS Blend	P3HT and PS are dissolved in chloroform/chlorobenzene mixture and electrospun	Field effect transistor	Hur et al. (2010)
Deoxyribonucleic acid (DNA)	Hexadecyltrimethylammonim chloride (CTMA), a surfactant which can dissolve DNA in organic solvents	Bio-FET	Bedford et al. (2008)
Barium titanate nanoparticles loaded polymer	Electrospun the solution containing, polymer, solvent, and barium titanate nanoparticles	Embedded capacitors	Carlberg et al. (2007)
PAN-derived carbon nanofiber	PAN fibers are electrostatically deposited from PAN/DMF solution and heat treated	Nanofibers with good conducting properties	Wang and Santiago-Aviles (2004)
CNT-loaded CA	CA, CNT and DMAc solution is electrospun	Thermal interface material (TIM)	Datsyuk et al. (2011)
PANI	PANI is doped with camphorsulfonic acid (HCSA) is dissolved in $CHCl_3$ along with PEO	Schottky nanodiode	Pinto et al. (2006)
$CoFe_2O_4$/PAN composite	$CoFe_2O_4$/PAN solution in DMF is electrospun	Electromagnetic interference (EMI) shielding	Chen et al. (2010)

Almost all the properties of the nanomaterials synthesized by any other method for the smart fabric can also be achieved by electospinning. There are potentially two methods to employ electrospinning in smart fabrics: one is to incorporate electrospun nanofibers to the fabric and the second is to develop the fabric from the electrospun fiber itself. Currently, electrospinning is exploited in the production of various kinds of smart fabrics. Electrospun fibers can be formed into traditional fiber yarns (Smit et al. 2005, Paneva et al. 2010), which is the raw material for the textile fabric production.

Electrospun solar cloth is composite nanofibers fabricated from P3HT (a conducting polymer)/PCBM (an electron conducting material) PCBM as the core and PVP as the shell to meet the energy requirements of personal electronic gadgets and indoor applications. The efficiency of the solar cloth is not pleasing, but the careful control of the various parameters will improve the efficiency to a required level. It has the potential application in many portable devices such as laptops, mobiles, and so on (Sundarrajan et al. 2010).

Electrospun fibers of piezoelectric and thermoelectric materials can be incorporated into fabrics to achieve sensing and actuation capabilities as discussed in the previous section. These fibers can also act as generators for low power production consistently. Electrospun PVDF fibers can be used as a nanogenerator (Chang et al. 2009) and actuators (Pu et al. 2010) for running miniaturized peltier coolers for mechanically driven cooling textiles (Fang et al. 2013). $CoFe_2O_4/BaTiO_3$ composite fibers (Zhou et al. 2010) also exhibit piezoelectric characteristics. It is also noted that these materials in their fibrous form exhibit superior properties than their film and bulk forms.

Protective clothing for laborers dealing with harmful atmosphere has been gaining lot of interest. The important feature of such a material is its impermeability to hazardous materials with comfort in hot and humid conditions and mechanical strength.

Polypropylene electrospun nanofibers can form a protective barrier against various pesticides with satisfactory air/water permeation (Lee and Obendorf 2006). Superhydrophobic nanofibers such as $PVDF/SiO_2$ composite (Wang et al. 2011) can be used in developing self-cleaning fabrics with the aspect of protection. PU nanofibers can be used to produce water proof fabric with breathability (Kang et al. 2007). A composite nanofiber containing PVA, nylon-6, PC, and PU shows excellent mechanical properties (Han et al. 2008b). PAN-derived carbon fiber yarns exhibit ultimate strength close to 1 GPa (Moon and Farris 2009). TiO_2 nanoparticle-loaded nylon-6 fibers exhibit hydrophilicity, mechanical strength, antimicrobial, and UV protecting abilities (Pant et al. 2011a). Electrospun composite nanofibers of PMMA/pyrene methanol (PM) (Wang et al. 2002b) and PAA/PM (Wang et al. 2002a) can be used in military garment for the detection of explosives such as DNT and TNT. PAN nanofibers are able to shield electromagnetic waves (Sonehara et al. 2008). PSU loaded with MgO nanofibers has the potential to protect against warfare chemicals (Sundarrajan and Ramakrishna 2007).

5.8.6 Catalysis

Catalysts in nanosize possess a high surface-to-volume ratio due to which it has a large number of favorable sites for catalyzing a reaction, leading to a significant improvement in the reaction rate, even at a miniscule quantity. However, for cost-effective production, the catalyst must be used repeatedly. The recovery and reuse of nanosized catalysts are extremely difficult, and the removal of the catalyst from the final product is always preferred; the best example is catalyst-assisted drinking water purification. To overcome such challenges, the nanosized catalysts must be immobilized so that they can be retained for the further reaction, meanwhile maintaining the threshold of the nanocatalyst in catalyzing the intended reaction. To achieve this, nanoparticle-loaded nanofibrous mats can be used. Electrospinning is the best method available for the synthesis of the nanofibrous mats. Electrospun polymer fibers or the polymer precursor-derived carbon nanofibers or ceramic fibers loaded with nanocatalyst are used in catalysis. Jia et al. (2002) reported that electrospun functionalized PS nanofibers loaded with α-chymotrypsin, a model protein, can be used as an active biocatalyst for genetic engineering applications. Nanofibrous enzymes exhibit a higher activity, three times in magnitude that of their native counterparts. Li et al. (2010) prepared nanofibers of SnO_2/Al_2O_3 catalyst for NO_x reduction. $PVP/SnCl_4 \cdot 5H_2O/AlCl_3$ electrospun composite fibers are calcined to get the resulting catalyst nanofibers. These fibers can be further doped with CeO_2 to enhance the reduction activity.

Im et al. (2008) studied the effect of metal/metal oxide catalyst-loaded electrospun-derived carbon nanofibers. The report says that the addition of catalyst nanoparticles increases the hydrogen storage capacity significantly. Copper oxide nanoparticle loading shows a higher storage capacity and pure metal fibers exhibit more significant properties than metal oxides. The catalyst-loaded electrospun nanofibers can have an outstanding role in direct methanol fuel cells. Electrospun carbon nanofibers loaded with platinum catalyst (Chang et al. 2011) have an increased fuel reduction rate leading to a high power density. Pt_xAu_{100-x} bimetallic catalyst loaded on carbon nanofibers (Huang et al. 2009) formed by electrospinning exhibited a remarkable increase in methanol reduction. The carbon nanofibers used in these applications have a high electrical conductivity; at the same time the fuel cross-overs across the electrodes are restricted. Sol–gel-assisted electrospun NiO nanofibers loaded on glass fiber mats are capable of reducing the CO and HC emissions from the exhaust of diesel engine (George and Anandhan 2014a).

Nanoporous zirconia electrospun fiber mats obtained by electrospinning (Yin et al. 2011) exhibited an improvement in photocatalysis as compared with the commercially available zirconia powder for the degradation of methyl orange in aqueous media. The increase in the catalysis can be attributed to the porous structure of the nanofibers. The electrospun fibrous membrane of PANi-PEO loaded with TiO_2, a known photocatalyst, can be used in catalytic filter membranes (Neubert et al. 2011).

5.8.7 Sensors

Sensor is a device that detects or measures a physical quantity of stimulus and converts it to a measurable form. Sensors are used in a wide range of applications such as sensing pressure, temperature, humidity, and hazardous gases, drinking water quality monitoring, weather monitoring, detection of explosives, medical diagnostics, and food inspection. There are a number of parameters that govern the quality of a sensor. A few essential sensor parameters are listed in Table 5.16.

TABLE 5.16 Sensor Parameters

Sensor Parameter	Description
Sensitivity	Sensor response with a unit of measurand.
Resolution	The smallest change in the measurand that can be detected.
Selectivity	Ability to detect the specific measurand from a mixture.
Linearity	Range over which the response is in direct proportion with the level of the measurand.
Limit of detection	Lowest value of the measurand that can be detected by the sensor.
Response time	Responding time required for a unit change in measurand.
Reversibility	Ability of the sensor to go back to the initial state after the measurement.
Reproducibility	Ability to reproduce the measurement.

The high specific surface area of nanostructured materials makes them the most conceivable candidates for sensor applications; nanowires and nanofibers are extensively used for the same. The challenging part is the fabrication of nanosized wires or fibers in a cost-effective way. Electrospinning is an economical and easy way for the fabrication of nanofibers as compared with drawing, template-based synthesis, lithography, electrochemical synthesis, and so on. Apart from their high specific surface area, electrospun nanofibers are featured with high porosity, interconnectivity, high axial strength, and flexibility, making them the right materials for sensor applications. By controlling the operating parameters of electrospinning various morphologies such as web, beaded, ribbon, hollow, multichannel tubular, multicore cable, and porous can be obtained (Ding et al. 2010). Sensors can be classified on the basis of their detection mechanisms. Electrospun nanofibers are used in sensors such as acoustic, resistive, photoelectric, optical, and amperometric. Sensors are also used in biomedical applications (Table 5.17). The applications include pathogen detection, glucose monitoring, temperature sensing, and so on.

5.8.8 Composites

The mechanical properties of continuous nanofibers differ from their counterparts with finite length and microsize. High aspect ratio (l/d), large surface area, and dispersion of nanofibers throughout the matrix make electrospun nanofibers a very good choice of fillers in composites. These nanofibers have the potential to improve the specific strength and modulus. Also, the addition of nanofibers does not alter the transparency of the matrix as the size of the nanofibers is less than the wavelength of the visible light (Huang et al. 2003).

Sun et al. (2010) used PAN–PMMA nanofibers in Bis-GMA dental restorative composite. The composite loaded with 0.6 wt.% of nanofibers showed an improvement in work of fracture by 30.4%. Chen et al. (2011a) reported that PI film reinforced with 2 wt.% CNT/PI nanofibers shows an increase in the tensile strength and elongation at break by 138% and 104%, respectively. Chen et al. (2011c) reported that the epoxy loaded with electrospun carbon nanofibrous mat shows an increase in shear strength by 86% and the out-of-plane thermal conductivity by 150%. The addition of electrospun nylon-66 nanofibers in small ratio to PE matrix improves the tensile properties of the composite tremendously (Lu et al. 2015).

TABLE 5.17　Electrospun Fiber in Sensor Applications with Different Approaches

Electrospun Fiber	Fabrication	Application	Reference
Resistive			
Measurand alters the resistance of the sensing element. The change in the resistance can occur due to the change in geometry or a change in the material.			
Lithium chloride doped TiO_2	Tetrabutyltitanate, lithium chloride, and PVP are mixed with acetic acid/ethanol. The mixture is electrospun, and the fiber is calcinated.	Humidity sensing	Li et al. (2008)
Campho sulfonic acid doped PANI/PS	Water/ethanol is used as a solvent.	Glucose sensing	Aussawasathien et al. (2005)
ZnO nanofibers	Zinc acetate and SAN is dissolved into DMSO and electrospun and the electrospun fiber are heat treated.	Ammonia sensing	Senthil and Anandhan (2014)
Tungsten oxide nanofibers	PVP and tungstic acid are dissolved into ethanol and electrospun.	Detection of NH_3 even at low concentrations	Nguyen et al. (2011)
Nickel oxide nanofibers	Poly(2-ethyl-2-oxazoline)/nickel acetate fibers are calcined.	Sensing of alcohol	George and Anandhan (2014b)
Nanofibers array of PEO, Poly(epichlorohydrin) (PECH), PVP, and poly(isobutylene) (PIB)	Electrospinning of PEO and PVP in dissolved in water, PIB is dissolved in toluene and PECH is dissolved in chloroform.	Array is sensible for volatile organic compound such as dichloropentane, methanol, toluene, and trichloroethylene	Kessick and Tepper (2006)
Nitrocellulose polymer	THF and DMF	Detection of *E. coli* bacteria.	Luo et al. (2010)
Photoelectric			
The change in the photoelectric response of the sensor with the analyte is taken into account.			
Cobalt doped ZnO nanofibers	Ethanol/water solution of PVP/zinc acetate/cobalt acetate is electrospun and annealed at 500°C.	Oxygen sensing	Yang et al. (2007)
ZnO nanowires	Gel of zinc acetate/PVA is used as pecursor. The electrospun fibers are calcined to get the ZnO fibers.	Detection of ethanol vapor	Wu et al. (2009)

(*Continued*)

TABLE 5.17 (Continued) Electrospun Fiber in Sensor Applications with Different Approaches

Electrospun Fiber	Fabrication	Application	Reference
	Optical		
It has three major components, optoelectronic system, an optical link, and a probe called optode. Optode is the sensing part and its changes in refractive index, optical absorbance, fluorescence, and intensity with respect to the analyte is measured using optoelectronic system. Electrospinning is used for the fabrication of optode.			
PAN-based nanocomposite	Nanocomposite fibers of PAN (PAN)/Fe_2O_3, PAN/ZnO, and PAN/Sb–SnO_2 were prepared using electrospinning.	PAN/Fe_2O_3 fiber shows good CO_2 sensing characteristics	Luoh and Hahn (2006)
PMMA–pyrene methanol (PM) nanofiber	The copolymer PMMA-PM is dissolved in propylene glycol/methyl ether acetate is used for electrospinning.	DNT detection.	Wang et al. (2007)
PVDF	DMF as solvent.	Temperature sensing	Morán et al. (2010)
Dimethylglyoxime/PCL nanofibers	PCL is dissolved into a mixture of DCM, and DMF; dimethylglyoxime (DMG) is added to the solution and electrospun.	Nickel ion sensing	Poltue et al. (2011)
PAM/aligned gold nanorods	Gold nanorods coated with poly(sodium 4-styrenesulfonate) is dispersed in PAM and electrospun.	Humidity sensing	Davis et al. (2014)
PAA	35 wt % solution in water is electrospun.	Humidity sensing.	Urrutia et al. (2013)
PAA–pyrene methanol(PM)	PAA-PM and cross-linkable PU latex was dissolved in DMF.	Detection if trinitro toluene and dinitro toluene	Wang et al. (2002b)
Polydiacetylene (PDA) embedded in PEO	DA monomer and PEO solution in ethanol/$CHCl_3$ was used for electrospinning.	Organic vapors and proteins sensing	Davis et al. (2014)
	Amperometric		
Sensors work on the basis of electrochemistry. Deposition of analyte on either anode or cathode changes the electrochemical properties, for example, current. The change in these properties is directly in relation with the analyte concentration.			
Hemoglobin (Hb)–collagen composite	Hemoglobin and collagen dissolved into HFIP is the precursor.	Hydrogen peroxide sensing	Guo et al. (2011)
Rhodium nanoparticle-loaded carbon nanofibers	Rhodium acetate, PAN, DMF, hydrazine and acetone were used as precursor.	Hydrazine sensing	Hu et al. (2010)
Mn_2O_3- Ag nanofibers	Prepared by calcining electrospun $Mn(NO_3)_2$–$AgNO_3$-PVP composite fibers.	Glucose monitoring	Huang et al. (2011b)
Ag/SnO_2	Ag/SnO_2 composite nanotubes were obtained by irradiating SnO_2 nanotubes in $AgNO_3$ solution with a mercury lamp.	Hydrogen peroxide sensing	Miao et al. (2013)

(Continued)

TABLE 5.17 (*Continued*) Electrospun Fiber in Sensor Applications with Different Approaches

Electrospun Fiber	Fabrication	Application	Reference
Acoustic Wave/Quartz Crystal Microbalance			
Sensing layers are deposited onto the active area of the SAW device. The change in electrical conductivity or mass of the sensing layer changes with the target materials the velocity of SAW due to mechanical and piezoelectric effects. The quartz crystal microbalance is an ultrasensitive mass sensor based on piezoelectric effect. The sensitivity depends on the absorbability of the coated material on the electrode of the sensor unit. The sensitivity of QCM gas sensors is highly increased in the presence of a coating that interacts with the target gas.			
PANI micro/nano dots	Electrospinning of Maleic acid doped PANI dissolved in NMP.	Humidity sensing at room temperature	Jaruwongrungsee et al. (2007)
PAA	Ethanol/water mixture is used as a solvent.	Water and ammonia sensing at room temperature	Ding et al. (2005)
PVP/MWCNTs	MWNT dispersed solution of PVP is made in deionized water is used for electrospinning.	Sensing of hydrogen	Chee et al. (2010)
The camphor sulfonic acid (CSA) doped PANI nanofibers	A solution containing aniline monomer, CSA, ammonium peroxydisuphate, and hexane is electrospun to obtain the nano fiber.	Humidity sensor	Wu et al. (2008b)
PLGA/Fullerene-C60 coated quartz crystal microbalance	PLGA is dissolved into DCM/THF and fullerene is dispersed into it.	Gluconic acid sensor	Şeker et al. (2010)

5.8.9 Miscellaneous Applications

There are plentiful applications of electrospinning that cannot be sorted into any of the categories we have been discussing in the previous sections and also innumerable applications of electrospinning are being explored day by day. Electrospun nanofibers can be used in specific applications such as sound absorption (Xiang et al. 2011), adsorbent for indigo carmine dye, a cancerous industrial waste contaminant, during water treatment (Teng et al. 2011), skin care products (Fan et al. 2010), and so on.

Salalha et al. (2006) used electrospinning to encapsulate biological materials, especially bacteria and viruses, to preserve their activity. To do so, at first bacteria or virus are dispersed in a dilute salt solution or LB media. This dispersion is mixed with an equal volume of 14% w/w aqueous solution of PVA and the electrospinning of this suspension is carried out. It is proved that this technique can be used for the efficient encapsulation of biological organism as well as biological materials such as DNA, proteins, and so on. Activated carbon nanofibers derived from electrospun PAN nanofibers is a strong absorbent of formaldehyde (Lee et al. 2010b), the most abundant airborne carbonyl chemical falling under the category of indoor volatile organic compounds. The abundant micropores on the surface of the activated carbon make its peculiar property suitable for an absorbent. Pant et al. (2011c) prepared electrospun composite nanofibers of TiO_2/nylon-6 containing silver nanoparticles. The antibacterial properties of silver and the photocatalytic activity of TiO_2 made it the effective filter membrane for the water purification applications.

Self-healing coatings are of substantial interest in coating systems because of their ability to repair automatically to protect the underlying substrates. Park and Braun (2009) introduced coaxially electrospun nanofibers in self-healing coating applications. Liquid materials, such as a healing agent(s) or catalyst–solvent mixture(s), are encapsulated into core–shell bead-on-string electrospun nanofibers and placed on top of the substrate. This technique is suitable for the large-scale production, and the size of the beads can be controlled by adjusting the process parameters. Presently, biologically active agents, vitamins, and so on have been added into cosmetics, which include cleansers, moisturizers, and UV-barrier creams. Vitamin C protects skin from damage by UV radiation and sunburn, photosensitive reactions, photoaging, and skin cancer, but is unstable in the presence of light, moisture, oxygen, and so on. Therefore, vitamin C loaded on silk fibers can effectively deliver the vitamins on skin at frequent intervals and at the same time it can attach to the cells to a larger extent because of its large surface area (Fan et al. 2010).

The addition of piezoelectric electrospun poly(γ-benzyl, L-glutamate) nanofibers to the mylar diaphragm of a microphone, as a transducer, improves the sensitivity of the microphone tremendously. These polymeric fibers are ideal transducing materials for high temperature and underwater applications (West et al. 2014).

ABBREVIATIONS

CA	Cellulose acetate
CNT	Carbon nanotube
DCM	Dichloromethane

DMAc	*N,N*-Dimethyl acetamide
DMF	*N,N*-Dimethylformamide
DMSO	Dimethyl sulfoxide
EC	Ethyl cellulose
EVA	Poly(ethylene-*co*-vinyl acetate)
EVOH	poly(ethylene-*co*-vinyl alcohol)
HFIP	1,1,1,3,3,3-Hexafluoro-2-propanol
MWCNT	Multiwalled carbon nanotube
NMP	*N*-Methylpyrrolidinone
P3HT	Poly(3-hexylthiophene-2,5-diyl)
PAA	Poly(acrylic acid)
PAAc	Poly(amic acid)
PAI	Poly(amide imide)
PAM	Poly(acrylamide)
PAN	Poly(acrylonitrile)
PANI	Poly(aniline)
PAN-MA	Poly(acrylonitrile-*co*-methyl acrylate)
PBI	Polybenzimidazole
PBS	Poly(1,4-butylene succinate)
PBT	Poly(butylene terephthalate)
PC	Poly(carbonate)
PCL	Poly(ε-caprolactone)
PDLLA	Poly(DL-lactide)
PEDOT	Poly(3,4-ethylenedioxythiophene)
PEG	Poly(ethylene glycol)
PEI	Poly(ethyleneimine)
PEO	Poly(ethylene oxide)
PES	Poly(ethersulfone)
PET	Poly(ethylene terephthalate)
PGA	Poly(glycolic acid)
PI	Poly(imide)
PLA	Poly(lactic acid)
PLGA	Poly(lactic-*co*-glycolic acid)
PLLA	Poly(L-lactic acid)
PMGI	Poly(methylglutarimide)
PMMA	Poly(methyl methacrylate)
PS	Poly(styrene)
PSU	Poly(sulfone)
PU	Poly(urethane)
PVA	Poly(vinyl alcohol)
PVAc	Poly(vinyl acetate)
PVC	Poly(vinyl chloride)
PVDF	Poly(vinylidene fluoride)

PVDF-HFP	Poly(vinylidenefluoride-*co*-hexafluoropropylene)
PVDF-TrFE	Poly(vinylidene fluoride-trifluoroethylene)
PVP	Poly(vinyl pyrrolidone)
SAN	Styrene-*co*-acrylonitrile polymer
TFA	Trifluoroacetic acid
TFE	2,2,2-Trifluoroethanol
THF	Tetrahydrofuran

REFERENCES

Abdelgawad, A.M., S.M. Hudson, and O.J. Rojas. 2014. Antimicrobial wound dressing nanofiber mats from multicomponent (chitosan/silver-NPs/polyvinyl alcohol) systems. *Carbohydrate Polym.* 100:166–178.

Adomavičiūtė, E. and R. Milašius. 2007. The influence of applied voltage on poly(vinyl alcohol) (PVA) nanofibre diameter. *Fibres Text. East. Eur.* 15:63.

Akhgari, A., Z. Heshmati, and B.S. Makhmalzadeh. 2013. Indomethacin electrospun nanofibers for colonic drug delivery: Preparation and characterization. *Adv. Pharm. Bull.* 3:85–90.

An, H., C. Shin, and G.G. Chase. 2006. Ion exchanger using electrospun polystyrene nanofibers. *J. Membr. Sci.* 283:84–87.

Asran, A.S., S. Henning, and G.H. Michler. 2010. Polyvinyl alcohol–collagen–hydroxyapatite bio-composite nanofibrous scaffold: Mimicking the key features of natural bone at the nanoscale level. *Polymer* 51:868–876.

Aussawasathien, D., J.-H. Dong, and L. Dai. 2005. Electrospun polymer nanofiber sensors. *Synth. Met.* 154:37–40.

Barakat, N.A.M., M.A. Kanjwal, F.A. Sheikh et al. 2009. Spider-net within the N6, PVA and PU electrospun nanofiber mats using salt addition: Novel strategy in the electrospinning process. *Polymer* 50:4389–4396.

Beachley, V. and X. Wen. 2009. Effect of electrospinning parameters on the nanofiber diameter and length. *Mater. Sci. Eng. C* 29:663–668.

Bedford, N., D. Han, and A.J. Steckl. 2008. Electrospun biopolymer-based micro/nanofibers. In *University/Government/Industry Micro/Nano Symposium, 2008. UGIM 2008. 17th Biennial*, Louisville, KY, pp. 139–141.

Bjorge, D., N. Daels, S. De Vrieze et al. 2009. Performance assessment of electrospun nanofibers for filter applications. *Desalination* 249:942–948.

Bölgen, N., İ. Vargel, P. Korkusuz et al. 2007. In vivo performance of antibiotic embedded electrospun PCL membranes for prevention of abdominal adhesions. *J. Biomed. Mater. Res., B.* 81B:530–543.

Bretcanu, O., S.K. Misra, D.M. Yunos et al. 2009. Electrospun nanofibrous biodegradable polyester coatings on Bioglass®-based glass-ceramics for tissue engineering. *Mater. Chem. Phys.* 118:420–426.

Camposeo, A., L. Persano, and D. Pisignano. 2013. Light-emitting electrospun nanofibers for nanophotonics and optoelectronics. *Macromol. Mater. Eng.* 298:487–503.

Carlberg, B., J. Norberg, and J. Liu. 2007. Electrospun nano-fibrous polymer films with barium titanate nanoparticles for embedded capacitor applications. In *Proceedings of 57th Electronic Components and Technology Conference, 2007. ECTC'07*, Reno, NV, pp. 1019–1026.

Chang, C., Y.-K. Fuh, and L. Lin. 2009. A direct-write piezoelectric PVDF nanogenerator. In *International Conference on Solid-State Sensors, Actuators and Microsystems, 2009. TRANSDUCERS 2009*, Denver, CO, pp. 1485–1488.

Chang, Y., G. Han, M. Li et al. 2011. Graphene-modified carbon fiber mats used to improve the activity and stability of Pt catalyst for methanol electrochemical oxidation. *Carbon* 49:5158–5165.

Chee, P.S., R. Arsat, X. He et al. 2010. Polyvinylpyrrolidone/multiwall carbon nanotube composite based 36°; YX LiTaO$_3$ surface acoustic wave H$_2$ gas sensor. In *2010 International Conference on Enabling Science and Nanotechnology (ESciNano)*, Kuala Lumpur, Malaysia, pp. 1–2.

Chen, D., R. Wang, W.W. Tjiu et al. 2011a. High performance polyimide composite films prepared by homogeneity reinforcement of electrospun nanofibers. *Compos. Sci. Technol.* 71:1556–1562.

Chen, I.-H., C.-C. Wang, and C.-Y. Chen. 2010a. Fabrication and characterization of magnetic cobalt ferrite/polyacrylonitrile and cobalt ferrite/carbon nanofibers by electrospinning. *Carbon* 48:604–611.

Chen, J.-P. and C.-H. Su. 2011. Surface modification of electrospun PLLA nanofibers by plasma treatment and cationized gelatin immobilization for cartilage tissue engineering. *Acta Biomater.* 7:234–243.

Chen, L.-J., J.-D. Liao, Y.-J. Chuang et al. 2011b. Polyvinylbutyral assisted synthesis and characterization of chalcopyrite quaternary semiconductor Cu(In$_x$Ga$_{1-x}$)Se$_2$ nanofibers by electrospinning route. *Polymer* 52:116–121.

Chen, M.-C., Y.-C. Sun, and Y.-H. Chen. 2013. Electrically conductive nanofibers with highly oriented structures and their potential application in skeletal muscle tissue engineering. *Acta Biomater.* 9:5562–5572.

Chen, Q., L. Zhang, A. Rahman et al. 2011c. Hybrid multi-scale epoxy composite made of conventional carbon fiber fabrics with interlaminar regions containing electrospun carbon nanofiber mats. *Composites Part A* 42:2036–2042.

Chen, S., G. He, A.A.C. Martinez et al. 2011d. Electrospun carbon fiber mat with layered architecture for anode in microbial fuel cells. *Electrochem. Commun.* 13:1026–1029.

Chen, Y., R.A. Pareta, and T.J. Webster. 2009. Self-assembling helical rosette nanotubes for cartilage tissue engineering. In *2009 IEEE 35th Annual Northeast Bioengineering Conference*, Boston, MA, pp. 1–2.

Chen, Z.G., P.W. Wang, B. Wei et al. 2010b. Electrospun collagen–chitosan nanofiber: A biomimetic extracellular matrix for endothelial cell and smooth muscle cell. *Acta Biomater.* 6:372–382.

Chiscan, O., I. Dumitru, P. Postolache et al. 2012. Electrospun PVC/Fe$_3$O$_4$ composite nanofibers for microwave absorption applications. *Mater. Lett.* 68:251–254.

Cho, T.-H., M. Tanaka, H. Onishi et al. 2008. Battery performances and thermal stability of polyacrylonitrile nano-fiber-based nonwoven separators for Li-ion battery. *J. Power Sources* 181:155–160.

Choi, J.S., S.J. Lee, G.J. Christ et al. 2008a. The influence of electrospun aligned poly(ε-caprolactone)/collagen nanofiber meshes on the formation of self-aligned skeletal muscle myotubes. *Biomaterials* 29:2899–2906.

Choi, J.S., K.W. Leong, and H.S. Yoo. 2008b. In vivo wound healing of diabetic ulcers using electrospun nanofibers immobilized with human epidermal growth factor (EGF). *Biomaterials* 29:587–596.

Choi, S.-S., Y.S. Lee, C.W. Joo et al. 2004. Electrospun PVDF nanofiber web as polymer electrolyte or separator. *Electrochim. Acta* 50:339–343.

Choi, S.W., Y.-Z. Fu, Y.R. Ahn et al. 2008c. Nafion-impregnated electrospun polyvinylidene fluoride composite membranes for direct methanol fuel cells. *J. Power Sources* 180:167–171.

Chong, E.J., T.T. Phan, I.J. Lim et al. 2007. Evaluation of electrospun PCL/gelatin nanofibrous scaffold for wound healing and layered dermal reconstitution. *Acta Biomater.* 3:321–330.

Christopherson, G.T., H. Song, and H.-Q. Mao. 2009. The influence of fiber diameter of electrospun substrates on neural stem cell differentiation and proliferation. *Biomaterials* 30:556–564.

Chuang, Y.-H., G.-B. Hong, and C.-T. Chang. 2014. Study on particulates and volatile organic compounds removal with TiO$_2$ nonwoven filter prepared by electrospinning. *J. Air Waste Manage. Assoc.* 64:738–742.

Cooley, J.F. 1902. Apparatus for electrically dispersing fluids. http://www.google.co.in/patents/US692631.

Corey, J.M., D.Y. Lin, K.B. Mycek et al. 2007. Aligned electrospun nanofibers specify the direction of dorsal root ganglia neurite growth. *J. Biomed. Mater. Res.* 83A:636–645.

Datsyuk, V., I. Firkowska, K.G. Hubmann et al. 2011. Carbon nanotubes based engineering materials for thermal management applications. In *2011 27th Annual IEEE Semiconductor Thermal Measurement and Management Symposium (SEMI-THERM)*, San Jose, CA, pp. 325–332.

Davis, B.W., A.J. Burris, N. Niamnont et al. 2014. Dual-mode optical sensing of organic vapors and proteins with polydiacetylene (PDA)-embedded electrospun nanofibers. *Langmuir* 30:9616–9622.

Deitzel, J.M., J. Kleinmeyer, D. Harris et al. 2001. The effect of processing variables on the morphology of electrospun nanofibers and textiles. *Polymer* 42:261–272.

Desai, K., K. Kit, J. Li et al. 2009. Nanofibrous chitosan non-wovens for filtration applications. *Polymer* 50:3661–3669.

Ding, B., H.-Y. Kim, S.-C. Lee et al. 2002. Preparation and characterization of a nanoscale poly(vinyl alcohol) fiber aggregate produced by an electrospinning method. *J. Polym. Sci. B Polym. Phys.* 40:1261–1268.

Ding, B., M. Wang, X. Wang et al. 2010. Electrospun nanomaterials for ultrasensitive sensors. *Mater. Today* 13:16–27.

Ding, B., M. Yamazaki, and S. Shiratori. 2005. Electrospun fibrous polyacrylic acid membrane-based gas sensors. *Sens. Actuators, B* 106:477–483.

Ding, Y., P. Zhang, Z. Long et al. 2009. The ionic conductivity and mechanical property of electrospun P(VDF-HFP)/PMMA membranes for lithium ion batteries. *J. Membr. Sci.* 329:56–59.

Dissanayake, M.A.K.L., H.K.D.W.M.N.R. Divarathne, C.A. Thotawatthage et al. 2014. Dye-sensitized solar cells based on electrospun polyacrylonitrile (PAN) nanofibre membrane gel electrolyte. *Electrochim. Acta* 130:76–81.

Dong, Z., S.J. Kennedy, and Y. Wu. 2011. Electrospinning materials for energy-related applications and devices. *J. Power Sources* 196:4886–4904.

Doshi, J. and D.H. Reneker. 1995. Electrospinning process and applications of electrospun fibers. *J. Electrostat.* 35:151–160.

Doyle, J.J., S. Choudhari, S. Ramakrishna et al. 2013. Electrospun nanomaterials: Biotechnology, food, water, environment, and energy. *Conf, Pap. Sci.* 2013:e269313.

Du, F., H. Wang, W. Zhao et al. 2012. Gradient nanofibrous chitosan/poly ε-caprolactone scaffolds as extracellular microenvironments for vascular tissue engineering. *Biomaterials* 33:762–770.

Duchoslav, J., L. Rubacek, L. Kavan et al. 2008. Electrospun TiO2 Fibers as a Material for Dye Sensitizied Solar Cells. In *Proceedings of NSTI Nanotech Conference*, Boston, MA, 88–90.

Dzenis, Y. 2004. Spinning continuous fibers for nanotechnology. *Science* 304:1917–1919.

Edmondson, D., A. Cooper, S. Jana et al. 2012. Centrifugal electrospinning of highly aligned polymer nanofibers over a large area. *J. Mater. Chem.* 22:18646–18652.

Fang, J., H. Niu, T. Lin et al. 2008. Applications of electrospun nanofibers. *Chin. Sci. Bull.* 53:2265–2286.

Fang, J., H. Niu, H. Wang et al. 2013. Enhanced mechanical energy harvesting using needleless electrospun poly(vinylidene fluoride) nanofibre webs. *Energy Environ. Sci.* 6:2196–2202.

Fan, L.-P., K.-H. Zhang, X.-Y. Sheng et al. 2010. A novel skin-care product based on silk fibroin fabricated by electrospinning. In *2010 Fourth International Conference on Bioinformatics and Biomedical Engineering (iCBBE)*, Chengdu, Sichuan, pp. 1–4.

Fong, H., I. Chun, and D.H. Reneker. 1999. Beaded nanofibers formed during electrospinning. *Polymer* 40:4585–4592.

Formhals, A. 1934. Process and apparatus for preparing artificial threads. http://www.google.co.in /patents/US1975504.

Formhals, A. 1940. Artificial thread and method of producing same. http://www.google.co.in /patents/US2187306.

Formhals, A. 1944. Method and apparatus for spinning. http://www.google.com/patents/US2349950.

Fridrikh, S.V., J.H. Yu, M.P. Brenner et al. 2003. Controlling the fiber diameter during electrospinning. *Phys. Rev. Lett.* 90:144502.

Frohbergh, M.E., A. Katsman, G.P. Botta et al. 2012. Electrospun hydroxyapatite-containing chitosan nanofibers crosslinked with genipin for bone tissue engineering. *Biomaterials* 33:9167–9178.

Gao, J. and B. Xu. 2009. Applications of nanomaterials inside cells. *Nano Today* 4:37–51.

George, G. and S. Anandhan. 2014a. Glass fiber–supported NiO nanofiber webs for reduction of CO and hydrocarbon emissions from diesel engine exhaust. *J. Mater. Res.* 29:2451–2465.

George, G. and S. Anandhan. 2014b. Synthesis and characterisation of nickel nanofibre webs with alcohol sensing characteristics. *RSC Adv.* 4:62009–62020.

Gestos, A., P.G. Whitten, G.M. Spinks et al. 2010. Crosslinking neat ultrathin films and nanofibres of pH-responsive poly(acrylic acid) by UV radiation. *Soft Matter.* 6:1045–1052.

Gibson, P., H.S. Gibson, and D. Rivin. 2001. Transport properties of porous membranes based on electrospun nanofibers. *Colloids Surf., A* 187–188:469–481.

González, R. and N.J. Pinto. 2005. Electrospun poly(3-hexylthiophene-2,5-diyl) fiber field effect transistor. *Synth. Met.* 151:275–278.

Gopalan, A.I., P. Santhosh, K.M. Manesh et al. 2008. Development of electrospun PVDF–PAN membrane-based polymer electrolytes for lithium batteries. *J. Membr. Sci.* 325:683–690.

Grätzel, M. 2003. Dye-sensitized solar cells. *J. Photochem. Photobiol., C* 4:145–153.

Guarino, V., F. Causa, P. Taddei et al. 2008. Polylactic acid fibre-reinforced polycaprolactone scaffolds for bone tissue engineering. *Biomaterials* 29:3662–3670.

Guo, F., X.X. Xu, Z.Z. Sun et al. 2011. A novel amperometric hydrogen peroxide biosensor based on electrospun Hb–collagen composite. *Colloids Surf., B* 86:140–145.

Hai-ling, H., S. Yu-qin, and S. Wen-gang. 2011. A biodegradable nanofiber by electrospinning and its cytocompatibility of polymer-coated sirolimus-eluting stents with cardiac muscle cell. In *2011 International Conference on Human Health and Biomedical Engineering (HHBE)*, Jilin, China, pp. 617–623.

Han, J., J. Zhang, R. Yin et al. 2011. Electrospinning of methoxy poly(ethylene glycol)-grafted chitosan and poly(ethylene oxide) blend aqueous solution. *Carbohydr. Polym.* 83:270–276.

Han, S.O., J.H. Youk, K.D. Min et al. 2008a. Electrospinning of cellulose acetate nanofibers using a mixed solvent of acetic acid/water: Effects of solvent composition on the fiber diameter. *Mater. Lett.* 62:759–762.

Han, X.-J., Z.-M. Huang, C.-L. He et al. 2008b. Coaxial electrospinning of PC(shell)/PU(core) composite nanofibers for textile application. *Polym. Compos.* 29:579–584.

Hardick, O., B. Stevens, and D.G. Bracewell. 2011. Nanofibre fabrication in a temperature and humidity controlled environment for improved fibre consistency. *J. Mater. Sci.* 46:3890–3898.

He, W., T. Yong, W.E. Teo et al. 2005. Fabrication and endothelialization of collagen-blended biodegradable polymer nanofibers: Potential vascular graft for blood vessel tissue engineering. *Tissue Eng.* 11:1574–1588.

Heikkilä, P. and A. Harlin. 2009. Electrospinning of polyacrylonitrile (PAN) solution: Effect of conductive additive and filler on the process. *Express Polym. Lett.* 3:437–445.

Hohman, M.M., M. Shin, G. Rutledge et al. 2001. Electrospinning and electrically forced jets. I. Stability theory. *Phys. Fluids* 13:2201–2220.

Homayoni, H., S.A.H. Ravandi, and M. Valizadeh. 2009. Electrospinning of chitosan nanofibers: Processing optimization. *Carbohydr. Polym.* 77:656–661.

Hong, K.H., J.L. Park, I.H. Sul et al. 2006. Preparation of antimicrobial poly(vinyl alcohol) nanofibers containing silver nanoparticles. *J. Polym. Sci. B Polym. Phys.* 44:2468–2474.

Huang, C., R. Chen, Q. Ke et al. 2011a. Electrospun collagen–chitosan–TPU nanofibrous scaffolds for tissue engineered tubular grafts. *Colloids Surf., B* 82:307–315.

Huang, F., Q. Wang, Q. Wei et al. 2010. Dynamic wettability and contact angles of poly (vinylidene fluoride) nanofiber membranes grafted with acrylic acid. *Express Polym. Lett.* 4:551–558.

Huang, J., H. Hou, and T. You. 2009. Highly efficient electrocatalytic oxidation of formic acid by electrospun carbon nanofiber-supported Pt_xAu_{100-x} bimetallic electrocatalyst. *Electrochem. Commun.* 11:1281–1284.

Huang, S., Y. Ding, and Y. Lei. 2011b. Enzymatic glucose sensor based on electrospun Mn_2O_3-Ag nanofibers. In *2011 IEEE 37th Annual Northeast Bioengineering Conference (NEBEC)*, Troy, NY, pp. 1–2.

Huang, Y., G. Zheng, X. Wang et al. 2011c. Fabrication of micro/nanometer-channel by near-field electrospinning. In *2011 IEEE International Conference on Nano/Micro Engineered and Molecular Systems (NEMS)*, Kaohsiung, Taiwan, pp. 877–880.

Huang, Z.-M., Y.-Z. Zhang, M. Kotaki et al. 2003. A review on polymer nanofibers by electrospinning and their applications in nanocomposites. *Compos. Sci. Technol.* 63:2223–2253.

Huang, Z.-M., Y.Z. Zhang, S. Ramakrishna et al. 2004. Electrospinning and mechanical characterization of gelatin nanofibers. *Polymer* 45:5361–5368.

Hu, G., Z. Zhou, Y. Guo et al. 2010. Electrospun rhodium nanoparticle-loaded carbon nanofibers for highly selective amperometric sensing of hydrazine. *Electrochem. Commun.* 12:422–426.

Hu, J., F. Zeng, J. Wei et al. 2014. Novel controlled drug delivery system for multiple drugs based on electrospun nanofibers containing nanomicelles. *J. Biomater. Sci., Polym. Ed.* 25:257–268.

Hur, J., S.-N. Cha, K. Im et al. 2010. P3HT-PS blend nanofiber FET based on electrospinning. In *2010 10th IEEE Conference on Nanotechnology (IEEE-NANO)*, Seoul, South Korea, pp. 533–536.

Hussain, A., G. Collins, and C.H. Cho. 2010. Electrospun chitosan-based nanofiber scaffolds for cardiac tissue engineering applications. In *Proceedings of the 2010 IEEE 36th Annual Northeast Bioengineering Conference*, New York, pp. 1–2.

Ignatious, F., L. Sun, C.-P. Lee et al. 2010. Electrospun nanofibers in oral drug delivery. *Pharm. Res.* 27:576–588.

Im, J.S., S.-J. Park, T.J. Kim et al. 2008. The study of controlling pore size on electrospun carbon nanofibers for hydrogen adsorption. *J. Colloid Interface Sci.* 318:42–49.

Im, J.S., S.-J. Park, T. Kim et al. 2009a. Hydrogen storage evaluation based on investigations of the catalytic properties of metal/metal oxides in electrospun carbon fibers. *Int. J. Hydrogen Energy* 34:3382–3388.

Im, J.S., S.-J. Park, and Y.-S. Lee. 2009b. Superior prospect of chemically activated electrospun carbon fibers for hydrogen storage. *Mater. Res. Bull.* 44:1871–1878.

Jaruwongrungsee, K., A. Tuantranont, Y. Wanna et al. 2007. Quartz crystal microbalance humidity sensor using electrospun PANI micro/nano dots. In *Seventh IEEE Conference on Nanotechnology, 2007 (IEEE-NANO 2007)*, Hong Kong, China, pp. 316–319.

Jeun, J.P., Y.M. Lim, J.H. Choi et al. 2007. Preparation of ethyl-cellulose nanofibers via an electrospinning. *Solid State Phenom.* 119:255–258.

Ji, L., Z. Lin, A.J. Medford et al. 2009. Porous carbon nanofibers from electrospun polyacrylonitrile/SiO_2 composites as an energy storage material. *Carbon* 47:3346–3354.

Ji, X., W. Yang, T. Wang et al. 2013. Coaxially electrospun core/shell structured poly(L-lactide) acid/chitosan nanofibers for potential drug carrier in tissue engineering. *J. Biomed. Nanotechnol.* 9:1672–1678.

Jia, H., G. Zhu, B. Vugrinovich et al. 2002. Enzyme-carrying polymeric nanofibers prepared via electrospinning for use as unique biocatalysts. *Biotechnol. Prog.* 18:1027–1032.

Jiang, S., H. Hou, A. Greiner et al. 2012. Tough and Transparent Nylon-6 Electrospun Nanofiber Reinforced Melamine–Formaldehyde Composites. *ACS Appl. Mater. Interfaces* 4:2597–2603.

Jiang, J., J. Xie, B. Ma et al. 2014. Mussel-inspired protein-mediated surface functionalization of electrospun nanofibers for pH-responsive drug delivery. *Acta Biomater.* 10:1324–1332.

Jin, G., M.P. Prabhakaran, and S. Ramakrishna. 2011. Stem cell differentiation to epidermal lineages on electrospun nanofibrous substrates for skin tissue engineering. *Acta Biomater.* 7:3113–3122.

Kakinoki, S., S. Uchida, T. Ehashi et al. 2011. Surface modification of poly(L-lactic acid) nanofiber with oligo(D-lactic acid) bioactive-peptide conjugates for peripheral nerve regeneration. *Polymers* 3:820–832.

Kalfa, D., A. Bel, A. Chen-Tournoux et al. 2010. A polydioxanone electrospun valved patch to replace the right ventricular outflow tract in a growing lamb model. *Biomaterials* 31:4056–4063.

Kanani, A.G. and S.H. Bahrami. 2011. Effect of changing solvents on poly(caprolactone) nanofibrous webs morphology. *J. Nanomater.* 2011:e724153.

Kang, Y.K., C.H. Park, J. Kim et al. 2007. Application of electrospun polyurethane web to breathable water-proof fabrics. *Fibers Polym.* 8:564–570.

Karatas, D., K.M. Sawicka, and S.R. Simon. 2010. Optimization of the electrospinning process parameters for a pandemic vaccine patch. In *Bioengineering Conference, Proceedings of the 2010 IEEE 36th Annual Northeast*, 1–2.

Kattamuri, N. and C. Sung. 2004. Uniform polycarbonate nanofibers produced by electrospinning. *Macromolecules* 3:425.

Kenawy, E.-R., F.I. Abdel-Hay, M.H. El-Newehy et al. 2007. Controlled release of ketoprofen from electrospun poly(vinyl alcohol) nanofibers. *Mater. Sci. Eng., A* 459:390–396.

Kenawy, E.-R., G.L. Bowlin, K. Mansfield et al. 2002. Release of tetracycline hydrochloride from electrospun poly(ethylene-*co*-vinylacetate), poly(lactic acid), and a blend. *J. Controlled Release* 81:57–64.

Kessick, R. and G. Tepper. 2006. Electrospun polymer composite fiber arrays for the detection and identification of volatile organic compounds. *Sens. Actuators, B* 117:205–210.

Khanam, N., C. Mikoryak, R.K. Draper et al. 2007. Electrospun linear polyethyleneimine scaffolds for cell growth. *Acta Biomater.* 3:1050–1059.

Kijeńska, E., M.P. Prabhakaran, W. Swieszkowski et al. 2014. Interaction of Schwann cells with laminin encapsulated PLCL core–shell nanofibers for nerve tissue engineering. *Eur. Polym. J.* 50:30–38.

Kim, C. 2005. Electrochemical characterization of electrospun activated carbon nanofibres as an electrode in supercapacitors. *J. Power Sources* 142:382–388.

Kim, C., Y.-O. Choi, W.-J. Lee et al. 2004. Supercapacitor performances of activated carbon fiber webs prepared by electrospinning of PMDA-ODA poly(amic acid) solutions. *Electrochim. Acta* 50:883–887.

Kim, H.I., R. Matsuno, J.-H. Seo et al. 2009. Preparation of electrospun poly(L-lactide-*co*-caprolactone-*co*-glycolide)/phospholipid polymer/rapamycin blended fibers for vascular application. *Curr. Appl. Phys.* 9:e249–e251.

Kim, H., D. Lee, and J. Moon. 2011. Co-electrospun Pd-coated porous carbon nanofibers for hydrogen storage applications. *Int. J. Hydrogen Energy* 36:3566–3573.

Kim, I.-D., J.-M. Hong, B.H. Lee et al. 2007. Dye-sensitized solar cells using network structure of electrospun ZnO nanofiber mats. *Appl. Phys. Lett.* 91:163109.

Kim, J.-S. and D.H. Reneker. 1999. Polybenzimidazole nanofiber produced by electrospinning. *Polym. Eng. Sci.* 39:849–854.

Kovacina, J.R., S.G. Wise, Z. Li et al. 2011. Tailoring the porosity and pore size of electrospun synthetic human elastin scaffolds for dermal tissue engineering. *Biomaterials* 32:6729–6736.

Kulkarni, A., V.A. Bambole, and P.A. Mahanwar. 2010. Electrospinning of polymers, their modeling and applications. *Polym. Plast. Technol. Eng.* 49:427–441.

Kumbar, S.G., S.P. Nukavarapu, R. James et al. 2008. Electrospun poly(lactic acid-co-glycolic acid) scaffolds for skin tissue engineering. *Biomaterials* 29:4100–4107.

Laforgue, A. 2011. All-textile flexible supercapacitors using electrospun poly(3,4-ethylenedioxythiophene) nanofibers. *J. Power Sources* 196:559–564.

Lee, B.H., M.Y. Song, S.-Y. Jang et al. 2009a. Charge transport characteristics of high efficiency dye-sensitized solar cells based on electrospun TiO$_2$ nanorod photoelectrodes. *J. Phys. Chem. C* 113:21453–21457.

Lee, D., K. Cho, J. Choi et al. 2015. Effect of mesoscale grains on thermoelectric characteristics of aligned ZnO/PVP composite nanofibers. *Mater. Lett.* 142:250–252.

Lee, J.B., S.E. Kim, D.N. Heo et al. 2010a. In vitro characterization of nanofibrous PLGA/gelatin/hydroxyapatite composite for bone tissue engineering. *Macromol. Res.* 18:1195–1202.

Lee, J.Y., C.A. Bashur, A.S. Goldstein et al. 2009b. Polypyrrole-coated electrospun PLGA nanofibers for neural tissue applications. *Biomaterials* 30:4325–4335.

Lee, K.J., N. Shiratori, G.H. Lee et al. 2010b. Activated carbon nanofiber produced from electrospun polyacrylonitrile nanofiber as a highly efficient formaldehyde adsorbent. *Carbon* 48:4248–4255.

Lee, S. and S.K. Obendorf. 2006. Developing protective textile materials as barriers to liquid penetration using melt-electrospinning. *J. Appl. Polym. Sci.* 102:3430–3437.

Lee, S.W., S.W. Choi, S.M. Jo et al. 2006. Electrochemical properties and cycle performance of electrospun poly(vinylidene fluoride)-based fibrous membrane electrolytes for Li-ion polymer battery. *J. Power Sources* 163:41–46.

Lee, Y.-S., G. Collins, and T.L. Arinzeh. 2011. Neurite extension of primary neurons on electrospun piezoelectric scaffolds. *Acta Biomater.* 7:3877–3886.

Lee, Y.-S., C. Ezebuiroh, C. Collins et al. 2009c. An electroactive conduit for spinal cord injury repair. In *2009 IEEE 35th Annual Northeast Bioengineering Conference*, Boston, MA, pp. 1–2.

Li, B., J. Zheng, and C. Xu. 2013a. Silver nanowire dopant enhancing piezoelectricity of electrospun PVDF nanofiber web. In *Fourth International Conference on Smart Materials and Nanotechnology in Engineering*. Gold Coast, Queensland, Australia, July 10, 2013 8793:879314–879314–7.

Li, J., X. Chen, N. Ai et al. 2011. Silver nanoparticle doped TiO_2 nanofiber dye sensitized solar cells. *Chem. Phys. Lett.* 514:141–145.

Li, J., E. Liu, W. Li et al. 2009. Nickel/carbon nanofibers composite electrodes as supercapacitors prepared by electrospinning. *J. Alloys Compd.* 478:371–374.

Li, Q., W. Kang, B. Cheng et al. 2010. Fabrication of the SnO_2/Al_2O_3 catalysts through electrospinning. In *2010 Third International Nanoelectronics Conference (INEC)*, Hong Kong, China, pp. 521–522.

Li, X., M.A. Kanjwal, L. Lin et al. 2013b. Electrospun polyvinyl-alcohol nanofibers as oral fast-dissolving delivery system of caffeine and riboflavin. *Colloids Surf., B* 103:182–188.

Li, Z., H. Zhang, W. Zheng et al. 2008. Highly sensitive and stable humidity nanosensors based on LiCl doped TiO_2 electrospun nanofibers. *J. Am. Chem. Soc.* 130:5036–5037.

Liao, Y., L. Zhang, Y. Gao et al. 2008. Preparation, characterization, and encapsulation/release studies of a composite nanofiber mat electrospun from an emulsion containing poly (lactic-co-glycolic acid). *Polymer* 49:5294–5299.

Lin, H.-L., S.-H. Wang, C.-K. Chiu et al. 2010a. Preparation of nafion/poly(vinyl alcohol) electrospun fiber composite membranes for direct methanol fuel cells. *J. Membr. Sci.* 365:114–122.

Lin, Z., L. Ji, M.D. Woodroof et al. 2010b. Electrodeposited MnO_x/carbon nanofiber composites for use as anode materials in rechargeable lithium-ion batteries. *J. Power Sources* 195:5025–5031.

Liu, L., W.W. Gu, W.T. Xv et al. 2009. Preparation of polyacrylamide nanofibers by electrospinning. *Adv. Mater. Res.* 87–88:433–438.

Liu, S.-J., Y.-C. Kau, C.-Y. Chou et al. 2010. Electrospun PLGA/collagen nanofibrous membrane as early-stage wound dressing. *J. Membr. Sci.* 355:53–59.

Liu, Y., J.-H. He, J. Yu et al. 2008. Controlling numbers and sizes of beads in electrospun nanofibers. *Polym. Int.* 57:632–636.

Liu, Z.H., C.T. Pan, L.W. Lin et al. 2013. Piezoelectric properties of PVDF/MWCNT Nanofiber using near-field electrospinning. *Sens. Actuators, A* 193:13–24.

Lu, B., G. Zheng, K. Dai et al. 2015. Enhanced mechanical properties of polyethylene composites with low content of electrospun nylon-66 nanofibers. *Mater. Lett.* 140:131–134.

Lu, X., L. Li, W. Zhang et al. 2005. Preparation and characterization of Ag_2S nanoparticles embedded in polymer fibre matrices by electrospinning. *Nanotechnology* 16:2233.

Luo, Y., S. Nartker, H. Miller et al. 2010. Surface functionalization of electrospun nanofibers for detecting *E. Coli* O_{157}:H_7 and BVDV cells in a direct-charge transfer biosensor. *Biosens. Bioelectron.* 26:1612–1617.

Luo, Y., S. Wang, M. Shen et al. 2013. Carbon nanotube-incorporated multilayered cellulose acetate nanofibers for tissue engineering applications. *Carbohydr. Polym.* 91:419–427.

Luoh, R. and H.T. Hahn. 2006. Electrospun nanocomposite fiber mats as gas sensors. *Compos. Sci. Technol.* 66:2436–2441.

Ma, G., D. Fang, Y. Liu et al. 2012. Electrospun sodium alginate/poly(ethylene oxide) core–shell nanofibers scaffolds potential for tissue engineering applications. *Carbohydr. Polym.* 87:737–743.

Maensiri, S. and W. Nuansing. 2006. Thermoelectric oxide $NaCo_2O_4$ nanofibers fabricated by electrospinning. *Mater. Chem. Phys.* 99:104–108.

Malakhov, S.N., A.Y. Khomenko, S.I. Belousov et al. 2009. Method of manufacturing nonwovens by electrospinning from polymer melts. *Fibre Chem.* 41:355–359.

Martinez, R.D., A.N.R. Da Silva, R. Furlan et al. 2004. Analysis of electrospinning of nanofibers as a function of polyacrylonitrile (PAN) concentration. In *International Symposium on Microelectronics Technology and Devices, Electrochemical Society*, Porto de Galinhas, Brazil, pp. 277–282.

Mathew, G., J.P. Hong, J.M. Rhee et al. 2005. Preparation and characterization of properties of electrospun poly(butylene terephthalate) nanofibers filled with carbon nanotubes. *Polym. Test.* 24:712–717.

Matthews, J.A., G.E. Wnek, D.G. Simpson et al. 2002. Electrospinning of collagen nanofibers. *Biomacromolecules* 3:232–238.

McCann, J.T., D. Li, and Y. Xia. 2005. Electrospinning of nanofibers with core-sheath, hollow, or porous structures. *J. Mater. Chem.* 15:735–738.

Meechaisue, C., P. Wutticharoenmongkol, R. Waraput et al. 2007. Preparation of electrospun silk fibroin fiber mats as bone scaffolds: A preliminary study. *Biomed. Mater.* 2:181.

Meng, Z.X., X.X. Xu, W. Zheng et al. 2011. Preparation and characterization of electrospun PLGA/gelatin nanofibers as a potential drug delivery system. *Colloids Surf., B* 84:97–102.

Miao, J., M. Miyauchi, T.J. Simmons et al. 2010. Electrospinning of nanomaterials and applications in electronic components and devices. *J. Nanosci. Nanotechnol.* 10:5507–5519.

Miao, J., R.C. Pangule, E.E. Paskaleva et al. 2011. Lysostaphin-functionalized cellulose fibers with antistaphylococcal activity for wound healing applications. *Biomaterials* 32:9557–9567.

Miao, Y.-E., S. He, Y. Zhong et al. 2013. A novel hydrogen peroxide sensor based on Ag/SnO_2 composite nanotubes by electrospinning. *Electrochim. Acta* 99:117–123.

Milleret, V., B. Simona, P. Neuenschwander et al. 2011. Tuning electrospinning parameters for production of 3D-fiber-fleeces with increased porosity for soft tissue engineering applications. *Eur. Cell. Mater.* 21:286–303.

Min, B.-M., G. Lee, S.H. Kim et al. 2004a. Electrospinning of silk fibroin nanofibers and its effect on the adhesion and spreading of normal human keratinocytes and fibroblasts in vitro. *Biomaterials* 25:1289–1297.

Min, B.-M., S.W. Lee, J.N. Lim et al. 2004b. Chitin and chitosan nanofibers: Electrospinning of chitin and deacetylation of chitin nanofibers. *Polymer* 45:7137–7142.

Mo, X.M., C.Y. Xu, M. Kotaki et al. 2004. Electrospun P(LLA-CL) nanofiber: A biomimetic extracellular matrix for smooth muscle cell and endothelial cell proliferation. *Biomaterials* 25:1883–1890.

Moon, S. and R.J. Farris. 2009. Strong electrospun nanometer-diameter polyacrylonitrile carbon fiber yarns. *Carbon* 47:2829–2839.

Morán, C.O.G., C.J. Rodriguez-Montoya, and E. Suaste-Gomez. 2010. Preparation of membranes of poly(vynilidene fluoride) as temperature sensors via electrospinning for biomedical applications. In *2010 Seventh International Conference on Electrical Engineering Computing Science and Automatic Control (CCE)*, Tuxtla Gutierrez, Mexico, pp. 261–264.

Morton, W.J. 1902. Method of dispersing fluids. http://www.google.co.in/patents/US705691.

Mukhatyar, V.J., M. Salmerón-Sánchez, S. Rudra et al. 2011. Role of fibronectin in topographical guidance of neurite extension on electrospun fibers. *Biomaterials* 32:3958–3968.

Neubert, S., D. Pliszka, V. Thavasi et al. 2011. Conductive electrospun PANi-PEO/TiO_2 fibrous membrane for photo catalysis. *Mater. Sci. Eng., B* 176:640–646.

Nguyen, T.-A., S. Park, J.B. Kim et al. 2011. Polycrystalline tungsten oxide nanofibers for gas-sensing applications. *Sens. Actuators, B* 160:549–554.

Nirmala, R., H.-M. Park, R. Navamathavan et al. 2011. Lecithin blended polyamide-6 high aspect ratio nanofiber scaffolds via electrospinning for human osteoblast cell culture. *Mater. Sci. Eng., C* 31:486–493.

Nisbet, D.R., A.E. Rodda, M.K. Horne et al. 2009. Neurite infiltration and cellular response to electrospun polycaprolactone scaffolds implanted into the brain. *Biomaterials* 30:4573–4580.

Norton, C.L. 1936. Method of and apparatus for producing fibrous or filamentary material. http://www.google.co.in/patents/US2048651.

Ojha, S.S., M. Afshari, R. Kotek et al. 2008. Morphology of electrospun nylon-6 nanofibers as a function of molecular weight and processing parameters. *J. Appl. Polym. Sci.* 108:308–319.

Orlova, Y., N. Magome, L. Liu et al. 2011. Electrospun nanofibers as a tool for architecture control in engineered cardiac tissue. *Biomaterials* 32:5615–5624.

Paneva, D., N. Manolova, I. Rashkov et al. 2010. Self-organization of fibers into yarns during electrospinning of polycation/polyanion polyelectrolyte pairs. *Digest J. Nanomater. Biostruct.* 5:811–819.

Pant, H.R., M.P. Bajgai, K.T. Nam et al. 2011a. Electrospun nylon-6 spider-net like nanofiber mat containing TiO_2 nanoparticles: A multifunctional nanocomposite textile material. *J. Hazard. Mater.* 185:124–130.

Pant, H.R., M.P. Neupane, B. Pant et al. 2011b. Fabrication of highly porous poly (ε-caprolactone) fibers for novel tissue scaffold via water-bath electrospinning. *Colloids Surf., B* 88:587–592.

Pant, H.R., D.R. Pandeya, K.T. Nam et al. 2011c. Photocatalytic and antibacterial properties of a TiO_2/nylon-6 electrospun nanocomposite mat containing silver nanoparticles. *J. Hazard. Mater.* 189:465–471.

Park, B.J., H.J. Seo, J. Kim et al. 2010. Cellular responses of vascular endothelial cells on surface modified polyurethane films grafted electrospun PLGA fiber with microwave-induced plasma at atmospheric pressure. *Surf. Coat. Technol.* 205:S222–S226.

Park, J.-H. and P.V. Braun. 2010. Coaxial electrospinning of self-healing coatings. *Adv. Mater.* 22:496–499.

Park, J.Y., I.H. Lee, and G.N. Bea. 2008. Optimization of the electrospinning conditions for preparation of nanofibers from polyvinylacetate (PVAc) in ethanol solvent. *J. Ind. Eng. Chem.* 14:707–713.

Park, S.-H., D.-H. Won, H.-J. Choi et al. 2011. Dye-sensitized solar cells based on electrospun polymer blends as electrolytes. *Sol. Energy Mater. Sol. Cells* 95:296–300.

Patlolla, A., G. Collins, and T.L. Arinzeh. 2009. A novel, composite scaffold for bone repair. In *2009 IEEE 35th Annual Northeast Bioengineering Conference*, Boston, MA, pp. 1–2.

Persano, L., C. Dagdeviren, Y. Su et al. 2013. High performance piezoelectric devices based on aligned arrays of nanofibers of poly(vinylidenefluoride-co-trifluoroethylene). *Nat. Commun.* 4:1633.

Pham, Q.P., U. Sharma, and A.G. Mikos. 2006. Electrospinning of polymeric nanofibers for tissue engineering applications: A review. *Tissue Eng.* 12:1197–1211.

Pinto, N.J., R. González, A.T.J. Jr, and A.G. MacDiarmid. 2006. Electrospun hybrid organic/inorganic semiconductor Schottky nanodiode. *Appl. Phys. Lett.* 89:033505.

Poltue, T., R. Rangkupan, S.T. Dubas et al. 2011. Nickel (II) ions sensing properties of dimethylglyoxime/poly(caprolactone) electrospun fibers. *Mater. Lett.* 65:2231–2234.

Prabhakaran, M.P., L. Ghasemi-Mobarakeh, G. Jin et al. 2011. Electrospun conducting polymer nanofibers and electrical stimulation of nerve stem cells. *J. Biosci. Bioeng.* 112:501–507.

Prabhakaran, M.P., E. Vatankhah, and S. Ramakrishna. 2013. Electrospun aligned PHBV/collagen nanofibers as substrates for nerve tissue engineering. *Biotechnol. Bioeng.* 110:2775–2784.

Prilutsky, S., P. Schechner, E. Bubis et al. 2010. Anodes for glucose fuel cells based on carbonized nanofibers with embedded carbon nanotubes. *Electrochim. Acta* 55:3694–3702.

Pu, J., X. Yan, Y. Jiang et al. 2010. Piezoelectric actuation of direct-write electrospun fibers. *Sens. Actuators, A* 164:131–136.

Qi, Y. and M.C. McAlpine. 2010. Nanotechnology-enabled flexible and biocompatible energy harvesting. *Energy Environ. Sci.* 3:1275–1285.

Qian, Y.-F., Y. Su, X.-Q. Li et al. 2010. Electrospinning of polymethyl methacrylate nanofibres in different solvents. *Iran Polym. J.* 19:123.

Qin, X.-H. and S.-Y. Wang. 2006. Filtration properties of electrospinning nanofibers. *J. Appl. Polym. Sci.* 102:1285–1290.

Qiu, Y., J. Yu, T. Shi et al. 2011. Nitrogen-doped ultrathin carbon nanofibers derived from electrospinning: Large-scale production, unique structure, and application as electrocatalysts for oxygen reduction. *J. Power Sources* 196:9862–9867.

Ravichandran, R., V. Seitz, J.R. Venugopal et al. 2013. Mimicking native extracellular matrix with phytic acid-crosslinked protein nanofibers for cardiac tissue engineering. *Macromol. Biosci.* 13:366–375.

Reneker, D.H. and I. Chun. 1996. Nanometre diameter fibres of polymer, produced by electrospinning. *Nanotechnology* 7:216.

Rockwood, D.N., R.E. Akins Jr., I.C. Parrag et al. 2008. Culture on electrospun polyurethane scaffolds decreases atrial natriuretic peptide expression by cardiomyocytes in vitro. *Biomaterials* 29:4783–4791.

Rubacek, L. and J. Duchoslav. 2008. Electrospun nanofiber layers for applications in electrochemical devices. In *Proceedings of NSTI Nanotech Conference*, Boston, MA, pp. 88–90.

Rujitanaroj, P., N. Pimpha, and P. Supaphol. 2007. Preparation of ultrafine poly(ethylene oxide)/poly(ethylene glycol) fibers containing silver nanoparticles as antibacterial coating. In *Second IEEE International Conference on Nano/Micro Engineered and Molecular Systems, 2007 (NEMS'07)*, Bangkok, Thailand, pp. 1065–1070.

Sadrabadi, M.M.H., I. Shabani, M. Soleimani et al. 2011. Novel nanofiber-based triple-layer proton exchange membranes for fuel cell applications. *J. Power Sources* 196:4599–4603.

Salalha, W., J. Kuhn, Y. Dror et al. 2006. Encapsulation of bacteria and viruses in electrospun nanofibres. *Nanotechnology* 17:4675.

Salata, O. 2005. Tools of nanotechnology: Electrospray. *Curr. Nanosci.* 1:25–33.

Sang, Y., F. Li, Q. Gu et al. 2008. Heavy metal-contaminated groundwater treatment by a novel nanofiber membrane. *Desalination* 223:349–360.

Schneider, A., X.Y. Wang, D.L. Kaplan et al. 2009. Biofunctionalized electrospun silk mats as a topical bioactive dressing for accelerated wound healing. *Acta Biomater.* 5:2570–2578.

Schnell, E., K. Klinkhammer, S. Balzer et al. 2007. Guidance of glial cell migration and axonal growth on electrospun nanofibers of poly-ε-caprolactone and a collagen/poly-ε-caprolactone blend. *Biomaterials* 28:3012–3025.

Schofer, M.D., P.P. Roessler, J. Schaefer et al. 2011. Electrospun PLLA nanofiber scaffolds and their use in combination with BMP-2 for reconstruction of bone defects. *PLoS ONE* 6:e25462.

Şeker, S., Y.E. Arslan, and Y.M. Elçin. 2010. Electrospun nanofibrous PLGA/fullerene-C60 coated quartz crystal microbalance for real-time gluconic acid monitoring. *IEEE Sens. J.* 10:1342–1348.

Senthil, T. and S. Anandhan. 2014. Structure–property relationship of sol–gel electrospun ZnO nanofibers developed for ammonia gas sensing. *J. Colloid Interface Sci.* 432:285–296.

Senthil, T., G. George, and S. Anandhan. 2013. Chemical-resistant ultrafine poly(styrene-co-acrylonitrile) fibers by electrospinning: Process optimization by design of experiment. *Polym. Plast. Technol. Eng.* 52:407–421.

Seo, M.-K. and S.-J. Park. 2009. Electrochemical characteristics of activated carbon nanofiber electrodes for supercapacitors. *Mater. Sci. Eng., B* 164:106–111.

Shalumon, K.T., S. Sowmya, D. Sathish et al. 2013. Effect of incorporation of nanoscale bioactive glass and hydroxyapatite in PCL/chitosan nanofibers for bone and periodontal tissue engineering. *J. Biomed. Nanotechnol.* 9:430–440.

Shawki, M., A. Hereba, and A. Ghazal. 2010. Formation and characterisation of antimicrobial dextran nanofibers. *Rom. J. Biophys.* 20:335–346.

Shin, M., O. Ishii, T. Sueda et al. 2004. Contractile cardiac grafts using a novel nanofibrous mesh. *Biomaterials* 25:3717–3723.

Shin, Y.M., M.M. Hohman, M.P. Brenner et al. 2001. Experimental characterization of electrospinning: The electrically forced jet and instabilities. *Polymer* 42:09955–09967.

Simons, H.L. 1968. Patterned non-woven fabrics comprising electrically-spun polymeric filaments. http://www.google.com/patents/US3413182.

Singh, G., D. Rana, T. Matsuura et al. 2010. Removal of disinfection byproducts from water by carbonized electrospun nanofibrous membranes. *Sep. Purif. Technol.* 74:202–212.

Smit, E., U. Büttner, and R.D. Sanderson. 2005. Continuous yarns from electrospun fibers. *Polymer* 46:2419–2423.

Soliman, S., S. Pagliari, A. Rinaldi et al. 2010. Multiscale three-dimensional scaffolds for soft tissue engineering via multimodal electrospinning. *Acta Biomater.* 6:1227–1237.

Sonehara, M., T. Sato, M. Takasaki, H. Konishi, K. Yamasawa, and Y. Miura. 2008. Preparation and characterization of nanofiber nonwoven textile for electromagnetic wave shielding. *IEEE Trans. Magn.* 44:3107–3110.

Srinivasan, S.S., R. Ratnadurai, M.U. Niemann et al. 2010. Reversible hydrogen storage in electrospun polyaniline fibers. *Int. J. Hydrogen Energy* 35:225–230.

Suarez-Garcia, F., E. Vilaplana-Ortego, M. Kunowsky et al. 2009. Activation of polymer blend carbon nanofibres by alkaline hydroxides and their hydrogen storage performances. *Int. J. Hydrogen Energy* 34:9141–9150.

Subbiah, T., G.S. Bhat, R.W. Tock et al. 2005. Electrospinning of nanofibers. *J. Appl. Polym. Sci.* 96:557–569.

Sun, D., C. Chang, S. Li et al. 2006. Near-field electrospinning. *Nano Lett.* 6:839–842.

Sun, W., Q. Cai, P. Li et al. 2010. Post-draw PAN–PMMA nanofiber reinforced and toughened bis-GMA dental restorative composite. *Dent. Mater.* 26:873–880.

Sundarrajan, S., R. Murugan, A.S. Nair et al. 2010. Fabrication of P3HT/PCBM solar cloth by electrospinning technique. *Mater. Lett.* 64:2369–2372.

Sundarrajan, S. and S. Ramakrishna. 2007. Fabrication of nanocomposite membranes from nanofibers and nanoparticles for protection against chemical warfare stimulants. *J. Mater. Sci.* 42:8400–8407.

Sundaray, B., V. Subramanian, T.S. Natarajan et al. 2004. Electrospinning of Continuous Aligned Polymer Fibers. *Appl. Phys. Lett.* 84:1222–1224.

Sutthiphong, S., P. Pavasant, and P. Supaphol. 2009. Electrospun 1,6-diisocyanatohexane-extended poly(1,4-butylene succinate) fiber mats and their potential for use as bone scaffolds. *Polymer* 50:1548–1558.

Taepaiboon, P., U. Rungsardthong, and P. Supaphol. 2006. Drug-loaded electrospun mats of poly(vinyl alcohol) fibres and their release characteristics of four model drugs. *Nanotechnology* 17:2317.

Takemori, R. and H. Kawakami. 2010. Electrospun nanofibrous blend membranes for fuel cell electrolytes. *J. Power Sources* 195:5957–5961.

Tao, J. and S. Shivkumar. 2007. Molecular weight dependent structural regimes during the electrospinning of PVA. *Mater. Lett.* 61:2325–2328.

Teng, M., F. Li, B. Zhang et al. 2011. Electrospun cyclodextrin-functionalized mesoporous polyvinyl alcohol/SiO$_2$ nanofiber membranes as a highly efficient adsorbent for indigo carmine dye. *Colloids Surf., A* 385:229–234.

Teo, W.-E., R. Inai, and S. Ramakrishna. 2011. Technological advances in electrospinning of nanofibers. *Sci. Technol. Adv. Mater.* 12:013002.

Theisen, C., S. Fuchs-Winkelmann, K. Knappstein et al. 2010. Influence of nanofibers on growth and gene expression of human tendon derived fibroblast. *Biomed. Eng. Online* 9:9.

Tong, H.-W. and M. Wang. 2010. Electrospinning of fibrous polymer scaffolds using positive voltage or negative voltage: A comparative study. *Biomed. Mater.* 5:054110.

Tsai, R.-Y., T.-Y. Kuo, S.-C. Hung et al. 2015. Use of gum arabic to improve the fabrication of chitosan-gelatin-based nanofibers for tissue engineering. *Carbohydr. Polym.* 115:525–532.

Tungprapa, S., I. Jangchud, and P. Supaphol. 2007. Release characteristics of four model drugs from drug-loaded electrospun cellulose acetate fiber mats. *Polymer* 48:5030–5041.

Urrutia, A., J. Goicoechea, P.J. Rivero et al. 2013. Electrospun nanofiber mats for evanescent optical fiber sensors. *Sens. Actuators, B* 176:569–576.

Uyar, T., R. Havelund, Y. Nur et al. 2009. Molecular filters based on cyclodextrin functionalized electrospun fibers. *J. Membr. Sci.* 332:129–137.

Uyar, T., R. Havelund, Y. Nur et al. 2010. Cyclodextrin functionalized poly(methyl methacrylate) (PMMA) electrospun nanofibers for organic vapors waste treatment. *J. Membr. Sci.* 365:409–417.

Vargas, E.A.T., N.C. do Vale Baracho, J. de Brito et al. 2010. Hyperbranched polyglycerol electrospun nanofibers for wound dressing applications. *Acta Biomater.* 6:1069–1078.

Vatankhah, E., M.P. Prabhakaran, D. Semnani et al. 2014a. Electrospun tecophilic/gelatin nanofibers with potential for small diameter blood vessel tissue engineering. *Biopolymers* 101:1165–1180.

Vatankhah, E., M.P. Prabhakaran, D. Semnani et al. 2014b. Phenotypic modulation of smooth muscle cells by chemical and mechanical cues of electrospun tecophilic/gelatin nanofibers. *ACS Appl. Mater. Interfaces* 6:4089–4101.

Vaz, C.M., S. van Tuijl, C.V.C. Bouten et al. 2005. Design of scaffolds for blood vessel tissue engineering using a multi-layering electrospinning technique. *Acta Biomater.* 1:575–582.

Veleirinho, B. and J.A. Lopes-da-Silva. 2009. Application of electrospun poly(ethylene terephthalate) nanofiber mat to apple juice clarification. *Process Biochem.* 44:353–356.

Wang, H., G. Zheng, and D. Sun. 2007. Electrospun nanofibrous membrane for air filtration. In *Seventh IEEE Conference on Nanotechnology, 2007 (IEEE-NANO 2007)*, Hong Kong, China, pp. 1244–1247.

Wang, H.B., M.E. Mullins, J.M. Cregg et al. 2009. Creation of highly aligned electrospun poly-L-lactic acid fibers for nerve regeneration applications. *J. Neural Eng.* 6:016001.

Wang, L., Y. Yu, P.-C. Chen et al. 2008a. Electrospun carbon–cobalt composite nanofiber as an anode material for lithium ion batteries. *Scr. Mater.* 58:405–408.

Wang, L., Y. Yu, P.C. Chen et al. 2008b. Electrospinning synthesis of C/Fe_3O_4 composite nanofibers and their application for high performance lithium-ion batteries. *J. Power Sources* 183:717–723.

Wang, Q., Q. Yao, J. Chang et al. 2012. Enhanced thermoelectric properties of CNT/PANI composite nanofibers by highly orienting the arrangement of polymer chains. *J. Mater. Chem.* 22:17612–17618.

Wang, S., Y. Li, X. Fei et al. 2011. Preparation of a durable superhydrophobic membrane by electrospinning poly (vinylidene fluoride) (PVDF) mixed with epoxy–siloxane modified SiO_2 nanoparticles: A possible route to superhydrophobic surfaces with low water sliding angle and high water contact angle. *J. Colloid Interface Sci.* 359:380–388.

Wang, X., C. Drew, S.-H. Lee et al. 2002a. Electrospun nanofibrous membranes for highly sensitive optical sensors. *Nano Lett.* 2:1273–1275.

Wang, X., S.-H. Lee, B.-C. Ku et al. 2002b. Synthesis and electrospinning of a novel fluorescent polymer PMMA-PM for quenching-based optical sensing. *J. Macromol. Sci. Part A Pure Appl. Chem.* 39:1241–1249.

Wang, Y., D.J. Blasioli, H.-J. Kim et al. 2006. Cartilage tissue engineering with silk scaffolds and human articular chondrocytes. *Biomaterials* 27:4434–4442.

Wang, Y. and J.J. Santiago-Aviles. 2004. Low-temperature electronic properties of electrospun PAN-derived carbon nanofiber. *IEEE Trans. Nanotechnol.* 3:221–224.

Wan, Y.-Q., J.-H. He, J.-Y. Yu et al. 2007. Electrospinning of high-molecule PEO solution. *J. Appl. Polym. Sci.* 103:3840–3843.

West, J.E., K. Ren, and M. Yu. 2014. Planar microphone based on piezoelectric electrospun poly(γ-benzyl-α,L-glutamate) nanofibers. *J. Acoust. Soc. Am.* 135:EL291.

Wu, H., L. Hu, M.W. Rowell et al. 2010. Electrospun metal nanofiber webs as high-performance transparent electrode. *Nano Lett.* 10:4242–4248.

Wu, H., D. Lin, R. Zhang et al. 2008a. ZnO nanofiber field-effect transistor assembled by electrospinning. *J. Am. Ceram. Soc.* 91:656–659.

Wu, T.-T., Y.-Y. Chen, and T.-H. Chou. 2008b. A high sensitivity nanomaterial based SAW humidity sensor. *J. Phys. D: Appl. Phys.* 41:085101.

Wu, W.-Y., J.-M. Ting, and P.-J. Huang. 2009. Electrospun ZnO nanowires as gas sensors for ethanol detection. *Nanoscale Res. Lett.* 4:513.

Wulkersdorfer, B., K.K. Kao, V.G. Agopian et al. 2010. Bimodal porous scaffolds by sequential electrospinning of poly(glycolic acid) with sucrose particles. *Int. J. Polym. Sci.* 2010:e436178.

Xiang, H., L. Zhang, Z. Wang et al. 2011. Multifunctional polymethylsilsesquioxane (PMSQ) surfaces prepared by electrospinning at the sol–gel transition: Superhydrophobicity, excellent solvent resistance, thermal stability and enhanced sound absorption property. *J. Colloid Interface Sci.* 359:296–303.

Xiao, S., H. Ma, M. Shen et al. 2011. Excellent copper(II) removal using zero-valent iron nanoparticle-immobilized hybrid electrospun polymer nanofibrous mats. *Colloids Surf., A* 381:48–54.

Xu, C., F. Xu, B. Wang et al. 2011. Electrospinning of poly(ethylene-*co*-vinyl alcohol) nanofibres encapsulated with Ag nanoparticles for skin wound healing. *J. Nanomater.* 2011:e201834.

Xu, L., C. Yue, J. Liu et al. 2008. Nano-thermal interface material with CNT nano-particles for heat dissipation application. In *International Conference on Electronic Packaging Technology High Density Packaging, 2008 (ICEPT-HDP 2008)*, Shanghai, China, pp. 1–4.

Yan, H., L. Liu, and Z. Zhang. 2011. Continually fabricating staple yarns with aligned electrospun polyacrylonitrile nanofibers. *Mater. Lett.* 65:2419–2421.

Yang, M., T. Xie, L. Peng et al. 2007. Fabrication and photoelectric oxygen sensing characteristics of electrospun Co doped ZnO nanofibres. *Appl. Phys. A* 89:427–430.

Yang, Q., Z. Li, Y. Hong et al. 2004a. Influence of solvents on the formation of ultrathin uniform poly(vinyl pyrrolidone) nanofibers with electrospinning. *J. Polym. Sci. B Polym. Phys.* 42:3721–3726.

Yang, X., C. Shao, H. Guan et al. 2004b. Preparation and characterization of ZnO nanofibers by using electrospun PVA/zinc acetate composite fiber as precursor. *Inorg. Chem. Commun.* 7:176–178.

Yang, Y., T. Xia, W. Zhi et al. 2011. Promotion of skin regeneration in diabetic rats by electrospun core-sheath fibers loaded with basic fibroblast growth factor. *Biomaterials* 32:4243–4254.

Yao, C., X. Li, and T. Song. 2007. Electrospinning and crosslinking of zein nanofiber mats. *J. Appl. Polym. Sci.* 103:380–385.

Yao, Y., P. Zhu, H. Ye et al. 2006. Polysulfone nanofibers prepared by electrospinning and gas/jet-electrospinning. *Front. Chem. China* 1:334–339.

Yarin, A.L., S. Koombhongse, and D.H. Reneker. 2001. Taylor cone and jetting from liquid droplets in electrospinning of nanofibers. *J. Appl. Phys.* 90:4836–4846.

Yarin, A.L., E. Zussman, J.H. Wendorff et al. 2007. Material encapsulation and transport in core–shell micro/nanofibers, polymer and carbon nanotubes and micro/nanochannels. *J. Mater. Chem.* 17:2585–2599.

Yin, L., J. Niu, Z. Shen et al. 2011. Preparation and photocatalytic activity of nanoporous zirconia electrospun fiber mats. *Mater. Lett.* 65:3131–3133.

Yin, Z., X. Chen, J.L. Chen et al. 2010. The regulation of tendon stem cell differentiation by the alignment of nanofibers. *Biomaterials* 31:2163–2175.

Yoon, K., B.S. Hsiao, and B. Chu. 2009. Formation of functional polyethersulfone electrospun membrane for water purification by mixed solvent and oxidation processes. *Polymer* 50:2893–2899.

Yoshimoto, H., Y.M. Shin, H. Terai et al. 2003. A biodegradable nanofiber scaffold by electrospinning and its potential for bone tissue engineering. *Biomaterials* 24:2077–2082.

Yun, K.M., C.J. Hogan Jr., Y. Matsubayashi et al. 2007. Nanoparticle filtration by electrospun polymer fibers. *Chem. Eng. Sci.* 62:4751–4759.

Zahedi, P., I. Rezaeian, S.-O. Ranaei-Siadat et al. 2010. A review on wound dressings with an emphasis on electrospun nanofibrous polymeric bandages. *Polym. Adv. Technol.* 21:77–95.

Zamarripa, N., S. Farboodmanesh, and C.K. Kuo. 2009. Novel biomimetic scaffold for tendon and ligament tissue engineering. In *2009 IEEE 35th Annual Northeast Bioengineering Conference*, Boston, MA, pp. 1–2.

Zeng, X., Y. Ou, Y. Yang et al. 2014. Janus-type thermoelectric Co_3O_4/TiO_2 nanofibers by electrospinning. *Nanosci. Nanotechnol. Lett.* 6:1075–1078.

Zhang, F., R. Liu, B.Q. Zuo et al. 2010. Electrospun silk fibroin nanofiber tubes for peripheral nerve regeneration. In *2010 Fourth International Conference on Bioinformatics and Biomedical Engineering (iCBBE)*, Chengdu, Sichuan, pp. 1–4.

Zhang, M., Z. Wang, and Y. Zhang. 2006. Preparation of polyimide nanofibers by electrospinning. In *The 2006 International Conference on MEMS, NANO and Smart Systems*, Cairo, Egypt, pp. 58–60.

Zhao, Y., X. Cao, and L. Jiang. 2007. Bio-mimic multichannel microtubes by a facile method. *J. Am. Chem. Soc.* 129:764–765.

Zhao, R., X. Li, B. Sun et al. 2014. Electrospun chitosan/sericin composite nanofibers with antibacterial property as potential wound dressings. *Int. J. Biol. Macromol.* 68:92–97.

Zheng, G., Y. Dai, L. Wang et al. 2007. Direct-write micro/nano-structure for flexible electronic manufacturing. In *Seventh IEEE Conference on Nanotechnology, 2007 (IEEE-NANO 2007)*, Hong Kong, China, pp. 791–794.

Zhong, Z., Q. Cao, B. Jing et al. 2012. Electrospun PVDF–PVC nanofibrous polymer electrolytes for polymer lithium-ion batteries. *Mater. Sci. Eng., B* 177:86–91.

Zhou, Z., X. Shen, F. Song et al. 2010. Structures and magnetic properties of nanocomposite $CoFe_2O_4$-$BaTiO_3$ fibers by organic gel-thermal decomposition process. *J. Cent. South Univ. Technol.* 17:1172–1176.

Zhuo, H., J. Hu, S. Chen et al. 2008. Preparation of polyurethane nanofibers by electrospinning. *J. Appl. Polym. Sci.* 109:406–411.

Zong, X., H. Bien, C.-Y. Chung et al. 2005. Electrospun fine-textured scaffolds for heart tissue constructs. *Biomaterials* 26:5330–5338.

CHAPTER 6

Electrospun PVDF-TrFE-Based Piezoelectric Sensors

D. Manjula Dhevi, Yu Jin Ahn, Kap Jin Kim, and A. Anand Prabu

CONTENTS

I̲N THIS CHAPTER, OUR aim is to explore the latest developments that use electrospun poly(vinylidene fluoride-trifluoroethylene) (PVDF-TrFE) nanoweb fibers in fabrication of stretchable/flexible piezoelectric sensors. The discussion will focus on the electrospinning parameters, morphological, spectral, and electrical properties of PVDF-TrFE nanoweb fiber and its applications as piezoelectric sensors. Moreover, some of the difficulties faced by electrospun fibers such as precise deposition, large-scale production, flexibility improvement, and subsequent integration into flexible/stretchable piezoelectric sensors are also discussed.

6.1 INTRODUCTION TO PIEZOELECTRICITY

Jacques and Pierre Currie discovered the piezoelectric effect in 1880 by some simple asymmetric crystals, and since then, this phenomenon has raised a lot of interest. The term *Piezoelectricity* is derived from Greek word *piezo* meaning pressure. It is the ability of some crystals and certain ceramics to generate an electrical potential under applied mechanical stress (direct piezoelectric effect). Conversely, when these materials are applied with an electric potential difference, mechanical deformation occurs (indirect or converse piezoelectric effect). This could result in the electrical charge separation across the crystal lattice, and if not shorted across the crystal lattice, the applied charge induces a voltage across the material (Kazmierski and Beeby 2011). By applying an external electric field upon a ferroelectric material such as lead zirconate titanate, its internal dipoles can be reoriented, thereby leaving a remnant polarization (P_r) at zero applied electric field. Moreover, with applied stress, there is a change in P_r, which leads to the understanding that every ferroelectric material is piezoelectric, but not every piezoelectric material is ferroelectric (Ramadan et al. 2014). Piezoelectric materials such as quartz, tourmaline, and ceramic are used for their various advantages, such as small size and light weight, broad frequency, wide dynamic range and temperature, ultra-low noise, simple signal conditioning, and cost-effective test implementation. Common uses for piezoelectric sensors include modal analysis, predictive/preventative maintenance, environmental stress screening, and health and usage monitoring systems.

Among the many different piezoelectric sensor designs, the common modes in use nowadays are (a) charge mode and (b) low impedance voltage mode (LIVM). Charge mode sensors are manufactured using ceramic and crystalline quartz piezoelectric elements and exhibit high charge output and internal capacitance, good stability, and relatively lower insulation resistance. Figure 6.1 illustrates the conventional charge mode accelerometer system.

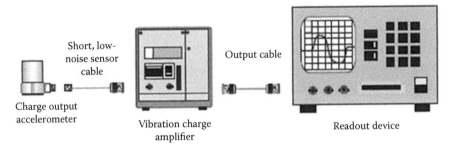

FIGURE 6.1   Conventional charge mode accelerometer system. (From www.pcb.com.)

The advantages of charge mode piezoelectric sensors are as follows:

1. The transfer function of the charge amplifier is dependent upon the feedback capacitor (C_f) value and is independent of input capacitance.

2. The charge mode sensors exhibit an upper temperature limit higher than the 121°C because no electronic components are housed within the sensor. Rather, it is only dependent on the Curie temperature (T_c) of the piezoelectric material or by the insulating material properties employed in that design.

3. System sensitivity is unaffected by changes in input cable length or type, which comes into play when interchanging the cables.

On the other hand, a LIVM transducer utilizes quartz/ceramic and has many features such as built-in electronics, low-cost signal conditioning, and stability over broad temperature range. LIVM accelerometer utilizes the voltage signal generated by the quartz element rather than the charge signal, which is the case in charge mode accelerometer. Figure 6.2 shows the schematic representation of the LIVM system.

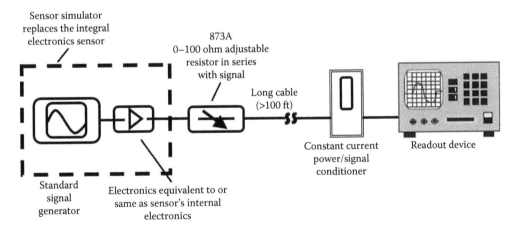

FIGURE 6.2   The LIVM system. (From www.pcb.com.)

The advantages of LIVM piezoelectric sensors are as follows:

1. LIVM sensor is independent of cable length because of its low output impedance (<100 ohms) within the limits of frequency response. Therefore, LIVM sensors can be driven using thousands of feet long cables.

2. Usage of expensive low noise cable is avoided and instead, inexpensive coaxial cable or twin-lead ribbon cable can be used to connect between the sensor and the power unit.

3. Sealed rugged construction, with high impedance connections contained within the sensor housing, makes LIVM sensors ideal for field use in dirty or moist environments.

4. At the time of assembly, sensitivity, and discharge time constant are fixed, which makes LIVM sensors ideal for dedicated healthcare monitoring applications (www.dytran.com).

6.2 PIEZOELECTRIC TRANSDUCERS

A piezoelectric transducer is a self-generating active sensor that does not require the assistance of an external power source. Figure 6.3a shows a typical piezoelectric transducer with a piezoelectric crystal inserted between the force summing member and a solid base. When applied with force on the pressure port, the exact force will fall on the force summing member, thereby generating a potential difference on the piezoelectric crystal, and the generated voltage will be proportional to the magnitude of applied force. It is essential to understand the basics of piezoelectric effect before going into details about the transducer (www.instrumentationtoday.com). There are only three or four piezoelectric coefficient elements in most of the piezoelectric materials, and d_{31} and d_{33} are the two common elements

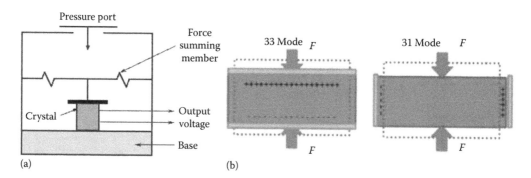

FIGURE 6.3 (a) Typical piezoelectric transducer. (From www.instrumentationtoday.com.) (b) Piezoelectric transduction modes, d_{33} and d_{31}. (Reprinted from Figure 1, Ramadan, K.S. et al., A review of piezoelectric polymers as functional materials for electromechanical transducers, *Smart Mater. Struct.*, 23, 033001, 2014, with permission from IOP Publishing Ltd.)

among them. The d_{31} coefficient, also known as the *transverse coefficient*, describes the electric polarization generated in a direction perpendicular to the direction of the applied stress, whereas the d_{33} coefficient, also called as the *longitudinal coefficient*, describes the electric polarization generated in the direction parallel to the applied stress direction. The 31 and 33 modes are commonly used in microelectromechanical systems (MEMS) to distinguish the two piezoelectric transduction mechanisms as illustrated graphically in Figure 6.3b (Kubba and Jiang 2014, Ramadan et al. 2014).

6.3 PIEZOELECTRIC POLYMERIC MATERIALS

When polymer films are subjected to an intense electric field, they exhibit piezoelectricity resulting from the orientation of their molecular dipoles along the electric field direction, also termed as *polarization* (Furukawa 1989). This polarization mainly arise from the spatial arrangement of macromolecular chains segments compared to the contribution of the injected charges. The piezoelectric activity in polymeric films is determined by the proportionality coefficients between the electrical effects and mechanical causes. The coefficient d is obtained by measuring the charge density (coulomb m^{-2}) appearing on the film surfaces (direction 3 = thickness) when a mechanical stress of 1 N m^{-2} is applied. Coefficient g is obtained if the electric field variation is measured per unit stress. These are connected with coefficient d by the correlation $g = d/\varepsilon$, where ε is the dielectric constant depending on the thickness of the film. Coefficient g is expressed in V m N^{-1}. Constants g and d are most widely used for the design of electromechanical transducers (www.piezotech.fr). The conversion of mechanical energy into electrical energy is represented by the electromechanical coupling coefficient K_T expressed in % as

$$K_T = \sqrt{\frac{\text{Converted energy}}{\text{Input energy}}} \qquad (6.1)$$

Figure 6.4 shows a graphical representation of the piezoelectric polymer types: (1) bulk piezo-polymers, further divided into amorphous dipolar and semicrystalline polymers, (2) piezoelectric composite polymer, and (3) voided charged polymer. Bulk piezo-polymers are solid polymer films and derive its piezoelectric mechanism through molecular arrangement. Piezoelectric composite polymers are embedded with piezoelectric particles, and combine the polymer's mechanical flexibility and the piezoelectric ceramic's high electromechanical coupling effect. Voided charged polymers are radically different from the first two categories such that gas voids are introduced in the polymer film and its surfaces are charged to form internal dipoles whose polarization is changed with the applied stress on the polymer film.

6.3.1 Bulk Piezoelectric Polymers

Ramadan et al. (2014) presented a detailed review of the theory and piezoelectric properties of bulk piezoelectric polymers. The requirements for a bulk polymer material to be piezoelectric are as follows: (1) the polymer's molecular structure should contain inherent

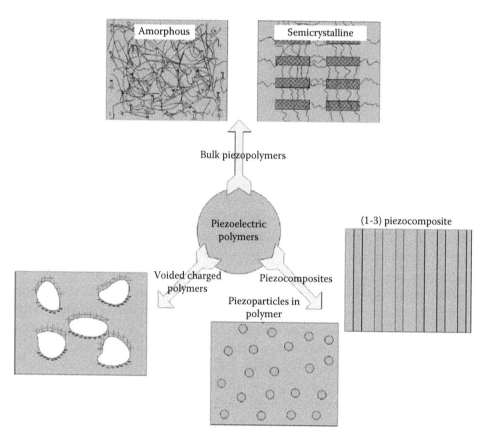

FIGURE 6.4 Schematic diagrams of piezoelectric polymer types: bulk piezo-polymers, piezoelectric/polymer composites, and voided charged polymers. (Reprinted from Figure 2, Ramadan, K.S. et al., A review of piezoelectric polymers as functional materials for electro-mechanical transducers, *Smart Mater. Struct.*, 23, 033001, 2014, with permission from IOP Publishing Ltd.)

molecular dipoles and (2) these dipoles can be reoriented through poling and kept in their preferred orientation state within the bulk material.

6.3.1.1 Semicrystalline Piezoelectric Polymers

This category includes PVDF, polyamides, liquid crystal polymers, and Parylene C. The polar groups (+ve and −ve charged ions) in this category are arranged in a crystalline form and cause polarization change with applied stress. The semicrystalline polymer's bulk structure is not a single crystal, but possesses randomly oriented microcrystals dispensed within an amorphous bulk. Poling is the effective way to get a piezoelectric response out of these polymers.

6.3.1.2 Amorphous Piezoelectric Polymers

Polyimide and polyvinylidene chloride belongs to this category because they exhibit non-crystalline behavior, even though their molecular structure contains dipoles. However, these dipoles can be aligned along the applied electric field by poling at a temperature few degrees above their T_g. After the poling process, P_r is linearly dependent on the poling electric field (E_p), and the resultant piezoelectric coefficient d_{31} is determined using the equation (Harrison and Ounaies 2002):

$$d_{31} = P_r(1-v)Y\left(\frac{\varepsilon_\infty}{3} + \frac{2}{3}\right) \qquad (6.2)$$

$$P_r = \Delta\varepsilon\varepsilon_0 E_p \qquad (6.3)$$

where:
   v is the mechanical Poisson's ratio
   Y is the Young's modulus
   ε_∞ is the material's permittivity at high frequencies
   The dielectric relaxation strength of a polymer $\Delta\varepsilon$ is equal to the change in the polymer's
       dielectric constant when heated from a temperature below T_g to a temperature
       higher than T_g

6.3.2 Polymer Piezoelectric Composites

The advantage of mixing polymers with piezoelectric ceramics include higher coupling factor and dielectric constant derived from the ceramics coupled with the mechanical flexibility of polymers. Because of the polymer's fewer spurious modes and low acoustic impedance, this category of materials is an ideal choice for acoustic devices. Ceramic/polymer composites are arranged as randomly scattered rods in polymer bulk films, also classified as (1–3) composites. Another arrangement is to impinge microscale/nanoscale particles inside a polymer matrix, and such a composite could be (0–3) if the particles are completely separated or it could be (3–3) if the particles are in contact.

6.3.3 Voided Charged Polymers

This category of polymers, also termed as *cellular polymers*, contain internal gas voids. The voided charged polymer's piezoelectric behavior arise from the following mechanism: When large electric field is applied across the polymer film, its surfaces surrounding the voids are charged resulting in the ionization of the gas molecules in the voids. Depending on the applied electric field direction, the opposite charges in the gas molecules are accelerated and get implanted on either side of the voids. These embedded dipoles respond to an applied electrical field or mechanical stress, resulting in the deformation of the charged voids which in turn, causes the piezoelectric effect. Aided by the coupling of electrical and mechanical energies, voided charged polymers exhibit a higher piezoelectric coefficient (d_{33}) up to 20,000 pC N^{-1}, which is comparably higher than that exhibited by piezoceramics (Hillenbrand and Sessler 2004, Ramadan et al. 2014).

6.3.4 Enhancement of Piezoelectric Effect in Polymers by Poling

Poling process results in the reorientation of the crystallites, that is, the molecular dipoles within the polymer bulk medium under higher applied electric field at an elevated temperature. Raising the temperature of the polymer close to its T_g before poling increases the mobility of the dipoles and allows rotation to occur during poling. Once oriented, the ordered state of the molecular dipoles can be maintained by cooling the material to room temperature in the presence of the electric field. Figure 6.5 illustrates the two methods: (a) electrode poling and (b) corona poling, which are commonly used in poling the polymeric materials. In the case of electrode poling as shown in Figure 6.5a, both sides of the polymer film are deposited with conductive electrodes and high voltage (AC sinusoidal, DC or triangular low-frequency wave forms) in the range from 5 to 100 MV m^{-1} is applied across the bulk piezoelectric polymer (Kim et al. 2009). The poling process is either carried out in a vacuum chamber (Park et al. 2004) or immersed inside an electrically insulating fluid (Li et al. 2008) to prevent the breakdown of the polymer under the strong electric field. The final quality of the crystallite alignment, and consequently, the piezoelectric coefficient d depends on the following factors: (1) the extent of contamination/voids between the polymer surface and the electrodes, (2) time and strength of the applied electrical field, and (3) degree of uniformity and value of the applied poling temperature. In the case of corona poling, as shown in Figure 6.5b, only one side of the polymer film is covered with an electrode, and a very high voltage V_c (8–20 kV) is applied to a conductive needle placed on top of a grid, which is subjected to a much lower DC voltage V_g (0.2–3 kV). The entire setup is enveloped in a dry air or argon medium. The applied electrical field, that is, the amount of deposited charges across the polymer surface is controlled by governing the grid position and applied voltage. The required temperature for both corona and electrode poling methods does not go beyond 300°C, which is much lower than the lead zirconate titanate (PZT)-processing temperatures (~1200°C). Although more complicated than electrode poling, the corona poling technique compensates for the surface roughness in polymer film and does not require electrode deposition (Tadigadapa and Materi 2009, Ramadan et al. 2014). Other less common methods of poling include electron beam poling,

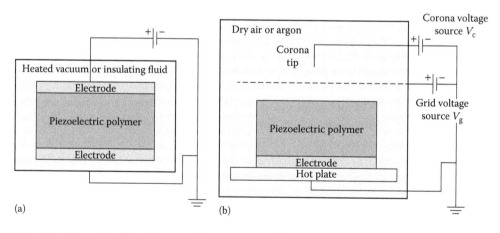

FIGURE 6.5 Piezoelectric polymer poling systems: (a) electrode poling and (b) corona poling. (Reprinted from Figure 4, Ramadan, K.S. et al., A review of piezoelectric polymers as functional materials for electromechanical transducers, *Smart Mater. Struct*, 23, 033001, 2014, with permission from IOP Publishing Ltd.)

in which irradiation from the electron beam can be focused locally to deposit electrons on the piezoelectric polymer surface (direct patterning method), which in turn reorient the dipoles. The drawbacks of this method include the chemical modification and material degradation under poling (Gross et al. 1987, Qiu 2010). Another ionization method is the use of soft X-rays for poling voided charged polymers.

6.3.5 Electrode Formation on Piezoelectric Films

Screen-printed silver, copper, and carbon electrodes can be used as electrodes on piezoelectric films, but the metal electrodes are rigid and hinder the device performance. Alternatively, conductive and flexible poly(3,4-ethylenedioxythiophene)/polystyrene sulfonate (PEDOT/PSS) greatly improves the device performance such as bimorphs and actuators. In the case of hydrophobic PVDF, the challenge comes in attaching the hydrophilic PEDOT, which does not stick well due to its bead formation on the PVDF surface, and the shrinking of PEDOT when dried, which deforms the thin PVDF sheets. PEDOT also becomes brittle and cracking when applied in large amounts. Until now, PEDOT can be applied to PVDF using three methods: (1) ion-assisted-reaction method to treat the PVDF surface followed by screen printing of PEDOT on it (Lee et al. 2003), (2) plasma etching method to treat the PVDF surface and then airbrushing of PEDOT on thin coats (Hugo et al. 2006), and (3) inkjet printing method to apply very small amounts of PEDOT to improve its adherence to PVDF. In this method, any electrode pattern can be created with excellent adherence without any prior treatment of the PVDF surface (Singh et al. 2010).

Among the known piezoelectric class of polymers, fluoropolymers such as PVDF and its copolymer with trifluoroethylene (TrFE) are the most promising because of their availability as flexible thin films with high piezoelectric activity. Toray® (Japan) and Piezotech® (France) are among the major companies focusing on the manufacturing of fluoropolymers based piezoelectric devices in the commercial market.

6.3.6 Poly(Vinylidene Fluoride)

PVDF has found wider use in human-related applications aided by its flexibility, light weight, and spinnable and biocompatible characteristics. Porous PVDF membrane exhibiting piezoelectric behavior is a better choice to be incorporated into clothing as the power source for wearable electronic devices owing to its good air and water vapor permeability compared to solid piezoelectric films (Swallow et al. 2008).

Although the piezoelectric effect on PVDF was first reported by Kawai (1969), its commercialized piezo films appeared on the world market only in 1981. Because of its molecular structure and its purity, PVDF has the ability to solidify in the crystalline form suitable for polarization. PVDF is a semicrystalline material with a nonpolar crystalline α- and δ-phases (*trans*-gauche conformation, TGTG′), crystalline γ- and ϵ-phases (T3GT3G′), and crystalline polar β-phase (all *trans*-conformation, TTTT) as shown in Figure 6.6a. The polarized electrets are thermodynamically stable up to about 90°C. In the case of PVDF film, its dipoles are still randomly oriented even after mechanical stretching and total summation of polarization is zero, thereby rendering PVDF

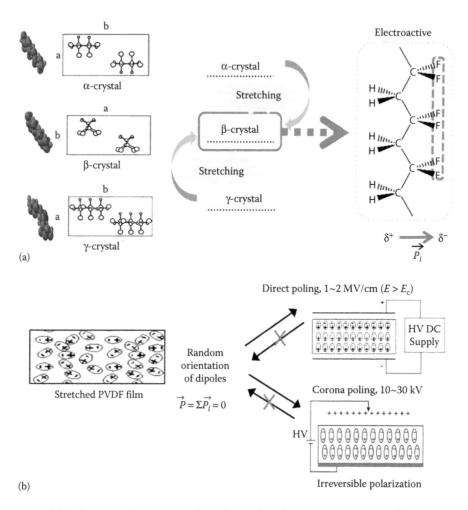

FIGURE 6.6 (a) Schematic illustration of α-, β-, and γ-phases in PVDF and (b) effect of direct poling and corona poling on stretched PVDF film.

as electrically inactive. Under electrical poling (direct or corona), the C-F dipoles are oriented along the poling direction as shown in Figure 6.6b. When an electric field (E) is applied across the sheets, they either contract in thickness and expand along the stretch direction (Figure 6.7a) or expand in thickness and contract along the stretch direction (Figure 6.7b), depending on which way the field is applied. Although the *trans*-gauche α-crystalline phase is the most stable and common phase, it is a known fact that the piezoelectric response of PVDF originates from the β-phase deformation resulting with the alignment of its dipoles in direction perpendicular to the polymer chains and parallel to the applied external electric field or applied pressure. However, the polymer chains in pristine/as-cast PVDF are entangled to form a random coil structure with the nonpolar α-crystalline (TGTG′) phase. Hence, it is imperative to transform the crystalline α-phase into crystalline β-phase by stretching the polymer chains, which is an additional process to be carried out (He et al. 2011).

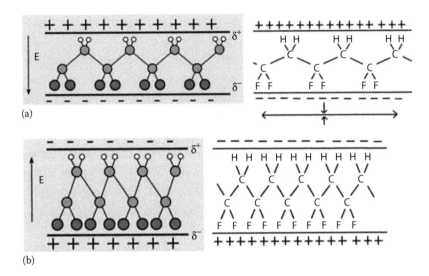

(a)

(b)

FIGURE 6.7 Effects of the two electric field directions (a) top-to-bottom and (b) bottom-to-top on a sheet of PVDF. (From www.physics.montana.edu.)

6.3.7 Poly(Vinylidene Fluoride-Trifluoroethylene)

Since the past decade, PVDF-TrFE has gained significant attention due to its advantages over PVDF such as unique ferroelectric ↔ paraelectric (FE ↔ PE) solid–solid crystalline phase transition at T_c, spontaneous polarization of dipoles at room temperature even for as-cast films and adequately large electromechanical coupling coefficient at room temperature. To induce a polar crystalline phase, stretching of the material is not required (Lovinger 1983). FE ↔ PE transition temperature in PVDF-TrFE is highly dependent upon its VDF content, thermal treatment and sample processing conditions, electrical poling, to name a few (Jin and Bum 1997, Tanaka et al. 1999). Its acoustic impedance (ρ_c) is close to water (used as immersion type transducer), piezoelectric constant (g) is very large (used as hydrophone), dielectric constant is small (ε) (used in high-frequency or large-area devices), electromechanical coupling factor (K_t) is the highest among piezoelectric polymers, and shock resistance as well as mechanical flexibility are superior. However, since the PVDF-TrFE piezoelectric charge constant is much smaller compared to piezoelectric ceramic materials, its practical application has been limited only to sensors and actuators, and no practical applications as a generator. Considering the limitations in the usage of thick/thin films (porosity, surface area, etc.) in wearable electronics, the following sections will focus more closely on the preparation of PVDF-TrFE electrospun nanoweb fibers for use as piezoelectric sensors, nanogenerators, and so on in wearable electronics.

6.4 ELECTROSPINNING: PRINCIPLES, WORKING PARAMETERS, AND NANOFIBER PROPERTIES

Although there are several high-volume and accurate methods such as fibrillation, island-in-sea, melt-blowing, nanolithography, and self-assembly systems for producing small diameter fibers, their usefulness is restricted by combinations of narrow material range and high cost. Electrospinning is a comparatively simple and low-cost process with an intermediate production rate. Electrospun nanoweb fibers exhibit many exceptional properties such as large surface area, high aspect ratio, flexible surface, superior mechanical properties, and tunable surface morphologies, and hence suitable for flexible and stretchable devices, which have generated more attention nowadays (Sun et al. 2013, Sharma et al. 2015a). Various polymers have been successfully electrospun into ultrafine fibers in recent years mostly in solvent solution and some in melt form.

Figure 6.8a shows a typical electrospinning setup consisting of three elements: a high voltage power supply, a capillary with a needle (jet source), and a cylindrical drum collector (target) wrapped with conductive fabric. In principle, electrospinning is a simple process of obtaining submicron scale polymer fibers with nanoscopic diameter (usually called *nanofibers*) by means of an electrostatic field. When a sufficiently high voltage is applied to a liquid droplet between the capillary and a grounded collection target, the body of the liquid becomes charged and electrostatic repulsion counteracts the surface tension, resulting in the stretching of the droplet (Dzenis 2004). The critical point at which the liquid stream is forced to elongate out of a capillary nozzle and forms a fine jet of atomized droplets is known as the *Taylor cone* (Taylor 1964). If the molecular cohesion of the liquid is sufficiently

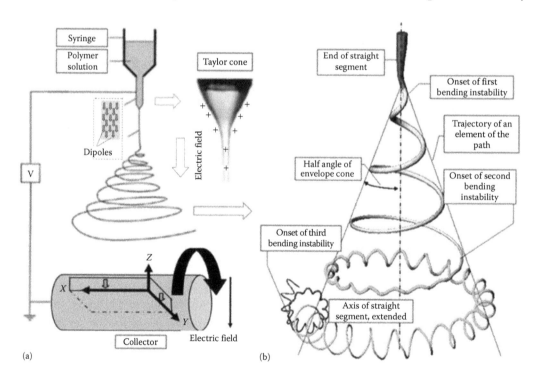

FIGURE 6.8 Schematic illustrations of (a) basic electrospinning setup showing the Taylor cone (*top right*), enlarged view of the induced dipoles in the polymer jet (*top left*), and the collector (*bottom*) showing the induced dipole direction and rotation direction and (b) prototypical instantaneous position of the electrospinning jet path that contained three successive electrical bending instabilities. (a and b readapted and reproduced, respectively from Figure 2, Ke, P. et al., *RSC Adv.* 4, 39704–39724, 2014. Reproduced by permission of The Royal Society of Chemistry.)

high, stream breakup does not occur and a charged liquid jet is formed. However, the jet that is stable near the spinneret tip soon enters a stage where it undergoes bending instability along with further stretching aided by the evaporation of the solvent.

Generally, the extension of the fluid is uniform at first followed by straight flow lines that undergo vigorous whipping and/or splitting motions caused by fluid instability as well as electrically driven bending instability as shown in Figure 6.8b. As the jet dries out in flight, the mode of current flow changes from ohmic to convective as the charge migrates to the surface of the fiber. Finally, the continuous as-spun fibers are deposited on sheets or other geometrical forms as a 2D nonwoven web (Sharma 2013, Ke et al. 2014).

The morphology and diameters of the electrospun fibers are highly influenced by parameters such as (1) intrinsic properties of the solution such as the polymer type, polymer chain conformation, elasticity, viscosity/concentration, electrical conductivity, and polarity of the solvent; (2) operational conditions such as the applied electric field strength, distance between spinneret and the collector, and the polymer solution feeding rate; and (3) humidity/temperature around the electrospinning setup. Among different nanoscale materials, nanofibers have been widely applied in industry because of the ease in production processes compared to other nanomaterials. Because the surface area is proportional to the fiber diameter and the volume is proportional to the square of the diameter, the specific surface area is inversely proportional to the fiber diameter, leading to high specific surface areas for nanofibers (Ellison et al. 2007). Nanofibers are usually fabricated by the electrospinning method as a nonwoven mat, and essentially consist of randomly oriented fibers connected together by physical entanglements or bonds between individual fibers, without any knitting or stitching. These nonwoven nanofibers are the primary alternative for traditional textiles such as tissue engineering (Bhattarai et al. 2004), reinforcement in composites (Chatterjee and Deopura 2002), micro–nano–electromechanical systems (Sundararajan et al. 2002), energy-storage media (Im et al. 2013), hygienic and health/personal care textiles (Lee 2009, Hu et al. 2012), and automotive textiles (Persano et al. 2013).

6.5 ELECTROSPINNING OF PVDF-TrFE COPOLYMER

Porous electrospun PVDF-TrFE nanoweb has stimulated much interest due to its high voltage sensitivity, low acoustic and mechanical impedance, and has potential applications in sensors, actuators, transducers, energy harvesters, and so on. Quite a few publications have adequately dealt with the processing methods and properties of electrospun nanofiber webs (Li et al. 2003, 2004). Many researchers have focused on the direct electrospinning of PVDF-TrFE for use in flexible/stretchable electronic sensor devices (Serrano and Pinto 2012), as nanogenerators/energy harvesters (Mandal et al. 2011), and as biosensors (Sencadas et al. 2012, Persano et al. 2013).

6.5.1 Electrospinning Solution Preparation and Nanoweb Formation

Neat solutions for electrospinning can be prepared by dissolving PVDF-TrFE (30–23 mol% TrFE) pellets (12–18 wt.% [w/v]) in any of the solvent systems: butan-2-one, DMF: acetone (8:2 v/v) or DMF: MEK (7:3 v/v) at 80°C for 3 h or by ultrasonication in THF at 60°C for 1 h. In some cases, nanoparticles such as barium titanate ($BaTiO_3$, average size 10–500 nm) were dispersed in the polymer solution, and the polymer to ceramic relative concentration ranged from 0% to 20% ceramic content (w/w). The electrospinning parameters are as follows: (1) spinning distance between needle tip and collector = 10–20 cm, (2) applied DC voltage = 10–25 kV, (3) solution flow rate = 1.0 mL·h^{-1}, (4) electrospinning time = approximately 30 min, and (5) cylindrical rotational speed = 1000 rpm (Mandal et al. 2011, Pereira et al. 2013).

6.5.2 Instrumentation

Electrospun nanoweb samples can be analyzed using varied instrumentation techniques for their morphological, spectral, and electrical characteristics to confirm their suitability for use as piezoelectric sensors. Nanoweb morphology is observed using scanning electron microscopy (SEM). Preferential chain- and dipole-orientation induced by electrospinning can be confirmed using FTIR spectrometer (number of scans: 500, resolution: 4 cm^{-1}). Piezoelectric signal from the nanofiber web is detected using a pressure sensor as shown in Figure 6.9. The circular bottom electrode, which is a conductive adhesive carbon tape of 2 cm diameter, is adhered to the outer surface of the nanofiber web. A Ni-Cu plated polyester fabric wrapped over the collector during electrospinning serves as the top electrode. One-sided adhesive transparent tape is used to encapsulate the whole sandwich structure (top electrode–nanowebs–bottom electrode). The piezoelectric signal is detected based on a simplified equivalent circuit diagram as shown in Figure 6.9. Sinusoidal pressure is imparted on the sensor to generate the output electrical signal, which is acquired in a voltage mode of the preamplifier with input impedance (R_{in}) of 1 GV through an NIDAQ board and saved on a PC. As shown in Figure 6.10a, the electrospun samples

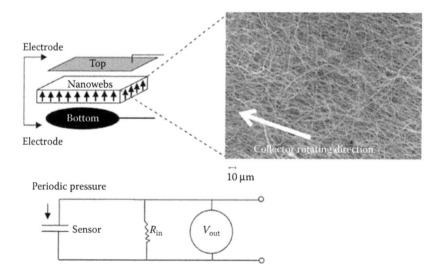

FIGURE 6.9 Schematic illustration of a nanofiber web-based pressure sensor (*left top*) with an SEM view of the nanofiber web (*right top*) and a simplified equivalent circuit (R_{in}: 1 GV) diagram for a detecting piezoelectric signal (*bottom*). (Reprinted from Figure 2, Mandal, D. et al.: Origin of piezoelectricity in an electrospun poly(vinylidene fluoride-trifluoroethylene) nanofiber web-based nanogenerator and nano-pressure sensor. *Macromol. Rapid Commun.* 2011. 32. 831–837. Copyright Wiley-VCH Verlag GmbH & Co. KGaA. Reproduced with permission.)

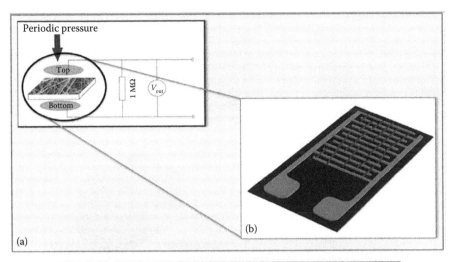

FIGURE 6.10 (a) Experimental setup used for the periodic bending tests, (b) inter-digitized electrode plate with electrospun fibers, and (c) electromechanical generator during energy harvesting bending test. (Reprinted from Figure 1, *Sensor. Actuat. A-Phys.*, 196, Pereira, J.N. et al., Energy harvesting performance of piezoelectric electrospun polymer fibers and polymer/ceramic composites, 55–62, Copyright 2013, with permission from Elsevier.)

were tested for their energy harvesting performance using low deformation-periodic bending measurement in an electromechanical generator excited from 1 Hz to 1 kHz using a signal generator having 1 MΩ internal resistance. Inter-digitized electrode plates were used as top and bottom electrodes as shown in Figure 6.10b. Measured from initial position (at 1 Hz), the maximum vertical displacement of the samples was ±7 mm as shown in Figure 6.10c. The voltage output data was collected using Picoscope software connected with an oscilloscope. Using the first 100 ms of vibration, the maximum average power was calculated (Pereira et al. 2013). In another study (Beringer et al. 2015), the piezoelectric response of PVDF-TrFE used as nanogenerator is tested using a piezoelectric cantilever (PEF). Application of a voltage across the top layer in the PEF generated an axial displacement at the cantilever tip due to the converse piezoelectric effect. The cantilever tip was slowly lowered to the surface of the nanogenerator until it made contact with the fibrous mat. Figure 6.11 shows the schematic illustration of this setup. Each nanogenerator was deformed at frequencies of 2 and 3 Hz with a force of approximately 8 mN.

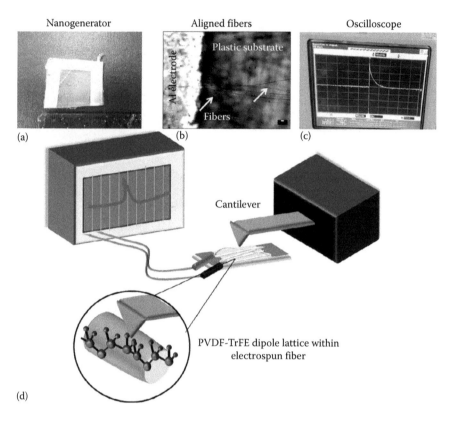

FIGURE 6.11   (a) Nanogenerator fabricated using aligned PVDF-TrFE nanoweb, (b) optical micrograph of nanogenerator revealing aluminum electrode with several aligned fibers across the flexible TOPAS substrate, (c) oscilloscope screen readout, and (d) illustration of cantilever deformation of aligned PVDF-TrFE nanogenerator. (Reprinted from Figure 1, *Sensor. Actuat. A-Phys.*, 222, Beringer, L.T. et al., An electrospun PVDF-TrFE fiber sensor platform for biological applications, 293–300, Copyright 2015, with permission from Elsevier.)

6.5.3 Effect of Electrospinning Conditions

The effect of varying parameters such as voltage, tip-to-collector distance, and solution concentration during the electrospinning of PVDF-TrFE copolymer is discussed to demonstrate the effect of spinning conditions on the final arrangement and quality of the fibers. Although the nanoweb with fiber diameters ranging from 2 to 3 μm could be obtained under favorable spinning conditions, minor changes in the spinning parameters will yield widely different mat quality as discussed elaborately by Pawlowski et al. (2003) and Sencadas et al. (2012).

6.5.3.1 Effect of Electrospinning Voltage

The applied voltage is the main driver behind the electrospinning process. In general, the fiber diameter gradually becomes smaller with increasing applied voltage, although it often enhances jet instability, which results in broader fiber diameter distribution. PVDF-TrFE fiber mean diameter varied between 518 ± 119 and 424 ± 72 nm for applied voltages of 15 and 30 kV, respectively (Sencadas et al. 2012). As shown in Figure 6.12a, under electrospinning

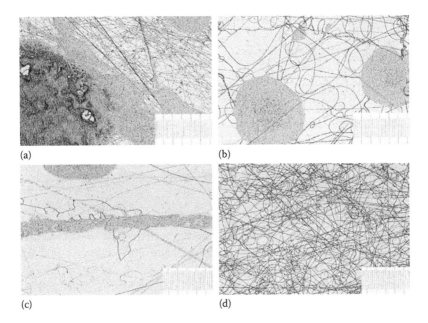

(a)

(b)

(c)

(d)

FIGURE 6.12   As voltage was increased from (a) 12 V to (b) 25 V, and as distance was increased from (c) 7.6 cm to (d) 25.4 cm, the proportion of fibers increased, were more defined and drier, and morphology and diameter were regular. Each division is 10 μm. (Reprinted from Figures 4 and 5, *Polymer*, 44, Pawlowski, K.J. et al., Electrospinning of a micro-air vehicle wing skin, 1309–1314, Copyright 2003, with permission from Elsevier.)

voltage of 12 V, a smaller proportion of fibers versus nonfibrous masses was evident, and the fibers were also not drier as desired. However, with increasing voltage to 25 V, the proportions of fibers increased and were better defined and drier as shown in Figure 6.12b.

6.5.3.2 Effect of Tip-to-Collector Distance
Studying the effect of tip-to-collector distance as shown in Figure 6.12c and d, the distance between the jet initiation point and target significantly influence the formation of fibers. At a distance of only 7.6 cm, it is too close for optimal fiber formation resulting in predominantly nonfibrous masses and fibers that did formed were wet. Under increased distance to 25.4 cm, the fibers were predominantly dry with uniform fiber morphology and diameter.

6.5.3.3 Effect of Needle Diameter and Flow Rate
Clogging and bead formation is prevented when using needle with smaller internal diameters, but the fiber size has a broader distribution. Surface tension of the droplet is increased with decreasing droplet size at the tip caused by the smaller needle inner diameters, and a larger Coulomb force is required to initiate the jet. The decreasing acceleration of the jet allows more time for the solution to be stretched/elongated, which in turn leads to broader distribution of the fiber diameter when used with smaller inner diameter needles for electrospinning. On the other hand, with increasing needle inner-diameter and feed-rate, the average fiber diameter was observed to be increased along with the presence of beads

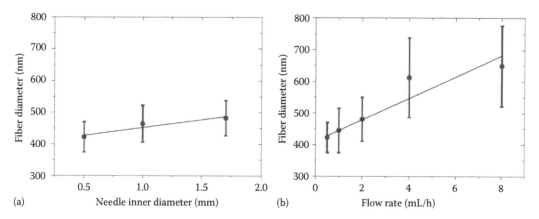

FIGURE 6.13 (a) Influence of tip inner diameter on the mean diameter of the electrospun PVDF-TrFE/BaTiO$_3$ fibers under constant flow rate (0.5 mL h^{-1}) and applied electric field (1.25 kV cm^{-1}) and (b) influence of flow rate on the mean diameter of the electrospun PVDF-TrFE/BaTiO$_3$ fibers under fixed tip inner diameter (0.5 mm) and applied electric field (1.25 kV cm^{-1}). Filler concentration in the electrospun solution is maintained at 5%. (Reprinted from Figure 3, *Sensor. Actuat. A-Phys.*, 196, Pereira, J.N. et al., Energy harvesting performance of piezoelectric electrospun polymer fibers and polymer/ceramic composites, 55–62, Copyright 2013, with permission from Elsevier.)

due to the larger solution volume as shown in Figure 6.13a. Mean fiber diameter increased from 422 ± 27 to 649 ± 127 nm with increasing flow rate from 0.5 to 8 mL h^{-1} as shown in Figure 6.13b. Increasing the drawn volume from the needle tip results in longer solvent evaporation time, and larger crystallization time for the polymer, thereby giving rise to broader fiber size distribution (Pereira et al. 2013).

6.5.3.4 *Effect of Solution Concentration and Aging*
At 10 wt.%, the collected sample was entirely nonfibrous mass with wet droplets. At 16 wt.%, tiny fiber formation was obvious to some extent with diameters of around 500 nm. At 20 wt.%, the fibers formed were more dry and defined with larger diameters. This sprayed wet droplets to spun dry fibers transition can typically help us to identify the optimal concentration for electrospinning (Pawlowski et al. 2003). In another study, two electrospun membranes from two different PVDF-TrFE concentrations (12 and 18 wt.% in 8:2 DMF/acetone) were used. SEM micrographs of the electrospun membrane prepared from 12 wt.% polymer concentration as shown in Figure 6.14a exhibit large number of polymeric beads. Low viscosity of the PVDF-TrFE solution made the electrospinning process unstable, thereby resulting in the dominant bead formation. With increasing the concentration to 18 wt.% as shown in Figure 6.14b, a fine nanofiber structure could be obtained along with complete absence of the beads. These two electrospun membranes were subjected to poling, but they exhibited high leakage current, probably caused by their high porosity. The porosity problem is reduced by hot pressing, which in-turn melted the electrospun membrane surface, and many of the nanofibers/beads merged together as observed from

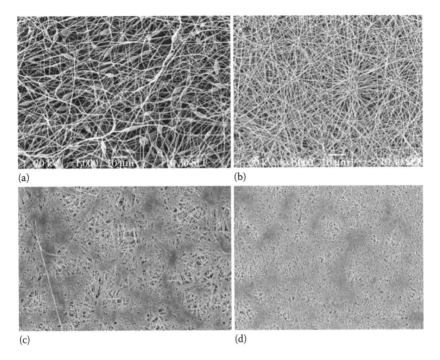

(a)   (b)

(c)   (d)

FIGURE 6.14   SEM morphology images of PVDF-TrFE electrospun membranes in (a) 12 wt.% and (b) 18 wt.% DMF/acetone (8/2) solvent mixture, (c) after hot-pressing (a) sample, and (d) after hot-pressing (b) sample to reduce the porosity. (Reprinted from Figure 2, *Polym. Test.*, 30, He, F. et al., Preparation and characterization of porous poly(vinylidene fluoride-trifluoroethylene) copolymer membranes via electrospinning and further hot pressing, 436–441, Copyright 2011, with permission from Elsevier.)

Figure 6.14c and d. The electrospun membrane density is also influenced significantly due to the hot pressing treatment as reported by He et al. (2011).

Sencadas et al (2012) attributed the slower evaporation rate of the DMF solvent (153°C) to the presence of high bead formation in PVDF and its related polymers. DMF:MEK solvent mixture is mostly preferred to dissolve PVDF-TrFE due to their thermal and electrical characteristics: DMF shows a high dielectric constant (38.2), whereas MEK has a lower boiling point (80°C), which allows faster polymer crystallization.

Beringer et al. (2015) observed an average nanofiber diameter of 340 and 360 nm using fresh and 25-year-old MEK solvents, respectively, for electrospinning PVDF-TrFE. FTIR data also showed no noticeable peak additions that would indicate chemical degradation when comparing new and old MEK solvents, and hence the aging of the MEK had no significant effect on the fiber diameter, but only on the surface morphology as observed from distinctively rough surface for old MEK case.

Sencadas et al. (2012) reported the optimum electrospinning conditions as follows: flow rate = 1 mL h⁻¹, applied voltage = 25 kV and tip-to-collector distance = 20 cm. Independent formation of fibers can be achieved when most of the solvent evaporation occur during the travel between tip to collector. With reduced distance, the solution jet travel distance is also reduced and consequently, the solvent has lesser time for its evaporation. In addition, the increasing electric field strength has a similar effect such as reducing the tip-to-collector distance and will accelerate the jet toward the collector leading to smaller fiber diameters.

6.6 SPECTRAL CHARACTERISTICS OF ELECTROSPUN PVDF-TrFE NANOWEB

Depending on the sample preparation conditions, PVDF exhibits at least four different crystalline forms at ambient temperature. Using FTIR spectroscopy, it is more difficult to find experimental evidence of crystalline modifications in electrospun PVDF as the IR spectrum is highly sensitive to chain and dipole orientation (Reynolds et al. 1989, Kim and Hsu 1994). On the other hand, PVDF-TrFE copolymer has a definite T_c and exhibit a single ferroelectric β-crystalline structure at ambient temperature far below its T_c.

A single-stage electrospinning process assists in preferentially orienting the C-F dipoles in PVDF-TrFE nanoweb as analyzed by researchers from our group using polarized FTIR spectroscopic techniques (Mandal et al. 2011). Figure 6.15a shows the IR optics and sample geometry. Figure 6.15b illustrates the as-electrospun nanofiber FTIR spectra measured with the parallel and perpendicular polarized IR beam, and it is clear that A_1 and B_2 band absorptions are prominent in the perpendicular polarized spectrum, whereas in the parallel polarized spectrum, B_1 band absorption is dominant. This is evidence toward the preferential orientations of electroactive CF_2 dipoles and *trans*-zigzag chains of PVDF-TrFE along the perpendicular and parallel directions, respectively, to the collector rotation direction as shown in Figure 6.15a.

It is well known that heating the poled copolymer sample above its T_c results in the depolarization of its dipoles, and hence the electrospun nanofiber is thermally treated to 130°C as shown in Figure 6.15c, which is well below its melting temperature ($T_m \approx 147°C$) but above its T_c ($\approx 115°C$), in order to obtain the nanofiber web with a randomized dipole orientation. Kim and Hsu (1994) reported the 1284, 1182, 884, and 845 cm^{-1} bands sensitivity toward changes in dipole orientation. In the IR optics scheme shown in Figure 6.15a, the IR beam's electric field vector ($\bar{E}$) is parallel to the nanofiber web surface. The $f_{dipolar}$ magnitude of the as-electrospun nanoweb is higher because of the preferential orientation of the CF_2 dipoles toward the nanoweb thickness direction during electrospinning. The magnitude of $f_{dipolar}$ is considerably reduced for the heat-treated sample due to the randomization of CF_2 dipoles above T_c. This is evidence toward the electric field-induced preferential orientation of CF_2 dipoles along the electric field direction during electrospinning.

In another study, the FTIR spectra of PVDF-TrFE film and electrospun melt-crystallized samples were compared. In PVDF-TrFE solid film, the peaks related to α-form crystallites were completely absent whereas strong peaks at 842 and 1290 cm^{-1} corresponding to β-crystalline phase were observed, which is an evidence that the PVDF-TrFE sample can directly crystallizes into the ferroelectric β-phase from its melt. In the electrospun PVDF-TrFE case with or without hot-press treatment, the spectra were found to be almost identical to that of the PVDF-TrFE solid film, which is an indication that the PVDF-TrFE crystallize in the β-phase from electrospinning and maintain its crystalline structure even after hot-pressing (Cebe and Runt 2004).

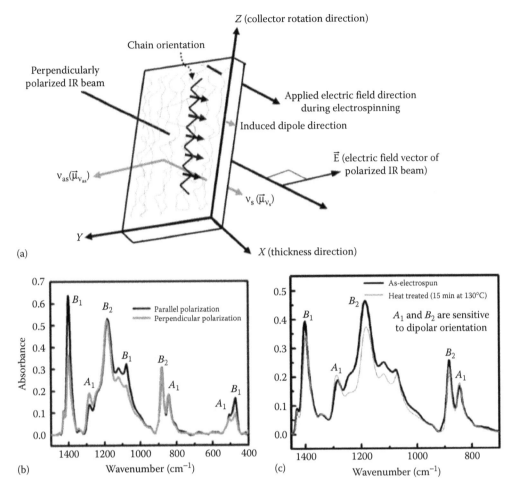

FIGURE 6.15 (a) Schematic drawing for the geometry of the PVDF-TrFE nanofiber web when the FTIR spectrum is measured using the perpendicularly polarized IR. The applied DC bias voltage direction is parallel to the incident IR beam direction, (b) FTIR spectra of as-electrospun PVDF-TrFE nanofiber web measured with parallel and perpendicular polarized IR beams, and (c) perpendicularly polarized FTIR spectra of PVDF-TrFE nanofiber web as a function of as-electrospun and heat-treated (at 130°C for 15 min). (Reprinted from Figure 3, Mandal, D. et al.: Origin of piezoelectricity in an electrospun poly(vinylidene fluoride-trifluoroethylene) nanofiber web-based nanogenerator and nano-pressure sensor. *Macromol. Rapid Commun.* 2011. 32. 831–837. Copyright Wiley-VCH Verlag GmbH & Co. KGaA. Reproduced with permission.)

6.7 PIEZOELECTRIC CHARACTERISTICS OF ELECTROSPUN PVDF-TrFE NANOWEB

Energy harvesting from mechanical vibrations has attracted enormous attention since the past decade aided by the development of smaller electronic components, which has a low operational power requirement. Basically, three mechanisms for vibration-to-electrical conversion is proposed: electromagnetic, electrostatic, and piezoelectric transductions (Anton and Sodano 2007), and among them, piezoelectric transduction has received the largest attention because of its high power density, ease of application, and the voltage output is directly obtained from the piezoelectric material itself, which in turn bypasses the need of an external voltage input. When a strain is applied, there is a change in the dipole moment of the piezoelectric material resulting in the formation of a potential difference which can be used to power devices. Commonly used ceramic piezoelectric materials for power harvesting such as PZT or $BaTiO_3$ is prone to fatigue and crack under high-frequency cyclic loading. Comparatively, PVDF and PVDF-TrFE copolymers are preferred because of their high chemical resistance, superior mechanical properties, and outstanding electroactive properties.

Fang et al. (2011) developed generators based on PVDF electrospun fibers, which achieved voltage output between 0.43 and 6.3 V when subjected to periodic oscillations between 1 and 10 Hz. Pereira et al. (2013) prepared electrospun composite fibers of PVDF-TrFE with $BaTiO_3$ on top of an interdigitated circuit in order to study the effect of ceramic filler on the energy harvesting efficiency of the polymer. Electrospun samples were subjected to mechanical excitations (periodic bending oscillation) in the frequency range between 1 Hz and 1 kHz. The polymer fiber is worked under alternating pressure in a longitudinal mode, which results in fiber surface charge variations due to the simultaneous tensile and bending stresses as shown in Figure 6.10c. Because of the induced voltage difference between the two adjacent interdigitated electrodes, which are connected in parallel to other units, the power output of the generator is significantly enhanced. When the external load is removed, the charge carriers flow in the reverse direction, thereby resulting in negative voltage signal as shown Figure 6.16a. Under fixed fiber loading, increasing the ceramic particle average diameter decreases the electrical output voltage measured for PVDF-TrFE/$BaTiO_3$ electrospun fibers as shown in Figure 6.16b. An average piezo-potential of ~100 mV could be measured for the fibers with $BaTiO_3$ particles of ~10 nm, which is higher than that achieved for the

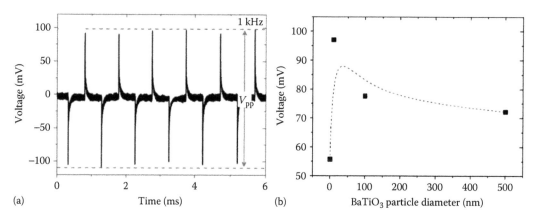

FIGURE 6.16 (a) Voltage generated during 6 ms at a frequency of 1 kHz, of an electrospun sample of PVDF-TrFE/$BaTiO_3$ with 20% of ceramic filler and (b) piezo-potential obtained at 1 kHz for PVDF-TrFE/$BaTiO_3$ with 20% ceramic filler. (Reprinted from Figure 6, *Sensor. Actuat. A-Phys.*, 196, Pereira, J.N. et al., Energy harvesting performance of piezoelectric electrospun polymer fibers and polymer/ceramic composites, 55–62, Copyright 2013, with permission from Elsevier.)

pure polymer matrix, along with a strong decrease in the piezo-potential with increasing filler size. Overall, power outputs in the range of 0.02 and 25 μW is obtained for pure PVDF fibers under low and high mechanical deformation, respectively, whereas PVDF-TrFE copolymer and its composites with ceramics exhibit reduced power output caused by enhanced mechanical stiffness.

The imparted pressure driven by AC signal served as the source of periodic pressure in pressure sensors resulting in a deformation that caused changes in the surface charge density which in turn, generated a potential difference between surfaces of the electrodes. The output signal amplitude is directly proportional to the maximum mechanical force (σ_{max}) exerted on the nanofiber web or the maximum nanofiber web deformation along the thickness direction. The output signal frequency is identical to the sinusoidal pressure imparted within the electromechanical coupling limit.

Figure 6.17a shows the two piezoelectric signals from as-electrospun (thick line) and heat-treated (thin line) nanofiber webs (active area diameter = 1.2 cm and applied frequency = 5.3 Hz). When the piezoelectric sensor using as-electrospun nanoweb is subjected to repetitive pressure and release, the output signal (current) with alternate signs opposite to each other, corresponding to a typical AC curve shape (thick line) is generated with a change in the total polarization as a function of time. Compared to the film and ceramic-based piezoelectric materials, the nanofiber web-based pressure sensor or generator shows higher piezoelectric effect at the identical pressure due to its higher compressibility associated with larger change in thickness. Figure 6.17a also shows the effect of heat-treatment (above T_c, 130°C for 15 min) on the as electrospun nanofiber web (thin line). The absence of an AC type signal (thin line) indicates the complete disappearance of piezoelectric coupling activity because of the complete randomization of the dipoles after the heat treatment above the T_c.

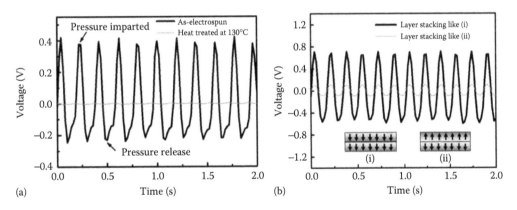

FIGURE 6.17 (a) Piezoelectric output signal from the PVDF-TrFE nanofiber web. The thick and thin line curves indicate the output signal from as-electrospun and heat treated (130°C for 15 min) nanofiber web-based sensors, respectively. (b) Piezoelectric output signals from structure (i) and (ii) as shown in the inset. (Reprinted from Figure 5, Mandal, D. et al.: Origin of piezoelectricity in an electrospun poly(vinylidene fluoride-trifluoroethylene) nanofiber web-based nanogenerator and nano-pressure sensor. *Macromol. Rapid Commun.* 2011. 32. 831–837. Copyright Wiley-VCH Verlag GmbH & Co. KGaA. Reproduced with permission.)

In another study, two types of sensors were fabricated as shown in the Figure 6.17b inset: (i) two nanofiber web layers stacked together with a (top-bottom)–(top-bottom) series structure and (ii) two nanofiber web layers stacked together with a (bottom-top)–(top-bottom) series structure. Compared with the output signal generated from the one layer sensor shown in Figure 6.17a, the output signal from structure (i) in Figure 6.17b was enhanced about two times (thick line), whereas the output signal from structure (ii) (thin line) is very much attenuated due to the opposite dipole arrangement in structure (ii). The cumulatively higher net dipole moment achieved using structure (i) is the result of higher density of the available charge carriers leading to enhanced output signal. The above results can be explained only if the dipoles are perpendicularly oriented to the surface plane (parallel to the thickness direction) of the as-electrospun nanofiber web.

6.8 APPLICATIONS OF ELECTROSPUN PVDF-TrFE NANOWEB

Flexible, self-powered materials are in demand for a multitude of applications such as energy harvesting, robotic devices, and lab-on-a chip medical diagnostics. However, simple flexibility allows only roll-to-roll production of these devices and does not guarantee their conformal bonding to nonplanar substrates. Therefore, it is significant to focus on electrospun nanofibers for use in highly stretchable devices that could withstand large deformations. Since the past decade, numerous polymeric electrospun nanofibers have been prepared for use in sensors, smart textiles, filters, fiber reinforcement, tissue engineering, drug delivery, and wound healing, and so on. In this section, the recent progresses achieved on the flexible and stretchable electronic devices based on electrospun PVDF-TrFE ultrathin fibers, including energy harvesters, biosensors, strain/pressure sensors, super capacitors, and organic field-effect transistors are summarized.

6.8.1 Energy Harvesters and Nanogenerators

Mandal et al. (2011) used PVDF-TrFE nanofiber web as flexible nanogenerator and nanopressure sensors. The nanofiber webs prepared through a simple and scalable electrospinning process can eliminate the need for conventional direct-contact/corona poling. Huang et al. (2003) studied the suitability of using the flexible and air permeable nanofiber webs in self-powered garments. Dey et al. (2012) studied the growth of electrospun PVDF-TrFE nanofibers and observed an increase in the duty cycle of response pulses as the vibration frequency is increased. The maximum output power is calculated to be around 0.6 nW. Pereira et al. (2013) investigated the energy harvesting efficiency of electrospun PVDF-TrFE/BaTiO$_3$ composites on interdigitated electrodes. Wang et al. (2015) proposed a flexible triboelectric and piezoelectric coupling nanogenerator with excellent flexibility. The sandwich-shaped nanogenerator delivered peak piezoelectric output voltage, energy power, and energy volume power density at 30 V, 9.74 μW, and 0.689 mW cm^{-3} under the pressure force of 5 N, respectively. The nanogenerator has some advantages such as flexibility, thickness controllability, and double coupling mechanisms.

6.8.2 Biological Sensors

Electrospun PVDF-TrFE has been reported to be biocompatible and has shown to be promising for tissue scaffolding and regenerative medicine, as well as for biological pressure sensor applications. Beringer et al. (2015) used PVDF-TrFE electrospun in an aligned format and interfaced with a flexible plastic substrate in order to create a platform for voltage response characterization. The biosensor is capable of producing −0.4 to 0.4 V when deformed by cantilever pressure of 8 mN at 2 and 3 Hz. This platform can be used for measurement and analysis of electromechanical behavior in cellular-powered nanodevices. Because of the ionic nature of cellular culture media, the baseline shift of the signal is to be expected as the oscilloscope is a tool for analyzing electrical signals and the culture media is conductive. The generated piezoelectric signal was not affected significantly by sterilization procedures, but showed dampened behavior when tested in cell culture media. The nanogenerators also demonstrated their biocompatibility, which is of importance for downstream cellular applications.

6.8.3 Strain and Pressure Sensors

Piezoelectric polymer-based sensors can exploit deformations induced by small forces, through pressure, mechanical vibration, elongation/compression, bending, or twisting, and PVDF-TrFE is an extremely robust and biocompatible material for strain and pressure sensing applications. For instance, Ren et al. (2013) developed flexible pressure sensors suitable for incorporation into fabrics using a custom-made setup, having a maximum sensitivity of 60.5 mV N^{-1}. The sensor output voltage was observed to be varied linearly with the applied forces, an evidence for linear piezoelectricity, and can measure dynamic force up to a frequency of 20 Hz. In another instance, Persano et al. (2013) prepared flexible piezoelectric sensors based on PVDF-TrFE electrospun aligned nanofibrous arrays giving larger response even under small applied pressures as shown in Figure 6.18. For a given pressure (10 Pa, L_{eff} = 3 mm), the output voltage does not change significantly with length of the fiber array over a range between 2 and 8 cm. Even without sensitive voltmeters, this level of response enables accurate measurement of compressive pressures as small as 0.1 Pa. Sharma et al. (2015b) developed pressure sensors utilizing PVDF-TrFE nanoweb fibers for endovascular applications and tested them *in vitro* under simulated physiological conditions. Using core–shell electrospun fibers, significant improvements in signal intensity gain (4.5×) was observed compared to PVDF nanofibers and nearly 40-fold higher sensitivity compared to PVDF-TrFE thin-film structures. These flexible nanofibers have a great potential for realizing more robust, reliable, and flexible pressure sensors on catheters in the field of minimally invasive surgeries.

6.8.4 Supercapacitors

Flexible/stretchable electronic devices can be powered using rechargeable and stretchable energy storage supercapacitors. Electrospun PVDF-TrFE perform well in this field because of its ability to form micro/nanofibers with larger voids interconnected between the fibers, thereby enabling the mobile ions in the electrolyte to move freely through the elastomeric separator during dynamic stretching/releasing or bending/releasing (Laforgue 2011, Wang et al. 2013). Dey et al. (2012) measured the capacitance value of

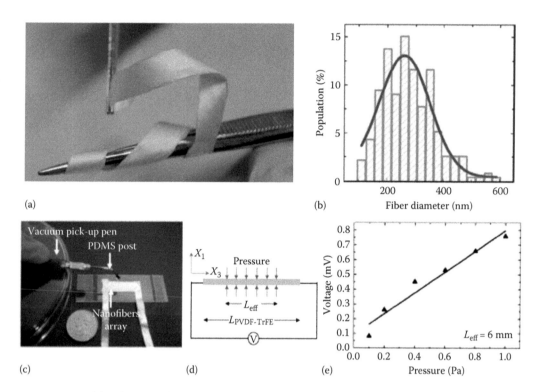

FIGURE 6.18 (a) A free-standing highly aligned piezoelectric PVDF-TrFE nanoweb fibers; (b) typical fiber diameter distribution and fit by a Gaussian curve (solid line); (c) manipulator used to apply pressures for the purpose of studying the voltage response; (d) illustration of an analytical model for the response of PVDF-TrFE nanofiber array under applied compression $-p$ along x_1 direction over the effective contact length (L_{eff}). $L_{PVDF-TrFE}$ is the total length of the PVDF-TrFE fiber array; and (e) experimental pressure response curve in the low-pressure regime (0.1–1 Pa) at $L_{eff} = 6$ mm. The line corresponds to a linear fit. (Reprinted by permission from Macmillan Publishers Ltd. *Nat. Commun.* Persano, L. et al., Copyright 2013.)

PVDF-TrFE nanoweb fiber as around 110 pF, which is always lower than the theoretical value of 1150 pF, and the measured internal resistance at around 13 MΩ.

6.8.5 OFETs

Organic field-effect transistors (OFETs) can be fabricated under reduced temperature and with reduced cost when compared to silicon thin-film transistors. The low fabrication temperature allows a wide range of substrates that can be used for fabricating OFETs, particularly RFID tags and chemical sensors, that require flexible polymeric substrates. Electrospun PVDF-TrFE nanofibers can promote charge transfer, thereby increasing the field effect mobility (Li et al. 2010).

6.9 SUMMARY AND FUTURE PROSPECTS

We have summarized the basic principles of piezoelectricity, types of piezoelectric polymers, mainly focusing on PVDF and PVDF-TrFE copolymer. Subsequently, the main approaches of electrospinning PVDF-TrFE ultrathin fibers and its use in flexible and stretchable electronic devices are discussed. It is interesting to note that PVDF-TrFE electrospun fibers exhibit high aspect ratio and flexibility, compatibility with many substrates, and scalability. Several strategies such as adding additives to tune the fiber conductivity have also been proposed. In this chapter, by utilizing polarized FTIR spectroscopic analysis (microscopically) and by detecting the piezoelectric signal (macroscopically), we have established a simple way to confirm the preferential orientation of CF_2 dipoles in the PVDF-TrFE nanofiber web. By changing the number of stacked nanoweb layers and other parameters such as electrode active area and frequency of the imparting pressure, there is a good possibility to obtain considerable amount of power output useful for operating portable electronic devices. Future research areas using PVDF-TrFE nanoweb could include developing a single-crystalline material, incorporating inorganics into the polymer to obtain higher electromechanical properties, enhancing electrospinning processability, and broadening its usable temperature range, so that the devices fabricated using the nanoweb can operate in extreme environments.

REFERENCES

Anton, S.R. and H.A. Sodano. 2007. A review of power harvesting using piezoelectric materials (2003–2006). *Smart Mater. Struct.* 16:R1.

Beringer, L.T., Xu, X., Shih, W., Shih, W.-H., Habas, R., and C.L. Schauer. 2015. An electrospun PVDF-TrFe fiber sensor platform for biological applications. *Sensor. Actuat. A-Phys.* 222:293–300.

Bhattarai, S.R., Bhattarai, N., Yi, H.K., Hwang, P.H., Cha, D.I., and H.Y. Kim. 2004. Novel biodegradable electrospun membrane: Scaffold for tissue engineering. *Biomaterials* 25:2595–2602.

Cebe, P. and J. Runt. 2004. P(VDF-TrFE)-layered silicate nanocomposites. Part 1. X-ray scattering and thermal analysis studies. *Polymer* 45:1923–1932.

Chatterjee, A. and B.L. Deopura. 2002. Carbon nanotubes and nanofibre: An overview. *Fiber Polym.* 3:134–139.

Dey, S., Purahmad, M., and S. Ray. 2012. Investigation of PVDF-TrFE nanofibers for energy harvesting. In *Proceedings of the IEEE Nanotechnology Materials and Devices Conference,* Waikiki Beach, HI, pp. 21–24.

Dzenis, Y. 2004. Spinning continuous fibers for nanotechnology. *Science* 304:1917–1919.

Ellison, C.J., Phatak, A., Giles, D.W., Macosko, C.W., and F.S. Bates. 2007. Melt blown nanofibers: Fiber diameter distributions and onset of fiber breakup. *Polymer* 48:3306–3316.

Fang, J., Wang, X., and T. Lin. 2011. Electrical power generator from randomly oriented electrospun poly(vinylidene fluoride) nanofibre membranes. *J. Mater. Chem.* 21:11088–11091.

Furukawa, T. 1989. Piezoelectricity and pyroelectricity in polymers. *IEEE Trans. Electr. Insul.* 24:375–394.

Gross, B., Gerhard-Multhaupt, R., Berraissoul, A., and G.M. Sessler. 1987. Electron-beam poling of piezoelectric polymer electrets. *J. Appl. Phys.* 62:1429–1432.

Harrison, J.S. and Z. Ounaies. 2002. Polymers, piezoelectric. In *Encyclopedia of Smart Materials,* ed. M. Schwartz, pp. 860–873. Hoboken, NJ: John Wiley & Sons.

He, F., Sarkar, M., Lau, S., Fan, J., and L.H. Chan. 2011. Preparation and characterization of porous poly(vinylidene fluoride-trifluoroethylene) copolymer membranes via electrospinning and further hot pressing. *Polym. Test.* 30:436–441.

Hillenbrand, J. and G.M. Sessler. 2004. Quasistatic and dynamic piezoelectric coefficients of polymer foams and polymer film systems. *IEEE Trans. Dielectr. Electr. Insul.* 11:72–79.

http://www.dytran.com/assets/PDF/Piezoelectric%20Measurement%20System%20 Comparison%20-%20Charge%20Mode%20vs.pdf. Accessed on February 27, 2016.

http://www.instrumentationtoday.com/piezoelectric-transducer/2011/07/. Accessed on February 27, 2016.

http://www.physics.montana.edu/eam/polymers/piezopoly.htm. Accessed on February 27, 2016.

http://www.piezotech.fr/image/documents/22-31-32-33-piezotech-piezoelectric-films-leaflet.pdf. Accessed on February 27, 2016.

http://www.pcb.com/techsupport/tech_accel. Accessed on February 27, 2016.

Hu, J., Meng, H., Li G., and S.I. Ibekwe. 2012. A review of stimuli-responsive polymers for smart textile applications. *Smart Mater. Struct.* 21:053001.

Huang, Z.M., Zhang, Y.Z., Kotaki, M., and S. Ramakrishna. 2003. A review on polymer nanofibers by electrospinning and their applications in nanocomposites. *Compos. Sci. Technol.* 63:2223–2253.

Hugo, S.V., Lediaev, L., Polasik, J., and J. Hallenberg. 2006. Piezoelectric actuators employing PVDF coated with flexible PEDOT-PSS polymer electrodes. *IEEE Trans. Dielectr. Electr. Insul.* 13:1140–1148.

Im, J.S., Yun, J., Kim, J.-G., and Y.-S. Lee. 2013. Preparation and applications of activated electrospun nanofibers for energy storage materials. *Curr. Org. Chem.* 17:1424–1433.

Jin, K.K. and K.G. Bum. 1997. Curie transition, ferroelectric crystal structure and ferroelectricity of a VDF/TrFE (7525) copolymer: 2. The effect of poling on curie transition and ferroelectric crystal structure. *Polymer* 38:4881–4889.

Kawai, H. 1969. The piezoelectricity in poly(vinylidene fluoride). *Jpn. J. Appl. Phys.* 8:975–976.

Kazmierski, T.J. and S. Beeby. 2011. *Energy Harvesting Systems: Principles, Modeling and Applications.* New York: Springer-Verlag.

Ke, P., Jiao, X.-N., Ge, X.-H., Xiao, W.-M., and B. Yu. 2014. From macro to micro: Structural biomimetic materials by electrospinning. *RSC Adv.* 4:39704–39724.

Kim, H., Tadesse, Y., and S. Priya. 2009. Piezoelectric energy harvesting. In *Energy Harvesting Technologies,* eds. S. Priya and D.J. Inman, pp. 3–36. New York: Springer.

Kim, K.J. and S.L. Hsu. 1994. An infra-red spectroscopic study of structural reorganization of a uniaxially drawn VDF/TrFE copolymer in an electric field. *Polymer* 35:3612–3618.

Kubba, A.E. and K. Jiang. 2014. A comprehensive study on technologies of tyre monitoring systems and possible energy solutions. *Sensors* 14:10306–10345.

Laforgue, A. 2011. All-textile flexible supercapacitors using electrospun poly(3,4-ethylenedioxythiophene) nanofibers. *J. Power Sources* 196:559–564.

Lee, C.S., Kim, J.Y., Lee, D.E. et al. 2003. Flexible and transparent organic film speaker by using highly conducting PEDOT/PSS as electrode. *Synt. Met.* 139:457–461.

Lee, S. 2009. Developing UV-protective textiles based on electrospun zinc oxide nanocomposite fibers. *Fiber Polym.* 10:295–301.

Li, C., Wu, P.M., Lee, S., Gorton, A., Schulz, M.J., and C.H. Ahn. 2008. Flexible dome and bump shape piezoelectric tactile sensors using PVDF-TrFE copolymer. *J. Microelectromech. Syst.* 17:334–341.

Li, D., Wang, Y., and Y. Xia. 2003. Electrospinning of polymeric and ceramic nanofibers as uniaxially aligned arrays. *Nano Lett.* 3:1167–1171.

Li, D., Wang, Y., and Y. Xia. 2004. Electrospinning nanofibers as uniaxially aligned arrays and layer-by-layer stacked films. *Adv. Mater.* 16:361–366.

Li, F., Zhao, Y., and Y. Song. 2010. Core-shell nanofibers: Nano channel and capsule by coaxial electrospinning. In *Nanofibers,* ed. A. Kumar, pp. 419–438. Croatia, Europe: InTech Europe.

Lovinger, A.J. 1983. Ferroelectric polymers. *Science* 220:1115–1121.

Mandal, D., Yoon, S., and K.J. Kim. 2011. Origin of piezoelectricity in an electrospun poly(vinylidene fluoride-trifluoroethylene) nanofiber web-based nanogenerator and nano-pressure sensor. *Macromol. Rapid Commun.* 32:831–837.

Park, C., Ounaies, Z., Wise, K.E., and J.S. Harrison. 2004. In situ poling and imidization of amorphous piezoelectric polyimides. *Polymer* 45:5417–5425.

Pawlowski, K.J., Belvin, H.L., Raney, D.L., Su, J., Harrison, J.S., and E.J. Siochi. 2003. Electrospinning of a micro-air vehicle wing skin. *Polymer* 44:1309–1314.

Pereira, J.N., Sencadas, V., Correia, V., Rocha, J.G., and S.L. Mendez. 2013. Energy harvesting performance of piezoelectric electrospun polymer fibers and polymer/ceramic composites. *Sensor. Actuat. A-Phys.* 196:55–62.

Persano, L., Dagdeviren, C., Su, Y. et al. 2013. High performance piezoelectric devices based on aligned arrays of nanofibers of poly(vinylidenefluoride-co-trifluoroethylene). *Nat. Commun.* 4:1633–1642.

Qiu, X. 2010. Patterned piezo, pyro, and ferroelectricity of poled polymer electrets. *J. Appl. Phys.* 108:011101.

Ramadan, K.S., Sameoto, D., and S. Evoy. 2014. A review of piezoelectric polymers as functional materials for electromechanical transducers. *Smart Mater. Struct.* 23:033001.

Ren, G., Cai, F., Li, B., Zheng, J., and C. Xu. 2013. Flexible pressure sensor based on a poly(VDF-TrFE) nanofiber web. *Macromol. Mater. Eng.* 298:541–546.

Reynolds, N.M., Kim, K.J., Chang, C., and S.L. Hsu. 1989. Spectroscopic analysis of the electric field induced structural changes in vinylidene fluoride/trifluoroethylene copolymers. *Macromolecules* 22:1092–1100.

Sencadas, V., Ribeiro, C., Pereira, J.N., Correia, V., and S.L. Mendez. 2012. Fiber average size and distribution dependence on the electrospinning parameters of poly(vinylidene fluoride–trifluoroethylene) membranes for biomedical applications. *Appl. Phys. A* 109:685–691.

Serrano, W. and N.J. Pinto. 2012. Electrospun fibers of poly(vinylidene fluoride-trifluoroethylene)/poly(3-hexylthiophene) blends from tetrahydrofuran. *Ferroelectrics* 432:41–48.

Sharma, J., Lizu, M., Stewart, M. et al. 2015a. Multifunctional nanofibers towards active biomedical therapeutics. *Polymers* 7:186–219.

Sharma, S. 2013. Ferroelectric nanofibers: Principle, processing and applications. *Adv. Mat. Lett.* 4:522–533.

Sharma, T., Naik, S., Langevine, J., Gill, B., and J.X.J. Zhang. 2015b. Aligned PVDF-TrFE nanofibers with high-density PVDF nanofibers and PVDF core–shell structures for endovascular pressure sensing. *IEEE Trans. Biomed. Eng.* 62:188–195.

Singh, M., Haverinen, H.M., Dhagat, P., and G.E. Jabbour. 2010. Inkjet printing-process and its applications. *Adv. Mater.* 22:673–685.

Sun, B., Long, Y.-Z., Liu, S.-L. et al. 2013. Fabrication of curled conducting polymer microfibrous arrays via a novel electrospinning method for stretchable strain sensors. *Nanoscale* 5:7041–7045.

Sundararajan, S., Bhushan, B., Namazu, T., and Y. Isono. 2002. Mechanical property measurements of nanoscale structures using an atomic force microscope. *Ultramicroscopy* 91:111–118.

Swallow, L.M., Luo, J.K., Siores, E., Patel, I., and D. Dodds. 2008. A piezoelectric fibre composite based energy harvesting device for potential wearable applications. *Smart Mater. Struct.* 17:025017.

Tadigadapa, S. and K. Materi. 2009. Piezoelectric MEMS sensors: State of the art and perspectives. *Meas. Sci. Technol.* 20:092001.

Tanaka, R., Tashiro, K., and M. Kobayashi. 1999. Annealing effect on the ferroelectric phase transition behavior and domain structure of vinylidene fluoride (VDF)-trifluoroethylene copolymers: A comparison between uniaxially oriented VDF 73 and 65% copolymers. *Polymer* 40:3855–3865.

Taylor, G. 1964. Disintegration of water droplets in an electric field. *Proc. R. Soc. A* 280:383–397.

Wang, X., Liu, B., Wang, Q. et al. 2013. Three-dimensional hierarchical $GeSe_2$ nanostructures for high performance flexible all-solid-state supercapacitors. *Adv. Mater.* 25:1479–1486.

Wang, X., Yang, B., Liu, J. et al. 2015. Flexible triboelectric and piezoelectric coupling nanogenerator based on electrospinning P(VDF-TrFE) nanowires. *Proceedings of the MEMS 2015, Estoril, Portugal*, pp. 110–113.

CHAPTER 7

Bespoke Superhydrophobic Materials

Role of Polymers and Polymer Nanocomposites

Sheila Devasahayam

CONTENTS

7.1 INTRODUCTION

"Biomimetics or biomimicry," under the industrial ecology is a form of engineering concerned with principles of the ecosphere. It captures the rules that govern the functioning of the ecosphere and uses biological analogy and ecosystem metaphor. Some of the nature-inspired innovations include manufacturing wet materials by imitating nature (e.g., spider silk); the surface of the gecko's foot; mussel adhesive which works underwater and sticks to everything, even without a primer; and beading of water on nonpolar surfaces such as waxy leaves (e.g., lotus effect) (Figure 7.1) [1].

Plant leaves show a wide variability in surface structures, such as various epidermis cell shapes (microstructure), epicuticular wax crystals (nanostructure), varying chemical composition of the wax, and coarse inhomogeneities such as leaf veins. The red rose "petal effect" on the other hand, a variant of lotus effect [2], demonstrates superhydrophobicity through a high adhesive force of droplets with the micro/nanostructures. All of these parameters seem to affect wettability [3]. Other plants have since been discovered to have a similar lotus leaf effect, including the *Nasturtium* spp., taro, and the prickly pear cactus. The self-cleaning effect of the cicada wings is attributed to the superhydrophobic cicada wings covered in rows of waxy cones about 200 nm or billionths of a meter high (Figure 7.2b). The rain droplets rolling or splashing off these wings can remove soil, dust, pollen, and microbes (http://www.insidescience.org/content/cicada-wings-are-self-cleaning/993). Another interesting example of a natural multiscale rough surface is the gecko lizard foot and legs of many insects (Figure 7.2c). The hierarchical roughness allows gecko to develop adaptive "smart" adhesion, which can be very high (allowing to climb a wall or even a ceiling) or small (when it needs to detach from the surface).

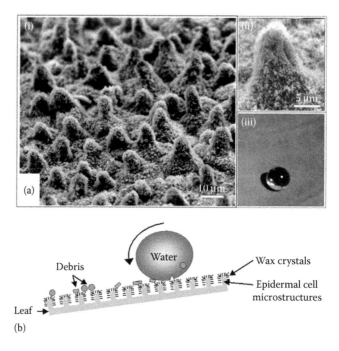

FIGURE 7.1 (a): (i) SEM image of the surface of a lotus leaf and (ii) a higher magnification with hierarchical structures clearly resolved. (iii) A water drop on the surface of the lotus leaf attains a nearly spherical shape. (From http://spie.org/documents/Newsroom/Imported/1441/1441_5423_0_2009-01-16.pdf.) (b) The lotus effect. Water forms droplets on the tips of the epidermal protrusions and collects pollutants, dirt, and small insects as it rolls off the leaf. (http://www.thenakedscientists.com/HTML/articles/article/biomimeticsborrowingfrombiology/.)

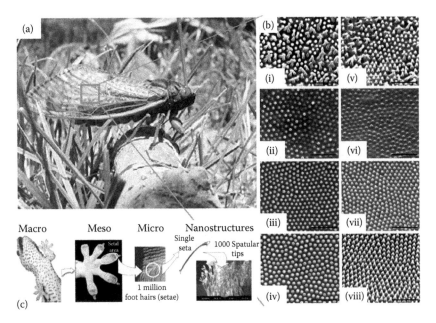

FIGURE 7.2   (a) Superhydrophobic wings of cicada. (b) SEM images of the four types of surfaces of cicada wings. i through iv: (i) Type A (*C. maculata*); (ii) Type B (*M. conica*); (iii) Type C (*M. microdon*); (iv) Type D (*T. jinpingensis*). The surfaces in v through viii were titled 30° from those in i through iv, respectively. Scale bars: 1 μm. (From Sun, M. et al., *J. Exp. Biol.*, 212, 3148, 2009. Reproduced with permission.) (c) The hierarchical structures of a gecko foot. (From https://www.google.com.au/ search?q=gecko's+foot&tbm=isch&tbo=u&source=univ&sa=X&ei=BM5HUvXAEonIkgXmuoD QBg&ved=0CDoQsAQ&biw=1920&bih=989&dpr=1#facrc=_&imgdii=_&imgrc=fIVSywVnKM IC6M%3A%3BG8fGSZwEQooCAM%3Bhttp%253A%252F%252Finfo.admet.com%252FPortals% 252F70514%252Fimages%252Fimage001-resized-600.png%3Bhttp%253A%252F%252Finfo.admet. com%252Findustry-news%252Fbid%252F74862%252FMIT-Researchers-Working-to-Develop-Biomimetic-Medical-Adhesive%3B600%3B235.)

The lotus effect refers to superhydrophobicity exhibited by the leaves of the lotus flower (Figure 7.1). Lotus leaf and other natural hydrophobic surfaces have a multiscale (or hierarchical) roughness structure, that is, nanoscale bumps superimposed over microscale asperities, minimizing the contact area between the leaf and the liquid. The lotus effect has led to several innovative applications in coatings and textiles to repel dirt, and in energy applications where superhydrophobic coating of the solar panel keeps the surface cleaner (Figure 7.3). Self-cleaning surfaces are desired for surfaces of satellite dishes, solar energy panels, photovoltaics, exterior architectural glass and green houses, and heat transfer surfaces in air-conditioning equipment. Nanoscale control of surface roughness results in nearly perfect optical clarity or transparency, which is a must for applications like self-cleaning, nonfogging displays, avoiding ice formation on optical elements, and protecting culturally important statuary from acid rain corrosion (http://www.phy.davidson.edu/ fachome/dmb/RESolGelGlass/SH/SuperhydrophobicCoatingsFinal.pdf). P2i's Aridion technology relies on superhydrophobic coatings to lower the surface energy of electronic devices such as the smartphones protecting the internal and external components

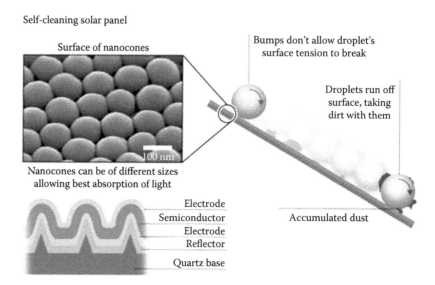

FIGURE 7.3 Self-cleaning solar panel. The pattern of tiny bumps lets the solar cell absorb more energy than standard cells while it to shed water and clean itself of dust. (From https://www.google.com.au/search?q=lotus+leaf+effect&tbm=isch&tbo=u&source=univ&sa=X&ei=YIk-UuvMHs O2kgW3h4Eo&ved=0CDEQsAQ&biw=1920&bih=946&dpr=1#facrc=_&imgdii=_&imgrc=gdj-EDk4TaslgDM%3A%3Bzh6pir2PKuHx-M%3Bhttp%253A%252F%252Fwww.newscientist.com%252 Fdata%252Fimages%252Farchive%252F2736%252F27366501.jpg%3Bhttp%253A%252F%252Fwww.democraticunderground.com%252Fdiscuss%252Fduboard.php%253Faz%253Dview_all%2526 address%253D115×218793%3B528%3B404. Copyrights NewScientist.)

from corrosion and water damage. While these applications involve solid surfaces, the emergence of flexible membrane has widened its uses in garments and barrier membranes, such as waterproofing textiles [4]. Other application of hydrophobic structures and materials include the development of micro fuel cell chips, where the selective removal of waste carbon dioxide (CO_2) from the liquid fuel is facilitated through the use of the hydrophobic membranes [5].

Drag reduction (DR) technologies vital for marine and defense industries rely on superhydrophobic coatings. They reduce "skin-friction drag for ships" hulls, allowing ships to increase their speed and range while reducing fuel costs. The HullKote Speed Polish used for smaller boats is a typical example of superhydrophobic coatings for surface protection [6]. They can also reduce corrosion and prevent biofouling of the ship's hull. Self-cleaning surfaces and hydrophobic coatings make it extremely difficult for a treated surface to harbor bacteria [7]. This allows tools and surfaces to remain sterile, even after contact with contaminating fluids. With bacteria unable to cling to tools, equipment, and cloth, they remain sterile for much longer without needing to constantly be cleaned or replaced. Hence, these coatings find application in surgical tools, medical equipments, and substrates to prepare spherical particles for drug release and also provide platforms for ex vivo high-throughput analysis of bio-materials–cells interactions [7].

It is anticipated that superhydrophobic coatings has potential energy savings up to 105 trillion Btu/year; potential cost savings of $1.74 billion/year; and potential reduction in CO_2 emissions of 6.16 million tons/year for U.S. economy (http://nanotech.ornl.gov/expertise/Qu%20Nano%20FS%20EERE.pdf).

The longevity of superhydrophobic coatings depends on the type of coating, environmental conditions, and the substrate on which the coating is applied. Friction and abrasion also determine the longevity of these coatings. A smooth surface subjected to a lot of abrasion may not usually hold up well. However, an application within a porous surface such as concrete or wood can act differently. Water hardness, chemicals, and other water impurities can also have an effect on the superhydrophobic properties of these coatings (https://waterbeader.com/).

Superhydrophobic surfaces are highly hydrophobic and extremely difficult to wet. They exhibit contact angles (CAs) of a water droplet exceeding 150° and the roll-off angle/CA hysteresis (CAH) is less than 10°. Typical features of superhydrophobic materials [3] include the following:

- When submerged in water, they are covered with a layer of air.

- After re-emerging from the water, the surface appears to be dry.

- Water drops on the surface roll off at slight tilting angles because of the weak adhesion.

- Sessile water drops have high CAs.

7.2 HYDROPHOBIC EFFECT

Hydrophobic effect is the observed tendency of nonpolar substances to aggregate in aqueous solution and to exclude water molecules. Kauzmann [8] discovered that nonpolar substances like fat molecules clump up together to minimize contact with water to reduce the interfacial tension (Figure 7.4a). The hydrophobic effect explains the separation of a mixture of oil and water into its two components, and the beading of water on nonpolar surfaces such as waxy leaves. Materials exhibiting this effect are known as hydrophobes. Surface chemistry of the material as well as the surface texture of the materials determine the degree of the hydrophobicity or the wetting behavior of materials, that is, the ability of the materials to react with water or the biological species (biofouling).

7.2.1 Thermodynamics of Hydrophobic Interactions

Hydrophobic interactions are spontaneous, relatively stronger than other weak intermolecular forces (i.e., Van der Waals interactions or hydrogen bonds). When the hydrogen bonds between the water molecules are broken due to the introduction of a hydrophobe it leads to an endothermic reaction. The distorted water molecules make new hydrogen bonds forming an ordered ice-like cage structure called a clathrate cage around the hydrophobe (Figure 7.4b). The hydrogen bonding is maintained near a small hydrophobic region and not maintained near a large hydrophobic region. This increases the entropy of the system; resulting in negative ΔS. The large ΔS values results in negative ΔG values making hydrophobic interactions spontaneous.

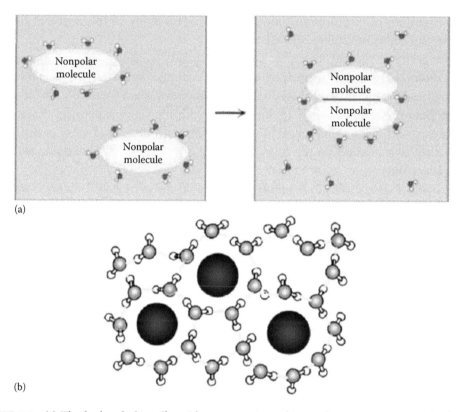

FIGURE 7.4   (a) The hydrophobic effect. The aggregation of nonpolar groups in water leads to an increase in entropy owing to the release of water molecules into bulk water. (From http://www. as.utexas.edu/astronomy/education/fall08/scalo/secure/309l_oct02_water.pdf.)   (b) Water molecules that are distorted by the presence of the hydrophobe will make new hydrogen bonds and form an ice-like cage structure called a clathrate cage around the hydrophobe. This orientation makes the system (hydrophobe) more ordered with an increase of the entropy of the system; therefore ΔS is negative. (From Molbert, S., The hydrophobic interaction, modeling hydrophobic interactions and aggregationof non-polar particles in aqueous solutions, PhD thesis, Université de Fribourg, Fribourg, Switzerland, 2003.)

The driving force for hydrophobic effect is the nucleation of an interface in water [9], which has significant length-scale dependence. A strong hydrophobic effect requires the presence of an oily surface in water extending about 1 (nm)^2. In this nanoscale regime, the strength of hydrophobic interactions scale linearly with oil–water surface area, whereas in the case of smaller separated oily components, hydrophobic interactions scale linearly with oily volume. The macroscopic effect is manifested by the separation of oil-rich and water-rich phases and the associated interfacial tension, γ. The microscopic effect is manifested by the relative insolubility of oil in water and its associated solvation free energy ΔG. Estimates of a microscopic surface tension from the variation of ΔG with molecular size yields values threefold smaller than γ. The discrepancy illustrates the difficulty in applying macroscopic concepts such as "interfacial surface" at the molecular level and can be formally resolved, qualitatively, by the predicted effect of surface curvature on surface tension [10]. For micro/nanoscale droplets, the value of the CA may be different than that for macroscale droplets. The effect of line tension is believed to be responsible for this phenomenon, since the energy of molecules in the vicinity of the triple line is different from that in the bulk or at the surface. However, this effect is significant only at nanometer range. Quere [11] and Checco et al. [12] attributed the scale dependence of the CA for nano/microsized droplets to the surface heterogeneity, assuming that tiny droplets first appear at the most wettable spots. At standard conditions, increasing temperature causes the interfacial tension to decrease while it causes the solvation free energy ΔG to increase. The order of factors affecting the strength of hydrophobic interactions depends on temperature > number of carbons on the hydrophobes > the shape of the hydrophobes (Justin Than, http://chemwiki.ucdavis.edu/Physical_Chemistry/Physical_Properties_of_Matter/Intermolecular_Forces/Hydrophobic_interactions).

7.2.1.1 Contact Angle

CA measurements often in combination with the tilting angle provides quantitative information on hydrophilic to superhydrophobic surfaces [3]. A CA less than 90° indicates that wetting of the surface is favorable, and the fluid will spread over a large area on the surface; while CAs greater than 90° generally means that wetting of the surface is unfavorable, so the fluid will minimize its contact with the surface and form a compact liquid droplet (Table 7.1). CA measures static hydrophobicity. The phenomenon of wetting is more than just a static state. CAs formed by expanding and contracting the liquid are referred to as the advancing CA θa and the receding CA θr, respectively. Dynamic CAs can be measured at various rates of speed and at a low speed; it is close or equal to static CA. The difference between the advancing angle and the receding angle is called the hysteresis (H):

$$H = \theta a - \theta r$$

CAH arises from surface roughness and/or heterogeneity. The actual microscopic variations of slope on the surface create barriers that pin the motion of the contact line and alter the macroscopic CAs [13,14]. Surfaces that are not homogeneous will have domains, which impede motion of the contact line. The CAH and slide angle are dynamic measures. Young [15] used the CA to describe the forces acting on a fluid droplet resting on a solid surface

TABLE 7.1   Surface Classification Based on Their Contact Angle

Contact Angle (degrees)	Type of Surface	Example
~0	Superhydrophilic	UV-irradiated TiO_2
<30	Hydrophilic	Glass
30–90	Intermediate	Aluminum
90–140	Hydrophobic	Plastic
>140	Superhydrophobic	Lotus leaf

Source: http://nanoyou.eu/attachments/502_EXPERIMENT%20D1_Teacher%20 document%2011-13.pdf.

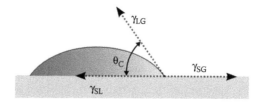

FIGURE 7.5 θ, is the angle formed by a liquid at the three phase boundary where the liquid, gas, and solid intersect CA and interphase energy between 3 phases (gas, liquid, solid).

surrounded by a gas based on surface energies between solid/liquid, solid/gas, and gas/liquid through CA as given by (Figure 7.5)

$$\gamma_{SG} = \gamma_{SL} + \gamma_{LG}\cos\theta$$

where:

γ_{SG} is the interfacial tension between the solid and gas
γ_{SL} is the interfacial tension between the solid and liquid
γ_{LG} is the interfacial tension between the liquid and gas

7.2.1.2 Microstructured Surface

Hydrophobic surfaces can be modified to superhydrophobic surfaces by a certain type of morphology. The thermodynamic equilibrium CAs on rough and heterogeneous surfaces are called Wenzel [16] and Cassie–Baxter angles, respectively. They are not equivalent to Young's CA. Wenzel [16] found that the CA of a liquid with a rough surface is different from that with a smooth surface. According to Wenzel, when the liquid is in intimate contact with a microstructured surface, θ changes to θ_w^*.

$$\cos\theta_w^* = r\cos\theta$$

where r is the ratio of the actual area to the projected area (Figure 7.6).

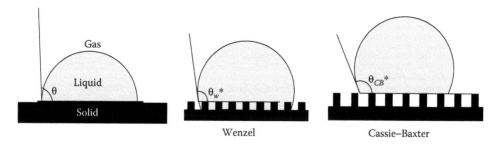

FIGURE 7.6 A droplet resting on a solid surface and surrounded by a gas forms a characteristic CA θ. If the solid surface is rough, and the liquid is in intimate contact with the solid asperities, the droplet is in the Wenzel state. If the liquid rests on the tops of the asperities, it is in the Cassie–Baxter state.

Wenzel's equation shows that microstructuring a surface amplifies the natural tendency of the surface [16]. In Wenzel's state, a hydrophobic surface (one that has an original CA greater than 90°) becomes more hydrophobic and a hydrophilic surface becomes more hydrophilic when microstructured [17]. Cassie and Baxter showed that air pockets may be trapped in the cavities of a rough surface, resulting in a composite solid–liquid–air interface, as opposed to the homogeneous solid–liquid interface. According to Cassie and Baxter the liquid suspended on the tops of microstructures, θ will change to θ_{CB}^*

$$\cos\theta_{CB}^* = \varphi\,(\cos\theta + 1) - 1$$

where φ is the area fraction of the solid that touches the liquid.

Liquids in the Cassie–Baxter state is more mobile than in the Wenzel state (Figure 7.6). For the Cassie–Baxter state to exist, the following inequality must be true [11]

$$\cos\theta < (\varphi - 1)/(r - \varphi)$$

Liquids in the Cassie–Baxter state generally exhibit lower slide angles and CAH than those in the Wenzel state (Figure 7.6). For the Cassie–Baxter state to exist the following two criteria need to be met: (1) contact line forces overcome body forces of unsupported droplet weight and (2) the microstructures are tall enough to prevent the liquid that bridges microstructures from touching the base of the microstructures [18]. Maintenance of the Cassie state requires that the Laplace pressure generated by the intruding interface balance the applied pressure. For roughness characterized by a single length scale δ, the requirement for water repellency is simply expressed as $\sigma/\delta > \Delta p$, where σ is the surface tension and Δp is the pressure difference [19].

7.2.1.3 Drag Reduction

There is a correlation between wetting properties of a liquid and the surface friction [20,21]. Various methods have been invented in the field, including riblets [22–24], compliant coatings [25], surfactants [26,27], polymer solutions [28–30], microbubbles [31–35], the Leidenfrost effect [36], and superhydrophobic coatings.

Among the above methods, the superhydrophobic coatings can yield "skin-friction DR for ships" hulls, thus increasing the fuel-efficiency, having vital applications in the maritime industry with a potential to save billions of dollars. They also reduce corrosion, and prevent marine organisms from growing on a ship's hull and can make removal of salt deposits possible without using freshwater. Superhydrophobic coatings also last longer than traditional coatings.

Drag increases manifold when flow transforms from laminar to turbulent condition. Under intense turbulent conditions, formation of pockets of air or vapor was reported by Vandogen and Roche Jr. [37]. Lumley [38] related DR to the viscosity ratio of the effective viscosity in the pipe core (which increased) to the viscosity adjacent to the wall, which did not change. A drag reducer will shift the transition from a laminar flow to a turbulent flow having higher flow velocity.

DR is given by Truong [39]:

$$\%DR = (\Delta Ps - \Delta Pp\} \times 100/\Delta Ps$$

where:

ΔPs is the pressure in a given length of tube for a pure solvent

ΔPp is the pressure drop for drag reducing solution with the same flow rate of liquid for both

The pressure loss in a pipe is due to fluid-frictional resistance, broadly classed in terms of laminar and turbulent flows by the fluid Reynolds number. Turbulent flow is defined here in the engineering sense of the flow exceeding a critical Reynolds number (Re) = 2000. In Newtonian fluids such as water, the friction drag is a function of the Reynolds (Re) number.

Re for pipes is given by

$$Re = VD/v > 2300$$

For an external flow, for example, over a ship hull

$$Re = VL/v > 500,000$$

where:

V is the flow velocity

L is the length

D is the pipe diameter

v is the kinematic viscosity of the drag reducing solution

The DR is related to the Darcy's frictional factor through pressure drop, which in turn is related to Re as

$$f_D = \frac{\Delta p}{\rho V^2} \frac{2D}{L}$$

$f_D = 64/Re$, for the laminar regime, and $1/(f_D)^{1/2} = 2\log_{10} Re\ (f_D\}^{1/2} - 0.8$ for the turbulent flow.

For pipes, the friction parameter is a function of Re and the relative roughness of the inner surface of the pipe. Friction factor is the ratio between the input of energy provided by an external pressure difference and the kinetic energy of the resulting mean flow in the pipe. It is a measure of the force that is required to sustain a certain mean flow (http://math.unice.fr/~musacchi/tesi/node45.html).

The effect of DR is to reduce the friction to a value considerably lower than the turbulent flow of the solvent, but not approaching that corresponding to laminar conditions

(Figure 7.7). In the laminar regime the friction drag decrease as Re^{-1} until the critical Re number (defined as 2000) is reached. Transition to turbulence causes a sudden increase of the friction drag which for fully developed turbulence reaches an almost constant value with only a weak logarithmic dependence on Re described by the Prandtl–Karman (P–K) law (a straight line in P–K coordinates) corresponding to the turbulent behavior of Newtonian fluids. The dependence of the friction drag on the Re number is conventionally shown in the P–K coordinates: $1/\sqrt{(f_D/4)}$ versus $\log(Re/\sqrt{(f_D/4)})$ (Figure 7.8).

British chemist Toms, in 1949, while performing experiments on the degradation of polymers observed that the addition of few parts per million of long chain polymers in turbulent flow produces a dramatic reduction of the friction drag, and polymer solutions are most often employed as drag reducing systems since then [40]. Experiments show that the higher the molecular weight (MW), the more effective a given polymer as a drag reducer for a given concentration and Re number [41]. Polymers with a MW below 100,000 seem to be ineffective [41].

Extension of the polymer chain is critical for DR. The longer polymer chain provides more chance for entanglement and interaction with the flow. For a given MW, polymer with linear structure such as poly(ethylene oxide), polyisobutylene, and polyacrylamide are most effective drag reducing polymers with maximum extensivity. Nonlinear polymers such as gum arabic and the dextrans are ineffective for DR [41].

Kim and coworkers [42–44] found that salt enhances the DR of water-soluble poly(acrylic acid) (PAA) because the salt molecules provides stabilizing mechanism and prevent the aggregation of PAA chains which lower the DR properties of the PAA solution.

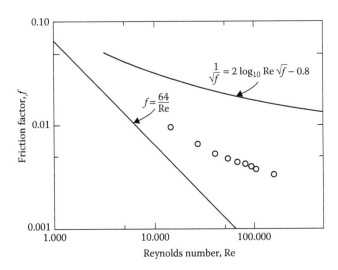

FIGURE 7.7 Typical data for drag reducing polymer solutions fall between the turbulent friction line for pipe flow, and the laminar line, 64/Re, extended beyond its usual limit of a Reynolds number of 2300, where f is the pipe friction coefficient in engineering terms, equal to pressure dropper length times the diameter, divided by $pV\,2$, and ρ is the fluid density. (From Hoyt, JW., Drag reduction, in *Encyclopedia of Polymer Science and Engineering*, 2nd ed., vol. 5, 1986, 129–151.)

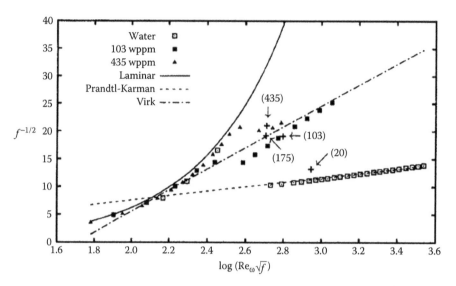

FIGURE 7.8 The relation between the friction factor and the (wall) Reynolds number in P–K coordinates. The data for water and for the 103 and 435 wppm solutions are presented at a number of (wall) Reynolds numbers as indicated for the symbols. (Reproduced with permission from Ptasinski, P.K. et al., *Turbul. Combust.*, 66, 159, 2001; http://mate.tue.nl/mate/pdfs/908.pdf.)

7.3 MECHANISMS: MOLECULAR EXTENSION

The polymer hydrodynamic coil interacts with and disrupts the eddies and microvortices present in turbulent flows. Tulin [45] and Berman [46] observed that the DR effect depends on the stretching of individual molecules by high strain rates in the flow. At high strain rates, the polymer chain tends to elongate along the principal strain rate axis, and large extensions result. At the same time, a form of strain rate hardening occurs in which the elongation viscosity becomes very high. As the elongation viscosity increases, the large-scale bursts and sweeps in the wall layer flows are inhibited, thus reducing the friction [45]. In order to observe significant DR, enough polymer molecules should be present so that the mean distance between stretched molecules is at least an order of the molecular size. However, it is not clear if it is a requirement for DR. It is argued that if individual fully elongated molecules can produce observable DR, then the effect should be observed for much lower concentrations than have been reported. The fact that large numbers of partially elongated molecules do give significant DR indicating that at low concentrations only a few molecules can be stretched by the flow.

DR increases as the flow strain rate increases for the randomly coiled molecules until a limit is reached. Both these molecules and extended polyelectrolytes show increased DR with concentration and MW up to a limit. Even the extended polyelectrolytes that have a constant level of DR over some range of Re eventually show a decline at high Re. The amount of DR depends on the diameter of the pipes: there is less DR for larger pipes even when the onset effect is taken into account [45,46].

7.3.1 DR with Surfactant Solutions

Surfactants are used as effective drag reducers owing to their chemical and mechanical stability. Surfactant systems exhibit DR at concentrations similar to dilute polymer solutions (<100 ppm) [39,49]. The main characteristic feature of the friction behavior of surfactant solutions is the disappearance of DR when a critical wall stress is reached [50].

The molecular structure of the surfactant has an important effect on its micelle size and shape which in turn profoundly influence the drag reducing ability [51,52]. Formation of micelles and their shapes are main factors influencing the DR ability of the surfactant solutions. The spherical micelle is generally conceived as a small ball-like particle of colloidal dimensions and fairly constant in size for a given surfactant. These spherical micelles exist only in relative dilute solutions. In concentrated solutions, however, the lamellar micelle is favored. Under the influence of an electrolyte, spherical micelles can rearrange into cylindrical or rod-like micelles (Figure 7.9) [49,53].

The shapes of micelles that are concentration-dependent play a vital role in DR. The shape of a surfactant molecule can be described by its surfactant packing parameter, N_s. The packing parameter takes into account the volume of the hydrophobic chain (V_c), the cross-sectional area of the hydrophobic core of the aggregate expressed per molecule in the aggregate (a), and the length of the hydrophobic chain (L_c) [54]:

$$N_s = \frac{V_c}{a * L_c}$$

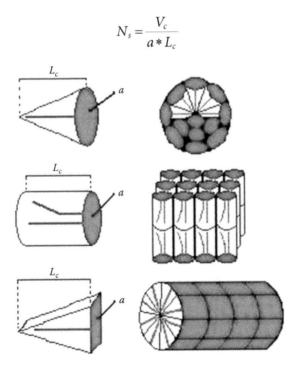

FIGURE 7.9 Shape of a surfactant molecule can be described by its surfactant packing parameter, N_s. The packing parameter takes into account the volume of the hydrophobic chain (V_c), the cross-sectional area of the hydrophobic core of the aggregate expressed per molecule in the aggregate (a) and the length of the hydrophobic chain.

The shape of a micelle is directly dependent on the packing parameter of the surfactant. Surfactants with a packing parameter of $N_s \leq 1/3$ are cone-shaped, packing together to form spherical micelles when in an aqueous environment (Figure 7.9) [54]. Surfactants with a packing parameter of $1/3 < N_s \leq 1/2$ have a wedge-like shape and will aggregate together in an aqueous environment to form cylindrical micelles (Figure 7.9) [54]. Surfactants with a packing parameter of $N_s > 1/2$ appear to have a cylindrical shape and pack together to form a bilayer in an aqueous environment (Figure 7.9) [54].

Savins [55] observed stress-controlled DR (30%) effect in the soap solutions of 0.2% sodium oleate, an anionic surfactant. The DR was observed to increase with increasing shear stress up to a critical value. Beyond the critical value, the DR of the soap solution became indistinguishable from that of the soap-free solution. This indicates that the network of micelles collapses if the shear stress exceeds a critical shear stress. This occurs because of a temporary disentanglement of the network induced by turbulent vortices and eddies in fully developed flow. If the wall shear stress is reduced from above to below the critical value, then the network bonds reform and the DR ability of the solution is restored [39].

Gadd [56] suggested the possibility of using the cetyltrimethylammonium bromide (CTAB), a cationic surfactant and naphthol mixture having shear-thinning characteristics to reduce turbulent friction. Similar to anionic surfactant solutions, the drag reducing ability of the CTAB–naphthol solution terminated at some upper Reynolds number corresponding to a critical shear stress where there was a scission of the micelles. Advantage of using cationic surfactants over the anionic ones is that these complex soaps do not precipitate in the presence of calcium ions. However, they are expensive and degrade chemically in aqueous solutions in a matter of a few days. Further, although they are mechanically stable, they are not thermally stable and thus limited in practical applications [39].

Zakin and Chang [51,52] reported the critical shear stress for mechanical degradation in the case of nonionic surfactant to be dependent on the surfactant concentration, electrolyte type and concentration, and on the temperature. Nonionic surfactants have an advantage over the anionic and cationic counterparts because they are both mechanically and chemically stable. They do not precipitate out in the presence of calcium ions and therefore can be used in impure waters, seawater, or concentrated brine solutions. Despite these merits, more studies are needed to exploit the potential of nonionic surfactants to their fullest extent as a drag reducer.

When one compares the data for surfactant solutions with that for polymer solutions, it becomes obvious that the DR behaviors in these two cases are different. The critical shear stress can be up to 100 Pa, which is quite high when compared to the wall shear stress where mechanical degradation starts in dilute polymer solutions [39].

While the soap solution exhibits DR low wall shear stress values, the polymer solutions show relatively small DR at low Reynolds numbers and increasingly large reduction at high Reynolds numbers, attributed to the morphological difference between micellar and polymeric structures [50]. The network of micelles collapses if the wall shear stress exceeds. However, if the wall shear stress is reduced from above to below the critical value, the network bonds reform and the DR ability of the surfactant solution is restored. In contrast, once the polymer chains are broken by high shear stress, the drag reducing ability of the polymer solution is permanently lost.

A flexible polymer molecule needs to be elongated by a large velocity gradient before its full drag reducing ability is developed, but the surfactant particles are oriented much more easily at lower velocity gradients, and the micelles collapse at high shear stresses associated with large velocity gradients [50]. In terms of equivalent MW, micelles are known to have larger values than polymers and therefore they would shift the onset of DR to a lower shear stress value [41,50]. Large extensional viscosity of wormlike micelles has led to significant DR in turbulent flows [57]. Although both wormlike micelle solutions and polymer solutions can be viscoelastic, wormlike micelles are physically quite different from polymers. Whereas the backbone of a polymer is covalently bonded and rigid, wormlike micelles are held together by relatively weak physical attractions and as a result are continuously breaking and reforming with time. In an entangled network, both individual polymer chains and wormlike micelles can relieve stress through reptation driven by Brownian motion. However, unlike polymeric fluids, wormlike micelle solutions have access to a number of stress relief mechanisms in addition to reptation. Wormlike micelles can relieve stress and eliminate entanglement points by either breaking or reforming in a lower stress state [58].

A remarkable aspect of polymers as a drag reducer is that the DR occurs at very low concentrations in the ppm region. According to Oliver and Bakhtiyarov [59], increasing the concentration beyond 30–40 ppm lowers DR for PEO in a small tube owing to increase of the viscosity with increasing concentration. Interestingly, DR can be observed in concentrations as low as 0.02 ppm [59]. Dilute polymer solutions deviate from the P–K law. For smaller Re, below the critical threshold, their behavior is similar to the Newtonian fluid, for larger Re numbers, the friction drag is drastically reduced with respect to the Newtonian case and it finally reaches a universal asymptote which is independent on the kind of polymers or the concentration of the solution, and is known in literature as the maximum DR asymptote (MDRA) (http://math.unice.fr/~musacchi/tesi/node45.html). A theory based on the elastic behavior of polymers was proposed by Tabor and de Gennes in 1986 [60] to explain both the onset of the DR and the presence of the universal upper bound. An overview of this theory can be found in Sreenivasan and White [61]. A similar MDRA exists for surfactant DR, however, even though the DR mechanism appears to be the same for polymers and wormlike micelles, the MDRA for surfactant drag reducers has been found to be between 5% and 15% higher depending on Reynolds number. Surfactant solutions are so much more efficient at reducing turbulent drag [57,62].

The shapes of the micelles and the polymers affect the DR. Main difference between flexible and rodlike polymers is that the former need a minimal value of Reynolds number in order to stretch (i.e., undergo the coiled-to-stretched transition, whereas rodlike polymers are "stretched" at any value of the Reynolds number). For a rodlike polymer at low Re, a drag enhancement is observed, which smoothly transforms into a DR at high Re. When the value of Re is small, the drag is enhanced due to the homogeneous increase in the effective viscosity. When Re is sufficiently large, due to the turbulent activity, the effective viscosity varies as a function of the distance to the wall. As a result, the amount of drag is reduced. It should be noted that only rodlike polymers exhibit drag enhancement for low Re. For flexible polymers, the drag is the same as that of the Newtonian flow for low Re, and only after a critical value of Re, the DR sets in. This difference in behavior from rodlike polymers is because the flexible polymers are coiled when Re is small. They do not affect the flow unless Re increases enough to allow the shear to develop to affect the coil-to-stretch transition in the flexible polymers. In contrast, rodlike polymers are always extended and, therefore, they can affect the flow for all Re [63].

Dong et al. [64] have demonstrated that while increase in the wetting area does affect DR, the plastron effect makes the major contribution to the drag reducing effect. With a plastron, the bubble layer hinders tight contact between the solid surface and the water, replacing the liquid/solid interface with a liquid/bubble/solid interface. (The gaseous plastron effect refers to the mechanism by which insects or water mammals form air bubbles on their skin in order to dive.) When such bubbles form on a coated surface the escaping gas is believed to reduce drag [65].

7.3.2 Biofouling

Biofouling control is a vital part of any DR projects. Surface wettability and microtopography can either enhance or deter larval settlement of many sessile marine organisms. On a tested towing tank, the fouling layer consisted of slimes and small barnacles, which were found to cause a fourfold increase in resistance compared to the original clean state [66]. Biofouling occurs when microorganisms, plants, algae, or animals accumulate on wetted surfaces and is most significant economically to the shipping industries, since high levels of fouling on a ship's hull significantly increases drag, reducing the overall hydrodynamic performance of the vessel increases the fuel consumption. Biofouling is also found in almost all circumstances where water-based liquids are in contact with other materials. Marine organisms are physically damaged or killed on contact with the wall made of nanofibers, preventing the settlement and growth of all organisms including bacteria. Nontarget organisms, not in contact, are unharmed. New generation composite materials are stronger and more rigid than steel while not corroding in the marine environment because the nanofiber wall in these structures provides a durable surface against the settling of fouling organisms.

Photoactive materials have the potential to prevent fouling through the generation of reactive oxygen species very close to a surface. The reactive oxygen species kills or repels the settling organisms, and is environmentally safe due to a very short half-life. Photoactive materials have the potential to prevent fouling through the generation of reactive oxygen species very close to a surface. The reactive oxygen species kills or repels the settling organisms, and is environmentally safe due to a very short half-life [67] http://blake.lsmc.u-bordeaux.fr/UFR_Chimie/licence_chimie/sujets/Pub_Faure.pdf.

7.4 SUPERHYDROPHOBIC COATINGS

Self-cleaning superhydrophobic coating on a ship has an important application in the DR technology. While the effects of surfactants and the polymers on the DR are discussed in the preceding section, in this section we focus on the superhydrophobic coatings for DR in reducing the friction, increasing the CA, and decreasing the CAH.

DR in laminar flow using hydrophobic surfaces fabricated by silicon wafers was carried out and pressure reductions up to 40% was reported by [68]. As discussed in the foregoing sections the key strategies to produce superhydrophoblic surfaces include reducing the surface free energy of the surface, roughening of the surface as well as imparting chemical heterogeneity to the surface [69]. The quality of superhydrophobic surfaces depends on various factors such as the apolar characteristic of the material, the surface topography (nano-, micro-, or hierarchical structure), and the occurrence of surface defects or discontinuities. There are many ways to make rough surfaces, including mechanical stretching, laser/plasma/chemical etching, lithography, sol–gel processing and solution casting, layer-by-layer and colloidal assembling, electrical/chemical reaction and deposition, electrospinning, and chemical vapor deposition [4].

7.4.1 Chemical Heterogeneity

Chemical heterogeneity [66] leads to fluctuations of the surface energy and thus has a similar effect upon surface energy as roughness. That is the atomic-scale energy barriers combined with the micron-scale roughness leads to the frictional dissipation. Absorption of hydrocarbon molecules at the surface results in chemical inhomogeneity providing energy barriers that leads to friction. Polymer interactions with the inorganic components, introduces heterogeneity within the polymer matrix on the nanoscale and deforming the polymer from its bulk conformation [70].

Polymeric nanocomposites are organic–inorganic hybrid materials, in which inorganic nanoscale building blocks are dispersed in a polymeric matrix. Thermoplastic and thermosetting matrices together with ceramic fillers provide the necessary platform for imparting chemical heterogeneity, and hierarchical structures. Filler addition causes structural variations in the polymeric network, significantly influencing the degree of crystallinity and producing defects in the polymeric matrix [71,72]. Incorporating nanoparticles significantly improves the mechanical properties of the polymer matrix. This is attributed to much higher specific surface area of submicron size particle promoting stress transfer from the matrix to submicron particles. This has relevance to the strength of hierarchical structures to retain the stabilities of the composite structures discussed under Section 7.6.

Random arrangements of nanoparticles within a nanocomposite polymer matrix will not provide optimized performance required for many potential high-technology applications. A hierarchy of morphology in a nanocomposite is more important than a unary (micrometer) scale filled polymers due to the extreme aspect ratio of many nanoparticles, for example, exfoliated clays and carbon nanotubes. Two general approaches to invoke hierarchical morphology control are external-in (directed patterning of nanoparticle dispersions) and internal-out (mesophase assembly of nanoparticles).

Materials like CNTs often exhibit hierarchical structures that are composed of several levels. Although CNT diameters are in the nanometer range, larger structures are formed because of re-agglomeration of the nanoparticles. The nanospheres can self-assemble in different ways, including as densely packed structures and linearly packed structures (necklace-like), which can further build up different configurations, such as spider-web-like networks or particulated structures. The resulting configuration tends to be far more complex for CNTs than for polymer chains because of the tubular structures of the former. This enables a larger number of possible morphologies [73]. Directed patterning of nanoparticle dispersions using external electric fields relies on the creation of a multi-dimensional morphology by an external means to transform a random distribution to a prescribed ordered construction. For example, particle–particle interactions, which arise from induced magnetic dipoles, lead to chaining and subsequently long range, periodic ordering of the particle chains parallel to the lines of magnetic flux [74].

Wu and Shi [75] reported the fabrication of a lotus-like micro–nanoscale binary structured surface with superhydrophobic behavior from copper phosphate dehydrate materials. Other materials include manganese oxide polystyrene (MnO_2/PS) nanocomposite, zinc oxide polystyrene (ZnO/PS) nanocomposite, precipitated calcium carbonate, and carbon nanotube structures (http://en.wikipedia.org/wiki/Superhydrophobic_coating#cite_note-1).

7.5 LOW SURFACE ENERGY MATERIALS

Fluorocarbons and silicones are well-known hydrophobic materials. The low surface energy of fluorinated polymers has prompted research in to creating superhydrophobic surfaces using these polymers (Table 7.2). Most artificial superhydrophobic surfaces are usually produced of silicon or silica using photolithography for micropatterning [76] or from polymers using evaporation [77]. The surface energy values of some typical materials are shown in Table 7.2. Materials like the fluorocarbons, anodized alumina, and mica exhibit low surface energies. However, the carbon-based surfaces including carbon nanotube, graphene, graphene oxide, and graphite flakes that exhibit very low surface energy are the ideal candidates as nanofillers in polymer matrix to produce superhydrophobic surfaces.

The polytetrafluoroethylene (PTFE), widely known as Teflon, is used as coatings for anti-adhesion agents and chemical insulators because of its very low surface energy. The hydrophobic properties of PTFE are caused by the fluorination of the carbon bonds. Fluorocarbon polymers have an inherently low friction resistance to all substrates. They are essentially nonbiodegradable, which would reduce growth of such sea organisms as barnacles and other related marine organisms [65]. Unfortunately, PTFE is difficult to handle because of its insolubility, viscosity, and melt strength. Zhang et al. [78] produced superhydrophobic film by stretching a PTFE, the extended film consisted of fibrous crystals with a large fraction of void space in the surface imparting superhydrophobicity to the film.

Fabrication of superhydrophobic coating by incorporating hydrophobically modified silica particles in PTFE matrix is described by Bharathibai and Dinesh Kumar [79]. They achieved superhydrophobic PTFE-based nanocomposite coatings by incorporating hydrophobically modified silica nanoparticles in PTFE emulsion. The effect of the concentration of hydrophobically modified silica particles on the hydrophobicity of PTFE–HMS composite coating revealed that the wetting CA (WCA) increased with increase in concentration of HMS nanoparticles in the coating. WCA of 165° and a low sliding angle >2° was achieved at hydrophobically modified silica nanoparticle concentrations >30%.

Research group of Karapanagiotis et al. [80,81] has produced polymer–particle composite films that have superhydrophobic properties when applied on surfaces of stone and

TABLE 7.2   Surface Energy Values of Solid Materials

Material Type	Surface Energy (mJ/m²), γS (Experimental, Estimate,* or Theoretical†)
Low Energy Surfaces—Plastics, Rubber, and Composites	
Polyhexafluoropropylene	12.4
Polytetrafluoroetylene—PTFE	19.1
Poly(vinylidene fluoride)—PVF	30.3
Poly(chlorotrifluoroethylene)	33.5
Polyethylene—PE	32.4
Polypropylene—PP	33*
Poly(methylmethacrylate)—PMMA	40.2
Polystyrene—PS	40.6
Polyamide—PA, Nylon-6,6	41.4
Poly(vinylchloride)—PVC	41.5
Poly(vinylidene chloride)	45.0
Poly(ethylene terephthalate)	45.1
Epoxy—typical rubber toughened	45.5
Epoxy—typical amine cured	46.2
Phenol–resorcinol resin	52*
Urea–formaldehyde resin	61*
Styrene–butadiene rubber	29.1
Acrylonitrile–butadiene rubber	36.0
Carbon fiber-reinforced plastic (CFRP) abraded	58.0
High Energy Surfaces—Metals, Oxides, and Ceramics	
Aluminum oxide (Al_2O_3)—anodized	169
Aluminum oxide (Al_2O_3)—sapphire	638
Berylium oxide (BeO)	1107†
Copper	1360†
Graphite	1250†
Iron oxide (Fe_2O_3)	1357†
Lead	442†
Mercury	319†
Mica	120
Nickel	1770†
Platinum	2672
Silicon dioxide—silica	287
Silver	890†
Graphene	46.7
Graphene oxide	62.1
Graphite flake	54.8
Carbon nanotubes	28–45

Source: Kinloch, A.J., *Adhesion and Adhesives: Science and Technology*, Chapman & Hall, London, 1987.

other building materials, such as glass or wood. Those films were created by spraying metal oxide nanoparticles dispersed in polymer (siloxane or acrylic) on the building substrate.

Facio and Mosquera [82] recommend a sol–gel process of mixing silica nanoparticles with a silica oligomer and a low-MW organic siloxane in the presence of a surfactant to produce an effective coating product and the nanoparticles to create a surface roughness that will enhance the hydrophobicity produced. These nanoparticles produce a densely packed coating in which the air is trapped. Thus, water droplets cannot penetrate into the coating, and the contact area between droplet and surface is significantly minimized. In addition, the organic component reduces the surface free energy. This results in a high static CA (~150°) and low CAH values (~7°). The surfactant plays a valuable role whereby it acts to coarsen the pore gel network, thus preventing cracking during drying.

A two-step process, involving creation of a rough surface (e.g., with a fractal structure) and modification of the surface with materials of low surface free energy, such as fluorinated or silicon compounds is reported by many researchers including Liu et al. [83]. The micro–nanoscale binary-structured composite particles of silica/fluoropolymer are prepared using an emulsion-mediated sol–gel process, and then these composite particles are applied to various substrates to mimic the surface microstructures of lotus leaves resulting in superhydrophobic surfaces with a water CA larger than 150°.

Yu and Son [84] reported on fabrication of a superhydrophobic surface with adjustable hydrophobicity and adhesivity based on a silica nanotube array. They reported the silica nanobundle structures showing hydrophobic properties without any surface modification to decrease the surface energy. The hydrophobicity is attributed to the irregularities of the pillar heights. The apparent CA increased gradually with increasing height of the silica nanopillars. This height-dependent hydrophobicity was attributed to the increasing frontal surface area of the silica layer with the nanobundle-array surface morphology.

Nature does not require the lower surface energy of –CH3 groups or fluorocarbons to achieve hydrophobic effects. That is, extremely low surface energy is not necessary to achieve nonwetting, rather, it is the ability to control the morphology of a surface on micron and nanometer length scales which is crucial. In nature paraffinic hydrocarbons are responsible for the wetting and self-cleaning lotus effect. The lotus plant achieves an apparent WCA >160° and nil sliding angle using paraffinic wax crystals containing predominantly $-CH_2-$ groups. Lu et al. [85] produced a highly porous superhydrophobic surface of polyethylene (PE), with WCA up to 173° simply by controlling its crystallization behavior to form nanostructured floral-like crystal structures. Jiang et al. [86] obtained a superhydrophobic film composed of porous microparticles and nanofibers by electrostaticspinning and spraying of a PS solution in dimethylformamide (DMF). Lee et al. [87] used nanoimprint pattern transfer process to produce a three-dimensionally rough surface with advancing and receding WCA of 155.8° and 147.6°, respectively using vertically aligned PS nanofibers. When the aspect ratio of the PS nanofibers increased, the nanofibers formed twisted bundles resulting in a three-dimensionally rough surface.

Polyamide [88], polycarbonate [89], and alkylketene dimer [90] have also been made into superhydrophobic surfaces. Using electrochemical polymerization, Yan et al. [91] synthesized the needle-like poly (alkylpyrrole) structures grown perpendicularly to the surface of the electrode to yield a stable superhydrophobic film.

The decoupling of wetting from simple surface energy opens up many possibilities for engineering surfaces [4], such as microfluidics, piping and boat hulls, surfaces for satellite dishes, solar energy panels, photovoltaics, exterior architectural glass and green houses, and the heat transfer surfaces in air-conditioning equipment.

7.6 HIERARCHICAL SURFACE STRUCTURES

Nosonovsky and Bhushan [92] have summarized the design requirements for artificial roughness-induced superhydrophobic surfaces as high CA, CAH, and high slip lengths, reflecting the surface friction that is able to support liquid pressure created by either liquid flow or liquid droplet weight. This requires the formation of a composite solid–liquid–air interface (air or gas pockets trapped in the cavities of a rough surface, results in a composite solid–liquid–air interface, as opposed to the homogeneous solid–liquid interface).

Hydrophobic properties increase as the adhesive and/or frictional forces decreases. Ratio of elastic to surface energy controls the friction. For a smooth surface, the atomic-level interfacial potential energy is much weaker than the elastic potential energy, so the contacting surfaces behave as rigid bodies, without being able to reach minima of the interfacial energy, leading to the superlubricity [93]. Graphite exhibits superhydrophobicity, that is, when two atomically smooth noncommensurate bodies are in contact, positions of the energy barriers and minima would not coincide, so virtually no friction is expected.

Hierarchical surfaces are related to the hierarchical friction mechanisms to deal with the multiscale dissipation mechanisms. Different mechanisms of friction acting simultaneously at different scales and hierarchy levels lead to hierarchical nature of friction. The transition to a higher hierarchy level is a result of instability, which, in turn, is a consequence of nonlinearity of the system. However, surface roughness leading to the multiple asperity contact can provide such a mechanism. For microscale asperities, the elastic energy is small due to large distances between the asperities, and multiple equilibrium states with energy barriers between them exist [92].

Enhanced superhydrophobic effect is reported when a superhydrophobic structure configured with a plurality of protrusions with different heights [86]. In order to minimize CAH, a stable composite interface should exist. In order for the composite to be stable, the roughness should be hierarchical with nanoasperities over microasperities mimicking the lotus leaf [92], since destabilizing factors for the composite interface, for example, hierarchical frictional forces have different length scales. These surface composite coatings of micro–nanoscale binary structured composite particles can mimic the surface microstructures of lotus leaves. Roughness-induced superhydrophobic surface can lead to a very low friction with flowing liquid due to a small liquid slip.

Nosonovsky and Bhushan [92] reported that hierarchical structure was crucial for a high CA and for the stability of the water–solid and water–air interfaces. To effectively resist destabilization of the composite interface, namely, the capillary waves, condensation and accumulation of nanodroplets; surface inhomogeneity, which are scale-dependent with different characteristic scale lengths; and a multiscale (hierarchical) roughness are required. High asperities resist the capillary waves, while nanobumps prevent nanodroplets from filling the valleys between asperities and pin the triple line in case of a hydrophilic spot [94]. If the amplitude of a standing capillary wave at the liquid–air interface formed due to an external perturbation is greater than the height of the asperity, the liquid can touch the valley between the asperities; and if the angle under which the liquid comes in contact with the solid is greater than θ, it is energetically profitable for the liquid to

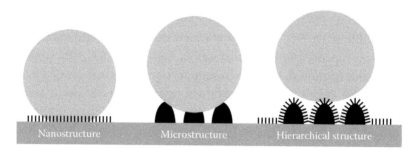

FIGURE 7.10 Effect of unitary (nonhierarchical) structures of micro- and nanoroughness, and hierarchical structures (microroughness covered with nanoroughness).

fill the valley. The effect of capillary waves is more pronounced for small asperities with heights comparable to the wave amplitude as in the case of unitary roughness, where the amplitude of asperity is very low. This is why the likelihood of instability of a unitary interface will be very high (Figure 7.10).

Nanoroughness is required to support nanodroplets, which may condensate in the valleys between large asperities. Both the condition of supporting the nanodroplets and condition of structural strength of the pillars result in smaller asperities having much higher aspect ratio, controlled by the asperities spacing factor [92].

On hierarchical structures when a water droplet falls from a high point to the superhydrophobic microstructure, the closed space formed by the water droplet and the protrusions can provide an air spring effect to bounce the water droplet away. The protrusions with different heights can disperse the impact of the falling water droplet, so that the superhydrophobic effect can be further enhanced [87]. Droplet's weight and curvature also contribute to the composite instabilities [90]. For small droplets, surface effects dominate over the gravity and the latter is not responsible for the transition. Instead, the curvature of the droplet may contribute to the composite instabilities. The curvature of a droplet is governed by the Laplace equation, which relates pressure inside the droplet to its curvature. The curvature is the same at the top and at the bottom of the droplet. For a patterned surface with pillars of height H and pitch P, maximum curvature with the radius on the order of P^2/H results. When actual radius of a droplet, R, exceeds this value, the transition to the homogeneous interface may occur. Therefore, the ratio RH/P^2 controls the transition due to the droplet's curvature [92].

Sun et al. [95] produced both a positive and a negative replica of a lotus leaf surface by nanocasting using poly(dimethylsiloxane), which has the CA with water of about 105°. This value is close to the CA of wax, which covers lotus leaves (about 103°). The positive and negative replicas have the same roughness factor and thus should produce the same CA in the case of a homogeneous interface; however, the values of surface curvature are opposite. The value of CA for the positive replica was found to be 160° (close to that of the lotus leaf), while for the negative replica it was only 110°. This result suggests that the sign of surface curvature plays a critical role in formation of the composite interface.

Convex nanoasperities can pin the liquid–air interface and thus prevent liquid from filling the valleys between asperities.

A high average roughness parameter resulting from a higher density of microcracks may reduce the water contact area, may also contribute to a greater CA (106.9°) [69]. Thus the surface cracks mainly determine the hydrophobicity. The wax particles tended to accumulate around the rims of the larger bumps increasing the surface roughness (1,326,102 nm) and thus increasing the CA from 71.3° to 112.4°. Holes and/or cracks may be related to the tensile strength of cellular solid materials.

Feng et al. [96] created superhydrophobic alumina surface based on rose petal effect through the manipulation of alumina's low surface free energy which results in low interaction force between the water and the alumina surface. Alumina gel is initially treated in the boiling water resulting in the roughened alumina coatings with porous structure. When stearic acid molecules are subsequently grafted onto the alumina surface, it is rendered hydrophobic. Both the roughened and porous structure and the hydrophobic materials at the alumina surface endue the alumina surface with outstanding superhydrophobic property and high adhesive force to water. Due to the hierarchical micro- and nanostructure at the alumina surface, namely, grooves–pillars and nipple-shaped protrusions, respectively, the water droplets cannot enter into the larger-scale grooves in the alumina surface, but water droplets can impregnate the nipple-shaped protrusions and plateaus. Consequently, the alumina surface shows super water repellence and strong adhesion to water when the surface is tilted in any angles (Figure 7.11) [96].

The adhesive force of the surface can affect the transition of a hydrophobic to a superhydrophobic surface [84]. It is reported that lotus leaves lose their superhydrophobic character ("lotus effect") when microscale water droplets are condensed on the leaf [97]. This is attributed to the change (increase) in the contact area between the water droplet and the object surface, thereby decreasing the CA θ between them.

7.7 REVERSIBLY SWITCHABLE WETTABILITY

UV patterning, exposure to short wavelength UV light through a photomask, allows the CA to be varied continuously from 170° to 0° depending on the exposure time. This ability to control the material's affinity for water in a spatially defined manner has promising applications in electrowetting, microfluidics, and nanodevice development (Table 7.3) (http://www.phy.davidson.edu/fachome/dmb/RESolGelGlass/SH/SuperhydrophobicCoatingsFinal.pdf).

The concept of spatially controlled wetting behavior is inspired by the patterned back of the Namib beetle. Exposure to high-intensity, low-wavelength UV light under special conditions allows to continuously vary the CA from over 170° to 0° and the light can control the path where the water droplets will roll. A superhydrophobic film patterned with a square array of hydrophilic patches can control whether water sticks to or slips at the surface within a microfluidic device. These patches grow into drops, which extend onto the surrounding superhydrophobic material. When the drops have grown to a sufficiently large size, gravity causes them to detach and roll, allowing collection (http://www.phy.davidson.edu/fachome/dmb/RESolGelGlass/SH/SuperhydrophobicCoatingsFinal.pdf).

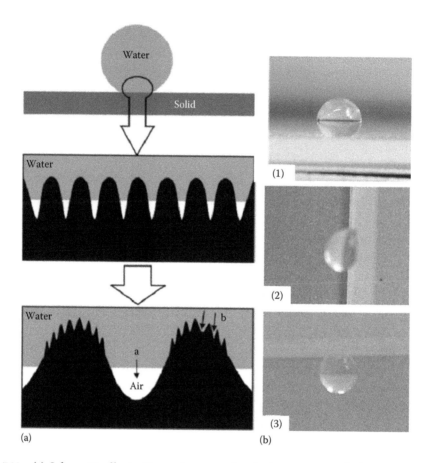

FIGURE 7.11 (a) Schematic illustrations of a water droplet in contact with the superhydrophobic alumina surface. (b) Photographs of a water droplet on the superhydrophobic alumina surface with different tilt angles: (1) 0°, (2) 90°, and (3) 180°. (Reproduced with permission from Feng, L. et al., *Colloid Surface.*, 410, 66, 2012.)

The strength of the microstructures also determines the switchable wettability. While a superhydrophobic microstructure that has higher structural strength can retain its superhydrophobicity, a pillar structure of insufficient strength can easily be broken when a slight lateral or vertical force is applied damaging the superhydrophobic effect [87]. A superhydrophobic nanopillar structure may lose its hydrophobic ability under some conditions. For example, a static water droplet standing on the rough or pillar structure surface may be hydrophobic due to the diminished contact area between the water droplet and the pillar structure surface. However, an open pillar structure can allow airflow in the pillars, and when the water droplet falling from a high point to the pillar structure, push the air between the pillars out, which can wet the pillar structure resulting in its loss of hydrophobicity.

A superhydrophobic ZnO nanorod films can be transformed to superhydrophilicity upon UV irradiation due to the generation of electron–hole pairs that absorb hydroxyl group on the ZnO surface. Subsequent dark storage of the UV-irradiated ZnO film for

TABLE 7.3 List of Superhydrophobic Process Parameters and the Characteristics

Coating Method	Contact Angle (deg)	Substrate Pretreatment	Deposition	Surface Modification	Tunable CA	Patternable	Transparent
Plastic transformation	160	Unknown	Dropwise/130°C vacuum oven	Rapid cooling from 130°C to 70°C	No	No	No
Carbon nanotube forest	170	Ni catalyst islands in PECVD@ 650°C	Plasma discharge acetylene/ammonia	HFCVD coating@ 500°C	No	No	No
Polyelectrolyte multilayers	172	Acidic soakings	100 + dip coatings	CVD@180°C/2h	No	Yes	No
Galvanic cell reaction	154	Ultrasonic washings in acetone and ethanol	$AgNO_3$ and HF immersion	Dodecanethiol soaking overnight	No	No	No
Nanosphere lithography	168	Unknown	Spin coating	Oxygen plasma/Ag deposition/octadecanethiol rinse	132°–170°	No	No
Sol–gel foam	160	No substrate	No deposition	Heated @ 300°C	No	No	No
Sol–gel alumina	160	Al_2O_3 heated to 400°C + immersion in boiling water	Dip coating	Heptadecafluorodecyl-trimethoxilane	5°–160°	Yes	No
Coating 1 (10/912.576)	172	Acidic soakings	10–100+ dip coatings w/Si particles 0.2–20 microns in diameter	CVD @ 180°C/2h	No	Yes	No
Coating 2 (PCT/AU2004/000462)	165	Unknown	Spray, spin, dip with hexane solvent	Room temp (days) or 150°C (hours) heating for crosslinking of the polymer strands	No	No	Depends on size of particulate material
Our simple sol–gel coating	172	None	Spray, spin, dip	None	0°–170°	Yes	Yes

Source: http://www.phy.davidson.edu/fachome/dmb/RESolGelGlass/SH/SuperhydrophobicCoatingsFinal.pdf.

7 days can render it superhydrophobic again. TiO$_2$ nanorod films also exhibit similar reversibly switchable wettability [98].

7.8 SYNERGISM OFFERED BY THE NANO AND MICROFILLERS

Friction and wear are related with each other and are controlled by roughness, hardness, ductility effect, oxide film effect, reaction layer effect, and transfer effect; the friction coefficient becoming smaller at smaller roughness extending the life of the coating longer [99]. Traditional fillers in a polymer system considerably improve the wear and friction coefficients. Addition of nano- and microfillers improves these properties in a synergistic fashion [100].

It is reported relative to the neat epoxy, the wear resistance could be improved by almost a factor of 3 after the addition of 4–6 vol.% of 300-nm-sized TiO$_2$ particles [100]. Adding traditional fillers, for example, short carbon fibers and graphite flakes, the wear improvement could be more effective than just in the case of the nanoparticles. Similar effects are observed for polyetheretherketone (PEEK), where the addition of nanoparticles led to an improved performance in wear and coefficient of friction at room temperature and at elevated temperatures [99,101]. Figure 7.12 shows the use of traditional fillers to improve the friction and wear properties of the neat polymer (polyamide 6, 6 [PA6, 6]) quite remarkably, and further topped by the additional incorporation of nanosized TiO$_2$ particles, especially

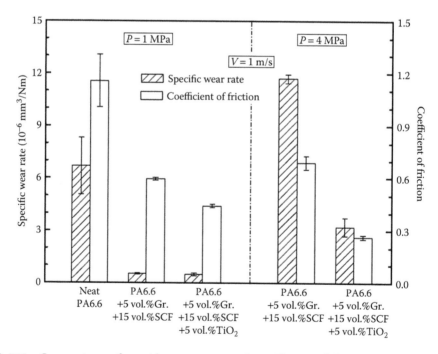

FIGURE 7.12 Comparison of specific wear rate and coefficient of friction of PA6, 6 matrix composites with and without 300 nm TiO$_2$ particles at various contact pressures. (Reproduced with permission from Friedrich, K. et al., Property improvements of polymer composites by spherical nanoparticles, in *Proceedings of ACUN-5, Developments in Composites, Advanced, Infrastructural, Natural and Nanocomposites*, Sri Bandyopadhyay et al., eds, Sydney, Australia, July 11–14, 2006, pp. 297–305.)

under higher apparent contact pressure. In particular, the coefficient of friction is appreciably reduced, when adding the nanoparticles to the tribologically modified PA6, 6 compound (containing 5 vol.% of graphite flakes and 15 vol.% of short carbon fibers) [100].

7.9 SUMMARY

Research in polymer nanocomposites has made it possible to realize many novel material properties [70], including superhydrophobicity. The key strategies to produce superhydrophobic surfaces based on biomimetics rely on the surface with chemical heterogeneity, low surface energy, and the hierarchical structures, which increase the CA and decrease the CAH and the hierarchical frictional forces. While the low surface energy materials such as silicon-based polymers are favored for producing superhydrophobic structures, nature relies on polymers like paraffins for superhydrophobic structures. The polymer nanocomposites in addition to contributing to the chemical heterogeneity also contribute to the hierarchical structures on the surface. Switchable and alternate patches of hydrophobic and hydrophilic domains also contribute to the superhydrophobic structures. The diminished WCA of the surface due to the plurality of roughness greatly affects the friction forces reducing them. The highly adhesive nature of the droplets sitting on the apex of the nanopillars is prevented from rolling over and hence the wettability is greatly diminished.

The synergy provided by the combination of micro- and nanoparticles in polymer composites also helps greatly to reduce the frictional forces which could lead to hydrophobicity due to the chemical heterogeneity of the matrix as well as the hierarchical structures provided by the fillers.

ACKNOWLEDGMENTS

The author is grateful to Bentham Science Publishers and the editors of the book for giving the opportunity to write this very important topic on superhydrophobicity—their fundamental and application-orientated aspects. Also thanks are due to the permissions provided by various authors to present their figures/diagrams and tables in the chapter to make the correlation clearly understandable. Most of all the CRN funding for the project by Federation University Australia is gratefully acknowledged.

REFERENCES

1. Benyus J. *Biomimicry: Innovation Inspired by Nature*. HarperCollins, New York, 1997.
2. Charles Q. Choi, Cicada Wings Are Self-Cleaning, INSIDE SCIENCE NEWS SERVICE, April 29, 2013.
3. Feng L, Zhang Y, Xi J, Zhu Y, Wang N. Petal effect: A superhydrophobic state with high adhesive force. *Langmuir* 2008, 24: 4114–4119. doi: 10.1021/la703821h.
4. Emmanuel Stratakis et al., Laser structuring of water-repellent biomimetic surfaces, SPIE Newsroom 10.1117/2.1200901.1441.
5. Ensikat HJ, Mayser M, Barthlott W. Superhydrophobic and adhesive properties of surfaces: Testing the quality by an elaborated scanning electron microscopy method. *Langmuir* 2012, 28: 14338–14346.
6. Ma M, Hill RM. Superhydrophobic surfaces. *Curr Opin Colloid Interface Sci* 2006, 11: 193–202.

7. Hur JI et al. *Membraneless Micro Fuel Cell Chip Enabled by Self-Pumping of Fuel-Oxidant Mixture*. Institute of Electrical and Electronics Engineers, Hong Kong, China, 2010.

8. Simpson T, John Dr, DeTrana A. Super-hydrophobic materials. Oak Ridge National Laboratory, Shawn Carson. n.d. Web. April 24, 2013. www.ornl.gov/adm/partnerships/events/b.

9. Privett BJ. Antibacterial fluorinated silica colloid super-hydrophobic surfaces. *Langmuir* n.p. June 30, 2011, Web. May 7, 2013 (ACS Publications). doi: 10.1021/la201801e.

10. Kauzmann W. Some factors in the interpretation of protein denaturation. *Adv Protein Chem* 1959, 14: 1–63.

11. Chandler D. Interfaces and the driving force of hydrophobic assembly. *Nature* insight review article. http://www.aathavan.org/chem130a/hydrophobic-chandler.pdf, 2003.

12. Tanford C. Interfacial free energy and the hydrophobic effect. *Proc Natl Acad Sci USA* 1979, 76: 4175–4176.

13. Quere D. Non-sticking drops. *Rep Prog Phys* 2005, 68(11): 2495–2532.

14. Checco A, Guenoun P, Daillant J. Nonlinear dependence of the contact angle of nanodroplets on contact line curvatures. *Phys Rev Lett* 2003, 91: 186101.

15. Johnson RE, Dettre, RH. Contact angle hysteresis. *J Phys Chem* 1964, 68(7): 1744–1750. doi: 10.1021/j100789a012.

16. Yuan Y, Randall Lee T, Bracco G, Holst B (eds). *Surface Science Techniques*. Springer Series in Surface Sciences, vol. 51. Springer-Verlag, Berlin, Germany, 2013.

17. Young T. An essay on the cohesion of fluids. *Phil Trans R Soc Lond* 1805, 95: 65–87. doi: 10.1098/rstl.1805.0005.

18. Wenzel RN. Resistance of solid surfaces to wetting by water. *Ind Eng Chem* 1936, 28(8): 988–994. doi: 10.1021/ie50320a024.

19. de Gennes, P-G. *Capillarity and Wetting Phenomena*. http://www.springer.com/materials /surfaces+interfaces/book/978-0-387-00592-8, 2004.

20. Extrand C. Criteria for ultralyophobic surfaces. *Langmuir* 2005, 68: 2495–2532.

21. Flynn MR, Bush JWM. Underwater breathing: The mechanics of plastron respiration. *J Fluid Mech* 2008, 608: 275–296.

22. Choi CH, Kim CJ. Large slip of aqueous liquid flow over a nanoengineered superhydrophobic surface. *Phys Rev Lett* 2006, 96: 066001.

23. De Gennes PG. On fluid/wall slippage. *Langmuir* 2002, 18: 3413.

24. Bechert DW, Bruse M, Hage W. Experiments with three dimensional riblets as an idealized model of shark skin. *Exp Fluids* 2000, 28: 403–412.

25. Zhao DY, Huang ZP, Wang MJ, Wang T, Jin YF. Vacuum casting replication of micro-riblets on shark skin for drag-reducing applications. *J Mater Process Technol* 2012, 212: 198–202.

26. Bechert DW, Bruse M, Hage W, Vander Hoeven JGT, Hoppe, G. Experiments on drag-reducing surfaces and their optimization with an adjustable geometry. *J Fluid Mech* 1997, 338: 59–87.

27. Gray J. Studies in animal locomotion, VI. The propulsive powers of the dolphin. *J Exp Biol* 1935, 13: 192–199.

28. Ferhat H, Sylvain G. Drag reduction by surfactant in closed turbulent flow. *Int J Eng Sci Tech* 2010, 2: 6876–6879.

29. Kawaguchi Y, Segawa T, Feng ZP, Li PW. Experimental study on drag reducing channel flow with surfactant additives-Spatial structure of turbulence investigated by PIV system. *Int J Heat Fluid Flow* 2002, 23: 700–709.

30. Virk PS. Drag reduction fundamentals. *AIChE J* 1975, 21: 625–656.

31. Vatankhah C, Jafargholinejad S, Mozaffarinia R. Experimental investigation on drag reduction performance of two kind of polymeric coatings with rotating disk apparatus. *Aust J Basic Appl Sci* 2011, 5: 143–148.

32. Anshuman R, Alexander M, Wim, VS, Larson RG. Mechanism of polymer drag reduction using a low-dimensional model. *Phys Rev Lett* 2006, 97: 234501.

33. McCormick M, Bhattacharyya R. Drag reduction of a submersible hull by electrolysis. *Nav Eng J* 1973, 85: 406–410.
34. Elbing, BR, Winkel ES, Lay KA, Ceccio SL, Dowling DR, Perlin MJ. Bubble-induced skin-friction drag reduction and the abrupt transition to air-layer drag reduction. *Fluid Mech* 2008, 612: 201–236.
35. Wu SJ, Hsu CH, Lin TT. Model test of the surface and submerged vehicles with the micro-bubble drag reduction. *Ocean Eng* 2007, 34: 83–93.
36. Gokcay S, Insel M, Odabasi AY. Revisiting artificial air cavity concept for high speed craft. *Ocean Eng* 2004, 31: 253–267.
37. Matveev KI, Burnett TJ, Ockfen AE. Study of air-ventilated cavity under model hull on water surface. *Ocean Eng* 2009, 36: 930–940.
38. Vakarelski IU, Marston JO, Chan DYC, Thoroddsen ST. Drag reduction by leidenfrost vapor layers. *Phys Rev Lett* 2011, 106: 214501.
39. Vandogen DB, Roche Jr. EC. Efflux time from tanks with exit pipes and fittings. *Int J Eng Educ* 1999, 15: 206–212.
40. Lumley JL. Drag reduction in turbulent flow by polymer additives. *J Polym Sci Macromol Rev* 1973, 7: 263–290.
41. Truong VT. *Drag Reduction Technologies.* DSTO Aeronautical and Maritime research laboratory, Melbourne, Australia, 2001.
42. Toms BA. *Proceedings of International Congress on Rheology,* vol. II, pp. 135, North Holland, Amsterdam, the Netherlands, 1949.
43. Hoyt JW. Drag reduction. In: *Encyclopedia of Polymer Science and Engineering,* 2nd ed., vol. 5, 1986, 129–151.
44. Kim OK, Choi LS, Long T, Yoon TH. *Polym Comm* 1988, 29: 168.
45. Kim OK, Choi LS, Long T, McGrath K, Armistead JP, Yoon TH. *Marcromolecules* 1993, 26: 379.
46. Kim OK, Choi LS. Drag-reducing polymers. In: *Polymeric Materials Encyclopedia,* Salomone JC (ed). CRC Press, Boca Raton, FL, 1996, 1937–1945.
47. Tulin MP. *Proceedings of the 3rd International Conference on Drag Reduction,* Sellin RHJ, Moses RT (eds). University of Bristol, Bristol, UK, 1984.
48. Berman NS. Drag reduction by polymers. *Annu Rev Fluid Mechanics* 1978, 10: 47–64.
49. Molbert S. The hydrophobic interaction, modeling hydrophobic interactions and aggregation of non-polar particles in aqueous solutions. PhD thesis, Université de Fribourg, Fribourg, Switzerland, 2003.
50. Ptasinski PK, Nieuwstadt FTM, Van Den Brule BHAA, Hulsen MA. Experiments in turbulent pipe flow with polymer additives at maximum drag reduction flow. *Turbul Combust* 2001, 66: 159–182.
51. Gyr A, Bewersdorff HW. *Drag Reduction of Turbulent Flows by Additives.* Kluwer Academic, Dordrecht, the Netherlands, 1995.
52. Shenoy AV. A review on drag reduction with special reference to micellar system. *Colloid Polym Sci* 1984, 262: 319.
53. Zakin JL, Chang JL. Non-ionic surfactants as drag reducing additives. *Nat Phys Sci* 1972, 239: 26.
54. Zakin JL, Chang JL, *Proceedings of the International Conference on Drag Reduction,* Cambridge, UK, 1974.
55. Everett DH. *Basic Principles of Colloid Science.* The Royal Society of Chemistry, London, UK, 1994.
56. Cullis P. Lipid polymorphism and the roles of lipids in membranes. *Chem Phy Lipids* 1986, 40: 127–144.
57. Savins JG. A stress–controlled drag reduction phenomenon. *Rheol Acta* 1967, 6: 323.
58. Gadd GE. Effects of long-chain molecule additives in water on vortex streets. *Nature* 1966, 212: 1348.
59. Rothstein JP. Strong flows of viscoelastic wormlike micelle solutions. *Rheol Rev* 2008, 1–46.

60. Larson RG. *The Structure and Rheology of Complex Fluids*. Oxford University Press, New York, 1999, 364p.

61. Oliver DR, Bakhtiyarov SIJ. Drag reduction in exceptionally dilute polymer solutions. *Non-Newtonian Fluid Mech* 1983, 12: 113.

62. Tabor M, de Gennes PG. A cascade theory of drag reduction. *Europhys Lett* 1986, 2: 519–522.

63. Sreenivasan KR, White CM. Mechanics and prediction of turbulent drag reduction with polymer additives. *J Fluid Mech* 2000, 409: 149–164.

64. Zakin JL, Myska J, Chara Z. New limiting drag reduction and velocity profile asymptotes for non-polymeric additives systems. *AIChE J* 1996, 42: 3544–3546.

65. Amarouchene Y, Bonn D, Kellay H, Lo TS, L'vov VS, Procaccia I. Reynolds number dependence of drag reduction by rodlike polymers. *Phys Fluids* 2008, 20: 065108. http://dx.doi.org/10.1063/1.2931576.

66. Dong H, Cheng M, Zhang Y, Wei H, Shi F. Extraordinary drag-reducing effect of a superhydrophobic coating on a macroscopic model ship at high speed. *J Mater Chem A* 2013, 1: 5886.

67. McCulloch CR, Gill RC. Submersible object having drag reduction and method. 1976, US 3973510.

68. Swain G. *Proceedings of the International Symposium on Seawater Drag Reduction*. Newport, RI, July 1998, p. 155.

69. Kinloch AJ. *Adhesion and Adhesives. Science and Technology*. Chapman & Hall, London, 1987. http://www.twi.co.uk/technical-knowledge/faqs/material-faqs/faq-what-are-the-typical-values-of-surface-energy-for-materials-and-adhesives/.

70. Jia OU, Perot B, Rothstein JP. Laminar drag reduction in micro channels using ultra hydrophobic surfaces. *Phys Fluids* 2004, 16(12): 4635–4644.

71. Sun M, Liang A, Watson GS, Watson JA, Zheng Y. Compound microstructures and wax layer of beetle elytral surfaces and their influence on wetting properties. *PLoS ONE* 2012, 7(10): 1–14: e46710. doi: 10.1371/journal.pone.0046710.

72. Devasahaym S, Bandyopadhyay S. Evolution of novel size-dependent properties in polymer-matrix composites due to polymer filler interactions (Chapter 1). In: *New Developments in Polymer Composites Research*, Laske S, Witschnigg A (eds). Nova Publication, 2013, pp. 1–32.

73. Abdelrazek EM, Abdelghany AM, Orabyand AH, Asnag GM. Investigation of mixed filler effect on optical and structural properties of PEMA films. *Int J Eng Technol IJET-IJENS* 2012, 12(04): 98–102.

74. Zidan HM. Filling level effect on the physical properties of $MgBr_2$ and $MgCl_2$ filled poly (vinyl acetate) films. *J Polym Sci: Part B: Polym Phys* 2003, 41: 112–119.

75. de Almeida Prado LAS. Morphological and structural aspects of polymer nanocomposites. 011 Society of Plastics Engineers (SPE). http://www.4spepro.org/view.php?source=003504-2011-01-29.

76. Vaia RA, Maguire JF. Polymer nanocomposites with prescribed morphology: Going beyond nanoparticle-filled polymers. *Chem Mater* 2007, 19(11): 2736–2751.

77. Wu X, Shi G. Fabrication of a lotus-like micro-nanoscale binary structured surface and wettability modulation from superhydrophilic to superhydrophobic. *Nanotechnology* 2005, 16: 2056.

78. Barbieri L, Wagner E, Hoffmann P. Water wetting transition parameters of perfluorinated substrates with periodically distributed flat-top microscale obstacles. *Langmuir* 2007, 23: 1723–1734.

79. Erbil HY, Demirel AL, Avci T. Transformation of a simple plastic into a superhydrophobic surface. *Science* 2003, 299: 1377–1380.

80. Zhang JL, Li JA, Fan Y, Han YC. Fabricating super-hydrophobic lotus-leaf-like surfaces through soft-lithographic imprinting. *Macromol Rapid Commun* 2004, 25: 1105–1108.

81. Bharathibai JB, Dinesh Kumar V. Fabrication of superhydrophobic nanocomposite coatings using polytetrafluoroethylene and silica nanoparticles. *ISRN Nanotechnol* 2011: Article ID 803910, 6p. doi: 10.5402/2011/803910.

82. Manoudis PN, Karapanagiotis I, Tsakalof A, Zuburtikudis I, Panayiotou C. Superhydrophobic composite films produced on various substrates. *Langmuir* 2008, 24: 11225–11232.
83. Karapanagiotis I, Manoudis PN, Savva A, Panayiotou C. Superhydrophobic polymer-particle composite films produced using various particle sizes. *Surf Interface Anal* 2012, 44: 870–875.
84. Facio DS, Mosquera MJ. Simple strategy for producing superhydrophobic nanocomposite coatings: In situ on a building substrate. *ACS Appl Mater Interfaces* 2013, 5(15): 7517–7526.
85. Liu Y, Chen X, Xin JH. Super-hydrophobic surfaces from a simple coating method: A bionic nanoengineering approach. *Nanotechnology* 2006, 17: 3259–3263.
86. Yu J, Son SJ. Fabrication of a superhydrophobic surface with adjustable hydrophobicity and adhesivity based on a silica nanotube array. *Bull Korean Chem Soc* 2012, 33(10): 3378–3382.
87. Lu XY, Zhang CC, Han YC. Low-density polyethylene superhydrophobic surface by control of its crystallization behavior. *Macromol Rapid Commun* 2004, 25:1606–1610.
88. Jiang L, Zhao Y, Zhai J. A lotus-leaf-like superhydrophobic surface: A porous microsphere/nanofiber composite film prepared by electrohydrodynamics. *Angew Chem Int Ed* 2004, 43: 4338–4341.
89. Lee SY, Liao SK, Chen YC. Super hydrophobic microstructure. 2012, US 20120177881. http://www.freepatentsonline.com/y2012/0177881.html.
90. Zhang J, Lu X, Huang W, Han Y. Reversible superhydrophobicity to superhydrophilicity transition by extending and unloading an elastic polyamide film. *Macromol Rapid Commun* 2005, 26: 477–480.
91. Zhao N, Xu J, Xie QD, Weng LH, Guo XL, Zhang XL. Fabrication of biomimetic superhydrophobic coating with a micro-nano-binary structure. *Macromol Rapid Commun* 2005, 26: 1075–1080.
92. Mohammadi R, Wassink J, Amirfazli A. Effect of surfactants on wetting of super-hydrophobic surfaces. *Langmuir* 2004, 20: 9657–9662.
93. Yan H, Kurogi K, Mayama H, Tsujii K. Environmentally stable super water-repellent poly(alkylpyrrole) films. *Angew Chem Int Ed* 2005, 44: 3453–3456.
94. Nosonovsky M, Bhushan B. Hierarchical roughness makes superhydrophobic states stable. *Microelectron Eng* 2007, 84(3): 382–386.
95. Sokoloff JB, Possible microscopic explanation of the virtually universal occurrence of static friction. *Phys Rev B* 2002, 65: 115415.
96. Nosonovsky M, Bhushan B. Hierarchical roughness optimization for biomimetic superhydrophobic surfaces. *Ultramicroscopy* 2007, 107: 969–979.
97. Sun M, Luo C, Xu L, Ji H, Ouyang Q, Yu D, Chen Y. Artificial lotus leaf by nanocasting. *Langmuir* 2005, 21: 8978–8981.
98. Feng L, Liu Y, Zhang H, Wang Y, Qiang X. Superhydrophobic alumina surface with high adhesive force and long-term stability. *Colloid Surface* 2012, 410: 66–71.
99. Cheng YT, Rodak DE. Is the lotus leaf superhydrophobic? *Appl Phys Lett* 2005, 86:144101.
100. Feng XJ, Feng L, Jin MH, Zhai J, Jiang L, Zhu DB. Reversible super-hydrophobicity to superhydrophilicity transition of aligned ZnO nanorod films. *J Am Chem Soc* 2004, 126: 62–63.
101. Kato K. Wear in relation to friction—A review. *Wear* 2000, 241: 151–157.
102. Friedrich K, Haupert F, Zhang Z. Property improvements of polymer composites by spherical nanoparticles. In: *Proceedings of ACUN-5, Developments in Composites, Advanced, Infrastructure, Natural and Nanocomposites*, Sri Bandyopadhyay et al., (eds). Sydney, Australia, July 11–14, 2006, pp. 297–305.
103. Oster F, Haupert F, Friedrich K, Bickle W, Müller M. Tribologische hochleistungsbeschichtungen aus neuartigen polyetheretherketon (PEEK)-compounds. *Tribologie und Schmierungstechnik*, 2004, 51(3): 17–24.
104. Sun M. et al., Wetting properties on nanostructured surfaces of cicada wings. *J. Exp. Biol.*, 2009, 212: 3148–3148.

CHAPTER 8

Polymer Coatings and Patterning Techniques

Sreeram K. Kalpathy

CONTENTS

8.1 INTRODUCTION

The world of coatings technology has come to rely extensively on polymeric materials today most academic universities or industrial firms dealing in coatings are invariably associated with a polymer division. Historically, the term coatings would have been reminiscent of the artist's paint, which originally used natural ingredients such as linseed oil or egg yolk extracts. Iskowitz (2006) provides an overview of the various stages of development of composition of the artist's paint. However, the gloss, durability, rheological specifications, and numerous other considerations have made polymers an essential ingredient in paints. Most of the binders used in paints happen to be polymeric, considering their good adhesion, drying, and chemical functionalities. Even in today's world of coatings which has broadened to applications ranging from art paints to photographic films to adhesives to metal corrosion prevention to printing inks to machine or wood finish, the use of polymers is inevitable. The final coated material happens to be polymeric either before the coating application or at the least, after the coating is cured.

In the context of smaller length scales being sought after everywhere by the present day scientific community, this chapter attempts to describe some of the latest developments in polymer coatings in the last decade or so, which have in turn strengthened tremendously the fields of nanotechnology and interfacial engineering. In Section 8.1 a review of materials, chemistry, and techniques used for polymer coatings is provided. Typical material choices based on suitability for industrial use are discussed here, followed by the preferred means of their application. With miniaturization of latest devices and displays being an endless effort, creation of small-scale patterns of polymeric surfaces have assumed importance, considering the flexible nature of such surfaces. Section 8.2 looks at the various patterning techniques for polymers, and Section 8.3 deals with a closely associated technique of liquid dewetting, which has also gained attention as a fabrication strategy for creating nanostructures. The focus in Section 8.3 will be on the theoretical framework and supporting mechanisms for such dewetting and pattern creation. The key features of all these latest developments are summarized in Section 8.4 and possible directions for future research in polymer coatings are suggested.

8.2 POLYMER-BASED COATINGS: MATERIALS AND TECHNIQUES

The utility of polymer coatings based on the coating composition has been so widely investigated that many polymers are almost invariably associated with a certain niche of applications. Tying a class of polymers in this manner is of course based on the specific property characteristics that the end-product demands. For example, an automotive coating would typically require gloss, exterior durability, and resistance to chemical and abrasive attack. These combined with cost, availability, and ease of quality control make acrylics, polyesters, and epoxy esters as suitable candidates. On the other hand, polyimides would better be utilized in microelectronics given their superior electrical properties. Similarly, polymer coatings for medical devices would seek a composition based on the toxicology, wettability, and microbial deterioration (Wright, 1986). Often, coating composition is also influenced by polymeric materials which are only additives (in the form of surfactants, dispersants, or pigments), but nevertheless play a significant role.

Just as diversity is involved in classification of polymeric coating materials, a similar challenge is in classification and optimization of the coating technique used for each material type. Polymer coatings could be solvent-based or waterborne (a more environmental-friendly alternative), or even more recently, vapor-deposition based (Martin et al., 2007). Conventional techniques such as slot-die (Lippert, 1991), wire-wound rod (Hanumanthu and Scriven, 1996), or gravure (Benkreira and Patel, 1993) may be used for solvent-based and waterborne coatings. However, a complex system like a PVC Plastisol (Krauskopf and Godwin, 2005) that switches between liquid and solid states during the manufacturing process may require a combination of hot and cold dipping. Alternatively, a spray coating or slush molding process (Krans, 1999) is used in which the molds could be heated.

8.2.1 Materials

Table 8.1 discusses 10 major classes of polymeric materials in use for industrial coatings, and the typical application ranges to which each is best suited for.

TABLE 8.1   Major Polymeric Material Classes Used for Coating Applications

Class	Forms in Which Used or Applied	Suggested Application Areas	Relevant Physical Properties
1. Acrylic polymers	Solid beads, solution polymers, emulsions (dominant among the three)	Regular coatings for automotive plastics/metal/wood	Hardness, chemical resistance, good adhesion to plastics
		Adhesives (for packaging, medical)	Wide range of glass transition temperatures rendering them useful for laminating adhesives to heat-sealable adhesives
		Formulating waterborne inks	Ready formation of emulsions, good film formation ability
2. Polyesters	Solvent-borne, waterborne, high solids paints, thermosetting powder coatings	Decorative coating on metal objects used in food packaging	Optimum elasticity (to permit can forming) and good surface hardness
		Paints for automotive parts vulnerable to stone chipping	Excellent elasticity, abrasion resistance
		Weld joints of metallic objects (Holscher, 1984)	Ready film formation of thermoplastic powders with little curing time
		Heat-sealable adhesives	Ready bonding with substrates upon heating
3. Vinyl ether	Solutions form, aqueous dispersions (especially for soft resins like Lutonal I 60 [Muller, 2006])	Starting materials for production of adhesives	Improvement in adhesion and anchorage upon blending with starch, dextrin
		Pressure-sensitive adhesives in the medicinal sector	Colorless, ready formation of secondary dispersions and resistance to hydrolysis
4. Alkyds (Solomon, 1977)	High solids coatings, emulsion-based coatings and powder coatings	Wall paints, marine parts, lacquers (especially those with high fatty acid content)	Good drying characteristics, flexible film formation, gloss
		Metal finishes (those with ~50% fatty acid content)	Durability and gloss, good drying characteristics
		Baking primers and enamels (those with low or zero fatty acid content)	Suitable for curing at elevated temperatures
5. Polyurea	Used as 100% solids, and may be directly sprayed on dry as well as damp surfaces, because they cure instantly	Metal structures in marine environments, roofing of steel or concrete, and secondary containment at cold temperatures	Near-instantaneous gel time, excellent chemical resistance (Technical Bulletin, 1996), film toughness at low temperatures, no catalyst needed for curing

(Continued)

TABLE 8.1 (*Continued*)   Major Polymeric Material Classes Used for Coating Applications

Class	Forms in Which Used or Applied	Suggested Application Areas	Relevant Physical Properties
6. Amino resins (Swaraj, 1997)	Solvent-based (butanol/xylene) or high solids (80%–100%) resins. Resins are mostly either urea–formaldehyde (UF) or melamine–formaldehyde (MF)	UF resins: Interior wood finishes MF resins: Automotive and appliance coatings	Poor exterior durability, can be cured athermally, less expensive Very glossy with distinct visibility, excellent mar resistance
7. Polyvinyl chloride (PVC)	Used as liquid (aqueous or solvent) dispersion systems of polymer an equivalent resin blend. Usually applied by spraying or dipping or roll-coating depending on viscosity	Household products—tool handles, dishwasher racks, garden gloves Encapsulation coatings for electrical parts Coating on porous substrates, clothing/fabric	Suitable for a single heavy coating with good viscosity control, pseudoplasticity, adhesion Viscosity control, good fusion characteristics, chemical stability Easy room temperature curing, easy to dispense small liquid volumes
8. Liquid polymers	Polymer binder precursors with functional end groups, blended with additives, are cured to form 100% solid coating	Diverse applications, depending on the specific functionality of the end group (Goethals, 1988). Furniture polish, textile coating Metallic coatings used as samples for mechanical testing, radiation curing.	Functionality of the end group perfluoroalkyl iodide (Thiokol Chemical Corporation, a, b.) Functionality of the acrylic or methacrylic end groups (Morris, 1984; Christmas, 1984)
9. Waxes	Used as pastes by dispersing waxes in solvents, dispersions of fine particles in water, emulsified in water	Several surface finish applications (Sayers, 1983)	Possible to obtain high gloss or high matte depending on wax solids content, Abrasion resistance
10. Elastomers	Extruded or blow molded to temperatures up to 220°C	Car hoods Seals, boots, and other joint applications for architectural use Food/water packaging, medicinal storage	Good heat resistance Elasticity Low toxicity

8.2.2 Coating Techniques

Table 8.1 reveals that polymers could be used in a variety of forms, which implies that the technique for their application also requires careful selection. For example, polyvinyl chloride dispersions with low blending resin contents can be quite viscous ruling out spray coating. In this case, heavy-walled and products for decorative applications are suitable for dip coating since a single heavy coat can be obtained. The use of all conventional mechanical coating techniques have been well investigated, including electrodeposition and extrusion coating with acidic polymers. This section focuses on two recent innovative coating techniques based on chemical vapor deposition (CVD) of polymers, both having promising implications for biomedical devices.

CVD methods have emerged as a major technique for polymer deposition, post the attention it had received for coating inorganic layers. First it is a convenient alternative as a solvent-free process. Other advantages include close control of composition, adaptability to complex polymers with functional groups, and excellent adhesion.

The review by Lahann (2006) describes free radical polymerization and deposition of poly (*p*-xylylene)s. The starting point is the synthesis of [2,2] paracyclophanes as precursors (Lahann et al., 2002). The subsequent CVD process is performed using a customized setup consisting first a zone for sublimation of the precursor, and then pyrolysis. Following, the monomers are transferred to a cooler deposition chamber which cause spontaneous polymerization. Working pressures of the order 0.05 mbar are used in the process. The resulting films have been shown to have good adhesion and robustness. More importantly, the bioactive nature of such coatings has also been investigated in light of biomedical applications.

More recently, Martin et al. (2007) have reported an improved version of CVD method, known as initiated CVD (iCVD). This method is particularly suited for coating fragile heat-sensitive substrates, since the formation of a polymeric film occurs directly on a cooled substrate (20°C–50°C). The work described pertains to deposition of poly (dimethylaminomethyl) styrene (PDMAMS). Initially, a free radical initiator species (here, di-*tert*-amyl peroxide [TAP]) is thermally cracked over a hot tungsten filament. The initiator steam is joined with the vaporized monomer stream of (dimethylaminomethyl) styrene, followed by direct deposition on the substrate. The method requires no solvent use, which renders it an ideal technique for medical device coatings. Combined with the idea that the coating can be closely applied along the contours of the substrate surface, it is possible to coat porous textiles using this technique. The work by Martin et al. (2007) precisely involves use of iCVD on a nylon fabric, followed by its antimicrobial testing, which gave effective results.

8.3 PATTERNED POLYMERIC SURFACES

Creation of chemically patterned surfaces with controlled wettability is an area that attracts considerable attention. Given that the range of commercial applications with heterogeneous surface functionalities at small length scales is extensive and increasing, newer methods are in demand to create patterned surfaces with tiny features. From fabric coatings to microlens array fabrication (Sadik et al., 2001) to microfluidics to electronic displays, the ability to coat or print on discrete areas of a substrate (which in many cases itself, is polymeric) is crucial. A particularly important advantage of polymeric substrates is with application to electronic devices and displays, where a lighter and robust device can be fabricated if the substrate is a patterned flexible web. This is true, for example, in manufacturing roll-to-roll LCDs (McCollough et al., 2006).

Patterned polymeric surfaces may be achieved through various routes, which may be broadly viewed as two major categories. The first is the class of techniques where the substrate could be a metal, ceramic, or a polymer which is patterned by soft lithographic (Xia and Whitesides, 1998) or contact printing methods. Second is a class of methods known as self-assembly where two or more components spontaneously organize into large aggregates in a templated fashion. Self-organization of polymers into patterned arrays can occur by dewetting of polymeric liquid onto a solid substrate. Since there is considerable recent literature on liquid dewetting and polymer flow based on theory and computational simulations, it is discussed separately in Section 8.4.

8.3.1 Soft Lithographic Techniques

Soft lithographic techniques were originally looked upon as convenient substitutes to photolithography, which is laborious and expensive. Although initially restricted largely to substrates like silicon, gold, and silver, soft lithographic approaches have hence been adapted to plastic and polymeric substrates too. For details of experimental techniques, the reader may refer to Fan et al. (2000) or Heule et al. (2004). In principle, a soft mold or stamp is made of an elastomer like poly(dimethylsiloxane) (PDMS) which contains the required pattern. Either a liquid polymer is cast against the mold, or the stamp is "inked" with the required solution, which is printed onto the substrate when the stamp makes contact with it. The former is known by terminologies such as contact molding or embossing, while the latter is referred to as microcontact printing or nanocontact printing. Micro- or nanocontact printing has particularly been explored for patterning self-assembled monolayers (SAMs). For instance, the stamp may be inked with long chain molecules of silanes (e.g., trichlorosilane) such that SAMs are patterned onto the substrate (Brzoska et al., 1994). This step can lead to the formation of a hydrophobic patterned surface. The substrate may then be coated with the target solution using standard coating/printing methods. Microcontact printing has also hence been used to pattern several materials other than SAMs: colloidal particles, polymers, and even biomolecules. While feature sizes of tens of micrometers are easily attainable, minimum feature sizes can be as small as 30 nm (Odom et al., 2002; Li et al., 2003).

In many cases, polymers have been surface modified using multistep synthesis techniques (Black et al., 1999; Patel et al., 2000; Yang and Chilikoti, 2000; Yang et al., 2000; Hyun et al., 2001) so that they could be used as substrates for microcontact printing. In a promising report of a simple fabrication method by Lee et al. (2008) a mold made of pentaerythritol propoxylate triacrylate (PPT) is coated with a thin PDMS layer, and then replicated into patterned templates. The oligomer PPT can easily be adhered to organic or inorganic substrates upon UV exposure. Besides, the strong bonding between the PPT and the thin PDMS layer assures durability of the mold, and many replicas of patterns (~200 per mold) can easily be fabricated.

In a comprehensive review by Gates et al. (2005), the authors discuss several conventional and innovative techniques for nanofabrication. In this section, two noteworthy contributions (Jacobs and Whitesides, 2001; Briseno et al., 2006) have been selected as advances or extensions to microcontact printing, used for patterning polymeric surfaces. In the work by Briseno et al. (2006), the authors have adapted microcontact printing for patterning organic semiconductors and fabricating devices. The resolutions reported here are of the order of 1 μm, but the technique seems to possess some distinct advantages. Instead of the usual practice of first "inking" the PDMS stamp with a solution, they have successfully printed high-resolution hydrophobic patterns onto hydrophilic substrates (here SiO_2, Si, flexible polyimide, or polyester) using a "dry" PDMS stamp. During contact, low molecular weight oligomers from the PDMS stamp are transferred onto the hydrophilic substrate surface. Later, a variety of organic semiconductors were coated onto the substrate, which deposit only on the contact-free regions (i.e., where PDMS oligomers are not present). This method is superior with respect to the following aspects. First, the ink-free process prevents potential defects that can occur due to ink smearing and diffusion on the substrate (Balmer et al. 2005). Besides, the oligomers transferred during contact will not interact with the "inks" and therefore, the quality of the hydrophobic surface created is superior. Finally, the process is easily scalable for applications where large area printing is required on various substrates.

Jacobs and Whitesides (2001) have demonstrated extension of the use of microcontact printing for patterning electrical charges. A thin dielectric film of material such as poly(methyl methacrylate) (PMMA), that can maintain an electrostatic potential is patterned using a PDMS stamp. The PDMS stamp, coated with electrode material (like a thin 100 nm gold layer) is actually used as a flexible electrode, and brought into contact with the dielectric film. Then a voltage pulse is applied between the PDMS and the dielectric film. When the electrode is removed, charge remains on the dielectric film in areas where the stamp contacted the film. The mechanism is therefore a deliberate injection and trapping of charges on the film, perhaps involving dipole interactions (Jacobs et al., 2002). While minimum feature sizes and charge densities that can be deposited remains to be known, Figure 8.1 reproduced from Gates et al. (2005) shows features obtained by Jacobs and Whitesides (2001) which are as small as 135 nm. Voltage pulses are of the order 10–30 V ranging for ~10 s, with current densities of ~10 mA/cm^2 (Gates et al., 2005). The electrical microcontact printing technique finds potential applications in high-density data storage and electrostatic printing of carbon toner.

Several modifications have also been tried for contact molding processes. A contribution that has received significant attention is the work by McClelland et al. (2002). Here, the authors report a novel nanocontact molding technique in which nanometer-scale features have been replicated on surfaces using functional cross-linked polymers. A polymeric mold is formed using a master pattern of SiO_2 pillars. This mold is then used to fabricate the stamp, out of a methacrylate photopolymer resist film. This resist film-based stamp is used to etch the target SiO_2 wafer, and obtain a pattern of SiO_2 pillars. A much greater control over the chemistry of the functional group can be achieved in this method. In a later work (Beinhoff et al., 2006) this method has been utilized for growth of fluorescent polyfluorene (PF) brushes on contact molded features. The PF brushes are grafted over cross-linked PMMA networks on silicon substrates, and have heights of the order 40 nm,

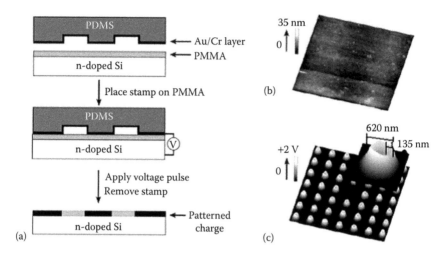

FIGURE 8.1 (a) Schematic of electrical microcontact printing. (b) Kelvin probe microscopy measurements of a PMMA thin film after patterning charge, as reported by Jacobs and Whitesides (2001) showing no change in surface topography. (c) Profile of a surface potential of a test pattern as reported by Jacobs and Whitesides (2001) showing features of size 135 nm. (Reprinted with permission from Gates, B.D. et al., *Chem. Rev.* 105, 1171–1196. Copyright 2005 American Chemical Society.)

spanning widths as small as 100 nm. Such precisely defined polymeric patterns find application in the manufacture of polymeric light emitting diodes (PLEDs).

8.3.2 Other Techniques

Two other important techniques for polymer patterning, especially useful for nanofabrication, are the breath-figure method (Widawski et al., 1994; Cui et al., 2005) and self-assembly (Jenekhe and Chen, 1999). The breath-figure method uses the idea similar to condensation of moist air into ordered patterns of water droplets on a cold surface. A typical experimental procedure for polymer patterning would involve the use of a dilute solution of a rodlike polymer (Song et al., 2004), from which the solvent is evaporated in a stream of moist air across the polymer–solution interface. The resulting ordered patterns are called "breath figures," which subsequently crystallize to form an array of holes on the polymer.

Self-assembly, where small components organize spontaneously in a predefined way could be carried out using an imposed spatial pattern or without it. The former, known as templated self-assembly is more commonly used for nanofabrication. Figure 8.2, taken from Kim et al. (2003), shows poly (styrene-*block*-methyl methacrylate) self-assembled onto a photopatterned template with a uniform spacing of 48 nm. The technique relies on a combination of conventional photolithography with patterned SAMs and self-assembly of the *block* copolymer by induced phase separation. Self-assembly has also been extended for patterning charged polymers, nanoparticles, and nanowires.

——— 50 nm

FIGURE 8.2 SEM image of poly(styrene-*block*-methyl methacrylate) self-assembled onto a photopatterned template of SAMs with periodicity of 48 nm. (Reprinted by permission from Macmillan Publishers Ltd. *Nature* 424, 411–414, Kim, S.O. et al., Copyright 2003.)

A relatively new method has been described by Wang et al. (2006) that results in patterns with wettability contrasts by micropore formation. In place of a regular polymer–solvent combination (here PMMA and polystyrene [PS], with solvents being tetrahydrofuran [THF], ethanol, or ethylene glycol), a small amount of nonsolvent (water) is used. The substrate material (glass or poly [tetrafluoroethylene] films) is immersed into this combination and removed. The solvent vaporizes on the thin polymer film formed on the substrate, and the nonsolvent droplets grow by phase separation. Depending on the growth rate of the droplets, this method leads to the formation of a patterned film with dense pores. Figure 8.3 (Wang et al., 2006) shows SEM images of surface patterns when different solvents were used. The simplicity of the method enables its adaption for patterning irregular or rough surfaces, and surfaces with large areas.

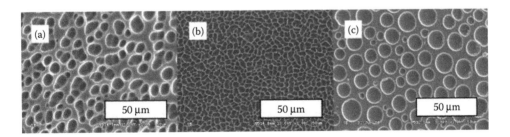

FIGURE 8.3 SEM micrographs of surface patterns obtained from evaporation at a temperature of 30°C of (a) 10 wt.% PMMA in acetonitrile solution with 1 wt.% water, (b) 10 wt.% PMMA in THF solution with 4 wt.% ethylene glycol, and (c) 10 wt.% PS in THF solution with 4 wt.% water. (Reprinted with permission from Wang, Y. et al., *Langmuir* 22, 1928–1931. Copyright 2006 American Chemical Society.)

8.4 DEWETTING OF THIN POLYMER FILMS

The most popular patterning techniques, based on contact molding or printing, rely more on deliberately confining the polymer to desired locations by use of an external pre-designed mold or stamp. However, it is possible to take advantage of the natural phenomenon that thin liquid films on nonwettable solid surfaces are unstable; and therefore they would "dewet" or uncover the substrate, forming disordered droplet arrays. Ordinarily, dewetting would be undesirable in preparation of a homogeneous polymer film. But it can be useful as a simple strategy for engineering patterns of polymers, thus avoiding the laborious lithographic processes. The process is largely governed by hydrodynamic forces such as capillary and viscous forces, and has been studied both experimentally and theoretically. One of the earliest papers that describe experiments in this field is of Reiter (1992), who observed development of holes in a thin film of PS. Since then other researchers have been able to demonstrate various morphological structures by controlled dewetting. In this section, first, some of the key features of the theoretical framework used for prediction of dewetting morphologies are reviewed. Then the relevant mechanistic explanations and experimental evidences are visited.

8.4.1 Theoretical Framework

A typical theoretical analysis of dewetting behavior of a liquid film on a solid substrate commences with the momentum balance (Navier–Stokes) and the mass balance (continuity) equations. For thin liquid films, with thicknesses of the order 100s of nm or smaller, the problem may be considerably simplified by invoking the lubrication approximation (Oron et al., 1997). As the aspect ratio of the film ε (i.e., ratio of the characteristic thickness to its horizontal extent) would be small in magnitude, only the leading order terms in ε would be retained in the dimensionless momentum balance and the continuity equations. Usually, this would lead to vanishing of inertial terms in the equations, while the dominant viscous terms would be retained. Combined with the appropriate boundary conditions, the spatial and temporal evolution of the film may be expressed as a single nonlinear partial differential equation for the film thickness H (Oron et al., 1997). A typical dimensionless evolution equation reads as (Kargupta and Sharma, 2002b)

$$\frac{\partial H}{\partial T} + \vec{\nabla}.\left[H^3\vec{\nabla}(\nabla^2 H)\right] - \vec{\nabla}.\left(H^3\nabla\phi\right) = 0. \tag{8.1}$$

where:
  T is the dimensionless time variable
  $\vec{\nabla}$ is the gradient vector operator

The first term in Equation 8.1 represents the local rate of film thinning. It also includes the viscous forces, which retard the growth of instability. The second term represents surface tension effects, and the third term is usually present to account for intermolecular interactions, such as van der Waals forces (Israelachvili, 1992). In Equation 8.1, ϕ denotes an intermolecular force potential, which may take various forms depending on the material properties of the system in consideration (Teletzke et al., 1987; Khanna et al., 1996; Schwartz and Eley, 1998; Sharma and Khanna, 1998; Kargupta and Sharma, 2002a). Evolution equations such as Equation 8.1 are solved using standard numerical integration techniques, with a finite difference or finite element method for space discretization.

8.4.2 Film Dynamics

The framework outlined in Section 8.3.1 has been used considerably to understand the dynamics of film dewetting. The occurrence of different dewetting modes in polymer films, morphological evolutions, and templating may all be predicted by this approach. Dewetting of a thin film may occur either by nucleation of holes (e.g., due to the presence of dust particles or substrate heterogeneities), or by a spinodal mechanism (driven by local thermal fluctuations). Seemann et al. (2001a, b, 2005) have published a series of work which illustrates these different dewetting modes, which are compared with a set of controlled experiments. They used thin PS films (2–80 nm) as a model system which was allowed to dewet silicon substrates with varying oxide thicknesses. The varying oxide thicknesses will in turn affect the substrate wettability, which may be modeled by varying the magnitude of the intermolecular force potentials (see Section 8.3.1). In Seemann et al. (2001a), the authors thus designed a stability map with oxide thicknesses and PS film thicknesses as the variables.

Theoretical studies on thin polymer liquid film dewetting have also been extended to cases where the substrate itself is pre-patterned, either chemically or topographically. Much of these is due to Kargupta and Sharma (2002a, b, 2003) who have adapted prior models to include spatially varying intermolecular force potentials. This is accomplished either by varying ϕ along the spatial coordinate as a step function (Kargupta and Sharma 2002a) or in a smooth continuous fashion, usually with trigonometric functions (Kargupta and Sharma 2002b). Kargupta and Sharma (2003) in particular identify conditions for nearly perfect templating with a variety of substrate patterns such as squares of less and more wettable regions, and checkerboard patterns. The characteristic length scales of dewetting have also been estimated. Of particular interest is the case where there is combined presence of chemical patterns (based on wettability gradients) and physical patterns (based on topographical differences). A much richer variety of architectures may be synthesized by this combination (Kargupta and Sharma 2003).

The theoretical predictions of film dynamics on pre-patterned substrates have also been observed experimentally. Julthongpiput et al. (2007) demonstrated experimentally using PS films that transition from isotropic dewetting to a pattern-directed dewetting occurs only if there is a minimum difference in surface energies of the less wettable and more wettable regions of the pattern. The features of the patterns were several 4 μm wide less wettable regions, separated by the reference SiO_2 matrix typically 20 μm wide. An experimental study using the same system was also reported by Luo et al. (2004) on both homogeneous and physically patterned substrates. In this work, PS films were allowed to dewet substrates having gratings in the form of stripes, with widths of the order 5–20 μm. The effect of annealing temperature on the final film morphology is probed, and is compared with the corresponding behavior on homogeneous flat substrates. Figure 8.4 is a set of images taken from their work,

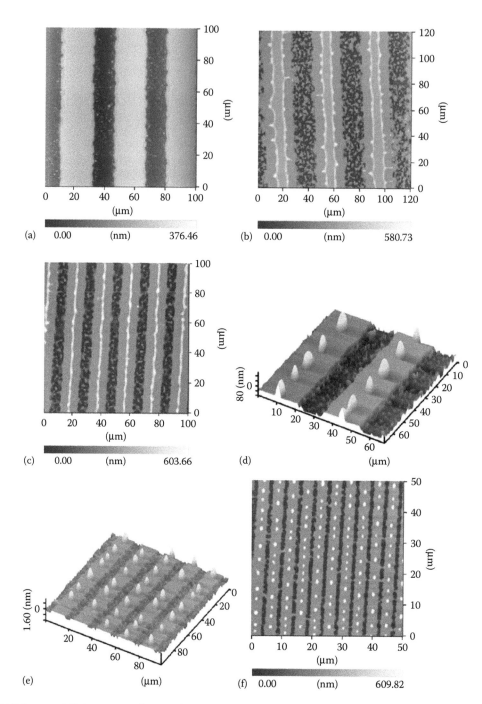

FIGURE 8.4 AFM images of typical morphological features observed for dewetting of thin PS films on physically striped patterns. (a) Initial thin PS films after spin casting. (b) The film moves to the center upon annealing at 167°C for 2 min. (c) Line formation upon annealing at 167°C for 8 min. (d) Ordered droplets with stripe widths being 20 μm. (e) Ordered droplets with stripe widths being 10 μm, upon annealing at 167°C for 3 h. (f) Ordered droplets with stripe widths being 5 μm, upon annealing at 167°C for 3 h. (Reprinted from *J. Colloid. Interface. Sci.*, 269, Luo, C., Xing, R., Zhang, Z., Fu, J., and Han, Y., pp. 158–163. Copyright 2004 with permission from Elsevier.)

which illustrates the various morphological features. The study also proposes mechanisms for the difference in dewetting behavior, and suggests conditions at which ordered droplet arrays can be obtained, In a more recent study, Yoon et al. (2008) have developed droplet arrays on much smaller topographic patterns, squares of size 200×200 nm^2, where heated PS films arrange into spherical caps of about 70 nm diameter. Such ordered droplet arrays are of great interest in micro- and nanofluidics (Squires and Quake, 2005), in applications where reproducible parcels of liquid are required.

8.5 CONCLUDING REMARKS

Research on polymeric materials in the present day scenario is motivated by their ubiquity. Coatings can no longer rely restrictively on natural ingredients and metals, polymers are an essential component, given their hybrid features. In spite of their widespread use, a lot remains to be explored in terms of structure–property correlations of many polymeric materials as a dried film. For example, the coefficient of thermal expansion, ultimate mechanical properties, creep and stress relaxation, and dielectric properties needs to be better correlated with the structure of thin polymeric coatings. Similarly, latex dispersions have come into use as a major coating material, but a lot more can be explored about their properties in terms of particle sizes, volatility, and composition. With regard to techniques for application also, innovations are needed, given the versatility in their flow properties, and in some cases, high sensitivity to temperature.

Patterning of polymers has also aroused great interest in the polymer research community, due mostly to nanofabrication needs in the electronics industry. While the techniques are by now well understood on a laboratory scale, their scale up for mass fabrication is still a challenge. The blends used for PDMS stamps need some modification. Some preliminary steps for scaling up the process were envisioned by Michel et al. (2001) at IBM Corporate Research, which have of course been built upon by future researchers. There have also been attempts to use stamps for continuous roll-to-roll patterning. However, direct printing processes like inkjet, gravure, and more recently aerosol jet printing (Renn, 2006) have taken over as more suitable candidate processes due to their efficient capability of integration into a roll-to-roll manufacturing line.

Finally, dewetting of thin polymer films has been recognized as a phenomenon of great practical relevance for patterning and templating. With many more technologies being available to create and observe such systems, the interest of theoretical physicists, mathematicians, and modeling engineers to study such phenomena has been aroused considerably. The fundamental aspects governing physical mechanisms of dewetting have been explained based on several models from hydrodynamics and statistical physics. A fairly detailed background about the field has been presented by Blossey (2012) in his recent book. There is still further scope to extend these models to more complex situations, such as presence of imposed external fields, effects of their non-Newtonian rheology, especially viscoelasticity, and influence of film thickness on the glass transition of polymeric films.

REFERENCES

Balmer, T. E., H. Schmid, R. Stutz et al. 2005. *Langmuir* 21:622–632.

Beinhoff, M., A. T. Appapillai, L. D. Underwood, J. E. Frommer, and K. R. Carter. 2006. *Langmuir* 22:2411–2414.

Benkreira, H. and R. Patel. 1993. *Chem. Eng. Sci.* 48:2329–2335.

Black, F. E., M. Hartshorne, M. C. Davies et al. 1999. *Langmuir* 15(9):3157–3161.

Blossey, R. 2012. *Thin Liquid Films.* New York: Springer.

Briseno, A. L., M. Roberts, M. Ling, H. Moon, E. J. Nemanick, and Z. Bao. 2006. *J. Am. Chem. Soc.* 128:3880–3881.

Brzoska, J. B., I. Benazouz, and F. Rondelez. 1994. *Langmuir* 10:4367–4373.

Christmas, B. K. 1984. *Mod. Paint Coat.* 74(10):152.

Cui, L., J. Peng, Y. Ding, X. Li, and Y. Han. 2005. *Polymer* 46(14):5334–5340.

Fan, H., Y. Lu, A. Stump et al. 2000. *Nature* 405:56–60.

Gates, B. D., Q. Xu, M. Stewart, D. Ryan, C. G. Willson, and G. M. Whitesides. 2005. *Chem. Rev.* 105:1171–1196.

Goethals, E. J. 1988. *Telechelic Polymers: Synthesis and Applications.* Boca Raton, FL: CRC Press.

Hanumanthu, R. and L. Scriven. 1996. *TAPPI J.* 79(5):126–138.

Heule, M., U. P. Schonholzer, and L. J. Gauckler. 2004. *J. Eur. Ceram. Soc.* 24:2733–2739.

Holscher, H. J. 1984. *Neue Verpack* 1:46.

Hyun, J., Y. Zhu, A. Liebmann-Vinson, T. P. Beebe, and A. Chilkoti. 2001. *Langmuir* 17(20):6358–6367.

Iskowitz, M. 2006. Artist's paints: Their composition and history. In *Coatings Technology Handbook*, A. A. Tracton (ed). Boca Raton, FL: CRC Press/Taylor & Francis Group, pp. 117-1–117-7.

Israelachvili, J. N. 1992. *Intermolecular and Surface Forces.* London: Academic Press.

Jacobs, H. O., S. A. Campbell, and M. G. Steward. 2002. *Adv. Mater.* 14(21):1553–1557.

Jacobs, H. O. and G. M. Whitesides. 2001. *Science* 291:1763–1766.

Jenekhe, S. A. and X. L. Chen. 1999. *Science* 283:372–375.

Julthongpiput, D., W. Zhang, J. F. Douglas, A. Karim, and M. J. Fasolka. 2007. *Soft Matter.* 3:613–618.

Kargupta, K. and A. Sharma. 2002a. *J. Chem. Phys.* 116(7):3042–3051.

Kargupta, K. and A. Sharma. 2002b. *Langmuir* 18(5):1893–1903.

Kargupta, K. and A. Sharma. 2003. *Langmuir* 19(12):5153–5163.

Khanna, R., A.T. Jameel, and A. Sharma. 1996. *Ind. Eng. Chem. Res.* 35:3081–3092.

Kim, S. O., H. H. Solak, M. P. Stoykovich, N. J. Ferrier, J. J. de Pablo, and P. F. Nealey. 2003. *Nature* 424:411–414.

Krans, D. E. 1999. *Met. Finish.* 97(5):135–138.

Krauskopf, L. G. and A. Godwin. 2005. Plasticizers. In *PVC Handbook*, C. E. Wilkes and J. W. Summers (eds). Munich, Germany: Hanser, pp. 173–198.

Lahann, J. 2006. *Polym. Int.* 55: 1361–1370.

Lahann, J., M. Balcells, T. Rodon et al. 2002. *Langmuir* 18(9):3632–3638.

Lee, M. J., J. Kim, and Y. S. Kim. 2008. *Nanotechnology* 19:355301-1–355301-16.

Li, H., B. V. O. Muir, G. Fichet, and W. T. S. Huck. 2003. *Langmuir* 19(6):1963–1965.

Lippert, H. G. 1991. Slot die coating for low viscosity fluids. In *Coatings Technology Handbook*, D. Satas and A. A. Tracton (eds). New York: Marcel Dekker, pp. 1559–1564.

Luo, C., R. Xing, Z. Zhang, J. Fu, and Y. Han. 2004. *J. Colloid. Interface. Sci.* 269:158–163.

Martin, T. P., S. E. Kooi, S. H. Chang, K. L. Sedransk, and K. K. Gleason. 2007. *Biomaterials* 28: 909–915.

McClelland, G. M., M. W. Hart, C. T. Rettner, M. E. Best, K. R. Carter, and B. D. Terris. 2002. *Appl. Phys. Lett.* 81(8):1483–1485.

McCollough, G. T., C. M. Rankin, and M. L. Weiner. 2006. *J. Soc. Inf. Display.* 14(1):25–30.

Michel, B., A. Bernard, A. Bietsch et al. 2001. *IBMJ. Res. Dev.* 45(5):697–719.

Morris, W. J. 1984. *J. Coat. Technol.* 56(715):49–56.

Muller, H. W. J. 2006. Vinyl ether polymers. In *Coatings Technology Handbook*, A. A. Tracton (ed). Boca Raton, FL: CRC Press/Taylor & Francis Group, pp. 117-1–117-7.

Odom, T. W., V. R. Thalladi, J. C. Love, and G. M. Whitesides. 2002. *J. Am. Chem. Soc.* 124:12112–12113.

Oron, A., S. H. Davis, and S. G. Bankoff. 1997. *Rev. Mod. Phys.* 69(3):931–980.

Patel, N., R. Bhandari, K. M. Shakesheff et al. 2000. *J. Biomater. Sci. Polym. Ed.* 11:319–331.

Reiter, G. 1992. *Phys. Rev. Lett.* 68(1):75–78.

Renn, M. J. 2006. U.S. Patent 7108894 B2.

Sadik, E., D. Hartmann, and O. Kibar. 2001. *World Patent* WO2001062400.

Sayers, R. 1983. *Wax: An introduction*. London: European Wax Federation and Gentry Books.

Schwartz, L. W. and R. R. Eley. 1998. *J. Colloid. Interface. Sci.* 202:173–188.

Seemann, R., S. Herminghaus, and K. Jacobs. 2001a. *J. Phys. Condens. Matter.* 13:4925–4938.

Seemann, R., S. Herminghaus, and K. Jacobs. 2001b. *Phys. Rev. Lett.* 86:5534–5537.

Seemann, R., S. Herminghaus, C. Neto et al. 2005. *J. Phys. Condens. Matter.* 17:S267–S290.

Sharma, A. and R. Khanna. 1998. *Phys. Rev. Lett.* 81(16):3463–3466.

Solomon, D.H. 1977. *The Chemistry of Organic Film Formers*. New York: Krieger.

Song, L., R. K. Bly, J. N. Wilson et al. 2004. *Adv. Mater.* 16(2):115–118.

Squires, T. M. and S. R. Quake. 2005. *Rev. Mod. Phys.* 77:977–1026.

Swaraj, P. 1997. *Surface Coatings: Science and Technology*, 2nd ed. Chichester, England: John Wiley.

Technical Bulletin, 0197. 1996. Lakewood, WA: Specialty products.

Teletzke, G. F., H. T. Davis, and L. E. Scriven. 1987. *Chem. Eng. Commun.* 55:41–81.

Thiokol Chemical Corporation. a. 1975. British Patent 1,403,649.

Thiokol Chemical Corporation. b. 1974. British Patent 3842053A.

Wang, Y., Z. Liu, Y. Huang, B. Han, and G. Yang. 2006. *Langmuir* 22:1928–1931.

Widawski, G., M. Rawiso, and B. Francois. 1994. *Nature* 369:387–389.

Wright, I. C. 1986. *Biodeterioration* VI:637–643.

Xia, Y. and G. M. Whitesides. 1998. *Angew. Chem. Int. Ed. Engl.* 37:550–575.

Yang, Z., A. M. Belu, A. Liebmann-Vinson, H. Sugg, and A. Chilkoti. 2000. *Langmuir* 16(19):7482–7492.

Yang, Z. and A. Chilikoti. 2000. *Adv. Mater.* 12(6):413–417.

Yoon, B., A. G. Lee, H. Kim, J. Huh, and C. Park. 2008. *Soft Matter.* 4:1467–1472.

III

Micro-, Macro-, Nanotesting and
Characterization of Polymers

Dynamic Mechanical Properties of Polymeric Materials Using Split Hopkinson Pressure Bar Apparatus

Mohd Firdaus Omar and Hazizan Md Akil

CONTENTS

9.1 POLYMERS AND SCOPE OF THIS CHAPTER

The demands of light material have increased continuously for the past few years. A huge number of funds and works have been invested to create lighter material without sacrificing its initial performances. To date, polymer is one of the promising materials that completely satisfied the lenient requirements with added excellent balance between impact resistance and weight. For these reasons, polymers have received remarkable attention from both industrial and educational sectors. Although polymers exhibit a lot of excellent abilities, their mechanical characteristics have become the primary criteria that determine the overall performances [1,2]. Recently, many sophisticated techniques exist to characterize the mechanical properties of polymers. Nevertheless, the techniques are totally different between static and dynamic assessments. As pointed out by Omar et al. [3], the universal testing machine used in static properties measurement would not be relevant for dynamic measurement because of the difficulties in providing high strain rate condition to the specimen. Therefore, a unique dynamic facility was first introduced by Kolsky [4] in 1949 to fulfill and satisfy the high strain rate testing requirements, which is known as the split Hopkinson pressure bar (SHPB) apparatus. Annually, the apparatus has experienced magnificent evolution by the following researchers [5–7], where currently, it has become a standard method of measuring material dynamic mechanical properties in the range of $10 \ s^2$ to $10 \ s^4$ of strain rates [8,9]. In polymer, factors such as chain structures, type of branching, and molecular weight might be key drivers that affect their mechanical characteristics [10–12]. Apart from internal issues, it is believed that external factor such as strain rate effect may also cause huge impacts to the mechanical behavior of polymers.

This chapter is concerned primarily on the mechanical properties of polymeric materials under high strain rate loading using an established dynamic testing apparatus, namely SHPB. It also covers some fundamental concepts of the SHPB technique, including its history, development, challenges, and solutions on polymeric specimens. The purpose of this chapter is also to describe and, when possible, to explain the effect of strain rates toward the mechanical proportions of various polymeric specimens. For the numerical approach, three material-constant constitutive equations are discussed to take into account the characteristics of strain rate sensitivity of polymeric materials at very high strain rates.

9.1.1 Importance of Dynamic Mechanical Properties Knowledge in Polymeric Materials

It is generally acknowledged that the applications of polymeric-based products have been extended from conservative to more challenging applications such as engineering components, constructions, load-bearing applications, and others. Hence, the strain rate effect should be the first priority factor to be investigated since almost all of the highlighted applications are mainly involved with both static and dynamic conditions. In addition, the knowledge of rate sensitivity is also important during material selection in order to estimate the magnitude of changes in the properties of materials. Without this knowledge, it is almost impossible to predict and prevent the unexpected failure during service. In order to simulate actual service environments, static and dynamic mechanical testing should be performed on polymeric-based products to fully characterize their overall mechanical performance. For static testing, the mechanical properties of materials can be easily characterized using conventional mechanical testing machine such as universal testing machine. However, it becomes more complicated with dynamic testing as it requires many additional requirements and assumptions that need to be satisfied before proceeding with the actual test. This is the main reason why scientists are more interested in the static behavior of material rather than their dynamic behavior recently. Although several previous studies have been carried out on the dynamic mechanical properties of polymeric materials and their composites, but the information and knowledge remain unclear because there is a limitation in terms of their experimental setup and parameter. Based on this limitation, it is believed that systematic works should be carried out in the future in order to gain reliable results, especially for engineering design and simulation purposes.

9.2 HIGH STRAIN RATE FACILITIES FOR DYNAMIC MECHANICAL TESTING

For the past few years, there has been various high strain rate facilities available such as the drop weight impact, Taylor impact, expanding ring, plate impact, and SHPB. Initially, the drop weight impact machine was purposefully designed to demonstrate the low-to-intermediate speed impact test. The fundamental of this machine is based on the dropped weight concept, where the impact velocity is dependent on the earth's gravity. For the machine design, a specimen is fixed on top of the steel base. After that, an impactor is elevated at a certain specific height before being released on the specimen's surface. During the collision between the specimen's surface and the impactor, the kinetic energy of the impactor is then absorbed by the progressive failure of the specimen. The kinetic energy absorption continues until the impactor has totally stopped. The load cell is held at the space between the specimen and the steel base in order to record the generated crushing force during the collision. The crushing force results are then captured and recorded using a data acquisition system. The main advantage offered by this machine is related to a cost effective solution as compared to other dynamic facilities that use a gas gun. Lin et al. [13] investigated the mechanical behavior of epoxy reinforced modified montmorillonite (cloisite 30B) and titanium dioxide nanocomposites at dynamic loadings using the falling mass impact tester, and they found prominent improvements in terms of impact strength by the filler weight contents. In a more recent study, Dhakal and his coworkers [14] investigated the effect of temperature, as well as the impact velocity toward the impact response of polymeric material using a drop weight impact tester. Interestingly, they found that the absorbed energy for the tested polymeric specimens tended to increase with increasing temperature. Although a falling mass impact test can give both impressive and convenient results, the tests are still limited by several factors, such as sensitivity toward the contact conditions between the impactor and the specimen [15]. In addition, the system is also restricted to lowered strain rate loading conditions (between 1 and 10 ms^{-1}) since the striking velocity of the impactor totally relies on the height of the machine [16].

Another famous dynamic facility that can be used to characterize the dynamic mechanical properties of materials is the Taylor impact test. This test was first developed by Taylor during the late 1940s [17]. His main purpose in demonstrating the technique was to investigate the dynamic strength of ductile materials under compression loading. For the experimental design, this method propels a cylindrical projectile toward the specimen, which is normally rigid and symmetrical in shape. From the measurement of the initial velocity of the projectile, the velocity of the target and the change of shape, as well as the dynamic behavior of the material can be obtained using the corresponding equations. More recently, Salisbury [18] speculated that computer simulation of the impact allows the selection of an appropriate material model by comparing the deformed shape to the simulation. Although the Taylor impact test can provide a higher strain rate condition than that of the drop weight impact test, somehow, it is only credible to certain specific specimen geometries. As revealed by Field et al. [8], the technique faces a difficulty in determining the dynamic deformation of a specimen in disc form. For this reason, the usage of the Taylor impact test is rarely implemented and slowly diminishing, especially for current dynamic characterizations.

Alternatively, the characterization of dynamic mechanical behavior of materials is based on the expanding ring as suggested by Hoggat and Recht [19]. The basic principle of this technique involves a hollow cylinder with an explosive core as a method of initiating a shock wave [20]. The velocity histories are then manipulated using a set of simple equations in order to calculate the stress–strain characteristic of the material. A variety of detonation products can be used to provide various strain rate conditions. The major advantage offered by this technique is the ability to produce an extremely high strain rate test, while the main drawback for this technique is the difficulty in the data reduction measurement that leads to inaccuracy in the stress measurement [21]. Daniel and LaBedz [22] performed the expanding ring for the unidirectional 0°–90° graphite/epoxy specimens of up to 500 s^{-1}. They found that the technique could give reliable results in which composites with 0° fibers showed some significant increment in modulus but no dramatic changes in strength. For composites with 90°, a much higher modulus and strength than the static value was exhibited.

A desire to scrutinize the characteristics of materials, at high strain rates, has revealed the most promising technique in studying the dynamic mechanical behavior of materials, namely the SHPB technique. The implementation of two Hopkinson bar was initiated by three great researchers in the early nineteenth century [23,24] to measure the dynamic properties of materials in compression [4,17,25]. Even though Kolsky introduced this technique almost five decades ago, it was only intensively used by researchers during the early 1970s. More recently, the SHPB technique has become the standard method of measuring material dynamic mechanical properties in the range of 10^2 to $10^4\,s^{-1}$ of strain rates [8,26]. As SHPB has increasingly become the standard method of measuring material dynamic mechanical properties in the strain rate range of 10^2 to $10^4\,s^{-1}$, tension [27] and torsion [28] versions have been developed.

9.2.1 Split Hopkinson Pressure Bar Apparatus

Typically, the SHPB apparatus consists of three separate bars, which are a striker bar, an incident bar, and a transmitter bar as shown in Figure 9.1. At the beginning stage of the SHPB test, the specimens are clamped between the incident and the transmitter bars (refer to Figure 9.1). The striker bar is accelerated by the pressure from an air gun and then launched through the gun barrel before colliding with the incident bar. During the collision, a compressive strain pulse (ε_i) is generated in the incident bar and travels toward the specimen. Because of the impedance mismatch between the bar and the specimen's surface, some of the generated pulse is reflected back (ε_r) to the incident bar and the remaining strain pulse (ε_t) will be transmitted through the specimen into the transmitter bar. The propagation of the strain pulse along the Hopkinson bars can be well understood by the Lagrangian x–t diagram, which is shown in Figure 9.2. The generated incident and the transmitted and reflected pulses are then captured by the piezoelectric strain gauges mounted on the incident and transmitter bars using a special adhesive. The output voltage of the Wheatstone circuit due to the change of resistance in the strain gauge when

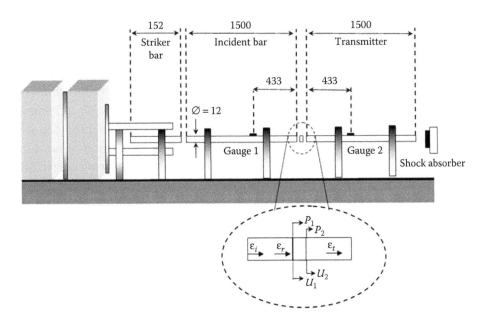

FIGURE 9.1 The schematic diagram of the SHPB apparatus. (From Omar, M.F. et al., *Mater. Design.*, 32[8–9], 4207–4215, 2011. With permission.)

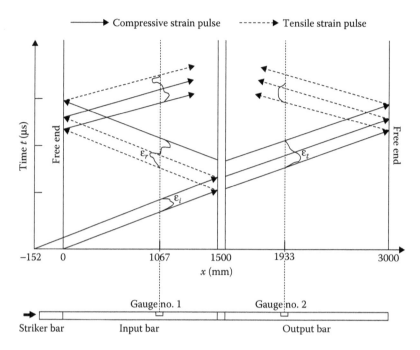

FIGURE 9.2 Lagrangian diagram illustrating wave movement in Silver steel Hopkinson Bars. (From Omar, M.F. et al., *Mater. Design.*, 31[9], 4209–4218, 2010. With permission.)

deformation occurred in the Hopkinson bars is then transferred to the transducer amplifier to amplify the voltages produced by the strain gauges. The signal is then captured using an oscilloscope and is saved in the computer for data processing.

9.2.1.1 The Advantages of SHPBA as a Dynamic Tester

The main element that is attributed to the high popularity of the SHPBA is related to its capability to easily obtain the stress/strain curve as the output result, which holds useful information to characterize materials. In addition, the versatility of the test configuration has also contributed to the high usage of the SHPB, where it is available in compression [30], tensile [31], and torsion [32]. However, there are various factors affecting the accuracy of the SHPB result, such as specimen–bar interface friction, extension–shear coupling, and the rise time of the loading pulse that needs to be addressed during the SHPBA test [33]. Based on previous studies, it can be concluded that the SHPB is suitable with almost all materials, including metals [34,35], ceramics [36,37], polymers [38,39], as well as composites [40,41]. Nevertheless, additional modifications and precautions should be taken into consideration during the SHPB test for soft specimens such as polymers, foam, and elastomers.

9.2.1.2 History and Development of SHPBA

The initial idea in manipulating a pressure bar to obtain the characteristics of a material was first introduced by Hopkinson [42]. He initiated a procedure that allowed the transaction of dropping weight energy to a wire and started to measure the deformation of the wire before failure. In 1914, his son Bertram Hopkinson continued his work by using a bar in order to obtain the pressures developed by the blast on impact from a bullet [23]. A schematic diagram of an orthodox SHPB device fabricated by Bertram Hopkinson is illustrated in Figure 9.3. As referred to in this figure, he used a bar (B) that was being suspended by two sets of wires. This bar was parallel to a box (D) that was also suspended. A secondary rod (C) was placed at the end of the main rod and held in place by a small magnetic force. A bullet was then shot at the end of the main rod (A). The collision induced a pressure pulse that was imparted into the main rod. The generated pulse traveled down the main rod into the secondary rod causing it to fly off and caught by the box. The displacement of the box and the secondary rod was measured with a basic measurement device, enabling the calculation of the momentum. Very simple measurement devices available at that time prohibited the accuracy of the result.

A very limited number of works were involved with this kind of research until 1948 when Davies [43] demonstrated a crucial study on the Hopkinson pressure bar. At that time, more accurate devices were used to precisely measure the displacement of the secondary bar (i.e., the end bar). This research was then continued by Kolsky [4] using a three-bar system, which contained a striker bar, an incident bar, and a transmitted bar. He mounted condenser units on both the incident and transmitter bars to gain knowledge on characterizing the mechanical properties of the tested specimen. It is also believed that a new promising era of the SHPBA had begun from this kind of research. Figure 9.4 shows the usage and development of the SHPB from 1940 to 1998. The histogram data was based on the published papers in any given year where the SHPB approach was used to determine the dynamic mechanical properties of various materials. From Figure 9.4, it is clearly seen that the usage of the SHPB apparatus for determining the dynamic mechanical properties of

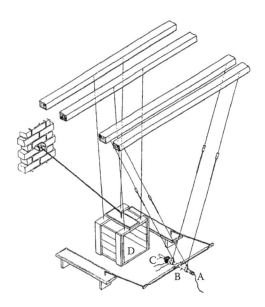

FIGURE 9.3 Schematic diagram of a traditional SHPB device in 1914. (From Hopkinson, B., *Proc. R. Soc. Lond. A*, 213[89], 437–456, 1914. With permission.)

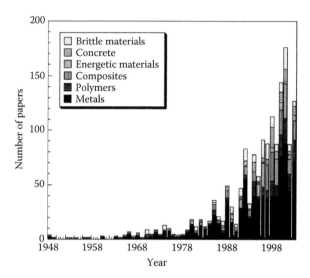

FIGURE 9.4 The usage and development of the SHPB apparatus from 1940 to 1998. (From Field, J. et al., *Int. J. Impact Eng.*, 30[7], 725–775, 2004. With permission.)

various materials started to become well known in the late 1970s. Additionally, the recent development and modification of the SHPB apparatus by various experts are combined together and summarized in Table 9.1. The rapid development of the SHPB is driven by the desire to scrutinize the characteristics of various materials at dynamic conditions for crucial applications.

TABLE 9.1   The Current Development of the SHPB Apparatus

Year	Development	References
1980	Gorham and field developed a miniaturized direct impact Hopkinson bar	[44]
1985	Albertini developed a large SHPB for testing structures and concrete	[45]
1991	Nemat–Nasser developed one pulse loading SHPBs (compression, tension, and torsion) and soft recovery techniques	[46]
1991–1993	Use of torsional SHPB for measurement of dynamic sliding friction and shearing properties of lubricants	[47]
1992–2003	Development of polymer SHPB for testing foams	[48,49]
1997–2002	Use of wave separation techniques to extend the effective length of a Hopkinson bar system	[50,51]
1998	Development of magnesium SHPB for soft materials	[52]
1998	Development of radiant methods for heating metallic SHPB specimens quickly	[53,54]
1998–2002	Analysis of wave propagation in non-uniform viscoelastic rods	[51]
1999	Development of one pulse torsion SHPB	[55]
2003	Extension of Hopkinson bar capability to intermediate strain rates	[56]
2004–2005	The extension of Kolsky bar techniques for tension and torsion tests based on very similar mechanisms but different loading and gripping methods	[8,57]
2009–2011	An innovative design of split Hopkinson bar that used the intense pressure created in a transient magnetic field formed by the passage of a pulse of electric current through a series of coils	[58]
2012–2014	Validation of material's model using SHPB apparatus	[59]

9.2.1.3 Theory Behind the Conventional SHPBA

In the conventional SHPBA, the behavior of materials is obtained from the difference in interface velocities (V_1, V_2 in Figure 9.5). As the elastic pulse deforms the sample length, the distance between the incident and transmitter bars decreases because $V_1 > V_2$. This deformation occurs over a period, which enables the calculation of the strain rate using the following equation:

$$\frac{d\varepsilon_s}{dt} = \frac{V_1 - V_2}{L_s} \tag{9.1}$$

Unfortunately, it is very difficult and almost impossible to measure the velocity at the end of each bar. Thus, an alternative approach using elastic wave propagation in the incident and transmitter bars is often adopted. Theoretically, the wave velocity in the material is defined as

$$C_o = \sqrt{\frac{E}{\rho}} \tag{9.2}$$

where C_o, E, and ρ are the wave speed, Young's modulus, and density, respectively. It is believed that the longitudinal wave propagates through the elastic media at this speed [18]. In order to determine the stress, strain, and strain rate history of the specimen, the incident strain ε_i (t), the reflected strain ε_r (t), and the transmitted strain ε_t (t) can be manipulated. The relationship between the velocities at the interface and the strain can be expressed by the following equations:

$$V_1 = C_o\varepsilon_i \quad \text{at} \quad (t = 0)$$
$$V_2 = C_o\varepsilon_t \tag{9.3}$$

At $t > 0$, the incident and reflected waves are overlaid and therefore reduce the velocity of V_1. As a result, V_1 becomes

$$V_1 = C_o(\varepsilon_i - \varepsilon_r) \tag{9.4}$$

Equations 9.3 and 9.4 are inserted into Equation 9.1 in order to calculate the strain rate. Thus, the strain rate equation can be summarized as follows:

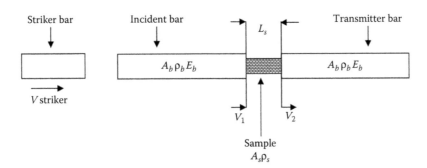

FIGURE 9.5   Typical diagram of the Hopkinson bars.

$$\dot{\varepsilon}_s = \frac{C_o}{L}\left(\varepsilon_i - \varepsilon_r - \varepsilon_t\right) \tag{9.5}$$

In our case, the Hopkinson bars are made from the same material and therefore the stress in the sample is obtained by using the following equation:

$$\sigma_s = \frac{F_1(t) + F_2(t)}{2A_s} \tag{9.6}$$

where, F_1 and F_2 are the applied force (i.e., by the bar) at the specimen surface. Meanwhile, A_s represents the cross-sectional area of the sample. As $E = \sigma/\varepsilon$ (stress/strain), the forces in the bar can be related to the strains in the bar by

$$F_1 = E_b A_b(\varepsilon_i + \varepsilon_r)$$
$$F_2 = E_b A_b(\varepsilon_t) \tag{9.7}$$

For the specimen's stress, Equation 9.7 is substituted into Equation 9.6 and then summarized as follows:

$$\sigma_s = \frac{E_b A_b}{2A_s}\left(\varepsilon_i + \varepsilon_r + \varepsilon_t\right) \tag{9.8}$$

where E_b and A_b are the Young's modulus and the cross-sectional area of the bar respectively. Generally, to achieve an equilibrium state, $F_1 = F_2$ and $\varepsilon_i + \varepsilon_r = \varepsilon_t$. Hence, the stress, strain, and strain rates can be simplified and summarized into the following equations:

$$\sigma_s(t) = E_b \frac{A_b}{A_s}\varepsilon_t(t) \tag{9.9}$$

$$\varepsilon_s = -2\frac{C_o}{L_s}\int_0^t \varepsilon_r(t)dt \tag{9.10}$$

$$\dot{\varepsilon}^s = \frac{d\varepsilon(t)}{dt} = -2\frac{C_o}{L_s}\varepsilon_r(t) \tag{9.11}$$

The derivations of Equations 9.9 through 9.11 are closely related with the following assumptions and ideas [60]:

1. The bars must remain elastic throughout the SHPB test.

2. The propagation of the wave in the Hopkinson bars is approximated by a one-dimensional theory, where the wave dispersion and attenuation are totally negligible.

3. The pulse is uniform and homogeneous over the cross section of the bar.

4. The friction and radial inertia effects are negligible, and the specimen remains in equilibrium throughout the test.

It is necessary to enforce the first assumption since the Hopkinson equations are mainly adopted from elastic wave equations. For the second assumption, it is clearly stated that the wave dispersion and attenuation are totally negligible, so that the strain measured by the strain gauges is assumed as similar with the strain experienced at the interface. Dispersion is a result of a bar's phase velocity dependence on the frequency, which in effect distorts the wave as it propagates [61]. The third assumption indicates that the generated pulse must be uniform and homogeneous over the cross section of the Hopkinson bars. This property must exist to prevent any needless pulse, such as a distortion pulse. Therefore, it is suggested that the pulse is fully developed in four [43] to ten [62] bar diameters from the interface. Ultimately, for the fourth assumption, it needs a detailed explanation on the specimen's geometry consideration and invites serious discussion among the experts. Hence, a typical discussion on the highlighted issue is made in the following section.

9.2.1.4 Specimen Geometry Considerations for Polymeric Specimens

The majority of SHPB experts agree that the geometry of the specimens will significantly influence the stress equilibrium within the specimen's body [4,63,64]. As pointed out by Davies [43], the equilibrium can be achieved if the back and forth pulses within the specimen is more than π times. Therefore, to achieve that condition, they suggested that the optimum slenderness ratio (length/diameter) for a polymeric specimen is 0.5. More recently, a study on the effect of the specimen's thickness on stress equilibrium was performed by Wu and Gorham [65]. They found that a thinner specimen would be able to achieve faster uniformity of deformation but would increase the effect of friction. Their finding almost agrees with the earlier result reported by Dioh et al. [64]. Dioh et al. [64] suggested that it is crucial to choose a suitable specimen's geometry in order to correctly determine the properties of materials, especially dynamic loading. An optimum specimen's geometry will reduce the effect of dispersive distortion and therefore, improve the uniformity of deformation.

9.2.1.5 Calibration and Verifications of the SHPB Results for Polymeric Specimens

In real practice, there are two common calibration methods (i.e., with and without specimens) that can be manipulated to verify the accuracy and reliability of the SHPB setup [66,67]. For the first calibration setup, two Hopkinson bars (elastic bars) were wrung together without a specimen between them, which was implemented by Naik et al. [9] and Omar et al. [68]. They also applied a thin (wax) lubricant at both ends of the Hopkinson bars in order to prevent any frictional effects. Therefore, both elastic bars can be treated together as a single bar, which was then impacted by a striker bar up to certain specific striking velocity. The oscilloscope readings captured during calibrations are presented in Figure 9.6 [68]. Gauge 1 represents the voltage measured by the strain gauge mounted on the incident bar. On the other hand, Gauge 2 represents the voltage measured by the strain gauge mounted on the transmitter bar. It can be observed that the amplitude and duration of both incident and the transmitted pulses are nearly similar. Moreover, no reflected pulses were recorded during this calibration. Apart from that, the stress histories from the strain gauge signals are presented in Figure 9.7. Typically, the stress history was based

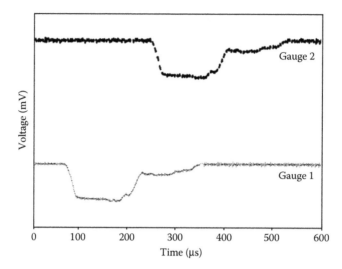

FIGURE 9.6   Strain gauge signals on the oscilloscope during calibration. (From Omar, M.F. et al., *Mater. Sci. Eng. A*, 528[3], 1567–1576, 2011. With permission.)

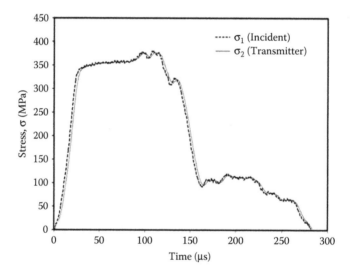

FIGURE 9.7   Comparison of stress vs. time characteristic derived from strain gauge signals during calibration. (From Omar, M.F. et al., *Mater. Sci. Eng. A*, 528[3], 1567–1576, 2011. With permission.)

on the signals captured from both gauges mounted on the Hopkinson bars, where σ_1 and σ_2 indicated the stresses within the incident bar and the transmitter bar, respectively. The stress history profile, shown in both Figures 9.6 and 9.7, is in good agreement with the profile presented by Naik et al. [9]. Based on the recorded stress history in Figure 9.7, it is clearly seen that the stress obtained in σ_1 and σ_2 matched very well. This indicates that the stress states within the incident bar and the transmitter bar are similar.

Another way of verifying and calibrating the SHPB setup is to closely relate the stress equilibrium within the specimen. Therefore, several previous researchers have done the dynamic stress equilibrium within soft specimens like polymers and polymer matrix

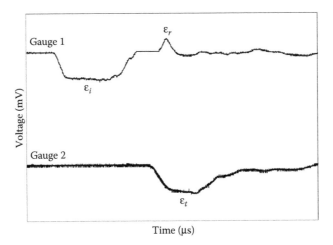

FIGURE 9.8 Oscilloscope traces from compression split Hopkinson pressure bar test on GFRC ($V_s = 16.9$ ms^{-1}). (From Omar, M.F. et al., *Mater. Design.*, 31[9], 4209–4218, 2010. With permission.)

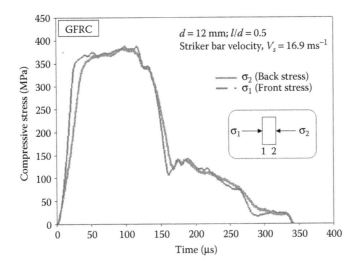

FIGURE 9.9 The applied stresses at each face of Hopkinson bars. (From Omar, M.F. et al., *Mater. Design.*, 31[9], 4209–4218, 2010. With permission.)

composites (PMCs) [3,69,70]. Omar et al. [3] clarified the stress equilibrium within their PMCs specimens (i.e., glass fiber reinforced composites) as shown in Figures 9.8 and 9.9. As can be seen in Figure 9.8, Gauge 1 represents the voltage measured by the strain gauge mounted on the incident bar, where it shows the incident strain pulse and reflected strain pulse (ε_i and ε_r). Conversely, Gauge 2 represents the voltage measured by the strain gauge attached on the transmitter bar, where it shows the transmitted strain pulse (ε_t). Meanwhile, Figure 9.9 illustrates the axial stress histories at the front and back ends of the specimen during the SHPB test with 16.9 ms^{-1} of striking velocity, V_s. From the graph in Figure 9.9, Omar and his coworkers [3] experimentally proved that the dynamic stress equilibrium within their PMCs specimens (i.e., GFRC) was approximately equilibrium, where both axial stress histories almost agreed with each other. Based on this calibration profile, it is convenient to say that the SHPB apparatus has fully satisfied the requirements of the high strain rate's testing and is also suitable for soft specimens such as polymeric specimens. Apart from that, it was experimentally proven that their SHPB setup is in perfect working condition (i.e., accurately aligned and friction free) and therefore, ready for further investigations.

9.2.2 Challenges and Solutions of SHPB Test on Polymeric Specimens

Soft specimens such as polymeric specimens might face difficulty in dealing with the conventional SHPB because the transmitted pulse is significantly decreased and consequently decreases the signal-to-noise ratio. Thus, it is necessary to boost the transmitted pulse in order to meet the SHPB requirements. Many studies have been performed to overcome this dilemma. As revealed by Chen et al. [71], the intensity of the transmitted pulse can be increased by reducing the cross-sectional area of the transmitter bar. This action will significantly reduce the impedance value of the transmitter bar and therefore increase the sensitivity of the transmitted pulse measurement. Besides, the method also prevents any needless effects such as dispersion and attenuation problems. A year later in 2000, Chen et al. [72] proposed another solution for increasing the signal in the transmitter bar. This time they embedded a small quartz crystal into the transmitter bar. Initially, the quartz crystal would replace the conventional strain gauges with three times more sensitivity than that of conventional strain gauges. The nature of quartz crystal with a similar shape and impedance also eliminates the undesired effects of placing this alternative gauge at the middle of the transmitter bar. However, problems such as reflections and refractions would occur when the generated pulse reaches the quartz crystal because of the discontinuities between the crystal-bar interfaces.

More recent studies by Frew et al. [73] and Vecchio and Jiang [74] suggested another alternative method that can be easily implemented during the SHPB test on soft materials. They believed that the use of a pulse shaper would increase the transmitter pulse, thus inducing faster stress equilibrium within the specimen body. On the other hand, Johnson et al. [75] proposed the idea of implementing a lower impedance bar, for example, a polymer bar, which has an almost similar impedance value to that of the tested specimen. By doing this, it would significantly boost the propagation of the transmitted pulse. As a pre-conclusion, with minor modifications, it is convenient to say that the SHPB apparatus is both reliable and consistent in determining the dynamic mechanical behavior of soft specimens, especially for polymeric materials.

9.3 SHPB TEST ON POLYMERIC MATERIALS

9.3.1 Effect of Strain Rates on Dynamic Mechanical Properties of Thermoplastic Specimens

Comparative studies between the static and dynamic mechanical properties of soft materials such as polymers were rarely reported in the past. However, a crucial preliminary investigation was initiated by several groups of researchers [29,70,76,77]. The most detailed and prominent finding was reported by Walley and Field [77]. They exposed the thermoplastic specimens to a wide range of investigated strain rates from 0.001 (static) to 10,000 s^{-1}

(dynamic). The stress/strain results for high-density polyethylene (HDPE), polypropylene (PP) and polycarbonate (PC) are shown in Figure 9.10a–c. From the stress–strain curves in Figure 9.10a–c, it is clearly indicated that PP and PC experienced pronounced load drops at all investigated strain rates. Conversely, HDPE exhibits some strain hardening before flowing at constant stress above a strain of about 0.3. Although the results might be useful to interpret the behavior of selected thermoplastic polymers under various loading conditions, somehow there is a limitation in terms of their experimental setup and parameters. For more conclusive findings, the average strain rate parameter should be constant from static to dynamic loading for all tested specimens. By implementing this approach, it is easier to compare materials from one to another. Based on this limitation, it is believed that systematic works should be carried out in the future in order to gain reliable results, especially for comparison purposes.

More recently, a study was performed by Nakai and Yokoyama [70], where they implemented the SHPB test on several commercial thermoplastic polymers, including ABS, PA-6, PA-66, and PC up to nearly 700 s^{-1} of strain rate. Meanwhile, the intermediate strain rate behavior was measured using a universal testing machine at a crosshead speed of 100 mm min^{-1}. They found that the Young's modulus of all polymeric specimens increased greatly

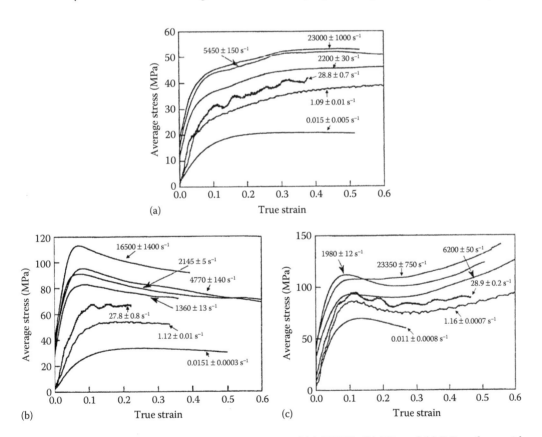

FIGURE 9.10 The compressive stress–strain curves of (a) HDPE, (b) PP, and (c) PC under a wide range of investigated strain rates. (From Walley, S.M. and Field, J.E., *DYMAT J.*, 1[3], 211–277, 1994. With permission.)

with increasing strain rates. Not only that further assessment observed that the effect of strain rate was further enhanced on several aspects of material proportions such as 2.5% of flow stress and strain energy, which can be seen in Figure 9.11a–c. From Figure 9.11a–c, both 2.5% of flow stress and the dissipation energy reacted similarly with the Young's modulus properties as the properties increased linearly with increasing strain rates for all polymeric specimens. Based on this observation, they believed that all tested thermoplastic polymers exhibit an inherent dynamic viscoelastic characteristic. The dissipation energy is mostly converted to heat during high rate deformation that causes the adiabatic temperature to rise within the specimen. Based on their finding, it can be pre-concluded that the viscoelastic properties of polymers are much affected by the strain rate where it may hugely influence the overall mechanical performance of polymeric specimens.

Li and Lambros [78] demonstrated both static and dynamic testing on PMMA specimens as illustrated in Figures 9.12 and 9.13. As for static range, PMMA shows obvious three regimes of stress–strain characteristic, which are an initial linear portion, a yielding and strain softening portion, and finally, a nonlinear strain hardening portion. This finding is consistent with the work reported by Arruda et al. [79]. In contrast to static characteristic, it was clearly observed that the dynamic response for PMMA did not exhibit a pronounced nonlinear regime (refer to Figure 9.13). They also mentioned that the ripples in the dynamic

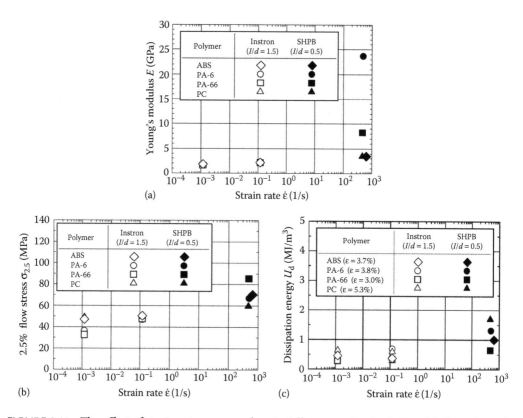

FIGURE 9.11 The effect of strain rate on several material's proportion including (a) Young's modulus, (b) 2.5% flow stress, and (c) dissipation energy. (From Nakai, K. and Yokoyama, T., *J. Solid Mech. Mater. Eng.*, 2[4], 557–566, 2008. With permission.)

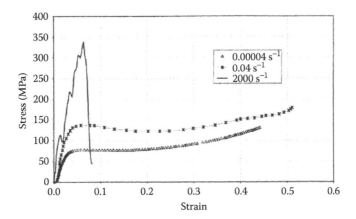

FIGURE 9.12   Compressive stress–strain curves for PMMA at strain rates 4×10^{-5}, 0.04, and 2000 s^{-1}. (From Li, Z. and Lambros, J., *Int. J. Solids Struct.*, 38[20], 3549–3562, 2001. With permission.)

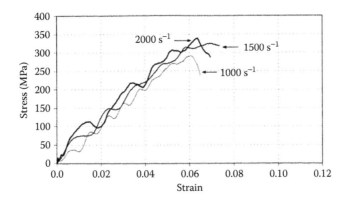

FIGURE 9.13   Dynamic compressive stress–strain curves for PMMA. (From Li, Z. and Lambros, J., *Int. J. Solids Struct.*, 38[20], 3549–3562, 2001. With permission.)

PMMA stress–strain curves did exist and might represent some localized shear yielding or failure, where the stress increased to its maximum value in a generally linear fashion and suddenly dropped to zero. This occurs considerably before the end of the applied loading pulse and corresponds to actual specimen failure. In fact, the PMMA specimens shattered completely from the impact event. Thus, there is a substantial difference in the response of PMMA over the strain rate range tested. This is best illustrated in the plots of Figure 9.12.

9.3.2 Effect of Strain Rates on Dynamic Mechanical Properties of Thermoset Specimens

Different to thermoplastic specimens, there are only a few works that are concerned on the effect of strain rates on dynamic mechanical properties of thermoset specimens using SHPB apparatus. This might be due to the nature ability of thermoset specimens that can be categorized as brittle materials. Among many types of thermoset specimens, epoxy resin has been widely used in many applications. Although a few works have been reported on epoxy resin and other thermoset polymers at high strain rates, but the works are generally restricted to only experimental data [80–82]. More recently, Gómez and Rodríguez [83] successfully investigated the compressive mechanical loading of the thermosetting system epoxy polymer-based on diglycidyl ether of bisphenol A (DGEBA) and an aromatic amine hardener, 4,4′-methylenedianiline (DDM) as a function of strain rate and temperature, ranging from 0.0025–2500 s^{-1} and from room temperature to 100°C, of which all below are the glass transition temperature. Their finding is summarized and portrayed in Figure 9.14. Figure 9.14 includes examples of stress–strain curves obtained at high strain rates in the Hopkinson bar. From Figure 9.14, it can be observed that the strain hardening phase was not reached during the high strain rate tests due to the limited compressive pulse length. They believed that the observation was not a consequence of any material response. Not only that, they also speculated that the increment in stress with increasing strain rate was attributed to the reduction of molecular mobility of the polymer chains due to the secondary relation processes.

9.3.3 Effect of Strain Rates on Dynamic Mechanical Properties of Polymer Matrix Reinforced Composites

It is generally acknowledged that huge attention is given on the high strain rate behavior of thermoset-based polymer matrix composites rather than their counterpart (i.e., thermoplastic-based polymer matrix composites), especially using the SHPB apparatus [9,69,84–86]. This can be clearly seen by the results reported by Guo and Li [86]. Initially, Gou and his coworkers had successfully obtained the compressive mechanical behavior of Epoxy/SiO$_2$ up to high levels of the strain rate investigated. Both low and high strain rate characteristics of Epoxy/SiO$_2$ composites were characterized using the universal testing machine and SHPB compression, respectively. Interestingly, they found that the yield stress, as well the

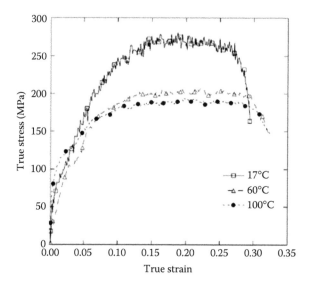

FIGURE 9.14 Compression true stress–strain curves for comparison of mechanical behavior at different strain rates for the epoxy between 1500 and 1700 s^{-1} over a wide range of temperatures. (From Gómez-del Río, T. and Rodríguez, J., *Mater. Design.*, 35, 369–373, 2012. With permission.)

flow stress of the Epoxy/SiO$_2$ nanocomposites, increased dramatically with increasing applied strain rate as depicted in Figure 9.15. A similar finding was also reported by Chen et al. [87] and Omar et al. [88], where they believed that the stress increment was closely related with the secondary molecular processes. Increasing the strain rate would significantly decrease the molecular structure of the polymer chains, thus making the material stiffer. Interestingly, they also found that in some cases, the strain-hardening effect had a secondary effect on the strain rate effect. This phenomenon can be clearly observed at curves (b) (at strain rate 0.01 s^{-1}) and overrun curve (c) (at strain rate 0.2 s^{-1}) at a strain of about 0.5 (refer to Figure 9.15). Based on this observation, the common sense deduction is that the positive rate sensitivity makes the material stronger is inappropriate in some cases. In this case, the strain hardening effect plays a primary role, where a decrease in the strain-hardening effect may be related to the micro-damage accumulated during loading within the composite body. At high loading rates, the mobility of the polymer chains decreases and thus, more ruptures have to take place to adapt to the deformation of the specimen. From a compressive strength point of view, there is a trend recorded for PMCs as a function of applied strain rates as revealed by both Guo and Li [86] and Omar et al. [88]. They found that at low strain rates, the compressive strength shows a linear relationship to the applied strain rate, whereas a sharp increase in the compressive strength value is observed at a high strain rate. Similar observation was also reported in the behavior of neat polymers such as PC [76] and PP [89]. Further explanation for this phenomenon is still unclear. However, a common interpretation of the phenomenon is closely related with changes in the deformation mechanism from a thermally activated flow mechanism at low rates to a viscous drag dominating flow process at high rates.

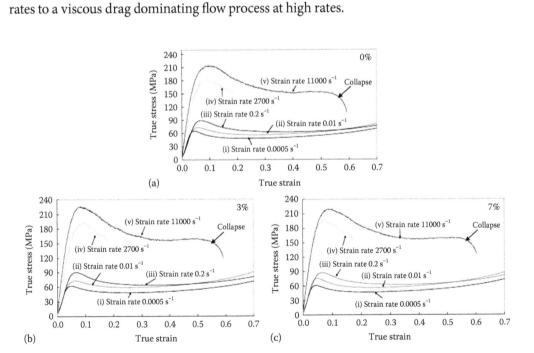

FIGURE 9.15 Typical stress–strain curves for epoxy–SiO$_2$ nanocomposites: (a) 0%, (b) 3%, and (c) 7% at room temperature. (From Guo, Y. and Li, Y., *Mater. Sci. Eng. A.*, 458[1–2], 330–335, 2007. With permission.)

Periasamy et al. [90] demonstrated the compression SHPB tests on a synthetic epoxy composite foam with a variety of hollow glass microballoon volume fractions (i.e., 10%, 20%, 30%, and 40%). The experimental strain rate was deliberately varied from a low to high level of strain rate in order to fully characterize the overall mechanical performance of the tested specimen (composite foam). Interestingly, they found that the stress–strain responses for low and high strain rates characteristic of all composite foams showed a significant variance from one to another as illustrated in Figure 9.16a and b. Specifically, from Figure 9.16a, it can be seen that the stress–strain responses under low strain rate region of composite foam portrayed four distinct regimes (elastic, softening, plateau, and densification regimes). Meanwhile, under high strain rate loading (refer to Figure 9.16b), the composite foams only showed two dominant regimes, which are a proportional loading zone up to a maximum stress and a monotonically softening zone subsequently. Apart from that the absorbed energy of all composite foams was also calculated up to 22% of true strain. It is interesting to note that all composite foam specimens absorbed higher energy under high strain rate loading than that of low strain rate loading. Instead of particulate and foam types of PMCs, the SHPB compression is also compatible with the fiber reinforcement of PMCs. As revealed by Ochola et al. [91], the SHPB compression has been successfully manipulated in order to characterize the mechanical performance of both glass and carbon fiber reinforced epoxy composites (i.e., GFRP and CFRP) under various level of strain rates investigated. They speculate that the ratio of strain to failure for CFRP was higher than that of GFRP ranging from low to high strain rate levels. However, they also found that the GFRC absorbed higher energy despite reduced strain to failure as compared to the CFRP system.

For tensile SHPB, a minor modification should be carried out, especially at the specimen clamping section and also the geometry of specimen in order to simulate dynamic tensile environment. However, PMCs specimens might face difficulty with this setup because it requires more complex machining of specimen's geometry. Based on this concern, other simpler tensile setups have been proposed by Fu and Wang [92] and Gómez-del Río et al. [93] in order to overcome this dilemma. Based on this concern, it is

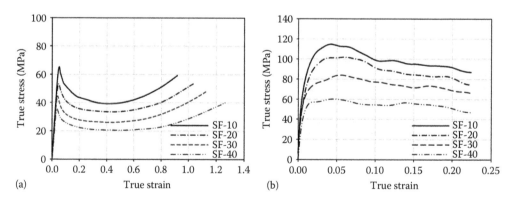

FIGURE 9.16 Effect of microballoon volume fraction on (a) static and (b) dynamic (strain rate ~1500 s^{-1}) stress–strain responses of SF samples. (From Periasamy, C. et al., *Mater. Sci. Eng. A.*, 527[12], 2845–2856, 2010. With permission.)

convenient to say that the SHPB test is also available and reliable in tensile configuration, especially for PMCs specimens.

Naik and his coworkers [69] experimentally proved that the tensile SHPB can be used to fully characterize the tensile mechanical properties of their PMCs specimens, which are woven fabric E-glass/epoxy composites. Based on their experimental analysis, it indicates that the thicker tensile strength is significantly lower than the tensile strength along the fiber direction. However, there is a significant increase in the thickness of tensile strength at high strain rate loading as compared to the low strain rate loading. Statistically, the increase was in the range of 75%–93% for the specimen and the range of strain rate considered. A similar approach has been demonstrated by Gómez-del Río and his coworkers [93] during their specimen's tensile properties evaluation of CFRP. Instead of just the strain rate effect, they also considered the effect of temperature toward dynamic tensile behavior of their CFRP specimens. The results of the finding can be found in Figure 9.17. From the bar chart in Figure 9.17, it can be summarized that there was little influence of temperature and strain rate on the tensile strength of a unidirectional laminate loaded in the fiber direction. In contrast, the strength increased appreciably in the transverse direction at low temperature and high strain rate. In addition, it also indicates that the temperature only attributed to having minor effect on the in-plane dynamic properties of the quasi-isotropic laminate considered in their study. For the effect of strain rate, a slender increase in tensile strength was observed under high strain rate loading. This was attributed to the rheology behavior of the epoxy matrix and nonsensitivity of the carbon fiber. The latest finding on the dynamic tensile behavior of the PMCs was demonstrated by Gan et al. [94] in their study regarding 3D braided E-glass/epoxy composites. The uniaxial tensile properties at high strain rates from 800 to 2100 s^{-1} were tested using the SHTB technique. The tensile properties at quasi-static strain rate were also tested and compared with those in high strain rates. Interestingly, they found that the stress–strain curves are rate sensitive. Besides, tensile modulus, maximum tensile stress, and corresponding tensile strain are sensitive to the strain rate as well. From visual observation, they found that with an increase in the strain rate (from 0.001 to 2100 s^{-1}), the tensile failure of the 3D-braided composite specimens has a tendency of transition from ductile failure to

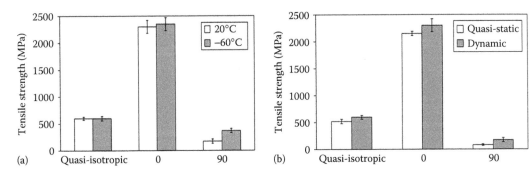

FIGURE 9.17 The effect of (a) temperature and (b) strain rates toward dynamic tensile properties of CFRP. (From Gómez-del Río, T. et al., *Compos. Sci. Technol.*, 65[1], 61–71, 2005. With permission.)

brittle failure. The magnitude response and phase response are very different in quasi-static loading with that in high-strain rate loading, as they conclude that the 3D-braided composite system is more stable at high strain rate than that of quasi-static loading.

Apart from compression and tensile configuration, the SHPB apparatus is also available in shear mode. However, this type of SHPB setup is not as simple as the other two SHPB setups. The example of shear SHPB test on PMCs specimens can be further understood through the work reported by Hosur et al. [95]. The results of the study exhibited an increment of dynamic shear stress with increasing of strain rate ranging from 1463 to 3478 s^{-1} for all tested temperature. Nevertheless, they also speculated that the severity of failure decreased with increasing temperature due to the increase of ductility and degradation of fiber-matrix interface. A similar shear SHPB setup was demonstrated by Sun and Gu [96] on 3D biaxial spacer weft knitted (E-glass) reinforced vinyl ester composites. They exposed the specimen in two directions of loading (illustrated in Figure 9.18). The overall results of this experiment are collected and summarized into Tables 9.2 and 9.3. From both Tables 9.2 and 9.3, it can be pre-concluded that the shear stiffness, failure stress, and strain exhibits strain-rate sensitivity under punch shear. The shear stiffness and the failure stress

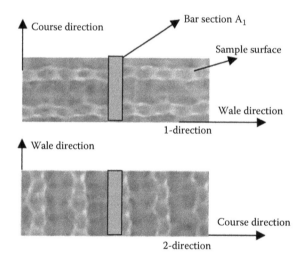

FIGURE 9.18 Schematic diagram of shear direction of 3D biaxial spacer weft knitted composites. (From Sun, B. and Gu, B., *J. Comp. Mater.*, 42[17], 1747–1762, 2008. With permission.)

TABLE 9.2 Mechanical Properties of the 3D Biaxial Spacer Weft Knitted Composite at Various Strain Rates in the 1-Direction

Shear Strain Rate (s^{-1})	Shear Stiffness (GPa)	Shear Peak Stress (MPa)	Failure Strain (%)
0.01	0.487	87.33	26.9
1200	0.631	97.64	17.7
2000	0.922	106.73	16.04
2700	1.383	111.92	15.2

Source: Sun, B. and Gu, B., *J. Comp. Mater.*, 42(17), 1747, 2008. With permission.

TABLE 9.3   Mechanical Properties of the 3D Biaxial Spacer Weft Knitted Composite at Various Strain Rates in the 2-Direction

Shear Strain Rate (s⁻¹)	Shear Stiffness (GPa)	Shear Peak Stress (MPa)	Failure Strain (%)
0.01	0.599	167.31	26.94
1200	0.740	175.87	23.29
2500	0.966	187.84	20.58
3200	1.230	195.62	19.91

Source: Sun, B. and Gu, B., *J. Comp. Mater.*, 42(17), 1747, 2008. With permission.

increased with the increase of shear strain rate, while the failure strain decreased with the increase in the strain rate.

9.4 RATE SENSITIVITY OF POLYMERIC MATERIALS UNDER VARIOUS LOADING RATES

The strain rate sensitivity of a material is referred to as the magnitude of changes of the properties of a material toward applied strain rates. This knowledge is essential during material selection, especially for extreme applications where a major change in material's properties is prohibited. Usually, this parameter is mostly presented in terms of the proportion of the material, such as Young's modulus, yield stress [97,98], flow stress [70,99], and others. In current practice, it is always beneficial if these characteristics are presented as a value in order to easily compare the magnitude of the strain rate sensitivity from one material to another. Recently, there have been several types of strain rate sensitivity parameters available that have been previously proposed [70,97,100]. However, in this case, the parameters suggested by Dasari and Misra [97] and Nakai and Yokoyama [70] are the most suitable parameters to be used for polymeric materials.

Zebarjad and Sajjadi [98] determined the strain rate sensitivity of polymer nanocomposites based on their yield stress value under various levels of strain rate investigated. Because the yield stress of polymer composites is thermally activated, their strain rate sensitivity can be described by using the well-known Eyring rate expression as follows:

$$\sigma_y = \sigma_y^0 + KLn\left(\frac{2\dot{\varepsilon}}{\dot{\varepsilon}_0}\right) \tag{9.12}$$

where σ_y, σ_y^0, $\dot{\varepsilon}_0$, $\dot{\varepsilon}$, and K are the yield stress at a specific temperature, the yield stress at a specified temperature when the strain rate is equal to $\dot{\varepsilon}_0$, the reference strain rate, and a measure of the strain rate dependency, respectively. A similar approach has been implemented by Sarang and Misra [101] where they also manipulated the Eyring equations to determine the strain rate sensitivity of the Wollastonite-reinforced ethylene–propylene copolymer composites. However, this strain rate sensitivity parameter is only suitable for measuring the magnitude of changes in the yield stress.

A more versatile strain rate sensitivity parameter was proposed by Malatynski and Klepaczko in the early 1980s [99]. The parameter was intensively used by Nakai and

Yokoyama [70], Gómez-del Río and Rodríguez [83], and Omar et al. [29] to measure the strain rate sensitivity of various polymeric specimens up to 2.5% of strain. The proposed parameter is as follows:

$$\beta = \frac{\sigma_2 - \sigma_1}{\ln(\dot{\varepsilon}_2/\dot{\varepsilon}_1)} \bigg|_{\varepsilon=0.025}^{\dot{\varepsilon}_2 > \dot{\varepsilon}_1} \tag{9.13}$$

where σ_1 and σ_2 is the flow stress at the fixed strain (in this case, the strain can be used up to fracture point) under different strain rates. Instead of versatility, the parameter also offers the simplest way to measure the strain rate sensitivity of materials from certain specific strains up to fracture point. Based on this concern, the parameter has been widely used in many studies to obtain the strain rate sensitivity of various polymeric specimens.

9.5 CONSTITUTIVE MODELING FOR POLYMERIC MATERIALS UNDER VARIOUS STRAIN RATES

9.5.1 Constitutive Equations

For numeral approach, there are three established equations that have been reported in the previous literature, which are the power law-based [102], the Eyring-based [103] and the cooperative [104] constitutive equations. However, recently, scientists claimed that the newly formulated model could be produced by the fact that the activation volume decreases with increasing strain rates. In this section, the power law-based equation was first introduced, followed by the Ree-Eyring and cooperative models; finally, the new model is presented in the final part of this section.

9.5.2 Power Law-Based Equations

The power law equation is an empirical model and assumes that the logarithm of the yield stress is linear in terms of the logarithm of the strain rate. Therefore, it is written as [102]:

$$\sigma_\pi = q\dot{\varepsilon}^m \tag{9.14}$$

where q and m are two material parameters. The subscript π holds for the power law formulation.

It has been proved that the Eyring model [103] that considers only one relaxation process fails to represent the true yield behavior at high strain rates. Therefore, two [105] and three [106] processes were proposed to better fit the polymer mechanical behavior. We can then assume that there are multiple relaxation processes, and each process is represented by an empirical power-law equation based on Equation 9.14. Therefore, the yield stress expression can be extended to include N mechanisms:

$$\sigma_{N\pi} = \sum_{i=1}^{N} q_i\dot{\varepsilon}^{m_i} \tag{9.15}$$

where q_i and m_i are $2N$ material constants and $N \geq 1$, and the subscript $N\pi$ holds for the extended power law formulation to include N relaxations processes, respectively.

9.5.3 Eyring-Based Equations

In contrast to the power law model, Eyring proposed a physically based model in which the yield stress can be described as follows:

$$\sigma_E = A_\alpha \left(\ln\left(2\, C_\alpha \dot{\varepsilon}\right) + \frac{Q_\alpha}{RT} \right) \tag{9.16}$$

where:

A_α and C_α are two material constants
Q_α depicts the activation energy
R is the universal gas constant
T is the absolute temperature
The subscript E holds for the Eyring model

Ree and Eyring [105] assumed the activation of two relaxation processes α and β. Accordingly, Equation 9.16 is modified as

$$\sigma_{RE} = A_\alpha T \left(\ln(2\, C_\alpha \dot{\varepsilon} + \frac{Q_\alpha}{RT}) + A_\beta T \sinh^{-1}\left(C_\beta \dot{\varepsilon} \exp\left(\frac{Q_\beta}{RT}\right) \right) \right) \tag{9.17}$$

where:

A_β and C_β denote two material constants
Q_β denotes the activation energy
The subscript RE holds for the Ree–Eyring approach

Ideally, there is no reason to stop the analysis at only two processes. Indeed, Safari et al. [106] integrated three processes to fit the data at strain rates higher than 5000 s⁻¹. The most general model including N processes leads to the following expression of the yield stress:

$$\sigma_{NRE} = A_1 T \left(\ln(2\, C_1 \dot{\varepsilon} + \frac{Q_1}{RT}) + \sum_{i=2}^{N} A_i T \sinh^{-1}\left(C_i \dot{\varepsilon} \exp\left(\frac{Q_i}{RT}\right) \right) \right) \tag{9.18}$$

$A_i \geq 1$ and $C_i \geq 1$ hold for $2N$ material constants and $Q_i \geq 1$ holds for the activation energy of each relaxation process respectively. The subscript NRE holds for the generalized Ree–Eyring approach that includes N relaxation processes. For an identification purpose, the equation above is simplified to

$$\sigma_{NRE} = a_1 + b_1 \ln\left(\frac{\dot{\varepsilon}}{\dot{\varepsilon}_0}\right) + \sum_{i=2}^{N} b_i \sinh^{-1}\left(c_i \dot{\varepsilon}\right) \tag{9.19}$$

where:

a_1, $b_i \geq 1$, and $C_i \geq 2$ are $2N$ material constants
$\dot{\varepsilon}_0$ is a normalizing strain rate chosen as equal to $1\ \text{s}^{-1}$ $(\dot{\varepsilon}_0 = 1\text{s}^{-1})$

9.5.4 Cooperative Constitutive Equation

In addition to the two constitutive equations presented above (i.e., Eyring-based and cooperative constitutive equations), there is another famous cooperative model, namely cooperative constitutive equation that was first proposed by Richeton et al. [107] and the following works by Fotheringham et al. [108] and Bauwens-Crowet [104]. In this case, the yield stress is given by the following equation:

$$\sigma_c = \sigma_i + \frac{2kT}{V}\sinh^{-1}\left(\frac{\dot{\varepsilon}}{\dot{\varepsilon}^*}\right)^{1/n} \tag{9.20}$$

where:
   σ_i, k, V, $\dot{\varepsilon}^*$, and n are the internal stress, the Boltzmann constant, the activation volume, the characteristic strain rate, and the material parameter, respectively
   σ_c denotes the yield stress obtained by the cooperative model

9.5.5 New Constitutive Equation

It is widely accepted that the most reported behavior refers to an increase of both yield stress and strain rate sensitivity with increasing strain rates. On the other hand, Omar et al. [29,109] and Gómez-del Río and Rodríguez [83,110] reported that the thermal activation volume is inversely proportional to the strain rate sensitivity. Therefore, this characteristic volume depends on the strain rate range. More precisely, the activation volume should decrease with increasing strain rate. Starting from the Eyring model, with the assumption that the plastic deformation is a stress-activated process, Walley et al. [77] wrote the activation volume V_0 as

$$V_0 = kT\frac{\partial \ln(\dot{\varepsilon}/\dot{\varepsilon}_{0c})}{\partial \sigma}\bigg|_{P.T} \tag{9.21}$$

where:
   P and $\dot{\varepsilon}_{0c}$ are the pressure and the characteristic strain rate, respectively
   σ and $\dot{\varepsilon}$ are the studied yield stress and strain rate, respectively
   T and K are the absolute temperature and the Boltzmann constant, respectively

Thus, the Eyring equation (Equation 9.16) can be rewritten as

$$\sigma_E = \sigma_0 + \frac{kT}{v_0}\ln\left(\frac{\dot{\varepsilon}}{\dot{\varepsilon}_0}\right) \tag{9.22}$$

where σ_0 is a characteristic stress and $\dot{\varepsilon}_0$ is a normalizing strain rate chosen as equal to $1s^{-1}$ $(\dot{\varepsilon}_0 = 1s^{-1})$.

In the latest equation (Equation 9.22), the activation volume V_0 is assumed to be constant and independent of strain rate. However, the strain rate sensitivity of the yield stress increases with increasing strain rate. This suggests that the activation volume decreases with increasing strain rate. In recent works, Gómez-del Río and Rodríguez [111] and

Omar et al. [112] suggest the integration of activation volume V^*, depending on the strain rate. If multiple processes should coexist, V^* is then interpreted as an apparent or equivalent activation volume including all processes. The expression of the apparent thermal activation volume is as follows:

$$V^* = V_0 e^{-\sqrt{\dot{\varepsilon}/\dot{\varepsilon}_c}} \tag{9.23}$$

where V_0 and $\dot{\varepsilon}_c$ are two material parameters. Actually, the authors have tried multiple equations to fit the variation of the activation volume in terms of the strain rate. It has come up that Equation 9.23 gives the best fit. Thus, the choice of Equation 9.23 is empirical. However, $\dot{\varepsilon}_c$ and V_0 have physical meanings. V_0 corresponds to the thermal activation volume at low strain rates that can be assimilated to the activation volume of the first Eyring process. Meanwhile, $\dot{\varepsilon}_c$ is interpreted as a critical strain rate. For strain rates that are insignificant compared to this critical value, that is, $\dot{\varepsilon} \ll \dot{\varepsilon}_c$, the equivalent activation volume can be assumed constant and is equal to V_0. Consequently, the first Eyring process dominates the other processes for $\dot{\varepsilon} \ll \dot{\varepsilon}_c$. Now, replacing V_0 in Equation 9.22 by the new strain rate-dependent activation volume V^* produces Equation 9.23, which yields

$$\sigma_{ME} = \sigma_0 + \frac{kT}{V_0} \exp\left(\sqrt{\frac{\dot{\varepsilon}}{\dot{\varepsilon}_c}}\right) \ln\left(\frac{\dot{\varepsilon}}{\dot{\varepsilon}_0}\right) \tag{9.24}$$

σ_{ME} holds for the yield stress predicted by the new modified Eyring constitutive equation. Similar to Equation 9.22, the normalizing strain rate $\dot{\varepsilon}_0$ is chosen as equal to 1 s^{-1}. This model depends only on three material constants; σ_0, V_0, and $\dot{\varepsilon}_0$. Each constant has a specific physical interpretation, giving a major advantage for their identification.

9.6 FUTURE RESEARCH INTEREST

The SHPB apparatus is one of the promising techniques among other dynamic facilities for several specific reasons that have been mentioned in the previous discussion. The most important reason SHPB receives such remarkable attention from scientists is related to its capability in generating the stress–strain curve that gives useful information to characterize the material mechanical performance. However, there is a very limited number of works that concerns high strain rate tensile and shear properties of PMCs, especially for thermoplastic reinforced polymer composites. In addition, numerical studies on the high strain rate properties of these composites are also infrequently reported and need additional effort to further clarify the relationship between the experimental and numerical results, which is important for engineering design and simulation purposes. Based on the highlighted issues, it is believed that a systematic study is necessary to fulfill the lack of information in this area. Apart from external factors such as the strain rate effect, it is also believed that the internal structures of polymer [113–115] and the filler-matrix-related characteristics of the PMCs (i.e., particle size, particle–matrix interface adhesion, particle shape, and geometry) may also influence the mechanical properties of the polymeric specimens. However, it has been recognized that a similar kind of study under a dynamic range of strain rates has never been reported in the past and remains a major challenge in the development of a better understanding on the mechanical behavior of polymeric-based products under various loading conditions.

9.7 CONCLUSION

The desire to scrutinize the characteristics of materials at very high strain rates leads to the most promising technique, namely the SHPB technique. However, soft specimens such as polymeric materials might face difficulty in dealing with the conventional SHPB approach because the transmitted pulse decreases significantly, and consequently decreases the signal-to-noise ratio. Thus, the mentioned topics aim to discuss the challenges and solutions that have been obtained by previous studies to ensure that the SHPB apparatus is also compatible and reliable for polymeric specimens. The implementation of SHPB on high strain rate behavior of neat polymer and PMCs is discussed along with a review of the previous studies. In addition, several established numerical equations related to the properties of polymeric materials under various strain rates have been covered. Unfortunately, it has been recognized that several critical aspects such as high strain rate tensile and shear properties of thermoplastic-based polymer matrix composites are still less reported and need extra attention from current scientists. Apart from that, the effect of filler-matrix-related characteristics on high strain rate of PMCs is also infrequently reported and remains as a major challenge in a dynamic perspective. Hopefully, there are huge efforts that can be coordinated in the future in order to fulfill the lack of knowledge in this dynamic area.

ACKNOWLEDGMENTS

The authors thank the Universiti Malaysia Perlis (Grant no: 9017-00014 and 9003-00390) and Cluster of Polymer Composites (CPCs), Universiti Sains Malaysia (USM) (Grant no: 1001/PBAHAN/8043057 and 811070) for sponsoring and bestowing financial assistance during this research work.

REFERENCES

1. Kuo H-C, Jeng M-C. Effects of part geometry and injection molding conditions on the tensile properties of ultra-high molecular weight polyethylene polymer. *Materials & Design.* 2010; 31(2):884–893.
2. Benkhenafou F, Chemingui M, Fayolle B, Verdu J, Ferouani AK, Lefebvre JM. Fracture behaviour of a polypropylene film. *Materials & Design.* 2011; 32(3):1515–1519.
3. Omar MF, Md Akil H, Ahmad ZA, Mazuki AAM, Yokoyama T. Dynamic properties of pultruded natural fibre reinforced composites using Split Hopkinson Pressure Bar technique. *Materials & Design.* 2010; 31(9):4209–4218.
4. Kolsky H. An investigation of the mechanical properties of materials at very high rates of loading. *Proceedings of the Physical Society Section B.* 1949; 62:676.
5. Haugou G, Markiewicz E, Fabis J. On the use of the non direct tensile loading on a classical split Hopkinson bar apparatus dedicated to sheet metal specimen characterisation. *International Journal of Impact Engineering.* 2006; 32(5):778–798.
6. Marais S, Tait R, Cloete T, Nurick G. Material testing at high strain rate using the split Hopkinson pressure bar. *Latin American Journal of Solids and Structures.* 2004; 1(3):219–339.

7. Sasso M, Newaz G, Amodio D. Material characterization at high strain rate by Hopkinson bar tests and finite element optimization. *Materials Science and Engineering: A*. 2008; 487(1–2):289–300.

8. Field J, Walley S, Proud W, Goldrein H, Siviour C. Review of experimental techniques for high rate deformation and shock studies. *International Journal of Impact Engineering*. 2004; 30(7):725–775.

9. Naik NK, Venkateswara Rao K, Veerraju C, Ravikumar G. Stress-strain behavior of composites under high strain rate compression along thickness direction: Effect of loading condition. *Materials & Design*. 2010; 31(1):396–401.

10. Tang L, Tam K, Yue C, Hu X, Lam Y, Li L. Influence of the molecular weight of ethylene vinyl acetate copolymers on the flow and mechanical properties of uncompatibilized polystyrene/ethylene–vinyl acetate copolymer blends. *Polymer International*. 2001; 50(1):95–106.

11. Antic V, Govedarica M, Djonlagic J. The effect of segment length on some properties of thermoplastic poly (ester–siloxane)s. *Polymer International*. 2003; 52(7):1188–1197.

12. Magniez K, De Lavigne C, Fox BL. The effects of molecular weight and polymorphism on the fracture and thermo-mechanical properties of a carbon-fibre composite modified by electrospun poly (vinylidene fluoride) membranes. *Polymer*. 2010; 51(12):2585–2596.

13. Lin J, Chang L, Nien M, Ho H. Mechanical behavior of various nanoparticle filled composites at low-velocity impact. *Composite Structures*. 2006; 74(1):30–36.

14. Dhakal HN, Arumugam V, Aswinraj A, Santulli C, Zhang ZY, Lopez-Arraiza A. Influence of temperature and impact velocity on the impact response of jute/UP composites. *Polymer Testing*. 2014; 35:10–19.

15. Hsiao H, Daniel I, Cordes R. Strain rate effects on the transverse compressive and shear behavior of unidirectional composites. *Journal of Composite Materials*. 1999; 33(17):1620–1642.

16. Richardson MOW, Wisheart MJ. Review of low-velocity impact properties of composite materials. *Composites Part A: Applied Science and Manufacturing*. 1996; 27(12):1123–1131.

17. Taylor G. The testing of materials at high rates of loading. *Journal of the Institution of Engineers*. 1946; 26(8):486–519.

18. Salisbury C. Spectral analysis of wave propagation through a polymeric hopkinson bar. M.Sc. thesis, University of Waterloo, Waterloo, Ontario, Canada, 2001.

19. Hoggatt C, Recht R. Stress-strain data obtained at high rates using an expanding ring. *Experimental Mechanics*. 1969; 9(10):441–448.

20. Meyers M. *Dynamic Behaviour of Materials*. New York: John Wiley & Son, 1994.

21. Hamouda A, Hashmi M. Testing of composite materials at high rates of strain: Advances and challenges. *Journal of Materials Processing Technology*. 1998; 77(1–3):327–336.

22. Daniel IM, LaBedz RH. Method for compression testing of composite materials at high strain rates. *Compression testing of homogeneous materials and composites, ASTM STP*. 1983; 808:121–139.

23. Hopkinson B. A method of measuring the pressure produced in the detonation of high explosive or by the impact of bullets. *Proceedings of the Royal Society of London A*. 1914; 213(89):437–456.

24. Landon J, Quinney H. Experiments with the Hopkinson pressure bar. *Proceedings of the Royal Society of London Series A*. 1923; 103(723):622–643.

25. Volterra E. Alcuni risultati di prove dinamichi sui materiali. *La Rivista del Nuovo Cimento*. 1948; 4:1–28.

26. Evora V, Shukla A. Fabrication, characterization, and dynamic behavior of polyester/TiO_2 nanocomposites. *Materials Science and Engineering: A*. 2003; 361(1–2):358–366.

27. Harding J, Wood E, Campbell J. Tensile testing of materials at impact rates of strain. *ARCHIVE: Journal of Mechanical Engineering Science*. (1959–1982[vols 1–23]). 1960; 2(2):88–96.

28. Duffy J, Campbell J, Hawley R. On the use of a torsional split Hopkinson bar to study rate effects in 1100-0 aluminum. *Journal of Applied Mechanics*. 1971; 38(1):83–91.

29. Omar MF, Akil HM, Ahmad ZA. Measurement and prediction of compressive properties of polymers at high strain rate loading. *Materials & Design*. 2011; 32(8–9):4207–4215.

30. Gowtham HL, Pothnis JR, Ravikumar G, Naik NK. High strain rate in-plane shear behavior of composites. *Polymer Testing*. 2013; 32(8):1334–1341.

31. Rong Z, Sun W. Experimental and numerical investigation on the dynamic tensile behavior of ultra-high performance cement based composites. *Construction and Building Materials*. 2012; 31:168–173.

32. Cao K, Wang Y, Wang Y. Experimental investigation and modeling of the tension behavior of polycarbonate with temperature effects from low to high strain rates. *International Journal of Solids and Structures*. 2014; 51(13):2539–2548.

33. Ninan L, Tsai J, Sun C. Use of split Hopkinson pressure bar for testing off-axis composites. *International Journal of Impact Engineering*. 2001; 25(3):291–313.

34. Mylonas G, Labeas G. Mechanical characterisation of aluminium alloy 7449-T7651 at high strain rates and elevated temperatures using split Hopkinson bar testing. *Experimental Techniques*. 2014; 38(2):26–34.

35. Mishra B, Mondal C, Goyal R, Ghosal P, Kumar KS, Madhu V. Plastic flow behavior of 7017 and 7055 aluminum alloys under different high strain rate test methods. *Materials Science and Engineering: A*. 2014; 612:343–353.

36. Chen S, Chen Y, Tsai L. High strain rate friction response of porcine molar teeth and temporary braces. In: *Dynamic Behavior of Materials*, Eds., Bo Song, Dan Casem, and Jamie Kimberle, volume 1. Springer, New York, London, 2014, pp. 29–35.

37. Ding Y, Tang W, Xu X, Ran X. Dynamic compressive response of unsaturated clay under confinements. In: *Dynamic Behavior of Materials*, Eds., Bo Song, Dan Casem, and Jamie Kimberle, volume 1. Springer, New York, London, 2014, pp. 479–487.

38. Kendall MJ, Siviour CR. Experimentally simulating adiabatic conditions: Approximating high rate polymer behavior using low rate experiments with temperature profiles. *Polymer*. 2013; 54(18):5058–5063.

39. Luong DD, Pinisetty D, Gupta N. Compressive properties of closed-cell polyvinyl chloride foams at low and high strain rates: Experimental investigation and critical review of state of the art. *Composites Part B: Engineering*. 2013; 44(1):403–416.

40. Jindal P, Pande S, Sharma P, Mangla V, Chaudhury A, Patel D et al. High strain rate behavior of multi-walled carbon nanotubes–polycarbonate composites. *Composites Part B: Engineering*. 2013; 45(1):417–422.

41. Hu Y, Liu T, Ding JL, Zhong WH. Behavior of high density polyethylene and its nano-composites under static and dynamic compression loadings. *Polymer Composites*. 2013; 34(3):417–425.

42. Hopkinson J. On the rupture of iron wire by a blow. *Proceeding of the Manchester Literary and Philosophical Society (2nd ed)*. 1872; 11:40–45.

43. Davies RM. *A Critical Study of the Hopkinson Pressure Bar*. London: Philosophical Transaction Royal Society London, 1948.

44. Gorham D. Measurement of stress-strain properties of strong metals at very high rates of strain. *Mechanical Properties at High Rates of Strain*. 1979; 1980:16–24.

45. Albertini C, Boone P, Montagnani M. Development of the Hopkinson bar for testing large specimens in tension. *Journal de Physique*. 1985; 46(C5):499–504.

46. Nemat-Nasser S, Isaacs JB, Starrett JE. Hopkinson techniques for dynamic recovery experiments. *Proceedings of the Royal Society of London Series A: Mathematical and Physical Sciences*. 1991; 435(1894):371–391.

47. Feng R, Ramesh K. Dynamic behavior of elastohydrodynamic lubricants in shearing and compression. *Journal de Physique IV*. 1991; 01(C3):69–76.

48. Zhao H. Testing of polymeric foams at high and medium strain rates. *Polymer Testing*. 1997; 16(5):507–516.

49. Wang L, Labibes K, Azari Z, Pluvinage G. Generalization of split Hopkinson bar technique to use viscoelastic bars. *International Journal of Impact Engineering*. 1994; 15(5):669–686.

50. Zhao H, Gary G. A new method for the separation of waves. Application to the SHPB technique for an unlimited duration of measurement. *Journal of the Mechanics and Physics of Solids*. 1997; 45(7):1185–1202.

51. Bacon C. Separation of waves propagating in an elastic or viscoelastic Hopkinson pressure bar with three-dimensional effects. *International Journal of Impact Engineering*. 1999; 22(1):55–69.

52. Gray III G, Idar D, Blumenthal W, Cady C, Peterson P. High-and low-strain rate compression properties of several energetic material composites as a function of strain rate and temperature. In: *11th International Detonation Symposium*, Snowmass Village, CO. Los Alamos, NM: Los Alamos National Laboratory, 1998.

53. Macdougall D. A radiant heating method for performing high-temperature high-strain-rate tests. *Measurement Science and Technology*. 1998; 9:1657.

54. Lennon A, Ramesh K. A technique for measuring the dynamic behavior of materials at high temperatures. *International Journal of Plasticity*. 1998; 14(12):1279–1292.

55. Chichili D, Ramesh K. Recovery experiments for adiabatic shear localization: A novel experimental technique. *Journal of Applied Mechanics*. 1999; 66:10.

56. Othman R, Bissac M, Collet P, Gary G. Testing with SHPB from quasi-static to dynamic strain rates. *Journal de Physique IV France*. 2003; 110:397–404.

57. Gama BA, Lopatnikov SL, Gillespie Jr JW. Hopkinson bar experimental technique: A critical review. *Applied Mechanics Reviews*. 2004; 57:223.

58. Silva CMA, Rosa PAR, Martins PAF. An innovative electromagnetic compressive split Hopkinson bar. *International Journal of Mechanics and Materials in Design*. 2009; 5(3):281–288.

59. Church P, Cornish R, Cullis I, Gould P, Lewtas I. Using the split Hopkinson pressure bar to validate material models. *Philosophical Transactions of the Royal Society A: Mathematical, Physical and Engineering Sciences*. 2014; 372(2023):20130294.

60. Li Z, Lambros J. Determination of the dynamic response of brittle composites by the use of the split Hopkinson pressure bar. *Composites Science and Technology*. 1999; 59(7):1097–1107.

61. Kaiser MA. *Advancements in the Split Hopkinson Bar Test*. Blacksburg, VA: Virginia Polytechnic Institute and State University, 1998.

62. Follansbee. *"The Hopkinson Bar," Mechanical Testing, Metals Handbook*. Metals Park, OH: American Society For Metal, 1985.

63. Davies E, Hunter S. The dynamic compression testing of solids by the method of the split Hopkinson pressure bar. *Journal of the Mechanics and Physics of Solids*. 1963; 11(3):155–179.

64. Dioh N, Leevers P, Williams J. Thickness effects in split Hopkinson pressure bar tests. *Polymer*. 1993; 34(20):4230–4234.

65. Wu X, Gorham D. Stress equilibrium in the split Hopkinson pressure bar test. *Le Journal de Physique IV*. 1997; 7(C3):3–3.

66. Naik N, Ch V, Kavala V. Hybrid composites under high strain rate compressive loading. *Materials Science and Engineering: A*. 2008; 498(1–2):87–99.

67. Yokoyama T. Impact tensile stress–strain characteristics of wrought magnesium alloys. *Strain*. 2003; 39(4):167–175.

68. Omar MF, Akil HM, Ahmad ZA. Static and dynamic compressive properties of mica/polypropylene composites. *Materials Science and Engineering: A*. 2011; 528(3):1567–1576.

69. Naik NK, Yernamma P, Thoram NM, Gadipatri R, Kavala VR. High strain rate tensile behavior of woven fabric E-glass/epoxy composite. *Polymer Testing*. 2010; 29(1):14–22.

70. Nakai K, Yokoyama T. Strain rate dependence of compressive stress-strain loops of several polymers. *Journal of Solid Mechanics and Materials Engineering*. 2008; 2(4):557–566.

71. Chen W, Zhang B, Forrestal M. A split Hopkinson bar technique for low-impedance materials. *Experimental Mechanics*. 1999; 39(2):81–85.

72. Chen W, Lu F, Zhou B. A quartz-crystal-embedded split Hopkinson pressure bar for soft materials. *Experimental Mechanics*. 2000; 40(1):1–6.

73. Frew D, Forrestal M, Chen W. Pulse shaping techniques for testing elastic-plastic materials with a split Hopkinson pressure bar. *Experimental Mechanics*. 2005; 45(2):186–195.

74. Vecchio KS, Jiang F. Improved pulse shaping to achieve constant strain rate and stress equilibrium in split-Hopkinson pressure bar testing. *Metallurgical and Materials Transactions A*. 2007; 38(11):2655–2665.

75. Johnson T, Sarva S, Socrate S. Comparison of low impedance split-Hopkinson pressure bar techniques in the characterization of polyurea. *Experimental Mechanics*. 2010; 50(7):931–940.

76. Rietsch F, Bouette B. The compression yield behaviour of polycarbonate over a wide range of strain rates and temperatures. *European Polymer Journal*. 1990; 26(10):1071–1075.

77. Walley SM, Field JE. Strain rate sensitivity of polymers in compression from low to high rates. *DYMAT Journal*. 1994; 1(3):211–277.

78. Li Z, Lambros J. Strain rate effects on the thermomechanical behavior of polymers. *International Journal of Solids and Structures*. 2001; 38(20):3549–3562.

79. Arruda EM, Boyce MC, Jayachandran R. Effects of strain rate, temperature and thermomechanical coupling on the finite strain deformation of glassy polymers. *Mechanics of Materials*. 1995; 19(2–3):193–212.

80. Mayr AE, Cook WD, Edward GH. Yielding behaviour in model epoxy thermosets—I. Effect of strain rate and composition. *Polymer*. 1998; 39(16):3719–3724.

81. Buckley C, Harding J, Hou J, Ruiz C, Trojanowski A. Deformation of thermosetting resins at impact rates of strain. Part I: Experimental study. *Journal of the Mechanics and Physics of Solids*. 2001; 49(7):1517–1538.

82. Naik NK, Gadipatri R, Thoram NM, Kavala VR. Shear properties of epoxy under high strain rate loading. *Polymer Engineering & Science*. 2010; 50(4):780–788.

83. Gómez-del Río T, Rodríguez J. Compression yielding of epoxy: Strain rate and temperature effect. *Materials & Design*. 2012; 35:369–373.

84. Kusaka T, Hojo M, Mai Y-W, Kurokawa T, Nojima T, Ochiai S. Rate dependence of mode I fracture behaviour in carbon-fibre/epoxy composite laminates. *Composites Science and Technology*. 1998; 58(3–4):591–602.

85. Hosur M, Alexander J, Vaidya U, Jeelani S. High strain rate compression response of carbon/epoxy laminate composites. *Composite Structures*. 2001; 52(3):405–417.

86. Guo Y, Li Y. Quasi-static/dynamic response of SiO_2-epoxy nanocomposites. *Materials Science and Engineering: A*. 2007; 458(1–2):330–335.

87. Chen LP, Yee AF, Moskala EJ. The molecular basis for the relationship between the secondary relaxation and mechanical properties of a series of polyester copolymer glasses. *Macromolecules*. 1999; 32(18):5944–5955.

88. Omar MF, Akil HM, Ahmad ZA. Mechanical properties of nanosilica/polypropylene composites under dynamic compression loading. *Polymer Composites*. 2011; 32(4):565–575.

89. Chou S, Robertson K, Rainey J. The effect of strain rate and heat developed during deformation on the stress-strain curve of plastics. *Experimental Mechanics*. 1973; 13(10):422–432.

90. Periasamy C, Jhaver R, Tippur H. Quasi-static and dynamic compression response of a lightweight interpenetrating phase composite foam. *Materials Science and Engineering: A*. 2010; 527(12):2845–2856.

91. Ochola R, Marcus K, Nurick G, Franz T. Mechanical behaviour of glass and carbon fibre reinforced composites at varying strain rates. *Composite Structures*. 2004; 63(3):455–467.

92. Fu S, Wang Y. Tension testing of polycarbonate at high strain rates. *Polymer Testing*. 2009; 28(7):724–729.

93. Gómez-del Río T, Barbero E, Zaera R, Navarro C. Dynamic tensile behaviour at low temperature of CFRP using a split Hopkinson pressure bar. *Composites Science and Technology*. 2005; 65(1):61–71.

94. Gan X, Yan J, Gu B, Sun B. Impact tensile behavior and frequency response of 3D braided composites. *Textile Research Journal.* 2012; 82(3):280–287.

95. Hosur M, Islam S, Vaidya U, Dutta P, Jeelani S. Experimental studies on the punch shear characterization of satin weave graphite/epoxy composites at room and elevated temperatures. *Materials Science and Engineering: A.* 2004; 368(1):269–279.

96. Sun B, Gu B. Shear behavior of 3-D biaxial spacer weft knitted composite under high strain rates. *Journal of Composite Materials.* 2008; 42(17):1747–1762.

97. Dasari A, Misra R. On the strain rate sensitivity of high density polyethylene and polypropylenes. *Materials Science and Engineering: A.* 2003; 358(1):356–371.

98. Zebarjad SM, Sajjadi SA. On the strain rate sensitivity of HDPE/$CaCO_3$ nanocomposites. *Materials Science and Engineering: A.* 2008; 475(1–2):365–367.

99. Malatynski M, Klepaczko J. Experimental investigation of plastic properties of lead over a wide range of strain rates. *International Journal of Mechanical Sciences.* 1980; 22(3):173–183.

100. Picu R, Vincze G, Ozturk F, Gracio J, Barlat F, Maniatty A. Strain rate sensitivity of the commercial aluminum alloy AA5182-O. *Materials Science and Engineering: A.* 2005; 390(1–2):334–343.

101. Sarang S, Misra R. Strain rate sensitive behavior of wollastonite-reinforced ethylene-propylene copolymer composites. *Materials Science and Engineering: A.* 2004; 381(1):259–272.

102. Acharya S, Mukhopadhyay A. High strain rate compressive behavior of PMMA. *Polymer Bulletin.* 2014; 71(1):133–149.

103. Eyring H. Viscosity, plasticity, and diffusion as examples of absolute reaction rates. *The Journal of Chemical Physics.* 1936; 4(4):283–291.

104. Bauwens-Crowet C. The compression yield behaviour of polymethyl methacrylate over a wide range of temperatures and strain-rates. *Journal of Materials Science.* 1973; 8(7):968–979.

105. Ree T, Eyring H. The relaxation theory of transport phenomena. *Rheology: Theory and Applications.* 1958; 2:83.

106. Safari KH, Zamani J, Ferreira FJ, Guedes RM. Constitutive modeling of polycarbonate during high strain rate deformation. *Polymer Engineering & Science.* 2013; 53(4):752–761.

107. Richeton J, Ahzi S, Vecchio K, Jiang F, Adharapurapu R. Influence of temperature and strain rate on the mechanical behavior of three amorphous polymers: Characterization and modeling of the compressive yield stress. *International Journal of Solids and Structures.* 2006; 43(7–8):2318–2335.

108. Fotheringham D, Cherry B. The role of recovery forces in the deformation of linear polyethylene. *Journal of Materials Science.* 1978; 13(5):951–964.

109. Omar MF, Akil HM, Ahmad ZA. Effect of molecular structures on dynamic compression properties of polyethylene. *Materials Science and Engineering: A.* 2012; 538:125–134.

110. Gómez-del Río T, Rodríguez J, Pearson RA. Compressive properties of nanoparticle modified epoxy resin at different strain rates. *Composites Part B: Engineering.* 2014; 57:173–179.

111. Gómez-del Río T, Rodríguez, J. Compression yielding of polypropylenes above glass transition temperature. *European Polymer Journal.* 2010; 46(6):1244–1250.

112. Omar MF, Akil HM, Ahmad ZA. Particle size—Dependent on the static and dynamic compression properties of polypropylene/silica composites. *Materials & Design.* 2013; 45:539–547.

113. Liu T, Baker W. The effect of the length of the short chain branch on the impact properties of linear low density polyethylene. *Polymer Engineering & Science.* 1992; 32(14):944–955.

114. Wood-Adams PM, Dealy JM, deGroot AW, Redwine OD. Effect of molecular structure on the linear viscoelastic behavior of polyethylene. *Macromolecules.* 2000; 33(20):7489–7499.

115. Wood-Adams PM. The effect of long chain branches on the shear flow behavior of polyethylene. *Journal of Rheology.* 2001; 45:203.

CHAPTER 10

Influence of Exfoliated Graphite Nanoplatelets on Properties of Polyethylene Terephthalate/Polypropylene Nanocomposites

I. M. Inuwa, Azman Hassan, and Reza Arjmandi

CONTENTS

10.1 INTRODUCTION

Engineering thermoplastics, such as polyamides and polycarbonates, possess superior mechanical and thermal properties, and hence are finding widespread use as structural materials in areas such as the automobile, aircraft, or electrical/electronic industries. Consequently, growing demand has led to a hike in prices of these thermoplastics. It is expected that the global revenue for engineering thermoplastics will hit 77 billion dollars by the year 2017 [1]. However, the increasing cost of engineering thermoplastics has motivated researchers both in academia and industry into focusing attention toward finding cheaper alternatives. Commodity thermoplastics (e.g., polyethylene terephthalate [PET] or polypropylene [PP]) are relatively inexpensive but have lower performance mechanical properties compared with engineering thermoplastics.

PET is a semicrystalline commodity thermoplastic with good mechanical properties, chemical resistance, thermal stability, low melt viscosity, and spinnability. PET has been used in several fields such as food packaging, film technology, automotive, electrical, beverages containers, and textile fibers. Despite its diversity of applications, PET is known to have poor impact properties, slow rate of crystallization, and moisture absorption, which tend to limit its use in engineering applications [2–4]. To overcome these drawbacks, PET was generally blended with other polyesters such as polybutylene terephthalate and olefinic polymers. PP is a linear olefinic commodity thermoplastic with good processability, light weight, low cost, and relatively better impact properties than PET. Its principal applications are in fiber and packaging industries. However, PP is characterized by poor flammability properties, low stiffness, low flexural modulus, and poor thermal properties that make it a poor candidate where these properties are required [5–7]. Blending PET and PP would offer an opportunity to combine the excellent properties of the two polymers due to synergistic effect and to overcome their individual shortcomings. For instance, the stiffness and heat deflection temperature of PP will be enhanced by PET, while the moisture absorption and impact strength of PET would be improved when blended with PP. Because PET and PP are thermodynamically incompatible due to differences in chemical structure and polarity, the use of suitable compatibilizers is then necessary to produce a material with desired properties. The use of elastomeric compatibilizers such as styrene-ethylene-butylene-styrene-*g*-maleic anhydride (SEBS-*g*-MAH) has led to a remarkable enhancement of the impact properties of PET/PP blends at the expense of other properties such as stiffness and strength [8].

Presently, research activities into the properties and structure of graphene have moved from curiosity oriented to application oriented [9]. Graphene is a monolayer carbon nanoparticle that consists of sp2-hybridized carbon atoms arranged in hexagonal planar structures. Properties that have endeared this unique material to diverse applications are its exceptional mechanical strength (Young's modulus of 1 TPa, tensile strength of 20 GPa), and excellent electrical (5000 S/m) and thermal conductivities (~3000 W/m K) [10–13]. Some of the areas in which research activity has blossomed include electrical and electronics devices, fuel, and solar cells and the development of multifunctional nanosized polymer composites (NPCs) [14–18]. Among the many types of graphitic nanofiller that have been employed in the development of NPCs, exfoliated graphite nanoplatelets (GNPs) have become a major focus as new reinforcing filler for the improvement of mechanical [19–21], thermal [22,23], and barrier properties [24]. In addition, GNP has been used to reduce the flammability of polymeric materials by inhibiting the vigorous bubbling process in the course of degradation during combustion. Thermally conductive polymer nanocomposites on the other hand offer new possibilities for replacing metal parts in several applications including power electronics, electric motors, generators, and heat exchangers due to their light weight, low cost, and ease of production [25–27]. Low cost of graphite, which is the precursor for GNP, is another factor for its increasing use in the fabrication of NPCs [28]. Studies have also shown that GNPs behaved as excellent conductive filler and lowered the percolation threshold of composites. GNP consists of short stacks of graphene sheets, which are characterized by high surface area, high aspect ratio, and platelet geometry. The platelet geometry can provide a tortuous path, which molecules have to follow to diffuse through composites. Additional advantages of GNPs over other types of fillers, such as carbon nanotubes, are its moderate cost compared with carbon nanotubes, and ease of processing in composite formulation [19].

Several methods have been used in the preparation of nanocomposites including melt intercalation, *in situ* exfoliation, and solution mixing. Melt intercalation has proven most convenient among the three methods from the industrial point of view, as it can be easily adapted to existing general plastic processing equipment such as extruders and injection-molding machines. The high amount of shear and heat required to exfoliate nanofillers in matrix can only be generated by melt blending. On the contrary, the *in situ* polymerization method is somehow complex and restricted to some polymer types, whereas the solution method requires large amounts of solvent, which is difficult to reclaim and with added cost. Many studies have been reported in which a direct melt-blending technique was used to disperse graphene in polymer matrix [29–31]. In a recent study, Fasihi et al. [32] reported that solid-state milling followed by low-temperature melt mixing has resulted in high-degree exfoliation of expanded graphite in PP composites. Melt mixing and solution methods have been compared with a coating technique in the preparation of polyethersulfone/exfoliated GNP nanocomposites where it is shown that the coating followed by the melt injection method is more effective than polymer solution or melt mixing in preserving the platelet morphology and increase in electrical conductivity of the prepared nanocomposites [33], although a more expensive and extra procedural step is required.

Furthermore, several studies have been conducted to assess the thermal conductivity of grapheme-reinforced polymer nanocomposites [34–37]. Min et al. [37] observed a 157% increase in thermal conductivity of GNPs/epoxy composites over that of pure epoxy. Teng et al. [35] studied the effect of functionalized graphene on epoxy composites and observed that the thermal conductivity was much higher than epoxy reinforced with multiwall carbon nanotubes. Steady-state thermal analysis is a method of measurement of the thermal conductivity of polymer composites and nanocomposites. This method is useful for specimens having thermal resistance in the range of $10–400 \times 10^{-4}\,m^2\,K/W$, which can be obtained from materials of thermal conductivity in the approximate range from 0.1 to 30 W/(m K) [38].

In this chapter, GNP-reinforced PET/PP blends were developed, and their morphology, structure, flexural, impact, and thermal properties were investigated. Conventional melt extrusion and injection-molding process were employed to fabricate the test samples; it is more attractive to industrialists because it is compatible with industrial processes and free of environmentally harmful or dangerous solvents. The use of a compatibilizer is expected to improve the compatibility of the PET/PP blends and aid the dispersion of GNP in the matrix. SEBS-g-MAH is expected to serve a dual function: to compatibilize PET/PP blends and to aid the dispersion of GNPs in the polymer matrix.

10.2 MATERIALS AND METHODS

10.2.1 Materials

Exfoliated GNPs, GNP-M-5 grade (99.5% carbon) consisting of graphene nanoplatelets of average diameter 5 mm and average thickness 6 nm was purchased as dry powder from XG Sciences, Inc. (East Lansing, MI, USA) and used as received. The BET surface area of the samples used in this experiment is 158 m^2/g measured in the laboratory. PET (grade M100) was obtained from Espet Extrusion Sdn Bhd (Malaysia) with an intrinsic viscosity of 0.82 g/dL. PP, a copolymer grade (SM240) with a density of 0.96 g/cm^3 and melt flow index (MFI) of 35 g/10 min (230°C and 2.16 kg load), was supplied by Titan Chemicals (Malaysia). Styrene-ethylene-butylene-styrene triblock copolymer grafted with 1.84 wt.% of maleic anhydride (SEBS-g-MAH) was supplied by Shell Chemical Company under the trade name of Kraton™ FG 1901X with weight ratio of styrene to ethylene/butylene in the triblock copolymer to be 30/70. MFI is 20 g/10 min (270°C, 5 kg); molecular weight of the styrene block is 7,000 g/mol and that of the ethylene/butylene block is 37,500 g/mol.

10.2.2 Sample Preparation

PET was predried in a vacuum oven at 100°C for 48 h, PP was dried at 80°C for 24 h, and SEBS-g-MAH was dried for 8 h at 60°C. PET, PP, SEBS-g-MAH, and GNP with various amounts of GNP, as summarized in Table 10.1, were melt blended using a counter-rotating twin screw extruder Plastic Corder, PL 2000. The temperature setting from the hopper to the die was 265°C/275°C/280°C/285°C and the screw speed was 60 rpm. The extruded material was pelletized and then dried at 80°C for 12 h before injection in an injection molding (JSW 100 Ton). The temperatures from the hopper to the nozzle were 225°C–270°C. Standard test

TABLE 10.1   Sample Formulation of GNP-Reinforced PET/PP Nanocomposites

Sample Name	PET (wt.%)	PP (wt.%)	SEBS-g-MA (phr)	GNP (phr)
GNP0	70	30	10	0
GNP0.5	70	30	10	0.5
GNP1	70	30	10	1
GNP2	70	30	10	2
GNP3	70	30	10	3
GNP4	70	30	10	4
GNP5	70	30	10	5
GNP7	70	30	10	7

samples (ASTM standards) were produced for tensile and impact tests. All tests were conducted more than 24 h after injection.

10.2.3 Characterizations

10.2.3.1 Mechanical Properties

10.2.3.1.1 Flexural Strength   Three-point loading utilizing a center loading technique was applied for the flexural test (ASTM D790). Samples for the flexural test were placed on a supported beam. The span distance was 100 mm and the strain rate was 3 mm/min. The average of five specimens was recorded.

10.2.3.1.2 Impact Strength   Impact testing (ASTM D256) was performed using an Izod Toyoseiki (11 J) impact tester at ambient temperature. Ten specimens were measured for each composition and the average values were determined.

10.2.3.2 Morphological Analysis

Dispersion of the graphene nanoplatelets was observed using field emission scanning electron microscopy (FESEM) and transmission electron microscopy (TEM). FESEM micrographs of fractured surfaces of the neat PET/PP blend and PET/PP/GNP nanocomposites were obtained using a Hitachi S-4800. The neat blend control and the nanocomposites were gold coated using a Balzers Union MED 010 coater. Thin sections (thickness of 70 nm) used for transmission imaging were microtomed using a Reichert Jung Ultracut E microtome. Transmission micrographs were collected using a JEOL JEM-2100 microscope, with an operating voltage of 200 kV.

10.2.3.3 X-Ray Diffraction

X-ray diffraction (XRD) patterns were collected using X'Pert, X-ray diffractometer (SIEMENS XRD D5000), and Ni-filtered CuKα radiation at an angular incidence of 0°–80° (2 h angle range). XRD scans of the GNP powder along with the nanocomposite samples were collected at 40 kV and 50 mA with an exposure time of 120 s.

10.2.3.4 Fourier Transform Infrared Spectroscopy

To study the interaction between GNP and the matrix, Fourier transform infrared spectroscopy (FTIR) was performed using a Perkin Elmer 1600 infrared spectrometer using the KBr method in the ratio of 1:100 and made to a thin pellet. FTIR spectra of the coated pellet were recorded using a Nicolet AVATAR 360 at 32 scans with a resolution of 4 cm⁻¹ and within the wave number range of 370–4000 cm⁻¹. The positions of significant transmittance peaks were determined by using the "find peak tool" provided by the Nicolet OMNIC 5.01 software.

10.2.3.5 Thermal Properties

10.2.3.5.1 Thermogravimetric Analysis   The thermal stability of samples was characterized using a thermogravimetric analyzer, model 2050 (TA Instruments, New Castle, DE). The specimens were scanned from 30°C to 600°C at a rate of 10°C/min under a nitrogen gas atmosphere.

10.2.3.5.2 Differential Scanning Calorimetry   The melting and crystallization behavior of PET/PP and PET/PP/GNP nanocomposites were characterized through differential scanning calorimetry (DSC) (Perkin Elmer DSC-6), using 5–10 mg samples sealed in aluminum pans. The temperature was raised from 30°C to 300°C at a heating rate of 10°C/min, and after a period of 1 min, it was swept back at the rate of 10°C/min. The fusion enthalpies, DHf (PET) and DHf (PP), were measured and the degree of crystallinity, X_c (PET) and X_c (PP) were calculated using Equations 10.1 and 10.2, respectively.

$$\% X_c(\text{PET}) = \frac{\Delta H_{f(\text{PET})}}{\Delta H^0_{f(\text{PET})}} \times \frac{1}{W(\text{PET})} \times 100 \tag{10.1}$$

$$\% X_c(\text{PP}) = \frac{\Delta H_{f(\text{PP})}}{\Delta H^0_{f(\text{PP})}} \times \frac{1}{W(\text{PP})} \times 100 \tag{10.2}$$

10.2.3.6 Fire and Flammability Tests

The heat release rate (HRR), mass loss (ML), and total heat release (THR) were determined using an ML calorimeter with a conical radiant heater and thermopile detector conforming to ISO5660 standards on samples measuring $100 \times 100 \times 3$ mm³. The tests were performed under 50 kW/m² external radiant heat flux that corresponds to a mild-to-intermediate fire scenario. Thermopile temperature and sample mass data were recorded throughout combustion using a data acquisition system. The macroscopic view of the char residue was imaged by a digital photograph.

The limiting oxygen index (LOI) test was conducted on specimens with dimensions $75 \times 6.5 \times 3$ mm³, which were enclosed in an open-vent chamber in a mixed oxygen/nitrogen environment. The specimen was ignited at its upper end with a methane flame. The specimens burnt from the top downwards. The atmosphere composition that permits steady burning is determined. The system was calibrated using a control sample with a defined LOI value in accordance with ASTM D-2863 standards.

Samples were also classified for their flammability resistance according to UL-94-V standards (ASTM D 3801). Specimens with dimensions of $127 \times 12.7 \times 3$ mm³ are exposed vertically to a methane gas burner flame for 10 s, as required by UL-94-V. The specimen was ignited at the bottom and burnt upward. The time required for the flame to self-extinguish after burner removal is measured, and the occurrence of dripping onto a piece of cotton placed underneath the sample is recorded. The test is repeated for five different samples. If any of the samples burns for more than 10 s and the drips do not ignite the cotton, the material is classified as V-0. If none burns for more than 30 s and the drips do not ignite the cotton, the material is classified as V-1, but if the cotton is ignited, the material is classified as V-2. If any of the samples burns for more than 30 s or if the entire sample is consumed, the material is classified as "burning."

10.2.3.7 Thermal Conductivity Test

A Cussons thermal conductivity analyzer was used to measure the effective thermal conductivity in accordance with ASTM E1530 standards. The specimens (Figure 10.1) were held under uniform compressive load between two polished surfaces, each controlled at a different temperature. The lower surface is part of a calibrated heat flow transducer. The heat flows from the upper surface through the sample to the lower surface establishing an axial temperature gradient in the stack. After reaching thermal equilibrium (steady state), the temperature difference across the sample was measured along with the output from the heat flow transducer.

The temperature difference and the sample thickness were then used to calculate the effective thermal conductivity using Equation 10.3. The temperature drop through the sample was measured with temperature sensors in the highly conductive metal surface layers on either side of the sample. By definition, thermal conductivity is the exchange of energy between adjacent molecules and electrons in a conducting medium; it is a material property that describes heat flow within a body for a given temperature difference per unit area. For one-dimensional heat flow, the equation is given as

$$Q = \frac{KA(T_1 - T_2)}{x} \tag{10.3}$$

FIGURE 10.1 Specimens for the thermal conductivity test.

10.3 RESULTS AND DISCUSSION

10.3.1 Mechanical Properties

10.3.1.1 Flexural Properties

In our recent study, flexural modulus and strength of PET/PP/GNP nanocomposites were investigated [39]. Figure 10.2 shows the effect of GNP loading on flexural modulus and strength of PET/PP/GNP nanocomposites. The flexural modulus increased steadily with GNP loading. A similar trend was observed previously [19]. The value of flexural modulus increased from 1.1 to 1.88 GPa at 5.0 phr loading (i.e., about 80% increase). These observations are attributed to the uniform dispersion of graphene nanoplatelets as evidenced by morphological studies. It is also evidence of the high modulus and crystallinity of graphene nanoplatelets [20]. Improvement in flexural modulus may also be attributed to the alignment of the nanofillers to the flow direction during extrusion and injection molding. Flexural strength shows a maximum at 3 phr corresponding to 38.7 MPa; on further addition of GNPs, the flexural decreased to 32.4 MPa at 5.0 phr. The decrease in flexural strength is due to agglomeration of the nanoplatelets at higher concentration leading to poor adhesion between the filler and matrix and the increase in filler–filler interaction due to the restacking of the nanoplatelets caused by van der Waals' forces.

10.3.1.2 Impact Strength

Figure 10.3 shows the effect of GNP loading on the impact strength of PET/PP/GNP nanocomposites that were investigated in our recent work [40]. Figure 10.3 shows the effect of GNP loading on the impact strength of PET/PP/GNP nanocomposites. A reduction in impact strength was observed in all nanocomposites compared with the neat blend. The

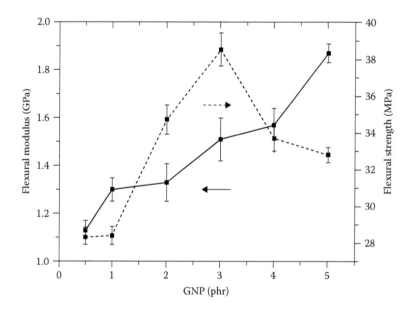

FIGURE 10.2 Effect of GNP loading on the flexural strength and flexural modulus of PET/PP/GNP nanocomposites.

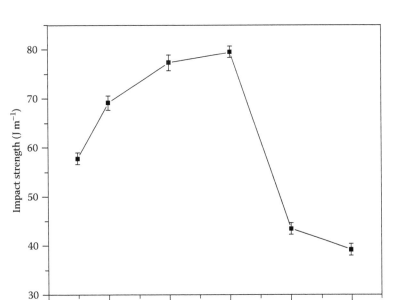

FIGURE 10.3   Effect of GNP loading on the impact strength of PET/PP/GNP nanocomposites.

decrease in the impact strength of the nanocomposites compared with the neat blend can be attributed to incompatibility between the matrix and filler and also the heterogeneous nature of the PET/PPGNP nanocomposites. A Similar trend is reported by Wang et al. [41] in a study to compare GNPs and carbon black on the mechanical properties of HDPE. Li and Chen [42] also reported a reduction of the impact strength of HDPE/expanded graphite nanocomposites prepared via a masterbatch process compared with pure HDPE due to the presence of the graphene layers in the matrix. However, as can be seen in the figure, PET/PP/GNP nanocomposites exhibit the least reduction of impact strength at 3 phr loading. This corresponds to the optimum filler concentration in which uniform dispersion of GNPs was achieved. Differences in dispersion situation among the different loadings may result in different energy-absorbing mechanisms at the impact fracture surface [30].

The sharp drop of impact strength after 3 phr is attributed to the restacking phenomenon. Figure 10.4 is a proposed scheme of GNP dispersion in polymer matrix as postulated by Zhao et al. [16]. It is proposed here that at low concentration, the GNP nanoparticles are individually dispersed in the polymer matrix at intervals. At higher concentration, the edges of the platelets just join together side by side. This condition is presumed to correspond with 3.0 phr GNP loading and is the ideal condition exhibiting ultimate contribution to the mechanical behavior with the greatest efficiency.

According to the experimental results, the optimum loading of GNP in PET/PP/blend is 3 phr. When the GNP concentration is increased beyond the optimum level (3 phr), the platelets begin to overlap on one another as illustrated in Figure 10.4. At higher loading, the platelets start restacking together in layers owing to strong van der Waals' forces and $\pi \rightarrow \pi$ attraction between GNP planes and the small distance between the graphene

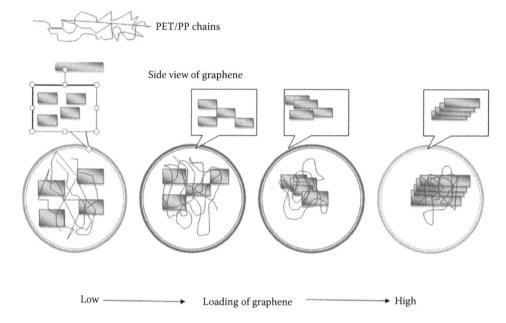

FIGURE 10.4   Illustrated scheme of GNP dispersion situations in PET/PP matrix.

sheets. Filler–filler interaction is presumed to be responsible for the observed decrease in the mechanical properties of the nanocomposites beyond the 3 phr concentration.

10.3.2 FESEM Analysis

Studies of the surface morphology and compositions of polymer nanocomposites by FESEM have been used extensively to characterize the dispersion of nanofillers in polymeric matrices [43,44]. In another study, the morphology of PET/PP blends and PET/PP/GNP nanocomposites was investigated using FESEM analysis [45]. Figure 10.5 is a typical FESEM of PET/PP blends and PET/PP/GNP nanocomposites. Figure 10.5a shows the FESEM micrograph of unreinforced PET/PP blend. It is interesting to note that the addition of SEBS-g-MAH has compatibilized the immiscible PET/PP blend (shown by arrow) by reducing the particle size of the dispersed PP phase and eliminating the voids. This resulted in the improved adhesion of the different phases and reduced interfacial tension. A similar finding was reported by Heino et al. [46]. The FESEM image of the impact fractured surface of PET/PP/GNP nanocomposites taken in the transverse direction is depicted in Figure 10.5b. The GNPs appear to be embedded and uniformly dispersed in the nanocomposites with the edges of the graphene sheets projecting out of the fractured surface. The uniform dispersion of GNPs in the polymer matrix is responsible for the observed increase in the properties. This observation is consistent with that of Ramanathan et al. [47]. Figure 10.5c shows the interaction of the GNP particles with the SEBS-g-MAH particles as shown by an arrow. The compatibilizer can be seen attached to the surface of GNP sheets. This indicates that the compatibilizer functions by aiding the dispersion of GNPs in the polymer matrix through encapsulation of GNPs by the rubber particles in addition to improving the adhesion between PET and PP particles. Similar findings were reported by Zhang and Deng [48].

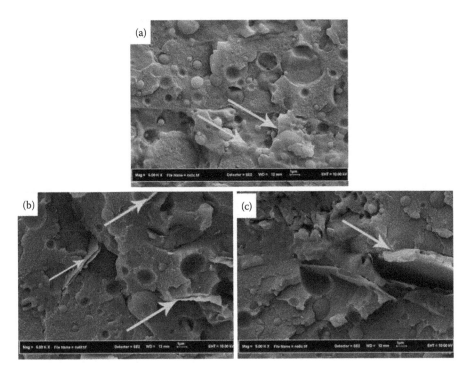

FIGURE 10.5 FESEM images of PET/PP/GNP nanocomposites at 3 phr showing (a) PET/PP blend, (b) uniform dispersion of GNP, and (c) attachment of GNP particles on the compatibilizer surface.

10.3.3 TEM Analysis

TEM analysis is a proven and reliable technique that has been used for investigating the dispersion of nanofillers in polymer matrix [45]. Figure 10.6 is the image analysis of GNP3. In Figure 10.6a, the interconnected GNP sheets can be seen within the matrix with maximum of three sheets overlapping on each other. Interconnected morphology is most suitable for the improvement of transport properties such as thermal conductivity. This confirms that the number of sheets was reduced from five to three originally supplied by the manufacturer, which can be attributed to shear mixing in the extruder equipment. Close examination of Figure 10.6b can reveal the interaction of GNP with the polymer matrix possibly SEBS-g-MAH. Folded graphene sheet off-plane can be seen surrounded by matrix that is represented by the blurred grey background with improved adhesion by the compatibilizer elastomer particle or absorption by the PET/PP chains [29]. The shear mixing also involves wrinkling (Figure 10.6c) and crumpling (Figure 10.6d) of exfoliated single layers due to the thin thickness of the graphene particles. It has been reported that the presence of the wrinkled and crumpled exfoliated graphene sheets may actually lead to nanoscale surface roughness that would likely produce an enhanced mechanical interlocking and adhesion with the polymer chains [49]. It is reported that SEBS-g-MAH improved the dispersion of fillers in the polymeric matrices thereby improving the toughness and wear resistance of the matrix due to interaction with fillers [48,50]. Overall, it is evident that good dispersion of GNP sheets has been achieved with isolated instances of exfoliation.

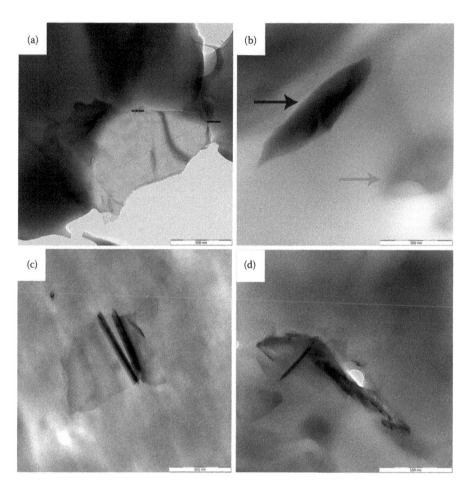

FIGURE 10.6 (a) Interconnected GNP sheets at 3 phr loading, (b) adhesion of GNP to polymer matrix promoted by SEBS-g-MAH, (c) wrinkled GNP sheets, and (d) crumpled GNP sheets.

10.3.4 X-Ray Diffraction

The XRD patterns of the pristine GNP, neat PET/PP blend, and PET/PP/GNP nanocomposites were also investigated and are shown in Figure 10.7 [45]. The characteristic peaks of the diffraction pattern for the graphene nanoplatelets show the graphene-2H characteristic peaks at 26.6° (d = 3.35 Å) and 54.7° (d = 1.68 Å) 2θ. The absence of the characteristic graphene peaks in the nanocomposites along with the observations by TEM and FESEM suggests a uniform dispersion of GNP sheets in the matrix. It may be concluded that GNPs in the nanocomposites may not have been substantially exfoliated but homogeneously distributed in the matrix thereby improving the properties. Bandla and Hanan [18] have reported similar observations.

10.3.5 Fourier Transform Infrared Spectroscopy

To observe any chemical changes occurring, the samples were analyzed using FTIR spectroscopy [40]. Figure 10.8 shows the FTIR spectra of GNP powder, neat blend, and PET/PP/GNP nanocomposites. No visible peaks were observed in the spectrum of GNPs.

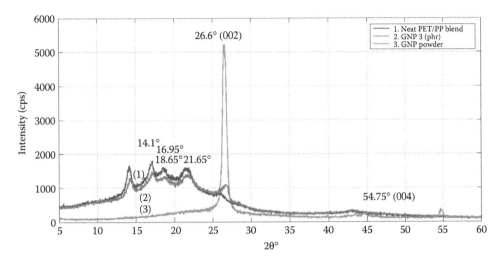

FIGURE 10.7 XRD features of the neat PET/PP blends GNP powder and PET/PP/GNP nanocomposites.

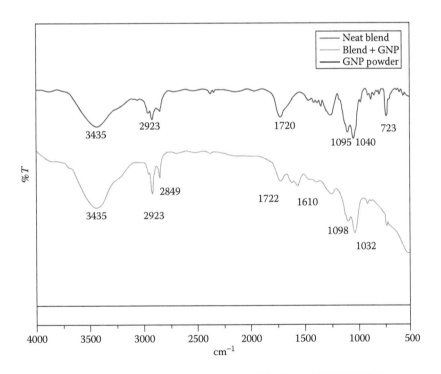

FIGURE 10.8 Typical FT-IR of the GNP powder, PET/PP blend, and PET/PP/GNP nanocomposites.

This is consistent with the findings of Geng et al. [51]. The absence of graphite and graphite oxide peaks confirms the purity of graphene sheets in the GNP powder. These peaks are at 3400 cm^{-1} (O–H stretching vibrations), 1720 cm^{-1} (C=O stretching vibrations), 1220 cm^{-1} (C–OH stretching), and 1060 cm^{-1} (C–O stretching) [52,53]. The broad absorption band at 3435 cm^{-1} in the neat blend and the nanocomposites is attributed to the hydroxyl group in PET [54]. The peaks at 2923 cm^{-1} appearing in both blend and nanocomposites are due to the SEBS group in the blend [2]. The peaks at 1610–1722 cm^{-1} characterize the C=O stretching vibrations existing in both PET and maleic anhydride groups. The overlapping of the C=O peaks existing both in PET and the anhydride group further promote the compatibility between the PET and the anhydride due to possible interaction of the polar groups of the anhydride with the ester group of PET. A similar observation was made by Chiu and Hsiao [55]. The peak at 723 cm^{-1} is attributed to unsaturated C–H stretching vibration. The foregoing discussion indicated that there is no chemical interaction between GNPs and blend matrix due to the absence of structural changes in the nanocomposites. Therefore, any property improvements of the nanocomposites are the result of the physical interactions (which improved the adhesion of GNPs to the matrix) only between the GNPs and the matrix. A similar conclusion was arrived at by Patole et al. [56]. They reported that there was no significant change in peak positions of graphene/polystyrene nanocomposites compared with pristine/polystyrene nanoparticles.

10.3.6 Thermogravimetric Analysis

The thermal stability of the neat blend and nanocomposites was also studied under nitrogen gas atmosphere [39]. The results of the measurements are shown in Figure 10.9a, b and Table 10.2. The thermal decomposition temperature of the blends and nanocomposites shows a single-step decomposition process indicating effective compatibilization of the blends. The results also show the effect of GNPs on the thermal stability of the neat PET/PP blends. As shown in Table 10.2, T_{10} and T_{50} as well as maximum decomposition temperatures (T_{max}) are characteristically high for all the nanocomposites compared with the neat blend. In particular, the 3 phr GNP concentration yielded highest thermal stability. The enhanced thermal stability of the nanocomposites was attributed to the high aspect ratio of GNPs that serve as a barrier and prevented the emission of gaseous molecules during thermal degradation. The dramatic improvement at 3.0 phr is attributed to the homogeneous dispersion of the GNPs in the matrix at this filler level. Homogeneously

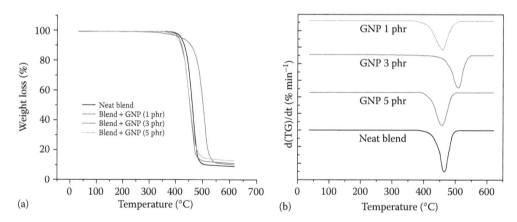

FIGURE 10.9 (a) TGA and (b) DTG curves of neat PET/PP and PET/PP/GNP composite showing higher thermal stability of 3 phr nanocomposites.

TABLE 10.2   TGA and DTG Data of Neat PET/PP Blend and PET/PP/GNP Nanocomposites

Sample	Degradation Temp. (°C)		DTG Peak Temp. (°C)	Residual Weight (%) at 600°C
	T_{10}	T_{50}	T_{max}	
Neat blend	242	454	466	8.774
GNP1	426	457	457	10.068
GNP3	458	502	505	11.808
GNP5	421	456	455	12.867

dispersed GNP disrupted the oxygen supply by forming charred layers on the surfaces of the nanocomposites thereby enhancing thermal stability [29]. However, the weight residue shows that highest amount of ash was obtained with the 5.0 phr GNP loading. This logically means that as the quantity of the GNP increased, the residue was also increased due to very high thermal stability of the graphene platelets. These observations are consistent with those made by Tantis et al. [57].

10.3.7 Differential Scanning Calorimetry

Figure 10.10a and b shows the heating and cooling scans of DSC traces for blends and nanocomposites [45]. The corresponding thermal data are provided in Table 10.3. From this table, it is clear that the T_c values of all nanocomposites are higher than those of the unreinforced PET/PP blend and increase with increasing GNP content. This phenomenon can be explained by the heterogeneous nucleation effect of the GNPs on the chain segments, which leads to the crystallization of PET at higher temperatures. Similar explanation can be made for T_{cPP}. This is consistent with a previous study [58].

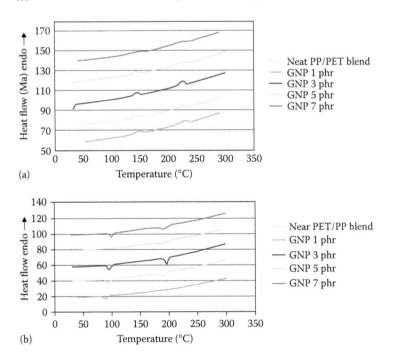

FIGURE 10.10   DSC plots: (a) heating scans and (b) cooling scans of PET/PP and PET/PP/GNP.

TABLE 10.3   DSC Data of PET, PET/PP Blend, and PET/PP/GNP Nanocomposites

Sample	Crystallization Temp. (°C)		Melting Temp. (°C)		Crystallinity (%)	
	PET	PP	PET	PP	PET	PP
GNP0	178.1	89.3	230.1	150.4	24.1	19.8
GNP1	189.7	93.3	230.0	140.6	21.6	17.6
GNP3	196.3	97.4	229.8	148.4	20.2	15.3
GNP5	186.5	99.5	228.7	148.3	19.5	13.4
GNP7	190.2	100.0	230.1	150.5	19.0	12.1

However, the crystallinity of PET and PP in PET/PP/GNP nanocomposites decreased with the addition of GNP. It can be thought that the presence of GNP has restricted the mobility of PET and PP chains during the crystallization process. The high surface area of the GNP sheets enables them to act as vast heterogeneous nucleation points that increased crystallization rates. Subsequently less crystalline regions are formed due to shorter time available. Additionally, due to the nanosized dimension of GNP particles, they can be present in both PET and PP phases resulting in the increased T_c for both PET and PP. The T_m of both PET and PP, however, remains essentially unaffected. Karevan et al. [59] observed that the decrease in the degree of crystallinity of PA12 with GNP content may be due to the decrease in the free volume and constraints of the polymer chains imposed by the rigid GNP that do not allow them to rearrange and form crystals.

10.3.8 Fire Properties

In our recent publication [60], the HRR and THR curves of unreinforced PET/PP blend and reinforced PET/PP/GNP nanocomposites with various GNPs contents are shown in Figure 10.11a and b, respectively. The corresponding data are shown in Table 10.4. The data in Table 10.4 show that the addition of GNPs to PET/PP blends, the HRR, and PHRR decreased continuously with the increase in the weight of GNPs in the matrix. A similar trend is observed for the THR (Figure 10.11b). Zhuge et al. [61] have observed that addition of graphitic nanofillers have remarkably improved fire resistance of PET copolymers. The broad plateau of the HRR curves at higher GNP loading (3–5 phr) is due to the formation of a protective char layer formed by the GNPs that indicate a substantial decrease in released heat due to the improved flame retardancy effect of GNPs in PET/PP matrix.

This is consistent with the report of a previous study on GNP-filled thermoplastic polyurethane nanocomposites [62]. A substantial decrease in HRR and PHRR was correlated with the extent of dispersion of nanofillers in polymer matrix [63]. Well-dispersed nanofillers significantly reduce the HRR and PHRR of the corresponding nanocomposites. Time-to-ignition (TTI) values of GNP0 and PET/PP/GNP nanocomposites are presented in Table 10.4. Notably, unreinforced PET/PP blend (GNP0) has early ignition and combustion compared with all the nanocomposites formulations. This is reflected in the highest values of HRR, THR, lowest values of TTI, and time-to-peak heat release rate (TPHRR) for the GNP0 samples. This behavior can be attributed to the inherent flammability of the polymeric matrices that formed the blend. The TTI of all the nanocomposites is higher than GNP0. The trend shows an increase in TTI with increasing GNP content; for instance, the TTI for GNP0 was 20 s, while that for

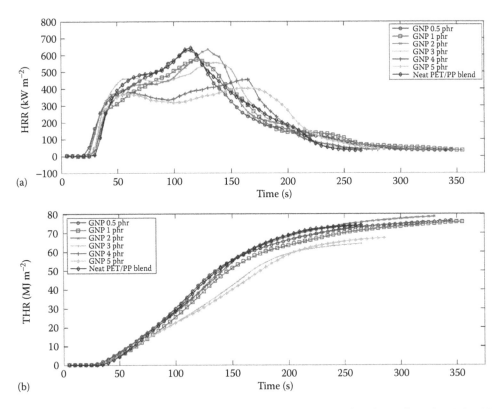

FIGURE 10.11   (a) HRR versus time for unreinforced PET/PP blend and reinforced PET/PP/GNP nanocomposites and (b) THR versus time for unreinforced PET/PP blend and reinforced PET/PP/GNP nanocomposites.

TABLE 10.4   Interpretation of Cone Calorimeter Data

Sample	PHRR (kW/m²)	TPHRR (s)	TTI (s)	Co (g/s)	FPI	THR (MJ/m²)	TSP (m²)
	±6%	±5%	±6%	±7%	±2%	±8%	±9%
GNP0	643	115	20	0.01182	0.0310	74.0	14.3
GNP0.5	627	115	22	0.00970	0.0351	76.3	14.0
GNP1	572	120	25	0.00994	0.0437	75.9	12.9
GNP2	553	130	27	0.00928	0.0488	78.6	14.0
GNP3	534	140	30	0.00732	0.0561	64.3	12.8
GNP4	453	165	33	0.00678	0.0728	74.0	10.9
GNP5	403	170	35	0.00591	0.0868	67.4	12.8

the GNP5 was 35 s, which is another proof of the flame retardancy of GNPs. This effect is more pronounced considering the time to peak release heat (TPHHR).

Peeterbroeck et al. [64] have studied the flame retardant effect of three types of multiwall carbon nanotubes in ethylene vinyl acetate copolymer. They showed that crushed multiwall carbon nanotubes remarkably improved the TTI from 91 to 170 s due to chemical reactivity of radical species present at the surface/extremities of crushed multiwalled carbon nanotubes during the combustion process. However, in recent detailed and comprehensive studies by Schartel et al. [65] and Dittrich et al. [66] using carbon-based fillers in the PP system, including functionalized graphene and multilayer graphene, they showed that the presence of layered carbon nanoparticles lowers the TTI. They attributed this observation to the difference in in-depth heat absorption between the PP and PP nanocomposites as the main reason for shorter times to ignition. Other parameters controlling TTI include the surface temperature of the specimen, thermal inertia, and the applied heat flux. The data in Table 10.4 show the FPI values, which is the TTI/PHRR ratio, for the blend and nano-composites. It can be clearly seen that FPI of GNP0 is 0.031, while that of GNP5 is 0.0868, an increase of about 160%. The FPI value is closely related to the real fire condition and is therefore used as a yardstick in fire service to design for the escape time for firemen in a real fire situation [26]. The longer TTI and lower PHRR result in higher FPI and better chance to reduce loss and casualty in real fire.

The carbon monoxide production (COP) rate and total smoke production (TSP) are shown in Figure 10.12a and b, respectively. The trend of COP and TSP plots can be compared with that of HRR and THR plots, respectively. The similarities of the plots indicate a similar pattern of behavior. It can be seen that just like HRR, the COP decreases with increasing GNP content; the COP of GNP0 (0.01182) is twice that of GNP5 (0.00928). A similar observation can be made for TSP and THR. This shows that the GNPs can effectively

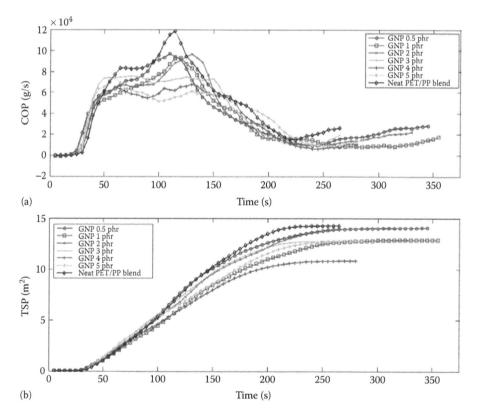

FIGURE 10.12 (a) COP versus time for unreinforced PET/PP blend and PET/PP/GNP nanocomposites and (b) TSP versus time for unreinforced PET/PP blend and PET/PP/GNP nanocomposites.

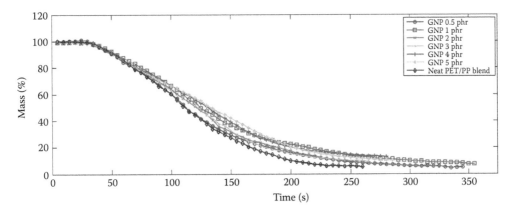

FIGURE 10.13 ML versus time for unreinforced PET/PP blend and reinforced PET/PP/GNP nanocomposites.

suppress smoke production. The decreasing ML rate with GNP loading (Figure 10.13) indicates an additive increase in the char level of the nanocomposites compared with the unreinforced PET/PP blend. This can be explained by the formation of a GNP network that occurred from 3 phr and that the network slowed down ML [67]. It is reported [25] that the mechanism of fire behavior of graphene is through the promotion of the formation of compact char layers in condensed phase during combustion in polymer matrix and that the char structure effectively prevents the inside thermal decomposition products into the flame zone and that of the oxygen into the underlying of polymer matrix.

However, in a study of GNP-filled thermoplastic polyurethane, Quan et al. [62] explained that the flame retardancy mechanisms observed in the nanocomposites were due to the intumescent flame retardancy of the GNPs. These combusted GNPs form an even layer of carbonaceous foams, thus separating the matrix from the incoming heat and oxygen, as well as cutting off the flaming path. The LOI values and the result of the UL-94-V flame test are two parameters that have been used to evaluate the flammability of flame retardant materials. The LOI and UL-94-V results of the samples are presented in Table 10.5. It can be seen that the addition of GNPs was beneficial to improve the flame retardancy of PET/PP/GNP nanocomposites. The LOI value increased from 21 for the GNP0 to 31 for GNP5, considering the results of the UL-94 V test; the sample reached the UL-94 V-0 rating when the content of GNPs was just 3 phr. The linear relationship of LOI values with GNP contents

TABLE 10.5   Interpretation of LOI and UL-94 Ratings

Sample	LOI (%)	UL-94 Ratings	
GNP0	21	Burning	Dripping
GNP0.5	24	V2	Moderate dripping
GNP1	26	V2	Low dripping
GNP2	27	V2	Less dripping
GNP3	28	V0	No dripping
GNP4	29	V0	No dripping
GNP5	31	V0	No ignited

fitted well with the trend for HRR and PHRR, i.e., increasing LOI corresponds to decreasing HRR and PHRR. The addition of GNPs has eliminated the severe dripping and the rapid burning that characterized unreinforced PET/PP blend.

Therefore, the incidence of increased fire spread caused by dripping in real fire has been eliminated by the addition of GNPs in PET/PP matrix. Figure 10.14a shows char formation in GNP3 nanocomposites due to the addition of GNPs. Figure 10.14b shows the self-extinguishing characteristics of other GNP nanocomposite formulations and the resistance to ignition by GNP5 compared with GNP0 that was totally consumed within the 3 min test period.

10.3.9 Proposed Flame Retardancy Mechanism

Based on the results of HRR, PHRR, and LOI in the above study [60], the flame retardant mechanism can be explained by the two competing factors [61,65]. The first is the formation of a thicker and more compact residue layer acting as an insulating barrier while reducing the escape of volatile decomposition byproducts to the flame. The formation of such a residue layer during burning is critical to obtain a low HRR. It should be noted that the thickness of the residue layer also increased with increasing GNP content as shown by a macroscopic view of the char residue (Figure 10.15). The second factor that explains the flame retardancy mechanism is the increase in the viscosity of the burning material; the effect of nanomaterials increasing the viscosity of polymer melt at low shear rates has been established before for different nanoparticle/polymer combinations [66]. It is observed that the dripping that has characterized PET/PP blends has been eliminated by the addition of GNPs to the blend. This observation was attributed to the increased viscosity of the burning material caused by the GNPs.

10.3.10 Thermal Conductivity

Thermal conductivity is determined primarily by the vibration of lattice phonons and also by the thermal motion of electrons [40]. The dependence of the thermal conductivity of PET/PP/GNP nanocomposites on GNP content is shown in Figure 10.16, as described

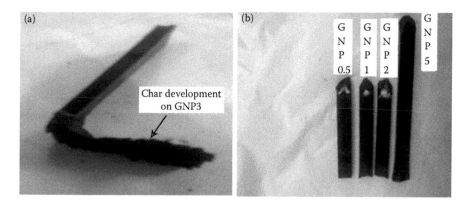

FIGURE 10.14 (a) Char formation and (b) self-extinguishing characteristics of GNP-filled PET/PP blends (LOI test specimens after test).

FIGURE 10.15   Macroscopic view of the char residue: (a) GNP0, (b) GNP1, (c) GNP3, and (d) GNP5.

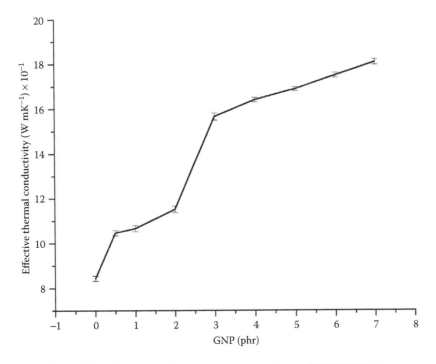

FIGURE 10.16   Effect of GNP loading on the thermal conductivity of PET/PP/GNP nanocomposites.

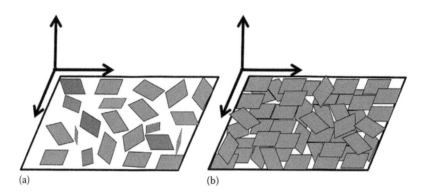

FIGURE 10.17   Typical distribution of GNPs in PET/PP showing: (a) random orientation (low GNP concentration) and (b) interconnected GNP sheets (higher GNP concentration at 3 phr and above).

in our recent investigation [60]. The thermal conductivity of PET/PP/GNP nanocomposites is strongly dependent on the GNPs loading, with higher weight fractions consistently resulting in higher thermal conductivities of the nanocomposites. A similar trend has been reported by Mamunya et al. [68].

This behavior is schematically depicted in Figure 10.17, in which at a lower concentration of GNPs (Figure 10.17a) (lower than 3 phr filler concentration), the GNPs are thought to be randomly distributed in the polymer matrix without interconnection. This results in a lower improvement in the values of the effective thermal conductivity of the nanocomposites. The platelets are randomly oriented in the matrix with each platelet standing isolated with little or no contact with each other, which led to low efficiency for thermal conductivity [38]. However, the development of an interconnected network led to a sharp enhancement of thermal conductivity starting from 3 phr, as illustrated in Figure 10.17b. This illustration is consistent with the morphological features of the TEM micrograph observed in Figure 10.6a.

Recently, it was found that fillers with a high aspect ratio, such as whiskers and platelets, can form a more continuous thermally conductive network in the polymer matrix and thus more effective in enhancing thermal transfer [37]. This factor combined with the high intrinsic thermal conductivity of GNPs offered reasonable explanations for the larger increments of thermal conductivity of PET/PP/GNP with increasing filler content. Because of Van der Waals' attraction, a homogeneous GNP network has been achieved under relatively high filler content as confirmed by TEM micrographs of GNP3 nanocomposites.

10.4 CONCLUSION

Exfoliated GNP-reinforced PET/PP nanocomposites were successfully prepared by the melt-blending technique. The structural and morphological studies show that the GNPs were well dispersed in the matrix; however, they may not have been substantially exfoliated. FTIR spectroscopy did not show any significant changes in the peak positions of PET/PP/GNP spectrum as compared with the neat PET/PP blend, which indicates the lack of significant chemical interaction between GNP and the blend matrix. Evidence from TEM,

FESEM, and XRD showed that the platelets remain intact and dispersed homogeneously in the polymer matrix without substantial exfoliation of GNPs in the polymer matrix. The maximum flexural strength and impact strength of the nanocomposites were obtained at 3 phr GNP loading. The improvements obtained for the mechanical properties are attributed to stiffness of the platelets, effective stress transfer between matrix and GNPs, and uniform dispersion of GNPs in the matrix. Thermogravimetric analysis results showed that GNPs have significantly enhanced the thermal stability of PET/PP/GNP nanocomposites with highest thermal stability at 3 phr loading. Data from cone calorimetry (HRR, PHRR, TTI, and ML), LOI, and UV94-V confirmed the enhancement of flame retardancy of PET/PP nanocomposites samples due to the addition of GNPs. The mechanism by which flame retardancy of PET/PP/GNP nanocomposites improved was through the formation of a uniform compact char layer in the condensed phase during decomposition of the polymer matrix. The char structure effectively prevents the inside thermal decomposition products into the flame zone and that of the oxygen into the underlying of polymer matrix. Thermal conductivity increased with increased GNP loading and exhibited a linear relationship with GNP content. This was attributed to the development of effective heat conduction bridges of interconnected GNPs that increased the efficiency for thermal conductivity. This study showed the potential of using GNPs to develop a flame retarded, thermally conductive polymer nanocomposites based on blends of PET/PP blends for applications in electrical/electronics industry where flame retardancy and heat dissipation in electrical components are essential.

ACKNOWLEDGMENTS

The authors acknowledge the Universiti Teknologi Malaysia (UTM) and Research University Grant 05H22, sub-code: Q.J130000.2509.05H22 for financial support.

REFERENCES

1. Kale, E. 2013. The eco-friendly car is driving engineering plastics market to $76 billion by 2017. http://www.tgdaily.com/general-sciences-features/78573-the-eco-friendly-car-is-driving-engineering-plastics-market-to-76-bi#VRkDTSm6J5LpjbcK.99. Accessed January 12, 2014.
2. Tanrattanakul, V., Hiltner, A., Baer, E., Perkins, W. G., Massey, F. L., and Moet, A. 1997. Toughening PET by blending with a functionalized SEBS block copolymer. *Polymer* 38: 2191–2200.
3. Tanrattanakul, V., Hiltner, A., Baer, E., Perkins, W. G., Massey, F. L., and Moet, A. 1997. Effect of elastomer functionality on toughened PET. *Polymer* 38: 4117–4125.
4. Loyens, W. and Groeninckx, G. 2002. Ultimate mechanical properties of rubber toughened semicrystalline PET at room temperature. *Polymer* 43: 5679–5691.
5. Smart, G., Kandola, B. K., Horrocks, A. R., Nazare, S., and Marney, D. 2008. Polypropylene fibers containing dispersed clays having improved fire performance. Part II: Characterization of fibers and fabrics from PP–nanoclay blends. *Polymers for Advanced Technologies* 19: 658–670.
6. Chan, C. M., Wu, J., Li, J. X., and Cheung, Y. K. 2002. Polypropylene/calcium carbonate nanocomposites. *Polymer* 43: 2981–2992.
7. Wen, X., Wang, Y., Gong, J., Liu, J., Tian, N., Wang, Y., Jianga, Z., Qiu, J., and Tang, T. 2012. Thermal and flammability properties of polypropylene/carbon black nanocomposites. *Polymer Degradation and Stability* 97: 793–801.

8. Abdul Razak, N. C., Inuwa, I. M., Hassan, A., and Samsudin, S. A. 2013. Effects of compatibilizers on mechanical properties of PET/PP blend. *Composite Interfaces* 20: 507–515.
9. Liu, X. G. 2013. How will carbon material lead to a Low Carbon Society–A review of Carbon 2012. *Carbon* 51: 438.
10. Geim, A. K., and Novoselov, K. S. 2007. The rise of graphene. *Nature Materials* 6: 183–191.
11. Lee, C., Wei, X., Kysar, J. W., and Hone, J. 2008. Measurement of the elastic properties and intrinsic strength of monolayer graphene. *Science* 321: 385–388.
12. Gómez-Navarro, C., Weitz, R. T., Bittner, A. M., Scolari, M., Mews, A., Burghard, M., and Kern, K. 2007. Electronic transport properties of individual chemically reduced graphene oxide sheets. *Nano Letters* 7: 3499–3503.
13. Balandin, A. A., Ghosh, S., Bao, W., Calizo, I., Teweldebrhan, D., Miao, F., and Lau, C. N. 2008. Superior thermal conductivity of single-layer graphene. *Nano Letters* 8: 902–907.
14. Kuila, T., Bose, S., Khanra, P., Kim, N. H., Rhee, K. Y., and Lee, J. H. 2011. Characterization and properties of *in situ* emulsion polymerized poly (methyl methacrylate)/graphene nanocomposites. *Composites Part A: Applied Science and Manufacturing* 42: 1856–1861.
15. Kuila, T., Bose, S., Mishra, A. K., Khanra, P., Kim, N. H., and Lee, J. H. 2012. Effect of functionalized graphene on the physical properties of linear low density polyethylene nanocomposites. *Polymer Testing* 31: 31–38.
16. Zhao, X., Zhang, Q., Chen, D., and Lu, P. 2010. Enhanced mechanical properties of graphene-based poly (vinyl alcohol) composites. *Macromolecules* 43: 2357–2363.
17. Rafiee, M. A., Rafiee, J., Wang, Z., Song, H., Yu, Z. Z., and Koratkar, N. 2009. Enhanced mechanical properties of nanocomposites at low graphene content. *ACS Nano* 3: 3884–3890.
18. Bandla, S. and Hanan, J. C. 2012. Microstructure and elastic tensile behavior of polyethylene terephthalate-exfoliated graphene nanocomposites. *Journal of Materials Science* 47: 876–882.
19. Kalaitzidou, K., Fukushima, H., Miyagawa, H., and Drzal, L. T. 2007. Flexural and tensile moduli of polypropylene nanocomposites and comparison of experimental data to Halpin-Tsai and Tandon-Weng models. *Polymer Engineering and Science* 47: 1796–1803.
20. Kalaitzidou, K., Fukushima, H., and Drzal, L. T. 2007. Mechanical properties and morphological characterization of exfoliated graphite–polypropylene nanocomposites. *Composites Part A: Applied Science and Manufacturing* 38: 1675–1682.
21. Bandla, S. and Hanan, J. C. 2011. Morphology and failure behavior of polyethyelene terephthalate-exfoliated graphene nanocomposites. In: *69th Annual Technical Conference of the Society of Plastics Engineers*, Boston, MA, 2011, 673p.
22. Kim, S., Do, I., and Drzal, L. T. 2010. Thermal stability and dynamic mechanical behavior of exfoliated graphite nanoplatelets-LLDPE nanocomposites. *Polymer Composites* 31: 755–761.
23. Kalaitzidou, K. 2006. Exfoliated graphite nanoplatelets as reinforcement for multifunctional polypropylene nanocomposites. PhD Thesis, Michigan State University, East Lansing, MI, 2006.
24. Al-Jabareen, A., Al-Bustami, H., Harel, H., and Marom, G. 2013. Improving the oxygen barrier properties of polyethylene terephthalate by graphite nanoplatelets. *Journal of Applied Polymer Science* 128: 1534–1539.
25. Huang, G., Gao, J., Wang, X., Liang, H., and Ge, C. 2012. How can graphene reduce the flammability of polymer nanocomposites? *Materials Letters* 66: 187–189.
26. Wu, X., Wang, L., Wu, C., Yu, J., Xie, L., Wang, G., and Jiang, P. 2012. Influence of char residues on flammability of EVA/EG, EVA/NG and EVA/GO composites. *Polymer Degradation and Stability* 97: 54–63.
27. Huang, G., Liang, H., Wang, Y., Wang, X., Gao, J., and Fei, Z. 2012. Combination effect of melamine polyphosphate and graphene on flame retardant properties of poly (vinyl alcohol). *Materials Chemistry and Physics* 132: 520–528.
28. Persson, H., Yao, Y., Klement, U., and Rychwalski, R. W. 2012. A simple way of improving graphite nanoplatelets (GNP) for their incorporation into a polymer matrix. *Express Polymer Letters* 6: 142–147.

29. Li, M. and Jeong, Y. G. 2011. Poly (ethylene terephthalate)/exfoliated graphite nanocomposites with improved thermal stability, mechanical and electrical properties. *Composites Part A: Applied Science and Manufacturing* 42: 560–566.
30. Jiang, X. and Drzal, L. T. 2010. Multifunctional high density polyethylene nanocomposites produced by incorporation of exfoliated graphite nanoplatelets 1: Morphology and mechanical properties. *Polymer Composites* 31: 1091–1098.
31. Istrate, O. M., Paton, K. R., Khan, U., O'Neill, A., Bell, A. P., and Coleman, J. N. 2014. Reinforcement in melt-processed polymer–graphene composites at extremely low graphene loading level. *Carbon* 78: 243–249.
32. Fasihi, M., Garmabi, H., Ghaffarian, S. R., and Ohshima, M. 2013. Preparation of highly dispersed expanded graphite/polypropylene nanocomposites via low temperature processing. *Journal of Applied Polymer Science* 130: 1834–1839.
33. Bian, J., Wei, X. W., Lin, H. L., Wang, L., and Guan, Z. P. 2012. Comparative study on the exfoliated expanded graphite nanosheet-PES composites prepared via different compounding method. *Journal of Applied Polymer Science* 124: 3547–3557.
34. Prusty, G. and Swain, S. K. 2013. Dispersion of expanded graphite as nanoplatelets in a copolymer matrix and its effect on thermal stability, electrical conductivity and permeability. *Carbon* 51: 436.
35. Teng, C. C., Ma, C. C. M., Lu, C. H., Yang, S. Y., Lee, S. H., Hsiao, M. C.,Yena, M.-Y., Chiou, K.-C., and Lee, T. M. 2011. Thermal conductivity and structure of non-covalent functionalized graphene/epoxy composites. *Carbon* 49: 5107–5116.
36. Jiang, X. and Drzal, L. T. 2012. Exploring the potential of exfoliated graphene nanoplatelets as the conductive filler in polymeric nanocomposites for bipolar plates. *Journal of Power Sources* 218: 297–306.
37. Min, C., Yu, D., Cao, J., Wang, G., and Feng, L. 2013. A graphite nanoplatelet/epoxy composite with high dielectric constant and high thermal conductivity. *Carbon* 55: 116–125.
38. Han, Z. and Fina, A. 2011. Thermal conductivity of carbon nanotubes and their polymer nanocomposites: A review. *Progress in Polymer Science* 36: 914–944.
39. Inuwa, I. M., Hassan, A., Samsudin, S. A., Kassim, M., Haafiz, M., and Jawaid, M. 2014. Mechanical and thermal properties of exfoliated graphite nanoplatelets reinforced polyethylene terephthalate/polypropylene composites. *Polymer Composites* 35(10): 2029–2035.
40. Inuwa, I. M., Hassan, A., Samsudin, S. A., Haafiz, M. K., Jawaid, M., Majeed, K., and Razak, N. C. 2014. Characterization and mechanical properties of exfoliated graphite nanoplatelets reinforced polyethylene terephthalate/polypropylene composites. *Journal of Applied Polymer Science* 131: 1–9.
41. Wang, L., Hong, J., and Chen, G. (2010). Comparison study of graphite nanosheets and carbon black as fillers for high density polyethylene. *Polymer Engineering and Science* 50: 2176–2181.
42. Li, Y. C. and Chen, G. H. 2007. HDPE/expanded graphite nanocomposites prepared via masterbatch process. *Polymer Engineering and Science* 47: 882–888.
43. Verdejo, R., Bernal, M. M., Romasanta, L. J., and Lopez-Manchado, M. A. 2011. Graphene filled polymer nanocomposites. *Journal of Materials Chemistry* 21: 3301–3310.
44. Wakabayashi, K., Brunner, P. J., Masuda, J. I., Hewlett, S. A., and Torkelson, J. M. 2010. Polypropylene-graphite nanocomposites made by solid-state shear pulverization: Effects of significantly exfoliated, unmodified graphite content on physical, mechanical and electrical properties. *Polymer* 51: 5525–5531.
45. Inuwa, I. M., Hassan, A., and Shamsudin, S. A. 2014. Thermal properties, structure and morphology of graphene reinforced polyethylene terephthalate/polypropylene nanocomposites. *The Malaysian Journal of Analytical Sciences* 18: 466–477.
46. Heino, M., Kirjava, J., and Hietaoja, P. 1997. Compatibilization of polyethylene terephthalate/polypropylene blends with styrene–ethylene/butylene–styrene (SEBS) block copolymers. *Journal of Applied Polymer Science* 65: 241–249.

47. Ramanathan, T., Stankovich, S., Dikin, D. A., Liu, H., Shen, H., Nguyen, S. T., and Brinson, L. C. 2007. Graphitic nanofillers in PMMA nanocomposites—An investigation of particle size and dispersion and their influence on nanocomposite properties. *Journal of Polymer Science Part B: Polymer Physics* 45: 2097–2112.

48. Zhang, J. G. and Deng, J. 2011. The effect of maleic anhydride grafted styrene-ethylene-butylene-styrene on the friction and wear properties of polyamide6/carbon nanotube composites. *Polymer-Plastics Technology and Engineering* 50: 1533–1536.

49. Ramanathan, T., Abdala, A. A., Stankovich, S., Dikin, D. A., Herrera-Alonso, M., Piner, R. D., Adamson, D. H. et al. 2008. Functionalized graphene sheets for polymer nanocomposites. *Nature Nanotechnology* 3: 327–331.

50. Lim, S. R. and Chow, W. S. 2012. Impact, thermal, and morphological properties of functionalized rubber toughened-poly (ethylene terephthalate) nanocomposites. *Journal of Applied Polymer Science* 123: 3173–3181.

51. Geng, Y., Wang, S. J., and Kim, J. K. 2009. Preparation of graphite nanoplatelets and graphene sheets. *Journal of Colloid and Interface Science* 336: 592–598.

52. Kuila, T., Bose, S., Hong, C. E., Uddin, M. E., Khanra, P., Kim, N. H., and Lee, J. H. 2011. Preparation of functionalized graphene/linear low density polyethylene composites by a solution mixing method. *Carbon* 49: 1033–1037.

53. Xu, Z. and Gao, C. 2010. In situ polymerization approach to graphene-reinforced nylon-6 composites. *Macromolecules* 43: 6716–6723.

54. Yu, Z. Z., Yang, M. S., Dai, S. C., and Mai, Y. W. 2004. Toughening of recycled poly (ethylene terephthalate) with a maleic anhydride grafted SEBS triblock copolymer. *Journal of Applied Polymer Science* 93: 1462–1472.

55. Chiu, H. T. and Hsiao, Y. K. 2006. Compatibilization of poly (ethylene terephthalate)/polypropylene blends with maleic anhydride grafted polyethylene-octene elastomer. *Journal of Polymer Research* 13: 153–160.

56. Patole, A. S., Patole, S. P., Kang, H., Yoo, J. B., Kim, T. H., and Ahn, J. H. 2010. A facile approach to the fabrication of graphene/polystyrene nanocomposite by *in situ* microemulsion polymerization. *Journal of Colloid and Interface Science* 350: 530–537.

57. Tantis, I., Psarras, G. C., and Tasis, D. L. 2012. Functionalized graphene–poly (vinyl alcohol) nanocomposites: Physical and dielectric properties. *Express Polymer Letter* 6: 283–292.

58. Akbari, M., Zadhoush, A., and Haghighat, M. 2007. PET/PP blending by using PP-g-MA synthesized by solid phase. *Journal of Applied Polymer Science* 104: 3986–3993.

59 Karevan, M., Eshraghi, S., Gerhardt, R., Das, S., and Kalaitzidou, K. 2013. Effect of processing method on the properties of multifunctional exfoliated graphite nanoplatelets/polyamide 12 composites. *Carbon* 64: 122–131.

60. Inuwa, I. M., Hassan, A., Wang, D. Y., Samsudin, S. A., Haafiz, M. M., Wong, S. L., and Jawaid, M. 2014. Influence of exfoliated graphite nanoplatelets on the flammability and thermal properties of polyethylene terephthalate/polypropylene nanocomposites. *Polymer Degradation and Stability* 110: 137–148.

61. Zhuge, J., Gou, J., and Ibeh, C. 2012. Flame resistant performance of nanocomposites coated with exfoliated graphite nanoplatelets/carbon nanofiber hybrid nanopapers. *Fire and Materials* 36: 241–253.

62. Quan, H., Zhang, B. Q., Zhao, Q., Yuen, R. K., and Li, R. K. 2009. Facile preparation and thermal degradation studies of graphite nanoplatelets (GNPs) filled thermoplastic polyurethane (TPU) nanocomposites. *Composites Part A: Applied Science and Manufacturing* 40: 1506–1513.

63. Uhl, F. M. and Wilkie, C. A. 2004. Preparation of nanocomposites from styrene and modified graphite oxides. *Polymer Degradation and Stability* 84: 215–226.

64. Peeterbroeck, S., Laoutid, F., Swoboda, B., Lopez-Cuesta, J. M., Moreau, N., Nagy, J. B., Alexandre, M., and Dubois, P. 2007. How carbon nanotube crushing can improve flame retardant behaviour in polymer nanocomposites? *Macromolecular Rapid Communications* 28: 260–264.

65. Schartel, B., Bartholmai, M., and Knoll, U. 2006. Some comments on the main fire retardancy mechanisms in polymer nanocomposites. *Polymers for Advanced Technologies* 17: 772–777.

66. Dittrich, B., Wartig, K. A., Hofmann, D., Mülhaupt, R., and Schartel, B. 2013. Flame retardancy through carbon nanomaterials: Carbon black, multiwall nanotubes, expanded graphite, multi-layer graphene and graphene in polypropylene. *Polymer Degradation and Stability* 98: 1495–1505.

67. Higginbotham, A. L., Lomeda, J. R., Morgan, A. B., and Tour, J. M. 2009. Graphite oxide flame-retardant polymer nanocomposites. *ACS applied materials and interfaces* 1: 2256–2261.

68. Mamunya, Y. P., Davydenko, V. V., Pissis, P., and Lebedev, E. V. 2002. Electrical and thermal conductivity of polymers filled with metal powders. *European Polymer Journal* 38: 1887–1897.

CHAPTER **11**

Influence of Organic Modifier on Miscibility and Chain Scission of the PA6 Phase and the Extent of Multiwalled Carbon Nanotubes Dispersion in Co-Continuous PA6/ABS Blends

Suryasarathi Bose, Petra Pötschke,
and Arup R. Bhattacharyya

CONTENTS

11.1 INTRODUCTION

It has been well established that both the extent and the stability of dispersion of carbon nanotubes in a polymer matrix are the key factors that dictate the properties of the nanocomposites. In this regard, various surface modification routes have been explored to ensure better interfacial interaction with the host matrix, and both opportunities and limitations of these routes have been thoroughly assessed (Bose et al. 2010).

In our earlier work, we have reported that by modifying multiwalled carbon nanotubes (MWNTs) by a simple yet interesting organic modifier, sodium salt of 6-amino hexanoic acid (Na-AHA), both the stability and the extent of dispersion of MWNTs in the blends of polyamide 6 (PA6) and acrylonitrile–butadiene–styrene copolymer (ABS) were greatly improved (Bose et al. 2008, 2009b, Kodgire et al. 2006b). This further manifested in a substantial reduction (by a factor of ca. 10 with respect to pristine MWNTs) in the electrical percolation threshold in 50/50 (wt/wt) PA6/ABS blends and also dependent very much on the fraction of Na-AHA with respect to the MWNTs in the blends. The melt-viscosity data have revealed that small amount of Na-AHA-modified MWNTs in the PA6 phase reduces its melt viscosity through plasticization albeit the dynamics was different in the low-frequency regime, which rather exhibited a *pseudo-solid like* behavior manifesting in a physical gel (Bose et al. 2009b). The intrinsic viscosity measurements and estimation of ~COOH end groups of PA6 by titration have revealed chain scission of the PA6 phase in the presence of Na-AHA (Bose et al. 2009b). If we exclude any possibility of polycondensation of Na-AHA at the mixing temperature (260 °C), considering its small size, the chain scission of PA6 could only be explained by coordination with water molecule. The latter is generated when the ~NH_2 functional moieties of Na-AHA react with the ~COOH end groups of PA6 via the formation of amide linkage. Transamidation of polyamides has also been reported by blending polyamides with other polyamides (Eersels and Groeninckx 1996). There is a possibility of transamidation of PA6 in the presence of Na-AHA during melt-mixing leading to small grafts of Na-AHA on the PA6 chain. In either case, chain scission may occur. It has been reported that blending maleic anhydride-containing polymers (viz.; styrene maleic anhydride [SMA]) with polyamides leads to polyamide graft on SMA via imide linkage (van Duin et al. 1998). The amide group can only take part once it is hydrolyzed by the water molecule, which is generated during the formation of imide linkage. The melting behavior of the polyamide phase typically remains unaltered on blending with MA-containing polymers though the crystallization kinetics change to a significant extent (van Duin et al. 1998). Interestingly, it has been observed that both melting and crystallization behaviors of the PA6 phase were influenced to a significant extent in the presence of Na-AHA, which may be due to the molecular-level miscibility in the amorphous phase of PA6 or due to the transamidation reaction of the PA6 chains leading to smaller imperfect crystals. The molecular-level miscibility may be associated with the enhanced chain mobility on account of interference with the hydrogen bonding. It has been well established that the superheated water disrupts the interchain hydrogen bonding as the O—H hydrogen is a strong electron acceptor than the N—H hydrogen; the carbonyl oxygen favors coordination to the O—H hydrogen (Stuart 1994).

These interesting results have prompted us to revisit and bring forward certain issues, viz.; the type, nature, and the degree of interactions that possibly exist between Na-AHA-modified MWNTs and the blends at the molecular level, which would allow a deeper understanding of the key role of Na-AHA toward improving the dispersion extent of MWNTs in the blends. At a given concentration of MWNTs (1 wt.%), the fraction of Na-AHA was varied to investigate the role of Na-AHA toward molecular-level miscibility by Fourier transform infrared (FTIR) spectroscopic analysis, melt-rheological investigation, differential scanning calorimetric (DSC) analysis, and dielectric spectroscopic analysis.

11.2 EXPERIMENTAL SECTION

11.2.1 Materials and Sample Preparation

PA6 was obtained from Gujarat State Fertilizer & Chemical corporation, Vadodara, India (Gujlon M28RC, relative viscosity 2.8, M_v is 38,642 in 85% formic acid, zero shear viscosity of 180 Pa.s at 260 °C). ABS copolymer (Absolac-120, with typical composition consisting of acrylonitrile: 24 wt.% rubber content: 16.5 wt.% and styrene: 59.5 wt.%) was supplied by Bayer India Ltd. Catalytic chemical vapour deposition (CCVD)-synthesized thin purified MWNTs (p-MWNTs) were obtained from Nanocyl SA, Sambreville, Belgium (NC 3100, L/D: 10–1000, purity > 95%) as per manufacturer specification. 6-Aminohexanoic acid (AHA) (Sigma Aldrich, $M_w = 132.18$; purity: 98%) was neutralized using sodium hydroxide to obtain Na-AHA. The detailed procedure to obtain Na-AHA and the solid mixtures of MWNTs and Na-AHA is described elsewhere (Kodgire et al. 2006a). Neat blends of PA6/ABS and blends with pristine MWNTs and Na-AHA-modified MWNTs were melt-mixed in a conical twin-screw micro-compounder (Micro 5, DSM Research, the Netherlands) at 260 °C with a rotational speed of 150 rpm for 15 min. All the experiments were performed under nitrogen atmosphere in order to prevent oxidative degradation. The following notation was used to represent the composition: *NxMyTzA*. For instance, N1M15T1A represents 50/50 (wt/wt) PA6/ABS blend with 1 wt.% Na-AHA-modified MWNTs, wherein MWNTs: Na-AHA ratio is 1:15 (*x:y*); N50T1A50 represents 50/50 (wt/wt) PA6/ABS blend with 1 wt.% MWNTs and N50S2A50 represent 50/50 (wt/wt) PA6/ABS blend with 2 wt.% Na-AHA. The above notation also represents that a two-step mixing protocol was employed, wherein various types of MWNTs (pristine versus modified) were initially mixed with the PA6 phase (for 10 min) followed by the addition of the ABS copolymer (for 5 min) without any interruption. Injection-molded samples (according to ASTM D 638, Type V, thickness: 3 mm; gauge length: 6.2 mm; width: 10 mm) were prepared using mini injection-molding machine from DSM Research, the Netherlands.

11.2.2 Characterization

Fourier transform infra-red (FTIR) spectroscopic analysis was carried out with a MAGNA 550 (Nicolet) spectrometer for composite powder samples using KBr pellets at room temperature in the scanning range of 400–4000 cm^{-1}.

Differential scanning calorimetric (DSC) measurements were carried out using a modulated DSC (Q1000, TA Instruments, New Castle, DE). The extruded samples of about 5 mg were dried in a vacuum oven prior to experiment. The heating–cooling–heating cycles were recorded in the temperature range from −60 °C to 260 °C at a scan rate of 10 K/min under nitrogen atmosphere. In the first heating run, all samples were annealed at the final temperature (260 °C) for 3 min to delete the previous thermal history.

The rheological measurements were performed using an ARES oscillatory rheometer (TA Instruments, New Castle, DE, USA) at 260 °C under nitrogen atmosphere with parallel plate geometry (plate diameter of 25 mm, gap of 1–2 mm). Frequency sweeps were carried out between 0.1 and 100 rad/s. The strains used were chosen in order to be within the linear viscoelastic range.

The dielectric spectroscopic measurements were performed on the injection-molded samples (across the thickness ~3 mm) in the frequency range between 10^{-2} and 10^{7} Hz using alpha high-resolution analyzer coupled with Novocontrol interface (broadband dielectric converter).

11.3 RESULTS AND DISCUSSION

11.3.1 Specific Interactions, Miscibility, and Chain Scission

The interchain hydrogen bonding may impart many intriguing properties in polyamides, viz.; high crystallinity and melting point, high specific strength, and extraordinary structural properties (Kohan 1995). However, it is envisaged that an increase in thermal energy may destroy hydrogen bonding (Stuart 1994). Hence, the degree of hydrogen bonding can be different under different processing conditions (melt vs. solution). Spectroscopic techniques have been found to be quite successful in the recent past in explaining the interchain hydrogen bonding as a function of temperature. In our previous work (Bose et al. 2009b), we argued that the amine (~NH$_2$) functional groups of Na-AHA may react with the terminal ~COOH end groups of PA6. This argument was also supported by FTIR analysis, wherein the intensity of the amide peak was increased as a function of Na-AHA concentration with respect to PA6 in the blends. In this context, few interesting observations in our previous report prompted us to revisit and take a deeper look into the finer details and are also the rationale behind this discussion. A significant debundling of MWNTs was observed in the presence of Na-AHA and the plausible mechanism was explained on the basis of moisture-induced plasticization and/or chain scission of the PA6 phase. In aqueous dispersion, the Na-AHA may get dissociated and the corresponding COO⁻ containing molecule may get adsorbed on the surface of MWNTs. As Na⁺ ions are mobile, an electrical *double layer* may form near the surface of MWNTs with net negative charge owing to gain in the entropy (Bose et al. 2011). The formation of net negative charge on the surface of MWNTs may help in debundling through electrostatic charge repulsion (Bose et al. 2011). In the melt, the loosely bound *agglomerates* of Na-AHA-modified MWNTs could further be dispersed by rupture and erosion mechanism (Khare et al. 2011). To simplify our discussion, let us consider the effect of Na-AHA alone on the local environmental changes in the functional groups involved in the reaction/interactions. A schematic representation of the possible interactions during melt-mixing of PA6/ABS blends in the presence of Na-AHA-modified MWNTs is illustrated in Figure 11.1 and the arguments are supported by the spectroscopic evidences. The changes in the local environment of the amide carbonyl group that occurred

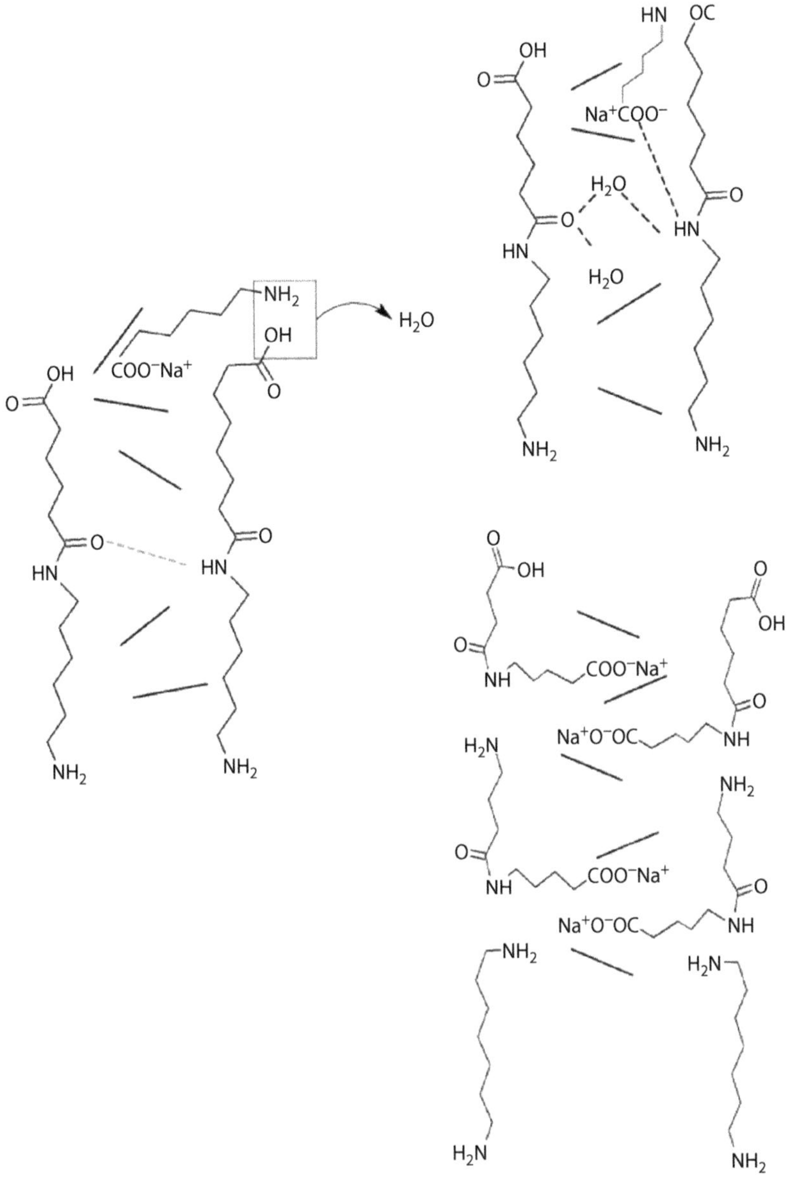

FIGURE 11.1 Schematic representation of two possible interactions between PA6 and Na-AHA-modified MWNTs (solid lines represent MWNTs).

upon blending PA6/ABS blends with Na-AHA-modified MWNTs are illustrated in Figure 11.2. As stated, the amine (~NH$_2$) functional groups of Na-AHA may react with the terminal ~COOH end groups of PA6 to form sort of Na-AHA graft on to PA6 via an amide linkage. The molecule of water during the amide bond formation is released in the melt at high temperature (260 °C) and possibly under these conditions may result in the hydrolysis of the PA chains. This argument is also supported by melt rheology (see subsequent sections), by intrinsic viscosity measurements, and by estimating the ~COOH end groups by

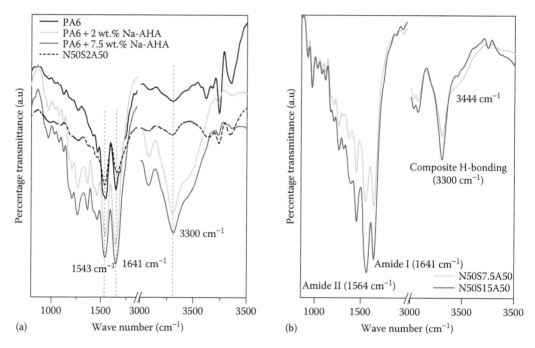

FIGURE 11.2   FTIR spectra for (a) PA6 and blends of PA6/Na-AHA and (b) 50/50 PA6/ABS blends with Na-AHA.

titration (Bose et al. 2009b). It has been well established that the superheated water disrupts the interchain hydrogen bonding as the O—H hydrogen is a strong electron acceptor than the N—H hydrogen; the carbonyl oxygen favors coordination to the O—H hydrogen (Stuart 1994). This in turn hydrolyzes the amide bonds and in this process disrupts the interchain hydrogen bonding. Interestingly, the water molecule establishes new hydrogen bonding as three molecules of water can be sorbed onto two neighboring amide groups, thus augmenting hydrogen bonding (Stuart 1994). Although the water molecule disrupts the existing hydrogen bonding in PA, it also augments it by forming different hydrogen bonding.

It has also been observed that the concentration of Na-AHA in the blends influenced significantly the crystallinity, miscibility, rheology, and the structural properties (Bose et al. 2009a), and hence, it is worth trying to understand the changes in the local environment of the carbonyl amide both at lower (Figure 11.2a) and at higher concentrations of Na-AHA (Figure 11.2b) in the blends. Figure 11.2a shows the FTIR spectra for PA6, PA6 with Na-AHA, and 50/50 blends of PA6/ABS with Na-AHA (2 wt.%). The characteristic peaks for PA6 are indicated in the FTIR spectra (Figure 11.2a). Peak at 1641 cm^{-1} represents carbonyl (C=O) stretching vibration ($v_{C=O}$), which is characteristic for amide I band. Peaks at 1543 and 1264 cm^{-1} suggest amide II band and amide III band, respectively, which are both assigned to nitrogen–hydrogen (N—H) bending vibration and carbon–nitrogen (C—N) stretching vibration (δN—H + vC—N); band at 3091 cm^{-1} is the characteristic double frequency absorption of amide II band. Peak at 690 cm^{-1} represents amide V band, which corresponds to bending vibration of amido-cyanogen (γN—H). Band at stretching frequency approximately 3301 cm^{-1} represents nitrogen–hydrogen (N—H) stretching vibration

(Chen et al. 2004). The frequency of the N—H stretching mode is strongly dependent on the strength of the hydrogen-bonded N—H groups (Skrovanek et al. 1985, 1986). For linear aliphatic polyamides, three N—H stretching bands occur at 3300, 3310, and 3447 cm^{-1}, which correspond to stretching vibrations of hydrogen-bonded N—H in the crystalline phase, hydrogen-bonded N—H in the amorphous phase, and free N—H groups, respectively. The spectrum of PA6 is characterized by an absorption band centered at 3300 cm^{-1}, which is believed to be a composite absorption consisting of hydrogen-bonded N—H stretching in the amorphous phase (at higher frequency) and that in the crystalline phase (at lower frequency) (Skrovanek et al. 1985, 1986). On increasing the concentration of Na-AHA in PA6 and in the blends, the peak intensities increased albeit there is no obvious shift in the peaks. Interestingly, at higher concentration of Na-AHA in the blends (>7.5 wt.%), a marked change in both the intensity and the peak position of the amide II is noticeable. It is worth noticing that because of the selective localization of Na-AHA in the PA6 phase in the blends, the effective concentration is much higher in contrast to PA6/Na-AHA blends. For instance, in a 50/50 PA6/ABS blends the effective concentration (assuming all the Na-AHA in the PA6 phase) would be higher by a factor of 2 with respect to neat PA6.

It is well documented that superheated water may plasticize PA6 phase mainly by hydrolysis of the amide groups further manifesting in chain scission (Marechal et al. 1995). In our case, the superheated water is released due to the melt interfacial reaction between the ~NH$_2$ functional groups of Na-AHA and the terminal ~COOH end groups of PA6 via an amide linkage. This process in turn leads to more Na-AHA grafts, where the existing amide bonds in the chains are replaced with the new amide bonds. This argument may possibly explain the reason behind observing reduced intrinsic viscosity as well as melt viscosity in PA6/ABS/Na-AHA blends. However, this argument is inexplicable for a shift in the amide II band in the blends. A plausible explanation to this could be stated as the superheated water hydrolyzes the amide and interferes with the hydrogen bonding; the N—H group changes its hydrogen bonding from N—H/O=C to N—H/COO$^-$. At a high temperature (260 °C) and under intense shear (during melt-mixing) COO$^-$Na$^+$ would not possibly disassociate but Na$^+$ would be in the vicinity of COO$^-$, thereby allowing hydrogen bonding with the polarized N—H bond. In the absence of COO$^-$Na$^+$, hydrolysis of amide would have resulted in ~COOH and ~NH$_2$ functional groups, which has been reported for PA6/SMA blends, wherein water molecule was generated via an imide linkage. The shift in amide II band in the blends at higher Na-AHA concentration can hence be attributed to the specific interactions. Transamidation reactions cannot be ruled out completely. At high temperature (260 °C), Na-AHA can take part in amide-interchange type of reaction, very similar to blending polyamides with polyamides, via transamidation (Eersels and Groeninckx 1996). This will cause chain scission and will lead to small grafts of Na-AHA on the PA6 chain (as shown in Figure 11.1) and in turn enhances the mobility of the chains, which otherwise was restricted due to interchain hydrogen bonding.

Another piece of information comes from the N—H stretching mode, which is again dependent on the strength of the hydrogen bonding in PA. The integral of this band has increased quite substantially in the case of PA6/Na-AHA and PA6/ABS/Na-AHA systems in contrast to PA6. This essentially indicates that both the degree and the nature of hydrogen bonding are altered to some extent or possibly to a large extent depending on the concentration of Na-AHA in the system. A detailed spectroscopic analysis of the changes occurring in the local environment of the hydrogen bonds in PA as a function of temperature would yield more information and is subjected to future investigations.

Figure 11.3 shows the melting endotherms of PA6, PA6/Na-AHA blend, and PA6/ABS blends with Na-AHA-modified MWNTs. Though blending maleic anhydride-containing polymer (like SMA copolymer) with PA6 results in graft copolymer formation and PA chain scission, it does not affect the melting point but noticeable change in the crystallization behavior has been reported (van Duin et al. 1998). In our case, blending Na-AHA with PA6 has resulted in an obvious depression in melting point essentially indicating miscibility in the amorphous segments. The latter may be as a result of small size of Na-AHA unlike SMA copolymer, which is typical of higher molecular weight. Another interesting observation is the *double melting endotherms* in both PA6 (215.9, 220.4 °C) and PA6/Na-AHA systems, but only a single melting endotherm blends with Na-AHA-modified MWNTs. Double melting endotherms can be due to *melting–recrystallization–melting* phenomenon or due to the presence of different crystal structures. The peaks at 930 and 973 cm^{-1} correspond to α- and γ-crystalline phases of PA6 (see Figure 11.2a). Both these peaks were present in the samples investigated here; however, the intensity of the latter crystalline phase was more prominent in the case of PA6/Na-AHA and blends with Na-AHA, which suggest that MWNTs favor a specific crystal structure. This was also supported by wide angle X-ray diffraction analysis (Bose et al. 2009a). The observation of the melting point depression was consistent with increasing concentration of Na-AHA in the blends as manifested from Figure 11.3b, which compares the melting endotherms at a given concentration of MWNTs (1 wt.%) but varying concentration of Na-AHA. It is envisaged that both chain rotation and mobility are enhanced if the interchain hydrogen bonding in PA is affected. Water molecule is only sorbed in the amorphous phase breaking the existing hydrogen bonding in PA6 and leads to

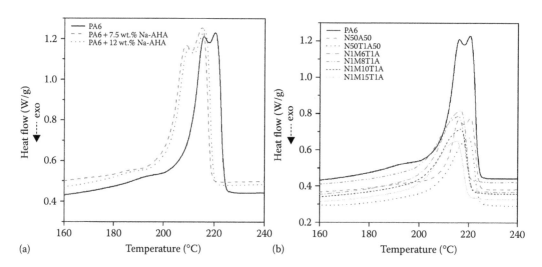

FIGURE 11.3   DSC second heating curves of (a) PA6 and blends of PA6/Na-AHA and (b) 50/50 PA6/ABS blends with Na-AHA-modified MWNTs.

plasticization without any associated complex formation, as in the case of PA/ionomer blends (Feng et al. 1996).

Figure 11.4a and b show the crystallization exotherm of the PA6 phase for pure PA6, PA6/Na-AHA, and blends with Na-AHA-modified MWNTs. A delay in crystallization can be observed in case of blending PA6 with Na-AHA. In the case of blends with MWNTs and Na-AHA-modified MWNTs, unusual crystallization behavior, i.e., two crystallization exotherms, could be observed and was discussed extensively (Bose et al. 2009a). Blends with MWNTs (1 wt.%) augmented the crystallization temperature by ca. 3 °C (196 °C in contrast to 193 °C in the case of PA6), whereas the crystallization temperature (i.e., the bulk crystallization temperature) was almost unaltered in blends with Na-AHA-modified MWNTs with respect to the bulk crystallization temperature of PA6 (except for Na-AHA concentration beyond 15 wt.%). These observations clearly prompt the expulsion of the grafted chains during crystallization; moreover, the population of these chains is more diffused in the amorphous regions. The latter effect favors miscibility and is also supported by the melting endotherm data.

11.3.2 Melt Rheology and Relaxations

The melt-rheological behavior of the blends in the presence and in the absence of Na-AHA-modified MWNTs has been studied using low amplitude oscillatory measurements in the linear viscoelastic regime (which was determined *a priori*). In our previous work we reported that blends with Na-AHA-modified MWNTs exhibited altogether a different rheological behavior with respect to the individual polymers and also with the blends with MWNTs. In the higher frequency regime, blends with Na-AHA-modified MWNTs exhibited significantly reduced melt-viscosity but interestingly, in the lower frequency regime, it showed a strong yield stress. The intrinsic viscosity measurements together with

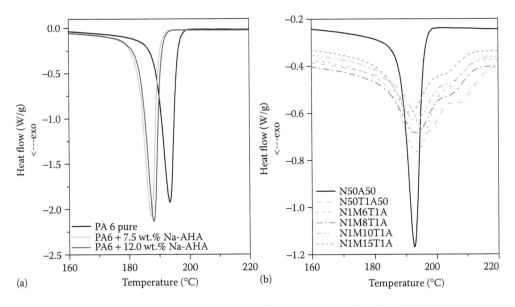

FIGURE 11.4   DSC cooling curves of (a) PA6 and PA6/Na-AHA blends and (b) 50/50 PA6/ABS blends with Na-AHA-modified MWNTs.

the estimation of ~COOH end groups by titration method suggest that chain scission of PA6 occurred by blending with Na-AHA during melt-mixing (Bose et al. 2009b). This can be attributed to the hydrolysis of amide by water molecule that was released due to melt-interfacial reaction between the ~NH_2 functional groups of Na-AHA and ~COOH end groups of PA6 or by amide interchange type of reactions via transamidation (as discussed). Even a small amount of Na-AHA in the PA6 phase reduces its viscosity through plasticization and accentuates further the difference in the rheological behavior of the two phases. This behavior is very much different from PA6/SMA blends, where a positive deviation in the melt viscosity has been noticed (van Duin et al. 1998). This could be attributed to longer grafts of PA6 on SMA unlike in our case where the size of Na-AHA molecule is smaller as compared to SMA. However, in the low frequency regime, the highly exfoliated network of MWNTs, aided by Na-AHA, seems to dominate further manifesting in *pseudo-solid like* behavior. This observation is further supported by the divergence in η^* versus G^* plot (Figure 11.5a), which essentially indicates a finite yield behavior due to *network-like* structure of MWNTs. This behavior is even consistent with increasing fraction of Na-AHA in the solid mixture of MWNTs and Na-AHA suggesting an increase in the phase miscibility, mostly in the amorphous phase of PA6, in the blends in the presence of Na-AHA. This argument is also supported by the DSC analysis, wherein the depression in melting point was observed in the blends in the presence of Na-AHA.

Figure 11.5b shows a plot of G' versus G'', analogous to Cole–Cole plot used in dielectric spectroscopy. In multi-component systems, such plot can elucidate structural changes between the matrix and the filled systems at a given temperature. For the blends with Na-AHA-modified MWNTs, the storage modulus, G' (for a given loss modulus, G''), increases significantly with increasing concentration of Na-AHA at a fixed loading of MWNTs especially at lower frequencies. This essentially suggests that at a given concentration of MWNTs, increasing Na-AHA concentration leads to a better exfoliation and also ensures enhanced interactions with the host matrix. For all the compositions investigated here, G' is higher than G'' and moreover, the slope of G' versus G'' decreases with increasing Na-AHA content (again at a given concentration of MWNTs).

Figure 11.6a shows the logarithmic plots of the real part of complex permittivity, ε', as a function of frequency. It appears that ε' increases with decreasing frequency especially for the blends with Na-AHA-modified MWNTs. It is worth noting that these compositions are well above the percolation threshold (Bose et al. 2009b). In contrast, blends with MWNTs though exhibit a similar behavior but to a moderate extent. It is envisaged that higher value of ε' at lower frequency originates from the contact regions. In the case of blends with Na-AHA-modified MWNTs, the contact region increases significantly due to specific interactions-mediated exfoliation of MWNTs. This leads to a decrease in the bulk capacitance. Figure 11.6b shows the logarithmic plots of imaginary part of complex permittivity, ε'', as a function of frequency. In case of blends with Na-AHA-modified MWNTs, the ε'' almost shows a linear behavior in a wide range of probed frequency. In general, in the case of MWNTs-based composites, above percolation threshold, the interfacial polarization is masked due to the DC electrical conductivity. In order to overcome the difficulty in evaluating the interfacial polarization, the electrical modulus formalism was

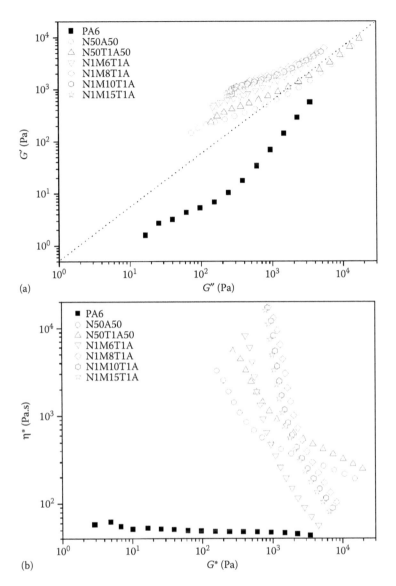

FIGURE 11.5   (a) Cole–Cole plots and (b) Complex viscosity (η^*) versus complex modulus (G^*) for 50/50 PA6/ABS blends with Na-AHA-modified MWNTs.

proposed by McCrum et al. (1967). One of the advantages of using electrical modulus data to interpret the bulk relaxation, in contrast to permittivity, is that the effect of DC electrical conductivity at low frequencies is eliminated, which otherwise masks other relaxations in the system. The frequency dependence of M'' at room temperature is described in Figure 11.6c. PA6 shows a loss peak at a certain frequency and the blends with Na-AHA-modified MWNTs exhibit a peak at a different frequency. Interestingly, the peak shifts to a higher frequency, especially for higher concentration of Na-AHA. Owing to the difference in conductive and dielectric differences, intrinsic immobilized charge can move freely at a certain applied field and then get jammed at the interface of the polymer and the conducting particle (Zhang et al. 2009). The latter results in interfacial polarization. It is evident from the frequency-dependent M'' that the relaxation time decreases (because the peak shifts to a higher frequency) as a function of Na-AHA concentration. This is attributed to the exfoliation of MWNTs, aided by Na-AHA, which further results in an increase in the interface and as a result more charges are accumulated at the interface between polymer and MWNTs. This is also supported by higher ε' values at lower frequencies, which also suggest charge accumulation at the bulk interface (interfacial polarization) and at the interface between the electrode and the sample (space charge polarization). Hence, it can be concluded that Na-AHA aids in significant exfoliation of MWNTs, which further

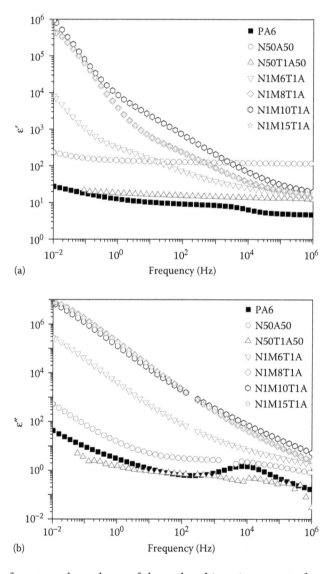

FIGURE 11.6 The frequency dependence of the real and imaginary part of complex permittivity (a and b) and the imaginary part of complex modulus for 50/50 PA6/ABS blends with Na-AHA-modified MWNTs (c). *(Continued)*

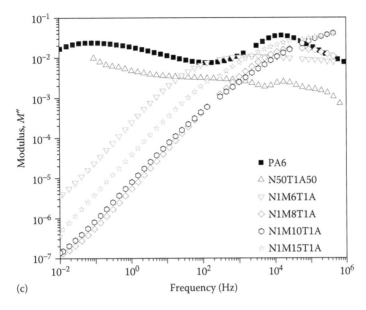

(c)

FIGURE 11.6 (Continued) The frequency dependence of the real and imaginary part of complex permittivity (a and b) and the imaginary part of complex modulus for 50/50 PA6/ABS blends with Na-AHA-modified MWNTs (c).

paved way for smaller conductive clusters of MWNTs in PA6. In solution, Na-AHA aids in dispersing MWNTs by electrical double layer mechanism and in the melt, the loosely bound Na-AHA-modified MWNTs, under shear, are finely dispersed. The latter effect is also mediated by enhanced chain mobility of PA6, mostly in the amorphous phases, on account of interference with the hydrogen bonding.

11.4 CONCLUSIONS

Melt blending 50/50 (wt/wt) PA6/ABS blends with Na-AHA-modified MWNTs has resulted in an enhanced chain mobility accompanied by the chain scission of the PA6 phase. Possible mechanisms of chain scission were discussed. At higher concentration of Na-AHA, a shift in the amide II peak of PA6 was observed from the FTIR spectra and could be attributed to the fact that the N—H group altered its hydrogen bonding from N—H/O=C to either N—H/ COO$^-$ or OH. The crystallization behavior of PA6 was greatly altered in the presence of Na-AHA-modified MWNTs suggesting an expulsion of the grafted chains during crystallization. Melting endotherm data suggest an enhanced miscibility in the amorphous phases of PA6. Interestingly, small amount of Na-AHA-modified MWNTs in the PA6 phase has reduced its melt-viscosity through plasticization and amplified further the difference in the melt-rheological behavior of the two phases. From electrical modulus data, it could be concluded that the relaxation time decreased as a function of Na-AHA concentration in the blends at a fixed concentration of MWNTs. This was attributed to the exfoliation of MWNTs aided by Na-AHA, which has resulted in an increase in the specific interfacial area, and as a result more charges could be accumulated at the interface between polymer and MWNTs.

ACKNOWLEDGMENTS

One of the authors (ARB) duly acknowledges the financial support from the Department of Science and Technology (DST), India (Project No. 08DS016). The authors acknowledge the support of "Microcompounder Central Facility" and CRNTS, IIT Bombay. One of the authors (ARB) would like to thank Leibniz Institute of Polymer Research Dresden, Germany, for providing the ARES rheometer facility. The authors also acknowledge Dr. Liane Häußler (IPF Dresden, Germany) for her assistance in DSC analysis. The authors also thank Professor Ajit R. Kulkarni for allowing us to use Dielectric Spectroscopy Facility.

REFERENCES

Bose, S., A. R. Bhattacharyya, L. Häusler, and P. Pötschke. 2009a. Influence of multiwall carbon nanotubes on the mechanical properties and unusual crystallization behavior in melt-mixed co-continuous blends of polyamide6 and acrylonitrile butadiene styrene. *Polymer Engineering and Science* 49(8):1533–1543. doi: 10.1002/pen.21392.

Bose, S., A. R. Bhattacharyya, R. A. Khare, S. S. Kamath, and A. R. Kulkarni. 2011. The role of specific interaction and selective localization of multiwall carbon nanotubes on the electrical conductivity and phase morphology of multicomponent polymer blends. *Polymer Engineering and Science* 51(10):1987–2000. doi: 10.1002/pen.21998.

Bose, S., A. R. Bhattacharyya, R. A. Khare, A. R. Kulkarni, and P. Pötschke. 2008. Specific interactions and reactive coupling induced dispersion of multiwall carbon nanotubes in co continuous poly-amide6/ionomer blends. *Macromolecular Symposia* 263(1):11–20. doi: 10.1002/masy.200850302.

Bose, S., A. R. Bhattacharyya, A. R. Kulkarni, and P. Pötschke. 2009b. Electrical, rheological and morphological studies in co-continuous blends of polyamide 6 and acrylonitrile-butadiene-styrene with multiwall carbon nanotubes prepared by melt blending. *Composites Science and Technology* 69(3–4):365–372. doi: 10.1016/j.compscitech.2008.10.024.

Bose, S., R. A. Khare, and P. Moldenaers. 2010. Assessing the strengths and weaknesses of various types of pre-treatments of carbon nanotubes on the properties of polymer/carbon nanotubes composites: A critical review. *Polymer* 51(5):975–993. doi: 10.1016/j.polymer.2010.01.044.

Chen, G., D. Shen, M. Feng, and M. Yang. 2004. An attenuated total reflection FT-IR spectroscopic study of polyamide 6/clay nanocomposite fibers. *Macromolecular Rapid Communications* 25:1121.

Eersels, K. L. L. and G. Groeninckx. 1996. Influence of interchange reactions on the crystallization and melting behaviour of polyamide blends as affected by the processing conditions. *Polymer* 37(6):983–989. doi: 10.1016/0032- 861(96)87281-8.

Feng, Y., A. Schmidt, and R. A. Weiss. 1996. Compatibilization of polymer blends by complexation.1. Spectroscopic characterization of ion-amide interactions in ionomer/polyamide blends. *Macromolecules* 29(11):3909–3917.

Khare, R. A., A. R. Bhattacharyya, A. S. Panwar, S. Bose, and A. R. Kulkarni. 2011. Dispersion of multiwall carbon nanotubes in blends of polypropylene and acrylonitrile butadiene styrene. *Polymer Engineering and Science* 51(9):1891–1905. doi: 10.1002/pen.21985.

Kodgire, P. V., A. R. Bhattacharyya, S. Bose, N. Gupta, A. R. Kulkarni, and A. Misra. 2006a. Control of multiwall carbon nanotubes dispersion in polyamide6 matrix: An assessment through electrical conductivity. *Chemical Physics Letters* 432(4–6):480–485. doi: 10.1016/j. cplett.2006.10.088.

Kohan, M.I., ed. 1995. *Nylon Plastics Handbook*. Cincinnati, OH: Hanser Gardner Publications.

Marechal, P., G. Coppens, R. Legras, and J. M. Dekoninck. 1995. Amine anhydride reaction versus amide anhydride reaction in polyamide anhydride carriers. *Journal of Polymer Science Part A—Polymer Chemistry* 33(5):757–766. doi: 10.1002/pola.1995.080330501.

McCrum, N. G., B. E. Read, and G. Williams. 1967. *Anelastic and Dielectric Effects in Polymeric Solids*. New York: John Wiley & Sons.

Skrovanek, D. J., S. E. Howe, P. C. Painter, and M. M. Coleman. 1985. Hydrogen bonding in polymers—Infrared temperature studies of an amorphous polyamide. *Macromolecules* 18(9):1676–1683.

Skrovanek, D. J., P. C. Painter, and M. M. Coleman. 1986. Hydrogen bonding in polymers—Infrared temperature studies of nylon-11. *Macromolecules* 19(3):699–705.

Stuart, B. H. 1994. A Fourier-transform Raman-study of water sorption by nylon-6. *Polymer Bulletin* 33(6):681–686. doi: 10.1007/bf00296082.

van Duin, M., M. Aussems, and R. J. M. Borggreve. 1998. Graft formation and chain scission in blends of polyamide-6 and -6.6 with maleic anhydride containing polymers. *Journal of Polymer Science Part A—Polymer Chemistry* 36(1):179–188. doi: 10.1002/(sici)1099-0518(19980115)36:1<179::aid-pola22>3.0.co;2-f.

Zhang, J., M. Mine, D. Zhu, and M. Matsuo. 2009. Electrical and dielectric behaviors and their origins in the three-dimensional polyvinyl alcohol/MWCNT composites with low percolation threshold. *Carbon* 47(5):1311–1320. doi: 10.1016/j.carbon.2009.01.014.

CHAPTER 12

Analysis of Strain Rate and Temperature Effects Data on Mechanical/Fracture Properties of Thermoset Resins and Their Composites

H. Pan and Sri Bandyopadhyay

CONTENTS

12.1 INTRODUCTION

Because of the viscous nature of thermoset resins, their mechanical properties are strongly dependent on strain rate and temperature (Coates and Ward 1978; George et al. 1999; Dasari and Misra 2003; Mae 2008; Plaseied and Fatemi 2008a, b). According to time–temperature physical equivalence (Van Krevelen and Nijenhuis 2009a, b), lower temperature should be comparable to higher strain rate, whereas higher temperature is equivalent to lower strain rate. Thus increasing strain rate or decreasing temperature would have similar effect on the mechanical properties of thermosets. In this review, strain rate and temperature effects on the yield strength (σ_Y), tensile strength (σ_T), and Young's modulus (E) of three types of thermosets are discussed: (a) epoxy, (b) vinyl ester, and (c) unsaturated polyester and some composites based on these materials.

12.2 TENSILE TEST

12.2.1 Test Specimen and Testing Parameters

The tensile test specimen has generally the shape of dumbbell as presented in ASTM D638 or ISO 527-2 standards. There are two common types of tension testing equipment: (a) hydraulic Instron testing machine and (b) spilt Hopkinson bar apparatus. At low-to-medium strain rate (up to 10^2 s^{-1}) (Shokrieh and Omidi 2011), hydraulic Instron testing machine is suitable, whereas spilt Hopkinson bar apparatus is designed for high strain rate (10^2–10^4 s^{-1}) (Wang and Sun 2008).

12.2.2 Mechanical Properties

As is well established, the tensile test results are recorded in the form of stress–strain (σ–ε) diagram (Askeland 1994). They can display the behavior of different types of mechanical characteristics of polymers as shown in Figure 12.1, where graphs 1, 2, 3, and 4 represent, respectively, rigid-brittle, rigid-strong, rigid-ductile, and soft-ductile behaviors (Elias 2003). Thermoset reins generally exhibit the rigid-strong behavior as shown in graph 1. For graphs 1 and 4, which have no yield point, yield strength (solid dot) is 0.2% offset yield

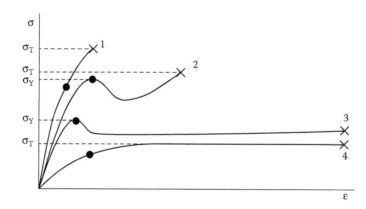

FIGURE 12.1 Schematic representation of stress–strain curves for different shapes with different mechanical properties. (Redrawn from Elias, H.-G., *An Introduction to Plastics*, Wiley-VCH, Weinheim, Germany, 2003.)

strength, whereas the yield strength is on the yield point such as graphs 2 and 3. Regarding tensile strength, it would be the stress at break (cross) or the yield strength (Elias 2003).

12.3 DATA FROM LITERATURE

12.3.1 Epoxy Resins (Neat and with Nanofillers)

12.3.1.1 Strain Rate Effect

Kody and Lesser (1997) performed tensile tests at low to moderate strain rate at 21°C and 40°C on a diglycidal ether of bis-phenol A (DGEBA) epoxy, cured with ethylene-di-amine (EDA) and methyl-ethylene-di-amine (MEDA)/N, N-dimethyl-ethylene-di-amine (DMEDA) with molecular weight M_c = 950 g mol^{-1} between cross-links. They found that at 21°C, as the strain rate increased from 3×10^{-3} to 4 s^{-1} the yield strength went up from 53 to 68 MPa. By contrast when testing was carried out under similar strain rates at 40°C, the yield strength increased from 44 to about 58 MPa. Zhou et al. (2007) observed that at constant room temperature (RT), both tensile strength and Young's modulus of neat epoxy and 1, 2, and 3 wt.% carbon nanofiber (CNF)-reinforced epoxy increased with increasing strain rate as shown in Figures 12.2 and 12.3, respectively.

As can be seen in Figure 12.2, the tensile strength of neat epoxy increased approximately by 6 Mpa (≈11%) from strain rate of 3.3×10^{-4} to 3.3×10^{-2} s^{-1}, whereas that of 2% CNF-reinforced epoxy increased by 12 MPa (≈21%) for the same increase in strain rate. In Figure 12.3 the Young's modulus of neat epoxy increased 0.47 GPa (≈20%) and that of 3% CNF-reinforced epoxy increased by 0.6 GPa (≈21%) under similar conditions. Gilat et al. (2002, 2007) observed that at RT under low strain rate (around 5×10^{-1} s^{-1}), epoxy resins are rather ductile (similar to graph 4 in Figure 12.1) while at medium (~1 s^{-1}) and high strain rate (360, 460, and 470 s^{-1}), specimens showed brittle behavior. However, at high

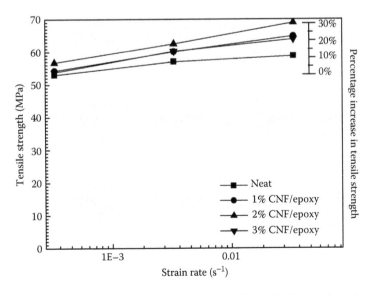

FIGURE 12.2 Tensile strength of neat and carbon nanofiber (CNF)-reinforced epoxy resins at strain rate of 0.00033, 0.0033, and 0.033 s^{-1}. (Data taken from Zhou, Y. X. et al., *Mater. Sci. Eng. a-Struct. Mater. Prop. Microstruct. Process.*, 465, 238, 2007.)

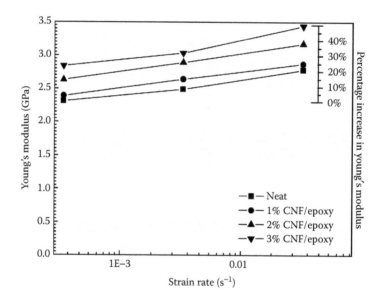

FIGURE 12.3 Young's modulus of neat and carbon nanofiber (CNF)-reinforced epoxy resins at strain rate of 0.00033, 0.0033, and 0.033 s^{-1}. (Data taken from Zhou, Y. X. et al., *Mater. Sci. Eng. a-Struct. Mater. Prop. Microstruct. Process.*, 465, 238, 2007.)

strain rate up to 470 s^{-1}, while Young's modulus continued to climb, there was no additional increase in tensile strength.

12.3.1.2 Temperature Effect

Kody and Lesser (1997) compared yield strength of DGEBA as evaluated at 40°C and 21°C at slow strain rate 3×10^{-3} s^{-1}. The strength was higher at 21°C by about 10 MPa ($\approx$21%), while it increased by 17% at 21°C at a faster strain rate of 4 s^{-1}. In addition, they observed that the specimens switched from ductile to brittle behavior with decreasing temperature. Zhao et al. (2010) performed tensile tests at a constant low strain rate, approximately 0.0033 s^{-1} on DGEBA epoxy resins cured by three different types of amine terminated polyesters (ATPE) curing agents, including α, ω-bis-amino-phenyl terminated poly (ethylene glycol) (BAPTPE), α, ω-bis-amino-methylene-phenyl terminated poly (ethylene glycol) (BAMPTPE), and α, ω-bis-amino-ethylene-phenyl terminated poly (ethylene glycol) (BAEPTPE). As can be seen in Table 12.1, tensile strength and Young's modulus of all the three types of ATPE-DGEBA increased significantly as the temperature was reduced from

TABLE 12.1   Tensile Strength and Young's Modulus of ATPE-DGEBA at Different Temperatures

ATPE	Tensile Strength (MPa)		Young's Modulus (MPa)	
	25°C	0°C	25°C	0°C
BAPTPE	57.55	64.36	408	471
BAMPTPE	15.59	50.02	163	413
BAEPTPE	14.32	48.14	131	307

Source: Zhao, L.Y. et al., *Chinese J. Polym. Sci.*, 28(6), 961, 2010.

25°C to 0°C, thereof the tensile strength rose by between 12% and up to 230%. Similarly, Kim and Mai (1991) observed that tensile strength of neat epoxy increased from around 19 MPa to above 70 MPa by decreasing temperature from 80°C to −50°C.

Yang et al. (2007) performed tensile tests on epoxy resins unmodified and modified by flexible diamine D-230 at RT and at −196°C at strain rates approximately 1.67×10^{-3} and $6.67 \times 10^{-4} \, s^{-1}$. As can be seen in Table 12.2, tensile strength of unmodified (0% D-230) epoxy resins increased by 30 MPa ($\approx$42%) and Young's modulus was elevated by 2.2 GPa ($\approx$75%) by lowering temperature from RT to −196°C. Similarly, modified (21% D230 and 49% D-230) epoxy resins increased around 20–30 MPa ($\approx$25%–37%) in tensile strength and 1.8–2.3 GPa ($\approx$57%–82%) in Young's modulus with lower temperature. Glass transition temperature of epoxy modified by 0%, 21%, and 49% are, respectively, 156.6°C, 110.9°C, and 99.9°C.

12.3.2 Vinyl Ester and Unsaturated Polyester

12.3.2.1 Strain Rate Effect

Plaseied and Fatemi (2008a, b) studied the mechanical properties of vinyl ester with a range of strain rates 1×10^{-4}, 1×10^{-3}, 0.01, 0.1, and 1 s^{-1} at 23°C. Their results, as shown in Figure 12.4, display an upward trend in both yield strength and Young's modulus with increased strain rate. The yield strength increased by 8%, while the Young's modulus increased by 12.5%.

12.3.2.2 Temperature Effect

Plaseied and Fatemi (2008a, b) studied the mechanical properties of vinyl ester over a range of temperature −35°C to 100°C at strain rate of 0.0001, 0.001, and 0.1 s^{-1}. They observed that the tensile strength and Young's modulus of vinyl ester, respectively, increased by 297% and 95% with decreasing temperature at strain rate of 0.0001 s^{-1}. While at strain rate of 0.001 s^{-1}, tensile strength and Young's modulus of vinyl ester, respectively, increased by 186% and 82% as temperature decreased from 100 to −35°C. They also pointed out that mechanical properties have a linear correlation to temperature, especially tensile strength and Young's modulus. As shown in Figures 12.5 through 12.7, the coefficient of linear regression, R^2, of each linear fitting is larger than 0.9, even up to 0.995, which means the linear relations between tensile strength and temperature, Young's modulus and temperature are reasonable if not good.

Escaig (1997) conducted tensile test on unsaturated polyester at temperature varying from 20°C to 80°C at strain rate of $3 \times 10^{-5} \, s^{-1}$, and they observed that the yield strength reduced

TABLE 12.2 Tensile Strength and Young's Modulus of Unmodified and Modified Epoxy Resins at −196°C and RT

D-230 (wt.%)	Tensile Strength (MPa)		Young's Modulus (GPa)	
	RT	−196°C	RT	−196°C
0	73.16 ± 3.16	103.55 ± 9.32	2.85 ± 0.03	5.00 ± 0.07
21	85.44 ± 2.76	106.44 ± 9.10	3.22 ± 0.21	5.06 ± 0.12
49	78.70 ± 1.90	107.60 ± 5.44	2.80 ± 0.17	5.11 ± 0.18

Source: Yang, G. et al., *Polymer*, 48(1), 302, 2007.

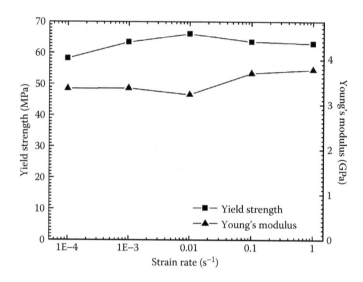

FIGURE 12.4 Yield strength and Young's modulus of vinyl ester at varying strain rates at 23°C. (Data taken from Plaseied and Fatemi, *Journal of Materials Science*, 43: 1191–1199, 2008a; Plaseied and Fatemi, *International Journal of Polymeric Materials*, 57: 463–479, 2008b.)

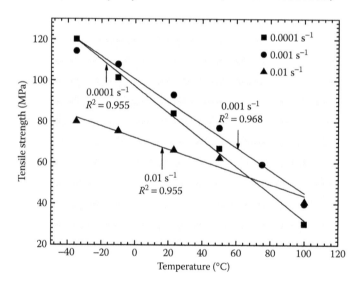

FIGURE 12.5 Linear correlation of tensile strength of vinyl ester to temperature at 0.0001, 0.001, and 0.01 s^{-1}. (Data taken from Plaseied and Fatemi, *Journal of Materials Science*, 43: 1191–1199, 2008a; Plaseied and Fatemi, *International Journal of Polymeric Materials*, 57: 463–479, 2008b.)

from 74 to 3.4 MPa ($\approx$95%). Banko et al. (1994) performed tensile tests on neat unsaturated polyester and 25 wt.% fly ash-filled composites at various temperatures at strain rate of approximately 0.002 s^{-1}. They observed that as temperature decreased from 50°C to 0°C, the Young's modulus of neat and 25 wt.% fly ash-filled unsaturated polyester rose, respectively, from 50 to 800 MPa ($\approx$1500%) and from 100 to 1250 MPa ($\approx$1150%), whereas the tensile strength increased, respectively, from 10 to 23 MPa ($\approx$130%) and from 5 to 14 MPa ($\approx$180%). However, the effect on strain to failure (measure of overall ductility) was not provided.

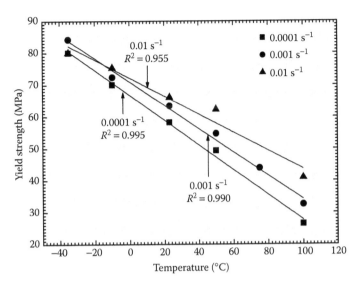

FIGURE 12.6   Linear correlation of yield strength of vinyl ester to temperature at 0.0001, 0.001, and 0.01 s⁻¹. (Data taken from Plaseied and Fatemi, *Journal of Materials Science*, 43: 1191–1199, 2008a; Plaseied and Fatemi, *International Journal of Polymeric Materials*, 57: 463–479, 2008b.)

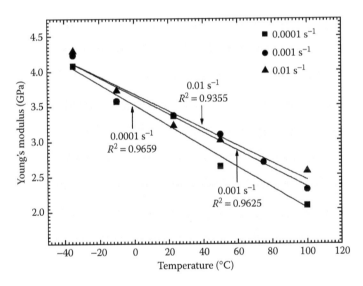

FIGURE 12.7   Linear correlation of Young's modulus of vinyl ester to temperature at 0.0001, 0.001, and 0.01 s⁻¹. (Data taken from Plaseied and Fatemi, *Journal of Materials Science*, 43: 1191–1199, 2008a; Plaseied and Fatemi, *International Journal of Polymeric Materials*, 57: 463–479, 2008b.)

12.3.3 Composites Using Fibers

12.3.3.1 Strain Rate Effect

Shokrieh and Omidi (2009) performed tensile tests on glass fiber-reinforced epoxy using 1-mm-thick laminate at low to intermediate strain rate, respectively, 1.7×10^{-3}, 0.55, 5.6, 46, and 85 s⁻¹, at RT. They found that the tensile strength increased from 783 to 1186 MPa (≈52%), while Young's modulus rose from 37 to 41 GPa (≈12%), as shown in Figure 12.8.

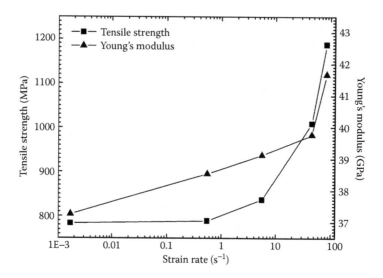

FIGURE 12.8 Tensile strength and Young's modulus of glass fiber-reinforced epoxy at different strain rates at room temperature. (Data taken from Shokrieh, M.M. and Omidi, M.J., *Compos. Struct.*, 88, 595, 2009.)

Okoli and Smith (2000) also achieved similar results from glass fiber-reinforced epoxy laminate composites (Table 12.3). It is clearly seen in Table 12.3 that tensile strength increased from 375 to 444 MPa ($\approx$18%) as strain rate rose higher from 0.0363 to 2.72 s^{-1}, while Young's modulus rose from 27 to 38 GPa ($\approx$41%). Al-Zubaidy et al. (2011) performed tensile test on carbon fiber-reinforced epoxy at strain rate varying from 2.42×10^{-4} up to 87.4 s^{-1} at RT and the tensile strength increased nearly 50%, while the Young's modulus increased about 25%.

12.3.3.2 Temperature Effect

Miwa and Endo (1994) observed the tensile strength of 9.9% carbon fiber-filled epoxy composites increased from around 8 to 112 MPa ($\approx$1300%) with a drop of temperature from 140°C to 20°C at a constant strain rate of 5×10^{-4} s^{-1}. Clearly this has some relationship to the ductile-brittle glass transition of the epoxy matrix. Cantwell et al. (1990) conducted tensile tests on epoxy composites filled by 42% volume fraction silica particles at 23°C, 50°C, 85°C, and 105°C at a range of low strain rate, respectively, 3.3×10^{-5}, 3.3×10^{-4}, 3.3×10^{-3}, and 0.0167 s^{-1}. They reported that the tensile strength increased more

TABLE 12.3 Tensile Strength and Young's Modulus of Glass Fiber-Reinforced Epoxy Composites at Varying Strain Rates

Strain Rate (s^{-1})	Tensile Strength (MPa)	Young's Modulus (GPa)
0.0363	375	27.2
0.281	403	28.0
1.52	412	35.6
2.72	444	38.4

Source: Okoli, O.I. and Smith, G.F., *Compos. Struct.*, 48(1–3), 157, 2000.

significantly as the temperature fell at lower strain rate, thereof an increase up to 1.3 times occurred at the lowest test strain rate of 3.3×10^{-5} s^{-1}. Similarly, Wang et al. (2002) found that Young's modulus of epoxy composites filled by 0%, 14%, 21%, 28%, 33%, and 39% silica, respectively, increased by 89%, 68%, 47%, 103%, 56%, and 79% as temperature decreased from 115°C, 100°C, 75°C, and 50°C to RT at the strain rate of 3.125×10^{-3} s^{-1}. The 0.2% offset yield strength increased by 148%, 87.5%, 117%, 1122%, 280%, and 247% with reduced temperature.

12.4 THEORETICAL ANALYSIS OF THE ABOVE

According to Ward (1971), strain rate (strain with time) represents the rate of deformation (Van Krevelen and Nijenhuis 2009), which means how quickly the fracture occurs during tension. Although the polymer molecules of thermosets are not as flexible as thermoplastics due to the thermoset's rigid curing network structure, they still have random coil behavior (Askeland 1994; Elias 2003) as other types of polymers.

If an external force such as tension is applied, polymer molecules will be disentangled along the tension direction. At lower strain rate, polymer molecules have sufficient time to disentangle and relax exerted stress (Plaseied and Fatemi 2008). This results in delayed fracture due to higher flow. At higher strain rate, the facture appears earlier because exerted stress has no time to relax and induces the polymer chains behave in a brittle manner.

Thus, extensional viscoelasticity diminishes as strain rate grows. If the growth of strain rate is more pronounced than reduced movement of molecular chains, tensile stress increases significantly. Young's modulus—which is used to evaluate stiffness of materials (Askeland 1994)—becomes higher as materials get *stiffer*. Polymers switch from ductility to brittleness by increasing strain rate (Gong and Bandyopadhyay 2007). Young's modulus therefore increases with increasing strain rate. Gilat et al. (2002, 2007) found that once strain rate reaches a very high level up to 360, 460, and 470 s^{-1}, tensile strength of epoxy resins tends to level off but Young's modulus still increases. In view of the present authors, perhaps, as strain rate increases to a certain high values, the decrease in viscoelasticity cancels the increasing strain rate.

The other parameter altering extensional molecular flow behavior (viscoelasticity) is temperature. Referring to the free volume theory (Ward 1971), viscoelasticity would preferably increase with increasing free volume. Free volume is a descriptive term for voids or holes where/through which polymer molecules or segments can move. Free volume is larger at higher temperature (particularly above glass transition) for the reason that (a) thermal expansion results in weaker van der Waals force, which holds polymer molecules together and the mobility of polymer molecules is then higher (George et al. 1999; Plaseied and Fatemi 2008), and (b) the thermal expansion of free volume with temperature results in enhanced free volume. Conversely, decreasing temperature contributes to lower possibility of chain flow or sliding. At a constant strain rate, tensile stress is therefore improved by reducing temperature. Regarding the strain rate effect on extensional viscoelasticity, the mechanical properties of thermosets should be most outstanding at low strain rate and low temperature theoretically.

12.4.1 Proposed Mathematical Equation Model from the Above Literature

The contributors of this chapter have come out with the following equations that can cover the results on strain rate and temperature effects in the thermoset resins and their composites using the published experimental data as used in this paper for strain rate

$$y = 8.55x + 34.9 \tag{12.1}$$

where:
  y is the percentage increase in strength or modulus
  x is the log of strain rate

For temperature

$$y = -0.78\exp(x/19.4) + 294 \tag{12.2}$$

where:
  y is the percentage increase in strength or modulus
  x is the temperature in °C

As can be seen in Figure 12.9, percentage increase in strength (upper figure), and percentage increase in modulus (the lower figure) show the similar trend lines as an increase in log of strain rate. There are two distinctly different trends (a) linear growth with

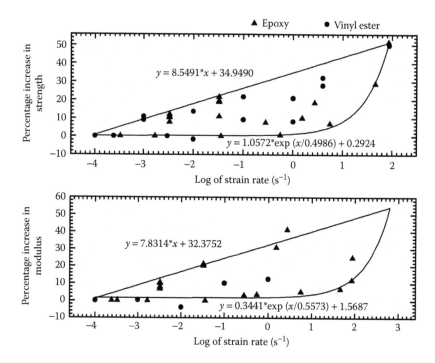

FIGURE 12.9   Strain rate effect on the mechanical properties of thermoset resins and their composites. (Data from Ward 1971; Van Krevelen and Nijenhuis 2009; Askeland 1994; Elias 2003; Plaseied and Fatemi 2008a, b; Gong and Bandyopadhyay 2007; Gilat et al. 2002, 2007; George et al. 1999.)

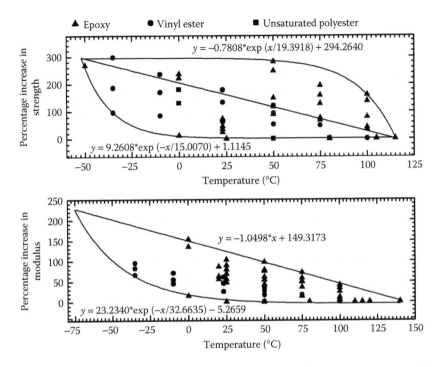

FIGURE 12.10 Temperature effect on the mechanical properties of thermoset resins and their composites. (Data from Ward 1971; Van Krevelen and Nijenhuis 2009; Askeland 1994; Elias 2003; Plaseied and Fatemi 2008a, b; Gong and Bandyopadhyay 2007; Gilat et al. 2002, 2007; George et al. 1999.)

equation $y = 8.55x + 34.9$ and (b) initial slow growth followed by a later sharp increase given by $y = 1.06\exp(x/0.50) + 0.29$.

For percentage increase in modulus, the upper limit equation is $y = 7.83x + 32.4$, while the lower limit equation is $y = 0.34\exp(x/0.56) + 1.57$.

Figure 12.10 shows the percentage increase in mechanical properties as temperature decreases. The lower limits for both percentage increase in strength and percentage increase in modulus are slow then exponential growth, but the upper limits are different. For percentage increase in strength, the upper limit equation is $y = -0.78\exp(x/19.4) + 294$, while the lower limit equation is $y = 9.26\exp(-x/15.0) + 1.11$. For percentage increase in modulus, the upper limit equation is $y = -0.05x + 149$, while the lower limit equation is $y = 23.2\exp(-x/32.7) - 5.27$.

12.5 CONCLUSIONS

1. The mechanical properties E, σ_T, and σ_Y of thermoset resins and composites increase with increasing strain rate or decreasing temperature.

2. According to the data from the literature, tensile test results compare well to theory at low to intermediate strain rate or at low to high temperature.

3. Some studies indicate that tensile strength of epoxy thermosets at high strain rate is likely to remain at the same level as that at medium rate.

4. These effects in the matrix resins can be translated into composites in similar or more complex manner.

5. The contributors of this chapters have transformed the numerous published data in form of simple equations covering the linear variations as well as the exponential changes as a function of log strain rate and temperature. Further work will continue.

ACKNOWLEDGMENT

The work was presented at ACUN-6 International Conference on *Composites and Nanocomposites in Civil, Offshore and Mining Infrastructure*, Monash University, Melbourne, Australia, November 14–16, 2012, and published in the ACUN-6 proceedings.

REFERENCES

Al-Zubaidy, H., X. L. Zhao et al., 2011. Mechanical behaviour of normal modulus carbon fibre reinforced polymer (CFRP) and epoxy under impact tensile loads. In *11th International Conference on the Mechanical Behavior of Materials*. Milano, Italy, June 5–9.

Askeland, D. R., 1994. *The Science and Engineering of Materials*. Boston, MA, PWS Publishing.

Banko, A. S., K. S. Rebeiz et al., 1994. Effect of fillers on the temperature relaxation behavior of unsaturated polyester based on recycled poly(ethylene-terephthalate). *Journal of Materials Science Letters*, 13(13): 934–936.

Cantwell, W. J., J. W. Smith et al., 1990. Examination of the processes of deformation and fracture in a silica-filled epoxy-resin. *Journal of Materials Science*, 25(1B): 633–648.

Coates, P. D. and I. M. Ward, 1978. The plastic deformation behaviour of linear polyethylene and polyoxymethylene. *Journal of Materials Science*, 13(9): 1957–1970.

Dasari, A. and R. D. K. Misra, 2003. On the strain rate sensitivity of high density polyethylene and polypropylenes. *Materials Science and Engineering: A—Structural Materials Properties Microstructure and Processing*, 358(1–2): 356–371.

Elias, H.-G., 2003. *An Introduction to Plastics*. Weinheim, Germany, Wiley-VCH.

Escaig, B., 1997. A physical model of the pressure dependence and biaxial mechanical properties of solid polymers. *Polymer Engineering and Science*, 37(10): 1641–1654.

George, J., S. Thomas et al., 1999. Effect of strain rate and temperature on the tensile failure of pineapple fibre reinforced polyethylene composites. *Journal of Thermoplastic Composite Materials*, 12(6): 443–464.

Gilat, A., R. K. Goldberg et al., 2002. Experimental study of strain-rate-dependent behavior of carbon/epoxy composite. *Composites Science and Technology*, 62(10–11): 1469–1476.

Gilat, A., R. K. Goldberg et al., 2007. Strain rate sensitivity of epoxy resin in tensile and shear loading. *Journal of Aerospace Engineering*, 20(2): 75–89.

Gong, S. Y. and S. Bandyopadhyay, 2007. Mechanical properties and fracture surface morphologies in unnotched specimens of rubber-PMMA composites. *Journal of Materials Engineering and Performance*, 16(5): 601–606.

Kim, J. K. and Y. W. Mai, 1991. Effects of interfacial coating and temperature on the fracture behaviors of unidirectional kevlar and carbon-fibre reinforced epoxy-resin composites. *Journal of Materials Science*, 26(17): 4702–4720.

Kody, R. S. and A. J. Lesser, 1997. Deformation and yield of epoxy networks in constrained states of stress. *Journal of Materials Science*, 32(21): 5637–5643.

Mae, H., 2008. Effects of local strain rate and micro-porous morphology on tensile mechanical properties in PP/EPR blend syntactic foams. *Materials Science and Engineering: A—Structural Materials Properties Microstructure and Processing*, 496(1–2): 455–463.

Miwa, M. and I. Endo, 1994. Critical fibre length and tensile-strength for carbon-fibre epoxy composites. *Journal of Materials Science*, 29(5): 1174–1178.

Okoli, O. I. and G. F. Smith, 2000. The effect of strain rate and fibre content on the Poisson's ratio of glass/epoxy composites. *Composite Structures*, 48(1–3): 157–161.

Plaseied, A. and A. Fatemi, 2008a. Deformation response and constitutive modeling of vinyl ester polymer including strain rate and temperature effects. *Journal of Materials Science*, 43(4): 1191–1199.

Plaseied, A. and A. Fatemi, 2008b. Strain rate and temperature effects on tensile properties and their representation in deformation modeling of vinyl ester polymer. *International Journal of Polymeric Materials*, 57(5): 463–479.

Shokrieh, M. M. and M. J. Omidi, 2009. Tension behavior of unidirectional glass/epoxy composites under different strain rates. *Composite Structures*, 88(4): 595–601.

Shokrieh, M. M. and M. J. Omidi, 2011. Investigating the transverse behavior of glass-epoxy composites under intermediate strain rates. *Composite Structures*, 93(2): 690–696.

Van Krevelen, D. K. and K. T. Nijenhuis, 2009. *Properties of Polymers: Their correlation with chemical structure*, Their Numerical Estimation and Prediction from Additive Group Contributions. 4th ed., Elsevier, Amsterdam, pp. 383–500, DOI:10.1016/B978-0-08-054819-7.00013-3.

Wang, H. Y., Y. L. Bai et al., 2002. Combined effects of silica filler and its interface in epoxy resin. *Acta Materialia*, 50(17): 4369–4377.

Wang, L. L. and Z. J. Sun, 2008. Studies on rate dependent damage evolution in polymer blends in high strain rates. *Plastics Rubber and Composites*, 37(5–6): 246–250.

Ward, I. M., 1971. *Mechanical Properties of Solids Polymers*. Bristol, England, John Wiley & Sons.

Yang, G., S. Y. Fu et al., 2007. Preparation and mechanical properties of modified epoxy resins with flexible diamines. *Polymer*, 48(1): 302–310.

Zhao, L. Y., J. G. Guan et al., 2010. Mechanical properties and curing kinetics of epoxy resins cured by various amino-terminated polyethers. *Chinese Journal of Polymer Science*, 28(6): 961–969.

Zhou, Y. X., S. R. Akanda et al., 2007. Nonlinear constitutive equation for vapor-grown carbon nanofibre-reinforced SC-15 epoxy at different strain rate. *Materials Science and Engineering: A—Structural Materials Properties Microstructure and Processing*, 465(1–2): 238–246.

IV

Specialty Polymers

CHAPTER 13

Recent Advances in Shape Memory Polymers

Shan Faiz, Mohammad Luqman, Arfat Anis, and Saeed M. Al-Zahrani

CONTENTS

13.1 INTRODUCTION

Shape memory materials (SMMs) belong to the family of smart materials, which is attributed to their ability to memorize the permanent shape after being deformed and fixed into a temporary one, and consequently, showing recovery by the application of an external stimulus. This phenomenon is termed *shape memory effect* (*SME*). A simple illustration is viewed in Figure 13.1.

The SME was observed first in the gold-cadmium (Au-Cd) alloy in 1951 [1]. It gained the commercial significance by its discovery in nickel-titanium (NiTi) alloy (NITINOL) [2]. A number of alloys were then studied and were deployed in various applications, including biomedical (mainly surgery, guide wire, and stents) systems, micro- and nano-electromechanical systems [3], thermal actuators, aerospace (antennas, shrouds, and solar array) [4,5], and automobiles [6]. The property associated with a large nonlinear

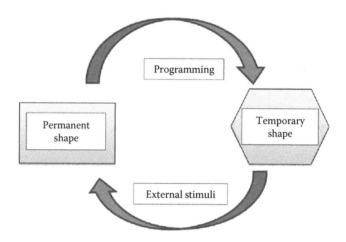

FIGURE 13.1   Schematic view of shape memory effect.

recoverable strain upon loading and unloading is termed *superelasticity* [7], which is mainly responsible for the SME.

Shape memory alloys (SMAs) have been serving over the past four decades, but there exist shortcomings in many of its properties: recovery volume, density, and high stiffness, to name a few (Table 13.1). This situation geared up the motivation for the alternative material which is none other than the shape memory polymers (SMPs).

The viscoelastic property is the key factor that enables the polymer network to recuperate the primary shape on the application of variety of physical and chemical stimuli. The maximum temperature of a polymeric material is responsible for the permanent shape. This temperature is often termed *highest thermal transition temperature* (T_{perm}). The region above this temperature is called *processing zone*, wherein a polymer can be melted and worked out by several processing techniques such as extrusion, molding, drawing, and calendaring. The other zone related to the temporary shape fixing is named *forming zone*. The modulation temperature is either glass transition (T_g) or melting temperature (T_m). The polymer can undergo through a programming method between the modulation temperature and T_{perm} to acquire a temporary shape, which is fixed by cooling below this zone. The shape can be retrieved by heating above the modulation temperature [10,20]. The phenomenon is illustrated in Figure 13.2.

It is essential to understand the variation between shape change effect and SME. In the former, a very high energy barrier between the two states exists. However, in the

TABLE 13.1   Characteristics of SMP Compared with SMA

Property	Shape Memory Alloys	Shape Memory Polymers
Recovery volume	6%–7%	400%–500%
Density	6.5 gm/cm³	1 gm/cm³
Workability	Good (thermoplastic)	Poor
Cost	$1000/kg	$50/kg

Source: Ishizawa, J., Research on application of shape memory polymers to space inflatable systems, Paper presented in the *7th International Symposium on Artificial Intelligence, Robotics and Automation in Space, SAIRAS 2003*, Nara, Japan, 2003.

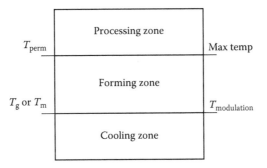

FIGURE 13.2 Different zones for a shape memory polymer.

latter, the elastic deformation of a material is the responsible characteristic, wherein stress is proportional to strain [9].

Several important reviews [10–16,20] have rendered the understanding of fundamental concepts, mechanisms, triggering stimuli, system of polymers being considered for SME and applications. We have covered all these areas to summarize the work done until now.

13.2 MECHANISMS

13.2.1 Shape Memory Cycle

Shape memory cycle (SMC) shows the correlation between the temperatures, stress, and strain. One complete cycle describes the thermomechanical procedure, that is, applying a programming technique to the sample to get it deformed into a temporary shape, and then redeeming its inaugural shape. The illustration of an SMC is shown in Figure 13.3.

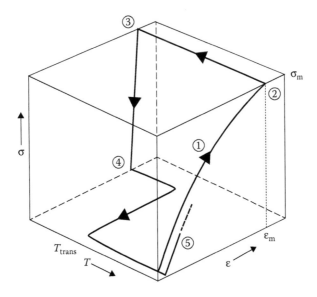

FIGURE 13.3 Schematic representation of shape memory cycle. (Reproduced from Lendlein, A. and Kelch, S., *Angew. Chem. Int. Ed.*, 41, 2034, 2002.)

The sample is heated at a temperature T (above T_g), inducing a maximum strain ε_m and stress σ_m. It must be kept in mind that the applied temperature must not exceed the highest transition temperature (T_{perm}), otherwise the polymer will melt. The sample is then cooled down below T_g keeping the stress constant. The load is then reduced to stress-free point and the sample is reheated up to T_{high} maintaining zero stress. Consequently, the inaugural shape is recovered. The cycle then begins again [12].

13.2.2 Mechanism of SMP

The polymer chains are at rest below the glass transition temperature. Their activation factor is the temperature that brought in a change from the glass transition phase to a rubbery elastic one. The thermodynamics studies reveal that all possible conformations of a polymer chain contain the same energy potential. Hence, the ascent of temperature will make the polymer chain to reassign into entropy-favored spherical conformation. For a precise time interval, there will be no movement or disentanglement of the chains, and the polymer returns to its original shape when the stress is released. However, a prolonged exposure to the external stress makes the polymer to relax and results in the disentanglement, which tends the polymer to result in plastic deformation [10,17].

On the molecular level, the physical net points and switch segments play vital role in the SME, which are irritable to an external stimuli. The net point that influences the permanent shape of SMP can be assembled by physical or chemical cross-linking, interlocking, or by interpenetrating segments [18]. Switch, on the other hand, provides shape fixity and shape retrievability control with the restoration of entropy upon external stimuli. The switches discovered by now include crystallization, glass transition, supramolecular hydrogen bonding, light-reduced reversible network, and percolating network (nanocomposites). The architecture of switch and net points segments are illustrated in Figure 13.4.

The literature review has revealed that the thermoresponsive SMPs show three basic mechanisms. The polymer is heated above its T_g, which takes the polymer chains into rubbery state, that is, chains set foot in motion. It is deformed by a programming technique and is cooled back below T_g to maintain the interim shape as the molecular motion is stopped. The reheating above T_g will revive this motion and polymer returns to its original shape. Polymers such as polymethyl methacrylate and silicone show such mechanism.

The other mechanism is observed in linear block copolymers such as polyurethane (PU) [21–30] and ethyl vinyl acetate [31] consisting of two segments. One is elastic with highest thermal transition T_{perm} acting as the physical cross-link and maintains the permanent shape. Above this temperature, the polymer melts and is processed by extrusion, drawing, and so on. The second part enables the fixation of temporary shape whose transition temperature is either T_g or T_m (melting temperature). The material is formed by heating above switching temperature but below T_{perm}, and temporary shape is fixed by cooling below T_g. Heating the material again above T_g splinters the cross-links into switching phase, and material returns to inaugural shape. The study has been advanced in recent years with the research on polymer blends of PU [32–38].

In the case of paraffin wax, the partial heating softens one component (inclusion), while the other remains as solid part. The solid part behaves as elastic component, while the

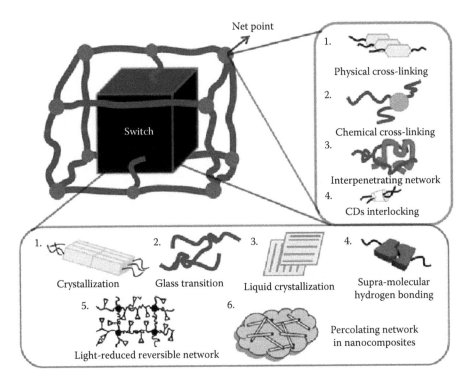

FIGURE 13.4 Overall architecture of SMP. (Reproduced from Hu, J. et al., *Prog. Polym. Sci.*, 37, 1720, 2012.)

melted part as transition component. The polymer upon cooling remains in deformed state because the inclusions are tough now. The reheating of polymer to transition temperature enables the elastic recovery of it. Huang [9] distinguishes these mechanisms as dual-state mechanism, dual component mechanism, and partial transition mechanism.

13.2.3 Quantitative Analysis

The strain recovery rate, R_r, represents the potential of material to memorize its permanent shape, and the strain fixity rate, R_f, is used to measure the ability of switching segments to fix the mechanical deformations. These are the two fundamental elements of quantitative analysis of SMPs, which are evaluated by the following equations [10].

$$R_r(N) = \frac{\varepsilon_m - \varepsilon_p(N)}{\varepsilon_m - \varepsilon_p(N-1)} \tag{13.1}$$

$$R_r(N) = \frac{\varepsilon_u(N)}{\varepsilon_m} \tag{13.2}$$

where:

ε_u, ε_p, and ε_m represent the fixed strain after unloading, the permanent strain after heat-induced recovery, and the temporal strain achieved by deformation, respectively

N represents number of cycles

The advancements in simulation techniques have enhanced the development of thermomechanical models, which has enabled the complex behavioral study of SMPs into more directional one. The contribution in this field [39–57] includes the uni- and multidimensional modeling, calibration, numerical implementations, optimization, and finite deformation response of SMPs.

Yipping [58] developed a 3D small strain internal state variable constitutive model. The model simplifies the quantitative study of storage and release of entropic deformation of a polymer during thermomechanical process. It helped to simplify pseudo-plastic and viscoelastic problems (mainly temperature and time effect) into a simple elastic one.

Baghani [60] has recently developed a 3D model for quantitative analysis in small strain regime. The strain is decomposed in six parts using first-order mixture rule. Moreover, the two boundary value problem is solved using finite element software. Two main constraints (1) time-discrete form of evolution equations and (2) integration scheme of tangent matrix in finite element modeling are also discussed. The model is claimed to be appropriate for SMP optimization.

13.2.4 SMP Systems

The SME is being studied for both a number of individual polymers and polymer systems inducing contrastive techniques (cross-linking, blending, etc.) [4]. The most common systems include segmented PU, segmented PU ionomers, cross-linked PE, cross-linked polycyclooctene, epoxy-based polymers, *trans*-polyisoprene, cross-linked ethylene vinyl-acetate copolymer, acrylate-based polymers, PE/nylon6 graft copolymers, styrene-based polymers, polynorbornene, and thioene-based polymers. Supramolecular SMP system is based on a partial cyclodextrin polyethylene glycol (PEG) inclusion complex.

13.2.5 Triggering Mechanisms

In the past few years, researchers have greatly focused for alternative triggering mechanisms to get benefitted with the increasing stimuli responsive polymer family. The most important ones include the irradiation (infrared, solar, or ultraviolet), electric current, and magnetic field. Small et al. employed the infrared activation on shape-memory PU (SMPU) doped with radiation absorbing dye (epolight 4121) [60–62]. A prototype device was formed consisting of SMP stent and SMP foam [61], which is now used for treating nonnecked fusiform aneurysms. They made a laser-activated SMP microactuator for the treatment of blood clot which gets deposited in arteries causing ischemic stroke. The devices are claimed to be compatible for *in vivo* deployment [63]. Yu and Ikeda [64] reviewed the photo-triggered materials, which include polymer gels, mono layers, solid films, and liquid crystalline elastomers (LCEs). They are termed *photo-deformable materials*.

Electric current as triggering mechanism has also been verified by the experiments conducted by Sahoo et al. [64–65]. Polypyrrole was used as a coating for SMPU, and the recovery of polymer was observed in approximately 25 s at 40 V.

Another alternative stimulus under research is based on water/solvent. These SMPs often come under the category of chemo-responsive SMPs. Research has been done on ether [67] and pyridine [68] based SMPUs with a conclusion that the response from this stimulus can be enhanced by using higher molecular weights of PEG. Chen [69] modified chitosan with PEG and cross-linked it with an epoxy material which was formulated into a hydrophilic biodegradable stent. A significant progress was made in polyethylene oxide (PEO)-based SMPUs [70,71].

The production of magnetically active SMPs (elastomers and gels) have gained significant attention in this expanding field of interest in recent years. The idea was implemented by the induction of magnetically responsive fillers such as Fe_3O_4 [72], Fe_2O_3/silica [73], and terfenol-D particles [74] in the SMP matrix. Recently, Chung [75] studied the behavior of magnetic polymer (composite of SMPU with Fe_3O_4). He found that the best procedure for preparing this composite is the melt-mixing compared with solvent blending and *in situ* reaction. The shape recovery maxima, up to 99%, were observed in repetitive cycles.

13.3 APPLICATIONS OF SMPs

Jung and Cho [76] prepared SMPU with poly cyclohexylmethacrylate (PCL) to form orthodontic wire, which is aesthetically more attractive than that with the commonly used metals and ceramics. They stated the valuable properties of SMPU for orthodontics application: transparency, low density, effective shape recovery (about 90%), and above all satisfactory aesthetic come out. They concluded that apart from usual elastomers, SMP wire can provide support to tooth for a longer time. The various applications of SMPs are summarized in Figure 13.5.

Ortega et al. [77] worked on the eradication of hemodynamic stress caused by needle flow in arteries. He used a deployable SMP to prepare tube-like adapter, which can be weaved through dialysis needle. The shape is triggered by the blood temperature. In this application, vascular wall shear stress is reduced, which is the responsible factor for intimal hypertension. The contributions in this field are incomplete without the discussion of prototype devices modeled by Small et al. [60,61]. SMP-based endovascular thrombectomy

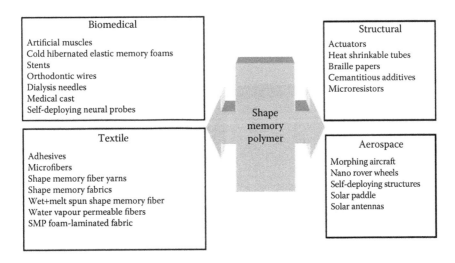

FIGURE 13.5　Application of SMPs in various fields.

device [61] was developed to bring operational simplification in the removal of thrombus (blood clot) in the brain. Another device [60] replaced the use of plasminogen activator (rt-PA) for curing ischemic stroke. The electromechanical device is triggered by laser heating, which captures thrombus and restores the flow of blood.

These days, the traditional invasive surgical techniques have been replaced by novel endovascular treatments such as aneurysm embolization with balloon-assisted coils, flow diversion devices, open- and closed-cell stents, and embolic materials. Most of these strategies depend on the total occlusion of the aneurysm by the formation of a dense interconnected thrombus matrix and scar tissue in the neck of the aneurysm [62]. The deployment of SMP coil for aneurysm occlusion is illustrated in Figure 13.6.

The application of SMP stent-foam device applied to *in vitro* model is another step ahead in medical sciences. This device obstructs the aneurysm and keeps lumen open in parent artery [60]. The use of SMP coil has also been reported [78] for the treatment of intracranial aneurysm. SMP neuronal probes are reported to be biocompatible with the brain tissues far better than the metallic and ceramic ones [63].

Rousseau et al. [79] researched on methods for using SMP as a medical cast and found it advantageous for this application owing to the low cost, easy processability, light weight, and recyclability.

Surgical stapler has become a widely used tool to cover surgical suture, particularly in minimally invasive surgery. Biodegradable staples are preferred since they disappear after few months and cancel the need for a second operation. However, unlike sutures, which are able to be tightened to close the wound, staples are lacking in this tightening function. Hence, as normally suggested, wound should be closed using forceps before firing a surgical stapler. After programming a commercial PLA-based staple, the staple has the ability to close the wound automatically by means of the thermally induced SME. The force generated by SME in polymers is compatible to human tissue, and also adjustable. The problem of overstressing, which may prevent good blood circulation and cause damage to the tissue, can be avoided. In addition, at present, staples of different sizes are required for tissue with different thicknesses which can be reduced to one by SME [9]. The images of a shape memory stent are shown in Figure 13.7 when it was deployed to a rabbit.

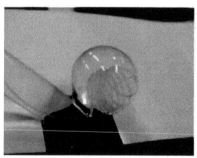

FIGURE 13.6 Deployment of two SMP coils for aneurysm occlusion under simulated flow conditions. (Reproduced from Hampikian, J.M. et al., *J. Mater. Sci. Eng. C*, 26, 1373, 2006. With permission.)

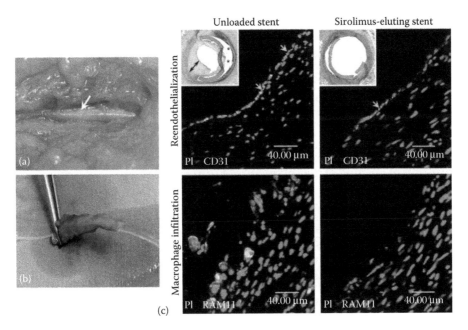

FIGURE 13.7 Images of a test stent after deployment in the rabbit infrarenal abdominal aorta (a) and patency test of the stent-implanted vessel at 4 weeks (b). Confocal fluorescence images show re-endothelialization (c, top) and macrophage infiltration after stent implantation (c, bottom). (Reproduced from Chen, M.C. et al., *Biomaterials*, 30, 79, 2009. With permission.)

Cold hibernated elastic memory of SMP has also been highlighted in scientific research in recent years. SMP foams due to their compressibility and high shape retrieve ratio have been implemented in hearing aids, micro tags, and microfoldable vehicles [81].

Noteworthy reviews in the textile and apparel field have revealed a revolutionary era of research especially for manufacturing of *smart fabrics* [13,19,82]. The area of interest gains attention by the induction of SMPs, which essentially supports the cause. Zhou et al. [83] prepared PU nanofibers by the electrospinning method using N, N-dimethylformamide as solvent. They concluded that concentration of the solution played the major role in the transformation of solution into fibers apart from other important parameters (voltage and feeding rate), which were set as 10–15 kV applied voltages, and 0.06–0.08 mm/min feeding rates.

Synthesized SMPU block copolymers were used to prepare electrospun nonwovens via electrospinning to study the mechanical properties. The electrospun SMPU nonwovens were prepared with hard-segment concentrations of 40 and 50 wt.% using N, N-dimethylformamide and tetrahydrofuran as solvent. An increase in the tensile strength was witnessed and the nonwoven were found to have a shape recovery >80% [84].

A similar approach was applied to study the thermal properties of SMPU nanofiber nonwoven. It was observed that the nonwoven properties are influenced by spinning and the recrystallization condition. It was concluded from temperature-dependent strain recovery curves that SMPU nonwoven showed lower recovery temperature than SMPU bulk film. The reason behind this behavior was the ultrafine diameter that made this happen [85]. Studies also produced tremendous results on nanofibers, micro- and nanofiber nonwovens [86–90].

Moreover, SMPs have been induced to manufacture laminations, [91,92] weavings [93], and coatings [94,95]. The researchers are now focusing on a separate field namely *shape memory fibers*. SMPUs, PEG-based copolymers with cotton, and LCEs are being used for primary shape memory fibers to ensure self-healing fabrics, enhancing washability, improving weavability of yarns, crease recoverability, and waterproofing. In a recent review, authors discussed the auto sleeve shortening shirt that was developed by CorpoNove which works on the principle of opening of microstructure of shirt with the change of temperature in the surroundings [13,84].

SMPs are capable of releasing their energy (entropy) in thermomechanical cycle when it is triggered by an external stimulus. Hence, their use as an actuator is very appropriate. LCEs with azobenzene [96,97] and laminated composites [98] are reported in this category. Commercial polymers such as PVC, FEP, and PTFE have been subjected to SME for the manufacturing of heat shrinkable tubings. These tubings are mainly used as insulator for cables [99]. PETG (glycol-modified polyethylene terephthalate) is a polyester and possesses high stiffness, hardness, and toughness as well as good impact strength. PETG is a commonly used material for transparency. Instead of using PU, PETG sheet, which is available off the shelf in almost any stationery shops, can be used as refreshable Braille paper for people with vision problem to easily remove any typographical errors by means of thermally induced shape recovery and retype just like writing with a pencil and eraser [100].

The implementation of SMP to create surface wrinkling is another interesting field of research. Wrinkles bring about unusual physical properties as self-cleaning, friction, adhesion, roughness, and hydrophobicity. The wrinkled materials can then be used as per the need of the situation [11]. Such type of fabric is illustrated in Figure 13.8.

The deployment of SMP in space application has also been done successfully [101–106]. Researchers successfully designed and developed morphing aircraft wing using SMP composite. Ishizawa [8] made an effort to replace SMA solar paddle with carbon fiber-reinforced SMP paddle which is light weight and more economically feasible than alloy. The team was successful to raise T_g of their paddle to an operable limit, and found good resistance of SMP in space environment. They also proposed SMPU-based solar antenna model.

13.4 CONCLUSION AND FUTURE PROSPECTS

13.4.1 Developing Areas

Self-healing in materials is a new and innovative approach to the extension of material lifetime and is highly dependable in its implementation. The damage is detected and repaired *in situ* by the incorporation of external healing agents. This approach is successful in healing damage at the structural length [11]. The concept originated when Xiao et al. [107] proposed that reverse plasticity SME is the basis of the self-healing of polymers. They fabricated epoxy-based SMP composites using free-standing nanolayered graphene as the filler, and found improved scratch resistance and thermal healing capability under stress-free conditions than the unfilled SMPs. They concluded that self-healing utilizes the

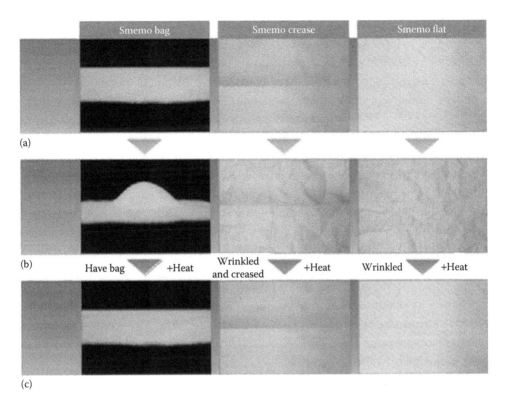

FIGURE 13.8   Typical shape memory recovery effects of shape memory-finished fabrics: (a) Original shape, (b) deformed shape, and (c) restored to its original shape. (Reproduced from Hu, J. et al., *Prog. Polym. Sci.*, 37, 1720, 2012. With permission.)

shape recovery of SMPs under unconfined and confined conditions which are applicable for curing external scratches, surfaces, internal cracks, and flaws of an object.

Rodriguez et al. [108] blended cross-linked PCL network with linear PCL; the interpenetrating network was termed *shape memory-assisted self-healing* because it was a combination of shape memory response (network) and self-healing capability (linear component). The results were impressive for the self-healing effect of films and led almost complete healing for linear PCL contents for 25 wt.%.

Another study was done to develop corrosion protecting film aluminum and its alloys by implementing segmented SMPUs as self-healing coating and results in the SMPU film recovery after heat treatment, and verified physically heal of the coating defects when it is damaged. It happened because the soft segments were relaxed with the temperature ascent, while hard segments preserved the properties [109,110].

13.4.2 Future Outlook

"This is not the end. It is not even the beginning of the end. But it is the end of the beginning." This famous quote of Sir Winston Churchill enlightens the versatility of SMPs, appealing research, and its promising future. Despite constraints in the applications and research of SMPs at present, we are very reasonable with our expectations in the near future. We seek more and more smart ideas, immense creativity, and advance technology to overcome the present limitations.

We look forward to the development of various effective materials with enhanced recovery rates, extensively biocompatible, drug elating, and adaptable metabolic conditions of the body in biomedical applications, and light weight, corrosion-resistant, tough, and scratch-resistant ones for space applications. Aesthetic principles such as color, softness, tactile, and comfort must dominate in SMPs for the textile applications.

Formulation techniques such as blending and interpenetrating networks lead to the formation of new and new SMPs, which offers elevated processability and operability. The world is looking forward to hybrid and energy conservative shape memory technology, which utilize the solar energy for heat and light responsive SMPs. How about developing an SMP car whose all exterior parts can be triggered in its original state even after an accident? We also suggest the portable raincoats, umbrellas, and tracksuits, which will retain their original shape on the impact of external weather conditions. A wide range of these materials are suggested to be applied in daily life home appliances where the wearing, friction, and impact disturb its exterior looks. This innovation will help a household person considerably to keep his or her home up to date and new. We believe a lot of advancements are currently in preliminary exploration stage, but the coming future is prospected to eradicate the shortcomings in the field.

REFERENCES

1. Chang, L.C. and Read, T.A. 1951. Plastic deformation and diffusionless phase changes in metals—The gold-cadmium beta phase. *Trans AIME* 189:47–52.
2. Buehler, W.J.,Gilfrich, J.W., and Wiley, R.C. 1963. Effect of low-temperature phase changes on the mechanical properties of alloys near composition TiNi. *J Appl Phys* 34:1475–1479.
3. Huang, W.M. 2012. The A to Z of materials: An introduction to shape memory materials. http://www.azom.com/article.aspx? Article ID: 5164 (Accessed November 28, 2012).
4. Schetky, M.D. 1991. Shape memory alloy applications in space systems. *Mater Des* 12:29–32.
5. Hartl, D.J. and Lagoudas, D.C. 2007. Aerospace applications of shape memory alloys. *Proc Int Mech Eng G: J Aerosp Eng* 221:535–552.
6. Stoeckel, D. 1990. Shape memory actuators for automotive applications. *Mater Des* 11:302–307.
7. Otsuka, K. 1998. *Shape Memory Materials*. 1st ed., Cambridge University Press, New York.
8. Ishizawa, J. 2003. Research on application of shape memory polymers to space inflatable systems. Paper Presented in the *7th International Symposium on Artificial Intelligence, Robotics and Automation in Space,* Nara, Japan.
9. Huang, W.M. 2012. Thermo/chemo-responsive shape memory effect in polymers: A sketch of working mechanisms, fundamentals and optimization. *J Polym Res* 19:9952.
10. Lendlein, A. and Kelch, S. 2002. Shape-memory polymers. *Angew Chem Int Ed* 41:2034–2057.
11. Hu, J., Zhu, Y., Huang, H., and Lu, J. 2012. Recent advances in shape–memory polymers: Structure, mechanism, functionality, modeling and applications. *Prog Polym Sci* 37:1720–1763.
12. Liu, C., Qin, H., and Mather, P.T. 2007. Review of progress in shape-memory polymers. *J Mater Chem* 17:1543–1558.
13. Dietsch, B. and Tong, T. 2007. A review—Features and benefits of shape memory polymers. *J Adv Mater* 39:3–12.
14. Anis, A., Faiz, S., Luqman, M. et al. 2013. Developments in shape memory polymeric materials. *Polym Plast Technol* 15:1574–1589.

15. Gunes, I.S. and Jana, S.C. 2008. Shape memory polymers and their nanocomposites: A review of science and technology of new multifunctional materials. *Nanosci Nanotechnol* 8:1616–1637.
16. Suna, L. and Huang, W.M. 2012. Stimulus-responsive shape memory materials: A review. *Mater Des* 33:577–640.
17. Mather, T. 2009. Shape memory polymer research. *Annu Rev Mater Res* 39:445–471.
18. Behl, M., Zotzmann, J., and Lendlein, A. 2010. Shape memory polymers and shape changing polymers. *Adv Polym Sci* 226:1–40.
19. Hu, J.L. and Chen, S.J. 2010. A review of actively moving polymers in textile applications. *J Mater Chem* 20:3346–3355.
20. Ratna, D. and Kocsis, J. 2008. Recent advances in shape memory polymers and composites: A review. *J Mater Sci* 43:254–269.
21. Hayashi, S. 1990. Technical report on preliminary investigation of shape memory polymers. Nagoya Research and Development Center. Mitsubishi Heavy Industries, Japan.
22. Kim, B.K., Lee, S.Y., and Xu, M. 1996. Polyurethanes having shape memory effects. *Polymer* 37:5781–5793.
23. Li, F.K., Zhang, X., Hou, J.N., and Xu, M. 1997. Studies on thermally stimulated shape memory effect of segmented polyurethanes. *J Appl Polym Sci* 64:1511–1516.
24. Wang, W.S, Ping, P., and Chen, X.S. 2006. Polylactide-based polyurethane and its shape-memory behavior. *J Eur Polym* 42:1240–1249.
25. Xu, J.W., Shi, W.F., and Pang, W.M. 2006. Synthesis and shape memory effects of Si–O–Si cross-linked hybrid polyurethanes. *Polymer* 457–465.
26. Hu, J. 2007. *Shape Memory Polymers and Textiles.* Woodhead Publishing Limited, Cambridge, UK.
27. Ping, P., Wang, W.S., and Chen, X.S. 2005. Poly(ε-caprolactone) polyurethane and its shape-memory property. *Biomacromolecules* 6:587–592.
28. Chung, Y.C., Choi, J.H., and Chun, B.C. 2008. Shape-memory effects of polyurethane copolymer cross-linked by dextrin. *J Mater Sci* 43:6366–6373.
29. Park, J.S., Chung, Y.C., Cho, J.W., and Chun, B.C. 2008. Shape memory effects of polyurethane block copolymers cross-linked by celite. *Fiber Polym* 9:661–666.
30. Jung, D.H., Jeong, H.M., and Kim, BK. 2010. Organic–inorganic chemical hybrids having shape memory effect. *J Mater Chem* 20:3458–3466.
31. Li, F.K., Zhu, W., Zhang, X., and Xu, M. 1999. Shape memory effect of ethylene–vinyl acetate copolymers. *J Appl Polym Sci* 71:1063–1070.
32. Behl, M., Ridder, U., Feng, Y., Kelch, S., and Lendlein, A. 2009. Shape-memory capability of binary multiblock copolymer blends with hard and switching domains provided by different components. *Soft Matter* 5:676–684.
33. Madbouly, S.A. and Lendlein, A. 2012. *Degradable Polyurethane/Soy Protein Shape-Memory Polymer Blends Prepared Via Environmentally-Friendly Aqueous Dispersions.* Wiley-VCH Verlag GmbH & Co. KGaA, Weinheim, Germany.
34. Ashton, J.H., Mertz, J.A.M., and Harper, J.L. 2011. Polymeric endoaortic paving: Mechanical, thermoforming, and degradation properties of polycaprolactone/polyurethane blends for cardiovascular applications. *Acta Biomater* 7:287–294.
35. Kurahashi, E., Sugimoto, H., and Nakanishi, E. 2012. Shape memory properties of polyurethane/poly(oxyethylene) blends. *Soft Matter* 8:496–403.
36. Luo, H., Hu, J., and Zhu, Y. 2012 Achieving shape memory: Reversible behaviors of cellulose-PU blends in wet–dry cycle. *J Appl Polym Sci* 125:657–665.
37. Chen, S.J. 2011. Study on the structure and morphology of supramolecular shape memory polyurethane containing pyridine moieties. *Smart Mater Struct* 20:065003.
38. Erden, N. and Sadhan, C.J. 2012. Synthesis and characterization of shape memory polyurethane–polybenzoxazine compounds. *MACP* 11:1225–1237.

39. Lin, J.R. and Chen, L.W. 1999. Shape-memorized cross linked ester-type polyurethane and its mechanical viscoelastic model. *J Appl Polym Sci* 73:1305–1319.
40. Conti, S., Lenz, M., and Rumpt, M. 2007 Modeling and simulation of magnetic-shape-memory polymer composites. *J Mech Phys Solid* 55:1462–1486.
41. Beloshenko, V.A., Varyukhin, V.N., and Voznyak, Y.V. 2005. Electrical properties of carbon-containing epoxy compositions under shape memory effect realization. *Compos Part A—Appl Sci* 36:65–70.
42. Sharma, B.K., Gajaev, I.Y., and Mattiasson, B. 2000. Thermosensitive, reversibly cross-linking gels with a shape memory. *Angew Chem Int Ed* 39:2364–2367.
43. Khan, A. and Lopez, O. 2002. Time and temperature dependent response and relaxation of a soft polymer. *Int J Plasticity* 18:1359–1372.
44. Reese, S. 2003. A micromechanically motivated material model for the thermo-viscoelastic material behavior of rubber-like polymers. *Int J Plasticity* 19:909–940.
45. Hasanpour, K., Ziaei-Rad, S., and Mahzoon, M. 2009. A large deformation framework for compressible viscoelastic materials: Constitutive equations and finite element implementation, *Int J Plasticity* 25:1154–1176.
46. Tobushi, H., Hashimoto, T., Ito, N., Hayashi, S., and Yamada, E. 1998. Shape fixity and shape recovery in a film of shape memory polymer of polyurethane series. *J Intel Mat Syst Str* 9:127–136.
47. Kolesov, I., Kratz, K., Lendlein, A., and Radusch, H. 2009. Kinetics and dynamics of thermally-induced shape-memory behavior of crosslinked short-chain branched polyethylenes. *Polymer* 50:5490–5498.
48. Kim, J., Kang, T., and Yu, W. 2010. Thermo-mechanical constitutive modeling of shape memory polyurethanes using a phenomenological approach. *Int J Plasticity* 26:204–218.
49. Volk, B., Lagoudas, D., Chen, Y., and Whitley, K. 2010. Analysis of the finite deformation response of shape memory polymers: I. Thermomechanical characterization, *Smart Mater Struct* 19:1–18.
50. Volk, B., Lagoudas, D., and Chen, Y. 2010. Analysis of the finite deformation response of shape memory polymers: II. 1-D calibration and numerical implementation of a finite deformation, thermoelastic model. *Smart Mater Struct* 19:094004-1–094004-18.
51. Barot, I.J. and Rajagopal. K.R. 2008. A thermodynamic framework for the modeling of crystallizable shape memory polymers *Int J Eng Sci* 46:325–335.
52. Srivastava, V., Chester, S., and Anand, L. 2010. Thermally-actuated shape-memory polymers: Experiments, theory, and numerical simulations. *J Mech Phys Solids* 58:1100–1124.
53. Atli, B., Gandhi, F., and Karst, G. 2009. Thermomechanical characterization of shape memory polymers. *J Intel Mat Syst Str* 20:87–95.
54. Crisfield, M. 1997. *Nonlinear Finite Element Analysis of Solids and Structures—Advanced Topics*, Wiley, New-York.
55. Gao, Z., Tuncer, A., and Cuitiño, A.M. 2011. Modeling and simulation of the coupled mechanical–electrical response of soft solids. *Int J Plasticity* 27:1459–1470.
56. Kafka, V. 2008. Shape memory polymers: A mesoscale model of the internal mechanism leading to the SM phenomena. *Int J Plasticity* 24:1533–1548.
57. Li, G. and Nettles, D. 2010. Thermomechanical characterization of a shape memory polymer based self-repairing syntactic foam. *Polymer* 51:755–762.
58. Liu, Y., Gall, K., Dunn, M.L., and Greenberg, A.R. 2006. Thermomechanics of shape memory polymers: Uniaxial experiments and constitutive modeling. *Int J Plasticity* 22:279–213.
59. Baghani, M. 2012. A thermodynamically-consistent 3-D constitutive model for shape memory polymers. *Int J Plasticity* 35:13–30.
60. Small, W., Buckley, P.R., Wilson, T.S., Benett, W.J., and Hartman, J. 2007. Shape memory polymer stent with expandable foam: A new concept for endovascular embolization of fusiform aneurysms. *IEEE Trans Biomed Eng* 54:1157–1160.

61. Small, W., Metzger, M.F., Wilson, T.S., and Maitland, D.J. 2005. Laser-activated shape memory polymer microactuator for thrombus removal following ischemic stroke: Preliminary in vitro analysis. *J Sel Top Quant Electron (IEEE)* 11:892–901.
62. Hampikian, J.M., Heaton, B.C., Tong, F.C., Zhang, Z., and Wong, C.P. 2006. Mechanical and radiographic properties of a shape memory polymer composite for intracranial aneurysm coils. *J Mater Sci Eng C* 26:1373–1379.
63. Yu, Y.L. and Ikeda, T. 2005. Photodeformable polymers: A new kind of promising smart material for micro-and nano-applications. *Macromol Chem Phys* 206:1705–1708.
64. Sahoo, N.G., Jung, Y.C., and Cho, J.W. 2007. Electroactive shape memory effect of polyurethane composites filled with carbon nanotubes and conducting polymer. *Mater Manuf Process* 22:419–423.
65. Sahoo, N.G., Jung, Y.C., and Cho, J.W. 2005. Conducting shape memory polyurethane-poly pyrrole composites for an electroactive actuator. *Macromol Mater Eng* 290:1049–1055.
66. Meng, H. and Hu, J. 2010. A brief review of stimulus-active polymers responsive to thermal, light, magnetic, electric, and water/solvent stimuli. *J Intel Mat Syst Str* 21:859–885.
67. Yang, B., Huang, W.M., Li, C., and Li, L. 2006. Effects of moisture on the thermo mechanical properties of a polyurethane shape memory polymer. *Polymer* 47:1348–1356.
68. Chen, S., Hu, J., Yuen, C.W., and Chan, L. 2009. Supramolecular polyurethane networks containing pyridine moieties for shape memory materials. *Mater Lett* 3:1462–1464.
69. Chen, M.C., Tsai, H.W., Chang, Y., and Lai, W.Y. 2007. Rapidly self-expandable polymeric stents with a shape-memory property. *Biomacromolecules* 8:2774–2780.
70. Jung, Y.C., So, H.H., and Cho, J.W. 2006. Water-responsive shape memory polyurethane block copolymer modified with polyhedral oligomeric silsesquioxane. *J Macromol Sci* 45:453–461.
71. Knight, P.T., Lee, K.M., Qin, H., and Mather, P.T. 2008. Biodegradable thermoplastic polyure-thanes incorporating polyhedral oligo silsesquioxane. *Biomacromolecules* 9:2458–2467.
72. Schmidt, A.M. 2006. Electromagnetic activation of shape memory polymer networks containing magnetic nanoparticles. *Macromol Rapid Commun* 27:1168–1172.
73. Mohr, R., Kratz, K., Weigel, T., Lucka-Gabor, M., Moneke, M., and Lendlein, A. 2006. Initiation of shape-memory effect by inductive heating of magnetic nanoparticles in thermoplastic polymers. *Proc Natl Acad Sci USA* 103:3540–3545.
74. Hazelton, C.S., Arzberger, S.C., Lake, M.S., and Munshi, N.A. 2007. RF actuation of a thermoset shape memory polymer with embedded magneto-electroelastic particles. *J Adv Mater* 39:35–39.
75. Chung, Y. 2012. Characterization of flexibly linked shape memory polyurethane composite with magnetic property. *J Thermoplast Compos* 25:283–303.
76. Jung, Y.C. and Cho, J.W. 2010. Application of shape memory polyurethane in orthodontic. *J Mater Sci Mater Med* 21:2881–2886.
77. Ortega, J.M., Small, W., Wilson, T.S. et al. 2007. A shape memory polymer dialysis needle adapter for the reduction of hemodynamic stress within arteriovenous grafts. *IEEE Trans Biomed Eng* 54:1722–1724.
78. Kang, S.M., Lee, S.J., and Kim, B.K. 2012. Shape memory polyurethane foams. *Exp Polym Lett* 6:63–69.
79. Rousseau, I.A., Berger, E.J., Owens, J.N., and Kia, H.G. 2010. Shape memory polymer medical cast. US 0249682.
80. Chen, M.C., Chang, Y., Liu, C.T. et al. 2009. The characteristics and in vivo suppression of neointimal formation with sirolimus-eluting polymeric stents. *Biomaterials* 30:79–88.
81. Havens, E., Snyder, E., and Tong, T. 2007. Light-activated shape memory polymers and associated applications. *Proc SPIE* 5762:48–54.
82. Vili, Y. 2007. Investigating smart textiles based on shape memory materials. *Text Res J* 77:290–300.

83. Zhuo, H., Hu. J.L., Chen. S.J., and Yeung, L.Y. 2008. Preparation of polyurethane nanofibers by electrospinning. *J Appl Polym Sci* 109:406–411.

84. Cha, D.I., Kim, H.Y., Lee, K.H. et al. 2005. Electrospun non-wovens of shape memory polyurethane block copolymers. *J Appl Polym Sci* 96:460–465.

85. Zhuo, H.T., Hu, J.L., and Chen, S.J. 2011. Study of the thermal properties of shape memory polyurethane nanofibrous nonwoven. *J Mater Sci* 46:3464–3469.

86. Zhuo, H.T., Hu, J.L., and Chen, S.J. 2011. Coaxial electrospun polyurethane core-shell nanofibers for shape memory and antibacterial nano-materials. *Exp Polym Lett* 5:182–187.

87. Chung, S.E., Park, C.H., Yu, W.R., and Kang, T.J. 2011. Thermoresponsive shape memory characteristics of polyurethane electrospun web. *J Appl Polym Sci* 120:492–500.

88. Zhang, J.N., Ma, Y.M, Zhang. J.J., Xu. D. et al. 2011. Microfiber SMPU film affords quicker shape recovery than the bulk one. *Mater Lett* 65:3639–3642.

89. McDowell, J.J., Zacharia, N.S., Puzzo, D., Manners, I., and Ozin, G.A. 2010. Electroactuation of alkoxysilane-functionalized poly ferrocenyl-silane microfibers. *J Am Chem Soc* 132:3236–3237.

90. Zhuo, H.T., Hu, J.L., Chen, S.J., and Zhu. Y. 2008. Study of shape memory nano fiber nonwoven fabrics. In *Proceedings of the International Conference on Advanced Textile Materials & Manufacturing Technology*, Hangzhou, China, pp. 463–466.

91. Ding, X.M., Hu, J.L., Tao, X.M., and Hu, C.R. 2006. Preparation of temperature-sensitive polyurethanes for smart textiles. *Text Res J* 76:406–413.

92. Hu, J.L., Zeng, Y.M., and Yan, H.J. 2003. Influence of processing conditions on the microstructure and properties of shape memory polyurethane membranes. *Text Res J* 73:172–178.

93. Liu, Y., Chung, A., Hu, J.L., and Lu, J. 2007. Shape memory behavior of SMPU knitted fabric. *J Zhejiang Univ Sci A* 8:830–834.

94. Mondal, S. and Hu, J.L. 2007. Water vapor permeability of cotton fabrics coated with shape memory polyurethane. *Carbohydr Polym* 67:282–287.

95. Bakhshi, R., Darbyshire, A., Evans, J.E., You, Z., and Lu, J. 2011. Polymeric coating of surface modified nitinol stent with POSS-nanocomposite polymer. *Colloids Surf B* 86:93–105.

96. Ikeda, T. and Ube, T. 2011. Photo mobile polymer materials: From nano to macro. *Mater Today* 14:480–487.

97. Yamada, M., Kondo, M., Miyasato, R. et al. 2009. Photomobile polymer materials-various three-dimensional movements. *J Mater Chem* 19:60–62.

98. Chen, S.J., Hu, J.L., Zhuo, H.T., and Zhu. Y. 2008. Two-way shape memory effect in polymer laminates. *Mater Lett* 62:408890.

99. Saito, T. 2003. System for holding a heat shrinkable tube during a heat shrinkage operation. US Patent 6531659.

100. Huang, W.M., Yang, B., Zhao, Y., and Ding, Z. 2010. Thermo-moisture responsive polyurethane shape memory polymer and composites: A review. *J Mater Chem* 20:3367–3381.

101. Keihl, M.M., Bortolin, R.S., Sanders, B., Joshi, S., and Tidwell, Z. 2005. Mechanical properties of shape memory polymers for morphing aircraft applications. *Proc SPIE* 5762:143–151.

102. Sofla, A.Y.N., Meguid, S.A., Tan, K.T., and Yeo, W.K. 2010. Shape morphing of aircraft wing: Status and challenges. *Mater Des* 31:1284–1292.

103. Yu, K., Yin, W., Sun, S., Liu, Y., and Leng, J. 2009. Design and analysis of morphing wing based on SMP composite. *Proc SPIE* 7290: doi:10.1117/12.815712.

104. Thill, C., Etches, J., Bond, I., Potter, K., and Weaver, P. 2008. Morphing skins. *Auronaut J* 112:117–139.

105. Yu, K., Sun, S., Liu, L., Zhang, Z., Liu, Y., and Leng, J. 2009. Novel deployable morphing wing based on SMP composite. *Proc SPIE* 7493: doi:10.1117/12.845408.

106. Yu, K., Yin, W., Liu, Y., and Leng, J. 2009. Application of SMP composite in designing a morphing wing. *Proc SPIE* 7375: doi:10.1117/12.839363.

107. Xiao, X.C., Xie, T., and Cheng, Y.T. 2010. Self-healable graphene polymer composites. *J Mater Chem* 20:3508–3514.
108. Rodriguez, E.D., Luo, X.F., and Mather, P.T. 2011. Linear/network poly(epsilon-caprolactone) blends exhibiting shape memory assisted self-healing (SMASH). *Appl Mater Inter* 3:152–161.
109. Jorcin, J.B., Scheltjens, G., Ingelgem, Y., Tourwé, E., and Van Assche, G. 2010. Investigation of the self-healing properties of shape memory polyurethane coatings with the odd random phase multisine electrochemical impedance spectroscopy. *Electrochimica Acta* 55:6195–6203.
110. Gonzalez-Garcia, Y., Muselle, T., De Graeve, I., and Terryn, H. 2011. SECM study of defect repair in self-healing polymer coatings on metals. *Electrochem Commun* 13:169–173.

CHAPTER 14

Thermoplastic Elastomers

Advances in Thermoplastic Vulcanizates

R. Rajesh Babu

CONTENTS

14.1 INTRODUCTION

The concept of polymer blending to develop new materials for scientific and commercial requirements has been growing exponentially during the last two decades. Polymer blends offer the possibility of combining the unique properties of individual polymers and that of producing new polymeric materials with tailor-made properties, which often have advantages over the development of a completely new polymeric material. The process of polymer blending is a simple, fast, effective, and economical method of producing high performance-to-cost ratio materials.

Polymer blends are broadly classified into three categories [1,2]:

1. Miscible blends, which are associated with a negative free energy change and exist in a single homogeneous phase and exhibit synergistic properties.

2. Immiscible blends, which are associated with a positive free energy change and exhibit multiphase morphology with distinct boundary layers and multiple glass transition temperatures (T_gs). They also have the less intense thermodynamic compatibility.

3. Technologically compatible blends or alloys exist in heterophase microscale; however, they show macroscopic properties like a single-phase material. The interfaces of these blends have generally been modified. They often exhibit enhanced physicomechanical properties as compared with the component polymers. A compatibilizer could be used to improve the performance properties.

14.2 THERMOPLASTIC ELASTOMERS

A thermoplastic elastomer (TPE) is a rubbery material or rubber-like material with properties and functional performance similar to those of a conventional vulcanized rubber; still it can be processed in a molten state as a thermoplastic polymer. Conventional rubbers are subjected to many processing steps, which require more processing equipment costing high capital investment. On the other hand, thermoplastic processing is simple to produce the final article and economically more advantageous. If rubber-like material is processed on plastic processing equipment similar to thermoplastics and still retains the characteristic properties of rubber/cross-linked rubber, it can become more economically feasible material for end use [3,4]. Such materials can be obtained by two main methods, essentially dissimilar methods, they are: synthesis of block copolymers having blocks of different chemical and physical characteristics, and blending of rubbers with plastics results in composite materials with new properties with an extended range of utilization. TPEs have been expanding the growth of usage in various fields and still experiencing a high growth rate in their commercial applications. TPEs may be broadly classified into the following classes according to their phase structure and physiochemical characteristics [5,6].

Block copolymers

1. Styrenic block copolymers (SBS)

2. Thermoplastic copolyesters (COPE)

3. Thermoplastic polyurethanes (TPU)

4. Thermoplastic polyamides (TPA)

Blends and alloys

1. Elastomeric rubber–plastic blends (TPOs)

2. Thermoplastic vulcanizates (TPVs)

Ionomers

1. Sulfonated EPDM rubber (S-EPDM)

2. Zn or Na salt of ethylene acrylic acids.

Broader families of these materials are becoming more familiar to a wider range of potential customers. The applications such as mobile phone grips, computer accessories, pen and toothbrush handles, knobs, and sports goods are targets for soft-touch feel using oil-extended styrene-(ethylene-butylene)-styrene block copolymers and TPUs. Similarly in the automotive sector, driver-side airbag covers and instrument panel skins are targets for thermoplastic polyolefin (TPO), TPV, and TPU materials, depending on the models, requirements, and philosophy toward design, style, and comfort. In this chapter, special attention has been given to the recent work on TPVs with an objective of not only updating the information but also appreciating the new developments in this field.

14.3 THERMOPLASTIC VULCANIZATES

TPEs based on polymer blends can be broadly classified as TPOs and TPVs. TPOs typically consist of a dispersion of noncross-linked elastomer particles in a semicrystalline thermoplastics matrix. The composition leads to a material of high toughness but limited elasticity. On the other hand, TPVs are prepared by a process called dynamic vulcanization. As the name implies, elastomer phase in the blend composition is cross-linked under a dynamic condition. The process was carried out at high shear rate and at elevated temperature to activate and complete the process of vulcanization. As a result of the operating condition, the cross-linked rubber particles are dispersed in the continuous thermoplastic matrix as microgels [7,8]. This morphology is permanent and does not change on further processing. Figure 14.1a and b shows the schematic representation of the TPV morphology and corresponding TEM micrograph of polypropylene/ethylene-propylene–diene terpolymer (PP/EPDM) TPVs. On cross-linking the elastomeric phase, TPVs exhibit a reduced set, better elastic recovery, improved mechanical, greater melt strength, and good fatigue resistance property compared with TPOs.

Historically, Uniroyal Chemical Corporation was the first company to introduce dynamically vulcanized EPDM/PP blends. This was based on the work of W.K. Fisher, who filed patent applications in 1971 on his discovery of partially cross-linking the EPDM phase of EPDM/PP blends with peroxide [9]. Greater commercial and industrial attention was gained only after the extensive work on TPVs based on various blend components by

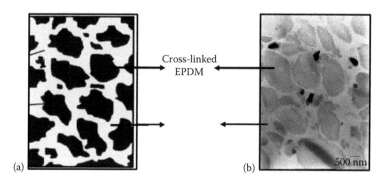

FIGURE 14.1 (a) Schematic representation of TPV morphology and (b) 2D-TEM image of PP/EPDM TPVs showing cross-linked EPDM dispersed in the PP matrix. (From Sengupta, P., PhD Thesis, Twente University, Enschede, the Netherlands, 2004. With permission.)

Coran and Patel in the 1980s. They proposed some critical criteria to judge the choice of the blend components and the processing attributes [10–12].

1. Low interfacial energy difference between the blend components.

2. Elastomers of high entanglement density are preferred for better elastic recovery properties.

3. Thermoplastic matrix phase should have high crystallinity.

4. Cross-linking or vulcanization process should not affect the thermoplastic matrix phase.

5. High shearing conditions are required to achieve fine particles of the cross-linked elastomer phase that in turn reflects in the mechanical and rheological properties.

The principles of dynamic vulcanization have gained interest in the reactive processing of polymer blends having large interfacial tension. Reactive compatibilization can be achieved either by the addition of block copolymers or *in situ* graft formation between the blend components. Earlier the subject had been reviewed by Abdou-Sabet [13], Coran [14], Karger [15], and Babu and Naskar [16]. Now, some TPVs have been commercialized with trade names Santoprene™ and Geolast™. They are widely used in automotives, household, wires/cables, biomedical and soft-touch applications, and so on. With the introduction of new polymers, a large number of rubber–thermoplastic combinations are possible, and the scope is extending to a greater level.

14.4 CHARACTERISTICS OF TPVs

1. Morphology

2. Rheology

3. Deformation behavior

4. Production and processing

5. Reinforcing stimulus

14.4.1 Morphology

Blend composition, interfacial tension, and processing condition play a critical role in the evolution of TPV morphology. It is generally accepted that with optimum cross-linking in the rubber phase reached and the critical stress achieved, the strands will break up into small particles under the applied deformation level. Typically, the particle size varies from 0.5 to 5 µm; the smaller the particles the better the mechanical properties. In the final morphology of TPVs, the cross-linked elastomer phase is always evenly dispersed and the volume ratio of the elastomer phase is 0.8. In other words, elliptical and/or spherical particles and their distribution of particle sizes lead to the closed packing of the cross-linked elastomer phase in the thermoplastic matrix [16–19].

One of the major advantages of the dynamically vulcanized blends over unvulcanized blends is that the morphology is fixed and is not altered by subsequent (re)processing. Radusch et al. [19] examined the morphology generation of phenolic resin cured PP/EPDM TPVs. The earlier studies by various authors on the development of morphology during dynamic vulcanization lead to the simplified representation as shown in Figure 14.2 [18]. In the early stage of mixing, two blend components are characterized in terms of non-molten thermoplastic materials dispersed in the elastomer matrix (a). As the melt mixing process progresses, two phases tend to form a cocontinuous phase morphology (b). The combined shear and elongational forces alter the state of cocontinuity (c) and lead to the refinement of cocontinuous strands as a function of mixing time (d). On the addition of a cross-linking agent, the viscosity ratio raises due to the cross-linking of the elastomer phase. As the elastomer becomes rigid, it causes an immobilization of the cross-linked elastomer particles, and therefore they break down to micron size smaller particles under

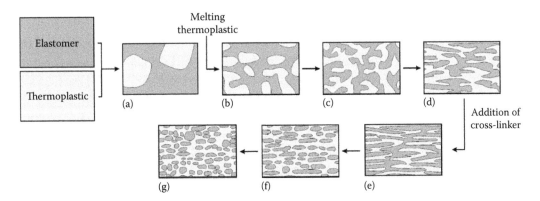

FIGURE 14.2 Schematic representation (a–g) of the morphology development during dynamic vulcanization of elastomer/thermoplastic blends. (From Abee, R.I., Thermoplastic vulcanizates—The rubber particle size to control the properties-processing balance, PhD Thesis, Technical University, Eindhoven, the Netherlands, 2009.)

the applied shear field (e). As the curing and mixing continues, phase inversion takes place and the elastomer becomes dispersed (f), even when its volume fraction is very high (~0.80). The dispersed elastomer phase will reach a final particle size (g) as a function of mixing time and also that depends on the deformation state and rate. Finally, at the end of the mixing process, the cross-linked rubber particle is finely dispersed in the thermoplastic matrix.

Sengupta and Noordermeer [20] studied the effect of PP/EPDM blend composition on the morphology evolution. Increasing the thermoplastic content results in a decrease of the elastomer particle size and the particles become more separated. The morphology of TPVs can be studied by several microscopic techniques such as scanning electron microscopy (SEM), transmitting electron microscopy (TEM), and atomic force microscopy (AFM). However, high volume fraction of the elastomer phase can cause misleading interpretation. The particles partially overlap and appear as a continuous phase in the microscopic images. The most suitable techniques to study TPVs were found to be TEM and low-voltage SEM. Furthermore, they reported the usage of electron tomography in reconstructing the 3D morphology in PP/EPDM TPV blends.

14.4.2 Rheology

TPVs to a good approximation, like vulcanized rubber at room temperature, undergo a thermoplastic melt process at temperatures above the melting point of thermoplastic matrix polymers. These materials are by nature complex polymer systems, that is, multiphase, heterogeneous, typically disordered materials for which a structure is as important as composition [21]. Correctly assessing the rheological properties of TPV is a challenging task for several reasons: their heterogeneous nature and their morphology-composition dependence, sensitivity to flow fields and to their influence in modifying the flow boundary conditions owing to the migration of small labile ingredients (e.g., oil, curative residue, etc.). The rheological properties of TPVs can be compared with those of highly filled polymers [22–24]. The flow characteristics of TPVs were first explored by Goettler et al. [22] and followed by several authors [25–27]. Han and White [23] described the comparative viscoelastic properties of uncross-linked and dynamically cross-linked PP/EPDM blends using various rheological instruments. Van Duin et al. [24] showed the variation of the viscoelastic property (viscosity and storage modulus) of the PP/EPDM TPVs as a function of frequency with special reference to the variation of the cross-linked EPDM particle size (Figure 14.3). Irrespective of the variation in particle size, TPVs' viscosity decreases with the increase in frequency, which clearly demonstrates the pseudoplastic behavior and significant non-Newtonian properties (shear thinning behavior). TPVs with smaller particle size show a plateau value of the storage modulus at low frequencies and higher moduli values in the entire range of frequency. Higher moduli in the low shear rate region and more shear thinning behavior with increasing shear rate can afford to give better processing characteristics. This can be attributed to the state and mode of a three-dimensional (3D) network structure formed by the dispersed cross-linked elastomer particles. As the frequency increases, the network structure, mainly aggregates/agglomerates, tends to collapse

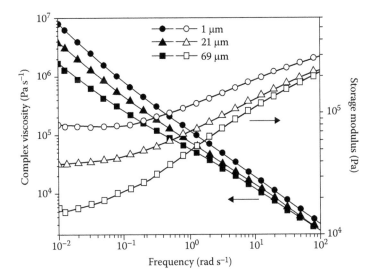

FIGURE 14.3 Variation of viscoelastic properties (complex viscosity and storage modulus) as a function frequency: Special reference to the rubber particle size variation. (From I'Abee, R.M.A. et al., *Soft Matter*, 6, 1758–1768, 2010.)

and deform to exhibit higher shear thinning behavior. It was generally accepted that TPVs have a yield stress for flow and that the value increases with the increase in the elastomer component in the TPV. The nonlinear viscoelastic property of the various commercially available TPVs with varying hardness was studied by Leblanc [28] using a Fourier transform rheometer (FT-rheo). In this study, the role of oil and plasticizer in the blend component and its influence on the morphology development were discussed in detail.

14.4.3 Deformation Behavior

As mentioned earlier, TPVs have a unique morphology due to the process of dynamic vulcanization. This morphology consists of a chemically cross-linked rubber phase embedded within a continuous semicrystalline thermoplastic matrix. Even though the matrix is a semicrystalline thermoplastic, dynamic vulcanizates (TPVs) behave elastically. The state of cross-linking in the elastomer phase results in better elastic recovery properties and a decrease in the permanent deformation after unloading (tension or compression set) [7,10]. The origins of this elasticity have been analyzed by several finite element studies and constitutive models have been proposed. Finite element modeling of TPVs during tensile deformation showed that the thermoplastic phase (that is PP) yielded partially in the equatorial region of the elastomeric particles and the rest of the PP phase remains unaffected [29]. The deformation mechanism of the phenolic resin cured PP/EPDM TPVs was proposed by Soliman et al. [30]. The proposal was based on the assumption that the interfacial adhesion between the PP and elastomer is adequate. During deformation, the PP phase acts as a glue between the EPDM particles and tends to yield partially. On unloading, the micromechanical model predicts that the elastic forces of the stretched elastomer particles pull back the highly plastically deformed (yielded) thin ligaments via buckling and bending of the ligaments. A schematic representation of the deformation and recovery of the dynamically vulcanized blends is shown in Figure 14.4 [31,32]. Such a buckling mechanism has been confirmed by the AFM studies of nylon-6/EPDM TPVs [32]. The plastic deformation is initiated in the region where the nylon matrix between the cross-linked elastomer particles is the thinnest. When the external force is removed, the elastic force of the stretched, dispersed rubber phase pulls back the plastically deformed nylon parts by either buckling or bending, as shown in Figure 14.5. This has been considered and believed to be the key mechanism for the elastic behavior of TPVs.

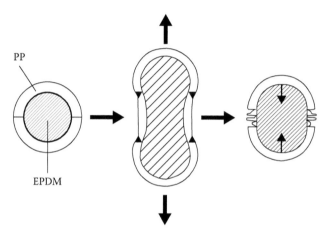

FIGURE 14.4   Sketch illustrating the deformation and recovery of dynamically vulcanized blends. (From Naskar, K., *Rubber Chem. Technol.*, 80, 504–526, 2007. With permission.)

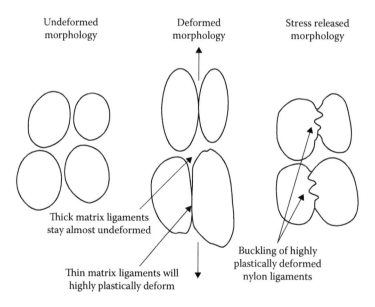

FIGURE 14.5   Sketch illustrating the deformation and recovery of dynamically vulcanized blends. (From Oderkerk, J. et al., *Macromolecules*, 35, 6623–6629, 2002. With permission.)

14.4.4 Production and Processing

TPVs are produced using either batch or continuous mixing techniques. On an industrial scale, TPVs are preferably produced in twin-screw extruders, which allow continuous processing with high throughputs and with a large degree of process flexibility. Dynamic vulcanization of TPVs on extruders is much more complex than that in batch kneaders, where the different steps of thermoplastic melting, rubber dispersion, and cross-linking are separated in time. In extruders, the elastomer and the thermoplastic are usually fed simultaneously through the hopper. Other ingredients, such as the cross-linking system, stabilizers, fillers, and oil, may be fed via the hopper and/or via side feeders. The thermoplastic has to melt; the elastomer and the thermoplastic have to mix, yielding a heterogeneous but finely dispersed blend; the extender oil that is often present has to redistribute over the two phases; and the cross-link system has to dissolve and cross-link the elastomer phase. These various processes will (partly) take place simultaneously and, in addition, will (partly) mutually interact. Furthermore, the temperature and the flow field of the melt fluctuate along the extruder axis, thereby affecting the viscosities, the cross-linking kinetics, and so on. At the end of the extruder, usually de-venting is applied for removing gaseous and volatile products. The final melt strand coming out at the die of the extruder is cooled and pelletized [33]. Because the curing process is very fast, the mixing speed should be high to obtain small rubber particles. The maximum shear rate should be higher than 2000 s^{-1} and the length of the extruder should be larger than 42 L/D.

Even now, there prevails a challenge on making a soft composition of TPVs. Soft compositions are typically prepared by increasing the rubber composition and the addition of oil or plasticizers. Such measures make the product less processable. In some cases, such compositions give poor extrudates and sometimes, cannot be extruded at all. The fabrication method and equipment for TPVs are essentially those of the thermoplastic in the material. Unlike thermoset rubber, thermoplastic injection molding, plastic extrusion, blow molding, thermoforming, and heat welding can also be performed for TPV-based materials [34].

14.4.5 Reinforcing Stimulus

Compatibilization of dissimilar polymers blends (high interfacial tension) is an efficient route to decrease the dispersed phase size in traditional polymer blends. The particle dispersions as small as 50–100 nm were reported for highly incompatible polymer pairs after compatibilization. Block or graft copolymers of the blend components are frequently used as compatibilizers, but the lack of economic viability limited the usage of the same. A good alternative is to generate such a component during melt blending, a process called *in situ* compatibilization or reactive blending. TPVs based on incompatible rubber/plastic blends can be prepared via the compatibilization technique. Assuming that the interfacial adhesion of the blends components and state of rubber dispersion are fair enough, two factors namely particle size of the cross-linked elastomer phase and the chain bridging nature of the matrix phase determine the reinforcing characteristics of TPVs.

14.4.5.1 Rubber Particle Size

In general, the rubber particle size in TPVs increases with the increase in the rubber content of the blend ratio. The balance between the particle break-up and coalescence is determinative for the particle size evolution. Once the rubber phase is cross-linked, the coalescence of the rubber particles is limited and the final morphology of the TPV is mainly determined by the break-up process. Rubber particle sizes in the range of 1–3 μm are typically observed for TPVs (PP/EPDM), and the preparation of sub-μm rubber particles is very challenging and has indeed not been achieved so far. This can be explained by the fact that further break-up of the μm-sized rubber particle is suppressed at a high viscosity ratio. The smaller the particle size, the higher the surface area and shorter the interaggregate distance, which results in exhibiting the best overall balance of mechanical properties and better elastic recovery properties. It was observed that both the elongation at break and the tensile strength increased by a factor of 5 on decreasing the rubber particle size (d_n) from 70 down to 5 μm (Figure 14.6). It is believed that cross-linked rubber domains are dispersed in the form of aggregates and agglomerates in the thermoplastic matrix. Furthermore, these aggregates and/or agglomerates are connected to form a 3D network structure. It is bit difficult to distinguish the network formation through the direct bonding mode or indirect mode (through polymer chains bridging on the different domains) [35].

14.4.5.2 Chain Bridging Nature

In TPVs, the difference between the surface energy of the thermoplastic matrix and the cross-linked elastomer particles leads to the adsorption of the matrix molecule onto the surface of the cross-linked elastomer particles through the segmental interfusion mechanism. Smaller cross-linked elastomer particles have a higher tendency to adsorb (physical and/or chemical adsorption) the matrix polymer chains on the surface and

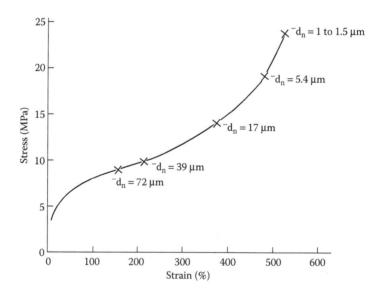

FIGURE 14.6 Influence of rubber particle size on tensile properties of PP/EPDM TPVs. (From Coran, A.Y. and Patel, R., *Rubber Chem. Technol.*, 53[1], 141–150, 1980. With permission.)

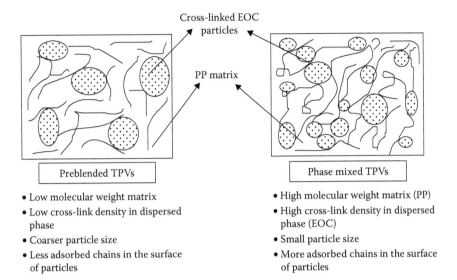

Cross-linked EOC particles

PP matrix

Preblended TPVs

- Low molecular weight matrix
- Low cross-link density in dispersed phase
- Coarser particle size
- Less adsorbed chains in the surface of particles

Phase mixed TPVs

- High molecular weight matrix (PP)
- High cross-link density in dispersed phase (EOC)
- Small particle size
- More adsorbed chains in the surface of particles

FIGURE 14.7 Schematic representation of possible structure developed by TPVs illustrating difference in binding the particles and the particle size of the TPVs prepared by preblended TPVs (conventional method) and phase mixed TPVs at 2 phr concentration of DCP. (From Babu, R.R. et al., *Polym. Eng. Sci.*, 50, 455–467, 2010. With permission.)

hence form more trapped polymeric chains by connecting the particles. The reinforcement mechanism of TPVs is also determined by a feature called the chain bridging effect, that is, the influence of matrix molecular weight on the adsorbing characteristics on the rubber particles. When the matrix chain length is sufficiently long, polymeric chains are immobilized by chain adsorption at one or several points along their chain lengths, which affects their stress response on the dynamic experiments. A schematic representation illustrating the difference in binding the particles and the particle size of the PP/ethylene–octene copolymer (EOC) TPVs prepared by the preblending and phase mixing method is shown in Figure 14.7. It is observed that TPVs prepared by adding an EOC curative master batch to the molten PP (phase mixing method) exhibit better mechanical properties and finer phase morphology than those prepared by the preblending method. The adsorbed chains on the surface of the cross-linked elastomer particles generally broaden the relaxation spectrum toward longer time, which has been reflected in the higher dynamic modulus and viscosity in the frequency and strain sweep experiments [36].

14.5 CATEGORIES OF TPVs

1. Nonpolar polymer-based TPVs

2. Polar polymer-based TPVs

3. Recycled polymer-based TPVs

4. TPVs by electron beam (EB) reactive processing.

14.5.1 Nonpolar Polymer-Based TPVs

14.5.1.1 PP/EPDM TPVs

Blends of PP- and EPDM-based TPVs are of most commercial importance. The elastomer component (EPDM) have saturated main chain, explaining the excellent stability against heat, oxygen, and ozone. Several cross-linking agents have been used in the preparation of TPVs. The most commonly used cross-linking systems are the sulfur vulcanization system, the peroxide system, and phenolic resin. However, EPDM/PP-based TPVs are conventionally cross-linked with acid-activated (resol phenolic resin). Van Duin et al. proposed the reaction mechanism of resol curing of ENB-containing EPDM using 2-ethylidene norbornene as a low-molecular-weight model compound [37]. The rubber-like properties of PP/EPDM or PP/NBR TPV could be improved by using a dimethylol-octyl phenol curing agent [38]. It was found that the *in situ* formation of a graft or block copolymer between the rubber particles and the PP matrix leads to the improvements in the compression set, oil resistance, and processing characteristics of the material.

The usage of a resol-based cure system leads to the formation of off-white color and black specks. These stimulate to explore the other cross-linking systems, such as platinum-catalyzed hydrosilane and peroxide/coagent. Naskar and Noordermeer [39] studied about the influence of the different peroxides on the PP/EPDM TPVs at fixed and varied blend ratio. The solubility parameter of peroxide relative to the polymers used (PP and EPDM), the decomposition mechanism of the peroxide, and the kinetic aspects of the peroxide fragmentation are found to govern the final mechanical properties of the TPV. Amongst various peroxides taken for the investigation, dicumyl peroxide (DCP) exhibits the best balance of the physicomechanical and elastic recovery property. It was found that the closer the solubility parameter of the peroxide to that of the EPDM, the higher is the tensile strength and the better the elastic recovery property. To overcome the blooming tendency of peroxide decomposition products, multifunctional peroxides were developed. These multifunctional peroxides contain functional unsaturated groups (coagent functionality) and the peroxide group in a same molecule. Particularly, 2,4-diallyloxy-6-*tert*-butylperoxy-1,3,5-triazine turns out to be a good alternative for the DCP/TAC combination [40].

Several patents were found in the literature describing hydrosilane as a curative for the preparation of TPV. PP/EPDM blends can also be cross-linked by silane grafting in the presence of a small amount of peroxide. Fritz et al. [41] studied the organosilane cross-linked PP/EPDM TPV. Grafting, hydrolysis, and condensation cross-linking reactions were carried out in a single-stage process. Lopez Manchado and Kenny [42] employed a special cross-linking agent such as benzene-1,3-bis(sulfonyl)azide for the preparation of TPV. They concluded that the sulfonyl-azide group can act as an effective cross-linking agent for the elastomeric phase and as a compatibilizing agent between the elastomeric and thermoplastic phase.

14.5.1.2 PP/Ethylene-Alpha Olefin-Based TPVs

Ethylene-alpha olefin or plastomers (or POEs) are relatively a family of polymers aligned with the development of single cited organometallic (metallocene catalyst) polymerization technique [43]. The uniform comonomer insertion (octene) resulted in a low-density product and the physical properties span the range between plastic and elastomeric behavior. These classes of polymers arise to make reduced crystallinity, unique miscibility, toughening, and elastic characteristics commercial reality. The physical properties of such polymers make them extremely useful for the automotive interior application. Because of the low level of unsaturation, these polymers exhibit outstanding heat, ultraviolet, and aging resistance. Their molecular structures enable these polymers to exhibit a low glass transition temperature (T_g), which in turn exhibits very good low temperature impact properties. Blends of POE and PP were surprisingly compatible and widely used as an impact modifier of PP [44]. The key features are the uniform intermolecular comonomer distribution, low extractable, comonomer-dependent melting point, and low ash content. A study on the effectiveness of the POE as impact modifiers for PP was carried out in relation to the traditional modifier EPDM. The results showed that PP/EPDM and PP/EOC represent a similar mechanical performance but the latter shows better processability.

Further attempts were made to prepare PP/EOC-based TPVs. To make TPVs based on a PP/EOC blend system, phenolic resin (resol) is ineffective, because a resol system needs an unsaturated cite to activate the cross-linking process to form a network structure. Peroxides can cross-link both the saturated and unsaturated polymers without any reversion characteristics. Babu et al. [45–47] made a detailed investigation on the coagent-assisted peroxide-cured PP/EOC TPVs with special focus on the influence of different peroxides, different coagents, and mixing protocol. The observed mechanical properties were determined by the ratio between the degree of degradation in the PP phase and the extent of cross-linking in the EOC phase. The addition of a coagent in the peroxide-cured PP/EOC TPVs increases the cross-linking efficiency in the EOC phase and decreases the degradation in the PP phase [46]. A one-to-one comparison of the dynamically vulcanized PP/EOC and PP/EPDM blends using the same type of peroxide/coagent combination was investigated. The results suggested that the uncross-linked and dynamically cross-linked blends of PP/EOC exhibit superior mechanical properties and lower viscosity in the melt state. Furthermore, PP/EOC blends show better cyclic loading properties over PP/EPDM-based blends but with marginal compromise in the set properties. It was demonstrated that the improved mechanical and dynamic properties of PP/EOC-based blends are due to the combined effect of the presence of smaller crystals and a better interfacial interaction of EOC with PP.

14.5.1.3 Natural Rubber-Based TPVs

Thermoplastic natural rubber (TPNR) blends were prepared by melt mixing of natural rubber (NR) and thermoplastic material via both simple blending and dynamic vulcanization process (i.e., TPVs). A thermoplastic matrix such as polypropylene (PP), low-density polyethylene, high-density polyethylene, linear low-density polyethylene, polystyrene, polyamide, poly(methyl methacrylate) (PMMA), and ethylene-vinyl acetate (EVA) copolymer. Modified forms of NR have also been used to prepare TPNRs [48]. Epoxidized NR (ENR) is one of the modified forms currently used to prepare TPNRs. Graft copolymers of NR with PMMA, polystyrene, and glycidyl methacrylate–styrene copolymer have also been used to prepare TPNRs. These materials were initially developed by the Malaysian

Rubber Producers' Research Association (MRPRA). It has been claimed that these TPNRs could replace vulcanized rubber in end products where high resilience and strength were not essential but good low-temperature performance was needed. They could also replace flexible plastics, such as plasticized PVC, EVA, and PP copolymers.

Varghese et al. [49] studied the influence of different cross-linking systems such as the sulfur and peroxide system in the preparation of NR/PP-based TPVs with special reference to reduce the incompatibility resulting from the molecular weight mismatch of NR/PP blends. Viscosity mismatch-derived incompatibility can be overcome by several methods such as improving the blending process, adjusting extender oil and filler concentrations in the dissimilar polymers, or adjusting the individual raw polymer viscosities. In general, the molecular weight (M_n) of NR is very high (~7×10^5) compared with that of PP (~2×10^5). One of the possible methods to improve the compatibility of NR with PP is to reduce the molecular weight of NR by masticating to a different level. It was found that mastication of NR for 10 min ($M_n \sim 4 \times 10^5$) is found to improve processability without significantly affecting the technological properties of the blend. However, reducing the molecular weight of NR below 3×10^5 adversely affected the technological properties. The processability and technological properties are best balanced in blends containing NR having a molecular weight (M_n) in the range of 4×10^5. The reduction in the molecular weight of NR by mastication and the consequent improvement in processability are more noticeable in the case of the peroxide-vulcanized blends compared with sulfur-vulcanized samples.

The effect of various types and concentrations of peroxides on the rheological, mechanical, and morphological properties of a 60/40 NR/PP TPVs was investigated by Nakason et al. [50]. They claimed that typical cross-linking temperature, the cross-link efficiency, and the relative amounts of decomposition products of each peroxide were the main factors that influenced the properties of TPVs. Among various peroxides studied, DCP and Di (tert-butyl peroxy isopropyl) benzene DTBPIB have typical cross-linking temperatures close to the mixing temperature, which provide high cross-link efficiency and highly reactive radicals.

14.5.2 Polar Polymer-Based TPVs

TPEs based on acrylonitrile–butadiene–styrene (ABS) and a variety of rubbers (including NBR) were first reported by Coran et al. in 1981 [11]. The isothermal/nonisothermal crystallization behavior of PP in acrylonitrile butadiene rubber (NBR)/PP TPVs prepared with three different processing methods was studied by Ming Tian et al. [51]. The vulcanized NBR particles in TPVs act as heterogeneous nucleation centers and increase the number of nuclei. The crystallization rate of PP thereby increases and the growth of PP spherulites is restrained because of the isolation of vulcanized NBR particles. Because the addition of compatibilizer improves the compatibility of NBR and PP, the smaller and uniformly dispersed NBR particles are obtained. The resulting system exhibits more and smaller PP crystals as well as a higher crystallization rate.

TPEs were developed from dynamically vulcanized blends of NBR/Poly(styrene-co-acrylonitrile) (SAN), NBR/waste NBR/SAN, and NBR/ABS-based scrap computer plastics (SCPs), and their properties were earlier reported by Anandhan et al. [52,53].

The aforementioned dynamically vulcanized blends exhibit a phase-separated microstructure in which dynamically cross-linked NBR particles are dispersed in SAN and ABS matrices, respectively, in the appropriate blends. Moreover, the blends dynamically vulcanized with the sulfur–accelerator system exhibit better mechanical properties than those dynamically vulcanized with DCP. Table 14.1 shows the mechanical properties of the uncross-linked and dynamically cross-linked blends of NBR/SAN with varying level of sulfur dose at different blend ratios. The TPE composition prepared by adding rubber-curative master batch to softened SAN gives higher mechanical properties than that prepared by adding curatives to the softened plastic–rubber preblend [52].

The investigation on the preparation and properties of TPVs based on ENR/PP blends using the sulfur cross-linking system was reported by Nakason et al. [54]. The effects of blend ratios of ENR/PP, types of compatibilizers, and reactive blending were investigated. Phenolic-modified PP (Ph-PP) and graft copolymer of maleic anhydride on PP molecules (PP-*g*-MA) were prepared and used as blend compatibilizers. It was found that a 60/40 blend ratio of ENR/PP TPV with Ph-PP as a blend compatibilizer showed the smallest dispersed rubber particles and the best overall balance of properties. A detailed investigation on NR and PMMA blends was studied by Oommen et al. [55], with reference to the effect of the blend ratio, cross-linking system, processing conditions, and graft copolymer concentration (NR-*g*-PMMA). Some polar polymer-based super-TPVs are shown in Table 14.2. As the name implies, these classes of TPVs were developed for special purpose applications, such as high-temperature utilization, oil-resistant condition, fatigue resistance, and commercial availability.

TABLE 14.1   Mechanical Properties of 60/40, 70/30, and 80/20 NBR/SAN Blends Dynamically Vulcanized at Different Sulfur Levels (Keeping MBT 1 phr; TMTD 0.5 phr as Constant)

Blend Details (NBR/SAN)	Sulfur Level	Tensile Strength (MPa)	Elongation at Break (%)	100% Modulus (MPa)	200% Modulus (MPa)	Tension set at 100% Elongation (%)	Hardness (Shore A)
60/40	—	6.2	150	5.3	—	50	64
	0.25	8.8	135	8.5	—	48	91
	0.50	11.9	213	9.4	11.7	46	89
	0.75	14.5	264	10.2	13.2	45	89
	1.0	14.6	199	11.6	—	47	90
70/30	—	2.2	191	2.1	—	44	38
	0.25	7.5	246	5.8	7.3	32	83
	0.50	9.5	268	6.2	8.6	24	81
	0.75	11.9	267	6.7	9.6	24	79
	1.0	13.4	243	8.3	12.1	28	81
80/20	—	1.5	280	1.3	1.4	16	25
	0.25	5.1	352	2.6	3.8	10	59
	0.50	6.6	345	3.0	4.6	8	61
	0.75	7.3	350	3.3	5.0	8	66
	1.0	9.3	360	3.8	5.9	10	66

Source: Anandhan, S. et al., *J. Appl. Polym. Sci.*, 88(8), 1976–1987, 2003. With permission.

TABLE 14.2   Commercially Available Super-TPVs

Grade Name	Elastomer Type	Matrix Phase	Supplier
TPSiV	Silicon	Polyamide (PA), thermoplastic polyurethane (TPU)	Dow corning
Zeotherm	Polyacrylate (ACM)	Polypropylene (PP), PA, polyester	Zeon
E-TPV	Ethylene acrylate	Thermoplastic copolyester (COPE)	Dupont
Uniprene-XL	Hydrogenated-styrene block copolymer (H-SBC)	PP	Teknor-apex
Serel	Styrene butadiene (SSBR)	PP	Goodyear
Septon V	H-SBC, reactive hard block	SBC triblock	Kuraray

Source: Robert Eller, Robert Eller Associates Inc., Intra-TPE competition in an expanding global market, TOPCON 2005, Akron, OH, 2004. With permission.

14.5.2.1 Nylon-Based TPVs

A very little information about TPVs based on nylon blends is available in the literature. Polyamides are defined as engineering pseudoductile polymers because of their high energy for craze initiation compared with the so-called brittle matrices. Generally, pseudoductile matrices deform under a uniaxial tensile stress mode mainly by shear yielding and exhibit a crazing phenomenon as the deformation mechanism. These possess limitations in their end use because of weak impact resistance particularly below their glass transition temperature. Blending is a well-known method for modifying polyamide properties. This induces a multiple craze irradiation and stopping throughout the material, which helps to avoid catastrophic crack failures. Blends of polyamides and unfunctionalized elastomers have low impact toughness because the rubber particles formed during melt blending are relatively large. Maleic anhydride grafted ethylene–propylene elastomers, EPR-*g*-MA, are frequently used for toughening polyamides.

Huang et al. [56] studied the effect of dynamic vulcanization on the crystallization behavior of an EPDM/nylon copolymer-based TPV with special reference to the compatibilizer. Morphological and thermal analyses were used to evaluate the effect of a compatibilizer on the size of dispersed rubber particles and its effect on the crystallization behavior of a nylon copolymer matrix in TPVs.

Oderkerk and Groeninckx [57] investigated the influence of the compatibilizer (MA-*g*-EPM) and EPDM-*g*-MA rubber/thermoplastic viscosity ratio on the blend phase morphology, mechanical properties, and deformation–recovery behavior by changing the molecular weight of nylon 6. It was found that the viscosity ratio plays a crucial role in achieving fine rubber dispersion in a nylon matrix. The viscosity of the nylon phase should be low enough to shift the phase inversion toward higher rubber content. On the other hand, if the viscosity of the nylon is too low, coarse blend morphology was observed resulting in poor mechanical and elastic recovery characteristics.

The effects of the vulcanizing agents, compatibilizer, PA content, aging, and reprocessing on the mechanical properties of EPDM/PA TPV were investigated by Liu et al. [58]. From the experimental results, it was found that chlorinated polyethylene (CPE) has a better effect in compatibilizing the EPDM/copolyamide blends compared with the other compatibilizers, including maleic anhydride grafted ethylene propylene rubber (MAHG-EPR), maleic anhydride grafted EPDM (MAH-G-EPDM), and epoxidized ethylene-propylene-diene rubber. Tensile strength and elongation at break go through a maximum value at a compatibilizer resin content (on total rubber content) of 20%. EPDM/PA TPV using sulfur as a vulcanizing agent has higher tensile strength and elongation than that of TPVs using a brominated *tert*-butyl phenolic resin or peroxide as a vulcanizing agent. The glass transition temperature (T_g) of PA moves toward a higher temperature with increasing CPE content, but the T_g of EPDM changes very slightly toward a lower temperature. A morphological study shows that the results show that the particles of the dispersed EPDM phase have an average size of 2 µm in the dynamically vulcanized EPDM/CPE/PA TPVs.

Kumar et al. [21] studied the rheological and morphological properties of nylon and nitrile rubber (NBR) as a function of the blend ratio, dynamic cross-linking, compatibilization, and processing temperature. The viscosity of the blends showed positive deviation from a linear rule of mixtures. Compatibilization using CPE increased the melt viscosity of the blends. The addition of the compatibilizer decreased the domain size of the dispersed phase, followed by an increase after a critical concentration of the compatibilizer, where the interface was saturated. The temperature dependence of the viscosity of different blend systems was studied using the Arrhenius equation and activation energy measurements. The compatibilized system showed higher values of activation energy compared with the uncompatibilized system, which means that the blends become more temperature sensitive in the presence of the compatibilizer. Among the vulcanized blends, the DCP-cross-linked blend showed a fairly stable morphology having a fine dispersion. Dynamic vulcanization reduced the die swell values of the nylon/NBR blends. The morphology of dynamically vulcanized systems is found to be highly stable. Nevertheless, the extrudate morphology of TPVs depended on the blend ratio, compatibilization, and shear rate.

14.5.3 Recycled Polymer-Based TPVs

One of the various problems that humankind faces as it enters into the twenty-first century is how to manage and dispose waste. Because polymeric materials do not decompose easily, disposal of waste polymers is a serious environmental problem. Reclaiming of scrap rubber products, for example, used automobile tires and tubes, hoses, conveyor belts, etc., is the conversion of a three-dimensionally interlinked, insoluble, and infusible strong thermoset polymer to a two-dimensional, soft, tacky, low modulus, processable, and vulcanizable polymer. The resultant mass of rubber is called the reclaim rubber or whole tire reclaim, or ground tire rubber (GTR). There are two approaches to the combination of scrap rubber and plastics. The initial interest was to use the crumb in minor proportions to toughen the plastics improving impact strength and to reduce the overall cost. A more recent interest is to develop a type of TPE wherein rubber is the major component bonded together by thermoplastics and which can be processed and recovered as thermoplastics. Deanin and Hashemiolya [59] studied the use of GTR as a filler in different thermoplastics. They found that the retention of the mechanical properties of the thermoplastic compounds containing GTR depends on the nature of the matrix polymer (low polarity and low crystallinity appear to favor compatibility). The loading of GTR and the compatibility between the GTR and the polymer matrix play a key role in the properties of the composite. Utilization of GTR in thermoplastic elastomeric blends as a partial replacement of the rubber phase provides an attractive alternative for the recycling of rubber waste. Osborn [60] studied the feasibility of activated tire rubber, a patented surface-modified crumb rubber product, as a modifying ingredient in TPEs such as Santoprene™. The most beneficial impact of GTR on TPE compounds may be economic, but a substantial drop in tensile strength and elongation was disappointing. Naskar et al. [61] observed that in TPE formulations based on the EPDM/poly(ethylene-*co*-acrylic acid) blend, 50% of EPDM could be substituted by GTR. Morphological studies show that virgin EPDM forms a coating over the GTR and the plastic phase forms the continuous matrix. Liu et al. [62] analyzed the feasibility of developing TPE from recycled rubbers such as EPDM, styrene butadiene rubber (SBR) and NR/SBR blends, and PP. Kumar et al. [63] used thermomechanically decomposed GTR to produce TPEs composed of low-density polyethylene (LDPE); virgin rubbers such as SBR, NR, and EPDM; and GTR with and without dynamic curing. Jacob et al. [64] replaced virgin rubber by ground EPDM vulcanizate powder in a PP/EPDM TPE and found that the mechanical properties can be improved by replacing EPDM with ground EPDM vulcanizate powder up to 45% loading, beyond which processing becomes difficult.

Recycling of plastics has been practiced since the invention of plastic processing methods. There are several methods available for the recycling of thermoplastics. But utilization of thermoplastics waste in making TPEs/TPVs is a newer method and widens new scope and opportunities in the field of recycling. Nevatia et al. [65] prepared TPEs from reclaimed rubber and scrap LDPE. They studied the physical, dynamic, and mechanical properties; rheological behavior; and morphology of these blends. A 50:50 rubber/plastic ratio was found to be the best in terms of processability, elongation, and set properties. A sulfur–accelerator system was found to be better than a peroxide system for dynamic cross-linking. Johnson [66] patented a process for the preparation of compatibilized melt processable TPO ground rubber compositions. Pulverized PP-based bumper waste and EPDM were melt blended in various ratios and dynamically vulcanized using DCP during twin-screw extrusion. The blends containing 40%–45% of EPDM have optimum thermoplastic elastomeric properties. Anandhan et al. [67] reported the development of TPEs from the blends of nitrile rubber (NBR)/SCP based on ABS terpolymer. ABS, one of the most important and high-volume engineering plastics, is mostly used for making housing of equipment. ABS is a polar thermoplastic and possesses very good oil resistance. NBR is a polar rubber, and hence it was blended with SCPs based on ABS with an objective of obtaining oil-resistant TPEs as well as a means of recycling ABS-based SCPs. Mechanical properties and morphology were reported, and it was found that the dynamically vulcanized 60/40, 70/30, and 80/20 NBR/SCP blends behave as TPEs. It was shown that dynamic cross-linking of NBR during its mixing with SCP improves its properties such as tensile strength and tension set. These blends have been found to exhibit excellent resistance to various polar solvents and IRM #903 oil. The mechanical properties of the dynamically vulcanized 60/40 NBR/SCP blend are comparable to those of Geolast™. The AFM amplitude and 3D height images of the 60/40 NBR/SCP uncross-linked and dynamically cross-linked blends are shown in Figure 14.8. The hills correspond to the NBR phase and the valleys to the ABS phase. The dynamically vulcanized blends have very uniformly dispersed NBR particles, whereas in the unvulcanized blends, dispersion is not so much uniform. Moreover, dynamically vulcanized NBR particles are finer than those of the unvulcanized blends. This is

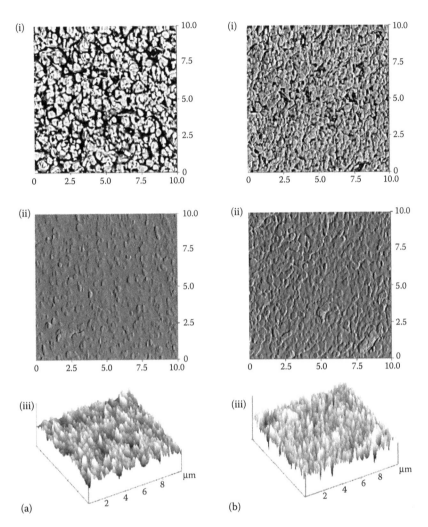

FIGURE 14.8 (a) (i) Phase, (ii) amplitude, and (iii) 3D height images of 60/40 NBR/SCP blends; (b) (i) Phase, (ii) amplitude, and (iii) 3D height images of dynamically vulcanized 60/40 NBR/SCP blends. (From Anandhan, S. et al., *Rubber Chem. Technol.*, 76, 1145–1163, 2003. With permission.)

proven from the surface area values obtained from roughness analysis. The surface area for the dynamically vulcanized blends is larger than that of the unvulcanized blends. These blends could be potential candidates for oil-resistant applications at a lower cost.

14.5.4 TPVs by Electron Beam Reactive Processing

EB processing utilizes the energy input of high-energy electrons to achieve the desired material changes via radical induced chemical reactions. Accelerated EBs have sufficient energy to attack the electrons in the atom shell, but not its nucleus, and can therefore initiate chemical reactions. The EB processing of polymers has gained special importance as it is extremely fast and completed in a fraction of seconds, improves state and rate of cure, involves no external crosslinking agent, and is versatile in application. Reactive processing of polymeric materials by an EB involves polymerization, cross-linking, modifications, grafting, and degradation, which were determined by the different reactive species formed initially by the EB.

In general, polymeric materials undergo cross-linking and chain scission when exposed to EB irradiation. It was suggested that polymers with a single or no side chain ($-CH_2-CR_1H-$ or $-CH_2-CH_2-$) are predominant to cross-linking, whereas those with two side chains attached to a single backbone carbon ($-CH_2-CR_1-R_2-$) are bound to main chain scission. Polyethylene, polystyrene, polyvinyl chloride, NR, and SBR are polymers that tend to cross-linking. In contrast, PMMA and butyl rubber (IIR) are some polymers that predominantly undergo chain scission [68]. The most relevant industrial applications are based on EB technology, such as pipes and cables, packing films, foams, as well as EB curing of large-scale surface finishing coatings, lacquers, and inks. Volume of information is available in the prevailing literature on the EB modification of polymers, and most of the irradiation techniques were practiced under stationary conditions [69].

A state-of-the art technology of the EB treatment of polymeric material at high shear and temperature (during melt mixing) was developed in Leibniz Institute for Polymer Research, Dresden, Germany, and termed as the EIReP technique [70–72]. The experimental setup of the EIReP process consists of an electron accelerator directly coupled with an internal mixer as shown in Figure 14.9. Such a setup aids to irradiate the polymers during the melt mixing even at elevated temperature. The penetration depth of electrons depends on acceleration voltage. As the voltage increases, the penetration depth also increases. The gas atmosphere also plays an important role in the modification of polymers, depending on the type of gas; functionalization changes for each specific polymer. In air, the amount of functional groups generated is more compared with nitrogen atmosphere. In addition, the amount of chain scission and cross-linking can also be controlled via gas atmosphere and additional use of cross-linking additives.

The important process control parameters involved in EIReP are listed below:

Dose: It is the amount of absorbed energy per unit of mass. It is measured in Gray (Gy). It can be controlled by total treatment time.

Kilogray is commonly used (kGy). 1 Gy = 1 J/kg.

Dose rate: It is defined as the absorbed dose per unit of time. It is measured in Gy/h and can be controlled by electron current at fixed treatment time.

Electron energy: It is controlled by acceleration voltage and is measured in Mega electron volts (MeV). It controls the treatment volume as well as the ratio of penetration depth to minimum rotor-to-rotor distance during EIReP.

Dose/rotation: It is defined as absorbed dose per rotation of rotors and is measured as kGy/rotation. It controls the absorbed dose during EIReP per single pass through the treatment zone.

Electron current: It is defined as the number of electron charges per unit of time. It is measured in milliamperes (mA).

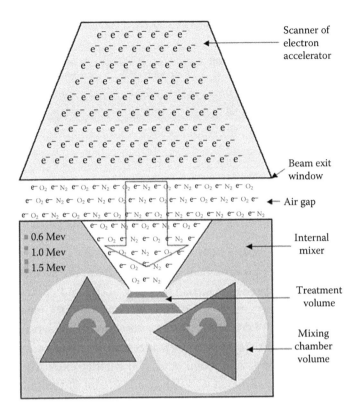

FIGURE 14.9   Schematic representation of the set-up: coupling of an electron accelerator with an internal mixer. (From Naskar, K. et al., *Exp. Polym. Lett.*, 3[11], 677–683, 2009.)

The dose per rotation can be controlled by the absorbed dose as well as rotor speed of the internal mixer. The total mixing volume is modified due to the change of polymer mass within the penetration depth of electrons during the mixing process. In the novel process, electron treatment time also influences the ratio of radical generation rate to mixing rate (dose per rotation) and electron energy controls the ratio of modified volume to total mixing chamber volume.

EIReP has several advantages compared with a conventional dynamic vulcanization technique. This can be applied to different polymeric systems that can bring about desired physical and chemical changes in the final state. This state-of-the-art polymer modification with high-energy electrons allows the control of

- Radical generation without any additives

- Intensive mixing at flexible reaction rate

- Customized reaction rate and diffusion rate

Naskar et al. [73] explored the use of an EB-induced reactive processing technique in the field of preparation of PP/EPDM TPVs. Table 14.3 summarizes the mechanical properties of the PP/EPDM TPVs prepared by EB reactive processing. It was reported that

TABLE 14.3 Comparative Physical Properties of Electron Induced Dynamic Vulcanization of 50/50 PP/EPDM Blend Ratio

Electron Energy (MeV)	Dose (kGy)	Treatment Time (s)	E Modulus (MPa)	Tensile Strength (MPa)	Elongation at Break (%)
0	0	0	112	4.7	46
1.5	25	60	159	7.0	135
1.5	50	60	156	9.2	298
1.5	100	60	173	9.8	282
1.5	100	30	191	9.5	215
1.5	100	15	176	14.7	624
0.6	50	60	150	8.2	168

Source: Naskar, K. et al., *Exp. Polym. Lett.*, 3(11), 677–683, 2009. With permission.

electron-induced reactive processing with 1.5 MeV electrons for 15 s at an absorbed dose of 50 kGy gives the best balance of mechanical properties within the accessible experimental window. On irradiation two processes simultaneously occur and contribute to enhancement in the mechanical properties: (a) cross-linking in the EPDM phase and (b) *in situ* compatibilization of PP and EPDM. These processes during EIReP depend mostly on electron treatment time as well as electron energy. Electron treatment time correlates with dose rate and radical generation rate. Babu et al. [74] further explored the use of EIReP technology in the preparation of PP/EOC TPVs as a function of treatment time, electron dosage, and electron energy. The mechanical, morphological, and rheological properties of PP/EOC TPVs were studied with special reference to the exposure time (16–64 s) keeping absorbed dose (100 kGy) and electron energy (1.5 MeV) invariable. Table 14.4 summarizes the physicomechanical properties of the PP/EOC TPVs prepared by the EIReP technique. It was found that chain scission dominates over chain cross-linking in both EOC and PP phases with the increase in exposure time. The primary factor is found to be the predominance of oxidative degradation during electron-induced reactive processing in air atmosphere. To extend the scope of the investigation, 30:70 blend ratios of PP and EOC are dynamically vulcanized using EIReP employing a range of absorbed doses (25, 50, and 100 kGy) while keeping the electron energy (1.5 MeV) and treatment time (16 s) as constant. Figure 14.10 shows the stress–strain behavior of the corresponding samples. The structure/property relationships of the prepared samples are studied using various characterization techniques such as DMA, DSC, SEM, and melt rheology. It was found that virgin EOC

TABLE 14.4 Comparative Physical Properties of Electron Induced Dynamic Vulcanization of 50/50 PP/EOC Blend Ratio

Electron Energy (MeV)	Dose (kGy)	Treatment Time (s)	Young's Modulus (MPa)	Tensile Strength (MPa)	Elongation at Break (%)
0	0	0	277	10.8 ± 0.6	147 ± 12
1.5	100	16	193	11.0 ± 0.4	208 ± 10
1.5	100	32	206	9.4 ± 0.2	58 ± 12
1.5	100	64	210	8.4 ± 0.5	23 ± 5

Source: Babu, R.R. et al., *Radiat. Phys. Chem.*, 80, 1398–1405, 2011. With permission.

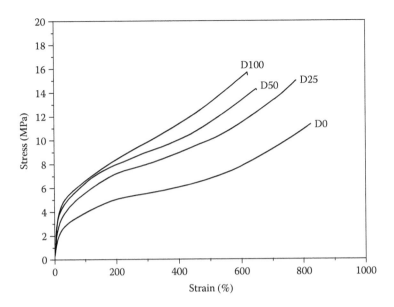

FIGURE 14.10 Tensile stress–strain curves of PP/EOC unmodified and EIReP modified samples (variation of absorbed doses [25, 50, and 100 kGy]; keeping the electron energy [1.5 MeV] and treatment time [16 s] as constant). (From Babu, R.R. et al., *Radiat. Phys. Chem.*, 80, 1398–1405, 2011. With permission.)

shows a prominent cross-linking characteristics when treated under a stationary condition and at ambient temperature in air (gel content: 77%), whereas the extent of cross-linking is severely restricted when treated under a dynamic condition at 180°C in air (gel content: 28%). The above observation clearly supports the interdiffusion of oxygen and subsequent oxidative degradation mechanism. Nevertheless, it is strongly expected that high degree of cross-linking can be promoted under the conditions where a coupling-agent and oxygen-free atmosphere is adapted during EIReP. Opportunities of exploring the EIReP technique pave a new technology and direction for the development of various types of TPVs in an ecofriendly fashion.

Mondal et al. [75,76] developed a range of TPEs and TPVs based on 50/50 wt.% PP/NR by the use of high-energy electrons. According to the mode of electron treatment, PP/NR-based TPEs and TPVs were prepared. PP/NR TPEs belong to the class of stationary mode of EB irradiation, whereas the EIReP technique was used to prepare PP/NR-based TPVs. It has been shown that the properties of these blends could be customized depending on the characteristics of raw materials and the process parameters of the electron treatment. In addition, the effect of polyfunctional monomers (such as trimethylol propane trimethacrylate (TMPTMA) and dipropylene glycol diacrylate (DPGDA)) on the properties of these blends was also investigated. It was reported that viscosity mismatch, structural incompatibility, and cure state mismatch need to be optimized by EIReP process parameters. The mastication was a very useful tool for lowering the viscosity of NR to match the viscosity of PP. Higher viscosity grade of PP can be selected to minimize the effect of thermal induced degradation during electron irradiation. Furthermore, before EIReP, a higher rotor speed led to an improved well-dispersed morphology due to the requirement of increased shear stress to break down highly viscous NR domains. On the contrary, this higher rotor speed during EIReP affected the level of tensile properties of PP/NR TPVs due to severe oxidative degradation tendency of blend components. The oxidative degradation of blend components (particularly PP) was limited by performing the reactive processing in nitrogen atmosphere.

14.6 SUMMARY AND CONCLUSIONS

This chapter represents a comprehensive and insight of the state of the art of the dynamic vulcanization process and preparation and properties of various TPVs. TPVs are a special class of TPEs, produced by a process called "dynamic vulcanization," which involves the cross-linking of a rubber phase while it is being mixed with a thermoplastic material at elevated temperature. The process is carried out at high shear and above the melting temperature of the thermoplastic component. Morphologically, TPV consists of cross-linked rubber particles dispersed in a continuous thermoplastic matrix, which explains both its elasticity and melt processability. The global annual growth rate of TPEs based on TPVs is about 15%. TPVs represent the second largest group of soft TPEs, after styrenic-based block copolymers. Several factors such as compatibility of blend components, molecular weight and its distribution of constituent polymers, state and mode of cross-linking nature, particle size and its distribution, and processing conditions have been found to be the deciding factors in framing the technological properties of TPVs. In principle, there is a wide variety of commercially available rubbers and plastics that can be considered for the preparation of TPVs. However, relatively few of them have been of technological importance, because most polymers are incompatible with each other. For TPVs, the preferred morphology is a fine and elastomeric dispersed phase and a thermoplastic continuous phase. The interfacial tension between the components should be relatively low to achieve a good and fine dispersion of the cross-linked rubber phase. A well-known scientific and commercial example of dynamically vulcanized TPE compositions is the ethylene propylene diene terpolymer/PP (EPDM/PP) TPVs. TPVs offer potential and proven applications in various sectors such as automotive, industrial hose, electrical applications, mechanical appliances, and soft touch applications. Recent reports show that the future of TPEs/TPVs is very bright, as a replacement for thermoset rubber in numerous applications and as new soft and flexible materials for new design and novel applications in industrial and health care products. In future, TPEs/TPVs will capture more ground in the biomedical sector and artificial organs market as replacement materials for thermoset rubber with all the performance advantages and low processing costs.

On industrial scale, TPVs are preferably produced in twin-screw extruders, which allow continuous processing with high throughputs and with a large degree of process flexibility. Dynamic vulcanization of TPVs on extruders is much more complex than that in batch kneaders. In continuous preparation of TPVs, different processes such as melt mixing of constitute polymers, yielding a homogenized fine dispersed blend; redistribution of extender oil or fillers; and finally the completion of cross-linking reaction will mutually interact and partly take place simultaneously. The challenging aspect lies in the fundamental understanding of the true insight in what is happening inside the extruder. Better understanding of the process will advance in a scientific approach in product development and process optimization.

Machado and Van Duin [37] studied the continuous preparation of TPVs in extruder, and their initial study shows that melting, dispersion, cross-linking, and phase inversion are complete, for most TPV compositions, in the first part of the extruder. However, further investigations are required to unwind the process flexibility and productivity.

REFERENCES

1. Paul, D.R. and S. Newman. 1978. *Polymer Blends.* New York: Academic Press.
2. Utracki, L.A. 1989. *Polymer Alloys and Blends.* New York: Hanser Publisher.
3. Legge, N.R. and G. Holden. 1987. *Thermoplastic Elastomers: A Comprehensive Review.* Munich, Germany: Hanser Publisher.
4. Morris, H.L. 1979. *Handbook of Thermoplastic Elastomers.* New York: Van Nostrand Reinhold.
5. White, J., S.K. De, and K. Naskar. 2009. *Rubber Technologists Handbook,* volume 2. Shawbury, UK: Smithers Rapra Technology.
6. Bhowmick, A.K. 2008. *Current Topics in Elastomer Research.* Boca Rotan, FL: CRC Press.
7. De, S.K. and A.K. Bhowmick. 1990. *Thermoplastic Elastomers from Rubber-Plastic Blends.* London: Ellis Horwood.
8. Sengupta, P. 2004. PhD Thesis. Morphology of olefinic thermoplastic elastomer blends : A comparative study into the structure-property relationship of EPDM/PP/oil based TPVs and SEBS/PP/oil blends. Enschede, the Netherlands: Twente University.
9. Fisher, W.K. 1973. Thermoplastic blend of partially cured monoolefin copolymer rubber and polyolefin plastic. U.S. Patent 3,758,643.
10. Coran, A.Y. and R. Patel. 1980. Rubber-thermoplastic compositions. Part I. EPDM-polypropylene thermoplastic vulcanizates. *Rubber Chemistry and Technology* 53(1):141–150.
11. Coran, A.Y., R. Patel, and D. Williams. 1982b. Rubber-thermoplastic compositions. Part V. Selecting polymers for thermoplastic vulcanizates. *Rubber Chemistry and Technology* 55(1):116–136.
12. Coran, A.Y. and R. Patel. 1983a. Rubber-thermoplastic compositions. Part VIII. Nitrile rubber-polyolefin blends with technological compatibilization. *Rubber Chemistry and Technology* 56(5):1045–1060.
13. Abdou-Sabet, S. and S. Datta. 2000. Thermoplastic vulcanizates. In *Polymer Blends,* eds. D.R. Paul and C.B. Bucknall. New York: John Wiley and Sons.
14. Coran, A.Y. 1987. Thermoplastic elastomers based on elastomer-thermoplastic dynamically vulcanized. In: *Thermoplastic Elastomers—A Comprehensive Review,* eds. N.R. Legge, and G. Holden. Munich, Germany: Hanser Publisher.
15. Karger, K.J. 1999. Thermoplastic rubbers via dynamic vulcanization. In *Polymer Blends and Alloys,* eds. G.O. Shonaike and G.P. Simon. New York: Marcel Dekker, pp. 125–153.
16. Babu, R.R. and K. Naskar. 2010. Recent developments in the thermoplastic vulcanizates. In *Advances in Polymer Science,* eds. G. Heinrich and B. Voit. Berlin, Germany: Springer, pp. 219–248.
17. Harrats, S., S. Thomas, and G. Groeninckx. 2006. *Micro and Nano Structured Multiphase Polymer Blend Systems—Phase Morphology and Interface.* Boca Raton, FL: CRC Press.
18. Abee, R.I. 2009. Thermoplastic vulcanizates—The rubber particle size to control the properties-processing balance. PhD Thesis. Eindhoven, the Netherlands: Technical University.
19. Radusch, H.J. and T. Pham. 1996. Morphologiebildung in dynamisch vulkanisierten PP/EPDM-Blends. *Kautsch Gummi Kunstst* 49:249–255.
20. Sengupta, P. and J.W.M. Noordermeer. 2005. Three-dimensional structure of olefinic thermoplastic elastomer blends using electron tomography. *Macromolecular Rapid Communications* 26(7):542–547.
21. Kumar, R.C., S.V. Nair, K.E. George, Z. Oommen, and S. Thomas. 2003. Blends of nylon/acrylonitrile butadiene rubber: Effects of blend ratio, dynamic vulcanization and reactive compatibilization on rheology and extrudate morphology. *Polymer Engineering and Science* 43(9):1555–1565.

22. Goettler, L.A., J.R. Richwine, and F.J. Wille. 1982. The rheology and processing of olefin-based thermoplastic vulcanizates. *Rubber Chemistry and Technology* 55(5):1448–1463.

23. Han, P.K. and J.L. White. 1995. Rheological studies of dynamically vulcanized and mechanical blends of polypropylene and ethylene-propylene rubber. *Rubber Chemistry and Technology* 68(5):728–738.

24. I'Abee, R.M.A., M. van Duin, A.B. Spoelstr, and J.G.P. Goossens. 2010. The rubber particle size to control the properties-processing balance of thermoplastic/cross-linked elastomer blends. *Soft Matters* 6:1758–1768.

25. Steeman, P. and W. Zoetelief. 2000. Rheology of TPVs. Paper presented at *ANTEC-2000*, Orland, CA.

26. Katbab, A.A., H. Nazockdast, and S. Bazgi. 2000. Carbon black-reinforced dynamically cured EPDM/PP thermoplastic elastomers. I. Morphology, rheology, and dynamic mechanical properties. *Journal of Applied Polymer Science* 75(9):1127–1137.

27. Sengers, W.G.F., P. Sengupta, J.W.M. Noordermeer, S.J. Picken, and A.D. Gotsis. 2004. Linear viscoelastic properties of olefinic thermoplastic elastomer blends: Melt state properties. *Polymer* 45(26):8881–8891.

28. Leblanc, J.L. 2006. Nonlinear viscoelastic properties of molten thermoplastic vulcanisates: An insight on their morphology. *Journal of Applied Polymer Science* 101(6):4193–4205.

29. Boyce, M.C., S. Socrate, O.C. Yeh, K. Kear, and K. Shaw. Micromechanisms of deformation and recovery in thermoplastic vulcanizates. *Journal of the Mechanics and Physics of Solids* 49:1323–1342.

30. Soliman, M., M.V. Dijk, M.V. Es, and V. Shulmeister. 1999. Deformation mechanism of thermoplastic vulcanisates investigated by combined FTIR and stress-strain measurements. In *Proceedings of ANTEC '99*, volume 2, pp. 1947–1954, New York.

31. Naskar, K. 2007. Thermoplastic elastomers based on PP/EPDM blends by dynamic vulcanization: A Review. *Rubber Chemistry and Technology* 80:504–526.

32. Oderkerk, J., G. De Schaetzen, B. Goderis, L. Hellemans, and G. Groeninckx. 2002. Micromechanical deformation and recovery processes of Nylon-6/rubber thermoplastic vulcanizates as studied by atomic force microscopy and transmission electron microscopy. *Macromolecules* 35:6623–6629.

33. Van, D.M. 2006. Recent developments for EPDM-based thermoplastic vulcanisates. *Macromolecular Symposia* 233:11–16.

34. Machado, A.V. and D.M. Van. 2005. Dynamic vulcanisation of EPDM/PE-based thermoplastic vulcanisates studied along the extruder axis. *Polymer* 46:6575–6586.

35. Goharpey, F., A. Katbab, and H. Nazockdast. 2003. Formation of rubber particle agglomerates during morphology development in dynamically crosslinked EPDM/PP thermoplastic elastomers. Part 1: Effects of processing and polymer structural parameters. *Rubber Chemistry and Technology* 76:239–252.

36. Babu, R.R., N.K. Singha, and K. Naskar. 2010. Melt viscoelastic properties of peroxide cured polypropylene-ethylene octene copolymer thermoplastic vulcanizates. *Polymer Engineering and Science* 50:455–467.

37. Van, D.M. and A.V. Machado. 2005. EPDM-based thermoplastic vulcanisates: Crosslinking chemistry and dynamic vulcanisation along the extruder axis. *Polymer Degradation and Stability* 90(2):340–345.

38. Abdou-Sabet, S., R.C. Puydak, and C.P. Rader. 1996. Dynamically vulcanized thermoplastic elastomers. *Rubber Chemistry and Technology* 69(3):476–494.

39. Naskar, K. and J.W.M. Noordermeer. 2003. Dynamically vulcanized PP/EPDM blends: Effects of different types of peroxide on properties. *Rubber Chemistry and Technology* 76:1001–1018.

40. Naskar, K. and J.W.M. Noordermeer. 2004b. Dynamically vulcanized PP/EPDM blends: Multifunctional peroxides as crosslinking agents—Part I. *Rubber Chemistry and Technology* 77(5):955–971.

41. Fritz, H.G., U. Bolz, and Q. Cai. 1999. Innovative thermoplastic vulcanizate (TPV) two-phase polymers: Formulation, morphology formation, property profiles and processing characteristics. *Polymer Engineering and Science* 39(6):1087–1099.
42. Lopez-Manchado, M.A. and J.M. Kenny. 2001a. Use of benzene-1,3-bis(sulfonyl)azide as crosslinking agent of TPVs based on EPDM rubber-polyolefin blends. *Rubber Chemistry and Technology* 74(2):198–210.
43. Walton, K.L. 2004. Metallocene catalyzed ethylene/alpha olefin copolymers used in thermoplastic elastomers. *Rubber Chemistry and Technology* 77(3):552–568.
44. Lai, S.M., F.C. Chiu, and T.Y. Chiu. 2005. Fracture behaviors of PP/mPE thermoplastic vulcanizate via peroxide crosslinking. *European Polymer Journal* 41(12):3031–3041.
45. Babu, R.R., N.K. Singha, and K. Naskar. 2009. Dynamically vulcanized blends of polypropylene and ethylene octene copolymer: Comparison of different peroxides on mechanical, thermal and morphological characteristics. *Journal of Applied Polymer Science* 113:1836–1852.
46. Babu, R.R., N.K. Singha, and K. Naskar. 2009. Dynamically vulcanized blends of polypropylene and ethylene octene copolymer: Influence of various coagents on mechanical and morphological characteristics. *Journal of Applied Polymer Science* 113:3207–3221.
47. Babu, R.R., N.K. Singha, and K. Naskar. 2010. Interrelationships of morphology, thermal and mechanical properties in uncrosslinked and dynamically crosslinked PP/EOC and PP/EPDM blends. *Express Polymer Letters* 4:197–209.
48. Ibrahim, A. and M. Dahlan. 1998. Thermoplastic natural rubber blends. *Progress in Polymer Science* 23(4):665–706.
49. Varghese, S., R. Alex, and B. Kuriakose. 2004. Natural rubber—Isotactic polypropylene thermoplastic blends. *Journal of Applied Polymer Science* 92:2063–2068.
50. Nakason, C., A. Thitithammawong, K. Sahakaro, and J. Noordermeer. 2007. Effect of different types of peroxides on rheological, mechanical, and morphological properties of thermoplastic vulcanizates based on natural rubber/polypropylene blends. *Polymer Testing* 26:537–546.
51. Tian, M., J. Han, H. Zou, H. Tian, H. Wu, Q. She, W. Chen, and L. Zhang. 2012. Dramatic influence of compatibility on crystallization behavior and morphology of polypropylene in NBR/PP thermoplastic vulcanizates. *Journal of Polymer Research* 19:9745.
52. Anandhan, S., P.P. De, S.K. De, and A.K. Bhowmick. 2003. Thermoplastic elastomeric blend of nitrile rubber and poly(styrene-co-acrylonitrile). I. Effect of mixing sequence and dynamic vulcanization on mechanical properties. *Journal of Applied Polymer Science* 88(8):1976–1987.
53. Anandhan, S., P.P. De, S.K. De, S. Swayajith, and A.K. Bhowmick. 2003. Thermorheological studies of novel thermoplastic elastomeric blends of nitrile rubber (NBR) and scrap computer plastics (SCP) based on acrylonitrile–butadiene–styrene terpolymer (ABS). *Plastics, Rubber and Composites* 32(8):377.
54. Nakason, C., P. Wannavilai, and A. Kaesaman. 2006. Effect of vulcanization system on properties of thermoplastic vulcanizates based on epoxidized natural rubber/polypropylene blends. *Polymer Testing* 25:34–41.
55. Oommen, Z., S. Thomas, C.K. Premalatha, and B. Kuriakose. 1997. Melt rheological behaviour of natural rubber/poly(methyl methacrylate)/natural rubber-g-poly(methyl methacrylate) blends. *Polymer* 38(22):5611–5621.
56. Huang, H., X. Liu, T. Ikehara, and T. Nishi. 2003. Investigation of crystallization behavior in dynamically vulcanized EPDM–nylon copolymer blends. *Journal of Applied Polymer Science* 90(3):824–829.
57. Oderkerk, J. and G. Groeninckx. 2002. Morphology development by reactive compatibilisation and dynamic vulcanization of nylon 6/EPDM blends with a high rubber fraction. *Polymer* 43:2219–2228.
58. Liu, X., H. Huang, Z.Y. Xie, Y. Zhang, Y. Zhang, K. Sun, and L. Min. 2003. EPDM/polyamide TPV compatibilized by chlorinated Polyethylene. *Polymer Testing* 22:9–16.

59. Deanin, R.D. and S.M. Hashemiolya. 1987. Polyblends of reclaimed rubber with eleven thermoplastics. *Polymer Materials Science and Engineering* 8:212–216.
60. Osborn, J.D. 1995. Reclaimed tire rubber in TPE compounds. *Rubber World* 212:34–35.
61. Naskar, A.K., A.K. Bhowmick, and S.K. De. 2001. Thermoplastic elastomeric composition based on ground rubber tire. *Polymer Engineering and Science* 41:1087–1098.
62. Liu, H.S., J.L. Mead, and R.G. Stacer. 2000. Thermoplastic Elastomers and Rubber-Toughened Plastics from Recycled Rubber and Plastics. Paper presented at the *Rubber Division Meeting of the ACS*, Dallas, TX.
63. Kumar, R.C., I. Fuhrmann, and J.K. Kocsis. 2002. LDPE-based thermoplastic elastomers containing ground tire rubber with and without dynamic curing. *Polymer Degradation and Stability*, 76:137–144.
64. Jacob, C., P.P. De, A.K. Bhowmick, and S.K. De. 2001. Recycling of EPDM waste. II. Replacement of virgin rubber by ground EPDM vulcanizate in EPDM/PP thermoplastic elastomeric composition. *Journal of Applied Polymer Science* 82:3304–3312.
65. Nevatia, P., T.S. Banerjee, B. Dutta, A. Jha, A.K. Naskar, and A.K. Bhowmick. 2002. Thermoplastic elastomers from reclaimed rubber and waste plastics. *Journal of Applied Polymer Science* 83:2035–2042.
66. Johnson, L.D. 1992. Includes functionalized olefin polymer. US patent 5157 082.
67. Anandhan, S., P.P. De, S.K. De, A.K. Bhowmick, and S. Bandyopadhyay. 2003. Novel thermoplastic elastomer based on acrylonitrile-butadiene-styrene terpolymer (ABS) from waste computer equipment and nitrile rubber. *Rubber Chemistry and Technology* 76:1145–1163.
68. Drobny, J.G. 2002. *Electron Beam Processes. Radiation Technology for Polymers*. Boca Raton, FL: CRC Press, pp. 63–104.
69. Bohm, G.G.A. and J.O. Tveekrem. 1982. The radiation chemistry of elastomers and its industrial applications. *Rubber Chemistry and Technology* 55:575–668.
70. Gohs, U., A. Leuteritz, K. Naskar, S. Volke, S. Wiessner, and G. Heinrich. 2010. Reactive EB processing of polymer compounds. *Macromolecular Symposia* 296:589–595.
71. Sritragool, K., H. Michael, M. Gehde, U. Gohs, and G. Heinrich. 2010. Polypropylene/rubber particle blends by electron induced reactive processing. *Kautschuk Gummi Kunststoffe* 63:554–558.
72. Thakur, V., U. Gohs, U. Wagenknecht, and G. Heinrich. 2012. Electron-induced reactive processing of thermoplastic vulcanizate based on polypropylene and ethylene propylene diene terpolymer rubber. *Polymer Journal* 44:439–448.
73. Naskar, K., U. Gohs, U. Wagenknecht, and G. Heinrich. 2009. PP-EPDM thermoplastic vulcanisates (TPVs) by electron induced reactive processing. *Express Polymer Letters* 3(11):677–683.
74. Babu, R.R., U. Gohs, K. Naskar, V. Thakur, U. Wagenknecht, and G. Heinrich. 2011. Preparation of polypropylene(PP)/ethyleneoctenecopolymer (EOC) thermoplastic vulcanizates (TPVs) by high energy electron reactive processing. *Radiation Physics and Chemistry* 80:1398–1405.
75. Mondal, M., U. Gohs, U. Wagenknecht, and G. Heinrich. 2013. Polypropylene/natural rubber thermoplastic vulcanizates by eco-friendly and sustainable electron induced reactive processing. *Radiation Physics and Chemistry* 88:74–81.
76. Mondal, M., U. Gohs, U. Wagenknecht, and G. Heinrich. 2013. Preparation of natural rubber based thermoplastic vulcanizates by electron induced reactive processing. *Kautschuk Gummi Kunststoffe* 66:47–50.
77. Eller, R. 2005. Intra-TPE Competition in an Expanding Global Market, TOPCON 2005, Robert Eller Associates Inc., Akron, OH.

CHAPTER 15

Multiferroic Composites Based on a Ferroelectric Polymer

Venimadhav Adyam

CONTENTS

15.1 MULTIFERROICS

Multiferroics are special class of materials in which at least two of the ferroic states such as magnetic, electric, or piezoelastic phases coexist [1–3]. Technologically, they are important materials because the coupling between the ferrophases and the piezoelastic properties facilitates a direct control of ferromagnetic (FM) and ferroelectric (FE) properties via externally mechanical stress. Alternately, external magnetic/electric fields can cause

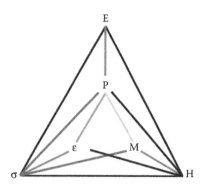

FIGURE 15.1 Concept of multiferroics and possible cross-couplings in multiferroics. H—magnetic field; M—magnetization; E—electric field; P—electric polarization; σ—applied mechanical stress; and ε—strain.

shape change due to induced mechanical stress. Coupling between various ferroic domains of a true multiferroic system is shown in Figure 15.1. Recently, a significant research is devoted on multiferroics, with coupling across magnetic and electric order parameters, called *magnetoelectric* (ME) *materials*. In fact, the coupling between magnetism and the motion charges was recognized in the nineteenth century, which was later combined into a common discipline with the culminating Maxwell equations. However, the long range ordering of electrical and magnetic dipoles in a solid was considered to be different, because displacement of ions and electrons are responsible for electric dipoles; on the other hand, electron spins govern magnetic properties.

Pierre was the first person to demonstrate the *magnetoelectric effect*. The term *multiferroic* was first coined in 1994 by Schmid [4]. The general usage of this term is not only restricted to spontaneous electric and magnetic polarizations; it even extended to any kind of magnetic ordering (FM, antiferromagnetic, ferrimagnetic, etc.) coupled with the FE or antiferrielectric properties [5]. In recent years, a great deal of research has been carried and summarized in many reports [6–8]. Such novel coupling scheme is likely to offer a whole range of new applications and phenomena [9]. The ability to couple either of electric or magnetic polarization allows an additional degree of freedom for designing novel spintronic devices. Other applications include multiple-state memory elements, in which data are stored both in the electric and magnetic polarizations, and novel memory media, which might allow the writing by FE data bit and reading by magnetic field. With piezo and magnetostriction effects, a multiferroic system can offer an undue contribution in harvesting clean energy. Solid-state refrigeration is a novel emerging areas of technology where multiferroics are expected to play a vital role [10].

In ferroelectromagnetic system, the free energy F is described by Landau theory and is given as

$$-F(E,H)=\frac{1}{2}\varepsilon_o\varepsilon_{ij}E_iE_j+\frac{1}{2}\mu_o\mu_{ij}H_iH_j+\alpha_{ij}E_iH_j+\frac{\beta_{ijk}}{2}E_iH_jH_K+\frac{\gamma_{ijk}}{2}H_iE_jE_k+\cdots \quad (15.1)$$

where:

ε_o, μ_o represent permittivity and permeability of the free space and ε, μ represent the same quantities in the material, respectively

α, β, and γ are the linear, first- and the second-order ME coefficients, respectively.

The first term on the right-hand side describes the contribution resulting from the electrical response to an electric field, the second term is magnetic equivalent of the first term, the third term describes the linear ME coupling, and the fourth and the higher terms describe the nonlinear ME coupling.

15.1.1 ME Effect

Linear ME is different from the multiferroics in the sense that they are also magnetically ordered oxides, but they do not posses spontaneous electric polarization [1]. However, applied magnetic fields drives the system to noncentrosymmetric space group and induce the electric polarization for a certain spin configuration (helical, spiral, and frustrated spin structures) by the Dzyaloshinskii–Moriya mechanism [11]. Since the presence/absence of the linear magnetic effect is a symmetry property, materials showing a linear ME effect can be distinguished using the magnetic point group of the material.

The ME coupling parameters can be obtained by differentiating the Equation 15.1 with respect to electric/magnetic field as

$$P_i = \frac{-\partial F}{\partial E_i} = \alpha_{ij} H_j + \frac{\beta_{ijk}}{2} H_j H_k + \cdots \tag{15.2}$$

$$M_j = \frac{-\partial F}{\partial H_j} = \alpha_{ij} E_j + \frac{\gamma_{ijk}}{2} E_j E_k + \cdots \tag{15.3}$$

where:

α_{ij} is called linear ME coefficient

β_{ijk} is called higher order ME coefficient. A ME material with linear ME effect has a relation

$$\alpha_{ij}^2 \le \varepsilon_o \mu_o \varepsilon_{ii} \mu_{jj} \tag{15.4}$$

From Equation 15.4, ME coupling is directly proportional to the relative permeability and permittivity of the materials [12], and by estimating or measuring this relation, one can conclude about the multiferroic or ME materials response [11]. It is important to note that all the ME systems need not be a multiferroic. The best example is Cr_2O_3 which is a well known ME material, but it does not show any spontaneous electrical polarization [13]. The converse is also true, that is, all the multiferroics need not exhibit the ME coupling.

Though having simultaneous FM and FE properties is rather unconventional, a few materials exhibit such properties intrinsically, for example, boracites ($Ni_3B_7O_{13}I$), $BiFeO_3$, and $YMnO_3$ [14,15]. The conditions required for the coexistence of ferroelectricity and ferromagnetism were thoroughly discussed by Hill, and it was concluded that such possibilities are rare [16]. Accordingly, several factors that inhibit the existence of such combinations were reported.

Conventional ferroelectricity was always associated with the structural distortion where the crystal structure changes from the high to low symmetry. This distortion removes the center of symmetry and the atomic displacements produce a dipole per unit cell and, therefore, a spontaneous electric polarization. According to crystal structure, only 13 point groups were identified to have the both properties. Apart from the rich physics and fascinating applications of multiferroic materials, there are several challenges to be solved before realizing them in applications such as (1) scarcity of materials that exhibit different ferroic properties near room temperature, (2) existing multiferroic have week coupling between them, and (3) many of the multiferroic are very leaky and fail in multiferroic electric characterization.

This has resulted in a concentrated effort of the academic and research communities to take innovative approaches and developed alternative multiferroic materials. An elegant solution to single phase multiferroics dilemma was the development of multiferroic composite materials. A schematic of combining magnetic and electric orders to make multiferroic/ME is shown in Figure 15.2 [17].

The composite ME effect depends on the magnetostrictive property and the piezoelectric property of the materials [16]. Then the ME effect can be written as

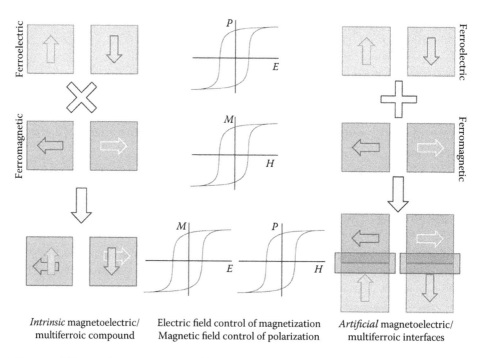

| Intrinsic magnetoelectric/ multiferroic compound | Electric field control of magnetization Magnetic field control of polarization | Artificial magnetoelectric/ multiferroic interfaces |

FIGURE 15.2 ME coupling consisting of both magnetic and electric field polarization and vice versa. (From Manuel Bibes, *Nature Materials* 11, 354–357, 2012. With permission.)

$$\text{Direct ME effect} = \frac{\text{Magnetic}}{\text{Mechanical}} \times \frac{\text{Mechanical}}{\text{Magnetic}}$$

$$\text{Converse ME effect} = \frac{\text{Electric}}{\text{Mechanical}} \times \frac{\text{Mechanical}}{\text{Magnetic}}$$

A strain interceded ME effect in bulk ME composites was found even above the room temperature.

15.2 IMPORTANT MATERIALS

15.2.1 Magnetic Materials

Magnetic materials are classified based on the magnetic moments of atoms (usually related to the spins of electrons) that align in a regular pattern with the neighboring spins (on same or different sublattices) pointing to either parallel or antiparallel direction [19,20]. Ferromagnets have parallel arrangement of neighboring atoms and particularly characterized by the presence of spontaneous magnetization. Antiferromagnetic or ferrimagnetic ordering is established depending on the strength of neighboring antiparallel spins.

15.2.2 FE Materials

Ferroelectricity in a material describes the presence of spontaneous polarization (P_s), the direction of which can be changed by the application of an external electric field [21]. FE materials retain the new state of polarization even after the removal of the external electric field. Another characteristic of FE materials is that the appearance of a hysteresis loop in polarization (P) versus electric field (E) similar to the magnetic hysteresis. The hysteresis disappears at temperatures known as Curie temperature. The polarizability of a dielectric materials show a linear dependence on the applied small an electric field. But in the case of FE material, the dielectric permittivity is high and shows strong nonlinearities with the applied electric field [22,23]. The distinction of various dielectric materials are graphically depicted in Figure 15.3.

Some of the well-known and extensively investigated FE materials are $CaTiO_3$, $BaTiO_3$, $PbTiO_3$, PZT, PMN-PT, $KNbO_3$, and $SrTiO_3$ [24,25].

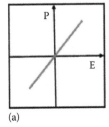
(a)
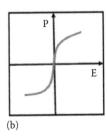
(b)
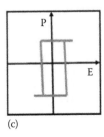
(c)

FIGURE 15.3 Polarization versus electric field loop of (a) linear, (b) nonlinear dielectric, and (c) ferroelectric materials.

15.3 SIGNIFICANCE OF POLY(VINYLIDENE FLUORIDE)

The discovery of the piezoelectric properties of poly(vinylidene fluoride) (PVDF) by Kawai [26], and the study of its pyroelectric and nonlinear optical properties [27] led to the discovery of its FE properties in the early 1970s. Since that time, considerable development and progress have been made on both materials and devices based on PVDF. Since the discovery of piezoelectricity in PVDF, ferroelectricity has been found only in a small number of crystalline polymers such as poly(vinylidene cyanide) copolymers, odd-numbered nylons, polyureas, and FE liquid crystal polymers. Among these, PVDF is the most developed and is applied in number of applications because of its chemical stability, highly compact structure, and large permanent dipole moment. Any variation in the structure of PVDF adversely affects the chemical stability and crystallinity, which therefore restrains ferroelectricity. PVDF exhibits various polymorphisms in three distinct conformations: TG_TG_ in α- and δ-phases, all trans (TTT) planar zigzag in β-phase, and T3G_T3G_ in γ- and δ-phases [28–31] (T and G represent trans and Gauche confirmations). Because of its specific chain conformation in crystal unit cell and providing the highest remnant polarization, β-phase has attracted more attention than the others in pyro- and piezoelectric applications. The ferroelectricity in PVDF is also found to be very stable even after preservation for several years [32]. Conversion of the β crystals to the nonpolar α form takes place thermally at the Curie temperature (T_C), which accordingly defines the upper limit of the polar phase. While the PVDF homopolymer needs to be mechanically oriented or poled in a electric field to achieve the polar β-phase [33].

Several researchers have proposed different methods to enhance the formation of β-phase. Most of the proposed techniques are based on a phase transformation mechanism in homopolymers and copolymers of PVDF involving a multi-stage process [34–36]. A common feature of various approaches for improving the electrostrictive properties of PVDF is reduction of the crystallite size by introducing defects. Smaller FE domains are more effectively oriented by an applied field, potentially yielding better electromechanical properties. Alternately, this objective is met through copolymers of VDF and trifluoroethylene (TrFE), where they spontaneously crystallize into the β-phase in PVDF [37–40]. For example, P(VDF-TrFE) is a random copolymer synthesized using two homopolymers: PVDF and poly(trifluoroethylene) (PTrFE).Figure 15.1 shows the chemical formula. PVDF is a crystalline polymer, has a monomer unit of —CH2—CF2—, in between polyethylene (PE) (—CH2—CH2—) and polytetrafluoroethylene (PTFE) (—CF2—CF2—) monomers. The similarity between PVDF and these two polymers gives rise to its physical strength, flexibility, and chemical stability. The FE property of PVDF copolymers originate from the large difference in electronegativity between fluorine (4.0), carbon (2.5), and hydrogen (2.1). Most of the electrons are attracted to the fluorine side of the polymer chain and polarization is created [41,42]. The copolymer crystal structure, phase transition behavior, and FE properties are affected by the ratio of VDF/TrFE content and the synthesizing conditions.

P(VDF-TrFE) thin films are prepared mostly in two methods. The first one is the melt and press method [43]. The copolymer crystallizes into β- or α-phases when it is slowly cooled to room temperature from the melt. The film has a high degree of crystallinity. Stretching or poling process is necessary to achieve the β-phase crystals. For the melt and press fabrication process, the film thickness is usually large and approximately 1 μm or more. Because of the relatively high coercive field of 50 MV m^{-1} of P(VDF-TrFE), sub-100 nm thick layers are required in order to arrive at a switching voltage below 10 V. Spin coating from solution is another common fabrication method. By changing the weight percentage of the polymer in solution, spin coating can be used to produce films with thickness ≤100 nm [44,45]. Different crystal phases can be achieved from polymer dissolved in different solvents. Spin coat from 2-butanone or cyclohexanone solutions allow the film to be crystallized into the β-phase directly. Another method of making ultra thin film reported by Bune et al. [46] is Langmuir–Blodgett deposition, which results in films that are a few monolayers thick and can be switched at 1 V. Unfortunately, these films are probably even less suitable for applications than the aforementioned spin-coated films because the switching times are longer by orders of magnitude [47]. This problem was solved with conductive contact layer coating with PEDOT:PSS [poly(3,4-ethylenedioxythiophene):poly(styrene sulfonic acid)] or Ppy:PSS [polypyrrole-poly(styrene sulfonic acid)] as a bottom and top interface [48,49]. Figure 15.4 shows a sub-10 V switching while retaining the remnant polarization and switching times of the bulk material.

The polymer chains in γ-PVDF have a conformation in between that of the α- and β-phases. The γ-phase is FE but experimentally hardly accessible. The δ-phase is a polar version of the α-phase. Its existence was proposed 30 years ago by electroforming from an originally α-phase bulk sample in a high electric field of about 170 MV m^{-1}. Recently, Li et al. have measured the FE properties of δ-phase and found that the remnant polarization and coercive field are comparable to those of the PVDF copolymers [50]. The high Curie temperature and stable remnant polarization up to 400 K (shown in Figure 15.5) finds this phase very attractive for high temperature ME applications.

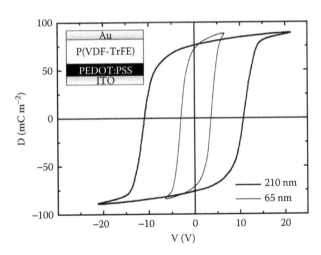

FIGURE 15.4 Displacement charge versus applied voltage V hysteresis loop measurements on thin PVDF capacitors. (From R. C. G. Naber, P.W.M. Blom, A. W. Marsman, D. M. de Leeuw, *Appl. Phys. Lett.* 2004, 85, 2032. With permission.)

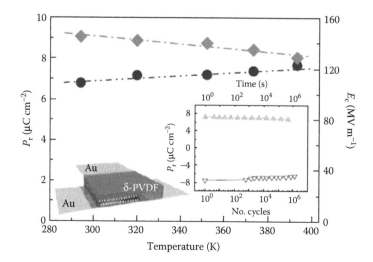

FIGURE 15.5 Remnant polarization, P_r (circles), and the corresponding coercive field, E_c (diamonds), as a function of temperature and cycle endurance of δ-PVDF. (From Mengyuan Li et al., *Nature Materials* 12, 433–438, 2013. With permission.)

15.4 COMPOSITE MULTIFERROICS

In composites, two ingredient phases in bulk/nanosystem are formed collectively by sintering or adhesive bonding. There are various ME composites, they are, particulate nanocomposite films (0-3), horizontal heterostructures (2-2), and vertical heterostructures (1-3), in which electrical polarization control by magnetic field or magnetic polarization control by the electric field. Particulate nanocomposite structures are shown in Figure 15.6a, where magnetic particles are dispersed in a matrix that could be inorganic or a polymer. The induced electrical polarization hysteresis loops did not show a saturated shape because of large leakage current [51]. Alternatively, thin film composites phases are combined at the atomic level to reduce the loss at the interface. Composite through, epitaxial or superlattice films has also been found to be a great method to obtain large ME voltages. Horizontal 2-2 heterostructures or laminars are shown in Figure 15.6b; they comprise alternating FE and magnetic layers. These heterostructures show very less ME coefficient because of large in-plane constraint due to the substrate [52]. This kind of structure is easy to process and substantially reduces the leakage current problem. In vertical 1-3 heterostructures, each

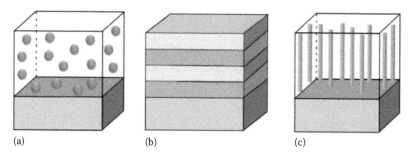

FIGURE 15.6 Various kinds of composites (a) 0–3 particulate, (b) 2–2 laminate, and (c) 1–3 fiber/rod.

nanopillar is surrounded by the matrix as shown in Figure 15.6c. In such system with CFO and BFO composite, CFO nanopillars are surrounded by a BFO matrix, where magnetization of the CFO pillars change by the electric field [53,54]. This phenomenon happens because of the electric field control of magnetization.

15.4.1 ME Effect in Bulk Composites

A variety of FE ceramics were combined with a number of magnetic metals, alloys, and oxides in various geometries and tested for the ME property. Their synthesis mainly involves high temperature sintering. It was discovered that industrially manufactured multilayer capacitors containing $BaTiO_3$ and FM nickel has displayed ME effect at a room temperature best suitable for magnetic field sensing with no electrical bias [55]. The sensitivity of these multilayer capacitors is in the order of 7 μV Oe^{-1}. Bulk ME ceramic composites have exhibited a larger ME effect.

In ME bulk composites, the magnetostrictive alloys such as Terfenol-D and Metglas-based composites have shown the strongest ME effect. Compared with the 0-3 particulate composite ceramics, the 2-2 type laminate composite ceramics, exhibit high ME coefficients as the laminates have smaller leakage. Giant ME effect was found in laminate composites with Terfenol-D and various piezoelectric materials. Well-designed Terfenol-D layers with PMN-PT layer has shown an ME voltage response of about 57 V cm Oe^{-1} at the resonance frequency [56]. For low-field applications, magnetostrictive materials with low coercivity and high permeability are desirable. Accordingly, Metglas win over Terfenol-D. In laminates of Metglas ribbon and PZT-fiber layer, greatly enhanced ME response up to about 10 V cm Oe^{-1} at low frequency and several hundred V cm Oe^{-1} at resonance under a low magnetic field [57]. Metglas has the advantage of significant magnetic flux concentration and a large planar aspect ratio markedly increase flux density at the central region [58]. The major advantage of ceramics is the availability of a large pool of FE and magnetic materials. Some of the short comings of ceramic composites are as follows: they are expensive, heavy, and brittle, and sometimes may fail during operation. Despite large ME coefficients, such composites may become *fragile* and are limited by deleterious reactions at the interface regions, leading to low electrical resistivity and high dielectric losses, which hinder their incorporation into devices. Adding flexibility to this structure using polymer has been found to be more attractive for applications. Some of the composites with high ME effect is shown in Table 15.1.

15.4.2 ME Effect in Polymer-Based Composites

Comparison with the ceramic and magnetic alloy-based ME composites, polymer-based composites, can be easily fabricated by low-temperature processing into a variety of forms such as thin sheets and molds with enriching mechanical properties. Polymers were included in three kinds of composites: nanocomposites, polymer as a binder, and laminated composites. Though a number of polymers can be used as binders, only PVDF and its copolymers stand unique for their FE/piezoelectric functionality. Most of the polymer-based composites use PVDF/copolymers in some means with inorganic magnetic materials. While ferromagnetic polymers are rare and challenging area in itself. Some of the polymer composites studies have also been made using polyurethane [67,68].

TABLE 15.1   Magnetoelectric Coefficient of Some of the Composite Systems

Materials	ME Voltage(V cm Oe^{-1})	Reference
(0-3) CFO/P(VDF-TrFE)	40 @ 5 kHz	[70]
(1-3) Terfenol-D in epoxy/PZT	18.2 @ f_r	[59]
(2-2) Terfenol-D/PMT-PT	10.3 @ 1 kHz	[60]
(2-2) Terfenol-D/PVDF	1.43	[61]
(2-1) FeBSiC/PZT-fiber [38]	750 @ f_r	[56]
(2-2) FeBSiC/PVDF	21.46 @ 20 Hz	[57]
(2-2) FeCoSiB/AlN	737 @ f_r	[62]
(2-2) NCZF/PZT/NCZF	0.782 @ 1 kHz	[63]
(2-2) PZT in PVDF/Terfenol-D in PVDF	3 @ f_r	[64]
(2-2) Gd crystal/P(VDF-TrFE)/silver conductive epoxy 500	500 @ f_r	[65]
(2-2) PVDF/Metglas unimorph	238 @ f_r	[74]
(2-2) PVDF/Metglas three-layer 310,000	310 @ f_r	[74]
(2-2) Cross-linked P(VDF-TrFE)/Metglas	383 @ f_r	[79]
(2-2) PU/PVDF/Fe$_3$O$_4$	753 @ f_r	[66]
(2-2) PVDF/Terfenol-D/PZT	6000 @ f_r	[87]

CFO: CoFe$_2$O$_4$, NZFO: Ni$_{0.8}$Zn$_{0.2}$Fe$_2$O$_4$, NCZF: Ni$_{0.6}$Cu$_{0.2}$Zn$_{0.2}$Fe$_2$O$_4$, PMN-PT: Pb(Mg,Nb)O$_3$-PbTiO$_3$, PVDF: polyvinylidene-fluoride, P(VDF-TrFE): poly(vinylidene fluoride-trifluoroethylene), PE: polyurethane [b], f_r: electromechanical resonance frequency.

Distinct methods of synthesis of 3-0 polymer nanocomposites have been followed. [69–72]. The key factor is the microstructure of the composite and the interface effect between the matrix and the filler. The effect depends on the surfaces and interfaces at the nanometer-scale and the effective properties of the nanocomposites can be either enhanced or reduced [73]. It was observed that such interface condition exerts a significant influence on the local and overall magnetoelectroelastic responses of MF composites, in particular when the fillers are at the nanometer-scale [74]. On the other hand, in Terfenol-D/polymer nanocomposites, interface layer induced by surfactant modifies the Terfenol-D particle surfaces. It was found that by adding a silane surfactant to the surface of the nanoparticles, the piezoelectricity of the composite is diminished because the surfactant modification leads to a reduction of the piezoelectric response [75]. Focusing on layered or laminate ME structures, those configurations are usually obtained by gluing together the electroactive polymer and the magnetostrictive phase using an epoxy binder [76–79] by the direct deposition of one phase on top of the other, [80,81] or by hot-molding techniques [82,83]. A number of ME studies were conducted with magnetic metal or other conducting magnetic materials in PVDF matrix [84,85]. By combining insulating piezo/FE materials with conducting materials have shown a huge interface states in case of inorganic ceramic composites. Theoretically, it was proposed that electronically assisted ME effect are present in addition to strain [82,86].

Using PVDF as a binder, a large ME has been found in PZT and Terfenol-D particulate composites [87]. While the three phase laminar structures have produced much larger ME coupling, where in these structures, Terfenol-D/PVDF particulate composites layers were sandwiched between PVDF/PZT particulate layers [85,88]. It was observed that optimal volume fraction of PVDF was necessary to achieve large ME. A maximum ME of 80 mV cm Oe^{-1} was found, which was higher than the three-phase particulate composites value of 42 mV cm Oe^{-1}. ME laminates of vinyl ester resin (VER)-bonded Terfenol-D magnetostrictive layer attached to PZT piezoelectric layer glued together with a conductive epoxy has shown increasing ME with loading of magnetostrictive amount and a maximum ME of 2.7 V mV cm Oe^{-1} was obtained at a magnetostrictive volume fraction of 0.48 [89]. ME interactions in laminar structures of magnetostrictive and piezoelectric phases are mediated by mechanical deformation. And the effect is large at the electromechanical resonance. Bichurin et al. have demonstrated this with both theory and experiments [90].

Flexible ME laminates composites have been realized by Zhai et al. is shown in Figure 15.7; these thin laminars were fabricated using Metglas and PVDF in bilayer and trilayer geometries [78]. The Metglas layer and PVDF layers were glued together using an epoxy and both laminate types have displayed strong ME with three-layer composite has showing $\alpha_{31} = 238$ mV cm Oe^{-1} and bilayer has showed $\alpha_{31} = 310$ mV cm Oe^{-1}, both near the longitudinal resonance frequency at 50 kHz as shown in the Figure 15.7. By taking advantage of the magnetic flux concentration effect of Metglas as a function of its sheet aspect ratio, the large the ME was further improved to 21.46 V cm Oe^{-1} in PVDF/Metglas 2605SA1 laminate at 20 Hz nonresonant frequency as depicted in Figure 15.8 [79]. Further improvement was made with the introduction of chain-end cross-linking and polysilsesquioxane structures into the P(VDF-TrFE) matrix, which led to the formation of larger crystalline samples and consequently better piezoelectric response in comparison to those of pristine P(VDF-TrFE) copolymers [91]. In these systems, ME value of 15.7 V cm Oe^{-1} was achieved under a direct current (DC) magnetic field of only 3.79 Oe at 20 Hz and a much higher value 374 V cm Oe^{-1} at the resonant frequency, and these values are indeed close to the maximum value found in inorganic laminars.

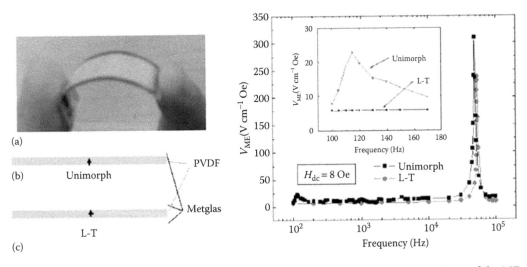

FIGURE 15.7   (a–c) shows Metglas/PVDF unimorph laminate; (d) frequency dependence of the ME voltage coefficient of both three-layer Metglas PVDF and unimorph and Metglas/PVDF laminates. (From Zhai, S. Dong, Z. Xing, J. Li, D. Viehland, *Appl. Phys. Lett.*, 89, 083507, 2006. With permission.)

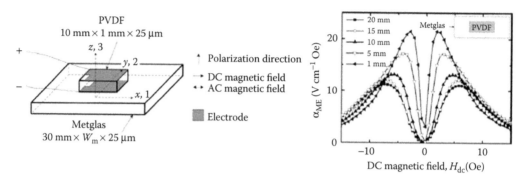

FIGURE 15.8 Schematic of PVDF/Metglas 2605SA1 laminate. α_{31} as a function of DC-biased magnetic field for Metglas/PVDF measured at 20 Hz and DC magnetic field of 0.38 Oe. (From Fang, S. G. Lu, F. Li, S. Datta, Q. M. Zhang, M. El Tahchi, *Appl. Phys. Lett.* 2009, 95, 112903. With permission.)

15.5 MAGNETODIELECTRIC EFFECT

In comparison to ME effect, magnetodielectric (MD) effect is less studied in composite materials. From measurement and application point of view, MD sensors are more easy to use as they simply measure capacitance. MD effect has been extensively studied in single-phase multiferroic systems as an alternative to ME effect. Because of the lack of the good insulating property of single-phase multiferroics, indirect method like MD is preferred to understand the dielectric coupling near the magnetic transition. A diverse family of materials display MD effect and indicates that the driving force of such a mechanism has no unique source, but it has different sources, and they may be intrinsic to the sample or extrinsic in nature.

Because of the nature of atoms and their arrangement in the crystal lattice (chemical bonds and atomic coordination number), the electrons in a dielectric medium cannot move over large distances. Their short hopping distance under the effect of an electric field gives rise to different types of polarization and can be expressed as [92]

$$\alpha = \alpha_e + \alpha_a + \alpha_o + \alpha_{sc}$$

where α_e, α_a, α_o, and α_{sc} represent the electronic, atomic or ionic, orientational, and space charge polarizations, respectively. The polarization spectrum for different kinds of polarization is shown in Figure 15.9.

Electronic polarization (α_e): All dielectric materials induce electronic polarization, which results from an oscillation of the electronic charge relative to the nucleus under the influence of external electric field. This deforms the originally symmetrical distribution of the electron cloud of atoms or molecules. Because of the low mass of electrons, these oscillations occur at very high frequencies, above 10^{14}–10^{16} Hz [93,94]. This contribution is dominating only in the optical region and follows the relation $n = \sqrt{\varepsilon}$, where n and ε are refractive index and polarization, respectively.

Atomic or ionic polarization (α_a): Only materials with ionic bonds may have ionic polarization, caused by the relative displacement of cations in one direction and anions in the

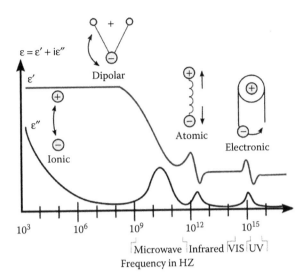

FIGURE 15.9 Spectrum of different kinds of dielectric polarizations with respect to frequency. (From C. Kittel, *Introduction to Solid State Physics*. Wiley, 1956. With permission.)

opposite direction. As a result, net dipole moments are induced on this new structure. The corresponding frequencies occur below the optical frequencies in the infrared range, about 10^{13} Hz. This type of polarization is characterized by vibrational spectroscopy.

Orientation (dipolar) polarization (α_o): This kind of polarization occurs in materials that contain the permanent dipoles. Applications of external electrical field reorient the dipoles along the field direction. This type of polarization is observed in the range of radio waves.

Interfacial (space charge) polarization: This polarization is because of accumulation of free charges at interfaces between different environments. This will be important in systems with high density of charge carriers where the application of external magnetic field migration of the charge carriers within the grains (hopping polarization) or relaxation of build of space charge carriers due to heterogeneous electrical properties at grain and grain boundaries [94,95].

15.5.1 Intrinsic Magnetodielectric Effect

Initially, Smolenskii et al. have discussed the possible origins of MD coupling in ferroelectromagnetic systems by using the order parameters of FE and magnetic sublattices [96]. All the multiferroic, ME materials exhibit the MD effect; however, the converse is not true. Such kind of MD coupling is reported in single-phase multiferroic $BiMnO_3$ system [96], and the dielectric anomaly observed around the magnetic transition at 100 K was correlated to spin pair correlation. On the other hand, it was discussed that the spin-phonon coupling is the one of the possible origin for MD in solids. Recently, Choudhury et al. have proposed spin-dependent asymmetric charge hopping in the partially disordered structure as the possible intrinsic origin of MD in La_2NiMnO_6 system [97]. For any MD material (including multiferroic, ME) the MD response can be correlated with magnetic moment of the system using a relation MD (%) = γM^2, where γ is the MD coupling coefficient. In some cases, MD was found to be varying with M^4 and this was found at the interface engineered superlattice structures [86]. Furthermore, this behavior was also noticed in La_2CoMnO_6 nanoparticles and PCMO nanoparticle dispersed in PVDF matrix [99,126].

15.5.2 Extrinsic Magnetodielectric Effect

Recent experimental and theoretical results clearly demonstrated that the MD effect may originate from the extrinsic contributions, even though the MD medium may or may not have magnetic ordering. Alternate situations are described here.

- Magnetoresistance combined with Maxwell–Wagner interfacial polarization
- Depletion of free charge carrier at the interfaces in the inhomogeneous materials

Catalan simulated and experimentally showed the MD behavior in electrically inhomogeneous materials, where the driving source for the MD behavior is MR combined with the Maxwell–Wagner interfacial polarization [100]. On the other hand, Maglione et al. have observed MD in a broad range of materials, having a short range magnetic ordering, and concluded that the free-charged carriers at the interfaces play a crucial role to invoke the MD behavior [101–106]. All the composites have Maxwell–Wagner interfacial polarization because of their microstructure.

There are several studies on ME behavior of FE polymer and magnetic oxide materials [84,85,90,107,108]. In particular, the high dielectric permittivity in the metal/semiconductor-polymer has been studied using the percolation theory [109–115]. Influence of several factors such as particle size, processing condition, and polar nature of the polymer matrix on the percolation threshold were studied [110,116–120]. On the other hand, large ME and MD behavior was observed in composite systems made of FE and FM materials [121].

In such composites, the FE polymers such as PVDF, P(VDF-TrFE), and PVDF-TrFE-CFE play a key role to achieve large interfacial strains, good connectivity between the constituting phases, and better mechanical flexibility. Moreover, the addition of high-insulating FE polymer prohibits the undesirable loss in the composite. A small MD was found in 30 wt.% $La_{0.7}Sr_{0.3}MnO_3$ in PVDF prepared by dip coating. $La_{0.7}Sr_{0.3}MnO_3$ nanoparticles were synthesized by the sol–gel route. A MD effect of 1.9% at 3.8 KOe was found at room temperature [122]. MD of different composite films of $BiFeO_3$/PVDF made by hot press technique with $BiFeO_3$ concentration of 70, 60, 50, and 40 wt.%, where 60 wt.% have shown maximum MD of approximately 0.37% at 8 kOe [123]. While $Ni_{0.5}Zn_{0.5}Fe_2O_4$ (NZFO)/PVDF shows a MD coefficient of 5.1% at room temperature under a magnetic field of 0.3 T [124].

Chandrashekar et al. have investigated the MD behavior in composites consisting of spin glass magnetic nanoparticles in δ-PVDF matrix, and the study has revealed a large MD of 30% at the spin glass transition (Figure 15.10a) [125,126]. With the increase of magnetic nanoparticle loading, dielectric property shows a percolation and at the percolation point (Figure 15.11), the MD showed a cross over from M^2 to M^4 behavior. With the help of

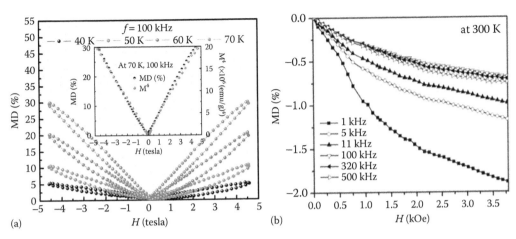

FIGURE 15.10   (a) Magnetic field variation of MD (%) of $Pr_{0.6}Ca_{0.4}MnO_3$/PVDF at 100 kHz. (From Chandrasekhar, K. D., Das, A. K.; Venimadhav, A., *Appl. Phys. Lett.* 2011, 98, 122908. With permission.) (b) Magnetic field variation of MD (%) of $La_{0.7}Sr_{0.3}MnO_3$/PVDF at 300 K is plotted at various frequencies. (From Ch. Thirmal, Chiranjib Nayek, P. Murugavel, and V. Subramanian, *AIP Advances* 3, 112109, 2013. With permission.)

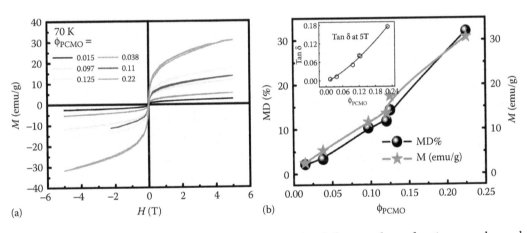

FIGURE 15.11   (a) Isothermal magnetization at 70 K for different volume fraction samples and (b) variation of MD (%) and magnetization at 4.6 T with respect to the volume fraction. (From K. D. Chandrasekhar, A. K. Das, and A. Venimadhav, *J Phys Chem C*, 2014, 118 27728. With permission.)

magnetic and microstructure analysis, it was suggested that above and below the percolation, inter- and intra-cluster magnetic interactions, respectively, contribute to the MD effect as shown in Figure 15.12. With the help of a temperature-dependent dielectric study, the contributions from PVDF, interfacial, and polaron relaxations were identified. Furthermore, the effect of magnetic filler on the glass temperature of the polymer was demonstrated [126]. A number of studies pertaining to MD effect [155–163] are shown in the Table 15.2.

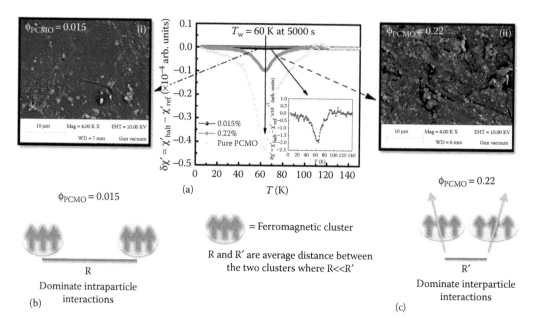

FIGURE 15.12 Magnetic memory measurements and the microstructure at different volume fractions of PCMO spin glass filler in δ-PVDF matrix. A model diagram for magnetic interaction of ferromagnetic clusters for 0.015 (below percolation) and 0.22 (above percolation) volume fraction samples, respectively. (From K. D. Chandrasekhar, A. K. Das, and A. Venimadhav, *J Phys Chem C*, 118, 27728, 2014. With permission.)

15.6 APPLICATIONS

Multiferroics have potential to cater a range of applications spanning to various industrial domains. Here, we discuss only a few important applications pertaining to electrical and electronic industries.

15.6.1 Magnetic Sensors

Magnetic field sensing has been of interest since the olden-day compass to present-day GPS sensors. Magnetic field sensors are widely used in a range of applications, such as automotive applications, mapping of the earth's magnetic field, GPS, security devices, magnetic data storage, biomedical applications, and metrologies, in our surrounding. Next generation magnetic field sensing demands additional features such as sensing without bias, small size, light weight, flexibility, and should be easy to use. Multiferroics detect alternate current (AC) magnetic fields with DC magnetic field bias. Laminar structures with large ME effect are ideal for this purpose [127]. ME elements can be designed with a sensitivity of 10^{-13} tesla [128–130]. The angle between the DC-biased magnetic field vector and AC excitation field is important for the operation of the multiferroic sensor [131]. Taking the advantage of this, Lage et al. have reported the vector field magnetometer based on two orthogonal self-biased multiferroic composite sensors that can be used for the detection of 2D magnetic field vectors [132].

TABLE 15.2   Magnetodielectric Study in Some of the Composite Systems

System	Preparation Technique	MD %	Reference
$Pr_{0.6}Ca_{0.4}MnO_3$ (22.5 %)/PVDF	Sol–gel	30 @ 4.6T 100 kHz, 70K	[126]
PVDF (70%)-$La_{0.7}Sr_{0.3}MnO_3$ (30%) film	Dip coating technique LSMO sol–gel	1.9% @ 3.8 kOe 300K	[122]
$BiFeO_3$-PVDF BFO 70, 60, 50 and 40 wt.% film	Hot press BFO sol–gel	60% BFO 0.37% at 8kOe	[123]
$Ni_{0.5}Zn_{0.5}Fe_2O_4$ (NZFO)-PVDF 0%, 5%, 10%, 20%, 30%, 40%	Modified polymer processing route	20% NZFO 5.1 @ .3T	[124]
$Ba_{0.7}Sr_{0.3}TiO_3$ (80%)— $Ni_{0.8}Zn_{0.2}Fe_2O_4$ (20%)	Electrospinning method	18.2% @1 kHz 6.3 kOe	[155]
0.7 BFO-0.3 BTO	Solid-state reaction	23 @ 9 KOe	[156]
$BiFeO_3$—$Na_{0.5}K_{0.5}NbO_3$ $x = 0.1$	Pechini method	1050 @ 6KOe 1 kHz	[157]
BSPT/LSMO/Si	Sol–gel	9.5 @ 7 T 52 kHz	[158]
0.5 $Co_{1.2}Fe_{1.8-x}Mn_xO_4$ (CFMO)—0.5 $Pb_{0.2}Ba_{0.8}TiO_3$(PBT)	Hydroxide coprecipitation	5.82 @ 6KOe 1 MHz	[159]
0.5 $Sr_{0.5}Ba_{0.5}Nb_2O_6$ (SBN)—0.5 $Co_{1.2-x}Mn_xFe_{1.8}O_4$ (CMFO)	Hydroxide coprecipitation	93 @ 4.5KOe 500 kHz	[160]
65 % $CoFe_2O_4$—35% $BaTiO_3$ Core shell	Coprecipitation followed by citrate-gel method	3.8 @ 7KOe 25 kHz	[161]
$(1-x)$PZT5/xNiFe$_2$O$_4$ $x = 0.2$	Solid-state reaction	10 @ 1T 1 kHz	[162]
Er_2O_3(0.5 mol %)—SiO_2	Sol–gel route	2.75 @ 9 T 2.5 kHz	[163]

15.6.2 ME Energy Harvesting

PVDF-based piezosensors are widely studied and many devices have been demonstrated. The ME composites have recently been evaluated for use in energy-harvesting applications [133,134]. As they have summation effect of both mechanical and magnetic energies, the stress could be converted into electric charges at the piezoelectric layer through a direct piezoelectric effect, and ME effect can also be used to convert the vibration and electromagnetic energies. By using an ME laminate attached to a cantilever beam with tip mass, a multimodal system was tested for energy harvesting from stray magnetic and mechanical energies. Dai et al. have demonstrated a prototype of vibration-energy harvester using a Terfenol-D/PZT/Terfenol-D laminate ME transducer, by which a load power of 1.05 mW was generated across a 564.7 k Ω resistor [134]. Recently Junqi Gao et al. have demonstrated charging of a battery using an ME element [130] (see Figure 15.13). By using PVDF composite ME element, one can use both vibration and magnetic fields to harvest energy.

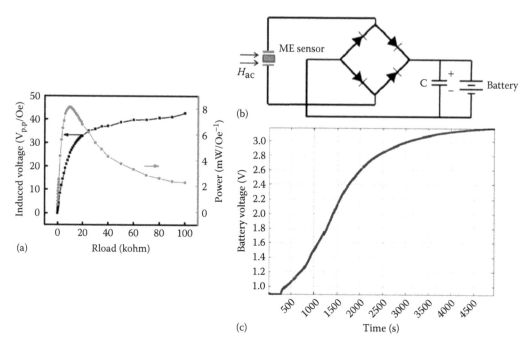

FIGURE 15.13 (a) Shows output voltage and power as a function of load resistance load for Metglas/ PMN–PT laminates. (b) and (c) show the experimental set up and testing of charging of a battery using an ME element. (From J. Gao, Z. Wang, Y. Shen, M. Li, Y. Wang, P. Finkel, J. Li, D. Viehland, *Materials Letters*, 82, 178–180, 2012. With permission.)

15.6.3 Random Access Memories

Continuous efforts are being made to develop storage technologies with higher storage speed and density. In the traditional two-state memories such as the magnetic tunnel junction, the resistance of the junctions depends on the relative orientation of the magnetic moments of the magnetic electrodes that determines the memory state (0 or 1) [135]. The bit writing is done via a magnetic field produced by carefully tuned currents passing through the bit and word line simultaneously. This is an inefficient, relatively slow, and energetically expensive, and limits the applicability of magnetic random access memory technology. While FE random access memory technology suffers from destructive read and reset operation. By using both magnetic- and electric-ordered system such as multiferroics, a four-state memory device can be constructed as depicted in Figure 15.14. The four-state memory has been demonstrated experimentally in a composite mutiferroic [136]. Many schemes of memory elements based on multiferroicity were proposed [137,138]. Changing magnetization state by current is energetically expensive; however, rotating the magnetization with voltage, which is generated by uniaxial strain/stress in a shape anisotropic magnetostrictive piezoelectric (multiferroic) nanomagnet, will be energetically more efficient. Ayan K. Biswas et al. have proposed a device using laminar multiferroic composite and magnetoresistive element as shown in the Figure 15.15 [139,140]. The strain or stress is generated by applying an electrical voltage across the piezoelectric layer that rotates the magnetization of the multiferroic nanomagnet. Importantly, they found that the write energy dissipation and

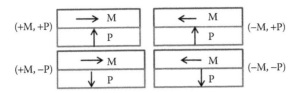

FIGURE 15.14 Schematic of a four-state memory.

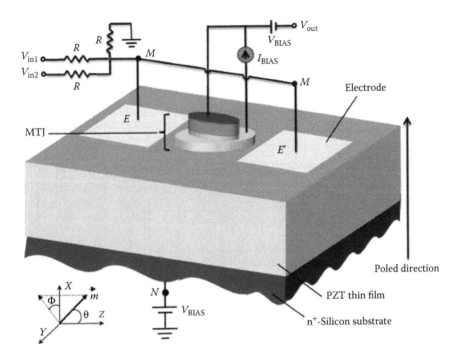

FIGURE 15.15 Schematic of memory element proposed. (From M. S. Fashami, J. Atulasimha, S. Bandyopadhyay, *Nature Scientific Reports*, 3, 3204, 2013. With permission.)

write error rate are reduced significantly. Such a scheme could be one of the most energy-efficient approaches for writing bits in magnetic nonvolatile memory. By using PVDF in this memory cells can be made flexible like paper memories.

15.6.4 Multiferroic for Refrigeration Application

Warburg first proposed the magnetocaloric effect (MCE) in 1881 and Giauque was awarded the Nobel Prize in 1949 for realizing this effect experimentally [141,142]. Over the past six decades, such magnetic refrigeration-based MCE has attracted considerable attention among the researchers because of its high energy efficiency and pollution-free environment [143,144]. This technology is an alternative to the conventional gas-compression refrigeration process. For the commercialization (in food preservation and air conditioning applications), it is important to obtain large MCE with relatively small magnetic fields at ambient temperatures.

In analogy to MCE, a change in temperature of electrically polarizable materials with a change in electric field is known as *electrocaloric effect* (ECE) [145–147]. Recent studies have explored ECE in organic and inorganic materials, especially in thin films [147]. For example, a higher ΔT_{ad} ~31 K was obtained in $PbZr_{0.95}Ti_{0.05}O_3$ thin films [148], and in lead-free materials such as $BaTiO_3$, a maximum ΔT_{ad} of 8–10 K was reported under the E ~50 kV mm^{-1} [149]. Furthermore, a significant value of ΔT_{ad} approximately 12.6 K at 353 K was also obtained in FE polymer [P(VDF-TrFE)] thin films [150,151].

In a given system, the refrigeration process is a result of thermal response to an adiabatic change in properties such as magnetization, electric polarization, volume, and strain. These changes can be induced by the application and removal of conjugate fields resulting in the well-known magnetocaloric, electrocaloric, barocaloric, and elastocaloric effects [152]. In this regard, MF materials are expected to play a vital role. Apart from usual MCE and ECE the coupling between ferroic-ordered phases can give an additional contribution to cooling. Recently, Vopson has given a general formalism for the resulting adiabatic temperature change in a strongly coupled MF system that consists of FM and FE phases [153] and Krishnamurthy et al. have experimentally tested the multicaloric effect in Y_2CoMnO_6 and signified the importance of sign of ME coupling [154]. PVDF being an electrocaloric material, combining this with the magnetocalorics can be a new system of materials for composite multicaloric system suitable for room temperature electronic cooling applications.

15.7 CONCLUSIONS

Polymers have proven its commercial technological merit after the success of polymer-OLEDs. There is a great deal of interest on polymer-based electronic devices. PVDF and its copolymers have great potential applications and multiferroic-based ME devices is one of the promising avenue. Their simplicity, good scalability, low-cost, 3D stacking capability, and large capacity for data-storage are important advantages of polymer devices. Furthermore, the synthesis of electroactive polymers by ink-jet printing, spin-coating, or vacuum evaporation on a variety of substrates can find them central status in the electronics market.

REFERENCES

1. J. Ma, J. Hu, Z. Li, C. W. Nan, Recent progress in Multiferroic magnetoelectric composites: from Bulk to films, *Adv. Mater.*, 2011, 23, 1062.
2. H. Schmid, On a magnetoelectric classification of materials, *Int. J. Magn.*, 1973, 4, 337–361.
3. M. Fiebig, Revival of the magnetoelectric effect, *J. Phys. D Appl. Phys.*, 2005, 38, R123.
4. H. Schmid, Multi-ferroic magnetoelectrics, *Ferroelectrics*, 1994, 162, 665–685.
5. M. E. Lines, A. M. Glass, *Principles and Applications of Ferroelectrics and Related Materials.* Clarendon Press, Oxford, 1977.
6. S. -W. Cheong, R. Ramesh, Multiferroics: Past, present, and future, *Phys. Today*, 2010, 63, 38–43.
7. W. Prellier, M. P. Singh, P. Murugavel, The single-phase multiferroic oxides: from bulk to thin film, *J. Phys. Condens. Matter*, 2005, 17, R803.
8. P. Martins, S. Lanceros-Méndez, Polymer-based magnetoelectric materials, *Adv. Funct. Mater.*, 2013, 23, 3371–3385.
9. T. Lottermoser et al., Magnetic phase control by an electric field, *Nature*, 2004, 430, 541–544.
10. M. V. Melvin, The multicaloric effect in multiferroic materials, *Solid State Commun.*, 2012, 152, 2067.

11. A. K. Agyei, J. L. Birman, On the linear magnetoelectric effect, *J. Phys. Condens. Matter*, 1990, 2, 3007.
12. W. Eerenstein, N. D. Mathur, J. F. Scott, Multiferroic and magnetoelectric materials, *Nature*, 2006, 442, 759.
13. I. E. Dzyaloshinskii, On the magneto-electrical effect in antiferromagnets, *Sov. Phys. JETP*, 1960, 10(3), 628–629 .
14. E. Ascher, H. Rieder, H. Schmid, H. Stossel, Some properties of ferromagnetoelectric Nickelae Iodine Boracite, $Ni_3B_7O_{13}I$, *J. Appl. Phys.*, 1966, 37, 1404–1405.
15. Z. J. Huang, Y. Cao, Y. Y. Sun, Y. Y. Xue, C. W. Chu, Coupling between the ferroelectric and antiferromagnetic orders in YMnO3, *Phys. Rev. B*, 1997, 56, 2623.
16. N. A. Hill, Why are there so few magnetic ferroelectrics? *J. Phy. Chem. B*, 2000, 104, 6694.
17. M. Bibes, Nanoferronics is a winning combination, *Nat. Mater.*, 2012, 11, 354–357.
18. J. van Suchtelen, Product properties: a new application of composite materials, *Phil. Res. Rep.*, 1972, 27, 28.
19. K. H. J. Buschow, *Handbook of Magnetic Materials*. Elsevier Science, Amsterdam, 2003.
20. D. C. Jiles, *Introduction to Magnetism and Magnetic Materials*, 2nd edn. Taylor & Francis Group, Boca Raton, FL, 1998.
21. F. Jona, G. Shirane, *Ferroelectric Crystals*. Pergamon Press, Oxford, 1962.
22. W. Känzig, Ferroelectrics and antiferroelectrics. In F. Seitz, T. P. Das, D. Turnbull, E. L. Hahn (eds.), *Solid State Physics*, vol. 4. Academic Press, New York, 1957.
23. K. M. Rabe, C. H. Ahn, J. -M. Triscone, *Physics of Ferroelectrics A Modern Perspective*. Springer, Berlin, 2007.
24. M. Dawber, K. M. Rabe, J. F. Scott, Physics of thin-film ferroelectric oxides, *Rev. Mod. Phys.*, 2005, 77, 1083–1130.
25. N. Setter, D. Damjanovic, L. Eng et al., Ferroelectric thin films: review of materials, properties, and applications, *J. Appl. Phys.*, 2006, 100, 051606.
26. H. Kawai, The piezoelectricity of poly(vinylidene Fluoride), *Jpn. J. Appl. Phys.*, 1969, 8, 975–976.
27. A. M. Glass, Pyroelectric properties of polyvinylidene fluoride and its use for infrared detection, *J. Appl. Phys.*, 1971, 42(13), 5219–5222.
28. E. Giannetti, Semi-crystalline fluorinated polymers, *Polym. Int.*, 2001, 50, 10.
29. M. G. Broadhurst, G. T. Davis, J. E. McKinney, R. E. Collins, Piezoelectricity and pyroelectricity in polyvinylidene fluoride—A model, *J. Appl. Phys.*, 1978, 49, 4992.
30. G. T. Davis, On structure morphology and models of polymer ferroelectrics. In T. T. Wang, J. M. Herbert, A. M. Glass (eds.), *The Applications of Ferroelectric Polymers*. Blackie & Sons Ltd., London, 1988, pp. 37–50.
31. A. J. Lovinger, Ferroelectric polymers, *Macromolecules*, 1982, 15, 40.
32. R. B. Symore, G. B. Kauffman, Pizoelectric polymers: Direct converters of work to electricity, *Journal of Chemical Education*, 990, 67(9), 763–765.
33. K. Tashiro, Crystal structure and phase transition of PVDF and related copolymers. In H. S. Nalwa (ed.), *Ferroelectric Polymers*. Marcel Dekker, Inc., New York, 1995, pp. 63–181.
34. R. Gregorio, Jr., M. Cestari, Effect of crystallization temperature on the crystalline phase content and morphology of poly (vinylidene fluoride), *J. Polym. Sci. Part B Polym. Phys.*, 1994, 32, 859.
35. T. C. Hsu, P. H. Geil, Deformation and transformation mechanisms of poly (vinylidene fluoride)(PVF2), *J. Mat. Sci.*, 1989, 24, 1219.
36. T. Kaura, R. Nath, M. M. Perlman, Simultaneous stretching and corona poling of PVDF films, *J. Phys. D Appl. Phys.*, 1991, 24, 1848.
37. X. Lu, A. Schirokauer, J. Scheinbeim, Giant electrostrictive response in poly (vinylidene fluoride-hexafluoropropylene) copolymers, *IEEE Trans. Ultrason. Ferroelectr. Freq. Control*, 2000, 47, 1291.

38. Y. Tajitsu, A. Hirooka, A. Yamagish, M. Date, Ferroelectric behavior of thin films of vinylidene fluoride/trifluoroethylene/hexafluoropropylene copolymer, *Jpn. J. Appl. Phys.*, 1997, 36, 6114.

39. A. Petchsuk, T. C. Chung, Synthesis and electric properties of VDF/TrFE/HFP terpolymer, *Polym. Prepr.*, 2000, 41(2), 1558.

40. H. Xu, Z. -Y. Cheng, D. Olson, T. Mai, Q. M. Zhang, G. Kavarnos, Ferroelectric and electrome-chanical properties of poly (vinylidene-fluoride–trifluoroethylene–chlorotrifluoroethylene) terpolymer, *Appl. Phys. Lett.*, 2001, 78, 2360.

41. A. Salimi, A. A. Yousefi, Conformational changes and phase transformation mechanisms in PVDF solution-cast films, *J. Polym. Sci. Part B Polym. Phys.*, 2004, 42(18), 3487–3495.

42. S. Fujisaki, H. Ishiwara, Y. Fujisaki, Low-voltage operation of ferroelectric poly(vinylidene fluoride-trifluoroethylene) copolymer capacitors and metalferroelectric-insulator-semiconductor diodes, *Appl. Phys. Lett.*, 2007, 90, 162902.

43. T. Yamada, T. Kitayama, Ferroelectric properties of vinylidene fluoride-trifluoroethylene copolymers, *J. Appl. Phys.*, 1981, 52(11), 6859–6863.

44. K. Kimura, H. Ohigashi, Polarization behavior in vinylidene fluoride-trifluoroethylene copolymer thin films, *Jpn. J. Appl. Phys.*, 1986, 25, 383.

45. Y. Tajitsu, Effects of thickness on ferroelectricity in vinylidene fluoride and trifluoroethylene copolymers, *Jpn. J. Appl. Phys.*, 1995, 34, 5418.

46. A. V. Bune, V. M. Fridkin, S. Ducharme, L. M. Blinov, S. P. Palto, A. V. Sorokin, S. G. Yudin, A. Zlatkin, Two-dimensional ferroelectric films, *Nature*, 1998, 391(6670), 874–877.

47. G. Vizdrik, S. Ducharme, V. M. Fridkin, S. G. Yudin, Kinetics of ferroelectric switching in ultrathin films, *Phys. Rev. B*, 2003, 68, 094113.

48. R. C. G. Naber, P. W. M. Blom, A. W. Marsman, D. M. de Leeuw, Low voltage switching of a spin cast ferroelectric polymer, *Appl. Phys. Lett.*, 2004, 85, 2032.

49. H. Xu, J. Zhong, X. Liu, J. Chen, D. Shen, Ferroelectric and switching behavior of poly (vinylidene fluoride-trifluoroethylene) copolymer ultrathin films with polypyrrole interface, *Appl. Phys. Lett.*, 2007, 90, 092903.

50. M. Li et al., Revisiting the δ-phase of poly (vinylidene fluoride) for solution-processed ferroelectric thin films, *Nat. Mater.*, 2013, 12, 433–438.

51. M. Liu et al., A modified sol-gel process for multiferroic nanocomposite films, *J. Appl. Phys.*, 2007, 102, 083911.

52. C. -W. Nan, G. Liu, Y. Lin, H. Chen, Magnetic-field-induced electric polarization in multiferroic nanostructures, *Phys. Rev. Lett.*, 2005, 94, 197203.

53. F. Zavaliche et al., Electric field-induced magnetization switching in epitaxial columnar nanostructures, *Nano Lett.*, 2005, 5, 1793.

54. H. Zheng et al., Controlling self-assembled perovskite-spinel nanostructures, *Nano Lett.*, 2006, 6, 1401.

55. C. Israel, N. D. Mathur, J. F. Scott, A one-cent room-temperature magnetoelectric sensor, *Nat. Mater.*, 2008, 7, 93.

56. Y. M. Jia, H. S. Luo, X. Y. Zhao, F. F. Wang, Giant magnetoelectric response from a piezoelectric/magnetostrictive laminated composite combined with a piezoelectric transformer, *Adv. Mater.*, 2008, 20, 4776.

57. S. X. Dong, J. Y. Zhai, J. F. Li, D. Viehland, Near-ideal magnetoelectricity in high-permeability magnetostrictive/piezofiber laminates with a (2-1) connectivity, *Appl. Phys. Lett.*, 2006, 89, 252904.

58. Z. Fang, S. G. Lu, F. Li, S. Datta, Q. M. Zhang, M. El Tahchi, Enhancing the magnetoelectric response of Metglas/polyvinylidene fluoride laminates by exploiting the flux concentration effect, *Appl. Phys. Lett.*, 2009, 95, 112903.

59. J. Ma, Z. Shi, C. W. Nan, Magnetoelectric properties of composites of single Pb (Zr, Ti) O3 Rods and Terfenol-D/Epoxy with a single-period of 1-3-type structure, *Adv. Mater.*, 2007, 19, 2571.

60. J. Ryu, K. Mori, M. Wuttig, Magnetoelectric coupling in terfenol-D/polyvinylidenedifluoride composites, *Appl. Phys. Lett.*, 2002, 81, 100.

61. S. Priya, K. Uchino, H. E. Kim, Magnetoelectric effect in composites of magnetostrictive and piezoelectric materials, *J. Electroceram.*, 2002, 8, 107.

62. H. Greve, E. Woltermann, H. J. Quenzer, B. Wagner, E. Quandt, Giant magnetoelectric coefficients in (Fe90Co10) 78Si12B10-AlN thin film composites, *Appl. Phys. Lett.*, 2010, 96, 182501.

63. R. A. Islam, Y. Ni, A. G. Khachaturyan, S. Priya, Giant magnetoelectric effect in sintered multilayered composite structures, *J. Appl. Phys.*, 2008, 104, 044103.

64. N. Cai, C. W. Nan, J. Y. Zhai, Y. H. Lin, Large high-frequency magnetoelectric response in laminated composites of piezoelectric ceramics, rare-earth iron alloys and polymer, *Appl. Phys. Lett.*, 2004, 84, 3516.

65. S. G. Lu, Z. Fang, E. Furman, Y. Wang, Q. M. Zhang, Y. Mudryk, K. A. Gschneidner, Jr., V. K. Pecharsky, C. W. Nan, Thermally mediated multiferroic composites for the magnetoelectric materials, *Appl. Phys. Lett.*, 2010, 96, 102902.

66. R. Belouadah, D. Guyomar, B. Guiffard, J. -W. Zhang, Phase switching phenomenon in magnetoelectric laminate polymer composites: experiments and modeling, *Phys. B Condens. Matter*, 2011, 406, 2821–2826.

67. D. Guyomar, D. F. Matei, B. Guiffard, Q. Le, R. Belouadah, Magnetoelectricity in polyurethane films loaded with different magnetic particles, *Mater. Lett.*, 2009, 63, 611.

68. D. Guyomar, B. Guiffard, R. Belouadah, L. Petit, Two-phase magnetoelectric nanopowder/polyurethane composites, *J. Appl. Phys.*, 2008, 104, 074902.

69. P. Martins, A. Lasheras, J. Gutierrez, J. M. Barandiaran, I. Orue, S. Lanceros-Mendez, Optimizing piezoelectric and magnetoelectric responses on CoFe2O4/P (VDF-TrFE) nanocomposites, *J. Phys. D Appl. Phys.*, 2011, 44, 495303.

70. P. Martins, X. Moya, L. C. Phillips, S. Kar-Narayan, N. D. Mathur, S. Lanceros-Mendez, Linear anhysteretic direct magnetoelectric effect in Ni0. 5Zn0. 5Fe2O4/poly (vinylidene fluoride-trifluoroethylene) 0-3 nanocomposites, *J. Phys. D Appl. Phys.*, 2011, 44, 482001.

71. J. X. Zhang, J. Y. Dai, L. C. So, C. L. Sun, C. Y. Lo, S. W. Or, H. L. W. Chan, The effect of magnetic nanoparticles on the morphology, ferroelectric, and magnetoelectric behaviors of CFO/P (VDF-TrFE) 0–3 nanocomposites, *J. Appl. Phys.*, 2009, 105, 054102.

72. G. Evans, G. V. Duong, M. J. Ingleson, Z. L. Xu, J. T. A. Jones,Y. Z. Khimyak, J. B. Claridge, M. J. Rosseinsky, Chemical bonding assembly of multifunctional oxide nanocomposites, *Adv. Funct. Mater.*, 2010, 20, 231.

73. R. C. Camrmarata, K. Sieradzki, Surface and interface stress, *Annu. Rev. Mater. Sci.*, 1994, 24, 215.

74. E. Pan, X. Wang, R. Wang, Enhancement of magnetoelectric effect in multiferroic fibrous nanocomposites via size-dependent material properties, *Appl. Phys. Lett.*, 2009, 95, 181904.

75. C. W. Nan, N., L. Liu, J. Zhai, Y. Ye, Y. Lin, Coupled magnetic–electric properties and critical behavior in multiferroic particulate composites, *J. Appl. Phys.*, 2003, 94, 5930.

76. J. Zhai, S. Dong, Z. Xing, J. Li, D. Viehland, Giant magnetoelectric effect in Metglas/polyvinylidenefluoride laminates, *Appl. Phys. Lett.*, 2006, 89, 083507.

77. Z. Fang, S. G. Lu, F. Li, S. Datta, Q. M. Zhang, M. El Tahchi, Enhancing the magnetoelectric response of Metglas/polyvinylidene fluoride laminates by exploiting the flux concentration effect, *Appl. Phys. Lett.*, 2009, 95, 112903.

78. M. Zeng, S. W. Or, H. L. W. Chan, Large magnetoelectric effect from mechanically mediated magnetic field-induced strain effect in Ni–Mn–Ga single crystal and piezoelectric effect in PVDF polymer, *J. Alloys Compd.*, 2010, 490, L5.

79. S. G. Lu, J. Z. Jin, X. Zhou, Z. Fang, Q. Wang, Q. M. Zhang, Large magnetoelectric coupling coefficient in poly (vinylidene fluoride-hexafluoropropylene)/Metglas laminates, *J. Appl. Phys.*, 2011, 110, 104103.

80. S. Zhao, J. -G. Wan, M. Yao, J. -M. Liu, F. Song, G. Wang, Flexible Sm–Fe/polyvinylidene fluoride heterostructural film with large magnetoelectric voltage output, *Appl. Phys. Lett.*, 2010, 97, 212902.

81. V. S. D. Voet, M. Tichelaar, S. Tanase, M. C. Mittelmeijer-Hazeleger, G. Ten Brinke, K. Loos, Poly (vinylidene fluoride)/nickel nanocomposites from semicrystalline block copolymer precursors, *Nanoscale*, 2012, 5, 184.

82. Y. H. Lin, N. Cai, J. Y. Zhai, G. Liu, C. W. Nan, Giant magnetoelectric effect in multiferroic laminated composites, *Phys. Rev. B*, 2005, 72, 012405.

83. N. Cai, J. Zhai, C. W. Nan, Y. Lin, Z. Shi, Dielectric, ferroelectric, magnetic, and magnetoelectric properties of multiferroic laminated composites, *Phys. Rev. B*, 2003, 68, 224103.

84. S. Valencia, A. Crassous, L. Bocher, V. Garcia, X. Moya, R. O. Cherifi, C. Deranlot et al., Interface-induced room-temperature multiferroicity in BaTiO3, *Nat. Mater.*, 2011, 10, 753.

85. J. -Q. Dai, H. Zhang, Y. -M. Song, Interfacial electronic structure and magnetoelectric effect in M/BaTiO3 (M = Ni, Fe) superlattices, *J. Magn. Magn. Mater.*, 2012, 324, 3937.

86. P. V. Lukashev, T. R. Paudel, J. M. López-Encarnación, S. Adenwalla, E. Y. Tsymbal, J. P. Velev, Ferroelectric control of magnetocrystalline anisotropy at cobalt/poly (vinylidene fluoride) interfaces, *ACS Nano*, 2012, 6, 9745.

87. C. W. Nan, L. Liu, N. Cai, J. Zhai, Y. Ye, Y. H. Lin, L. J. Dong, C. X. Xiong, A three-phase magnetoelectric composite of piezoelectric ceramics, rare-earth iron alloys, and polymer, *Appl. Phys. Lett.*, 2002, 81, 3831.

88. C. W. Nan, N. Cai, Z. Shi, J. Zhai, G. Liu, Y. Lin, Large magnetoelectric response in multiferroic polymer-based composites, *Phys. Rev. B*, 2005, 71, 014102.

89. N. Nersessian, S. W. Or, G. P. Carman, Magnetoelectric behavior of Terfenol-D composite and lead zirconate titanate ceramic laminates, *IEEE Trans. Magn.*, 2004, 40, 2646.

90. M. I. Bichurin, D. A. Filippov, V. M. Petrov, V. M. Laletsin, N. Paddubnaya, G. Srinivasan, Resonance magnetoelectric effects in layered magnetostrictive-piezoelectric composites, *Phys. Rev. B*, 2003, 68, 132408.

91. J. Jin, S. -G. Lu, C. Chanthad, Q. Zhang, M. A. Hague, Q. Wang, Multiferroic polymer composites with greatly enhanced magnetoelectric effect under a low magnetic bias, *Adv. Mater.*, 2011, 23, 3853.

92. K. C. Kao, *Dielectric Phenomenon in Solids with Emphasis on with Physical Concepts and Electronic Processes*. Academic Press, Amsterdam, 2004.

93. C. Kittel, *Introduction to Solid State Physics*. Wiley, New York, 1956.

94. A. K. Jonscher, *Dielectric Relaxation in Solids*. Chelsea Dielectric Press, London, 1983.

95. G. A. Smolenskiĭ, I. E. Chupis, Ferroelectromagnets, *Sov. Phys. Uspekhi*, 1982, 25 475.

96. R. Schmidt, W. Eerenstein, P. A. Midgley, Large dielectric response to the paramagnetic-ferromagnetic transition (TC< 100 K) in multiferroic BiMnO3 epitaxial thin films, *Phys. Rev. B*, 2009, 79, 214107.

97. D. Choudhury, P. Mandal, R. Mathieu, A. Hazarika, S. Rajan, A. Sundaresan, U. V. Waghmare et al., Near-room-temperature colossal magnetodielectricity and multiglass properties in partially disordered La2NiMnO6, *Phys. Rev. Lett.*, 2012, 108, 127201.

98. P. Padhan, P. LeClair, A. Gupta, M. A. Subramanian, G. Srinivasan, Magnetodielectric effect in Bi2NiMnO6–La2NiMnO6 superlattices, *J. Phys. Condens. Matter*, 2009, 21, 306004.

99. J. Krishna Murthy, A. Venimadhav, Magnetodielectric behavior in La2CoMnO6 nanoparticles, *J. Appl. Phys.*, 2012, 111, 024102.

100. G. Catalan, Magnetocapacitance without magnetoelectric coupling, *Appl. Phys. Lett.*, 2006, 88, 102902.

101. M. Mario, Interface-driven magnetocapacitance in a broad range of materials, *J. Phys. Condens. Matter*, 2008, 20, 322202.

102. C. -C. Chang, L. Zhao, M. -K. Wu, Magnetodielectric study in SiO2-coated Fe3O4 nanoparticle compacts, *J. Appl. Phys.*, 2010, 108, 094105.

103. D. Niermann, F. Waschkowski, J. de Groot, M. Angst, J. Hemberger, Dielectric properties of charge-ordered LuFe2O4 revisited: the apparent influence of contacts, *Phys. Rev. Lett.*, 2012, 109, 016405.

104. P. Ren, Z. Yang, W. G. Zhu, C. H. A. Huan, L. Wang, Origin of the colossal dielectric permittivity and magnetocapacitance in LuFe2O4, *J. Appl. Phys.*, 2011, 109, 074109.

105. T. Bonaedy, Y. S. Koo, K. D. Sung, J. H. Jung, Resistive magnetodielectric property of polycrystalline g-Fe2O3, *Appl. Phys. Lett.*, 2007, 91, 132901.

106. S. Zhang, X. Dong, F. Gao, Y. Chen, F. Cao, J. Zhu, X. Tang, G. Wang, Magnetodielectric response in 0.36 BiScO3-0.64 PbTiO3/La0. 7Sr0. 3MnO3 thin films and the corresponding model modifications, *J. Appl. Phys.*, 2011, 110, 046103.

107. D. Bhadra, M. G. Masud, S. Sarkar, J. Sannigrahi, S. K. De, B. K. Chaudhuri, Synthesis of PVDF/BiFeO3 nanocomposite and observation of enhanced electrical conductivity and low-loss dielectric permittivity at percolation threshold, *Polym. Phys.*, 2012, 50, 572.

108. J. -K. Yuan, W. -L. Li, S. -H. Yao, Y. -Q. Lin, A. Sylvestre, J. Bai, High dielectric permittivity and low percolation threshold in polymer composites based on SiC-carbon nanotubes micro/nano hybrid, *Appl. Phys. Lett.*, 2011, 98, 032901.

109. M. Panda, V. Srinivas, A. K. Thakur, On the question of percolation threshold in polyvinylidene fluoride/nanocrystalline nickel composites, *Appl. Phys. Lett.*, 2008, 92, 132905.

110. L. Wang, Z. -M. Dang, Carbon nanotube composites with high dielectric constant at low percolation threshold, *Appl. Phys. Lett.*, 2005, 87, 042903.

111. C. Huang, Q. Zhang, Enhanced dielectric and electromechanical responses in high dielectric constant all-polymer percolative composites, *Adv. Funct. Mater.*, 2004, 14, 501.

112. S. Kirkpatrick, Percolation and conduction, *Rev. Mod. Phys.*, 1973, 45, 574.

113. C. W. Nan, Y. Shen, J. Ma., *Annual Review of Materials Research*, Physical Properties of Composites Near Percolation vol 40. Annual Reviews, Palo Alto, 2010, 131–151.

114. L. Qi, B. I. Lee, S. Chen, W. D. Samuels, G. J. Exarhos, High-dielectric-constant silver–epoxy composites as embedded dielectrics, *Adv. Mater.*, 2005, 17, 1777.

115. Z. M. Dang, Y. H. Lin, C. W. Nan, Novel ferroelectric polymer composites with high dielectric constants, *Adv. Mater.*, 2003, 15, 1625.

116. M. Panda, V. Srinivas, A. Venimadhav, Synthesis and characterization of Ni-PVDF nanocomposites, *AIP, Conf. Proc.*, 2008, 99, 042905.

117. K. S. Deepa, M. T. Sebastian, J. James, Effect of interparticle distance and interfacial area on the properties of insulator-conductor composites, *Appl. Phys. Lett.*, 2007, 91, 202904.

118. Y. R. Hernandez, A. Gryson, F. M. Blighe, M. Cadek, V. Nicolosi, W. J. Blau, Y. K. Gun'ko, J. N. Coleman, Comparison of carbon nanotubes and nanodisks as percolative fillers in electrically conductive composites, *Scripta Mater.*, 2008, 58, 69.

119. C. J. Kerr, Y. Y. Huang, J. E. Marshall, E. M. Terentjev, Comparison of carbon nanotubes and nanodisks as percolative fillers in electrically conductive composites, *J. Appl. Phys.*, 2011, 109, 094109.

120. S. -H. Yao, Z. -M. Dang, M. -J. Jiang, H. -P. Xu, J. Bai, Effect of filament aspect ratio on the dielectric response of multiwalled carbon nanotube composites, *Appl. Phys. Lett.*, 2007, 91, 212901.

121. C. -W. Nan, M. I. Bichurin, S. Dong, D. Viehland, G. Srinivasan, Influence of aspect ratio of carbon nanotube on percolation threshold in ferroelectric polymer nanocomposite, *J. Appl. Phys.*, 2008, 103, 031101.

122. C. Thirmal, C. Nayek, P. Murugavel, V. Subramanian, Magnetic, dielectric and magneto-dielectric properties of PVDF-La$_{0.7}$Sr$_{0.3}$MnO$_3$ polymer nanocomposite film, *AIP Advances*, 2013, 3, 112109.

123. A. Kumar, K. L. Yadav, Temperature dependence of the crystal and magnetic structures of BiFeO$_3$, *J. Alloys Compd.*, 2012, 528, 16–19.

124. Y. P. Guo, Y. Liu, J. Wang, R. L. Withers, H. Chen, L. Jin, P. Smith, Giant Magnetodielectric Effect in 0-3 Ni0.5Zn0.5Fe2O4-Poly(vinylidene-fluoride) Nanocomposite Films, *J. Phys. Chem. C*, 2010, 114, 13861–13866.

125. K. D. Chandrasekhar, A. K. Das, A. Venimadhav, Large magnetodielectric response in Pr0.6Ca0.4MnO3/polyvinylidenePr0.6Ca0.4MnO3/polyvinylidenefluoride nanocomposites, *Appl. Phys. Lett.*, 2011, 98, 122908.

126. K. Devi Chandrasekhar, A. K. Das, A. Venimadhav, Magnetic glassy behavior of Pr0.6Ca0.4MnO3 nanoparticles: effect of intra and interparticle magnetic interactions on magnetodielectric property, *J. Phys. Chem. C*, 2014, 118, 27728.

127. M. Li, Y. Wang, J. Gao, D. Gray, J. Li, D. Viehland, Dependence of magnetic field sensitivity of a magnetoelectric laminate sensor pair on separation distance: effect of mutual inductance, *J. Appl. Phys.*, 2012, 111, 033923.

128. U. Laletsin, N. Padubnaya, G. Srinivasan, C. P. Devreugd, Frequency dependence of magnetoelectric interactions in layered structures of ferromagnetic alloys and piezoelectric oxides, *Appl. Phys. A*, 2004, 78, 33.

129. S. Dong, J. –F. Li, D. Viehland, J. Cheng, L. E. Cross, A strong magnetoelectric voltage gain effect in magnetostrictive-piezoelectric composite, *Appl. Phys. Lett.*, 2004, 85, 3534.

130. J. Gao, Z. Wang, Y. Shen, M. Li, Y. Wang, P. Finkel, J. Li, D. Viehland, Self-powered low noise magnetic sensor, *Mater. Lett.*, 2012, 82, 178–180.

131. Y. K. Fetisov, A. A. Bush, K. E. Kamentsev, A. Y. Ostashchenko, G. Srinivasan, Ferrite-piezoelectric multilayers for magnetic field sensors, *IEEE Sensors J.*, 2006, 6, 935.

132. E. Lage, F. Woltering, E. Quandt, D. Mayners, Fundamentals of multiferroic materials and their possible applications, *Appl. Phys. Lett.*, 2013, 113, 17C725.

133. S. X. Dong, J. Y. Zhai, J. F. Li, D. Viehland, S. Priya, Multimodal system for harvesting magnetic and mechanical energy, *Appl. Phys. Lett.*, 2008, 93, 103511.

134. X. Z. Dai, Y. M. Wen, P. Li, J. Yang, G. Y. Zhang, Modeling, characterization and fabrication of vibration energy harvester using Terfenol-D/PZT/Terfenol-D composite transducer, *Sens. Actuators A*, 2009, 156, 350.

135. E. Y. Tsymbal, O. N. Mryasov, P. R. LeClair, Spin-dependent tunnelling in magnetic tunnel junctions, *J. Phys. Condens. Matter*, 2003, 15, R109–R142.

136. Z. Shi, C. P. Wang, X. J. Liu, C. W. Nan, A four-state memory cell based on magnetoelectric composite, *Chin. Sci. Bull.*, 2008, 53, 2135–2138.

137. A. Gruverman, D. Wu, H. Lu, Y. Wang, H. W. Jang, C. M. Folkman, M. Ye. Zhuravlev et al., Tunneling electroresistance effect in ferroelectric tunnel junctions at the Nanoscale, *Nano Lett.*, 2009, 9, 3539.

138. M. Bibes, A. Barthelemy, Towards a magnetoelectric memory, *Nat. Mater.*, 2008, 7, 425–426.

139. A. K. Biswas et al., An error-resilient non-volatile magneto-elastic universal logic gate with ultralow energy-delay product, *Nature Sci. Rep.*, 2014, 4, 7553.

140. A. K. Biswas, S. Bandyopadhyay, J. Atulasimha, Complete magnetization reversal in a magnetostrictive nanomagnet with voltage-generated stress: A reliable energy-efficient non-volatile magneto-elastic memory, *Appl. Phys. Lett.*, 2014, 105, 072408.

141. E. Warburg, Magnetische untersuchungen, *Ann. Phys.*, 1881, 249 141.

142. W. F. Giauque, D. P. MacDougall, Attainment of Temperatures Below 1° Absolute by Demagnetization of $Gd_2(SO_4)_3 \cdot 8H_2O$, *Phys. Rev.*, 1933, 43(9), 768.

143. A. M. Tishin, Y. I. Spichkin, *The Magnetocaloric Effect and its Applications*. Taylor & Francis Group, Boca Raton, FL, 2003.

144. K. A. Gschneidner, Jr, V. K. Pecharsky, A. O. Tsokol, Recent developments in magnetocaloric materials, *Rep. Prog. Phys.*, 2005, 68, 1479.

145. X. Moya, S. Kar-Narayan, N. D. Mathur, Caloric materials near ferroic phase transitions, *Nat. Mater.*, 2014, 13, 439.

146. H. Meng, B. Li, W. Ren, Z. Zhang, Coupled caloric effects in multiferroics, *Phys. Lett. A*, 2013, 377, 567.
147. J. F. Scott, Electrocaloric materials, *Annu. Rev. Mater. Res.*, 2011, 41, 229.
148. A. S. Mischenko, Q. Zhang, J. F. Scott, R. W. Whatmore, N. D. Mathur, Giant electrocaloric effect in thin-film PbZr0.95Ti0.05O3, *Science*, 2006, 311, 1270.
149. X. Moya, E. Stern-Taulats, S. Crossley, D. González-Alonso, S. Kar-Narayan, A. Planes, L. Mañosa, N. D. Mathur, Giant electrocaloric strength in single-crystal BaTiO3, *Adv. Mater.*, 2013, 25, 1360.
150. S. G. Lu, B. Rozic, Q. M. Zhang, Z. Kutnjak, X. Li, E. Furman, L. J. Gorny, Organic and inorganic relaxor ferroelectrics with giant electrocaloric effect, *Appl. Phys. Lett.*, 2010, 97, 162904.
151. B. Neese, B. Chu, S. -G. Lu, Y. Wang, E. Furman, Q. M. Zhang, Large electrocaloric effect in ferroelectric polymers near room temperature, *Science*, 2008, 321, 821.
152. S. Lisenkov, B. K. Mani, C. M. Chang, J. Almand, I. Ponomareva, Multicaloric effect in ferroelectric PbTiO3 from first principles, *Phys. Rev. B*, 2013, 87, 224101.
153. M. M. Vopson, The multicaloric effect in multiferroic materials, *Solid State Comm.*, 2012, 152, 2067.
154. J. Krishna Murthy, A. Venimadhav, Multicaloric effect in multiferroic Y2CoMnO6, *J. Phys. D Appl. Phys.*, 2014, 47, 445002.
155. B. Li, C. Wang, W. Zhang, C. Hang, J. Fei, H. Wang, Fabrication of multiferroic Ba 0.7 Sr 0.3 TiO3–Ni0.8Zn0.2Fe2O4 composite nanofibers by electrospinning, *Mater. Lett.*, 2013, 91, 55–58.
156. H. Singh, A. Kumar, K. L. Yadav, Structural, dielectric, magnetic, magnetodielectric and impedance spectroscopic studies of multiferroic BiFeO3 –BaTiO3 ceramics, *Mater. Sci. Eng. B*, 2011, 176, 540–547.
157. S. X. Huo, S. L. Yuan, Y. Qiu, Z. Z. Ma, C. H. Wang, Crystal structure and multiferroic properties of BiFeO3–Na0.5K0.5NbO3 solid solution ceramics prepared by Pechini method, *Mater. Lett.*, 2012, 68, 8–10.
158. S. Zhang, X. Dong, Y. Chena, G. Wang, J. Zhuc, X. Tang, Co-contributions of the magnetostriction and magnetoresistance to the giant room temperature magnetodielectric response in multiferroic composite thin films, *Solid State Comm.*, 2011, 151, 982–984.
159. S. M. Salunkhe, S. R. Jigajeni, M. M. Sutar, A. N. Tarale, P. B. Joshi, Magnetoelectric and magnetodielectric effect in CFMO-PBT nanocomposites, *J. Phys. Chem. Solids*, 2013, 74, 388–394.
160. S. R. Jigajeni, A. N. Tarale, D. J. Salunkhe, P. B. Joshi, S. B. Kulkarni, Dielectric, magnetoelectric and magnetodielectric properties in CMFO-SBN composites, *Ceram. Int.*, 2013, 39, 2331–2341.
161. M. Malar Selvi, P. Manimuthu, K. Saravana Kumar, C. Venkateswaran, Magnetodielectric properties of CoFe2O4–BaTiO3 core–shell nanocomposite, *J. Magn. Magn. Mater.*, 2014, 369, 155–161.
162. T. Cheng, L. F. Xu, P. B. Qi, C. P. Yang, R. L. Wang, H. B. Xiao, Tunable dielectric behaviors of magnetic field of PZT5/NiFe2O4 ceramic particle magnetoelectric composites at room temperature, *J. Alloys Compd.*, 2014, 602, 269–274.
163. S. Mukherjee, C. H. Chen, C. C. Chou, K. F. Tseng, B. K. Chaudhuri, H. D. Yang, Colossal dielectric and magnetodielectric effect in Er2O3 nanoparticles embedded in a SiO2 glass matrix, *Phys. Rev. B*, 2010, 82, 104107.

CHAPTER 16

Thiophene-Based Conjugated Polymers

Synthesis, Linear, and Third-Order Nonlinear Optical Properties

M.G. Murali and D. Udayakumar

CONTENTS

16.1 INTRODUCTION

Conjugated polymers (CPs), which are considered as intrinsic semiconductors due to their delocalized π-electrons, have attracted much research attention in the past three decades. CPs are organic materials and are composed of carbon, hydrogen, and simple heteroatoms such as nitrogen, oxygen, and sulfur. These polymers generally have a regular alternation of single and double bonds along the polymer chain. Conductivity in these materials arises uniquely from the extended and delocalized π-conjugation. Therefore, the polymer that possesses the electrical, electronic, magnetic, and optical properties of a metal while retaining the mechanical properties and processability commonly associated with a conventional polymer, is termed as an "intrinsically conducting polymer," which is more commonly known as a "synthetic metal." A schematic representation of π-conjugation in poly(acetylene) is shown in Figure 16.1.

FIGURE 16.1   Schematic representation of π-conjugation in poly(acetylene).

Extensive research work in the field of CPs has started in late 1970s with the discovery of electrical conductivity in oxidatively doped poly(acetylene) [1]. This important discovery led to the award of Nobel Prize in chemistry in the year 2000 to three scientists: Hideki Shirakawa, Alan G MacDiarmid, and Alan J. Heeger. This pioneering work motivated the researchers to synthesize and to study the properties of other new classes of π-CPs based on aromatic precursors such as p-phenylenevinylene, thiophene, pyrrole, carbazole, fluorene, and their derivatives (Figure 16.2) [2–5]. Early work in this field held the hope that these types of polymer systems would serve as replacements for highly conductive metals, such as copper and aluminum for electrical transport or related applications. However, due to the instability of these systems when highly doped, other more practical uses have been realized such as polymer light-emitting diodes (PLEDs) [6–11]. Moreover, in order to tune the specific properties desired for end-user applications, a proper modification of the monomer structure is essential that manipulates both the electronic and optical properties of these polymer systems.

In CPs, the highest occupied band (analogous to the valence band in inorganic semiconductors) originates from the highest occupied molecular orbital (HOMO) of the monomer unit, whereas the lowest unoccupied band (the conduction band) originates from the lowest unoccupied molecular orbital (LUMO) of the monomer unit. The difference in energy between these levels is called the band gap (E_g). The band structure of a CP originates from the interaction of the π-orbitals of the repeating units (monomers) throughout the chain.

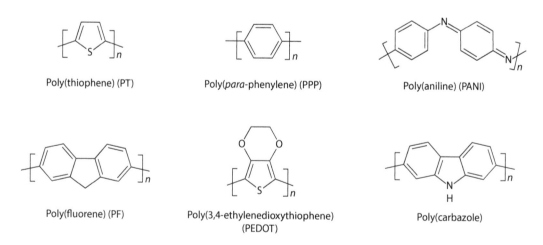

Poly(thiophene) (PT)   Poly(para-phenylene) (PPP)   Poly(aniline) (PANI)

Poly(fluorene) (PF)   Poly(3,4-ethylenedioxythiophene)   Poly(carbazole)
                                         (PEDOT)

FIGURE 16.2   Some important classes of π-conjugated polymers.

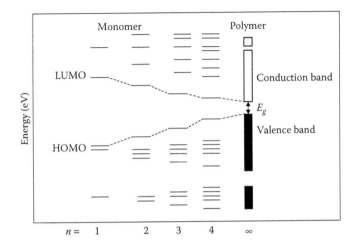

FIGURE 16.3 Schematic representation of band formation during the polymerization of a conjugated monomer into a π-conjugated polymer.

During the conversion of monomers to a polymer in the polymerization reaction, the HOMO and LUMO levels of the repeating unit disperse into the valence and conduction bands. A schematic representation of the band structure of a CP is illustrated in Figure 16.3.

16.1.1 Donor–Acceptor (D–A) CPs

CPs with carbon and heteroatom frameworks have attracted considerable attention due to their structural diversity and fascinating optoelectronic properties. However, design and synthesis of π-CPs with a low oxidation potential, broad absorption spectrum, low band gap, efficient photoinduced charge transfer and separation, and ambipolar charge transport with high mobilities are crucial for the fabrication of optoelectronic devices. In particular, tuning of the HOMO–LUMO energy levels and hence the band gap energy are of great importance for directing material properties toward the desired applications.

In this direction, the donor–acceptor (D–A) approach, introduced by Havinga et al. [12] in macromolecular systems via alternating electron-rich (D) and electron-deficient (A) substituents along the conjugated backbone, has attracted a good deal of attention in recent years. The D–A CPs absorb light at longer wavelengths than the wide-band-gap CPs made of all donor units. Judiciously chosen D and A groups are particularly desirable for low band gap polymers due to the significant enhancement of intramolecular charge-transfer (ICT) interaction and conjugation length, which leads to extended absorption and a higher absorption coefficient. Also, an extended, rigid π-conjugation with the quinonoid character in the polymer backbone facilitates intermolecular interactions between the polymer chains and increases the charge mobility of the polymer [13]. Hence, a CP with an alternating sequence of appropriate donor and acceptor units in the main chain can induce a reduction in its band gap energy. Further molecular orbital calculations have shown that the hybridization of the energy levels of the donor and the acceptor moieties results in D–A systems with an unusually low HOMO–LUMO separation [14]. If the HOMO levels of the donor

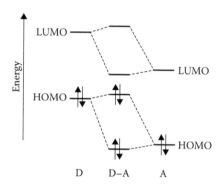

FIGURE 16.4   Molecular orbital interactions between donor (D) and acceptor (A) moieties.

and the LUMO levels of the acceptor moiety are close in energy, the resulting band structure will show a low energy gap [15,16]. In such systems, the HOMO energy level is raised and the LUMO energy level is lowered by the interaction of molecular orbitals of D and A groups as shown in Figure 16.4.

In D−A systems, the introduction of electron withdrawing groups reduces E_g by lowering the LUMO levels, whereas the introduction of electron-donating groups reduces E_g by raising the HOMO levels [17]. Therefore, the introduction of strong electron donors and acceptors in the CP backbone is one of the design criteria to obtain extremely low band gap polymers. Also, the electron or hole affinity can be enhanced simultaneously or controlled independently in a D−A system by modifying the chemical structure of the electron-rich or the electron-deficient substituents. In other words, CPs with desired HOMO and LUMO energy levels could be obtained by the proper selection of donor and acceptor units. Because of this reason, D−A CPs are emerging as promising materials for PLEDs and polymer solar cells (PSCs). In addition, the increase in effective π-electron delocalization in D−A polymers results in large third-order nonlinear optical (NLO) susceptibilities, and hence these polymers are also considered as potential candidates in the area of nonlinear optics [18,19].

In this context, a number of research groups have designed and synthesized various D−A CPs for optoelectronic applications. The most commonly used electron-donating moieties in D−A CPs are thiophene, fluorene, carbazole, and pyrrole with various substitution patterns. The substituent groups play an important role in controlling the polymer properties such as chemical, thermal, and electronic properties. The most widely used electron withdrawing groups are cyanovinylene, 1,3,4-oxadiazole, 2,1,3-benzothiadiazole, perylene diimides, 2,6-dimethyl-4H-pyran-4-(ylidene)propanedinitrile, and so on. The D−A CPs can be synthesized by reacting two monomeric units containing proper reactive functionalities, using various reaction methodologies such as condensation reaction and coupling reaction. Alternatively, suitable D−A type monomers can be synthesized first; then they undergo facile chemical or electrochemical polymerization leading to the formation of D−A polymers.

In this view, we have designed and synthesized a new set of D−A CPs (P1–P12) for optoelectronic applications. The chemical structure of D−A CPs is designed by the selection of proper electron donor and acceptor units, and the polymers are synthesized using multi-step synthetic routes. 3,4-Dialkoxythiophene and/or 9,9′-dialkylfluorene units are used as the electron donor moieties in the polymeric structures. The electron withdrawing units such as 1,3,4-oxadiazole, cyanovinylene, or cyanophenylenevinylene are used as electron acceptor segments.

16.2 DESIGN CRITERIA, STRUCTURAL FEATURES, AND SYNTHESIS OF POLYMERS P1–P12

In order to achieve desired properties in D–A CPs, the following important factors should be considered during the molecular design of these polymers: (1) presence of extended conjugation length, (2) change in the identity of the conjugation bridge, (3) increase in the π-donor and π-acceptor strength, (4) incorporation of more polarizable double bonds, (5) increase in the planarity of the polymer backbone, and (6) good solution process-ability of the polymers. We have considered these factors for the design of a new set of thiophene-based D–A type CPs (P1–P12). The core structure of the polymers consists of 3,4-didodecyloxythiophene as an electron donor unit and is attached to various electron withdrawing units, namely, 1,3,4-oxadiazole, cyanovinylene, and cyanophenylenevinylene moieties along with conjugative spacers like phenylenevinylene, thiophenevinylene, and biphenyl units. The chemical structure of the polymers are designed in such a way as to obtain polymers having a low band gap, high thermal stability, good solvent processability, and film-forming property with desired optical and electrochemical properties for opto-electronic and photonic applications. The reason for choosing 3,4-dialkoxythiophene unit as the core donor segment is that poly(3,4-dialkoxythiophene)s show facile dopability and a lower band gap due to the electron-donating nature of the alkoxy moiety. Further intro-duction of long alkoxy pendants at 3- and 4-positions of the thiophene ring facilitates the solvent processability of the corresponding polymer. Also, to overcome the possibility of steric interactions between the alkoxy groups of adjacent thiophene rings, which reduce the coplanarity and thus decrease the effective conjugation length of the polymer, conju-gated spacers are introduced between two thiophene rings in the polymer chain.

The (1,3,4-oxadiazolyl)benzene unit in P1 and (1,3,4-oxadiazolyl)-3,4-propelene-dioxythiophene unit in P2 are incorporated between two alkoxy-substituted thiophene rings, so as to minimize steric interactions between the alkoxy groups. In the case of P3, 3,4-dialkoxythiophene rings are separated by a spacer, 2-(ethenylphenyl)-1,3 -4-oxadiazole moiety. The chemical structure of the repeating unit in P4 is almost similar to that of P3, except that one of the 3,4-dialkoxythiophene rings in P3 is replaced by an unsubstituted thiophene ring in P4. Such a simple structural modification in the polymer chain allows studying the effect of alkoxy pendant groups on optical and electrochemical properties of the CPs. In the case of polymers P5–P8, cyanovinylene group was introduced along the polymer chain as cyano-containing polymers are expected to show a low band gap and extended absorption in the visible region. The repeated units in P6–P8 have an extended conjugation length as compared to that in P5, with the presence of an electron-donating dialkylfluorene unit in P6, an electron-withdrawing cyanophenylenevinylene unit in P7, and a D–A conjugated system in P8. The incorporation of these units along the polymer chain hence affects optical and electrochemical properties of the polymers. As compared with P6–P8, the number of vinylic linkages in polymers P9 and P10 chains is reduced so as to obtain chemically stable polymers. The chain structures of P11 and P12 are comparable with those of P9 and P10, respectively. The electron-donating dilkylfluorene rings in P9 and P10 are replaced with a strong electron-withdrawing cyanophenylenevinylene unit in P11 and P12. This structural modification is expected to enhance electron-transporting and hole-blocking properties of the polymers. Further, it is envisaged that such structural modifications would help understand the structure–property relationship in D–A CPs.

The five series of thiophene-based D–A type CPs (P1–P12) were synthesized from the corresponding monomers using a proper reaction methodology. The chemical structure of building block monomers, M1–M10, is shown in Figure 16.5.

The synthetic methodology followed for the synthesis of polymers P1–P12 is summarized in Scheme 16.1. The polycondensation reaction technique was used for the synthesis of polymers P1 and P2, wherein the monomer dihydrazide M1 was reacted with the corresponding diacid chloride monomer (M3 for P1 and M2 for P2) followed by a cyclodehydration reaction. To synthesize polymers P3 and P4, the well-known Wittig reaction methodology was employed, wherein the Wittig salt monomer M4 was treated with the appropriate thiophene

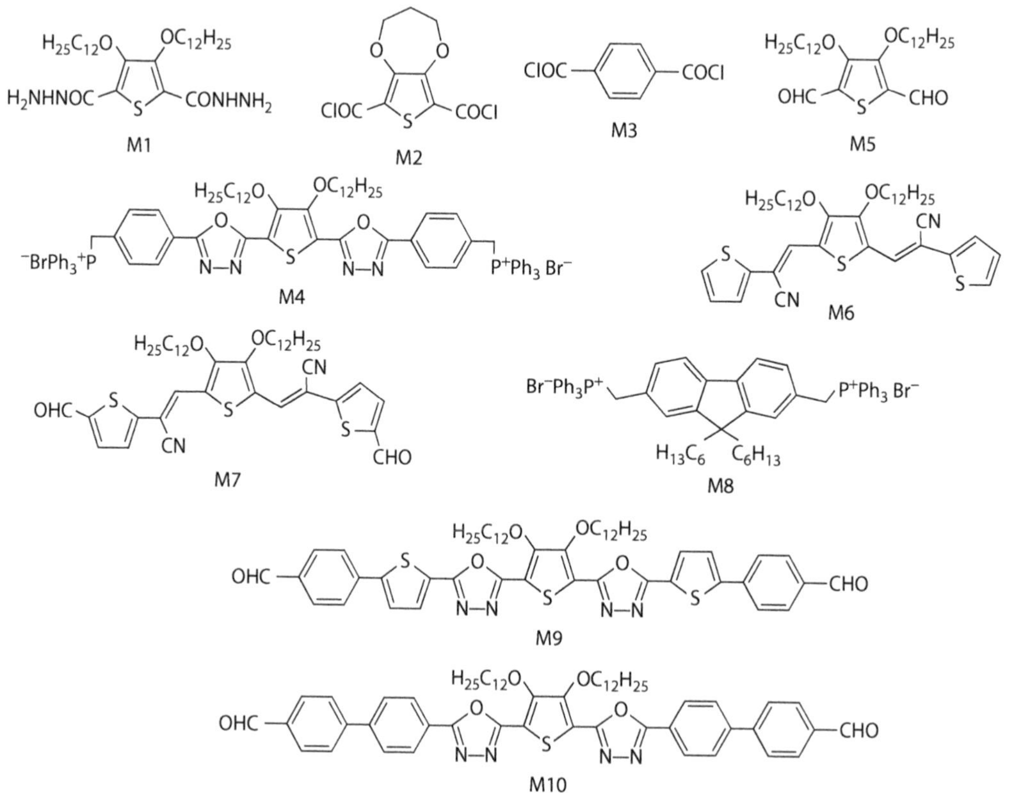

FIGURE 16.5 Chemical structures of synthesized monomers (M1–M10).

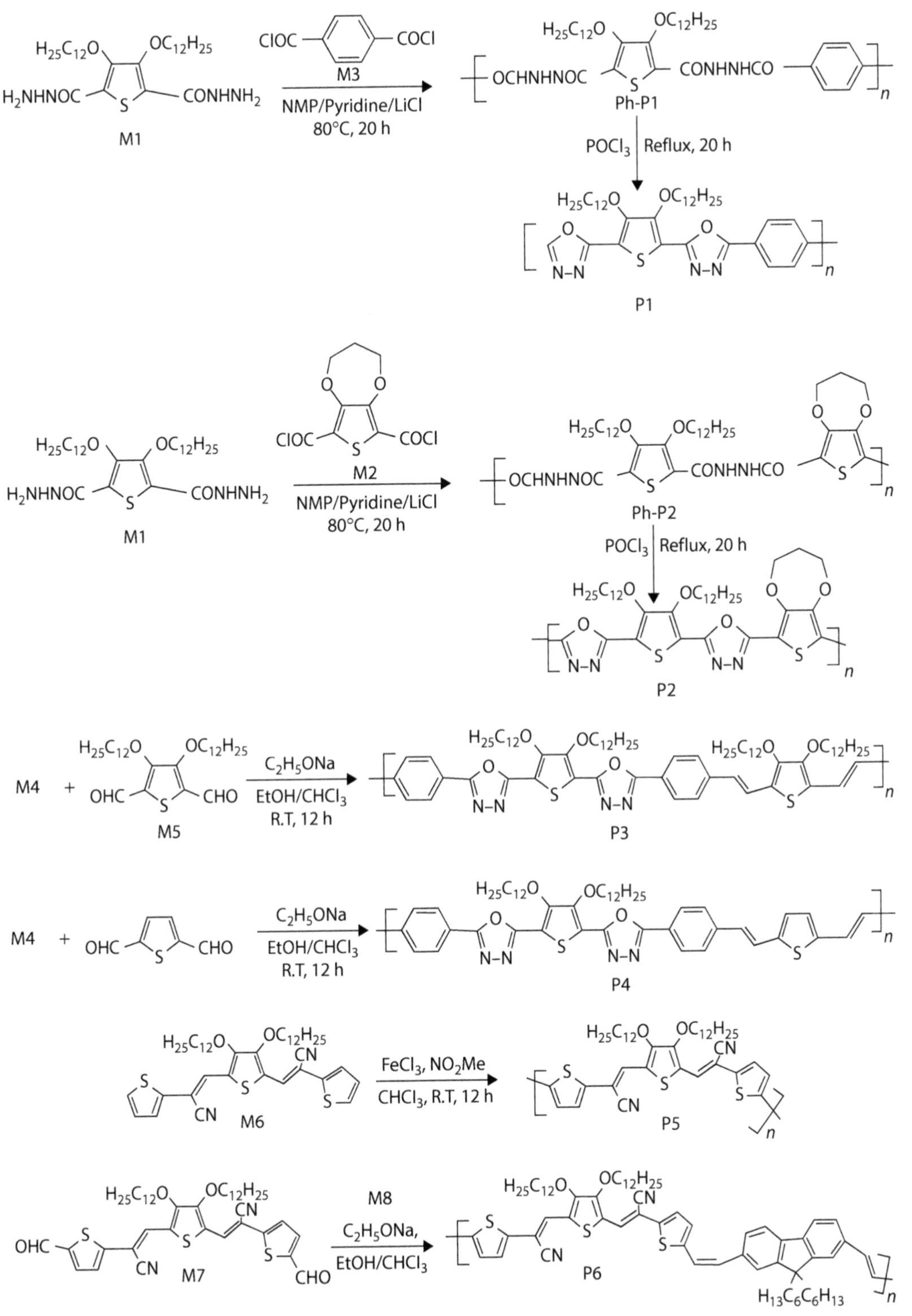

SCHEME 16.1 Synthesis of polymers P1–P12.

(Continued)

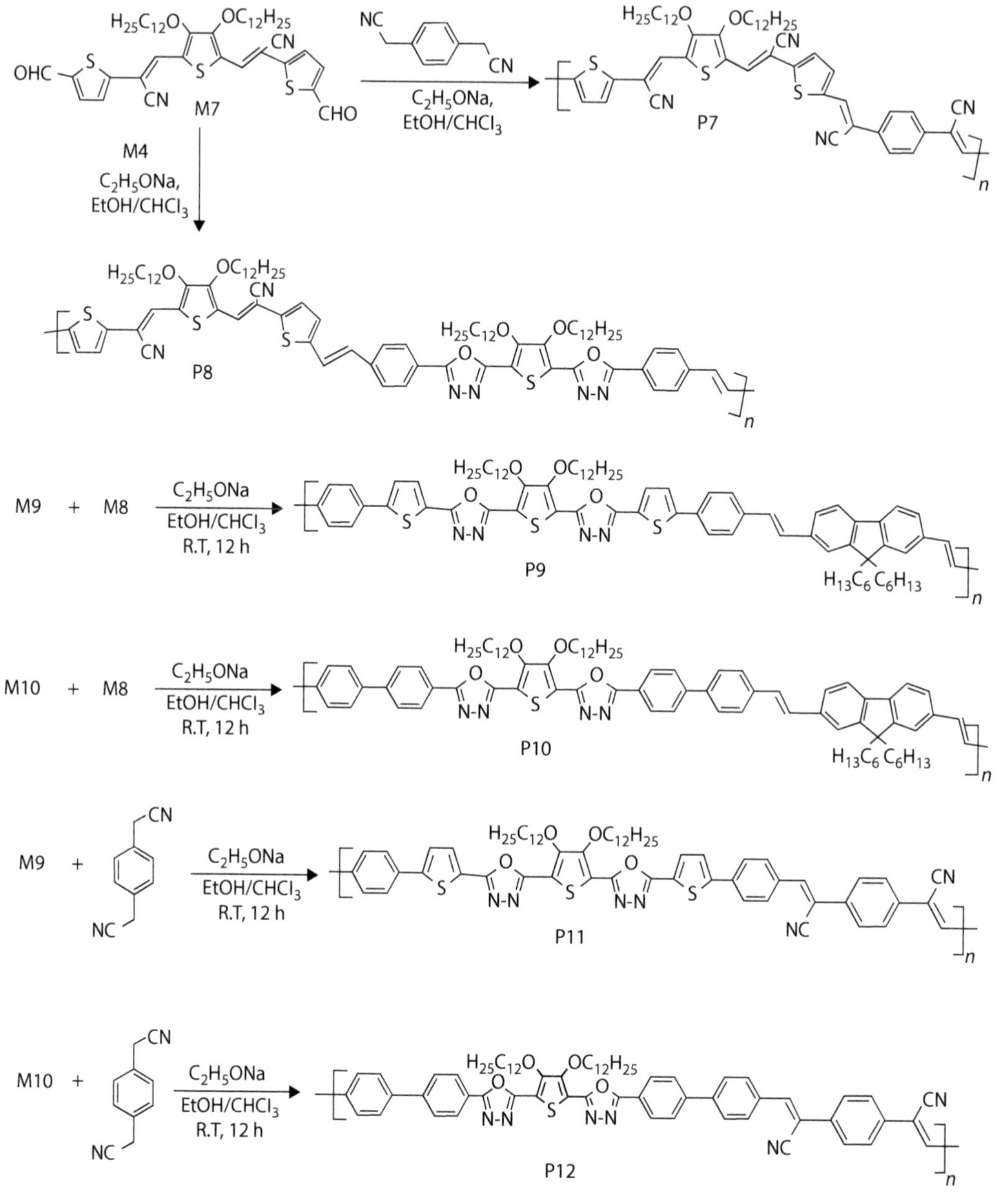

SCHEME 16.1 (Continued) Synthesis of polymers P1–P12.

dialdehyde monomer (dicarbaldehyde M5 for P3 and thiophene-2,5-dialdehyde for P4). The chemical polymerization reaction was performed to synthesize polymer P5 from its monomer M6. The polymers P6 and P8 were obtained through the Wittig reaction methodology, by reacting cyanodialdehyde monomer M7 with the corresponding Wittig salt monomer (M8 for P6 and M4 for P8). Polymer P7 was prepared by using a Knoevenagel condensation reaction between monomer M7 and 1,4-phenyldiacetonitrile. The Wittig reaction methodology was employed to prepare polymers P9 and P10, by reacting fluorene Wittig salt monomer M8 with the corresponding dialdehyde monomer (M9 for P9 and M10 for P10). To synthesize polymers P11 and P12, the corresponding dialdehyde monomer (M9 for P11 and M10 for P12) was condensed with 1,4-phenylenediacetonitrile using a Knoevenagel condensation methodology. All the polymers showed good solubility in common organic solvents due to the presence of long alkoxy pendants at the 3- and 4-positions of the thiophene ring.

16.3 LINEAR OPTICAL PROPERTIES OF THE POLYMERS (P1–P12)

The basic electronic structure of the CPs can be evaluated by investigating their photophysical processes that occur in the electronic excited states. UV–Vis absorption and fluorescence emission spectroscopies are the two techniques used to study the linear optical properties of the polymers. The UV–Vis absorption spectroscopy deals with the absorption of a sufficiently energetic photon ($h\nu$) in the UV–Vis region by the polymer and measures the electronic transition from the ground state to the excited state. Generally, the observed transitions in the absorption spectroscopy of CPs are attributed to the electronic transition from the π to π* state. The fluorescence emission spectroscopy deals with the transitions from the excited state to the ground state, as the electronic transition from the π* to π state. The measured data in linear optical studies of the polymers are essential, and one can evaluate the optical band gap and fluorescence quantum yield of the polymers. These results are very useful in selecting such polymers for device applications such as PLEDs and PSCs. In the following section, the UV–Vis absorption and fluorescence emission characteristics of these polymers have been discussed.

The UV–Vis absorption and the fluorescence emission data of the polymers are summarized in Table 16.1. The normalized UV–Vis absorption spectra of polymers P1 and P2 in a dilute chloroform solution (10^{-4} g/mL) and in the film state are given in Figure 16.6. The polymer solutions of P1 and P2 showed absorption maxima at 378 and 418 nm respectively,

TABLE 16.1   UV–Vis Absorption and Fluorescence Emission Spectral Data of the Polymers

Polymer	Λ_{max}^{a} (nm)	Λ_{max}^{b} (nm)	Λ_{em}^{a} (nm)	Λ_{em}^{b} (nm)	E_g^{opt} (eV)	ϕ_{fl} (%)
P1	378	394	506	516	2.28	35
P2	418	430	481	500	2.3	32
P3	395	395	547	565	2.29	40
P4	395	416	554	569	2.25	42
P5	550	550	657	—	1.75	38
P6	548	550	655	—	1.77	42
P7	520	536	590	—	1.70	38
P8	527	540	623	—	1.71	40
P9	427	443	506	513	2.42	40
P10	388	398	457	490	2.63	42
P11	411	458	536	542	2.3	35
P12	360	386	440	514	2.47	38

a Measured in CHCl₃ solution.
b Cast from CHCl₃ solution on glass substrates.
E_g^{opt} Optical band gap calculated from the onset absorption edge of the polymer film.

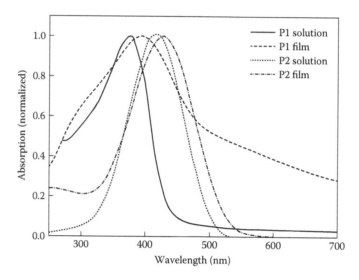

FIGURE 16.6   UV–Vis absorption spectra of P1 and P2.

which correspond to the π–π* transition in the polymer. The absorption maximum of P2 in the solution state showed a redshift of about 40 nm in comparison with that of P1. This redshift can be explained in terms of a strong ICT interaction between strong electron-donating 3,4-propylenedioxythiophene and electron-accepting 1,3,4-oxadiazole units in the polymer chain [20–22]. Smooth and optically clear thin polymer films on glass substrates were obtained by spin-coating the chloroform solutions of the polymer (1 mg mL⁻¹) at a spin rate of 1500 rpm. The polymer films displayed absorption peaks at 394 and 430 nm respectively for P1 and P2. The absorption spectrum of P1 in the film state was broadened, and the absorption maximum was redshifted by about 16 nm as compared to its spectrum in the solution state. A similar redshift in the absorption maximum was also observed for P2 in the film state. The observed redshift and broadening in the absorption spectra could be attributed to the increased extent of the π-stacking in the film state as well as to the increased polarizability of the polymer film [23,24]. The optical band gaps (E_g^{opt}), defined by the onset absorption (λ_{onset}) of the polymer in the film state, are 2.28 eV for P1 and 2.3 eV for P2, and these were calculated using the equation $E_g^{opt} = hc/\lambda_{onset}$, where h is the Planck constant and c is the velocity of light.

The normalized fluorescence emission spectra of P1 and P2 in a dilute chloroform solution (10⁻⁴ g/mL) and in the thin film state are given in Figure 16.7. The emission maxima of P1 and P2 were observed at 506 and 481 nm, which correspond to bluish green and blue, respectively. The emission maxima of P1 and P2 in the film state exhibited slight redshifts of about 10 and 19 nm, respectively, in comparison with those of their solutions. This redshift can be attributed to the interchain or/and intrachain mobility of the excitons and excimers generated in the polymer solid state [25]. The polymers emitted green (P1) and bluish green (P2) light in their film state. Based on the absorption maxima of the polymers, polymer P2 was expected to show a redshift in the emission maxima as compared to that of polymer P1. However, a blueshift was observed in the emission maxima of P2 as compared to that of P1. These results suggest that the polymer backbone in the excited state is more

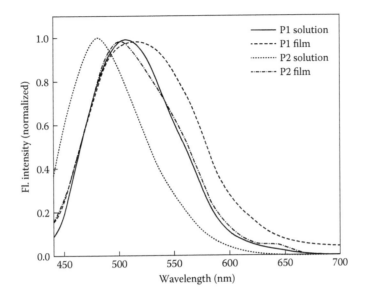

FIGURE 16.7   Fluorescence emission spectra of P1 and P2.

planar in P1 than that in polymer P2, leading to the bathchromic shift in its emission [26]. The fluorescence quantum yield (φ_{fl}) of the polymers in solution were calculated by comparing with the standard of quinine sulfate (ca. 1×10^{-5} M solution in 0.1 M H_2SO_4 having φ_{fl} of 55%), using the following equation:

$$\Phi_S = \Phi_r \left(\frac{A_r}{A_S} \cdot \frac{I_S}{I_r} \right) \qquad (16.1)$$

where:
  Φ represents the quantum yield of luminescence
  A is the absorbance at the excitation wavelength
  I is the corresponding relative integrated fluorescence intensity
  Subscript s represents an unknown sample
  Subscript r represents a standard

The quantum yields of P1 and P2 were found to be 35% and 32%, respectively.

The normalized absorption spectra of polymers P3 and P4 in a dilute $CHCl_3$ solution and in the thin film state are given in Figure 16.8. Polymer P3 in solution displayed an absorption maximum at 395 nm, which is due to the π–π^* transition. The polymer film displayed a similar absorption spectrum with no significant shift in the absorption maximum. The observed result could be attributed to the larger interchain distance imposed by the dodecyl chains attached to two thiophene moieties in the polymer backbone, which decreases the π-stacking interactions in the solid state [27]. The absorption spectrum of P4 in the solution state displayed an optical absorption maximum at 395 nm, while in the solid state the π–π^* transitions showed a redshift of about 21 nm with a shoulder in the

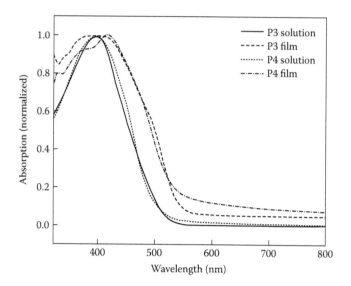

FIGURE 16.8   UV–Vis absorption spectra of P3 and P4.

shorter wavelength region at 368 nm. The redshift in the absorption maximum of P4 in the film state could be due to the π-stacking effect. Unlike in P3, the repeating unit in P4 contains pendant alkoxy groups only on one of the thiophene rings resulting in a lesser interchain distance in P4 as compared to that in P3. Hence, π-stacking interactions could be operative in P4 in the solid state. The optical band gaps of P3 and P4 were determined in the film state and were found to be 2.29 and 2.25 eV, respectively. The normalized fluorescence emission spectra of polymers P3 and P4 in dilute solution and the thin film state are shown in Figure 16.9. The emission maxima for P3 and P4 in the solution were observed at 547 and 554 nm, respectively. In comparison with their solution state, the emission

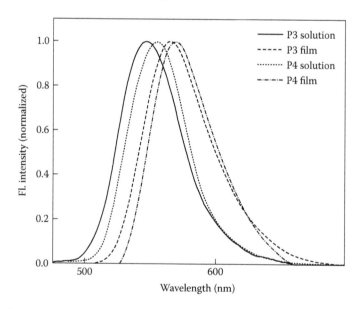

FIGURE 16.9   Fluorescence emission spectra of P3 and P4.

maxima of the polymer thin films showed bathochromic shifts of about 18 and 15 nm respectively for P3 and P4. The polymers emit a greenish yellow light in their film state.

The normalized absorption spectra of polymers P5–P8 in a dilute chloroform solution and in the thin film state are given in Figures 16.10 and 16.11, respectively. The absorption maxima of P5, P6, P7, and P8 in the solution state were 550, 548, 520, and 527 nm, respectively. In addition, the absorption spectrum of P6 and P8 showed peaks in the shorter wavelength region at 351 and 361 nm, respectively. The observed absorption maxima of the polymers in the longer wavelength region are due to the effective conjugation length with an efficient ICT through D–A units in the polymers. The incorporation of strong electron-donating 3,4-didodecyloxythiophene and electron-accepting cyanovinylene units in the polymer backbone resulted in an enhanced ICT between donor and acceptor segments,

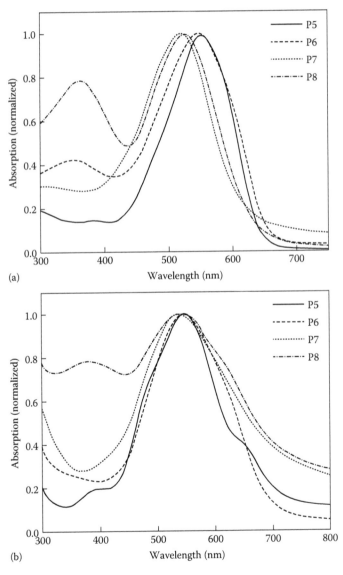

FIGURE 16.10 UV–Vis absorption spectra of P5–P8 (a) in solution and (b) in thin film.

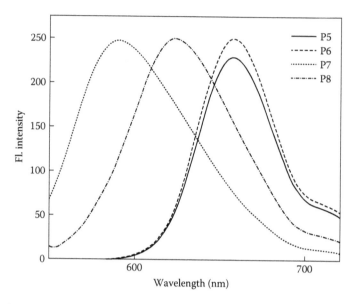

FIGURE 16.11 Fluorescence emission spectra of P5–P8 in solution.

causing an effective electronic delocalization along the polymer backbone. This further improves the effective conjugation length and leads to an extension of absorption [28]. Furthermore, the presence of a strong electron-withdrawing cyanovinylene group in these polymers extends the absorption to a longer wavelength region. However, the observed variation in the absorption maxima of the polymers P6–P8 is due to the difference in the chemical structures of the polymer chains. The absorption maximum of P6 was redshifted in comparison with those of P7 and P8. This can be understood in terms of a more planar backbone and the increased conjugation along the polymer backbone by the introduction of electron-donating fluorenevinylene unit in P6.

Polymer P5 in the film state showed an absorption maximum at the same wavelength as that of its solution state, but with a redshift of about 68 nm in the onset absorption edge in the film state. In addition, the spectrum of the polymer film showed a peak in the shorter wavelength region at 390 nm and a shoulder at 661 nm. The absorption spectrum of P6 in the thin film state is almost identical to that of its solution state, with no prominent shift in the absorption maximum. The observed behavior could be attributed to the presence of alkyl chains on the fluorene unit that reduce the intermolecular interaction and thus lead to poorer π-stacking in the film state [29]. The absorption spectra of P7 and P8 in the film state displayed a redshift and broadening in the absorption maxima as compared to those of their solutions. The observed redshifts for P7 and P8 films were about 16 and 13 nm, respectively, which could be due to the π-stacking effect in the film state. The optical band gaps defined by the onset absorption of the polymers in the film state were 1.75, 1.77, 1.70, and 1.71 eV, respectively, for P5–P8. Hence these polymers, with low band gaps as well as broad absorption bands, could be promising candidates for high efficiency PSCs. The polymers, as expected from the D–A arrangement along the polymer backbone, show narrow band gap energies. The band gaps of the polymers are lower than those of some important CPs reported in the literature like thiophene-based polymers containing

2,1,3-benzothiadiazole segments [30], *bis*(1-cyano-2-thienylvinylene) phenylene-based D–A type polymers [27], fluorene- and thiophene-based polymer [31], PFV [32], and some fluorene-based CPs containing oxadiazole pendants [33], which showed improved device efficiencies in PLEDs and PSCs.

The fluorescence emission spectra of P5–P8 in dilute solution are given in Figure 16.11. The polymers in the solution emit yellowish orange to red light with broad emission spectra. Polymer P5 showed an emission peak at 657 nm, whereas polymers P6–P8 showed emission peaks at 655, 590, and 623 nm, respectively. The fluorescence quantum yields (φ_{fl}) of the polymers in a chloroform solution were found to be 38%–42%. The photophysical data suggest that the polymers have low band gaps with broad absorption bands and are highly fluorescent with reasonably good fluorescence quantum yields. Hence, these polymers could be promising candidates for the fabrication of polymer-based optoelectronic devices such as PLEDs and PSCs.

The normalized UV–Vis absorption spectra of P9 and P10 in a dilute chloroform solution and in the thin film state are given in Figure 16.12a. The absorption maxima of polymers in the solution state were observed at 427 and 388 nm, respectively, for P9 and P10, which are attributed to the π–π^* transition in the polymer. The polymers in the film state showed redshifts of about 16 and 10 nm, respectively, for P9 and P10 due to the π-stacking effect in the film state. It was observed that P9 exhibited redshifts of about 39 and 45 nm in the absorption maxima in solution and the film state, respectively, as compared to those of P10. The observed behavior is a result of the strong electron-releasing effect of the conjugative spacer thiophene unit attached to 1,3,4-oxadiazole ring in P9 (as compared to the spacer phenyl ring in P10), which improves the ICT between the donor and the acceptor units. The optical band gaps defined by the onset absorption in the film state were 2.42 and 2.63 eV, respectively, for P9 and P10. The thin films of P9 and P10 emit green and bluish green light, respectively, under the irradiation of UV light. The fluorescence emission maxima of polymer solutions were observed at 506 and 457 nm, respectively, for P9 and P10 (Figure 16.12b). In the film state, bathchromic shifts of about 7 and 33 nm were observed for P9 and P10, respectively. The fluorescence quantum yield of the polymers in a chloroform solution was in the range 40%–42%.

The normalized absorption spectra of P11 and P12 are shown in Figure 16.13a. The absorption maxima of the polymers in solution were observed at 411 nm for P11 and at 360 nm for P12, which are due to the π–π^* transition in the polymer. The polymer films showed a redshift in the absorption maximum in comparison to those of their solutions. A redshift of about 47 nm with a broad absorption spectrum was observed for P11 thin film, whereas P12 thin film showed a redshift of about 26 nm, which are attributed to the π-stacking effect in the solid state. However, the absorption maxima of P11 in solution and the film state were redshifted about 51 and 72 nm, respectively, as compared with those of P12. This is due to the electron releasing effect of the thiophene moiety attached to 1,3,4-oxadiazole ring in P11, which resulted in the extended conjugation along the polymer backbone. The optical band gaps of the polymers in the thin film state were calculated from their onset absorption edge and were found to be 2.3 and 2.47 eV, respectively, for P11 and P12.

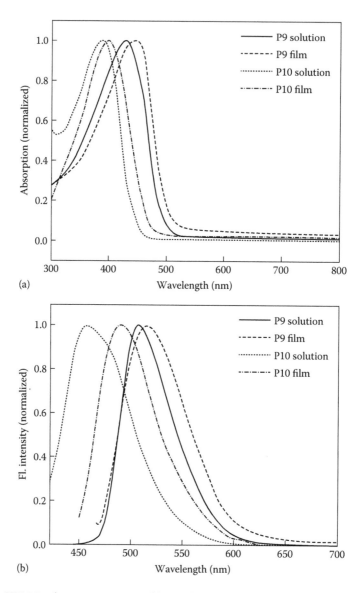

FIGURE 16.12   UV–Vis absorption spectra (a) and fluorescence emission spectra (b) of P9 and P10.

The fluorescence emission spectra of P11 and P12 are given in Figure 16.13b. The polymers in the solution state displayed emission peaks at 536 and 430 nm, respectively, for P11 and P12. The solid-state emission maxima of the polymers were redshifted about 6 and 74 nm, respectively, for P11 and P12, from the corresponding emission maxima in the solution. The polymer films emit green light under the irradiation of UV light. The quantum yield of the polymers in solution is found to be 35% for P11 and 38% for P12. Photographs of polymer solutions (P1–P12) under day light are given in Figure 16.14.

The presence of biphenyl ring along the polymer backbone was found to affect the photophysical properties of the polymers. The 2-phenylthiophene unit in P9 was replaced by a biphenyl ring in P10. A similar structural change was also made between polymers P11 and P12.

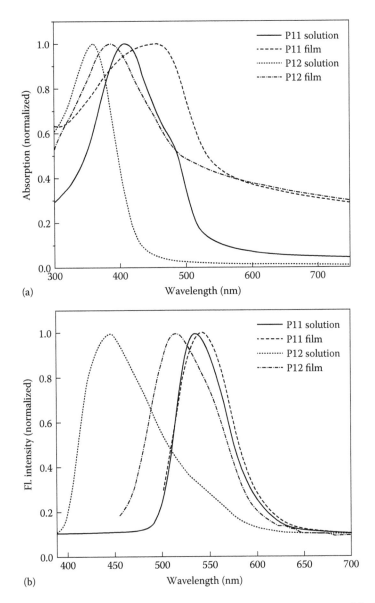

(a)

(b)

FIGURE 16.13   UV–Vis absorption spectra (a) and fluorescence emission spectra (b) of P11 and P12.

FIGURE 16.14   Photographs of polymer solutions (P1–P12) under day light.

It is well known that the ground-state geometry of the biphenyl molecule has two phenyl rings bent ~43° out of the plane. This lack of planarity causes a significant conjugation break within the backbone of the polymer. Therefore, placing biphenyl conjugative spacers within the backbone of polymers P10 and P12 causes a drop off in further conjugation enhancement. The lower the conjugation length, the higher is the energy required for the π–π^* transition. As a result, P10 absorbs light of a lower wavelength than that absorbed by P9. Similarly, the absorption maximum of P12 was blueshifted compared to that of P11. The nonplanarity caused by the introduction of biphenyl units also affects the band gap energy of the polymers. As a result, P10 showed a higher band gap energy as compared to that of P9. A similar trend was also observed between polymers P11 and P12. In addition, the twisted structure of the biphenyl unit not only decreases the effective conjugation length of the polymer but also limits the interchain interactions, enhancing its fluorescence quantum efficiencies. Because of this effect, P10 and P12 showed slightly higher fluorescence quantum yields when compared to those of P9 and P11, respectively.

16.4 NLO PROPERTIES OF THE POLYMERS (P1–P12)

Conjugated organic molecules and polymers offer exciting opportunities in the area of nonlinear optics due to their possible applications in all optical devices such as optical limiters, optical switches, and optical modulators [34]. As a result, in the past three decades, there has been a great interest in the synthesis of novel monomers and polymers with the intention of producing materials having large NLO properties. The study has been focused mainly on finding materials with large NLO effects, fast responses, good solubility, processability, and high durability. In this direction, CPs are a promising class of third-order nonlinear materials because of their potentially large third-order susceptibilities associated with a fast response time in addition to their variety and processability. Further, the strong delocalization of π-electrons in a conjugated polymeric backbone determines a very high molecular polarizability and thus remarkable third-order optical nonlinearities [35]. Moreover, these macromolecules offer good flexibility at both molecular and bulk levels so that structural modifications necessary to optimize them for specific device applications are possible. Further, these materials possess a high mechanical strength as well as excellent environmental and thermal stability. In contrast to misconceptions about the frailty of simple organic molecules, the optical damage threshold for polymeric materials can be greater than 10 GW/cm². As the nonlinear response of these systems is determined primarily by their chemical structure, one can design unique molecular structures and synthesize compounds with an enhanced nonlinear response by introducing suitable substituent groups. In this regard, a deeper understanding of the structure–property relationship would help design new organic conjugated molecules and polymers by the judicious choice of functional substituents, and thus to tune their optical properties for photonic applications.

In this direction, design of D–A CPs containing proper electron-rich and electron-deficient moieties along the polymer backbone is emerging as a promising strategy for tuning the linear and nonlinear optical properties of CPs [36–38]. Here, the presence of alternate electron donor and electron acceptor units in the polymer backbone would enhance the NLO properties of the polymer, mainly due to the increase in effective π-electron delocalization. Among various π-conjugated materials, thiophene-based polymers are currently under intensive investigation as materials for NLO because of their large third-order response, chemical stability, readiness of functionalization, and good film forming characteristics [39–41]. Additionally, the NLO properties in poly(thiophene)s can be synthetically tuned by introducing electron-releasing and electron-accepting segments in the polymer chain [42–44].

There are various phenomena responsible for the nonlinear absorption property exhibited by the optical materials. The materials with different nonlinear absorption processes such as saturable absorption (SA), reverse SA (RSA), two-photon absorption (2PA), and multiphoton absorption (MPA) are promising in various applications of science and technology [45,46]. For example, SA materials (the transmittance increases with the increase in optical intensity) have been used in lasers as Q-switching elements, and 2PA, 3PA, MPA, and RSA materials (their transmittances decrease with the increase of optical intensity) have been used in two-photon microscopy and optical limiters. Therefore, in cases of materials showing nonlinear absorption properties, it is necessary to identify their nonlinear absorption effects and to determine their nonlinear absorption parameters, such as the saturable intensity for a saturable absorber and the MPA coefficient for multiphoton absorbing material. The mechanism of number of nonlinear absorption processes can be represented in an energy level diagram as shown in Figure 16.15. The diagram shows the different energy levels of a molecule, the singlet ground state S_0, the excited singlet states S_1 and S_2, as well as the triplet excited states T_1 and T_2. It also displays the different transitions between the energy levels.

When two photons of the same or different energy levels are simultaneously absorbed from the ground state to a higher excited state (S_0–S_1), it is denoted as 2PA. When the excited state absorption (ESA) occurs, molecules are excited from an already excited state to a higher excited state (e.g., S_1–S_2 and/or T_1–T_2). For this to happen, the population of the excited states (S_1 and/or T_1) needs to be high so that the probability of photon absorption from that state is high. Therefore, the high-intensity light is needed to pump up the molecules to the excited state before a substantial amount of ESA takes place. The ESA could be enhanced if the molecules would undergo intersystem crossing (ISC) to the triplet state. If more absorption occurs from the excited state than from the ground state, it is usually called reverse saturable absorption. The triplet ESA may result in RSA if the absorption cross section of the triplet excited state is greater than that of the singlet excited state.

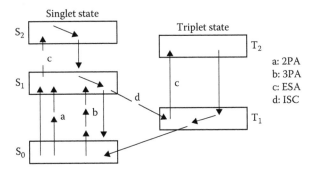

FIGURE 16.15   Energy level diagram showing 2PA, 3PA, ESA, and ISC.

The "open aperture" z-scan technique has been a widely used method [47] to measure the nonlinear absorption properties of various materials. In this method, a laser beam is used for optically exciting the sample, and its propagation direction is considered as the z-axis. The beam is focused using a convex lens and the focal point is taken as $z = 0$. Obviously, the beam will have maximum energy density at the focal point, which will symmetrically reduce toward either side of it, for the positive and negative values of z. The schematic representation of the open aperture z-scan experimental set up is given in Figure 16.16. In the experiment, the sample was placed in the beam at different positions with respect to the focus (i.e., at different values of z), and the corresponding optical transmission values were measured. Then a graph was plotted between z and the measured sample transmission (normalized to its linear transmission), which is known as the z-scan curve. The shape of the z-scan curve will provide information on the nature of the nonlinearity: reverse saturable, 2PA, three photon absorption, SA, and so on. By fitting the experimental data to theory, the parameters such as nonlinear absorption coefficient (β) and saturation intensity (I_s) of the sample can be calculated. The experimental data are very useful in assessing the NLO behavior of the sample under study.

In the z-scan experiment, a stepper motor-controlled linear translation stage was used to move the sample through the beam in precise steps. Two pyroelectric energy probes (Rj7620, Laser Probe Inc.) were used to measure the transmission of the sample at each point. One energy probe monitors the input energy, whereas the other monitors the transmitted energy through the sample. The second harmonic output (532 nm) of a Q-switched Nd:YAG laser (Minilite-Continuum, 5 ns full width at half maximum laser pulses) was used for excitation of samples. The experiments were carried out in the "single shot" mode, allowing sufficient time between successive pulses to avoid accumulative thermal effects in the sample. The experiment was automated, controlled by a data acquisition program written in Lab VIEW.

In the z-scan experiment, the linear transmittance of all the samples studied was fixed between 50% and 60%. The laser pulse energies used to excite the samples were between 75 and 100 μJ. The open aperture z-scan curves of polymers P1–P4 and P9–P12 in a chloroform solution are given in Figures 16.17 and 16.18, respectively. The samples show a strong optical limiting behavior, that is, the transmission of the sample decreases as the input light intensity increases (intensity is maximum at $z = 0$). In polymer systems under resonant

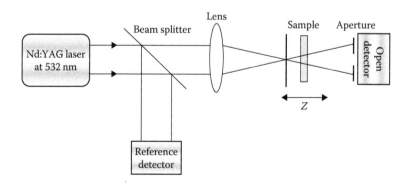

FIGURE 16.16 The open aperture z-scan setup.

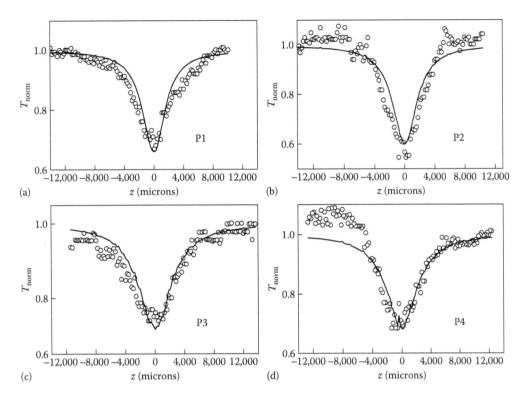

FIGURE 16.17 Z-scan curves (a–d) of polymers P1–P4.

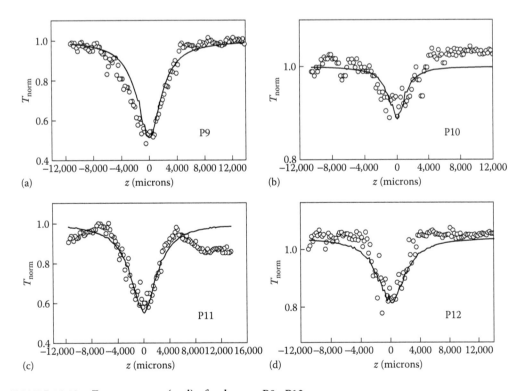

FIGURE 16.18 Z-scan curves (a–d) of polymers P9–P12.

excitation conditions, an optical limiting behavior can be attributed to effects such as ESA (excited singlet and/or triplet absorption), two- or three-photon absorption (2PA or 3PA), and self-focusing/defocusing. Of these, 2PA or 3PA and self-focusing/defocusing are electronic nonlinearities that require high laser intensities usually available only from pulsed picosecond or femtosecond lasers. Therefore, the cause of the observed optical limiting turns out to be ESA. The net effect is then known as "effective" 2PA process. The nonlinear transmission behavior of the present samples can be modeled by defining an effective nonlinear absorption coefficient of the form

$$\alpha(I) = \frac{\alpha_0}{1 + (I/I_s)} + \beta I \qquad (16.2)$$

where:

α_0 is the unsaturated linear absorption coefficient at the wavelength of excitation
I is the input laser intensity
I_s is the saturation intensity (intensity at which the linear absorption drops to half of its original value)
β is the effective 2PA coefficient

For calculating the transmitted intensity for a given input intensity, the propagation Equation 16.3 was numerically solved:

$$\frac{dI}{dz'} = -\left[\left(\frac{\alpha_0}{[1 + (I/I_s)]}\right) + \beta I\right] I \qquad (16.3)$$

Here z' indicates the propagation distance within the sample. By determining the best-fit curves for the experimental data, the nonlinear parameters could be calculated. In the z-scan curves (Figures 16.17 and 16.18), circles are data points, whereas the solid curve is a numerical fit according to Equation 16.3.

The effective 2PA coefficient (β) of the polymers was found to be of the order of 10^{-10}–10^{-12} m/W. The estimated β values of the polymers are summarized in Table 16.2. The higher β values observed for these polymers could be attributed to the effective π-electron delocalization along the polymer backbone due to the D–A arrangement. The β values of P1 and P2 were found to be of the order 10^{-10} m/W. The chemical structures of polymers P1 and P2 were almost similar and hence no significant difference in their β value was observed. The observed improvement in the NLO behavior of these polymers could be attributed to the regular (alternating) D–A structure of the polymer chains. For P3 and P4, the β value was found to be of the order 10^{-11} m/W. The β value for P9 was determined to be 3.5×10^{-10} m/W, whereas for P10 it was found to be 3.3×10^{-12} m/W.

TABLE 16.2   Effective 2PA Coefficient (β) of the Polymers

Polymer	P1	P2	P3	P4	P9	P10	P11	P12
β (m/W)	4×10^{-10}	3×10^{-10}	4.4×10^{-11}	3.1×10^{-11}	3.5×10^{-10}	3.3×10^{-12}	5.3×10^{-11}	1.0×10^{-11}

The observed β value of P9 is almost 100 times higher than that of P10. The enhancement of the NLO response in P9 could be due to the presence of strong electron-donating fluorenevinylene unit in the polymer backbone that extends the π-electron delocalization in the polymer, whereas the presence of nonplanar biphenyl ring attached to the 1,3,4-oxadiazole ring in P10 limits the π-electron delocalization along the polymer chain. Polymers P11 and P12 showed β values of the order 10^{-11} m/W. The β value of P11 is slightly higher than that of P12, which could be attributed to the presence of an electron-rich thiophene unit in P11.

Overall, in this study, polymers P1, P2, and P9 showed the highest nonlinearity with promising optical limiting behavior. The β value was found to be 4×10^{-10} m/W, 3×10^{-10} m/W, and 3.5×10^{-10} m/W, respectively, for P1, P2, and P9. For comparison, under similar excitation conditions, the effective 2PA coefficient values of some materials are 10^{-10}–10^{-12} m/W for Cu nanocomposite glasses [48], 3×10^{-11} m/W for functionalized carbon nanotubes [49], and 0.53×10^{-10} m/W for bismuth (Bi) nanorods [50]. Further, the obtained β values are comparable with some important D−A CPs reported in the literature [51,52]. So, the observed results indicate that the present samples exhibit an improved optical nonlinearity and hence expected to be promising candidates for optical limiting applications.

The z-scan profile usually shows a valley with a maximum and a minimum on each side of the focal point. However, the open-aperture z-scan profile of polymers P5−P8 showed a typical peak, symmetric about the focus, which is known to be the signature of a SA phenomenon (Figure 16.19). The peak appears at the focal point where the laser pulse has the strongest fluence. The linear absorption spectra of the P5−P8 further confirm the observed

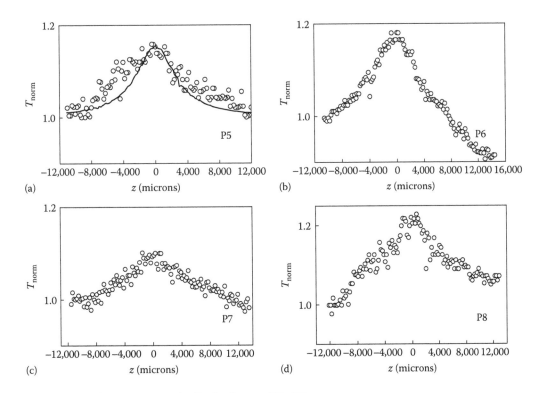

FIGURE 16.19   Z-scan curves (a-d) of polymers P5−P8.

saturation behavior, where the excitation wavelength of 532 nm is close to their absorption peak, which is a favorable situation for the absorption saturation. SA is a property of a material where the absorption of light decreases with the increase in light intensity. At sufficiently high-incident light intensity, atoms or molecules in the ground state of a saturable absorber material become excited into an upper energy state at such a rate that there is insufficient time for them to decay back to the ground state before the ground state becomes depleted and the absorption subsequently saturates. One consequence of SA is optical bistability. Certain NLO systems can possess more than one output state for a given input state. The term optical bistability refers to the situation in which two different output intensities are possible for a given input intensity. Also, the term optical multistability is used to describe the circumstance in which two or more stable output states are possible. A similar type of SA behavior in the near-field transmission was observed for inorganic materials [53].

The z-scan curves (Figure 16.19) obtained were fitted with numerically simulated results using Equation 16.4 [54],

$$\alpha = \alpha_0 \frac{1}{1+\left(I/I_s\right)} \tag{16.4}$$

where:

α_0 is the linear absorption coefficient at the wavelength of excitation

I is the incident intensity

I_s is the saturation intensity (intensity at which the absorption becomes half of the linear absorption)

By determining the best-fit curves for the experimental data, the nonlinear parameters could be calculated. In Figure 16.19, circles denote the experimental data points and the solid line is a theoretical fit with Equation 16.4. For P5 solution, I_s was found to be 2×10^{12} W/m^2.

Generally, RSA and SA are the two basic ESA processes taking place in organic materials. If the excited state has strong absorption compared to the ground state, the transmission can exhibit the RSA behavior. These nonlinear absorption processes are highly dependent on the excitation wavelengths, intensities, and excited state lifetime of the molecules. The most important application of RSA materials is in optical limiting devices used for the protection of sensitive optical devices. Optical limiting is defined as any process that limits the amount of light that can get through a material at a high intensity. Moreover, optical limiters for low-power continuous-wave (cw) lasers are very important in applications such as optical damage protection of extremely sensitive sensors including human eyes, which are vulnerable to damage even with a few mW of radiation from a laser pointer. Also, they find applications in pulse smoothing, pulse shortening, mode locking, spatial light modulation, and so on. Further, when the frequency of the incident light is near to the absorption resonance of the material, the absorption may saturate as the intensity increases.

The SA materials are mainly useful in the areas of optical communication and optical computing, particularly for Q-switching and mode-locking lasers. Because the lowest loss occurs when the laser modes are locked together into pulses, introduction of a saturable absorber into a laser cavity enables it to passively mode-lock, an important way to generate ultrashort pulses. Saturable absorbers are also useful for nonlinear filtering outside laser resonators, which can clean up pulse shapes. SA is particularly strong in semiconductor lasers at wavelengths just above the band edge.

Polymer P1, which showed strong optical limiting behavior, and P5, which showed interesting SA behavior, along with good film forming properties were used to prepare polymer/TiO$_2$ nanocomposite films. The prepared nanocomposites were characterized, and the z-scan method was employed to study their NLO properties.

16.5 POLYMER/TiO$_2$ NANOCOMPOSITES: PREPARATION AND CHARACTERIZATION

Metal and semiconductor nanoparticles exhibit a characteristic size and shape-dependent electronic structure leading to unique optical and NLO properties [55,56]. These attributes find application in a wide range of fields, including electronics, photonics, plasmonics, and sensing. Further, the plasmon absorption of metal nanoparticles such as copper, silver, gold, and palladium occurs in the UV–Vis region of the spectrum. Multiphoton excitations in this energy range and further ESAs can be exploited to elicit NLO responses from these nanomaterials. For instance, a third-order NLO susceptibility ($\chi^{(3)}$) value of 0.8×10^{-12} esu has been observed for yellow Ag colloidal nanoparticles [57]. A large third-order NLO susceptibility value as high as 2×10^{-5} esu has been observed for nanoporous layers of titanium dioxide (TiO$_2$) [55].

From a material point of view, it is advantageous to embed metal/semiconductor nanoparticles in thin polymer films for optical applications. The polymer matrix serves not only as a medium to assemble the nanoparticles and stabilize them against aggregation but also provides characteristic mechanical properties which are suitable for device applications. Nanocomposites wherein materials are mixed on the nanoscale are of particular interest as they combine the properties of two or more different materials with the possibility of observing novel mechanical, electronic, or chemical behaviors [58]. Further, nanocomposite structures are also known to enhance optical nonlinearities substantially. According to the local field enhancement under the surface plasmon resonance condition, larger third-order nonlinearity of composite films has been observed in the presence of metal nanoparticles. These high nonlinear materials have been thought to be good candidates for new optical and electronic devices. As a result, several CP–metal/semiconductor nanocomposites have been prepared from a range of different metals and with different types of CPs. The effects of this on the optical properties of the nanoparticles, electronic behavior of both the nanoparticles and conjugated materials, and some applications have been investigated.

For instance, a larger third-order nonlinearity of polydiacetylene (PDA) composite films has been observed in the presence of metal/semiconductor nanoparticles [59]. In the wake of this observation, the third-order NLO optical properties of other such metal/ semiconductor polymer nanocomposites have been investigated later on. Chen et al. [57] reported the synthesis of nanometer-size silver-coated PDA composites. NLO properties of these PDA/Ag nanocomposite vesicles were measured by the z-scan technique. The value of nonlinear refractive index (n_2) for pure PDA vesicles was 1.2×10^{-14} cm^2 W^{-1} and for PDA/Ag nanocomposite vesicles (on the outer surface) was 7.3×10^{-14} cm^2 W^{-1}. Nearly a seven times enhancement of the n_2 value was observed as a result of local-field enhancement under the surface plasmon resonance of silver nanoparticles at the interface. The 2PA of poly(styrene maleic anhydride)/TiO$_2$ nanocomposites was studied by Wang et al. [60] by the z-scan technique. Based on the 2PA, they have investigated the optical limiting behavior of the nanocomposites. Sezer et al. [61] reported the synthesis and NLO properties of poly(aniline) and poly(aniline) silver nanocomposite thin films. Chen et al. [62] reported the synthesis of poly(substituted diacetylene) (PNADA)/silver nanocomposites. The silver nanoparticles were dispersed in the polymer films. The introduction of silver nanoparticles into the polymer films led to the enhancement of NLO properties. The value of nonlinear refractive index for PNADA/Ag nanocomposite films was 11.6×10^{-15} cm^2 W^{-1}.

Thus it is evident that a combination of CP and metal/semiconductor nanoparticles in the form of nanocomposites show improved optical and NLO properties in comparison with those of CPs alone. In this regard, TiO$_2$ nanomaterials have attracted much attention, and their optical nonlinearities have been extensively investigated. Incorporation of TiO$_2$ nanoparticles into CPs could improve the mechanical, electrical, and optical properties of the nanocomposite. These composite materials have attracted interest in both fundamental studies and applications: inorganic semiconducting nanoparticles for their small size and novel properties, and CPs for their attributes as easily processed semiconductor materials with potential for optoelectronic applications. However so far, there is no report available in the literature on NLO studies of thiophene-based D−A CP/ TiO$_2$ nanocomposites.

For the preparation of polymer (P1 or P5)/TiO$_2$ nanocomposites, 10 wt.% TiO$_2$ nanoparticles were dispersed in the polymer using chloroform/chlorobenzene solvent system (10:1 volume ratio) and sonicated for 2 h. Smooth and optically clear thin polymer films on glass substrates were obtained by spin-coating the chloroform solutions of the polymer (1 mg mL^{-1}) at a spin rate of 1500 rpm. In a similar way, nanocomposite films were also prepared. The polymer and the nanocomposite films were dried under vacuum for 1 h. Figure 16.20 shows the FESEM images of P1/TiO$_2$ and P5/TiO$_2$ nanocomposites. A moderately uniform distribution of TiO$_2$ nanoparticles was observed with average particle sizes ranging from 25 to 50 nm. The thickness of polymer films and nanocomposite films were determined by SEM cross section and was found to be in the range 0.9–1 μm.

The FTIR spectra of P1, TiO$_2$ nanoparticles, and P1/TiO$_2$ nanocomposite are shown in Figure 16.21a. The strong bands at 2919 and 2851 cm^{-1} in the FTIR spectrum of P1 are due to (−C−H) stretching vibrations of the alkyl chains. The band at 1572 cm^{-1} is assigned to the imine >C=N in an oxadiazole ring. The band at 1043 cm^{-1} is due to the oxadiazole =C−O−C= stretching vibration. The FTIR spectrum of TiO$_2$ nanoparticles exhibits a band at 3395 cm^{-1} due to the O−H stretching mode of Ti–OH. The characteristic

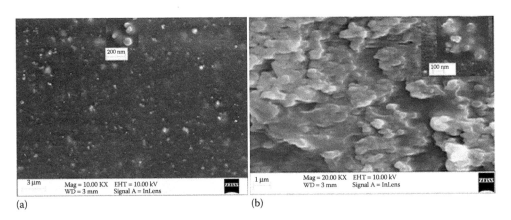

FIGURE 16.20 FESEM images of (a) P1/TiO$_2$ nanocomposite and (b) P5/TiO$_2$ nanocomposite film (inset: magnified image, Mag = 100 KX).

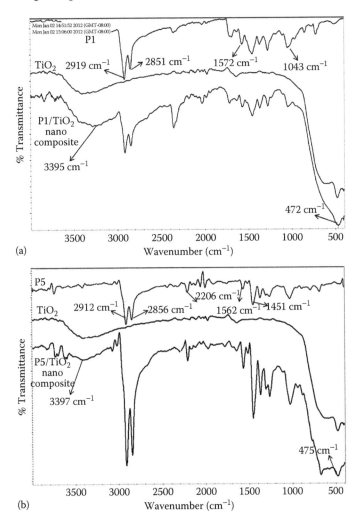

FIGURE 16.21 FTIR spectra of (a) P1, pure TiO$_2$ nanoparticles, and P1/TiO$_2$ nanocomposite and (b) P5, pure TiO$_2$ nanoparticles, and P5/TiO$_2$ nanocomposite.

absorption band of Ti–O–Ti was observed at 472 cm^{-1}. All these characteristic bands arising from both P1 and TiO$_2$ nanoparticles were observed in the FTIR spectrum of P1/TiO$_2$ nanocomposite, confirming the incorporation of TiO$_2$ nanoparticles in the nanocomposite. A similar spectral trend was also observed for P5/TiO$_2$ nanocomposite and its FTIR spectrum is given in Figure 16.21b.

The UV–Vis absorption spectra of polymer P1 in a chloroform solution, polymer thin film, and P1/TiO$_2$ nanocomposite film are shown in Figure 16.22a. Polymer P1 in solution displayed an absorption maximum at 378 nm, while its film showed a redshift of 16 nm in the absorption spectrum. The polymer P1/TiO$_2$ nanocomposite film showed absorption peaks at 412 nm ($\pi \rightarrow \pi^*$ of polymer) and at 310 nm and a shoulder at 250 nm (characteristic

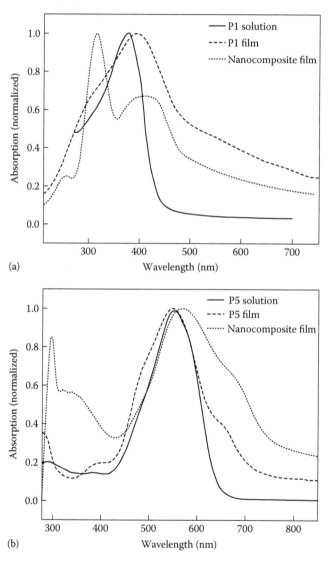

FIGURE 16.22 (a) UV–Vis absorption spectra of P1 in a chloroform solution, P1 thin film, and P1/TiO$_2$ nanocomposite thin film. (b) UV–Vis absorption spectra of P5 in a chloroform solution, P5 thin film, and P5/TiO$_2$ nanocomposite thin film.

absorptions of TiO_2). The incorporation of TiO_2 nanoparticles causes a slight redshift in the absorption maximum. The absorption maximum of the nanocomposite film shifts about 18 nm as compared to that of the polymer film.

Similarly, a redshift of about 18 nm in the absorption maximum was observed for P5/TiO_2 nanocomposite film as compared to that of the P5 film (Figure 16.22b). The observed redshifts in the absorption maxima can be understood in terms of effective π-conjugation length in polymer/TiO_2 nanocomposite films, which is longer than that in pure polymer films. A similar redshift in the absorption maximum was observed for MEH-PPV/TiO_2 nanocomposite films [63]. These optical results indicate that some interactions occur between the CP chains and TiO_2 nanoparticles in the nanocomposite structure.

Thermogravimetric analyses of polymers P1 and P5 and their nanocomposites were carried out under nitrogen atmosphere at a heating rate of 5°C/min. As shown in Figure 16.23a, polymer P1 decomposes slowly in the temperature region 180°C–310°C, whereas the onset decomposition temperature of P1/TiO_2 nanocomposite was observed at 300°C, indicating that the nanocomposite is thermally more stable than the polymer. A similar trend in the thermal stability was also observed for P5/TiO_2 nanocomposite. It is evident from the TGA traces given in Figure 16.23b that the nanocomposite is thermally stable up to 300°C. When the temperature was increased beyond 300°C, there was a sharp weight loss in the temperature range of 350°C–650°C. These results indicate that a strong interaction exists at the interface of the polymer and TiO_2 nanoparticles in the nanocomposite. A similar trend in the thermal behavior was observed for polymer nanocomposites reported in the literature [64]. The improved thermal stability in polymer/TiO_2 nanocomposites could be explained through the reduced mobility of the polymer chains in the nanocomposite. Consequently, the degradation process will be slowed and hence decomposition will take place at a higher temperature in the nanocomposite structure.

16.6 NLO PROPERTIES OF POLYMER FILMS AND POLYMER/TiO_2 NANOCOMPOSITE FILMS

In the z-scan experiment, the linear transmittances of the film samples studied were between 50% and 60%. The laser pulse energies used for P1 and its nanocomposites were between 75 and 100 μJ. The open aperture z-scan curves obtained for P1 and P1/TiO_2 films are given in Figure 16.24. Both films show a strong optical limiting behavior, that is, the transmission of the sample gets decreased as the input light intensity increases. The effect is quite strong because the normalized transmission gets decreased to values such as 0.1. The observed optical limiting behavior under the present experimental conditions can be attributed to effects such as ESA, thermal blooming, and induced thermal scattering. However, no induced scattering was visually observed during the experiment. Further, the numerical aperture of the detector was large enough to accommodate the transmitted beam fully even if moderate thermal blooming were to happen. Hence, the cause of the observed optical limiting turns out to be the ESA. The "effective" 2PA coefficient was calculated by fitting the experimental data to the standard nonlinear transmission equation.

The effective nonlinear absorption coefficient was estimated by determining the best-fit curves for the experimental data according to the numerical Equation 16.3. For example,

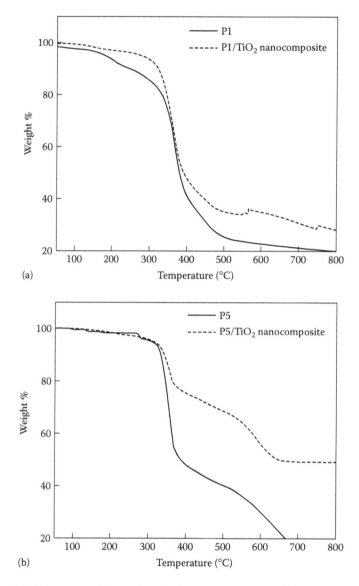

FIGURE 16.23 (a) TGA curves of P1 and P1/TiO$_2$ nanocomposite. (b) TGA curves of P5 and P5/ TiO$_2$ nanocomposite.

polymer P1 film effective 2PA coefficient was found to be 1×10^{-7} m/W, and for the P1/TiO$_2$ nanocomposite film it was 2×10^{-7} m/W. Obviously there is an enhancement of nonlinearity in P1/TiO$_2$ nanocomposite film compared to the pure polymer film. This is not substantial though, the reason being that the polymer films themselves are highly nonlinear in nature. From a device point of view, both P1 and P1/TiO$_2$ nanocomposite films are equally useful because the optical limiting efficiency exhibited by them is high. To put the aforementioned β values in perspective, the values obtained in similar systems under similar excitation conditions are 6.0×10^{-8} m/W in p-(N,N-dimethylamino)-dibenzylideneacetone in the PMMA matrix [65], 10^{-7}–10^{-9} m/W in

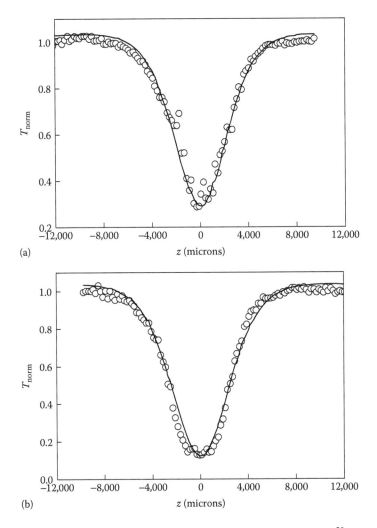

FIGURE 16.24 Z-scan curves of (a) P1 film and (b) P1/TiO$_2$ nanocomposite film.

Au:Ag-PVA nanocomposite films [66], and 6.8 × 10^{-7} m/W in a ZnO/PMMA nano-composite [67]. Obviously, the present films are potentially suited for fabricating optical limiters that can protect sensitive light detectors and also human/animal eyes from accidental exposure to high levels of optical radiation while maintaining normal transparency for safe low level inputs.

The open-aperture z-scan curves obtained for the P5 film and P5/TiO$_2$ nanocomposite film are shown in Figure 16.25. The laser pulse energies used to excite the samples were between 10 and 100 μJ. The z-scan curves obtained were fitted with numerically simulated results using Equation 16.4. By determining the best-fit curves for the experimental data, the nonlinear parameters were calculated. For the P5 film, I_s was found to be 6 × 10^{11} W/m^2, and for P5/TiO$_2$ nanocomposite film it was found to be 9 × 10^{11} W/m^2. There is an enhancement of the nonlinearity in P5/TiO$_2$ nanocomposite film compared to a pure polymer film.

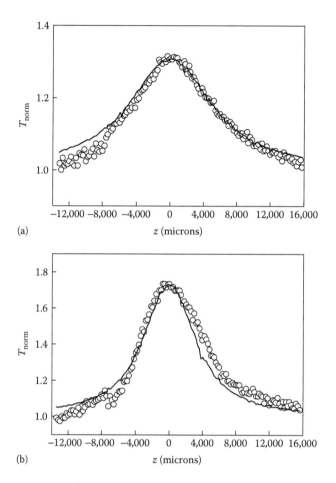

FIGURE 16.25   *Z*-scan curves of (a) P5 film and (b) P5/TiO$_2$ nanocomposite film.

For comparison, a similar SA behavior was observed for some organic materials reported in the literature. The values obtained are 1.5–4.5 × 10^{13} W/m^2 in phthalocyanines [68], 1–4 × 10^{10} W/m^2 in poly(indenofluorene) [69], 10^{10}–10^{11} W/m^2 for Rhodamine B [70], and 1.4–6.8 × 10^{13} W/m^2 for thiophene-based CPs [71]. The I_s obtained for the P5 film is almost 100 times lower than the values obtained for thiophene-based D–A polymers [71], indicating a better NLO response in P5. This could be due to the presence of stronger electron-withdrawing cyanovinylene group in P5 in comparison with 1,3,4-oxadiazole groups in the reported polymers. Because of their large SA, both P5 and P5/TiO$_2$ nanocomposite films are expected to be useful candidates for photonic applications mainly in areas such as Q-switching and mode-locking of lasers, pulse shaping, and optical switching.

16.7 CONCLUSIONS

A series of thiophene-based D–A CPs (P1–P12) are synthesized using multistep synthetic routes. All the polymers showed good solubility in common organic solvents due to the presence of long alkoxy pendants at the 3- and 4-positions of the thiophene ring. The linear optical properties of these polymers were investigated by UV–Vis absorption and fluorescence emission spectroscopic techniques. All the polymer films, except films of P3 and P5, showed a redshift in the absorption maxima as compared to those of their solutions due to the π-stacking effect in polymer films. In the P3 film, the larger interchain distance imposed by the additional dodecyl chains is attached to the thiophene ring, whereas the closer proximity of the alkoxy chains in P5 decreases the π-stacking interactions in the solid state (polymer films). The emission maxima of the polymers in the film state exhibited a redshift in comparison to those of their solutions. This redshift can be attributed to the interchain or/and intrachain mobility of the excitons and excimers generated in the polymer solid state. The polymers emitted blue to red light when irradiated with UV light, depending upon the chemical structure of the polymer backbone. Among these polymers, polymer P5−P8 showed a low band gap mainly due to the presence of cyanovinylene groups and also because of the D−A structure of the polymer chain. Polymer P7 containing cyanovinylene group and a regular D−A arrangement was found to show the lowest band gap. Because of the same reason, P7 also exhibited the lowest LUMO energy level among these polymers. Thus, introduction of cyanovinylene groups and maintaining a strict alternation of D and A units along the conjugated chain could be a promising molecular design to obtain low band gap polymers with a high electron affinity.

The NLO properties of P1−P12 were studied using the z-scan technique. The polymer samples showed improved nonlinearity as compared to some important NLO materials reported in the literature. Among these polymers, P1, P2, and P9 showed the highest optical nonlinearity with promising optical limiting behavior. The regular D−A structure in P1 and P2 and the presence of electron-donating dialkylfluorene unit with extended conjugation in P9 could be responsible for the observed trend. Interestingly, polymers P5−P8 showed SA behavior at the measured wavelength whose linear absorption peak is close to the excitation wavelength of 532 nm. Polymers P1 and P5 were used to prepare polymer/TiO$_2$ nanocomposites. Incorporation of TiO$_2$ nanoparticles into the polymer matrix was found to improve the thermal stability of the polymer. The polymer nanocomposite films showed a redshift in the absorption maxima as compared to those of polymer films. These results indicate the presence of some interactions between the polymer and TiO$_2$ nanoparticles in the nanocomposite structures. The NLO properties of the P1 film and P1/TiO$_2$ nanocomposite film were measured by the open aperture z-scan technique. All film samples showed strong optical limiting behavior, and the incorporation of TiO$_2$ nanoparticles marginally enhances the nonlinear absorption coefficient value of the polymer. These results suggest that P1 and P1/TiO$_2$ nanocomposite films are expected to be potential candidates for fabricating efficient optical limiters. Polymer P5 and P5/TiO$_2$ nanocomposite films exhibited strong SA behavior and the I_s was found to be of the order of 10^{11} W/m^2. These NLO results signify that P5 and its nanocomposite films are promising materials for applications in photonic switching devices. Hence, D−D CPs containing strong electron-donating and electron-withdrawing units arranged alternatively along the polymer chain could be a promising class of molecular materials to achieve high NLO responses. Further, enhancement in the NLO properties in these systems could be achieved by incorporating TiO$_2$ nanoparticles into the polymer matrix in the form of nanocomposite structures.

REFERENCES

1. Chiang, C. K.; Fincher, Jr C. R.; Park, Y. W.; Heeger, A. J.; Shirakawa, H.; Louis, E. J.; Gau, S. C.; MacDiarmid, A. G. Electrical conductivity in doped polyacetylene. *Phys. Rev. Lett.*, 1977, *39*, 1098–1101.

2. Roncali, J. Conjugated poly(thiophenes): Synthesis, functionalization and applications. *Chem. Rev.*, 1992, *92*, 711–738.

3 McCullough, R. D. The chemistry of conducting polythiophenes. *Adv. Mater.*, 1998, *10*, 93–116.

4. Skotheim, T. A. and Reynolds, J. R. *Handbook of Conducting Polymers*. 3rd ed.; CRC Press/ Taylor & Francis Group: Boca Raton, FL, 2007.

5. Sahin, O.; Osken, I.; Ozturk, T. Investigation of electrochromic properties of poly(3,5-bis(4-methoxyphenyl)dithieno[3,2-b;2′,3′-d]thiophene). *Synth. Met.*, 2011, *161*, 183–187.

6. Kraft, A.; Grimsdale, A. C.; Holmes, A. B. Electroluminescent conjugated polymers-seeing polymers in a new light. *Angew. Chem. Int. Ed.*, 1998, *37*, 402–428.

7. Brabec, C. J.; Sariciftci, N. S.; Hummelen, J. C. Plastic solar cells. *Adv. Funct. Mater.*, 2001, *1*, 15–26.

8. Halls, J. J. M.; Walsh, C. A.; Greenham, N. C.; Marseglia, E. A.; Friend, R. H.; Moratti, S. C.; Holmes, A. B. Efficient photodiodes from interpenetrating polymer networks. *Nature*, 1995, *376*, 498–500.

9. McQuade, D. T.; Pullen, A. E.; Swager, T. M. Conjugated polymer-based chemical sensors. *Chem. Rev.*, 2000, *100*, 2537–2574.

10. Prasad, P. N.; David, J. W. *Introduction to Nonlinear Optical Effects in Molecules and Polymers*. Wiley-Interscience: New York, 1991.

11. Invernale, M. A.; Acik, M.; Sotzing, G. A. *Thiophene-Based Electrochromic Materials, Handbook of Thiophene-Based Materials: Applications in Organic Electronics and Photonics*. John Wiley & Sons: Chichester, UK, 2009.

12. Havinga, E. E.; Hoeve, W.; Wynberg, H. Alternate donor-acceptor small-band-gap semiconducting polymers: polysquaraines and polycroconaines. *Synth. Met.*, 1993, *55*, 299–306.

13. Xiao, S.; Stuart, A. C.; Liu, S.; Zhou, H.; You, W. Conjugated polymer based on polycyclic aromatics for bulk heterojunction organic solar cells: A case study of quadrathienonaphthalene polymers with 2% efficiency. *Adv. Funct. Mater.*, 2010, *20*, 635–643.

14. Brocks, G.; Tol, A. Small band gap semiconducting polymers made from dye molecules: Polysquaraines. *J. Phy. Chem.*, 1996, *100*, 1838–1846.

15. Zotti, G.; Zecchin, S.; Schiavon, G.; Berlin, A.; Penso, M. Ionochromic and potentiometric properties of the novel polyconjugated polymer from anodic coupling of 5,5′-bis(3,4-(ethylenedioxy)thien-2-yl)-2,2′-bipyridine. *Chem. Mater.*, 1999, *11*, 3342–3351.

16. Balan, A.; Gunbas, G.; Durmus, A.; Toppare, L. Donor-acceptor polymer with benzotriazole moiety: Enhancing the electrochromic properties of the donor unit. *Chem. Mater.*, 2008, *20*, 7510–7513.

17. Ajayaghosh, A. Donor-acceptor type low band gap polymers: Polysquaraines and related systems. *Chem. Soc. Rev.*, 2003, *32*, 181–191.

18. Albota, M.; Beljonne, D.; Bredas, J. L.; Ehrlich, J. E.; Fu, J. Y.; Heikal, A. A.; Hess, S. E. et al. Design of organic molecules with large two-photon absorption cross sections. *Science*, 1998, *281*, 1653–1656.

19. Huang, C.; Sartin, M. M.; Cozzuol, M.; Siegel, N.; Barlow, S.; Perry, J. W.; Marder, S. R. Photoinduced electron transfer and nonlinear absorption in poly(carbazole-alt-2,7-fluorene)s bearing perylene diimides as pendant acceptors. *J. Phys. Chem. A*, 2012, *116*, 4305–4317.

20. Yu, G.; Gao, J.; Hummelen, J. C.; Wudl, F; Heeger, A. J. Polymer photovoltaic cells: Enhanced efficiencies via a network of internal donor-acceptor heterojunctions. *Science*, 1995, *270*, 1789–1791.

21. Jenekhe, S. A; Yi, S. Efficient photovoltaic cells from semiconducting polymer heterojunctions. *Appl. Phys. Lett.*, 2000, *77*, 2635–2638.

22. Campos, L. M.; Tontcheva, A.; Gunes, S.; Sonmez, G.; Neugebauer, H.; Sariciftci, N. S.; Wudl, F. Extended photocurrent spectrum of a low band gap polymer in a bulk heterojunction solar cell. *Chem. Mater.*, 2005, *17*, 4031–4033.

23. Shang, Y. L.; Wen, Y. Q.; Li, S. L.; Du, S. X.; He, X. B.; Cai, L.; Li, Y. F.; Yang, L. M.; Gao, H. J.; Song, Y. A triphenylamine-containing donor–acceptor molecule for stable, reversible, ultra-high density data storage. *J. Am. Chem. Soc.*, 2007, *129*, 11674–11675.

24. Surin, M.; Sonar, P.; Grimsdale, A. C.; Mullen, K.; De Feyter, S.; Habuchi, S.; Sarzi, S. et al. Solid-state assemblies and optical properties of conjugated oligomers combining fluorene and thiophene units. *J. Mater. Chem.*, 2007, *17*, 728–735.

25. Harrison, N. T.; Baigent, D. R.; Samuel, I. D. W.; Friend, R. H.; Grimsdale, A. C.; Moratti, S. C. Holmes, A. B. Site-selective fluorescence studies of poly(p-phenylene vinylene) and its derivatives. *Phys. Rev. B*, 1996, *53*, 15815–15822.

26. Karastatiris, P.; Mikroyannidis, J. A.; Spiliopoulos, I. K.; Kulkarni, A. P.; Jenekhe, S. A. Synthesis, photophysics, and electroluminescence of new quinoxaline-containing poly(p-phenylenevinylene)s. *Macromolecules*, 2004, *37*, 7867–7878.

27. Colladet, K.; Fourier, S.; Cleji, T. J.; Lutsen, L.; Gelan, J.; Vanderzande, D. Low band gap donor–acceptor conjugated polymers toward organic solar cells applications. *Macromolecules*, 2007, *40*, 65–72.

28. Wu, P.T.; Bull, T.; Kim, F. S.; Luscombe, C. K.; Jenekhe, S. A. Organometallic donor–acceptor conjugated polymer semiconductors: Tunable optical, electrochemical, charge transport, and photovoltaic properties. *Macromolecules*, 2009, *42*, 671–681.

29. Rodrigues, R. F. A.; Charas, A.; Morgado, J.; Macanita, A. Self-organization and excited-state dynamics of a fluorene–bithiophene copolymer (F8T2) in solution. *Macromolecules*, 2010, *43*, 765–771.

30. Anant, P.; Mangold, H.; Lucas, N. T.; Laquai, F.; Jacob, J. Synthesis and characterization of donor–acceptor type 4,4′-bis(2,1,3-benzothiadiazole)-based copolymers. *Polymer*, 2011, *52*, 4442–4450.

31. Lim, E.; Jung, B. J.; Shim, H. K. Synthesis and characterization of a new light-emitting fluorene–thieno[3,2-b]thiophene based conjugated copolymer. *Macromolecules*, 2003, *36*, 4288–4293.

32. Jin, S. H.; Park, H. J.; Kim, J. Y.; Lee, K.; Lee, S. P.; Moon, D. K.; Lee, H. J.; Gal, Y.S. Poly(fluorenevinylene) derivative by gilch polymerization for light-emitting diode applications. *Macromolecules*, 2002, *35*, 7532–7534.

33. Sung, H. H.; Lin, H. C. Novel alternating fluorene-based conjugated polymers containing oxadiazole pendants with various terminal groups. *Macromolecules*, 2004, *37*, 7945–7954.

34. Zyss, J. *Molecular Nonlinear Optics: Materials, Physics and Devices*. Academic Press: Boston, MA, 1994.

35. Udayakumar, D.; Kiran, J. A.; Adhikari, A.V.; Chandrasekharan, K.; Umesh. G.; Shashikala, H. D. Third order nonlinear optical studies of newly synthesized polyoxadiazoles containing 3,4-dialkoxythiophenes using z-scan and degenerate four wave mixing methods. *Chem. Phys.*, 2006, *331*, 125–130.

36. Cassano, T.; Tommasi, R.; Babudri, F.; Cardone, A.; Farinola, G. M.; Naso, F. High third-order nonlinear optical susceptibility in new fluorinated poly(p-phenylenevinylene) copolymers measured with the z-scan technique. *Opt. Lett.*, 2002, *27*, 2176–2178.

37. Kiran, A. J.; Udayakumar, D.; Chandrasekharan, K.; Adhikari, A. V.; Shashikala, H. D. Z-scan and degenerate four wave mixing studies on newly synthesized copolymers containing alternating substituted thiophene and 1,3,4-oxadiazole units. *J. Phys. B: Atom. Mol. Opt. Phys.*, 2006, *39*, 3747.

38. Ramos-Ortiz, G.; Maldonado, J. L.; Hernández, M. C. G.; Zolotukhin, M. G.; Fomine, S.; Fröhlich, N.; Scherf, U. et al. Synthesis, characterization and third-order non-linear optical properties of novel fluorene monomers and their cross-conjugated polymers. *Polymer*, 2010, *51*, 2351–2359.

39. Nisoli, M.; Cybo-Ottone, A.; De Silvestri, S.; Magni, V.; Tubino, R.; Botta, C.; Musco, A. Femtosecond transient absorption saturation in poly(alkyl-thiophene-vinylene)s. *Phys. Rev. B*, 1993, *47*, 10881–10884.

40. Kishino, S.; Ueno, Y.; Ochiai, K.; Rikukawa, M.; Sanui, K.; Kobayashi, T.; Kunugita, H.; Ema, K. Estimate of the effective conjugation length of polythiophene from its $|\chi(3)(\omega;\omega,\omega,-\omega)|$ spectrum at excitonic resonance. *Phys. Rev. B*, 1998, *58*, R13430–R13433.

41. Hegde, P.K.; Adhikari, A. V.; Manjunatha, M. G.; Sandeep, C. S. S.; Philip, R. Nonlinear optical studies on new conjugated poly{2,2l-(3,4- dialkoxythiophene-2,5-diyl) bis[5-(2-thienyl)-1,3,4-oxadiazole]}s. *J. Appl. Polym. Sci.*, 2010, *117*, 2641–2650.

42. Gubler, U.; Concilio, S.; Bosshard, C.; Biaggio, I.; Gunter, P.; Martin, R. E.; Edelmann, M. J.; Wykto, J. A.; Diederich, F. Third-order nonlinear optical properties of in-backbone substituted conjugated polymers. *Appl. Phys. Lett.*, 2002, *82*, 2322–2325.

43. Ronchi, A.; Cassano, T.; Tommasi, R.; Babudri, F.; Cardone, A.; Farinola, G. M.; Naso F. $\chi(3)$ measurements in novel poly(2′,5′-dioctyloxy-4,4′,4″-terphenylenevinylene) using the Z-scan technique. *Synth. Met.*, 2003, *139*, 831–834.

44. Ellinger, S.; Graham, K. R.; Shi, P.; Farley, R. T.; Steckler, T. T.; Brookins, R. N.; Taranekar, P. et al. Donor-acceptor-donor-based π-conjugated oligomers for nonlinear optics and near-IR emission. *Chem. Mater.*, 2011, *23*, 3805–3817.

45. Sutherland, R. L. *Handbook of Nonlinear Optics*. Marcel Dekker: New York, 1996.

46. He, G. S.; Yong, K. T.; Zheng, Q.; Sahoo, Y.; Baev, A.; Ryasnyanskiy, A. I.; Prasad, P. N. Multiphoton excitation properties of CdSe quantum dots solutions and optical limiting behavior in infrared range. *Opt. Express*, 2007, *15*, 12818–12833.

47. Sheik-Bahae, M.; Said, A. A.; Wei, T.-H.; Hagan, D. J.; Van Stryland, E. W. Sensitive measurement of optical nonlinearities using a single beam. *IEEE J.Quantum Electron.*, 1990, *26*, 760–769.

48. Karthikeyan, B.; Anija, M.; Sandeep, C. S. S.; Muhammad, N. T. M.; Philip, R. Optical and nonlinear optical properties of copper nanocomposite glasses annealed near the glass softening temperature. *Opt. Commun.*, 2008, *281*, 2933–2937.

49. He, N.; Chen, Y.; Bai, J.; Wang, J.; Blau, W. J.; Zhu, J. Preparation and optical limiting properties of multiwalled carbon nanotubes with π-conjugated metal-free phthalocyanine moieties. *J. Phys. Chem. C*, 2009, *113*, 13029–13035.

50. Sivaramakrishnan, S.; Muthukumar V. S.; Sivasankara, S. S.; Venkataramanaiah, K.; Reppert, J.; Rao, A. M.; Anija, M.; Philip, R.; Kuthirummal, N. Nonlinear optical scattering and absorption in bismuth nanorod suspensions. *Appl. Phys. Lett.*, 2007, *91*, 093104–093107.

51. Hua, J. L.; Li, B.; Meng, F. S.; Ding, F.; Qian, S. X.; Tian, H. Two-photon absorption properties of hyperbranched conjugated polymers with triphenylamine as the core. *Polymer*, 2004, *45*, 7143–7149.

52. Qian, Y.; Meng, K.; Lu, C. G.; Lin, B. P.; Huang, W.; Cui, Y. P. The synthesis, photophysical properties and two-photon absorption of triphenylamine multipolar chromophores. *Dyes Pigm.*, 2009, *80*, 174–180.

53. Cassano, T.; Tommasi, R.; Tassara, M.; Babudri, F.; Cardone, A.; Farinola, G. M.; Naso, F. Substituent-dependence of the optical nonlinearities in poly(2,5-dialkoxy-p-phenylenevinylene) polymers investigated by the z-scan technique. *Chem. Phys.*, 2001, *272*, 111–118.

54. Boyd, R. W. *Nonlinear Optics*. 3rd ed.; Academic Press: Millbrae, CA, 2008.

55. Gayvoronsky, V.; Galas, A.; Shepelyavyy, E.; Dittrich, T.; Timoshenko, V. Y.; Nepijko, S. A.; Brodyn, M. S.; Koch, F. Giant nonlinear optical response of nanoporous anatase layers. *Appl. Phys. B*, 2005, *80*, 97–100.

56. Porel, S.; Venkatram, N.; Rao, D. N.; Radhakrishnan, T. P. Optical power limiting in the femtosecond regime by silver nanoparticle–embedded polymer film. *J. Appl. Phys.,* 2007, *102,* 33107–33113.

57. Chen, X.; Zou, G.; Deng, Y.; Zhang, Q. Synthesis and nonlinear optical properties of nanometer-size silver-coated polydiacetylene composite vesicles. *Nanotechnology,* 2008, *19,* 195703–195711.

58. Sih, B. C.; Wolf, M. O. Metal nanoparticle conjugated polymer nanocomposites. *Chem. Commun.,* 2005, 3375–3384.

59. Masuhara, A.; Kasai, H.; Kato, T.; Okada, S.; Oikawa, H.; Nozue, Y.; Tripathy, S. K.; Nakanishi, H. Hetero multilayered thin films made up of polydiacetylene microcrystals and metal fine particles. *J. Macromol. Sci. Part A Pure Appl. Chem.,* 2001, *38,* 1371–1382.

60. Wang, S. X.; Zhanga, L. D.; Sub, H.; Zhanga, Z. P.; Li, G. H.; Menga, G. W.; Zhanga, J.; Wang, Y. W.; Fana, J. C.; Gao, T. Two-photon absorption and optical limiting in poly(styrene maleic anhydride)/TiO$_2$ nanocomposites. *Phy. Lett. A,* 2001, *281,* 59–63.

61. Sezer, A.; Gurudas, U.; Collins, B.; Mckinlay, A.; Bubb, D. M. Nonlinear optical properties of conducting polyaniline and polyaniline–Ag composite thin films. *Chem. Phys. Lett.,* 2009, *477,* 164–168.

62. Chen, X.; Tao, J.; Zou, G.; Zhang, Q.; Wang, P. Nonlinear optical properties of nanometer-size silver composite azobenzene containing polydiacetylene film. *Appl. Phys. A,* 2010, *100,* 223–230.

63. Yang, S.-H.; Rendu, P. L.; Nguyen, T.-P.; Hsu, C.-S. Fabrication of MEH-PPV/SiO$_2$ and MEH-PPV/TiO$_2$ nanocomposites with enhanced luminescent stabilities. *Rev. Adv. Mater. Sci.,* 2007, *15,* 144–149.

64. Zhu, Y.; Xu, S.; Jiang, L.; Pan, K.; Dan, Y. Synthesis and characterization of polythiophene/Titanium dioxide composites. *React. Funct. Polym.,* 2008, *68,* 1492–1498.

65. Kiran, A. J.; Rai, N. S.; Udayakumar, D.; Chandrasekharan, K.; Kalluraya, B.; Philip, R.; Shashikala, H. D.; Adhikari, A. V. Nonlinear optical properties of *p*-(*N, N*-dimethylamino) dibenzylideneacetone doped polymer. *Mater. Res. Bull.,* 2008, *43,* 707–713.

66. Karthikeyan, B.; Anija, M.; Philip, R. In situ synthesis and nonlinear optical properties of Au:Ag nanocomposite polymer films. *Appl. Phys. Lett.,* 2006, *88,* 053104–053107.

67. Sreeja, R.; John, J.; Aneesh, P. M.; Jayaraj, M. K. Linear and nonlinear optical properties of luminescent ZnO nanoparticles embedded in PMMA matrix. *Opt. Commun.,* 2010, *283,* 2908–2913.

68. Venkatram, N.; Rao, D. N.; Giribabu, L.; Rao, S. V. Femtosecond nonlinear optical properties of alkoxy phthalocyanines at 800 nm studied using z-scan technique. *Chem. Phys. Lett.,* 2008, *464,* 211–215.

69. Samoc, M.; Samoc, A.; Davies, B. L.; Reish, H.; Scherf, U. Saturable absorption in poly(indenofluorene): A picketfence polymer. *Opt. Lett.,* 1998, *23,* 1295–1297.

70. Venkatram, N.; Naga Srinivas, N. K. M.; Rao, D. N. Nonlinear absorption and excited state dynamics in Rhodamine B studied using z-scan and degenerate four wave mixing techniques. *Chem. Phys. Lett.,* 2002, *361,* 439–445.

71. Hegde, P. K.; Adhikari, A. V.; Manjunatha, M. G.; Suchand C. S. S; Philip, R. Nonlinear optical characterization of new thiophene-based conjugated polymers for photonic switching applications. *Adv. Polym. Tech.,* 2011, *30,* 312–321.

CHAPTER 17

Properties and Processing of Polymer-Based Dielectrics for Capacitor Applications

Amoghavarsha Mahadevegowda and Patrick S. Grant

CONTENTS

17.1 INTRODUCTION

This chapter gives a background to polymer-based dielectrics, the requirements for short-term energy storage and associated challenges, and will familiarize the reader with the basic principles and recent developments in the field of electrostatic capacitors. The various material systems comprising of polymer matrices and fillers, factors that influence the energy storage capability and processing techniques for polymer nanocomposites (PNCs; as dielectrics for capacitors), are also discussed.

Although high dielectric permittivity (high capacity for energy storage) fillers play an important role in determining the overall dielectric properties of PNC-based capacitors, the techniques (mostly chemical) to produce fillers themselves are not discussed in this chapter, unless they are formed *in situ* during the processing of the PNC. Alternatively, more emphasis is laid on cost-effective scalable processing techniques for dielectric composites.

17.1.1 Polymers as Dielectrics

Polymers are used widely as "dielectrics" in capacitors with a simple parallel-plate configuration in microelectronic circuits at low voltage and current, across a wide range of frequencies. Polymers are attractive dielectric materials for electrostatic capacitors because they offer low cost, low density, and ease of processing to form large area and thin films, with comparatively high breakdown strengths of $\sim 10^5$ kV/m. These attributes provide a different balance of properties when compared to electrolytic capacitor counterparts, which are usually based on a thin Ta-based oxide film mounted on mica or paper, rolled, and immersed in a liquid. These electrolytic capacitors provide very high absolute capacitances (of up to several thousand farads, F) but tend to have a comparatively high mass and require the use of liquid electrolytes that should neither freeze nor boil during operation, which places awkward limitations on design flexibility, particularly in some mobile electrical systems such as aircrafts and electric vehicles.

Although polymer-only capacitors are popular and reliable, polymers have a low dielectric permittivity of usually 3–5, compared with 10^3 for certain ceramics (such as barium titanate, $BaTiO_3$), which limits the absolute capacitances that can be achieved and hence their applicability. So, a classic composite approach wherein polymers loaded with a minority fraction of fillers that enhance the effective dielectric permittivity of the material between the electrodes has been extensively researched for higher energy density applications, for example, in advanced electronic devices, hybrid electric vehicles and pulse power applications.[1]

17.1.2 Electrostatic Capacitors

Electrostatic capacitors are one of the most fundamental devices in electronic circuits and consist of two conductive electrodes separated by an insulator, called the dielectric. A capacitor stores electrical energy in the electric field in the dielectric due to the electrical charges segregated on the conductive electrodes. The ability of a capacitor to store electrical energy is measured by its capacitance. For a parallel-plate capacitor, capacitance C is given by

$$C = \frac{\epsilon_0 kA}{d} \tag{17.1}$$

where:
  A is the area of the dielectric
  k is the dielectric constant (also known as relative permittivity) of the material between
    the two plates
  $\epsilon_0 = 8.854 \times 10^{-12}$ Fm^{-1} is the permittivity of free space
  d is the separation between the plates

Figure 17.1 shows a simple parallel-plate capacitor with a dielectric material in between the conductive plates (each plate of area A). The greater the area A, the greater the number of charges that can be stored on each plate. The smaller the distance d between the plates, the better the interaction between the positive and negative charges on the opposite plates. This interaction is important in keeping the charges firmly tethered to the plates, because the greater the number of charges remaining on plates, then the higher the capacitance. The dielectric constant k can be increased by introducing a dielectric material that has a high intrinsic polarizability. As a result of polarization, an internal opposing electric field develops in the dielectric which decreases the effective electric field across the capacitor. So, in order to increase the effective electric field across the capacitor to match the applied electric field, more charges have to be stored on the plates, resulting in enhanced capacitance.

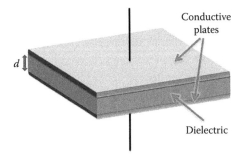

FIGURE 17.1 A parallel-plate capacitor with a dielectric material in between the conductive plates.

The electrical energy E_s stored in a parallel-plate capacitor is given by

$$E_s = \frac{1}{2}CV^2 \tag{17.2}$$

where V is the potential difference across the plates. The energy stored in a real capacitor is dissipated due to dielectric losses (charge migration resulting in current leakage and molecular vibrations leading to conversion of electrical energy to thermal energy) and resistive losses in electrodes. These losses are accounted for by a term called dissipation factor or loss tangent tan δ. The dielectric loss is given by the imaginary part of the permittivity, ϵ''. Permittivity ϵ^*, which is a complex entity, is given by

$$\epsilon^* = \epsilon' - j\epsilon'' \tag{17.3}$$

where:

$j = \sqrt{-1}$

ϵ' and ϵ'' are the real and imaginary parts of complex permittivity, respectively

The real part of the permittivity ϵ' is equal to $\epsilon_0 k$, which was introduced earlier in Equation 17.1.

The dissipation factor tan δ is defined as

$$\tan \delta = \frac{\epsilon''}{\epsilon'} \tag{17.4}$$

The energy density W of a dielectric capacitor is given by

$$W = \frac{1}{2}kE^2 \tag{17.5}$$

where E is the electric field strength. From Equation 17.5, it is apparent that by increasing the electric field strength, the energy density of the capacitor can be increased, but will be limited by the intrinsic dielectric breakdown strength of the material E_d. So, maximum energy density W_{max} is attained when E equals E_d.

17.1.2.1 Frequency Response of Permittivity

Permittivity is complex in nature because the response of polarization that occurs in a material under the influence of an external field is not instantaneous. The magnitude and phase difference of this response is accounted for by expressing permittivity as a complex entity. Hence, the dielectric constant of a material is dependent on the frequency of the applied electric field. Various polarization mechanisms[2] gain dominance at different ranges of frequency as shown in Figure 17.2a. The extent of polarization that the dielectric undergoes decides the energy density attainable for a particular frequency.

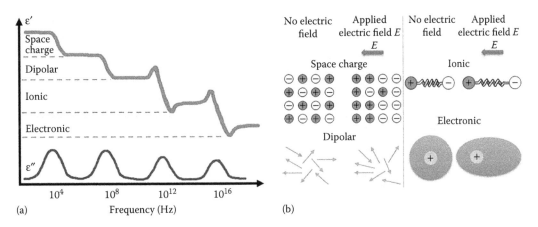

FIGURE 17.2 (a) Dominance of various polarization mechanisms at different ranges of frequency, and (b) schematic representation of different mechanisms of polarization.

Space charge polarization, also known as Maxwell–Wagner interfacial polarization, arises due to moving and piling up of the charge carriers at the interfaces of grains or phases under the influence of an electric field below 1 kHz.

A molecule can be polar or nonpolar depending on the distribution of the electron cloud between the constituting atoms. This polarity can be permanent or induced. In the absence of an electric field, the dipole moments of molecules are randomly oriented. On the application of an electric field, the dipoles rotate and align themselves in the direction of the applied field, giving rise to dipolar polarization observed in the frequency range 10^3–10^9 Hz. The rotation of dipoles gives rise to dielectric loss due to friction during orientation. This loss can be so significant that it has found practical application in heating food in a microwave oven where the heat is produced due to rotation of permanent dipoles of water molecules (here, the dipoles rotate to keep up with the alternating electric field).

When the applied electric field is reversed, the space charge and dipolar polarization mechanisms also change direction, which involves the movement of space charges and the rotation of dipoles. However, this change is not instantaneous and the time taken is known as the relaxation time. So relaxation effects are seen in space charge and dipolar polarization mechanisms. Ionic polarization, which is dominant in the range 10^{10}–10^{13} Hz, occurs due to the displacement between the positive and negative ions in the presence of an electric field. Electronic polarization is caused in neutral atoms due to an induced shift of electron clouds with respect to the nucleus under the influence of an electric field. Electronic polarization occurs in all dry solids up to 10^{16} Hz, and it is the sole contribution to loss above 10^{14} Hz.

The frequency dependence of ionic and the electronic polarization mechanisms is governed by resonance phenomena, similar to driven damped harmonic oscillators. This leads to peaks at the resonance frequencies of the ionic and electronic polarization modes as seen in Figure 17.2a.[2] A dip can be seen at frequencies just above the resonance peaks, corresponding to the response of the oscillating system (ionic or electronic) being out of phase with the driving force (applied AC field). A schematic representation of the polarization mechanisms is shown in Figure 17.2b.

The dielectric loss is especially high in the vicinity of the relaxation or resonance frequencies of the polarization mechanisms, as seen in Figure 17.2a, because the polarization lags behind the applied electric field, giving rise to an interaction between the field and the dielectric medium's polarization that results in heating. Apart from these polarization mechanisms, dielectric loss also depends on a number of other factors such as ambient temperature and humidity (which affect the conductance and polarization loss in the dielectric), electromagnetic interference from other operating devices in the vicinity of the capacitor and channel noise during transmission of a monitored signal.[3]

17.2 POLYMER CAPACITORS

17.2.1 Polymer-Only Dielectrics

High dielectric constant polymers find applications in artificial muscles, actuators, and charge storage devices.[4] From Equations 17.1 and 17.5, the main advantages of using polymers for capacitors can be understood as follows:

- Easily processed into thin films (low d) with large area (high A)

- Relatively high electric breakdown strength E_d

In the initial stages of research in the field of dielectrics, ferroelectric polymers were investigated and the first organic material that was found to have appreciable ferroelectric and piezoelectric properties was poly(vinylidene fluoride) (PVDF). Among polymers, PVDF has a relatively high dielectric constant of ~10 and remarkable mechanical and chemical properties.[5] The high dielectric constant is due to the presence of C–F dipoles that reorientate in the crystalline phase under the influence of an applied electric field.[6] Similar to PVDF, all polymers that have strong electronegative groups can spontaneously align in the presence of the electric field to some extent, and so possess a comparatively high dielectric constant. However, PVDF and similar polymers are not suitable for *high-energy density* capacitors due to their low dielectric permittivities when compared with certain ceramics. PVDF also tends to exhibit relatively high dielectric loss with tan δ ~ 0.02.

The focus on polymer dielectrics then moved to lower loss polymers such as silicone, benzocyclobutene, polynorbornene, and polyimide.[7] Table 17.1 gives a list of most commonly

TABLE 17.1 Dielectric Permittivities of Commonly Used Polymer Dielectrics

Polymer	Dielectric Permittivity
Polytetrafluoroethylene	1.7–2
Polypropylene	~2.2
Polystyrene	~2.6
Poly(methyl methacrylate)	~2.7
Polyester	~3
Polycaprolactam	~4
Poly(vinylidene fluoride)	~10

Source: Fujisaki, S. et al., *Appl. Phys. Lett.*, 90, 162902, 2007; Takele, H. et al., *Nanotechnology*, 17, 3499, 2006; Barber, P. et al., *Materials*, 2, 1697, 2009; Mahadevegowda, A. et al., *J. Phys. Conf. Ser*, 522, 012041, 2014.

used polymer dielectrics.[5,8–10] Due to their low dielectric permittivities, which are apparent from Table 17.1, none of them are generally considered ideal for high-energy applications, but many are very widely used as cheap and effective lightweight capacitors in mass mobile electronics market such as laptops and mobile phones.

17.2.2 Polymer Composite Dielectrics

Ceramics offer high dielectric constant (e.g., up to ~100,000 for $La_{1.8}Sr_{0.2}NiO_4$) but have low breakdown strength (~10^4 kV/m). Also, processing of ceramics to manufacture capacitors is difficult, especially in thin layers, and high dielectric permittivity ceramics are often dense, for example, high permittivity $BaTiO_3$ has a density of ~6.08 g/cc, which undermines specific energy density. Therefore, the polymer-based composite approach is to add a high permittivity ceramic as a minority, second phase "filler" in the polymer "matrix," to promote polarizability, without significantly undermining the intrinsic polymer properties. The filler phase should be sufficiently of small length scale, so that significant volume fraction can be incorporated without compromising film thicknesses and well dispersed to avoid percolation of current loss. High filler volume fractions will also decrease the breakdown strength, and particle clustering can reduce overall dielectric permittivity due to the promotion of the formation of voids.[1] Thus, using polymer composites as dielectrics involves a trade-off between high dielectric constant, processability, high breakdown strength, and low dielectric loss.

The increase in the dielectric constant of these composites is due to the increase in the average electric field within the polymer matrix. However, the internal electric field will be inhomogeneous within the polymer layer because of the large difference between the dielectric constants of the constituent phases, concentrating electric field lines at or around microstructural features. These field concentration effects can promote localized breakdown, excessive ohmic heating in an oscillating field, and cascade breakdown where local breakdown then concentrates the field elsewhere, leading to further breakdown, and so on until the whole film breaks down at lower fields than expected.

The distortion of electric field can be very significant in systems where small particles agglomerate to form larger particles[9,11] or where particles are angular or even needle-shaped. So, using near spherical nanosized fillers in order to minimize electric field distortion and facilitate fabrication of thin films can be advantageous, but agglomeration must be avoided.

17.2.3 PNC Dielectrics

Adding nanosized fillers to polymers increases the dielectric constant of the composite more than microsized fillers.[12] For example, at a filler loading of 50 wt.% of 100 nm sized Al particles, epoxy-based nanocomposites exhibited a dielectric constant of ~60, which was more than twice that of Al-epoxy microcomposites with a dielectric constant of ~30 and ~25 for 3 and 10 µm sized Al particles, respectively.[13] Among the reasons for this, the higher surface area to volume ratio of a nanoparticle provides opportunities to functionalize the surface, which can be exploited to improve the compatibility between the polymer and the nanoparticle[1] and can also be used to reduce internal fields[11] (and hence improve the breakdown strength of the resulting nanocomposite).

PNC systems, such as those popular systems containing $BaTiO_3$[14] and Al_2O_3 nanoparticles[13,15] in epoxy-like matrices, have been studied widely (particularly as a function of filler volume fraction) for embedded capacitor applications. For example, an increase in the volume fraction of $BaTiO_3$ by two times (from 0.25 to 0.5) resulted in a three times (from 15 to 45) increase in the effective dielectric constant of the nanocomposite.[14] These so-called embedded passives reduce internal parasitic resistance and inductance associated with surface-mounted discreet components.[13]

Until recently, the concept that enhanced dielectric properties might also be contrived by *interfacial* mechanisms operating between the two phases were generally overlooked, although as the size of the filler particles reduces to the nanometer scale, the dielectric properties of the interface between the polymer and filler might be expected to play a progressively more important role in the effective dielectric behavior.[16,17] The role of the interface in effective dielectric behavior is now gaining prominence and it has been suggested, for example, that large interfacial areas that facilitate interface exchange coupling through a dipolar interface layer lead to enhanced polarization at the interface and hence a high drop in the local electric field,[18–20] ultimately resulting in an enhanced effective dielectric constant. The effect of the interface will be discussed in Section 17.4.4.

To predict the effective dielectric constant and understand the behavior of the composite as an effective single medium dielectric, many analyses and models have been proposed. In the simplest model, the effective dielectric permittivity (ϵ_{eff}) is described by a rule of mixtures approach:

$$\epsilon_{eff} = \upsilon_{f_1}\epsilon_1 + \upsilon_{f_2}\epsilon_2$$

(17.6)

where the subscripts 1 and 2 represent the two phases, ϵ and υ_f represent dielectric constant and volume fraction, respectively. Equation 17.6 is also known as Lichtenecker's parallel mixing rule,[21] which gives the upper limit of the effective dielectric constant. The ϵ_{eff} can also be estimated by Lichtenecker logarithmic mixing rule:

$$\log(\epsilon_{eff}) = \upsilon_{f_1}\left[\log(\epsilon_1)\right] + \upsilon_{f_2}\left[\log(\epsilon_2)\right]$$

(17.7)

Equations 17.6 and 17.7 do not predict reliably the effective dielectric behavior of many composite systems, especially at low filler volume fractions of real dielectric materials.[21–25]

Other models based on mean field theory have been developed. For filler volume fraction in the range of 10%–50%, Maxwell's equation has been generally shown to give a good approximation of the effective dielectric constant[26]:

$$\epsilon_{eff} = \epsilon_1 \left(\frac{\epsilon_1 + 2\epsilon_1 - 2\left(1 - \upsilon_{f1}\right)\left(\epsilon_1 - \epsilon_2\right)}{\epsilon_2 + 2\epsilon_1 + \left(1 - \upsilon_{f1}\right)\left(\epsilon_1 - \epsilon_2\right)} \right)$$

(17.8)

Piezoelectric investigations of PVDF and lead zirconate titanate composites[27] lead to an alternative theoretical approach to the calculation of dielectric constant, which is given by Yamada equation:

$$\epsilon_{\text{eff}} = \epsilon_p \left[1 + \frac{n \upsilon_{fc} \left(\epsilon_c - \epsilon_p \right)}{n \epsilon_p + \left(\epsilon_c - \epsilon_p \right) \left(1 - \upsilon_{fc} \right)} \right] \tag{17.9}$$

where:
ϵ_p and ϵ_c are the dielectric constants of polymer and ceramic, respectively
n is a parameter attributed to the shape of ellipsoidal particles
υ_{fc} is the volume fraction of the ceramic

The Yamada equation has become perhaps the most widely fitted equation to experimental data, because it offers an extra degree of freedom through the shape factor n, although physical interpretation of n obtained by experiment is difficult.

Predictions based on effective medium theory (EMT) are not applicable to *conductive* filler–polymer composite systems where near percolation effects have a significant influence on the dielectric properties. Very high values of ϵ_{eff} (~10^7) have been reported just below the critical volume fraction of the minority phase at which significant current percolation occurs. This phenomenon in the vicinity of the percolation threshold follows a power law[28,29]:

$$\epsilon_{\text{eff}} \propto \sigma \propto \left(f_c - f \right)^{-t} \left(f < f_c \right) \tag{17.10}$$

where:
σ is the AC conductivity
t is a scaling exponent
f is the filler fraction
f_c is the percolation threshold fraction

When f approaches f_c, the filler particles are on average very close to one another, but remain separated by a thin insulating layer of polymer, thereby forming a microcapacitor network,[29] resulting in very high dielectric permittivities.

Although such a high dielectric constant appears attractive, losses also dramatically increase near the percolation threshold. Further, because the increase in dielectric constant is often strongly nonlinear, see Equation 17.10, in practice the manufacture of these films at an industrial scale is problematical: too little filler and there is barely an increase in dielectric behavior, slightly too much and the capacitor is highly lossy, leaks excessive current, and may breakdown completely. The formation of the microcapacitor networks is discussed in Section 17.3.2.1.

17.3 TYPES OF FILLERS

17.3.1 Ceramic Fillers

Generally, ceramics have a high dielectric constant, especially those with perovskite crystal structure, and often highly temperature sensitive. Table 17.2 gives the dielectric constants of commonly used ceramics for capacitors.[9,19,30]

Ceramic powders such as $BaTiO_3$, $CaCu_3Ti_4O_{12}$,[31] $SrTiO_3$,[32] Li_2MgSiO_4,[33] and TiO_2[34] have been investigated as fillers in polymers to obtain high-energy density capacitors, and all contribute to enhanced performance, which can usually be best fitted to one of the expressions in Equations 17.6 through 17.9. However, problems arising due to agglomeration of particles have necessitated the addition of surfactants, which often leads to an increase in the dielectric loss and impurity content in the resulting composite.[30,35,36]

17.3.2 Conductive Fillers

As previously described, conductive filler–polymer composites can have high dielectric constant at very low volume fraction of filler. The effective dielectric constant of composites with conductive fillers follows the power law in Equation 17.10 and enhancement is due to the microcapacitor network effect.

17.3.2.1 Formation of the Micro-Capacitor Network

Figure 17.3 shows the effective dielectric constant of PVDF-based nanocomposites as a function of exfoliated graphite nanoplate (xGnP) filler volume fraction f, measured at 1 kHz and room temperature.[37] The variation in the microstructure of the nanocomposite film at $f < 0.75$, 0.75–2.25, and >2.25 is shown in the schematics in Figures 17.4a–c, respectively.

In Figure 17.4a, when $f < 0.75$, the fillers were too far from each other to form an electrically conductive network. So the nanocomposite film remained largely insulating. When $f = 0.75$–2.25, the fillers were separated by a very thin layer of polymer, as shown in Figure 17.4b, but were still not sufficiently close to form a percolative network. The nanocomposite is then considered to be a near-percolating medium in which two adjacent conductive fillers separated by a thin dielectric layer of thickness d' can be thought of as two electrodes of a miniature capacitor.

TABLE 17.2   Dielectric Permittivities of Commonly Used Ceramics in Capacitors (Room Temperature Properties)

Composition	Dielectric Permittivity
SiO_2	3.9
Al_2O_3	9
Ta_2O_5	22
TiO_2	80
$PbNb_2O_6$	225
$BaTiO_3$	1,700
$CaCu_3Ti_4O_{12}$	~60,000
$La_{1.8}Sr_{0.2}NiO_4$	~100,000

Source: Barber, P. et al., *Materials*, 2, 1697, 2009; Mahadevegowda, A. et al., *Nanotechnology* 25, 475706, 2014; Rao, Y. and Wong, C.P., *J. Appl. Polym. Sci.*, 92, 2228, 2004.

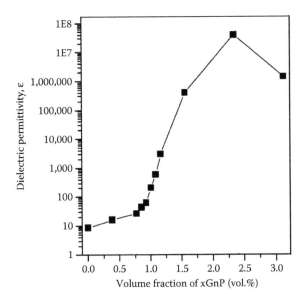

FIGURE 17.3 Dielectric permittivity of PVDF-based nanocomposites as a function of exfoliated graphite nanoplate (xGnP) filler volume fraction, measured at 1 kHz and room temperature. (He, F., Lau, S., Chan, H.L., and Fan, J.: High dielectric permittivity and low percolation threshold in nanocomposites based on poly[vinylidene fluoride] and exfoliated graphite nanoplates. *Adv. Mater.* 2008. 21. 710. Copyright Wiley-VCH Verlag GmbH & Co. KGaA. Reproduced with permission.)

Such a miniature capacitor has a relatively high capacitance due to the ultra-thin nature of the dielectric film between the fillers. A number of such miniature capacitors in random series and parallel combinations formed a high-*k* micro/nanocapacitor network that leads to an enhancement in the overall dielectric permittivity of the nanocomposite film. At *f* > 2.25, the filler fraction was sufficiently high to form a continuous percolating network throughout the film, as shown in Figure 17.4c, resulting in a relatively low-*k* film.

Conducting carbon fillers (carbon black, graphite nanoplates, and carbon nanotubes), metallic fillers (Ni, Ag, Mo), and nonmetallic conductive polymer particles (e.g., polyaniline) have all been used as fillers in conductor–insulator composites.[29,37–45]

In these composites, unlike composites containing ceramics, an increase in filler fraction does not always ensure a corresponding increase in dielectric constant, and beyond the percolation threshold, the composite becomes conductive as seen in Figure 17.4. The main disadvantages of using a conductive filler is the control of the volume fraction at close to the critical point during large-scale production, and near-percolating composite films tend to suffer from unusable high loss, especially at low frequencies (<10 kHz).[45] Moreover, operational parameters such as temperature can bring about microstructural changes that can alter the interparticle spacing and "convert" these dielectrics into lossy, *conducting* materials.

17.3.3 Ceramic and Conductive Fillers

A polymer film may contain two different types of ceramic/conductive fillers at the same time: (1) core–shell particles, where the ceramic and conductive phases are closely associated with each other; or (2) two nonconnected ceramic and conductive phases.

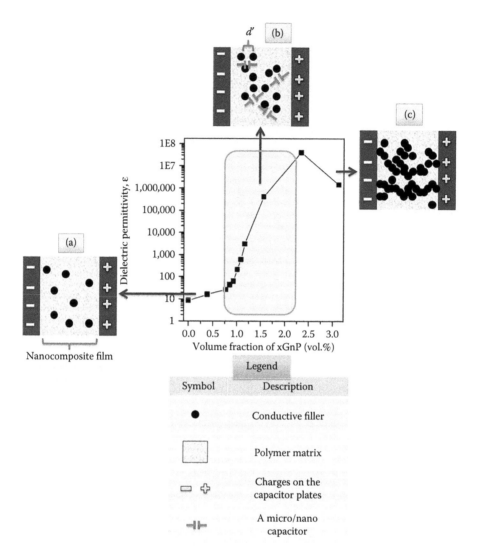

FIGURE 17.4 Dielectric permittivity of PVDF-based nanocomposites as a function of exfoliated graphite nanoplate (xGnP) filler volume fraction, measured at 1 kHz and room temperature. The variation in the microstructure of the nanocomposite film at filler volume fraction $f < 0.75$, 0.75–2.25 and >2.25 is shown in the schematics (a–c), respectively. (He, F., Lau, S., Chan, H.L., and Fan, J.: High dielectric permittivity and low percolation threshold in nanocomposites based on poly(vinylidene fluoride) and exfoliated graphite nanoplates. *Adv. Mater.* 2008. 21. 710. Copyright Wiley-VCH Verlag GmbH & Co. KGaA. Reproduced with permission.)

17.3.3.1 Core–Shell Particles

Core–shell particles with metallic cores and ceramic shells can enhance the dielectric constant of composite films.[7,46] The shell can be an oxide of the metal constituting the core or an oxide of another metal. For example, Co particles coated with ZnO were used as core–shell fillers in PVDF to enhance the dielectric constant to 51 at 100 Hz and 10% filler loading.[46] In the case of composite films with Al filler particles, post-deposition reaction

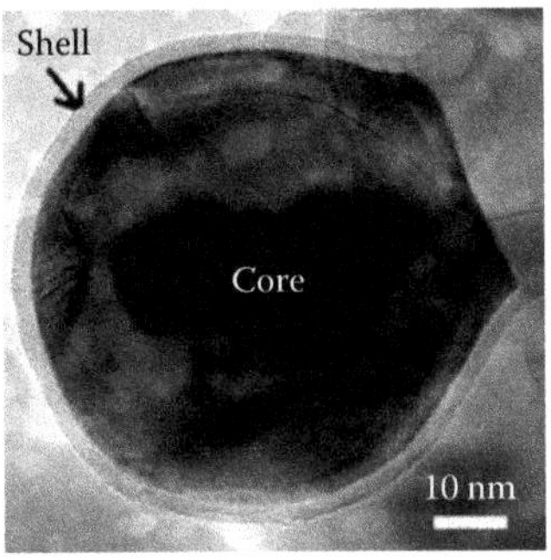

FIGURE 17.5 TEM image of a 100 nm aluminum particle with Al_2O_3 shell of thickness 2.8 nm. (Reprinted with permission from Xu, J.W. and Wong, C.P., *J. Electr. Mater.*, 35, 1087, 2006; Xu, J. and Wong, C.P., *Compos. Pt. A Appl. Sci. Manufact.*, 38, 13, 2007; Xu, J.W. and Wong, C.P., *Appl. Phys. Lett.*, 87, 082907. Copyright 2005, American Institute of Physics.)

of isolated Al regions may lead to the formation of core–shell particles, with Al as the core with an Al_2O_3 shell.[13,15] Figure 17.5 is a TEM image of a 100 nm aluminum particle with Al_2O_3 shell of thickness 2.8 nm.[7,15,47] Composites containing 80 wt.% Al-Al_2O_3 core–shell particles have shown dielectric constants as high as ~10^2 and low dissipation factor of ~0.02.[48] This effective dielectric constant was ~30 times higher than the dielectric constant of the epoxy matrix and ~10 times greater than bulk Al_2O_3, and therefore could not be explained by EMT models (because the effective dielectric constant of the composite was more than dielectric constant of either of the constituent phases). This "anomalous" increase in dielectric constant was attributed instead to near-percolation effects and interfacial polarization due to spatial or space charges (electrons) that tunnel out from the metallic Al core to the Al_2O_3 shell. Although a detailed model of the tunneling phenomena occurring in the vicinity of the core–shell particles relating to capacitance is generally unavailable, it can be implied that the role played by the tunneling process in enhancing polarization is to contribute spatial charges that migrate to and accumulate at the interfaces of dielectrically different phases, such as oxide/polymer interfaces. This build-up of charge at heterogeneous interfaces leads to so-called interfacial polarization, resulting in an elevation of the overall or effective dielectric response of the composite.

Returning to the aspect of space charge and tunneling effects, researchers in the mid-twentieth century showed that electron tunneling through 3–5 nm thick layers of Al_2O_3 was possible,[49,50] because the wave function of a free electron ψe in the metal (here Al) remains finite across the thickness of the insulator (here Al_2O_3), with a measurable probability of electron passage from one side of the insulator to the other.[51] These reports on electron tunneling stress the importance of the thickness of the oxide shell. If the shell is too thick (tens of nm), electron tunneling is suppressed and the core–shell particles' contributions to dielectric enhancement is limited. It is important to note that the effect of a number of defect energy levels that might be present in an oxide around a tiny, high curvature metallic particle with highly localized chemical and structural variations were not considered.

There is also some uncertainty[52–56] on the timescale involved in the tunneling process. However, in terms of the practical range of frequencies for capacitors (Hz–MHz), the near-instantaneous nature of tunneling rules out the possibility of any considerable phase difference between the tunneling current and the applied electric field. Nevertheless, under the influence of an AC field, the physical processes involving migration, accumulation, and reordering of the charged species at interfaces, which might "trap" charges, might result in a phase shift and make them capacitive in nature.

A recent in-depth study of the structure and chemistry of core–shell particles on a nanoscale together with systematic measurements of dielectric response has provided an insight into the physical processes governing the "super-k" behavior in composites containing core–shell particles.[19] The study suggested that the presence of a thin and highly defective oxide shell with a range of intermediate energy levels between that of the metallic core and the polymer matrix contributed positively to interfacial polarizability.

The development of an out-of-phase dipole moment at the interfaces, particularly at the poles of the particles in nanocomposite films, due to the accumulation of charges at relatively low frequencies, has been suggested to be responsible for an enhancement in the dielectric properties beyond the predictions of the Maxwell–Wagner theory.[57,58] So core–shell particles, which inherently have multiple interfaces, represent a microstructural condition that will tend to promote superimposed enhanced interfacial polarization effects.

17.3.3.2 Nonconnected Ceramic and Conductive Particles

Composite systems consisting of both ceramic and conductive filler particles can exhibit an increase in dielectric constant by superimposing the effects of the relatively polarizable ceramics and the microcapacitor network effect associated with conductive fillers. For example, $BaTiO_3$ has been investigated in combination with Al,[59] Ag,[60,61] Ni,[62] and carbon fiber.[63] These systems, although believed to show near-percolative effects have so far not exhibited enhancement in dielectric properties beyond those already reported conductive filler–polymer composites and their distinct benefits (if any) have not been demonstrated.

17.3.4 Surface-Modified Fillers

Surface functionalization has been proved to improve the mechanical flexibility of composite films, compatibility between filler and matrix, and also has resulted in better dispersion of nanofillers in the matrix leading to enhanced energy storage.[18,34,46,64–66] For example, the compatibility between $CaCO_3$ and nylon-6 was improved by surface modification with oleic acid[67] and $BaTiO_3$-based core–shell particles functionalized by methacrylate groups showed reduced filler coalescence and loss.[68] Coating metal particles with organic dielectrics which form an insulating high dielectric constant shell has also been used as a technique to enhance the dielectric properties of composites.[44] Such core–shell particles showed percolative-like enhancement in dielectric constant, which can be increased by as much as two orders of magnitude.

However, surface modification has also been shown to have some detrimental effects on the physical properties of the nanocomposites. Although an increase in the dielectric constant and a reduction in the particle aggregation was achieved by the use of surface-modified $BaTiO_3$ nanoparticles, a decrease in the breakdown strength was also observed.[65] In contrast, introduction of surface-modified TiO_2 nanoparticles into low-density polyethylene increased the breakdown strength by 40% when compared with composites containing untreated TiO_2 nanoparticles.[11] So, surface modification of fillers can have contrasting effects on the dielectric properties of the composites and the effects are highly dependent on the specific nature of modification, the nature of both filler and matrix.

17.4 FACTORS INFLUENCING ENERGY DENSITY

17.4.1 Dielectric Breakdown Strength

Polymers are excellent insulators because the band gap between valence and conduction bands is >2 eV[69] and typically ~7 eV,[70] as shown in Table 17.3.[71,72] Due to the high band gap energy, the electric field needs to be >10^7 kV/m for tunneling of electrons from the valence to the conduction band, which might lead to breakdown of the polymer. However, polymers can breakdown at lower fields of ~10^5 kV/m by other mechanisms.

Avalanche breakdown (also known as ionization breakdown) is characterized by conduction and multiplication of a large number of electrons (and other charge carriers) resulting from multiple high-energy collisions between the electrons in the polymer matrix. This multiplication of electrons occurs when free electrons in polymers acquire sufficient kinetic energy by accelerating (under the influence of an external field) along the mean free path and knock out other electrons. Polymers are good hosts for this mechanism because they often contain regions of amorphous phases. These amorphous phases have unoccupied volume consisting of voids/cavities which are filled with gases of dielectric strength ~10^3 kV/m. The differences between the dielectric properties of the polymer and the gases in the voids, and a number of other factors (such as gas pressure, void size, and shape) might lead to a local intensification of field resulting in an additional breakdown within the void. Since such breakdown occurs only inside the void and not necessarily in the whole polymer, such a breakdown is known as partial discharge breakdown, which can be supported by relatively large voids of dimensions of a few microns. The electrons that are generated by breakdown are accelerated along the length of the void and they might gain sufficient energy to knock off more electrons on the other side of the void and erode the matrix. The distribution of such voids throughout the matrix might lead to the overall breakdown of the polymer.

In relatively small voids of molecular order, which are referred to as holes,[73] the void size is not sufficient to support partial discharge breakdown because the voltage across the voids is not sufficient to cause breakdown. However, these holes provide longer mean

TABLE 17.3 Band Gap Energies of Some Polymers

Polymer	Band Gap Energy (eV)
Polypropylene	7
Polyethylene	6.9
Polyamide-6 (Nylon-6)	4.9
Polystyrene	4.4

Source: Ohki, Y. et al., *Annual Report Conference on Electrical Insulation and Dielectric Phenomena*, 1, West Lafayette, IN, 2010; Sreelatha, K. and Predeep, P., *J. Plastic Film Sheet.*, 29, 127 2013.

free paths than the polymer for the *free electrons* (not the electrons produced by ionization because the breakdown has not occurred yet) to accelerate and gain sufficient energy to *initiate* breakdown.[9,73]

A simple solution to the problem caused by voids is to fill them, preferably with nonconducting material. When the voids of poly(ethylene terephthalate) were filled with polyaniline, which is nonconductive in the emeraldine base form, the dielectric breakdown strength increased by 30%.[74] More complex solutions, such as surface modification are employed in composite films to improve their breakdown strength, as discussed in Section 17.3.

In polymers that have short free mean paths, conductive paths are formed as a result of polymer degeneration. Since the life of a polymer decreases exponentially with the applied electric field, the formation of conducting channels depends on the intensity and duration of the external applied field.[75] Also, high electric fields can ionize molecules due to tunneling of electrons through a reduced potential barrier. Due to this ionizing effect, the electric field intensity at which ionization of macromolecules starts can be considered as the theoretical maximum breakdown strength of polymers. Localized high electric fields can form in a polymer due to structural differences and defects. These local fields can also lead to degeneration of polymer. Degeneration leading to premature dielectric breakdown not only limits the energy storage capability of the capacitor, but also makes the capacitor less reliable.

17.4.2 Operating Temperature

The degradation temperature of polymer matrices limits the temperature at which the capacitors can be used. Also, the mismatch between the coefficients of thermal expansion of the passive components in an embedded system may lead to delamination and system failure.[7] So, choosing the right materials for the components gains importance over wide operating temperatures (such as might be expected in electric vehicles and aircraft). Incorporation of nanoparticles generally increases the crystallization temperature of the polymer.[18] For example, a 5% addition of $BaTiO_3$ increased the crystallization temperature of poly(vinylidene fluoride-*ter*-trifluoroethylene-*ter*-chlorotrifluoroethylene) from ~100°C to ~106°C. The presence of crystalline phases then can enhance charge transport due to the more ordered structure. In a crystalline polymer, fillers preferentially occupy the noncrystalline spaces (boundaries) and this can form a continuous network that facilitates charge transport (and thus make the capacitor more lossy).

In a composite, the dielectric constant generally increases with temperature in the range of 25°C–150°C. This increase can be due to an increase in the mobility of polymer chains, any increase in the permittivity of the filler (in the case of semiconducting fillers) and enhanced charge carrier diffusion with temperature.[34,45,76,77] However, this increase is not always useful or significant: for example, in the temperature range of 25°C–150°C, the dielectric constant of polymerized tripropylene glycol diacrylate-$BaTiO_3$ films ($BaTiO_3$ volume fraction ~0.3) increased from 39 to 45[76] and in composites containing ZnO fillers (volume fraction ~0.4) in low-density polyethylene, there was a slight increase in dielectric permittivity from ~5 at 20°C to ~7.5 at 100°C.[77]

A change in temperature can also lead to changes in the allotropic forms of the ceramic filler. If this change is exhibited in bulk properties, then the dielectric constant of the whole composite can be altered. For example, the dielectric constant of $BaTiO_3$ increases toward the Curie temperature of 132°C when it changes its crystal structure from cubic to tetragonal,[78] which results in progressive decrease in dielectric constant at higher temperatures.

The study of variation in dielectric constant at high temperatures is important in the context of industrial applications because capacitors that have relatively stable dielectric constant can be placed much closer to other heat-generating components such as generators or combustors, and may require less active temperature management or thermal proof packaging.

17.4.3 Size, Shape, and Distribution of Filler

London–van der Waals forces are high at the nanoscale as these forces are inversely proportional to the sixth power of the particle radius. So, nanoparticles have a strong tendency to aggregate and form clusters.[38,79] At high filler volume fraction, the particle size limits the minimum thickness of the film that avoids percolation, and the film will generally have a roughness of at least the average particle size. Therefore, beneficial increases in effective dielectric constant due to increased filler fraction might be offset by unhelpful increases in film thickness.

For a given filler fraction, nanocomposites containing two different filler particle sizes (bimodal) have been shown to have higher capacitances than unimodal nanocomposites.[14,80] Figure 17.6 shows the variation of capacitance with frequency of $BaTiO_3$–epoxy bimodal and unimodal nanocomposites for the same thickness (7 μm) capacitors at a

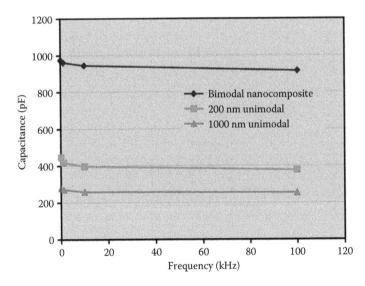

FIGURE 17.6   Variation of capacitance with frequency of $BaTiO_3$–epoxy bimodal and unimodal nanocomposites for the same thickness (7 μm) capacitors at a constant $BaTiO_3$ volume fraction of 0.6. (Reprinted from *Thin Solid Films*, 518, Rasul et al., Flexible high capacitance nanocomposite gate insulator for printed organic field-effect transistors, 7024, Copyright 2010, with permission from Elsevier.)

BaTiO$_3$ volume fraction of 0.6.[80] At 1 kHz, the capacitance of bimodal nanocomposite was ~1000 pF compared with ~300 pF and ~400 pF for unimodal films with 1000 and 200 nm BaTiO$_3$ particles, respectively. This was attributed to the high packing efficiency (the small particles located at the interstices between large particles) and increased particle surface coverage in the bimodal nanocomposites.

The effect of filler shape should be considered seriously when the particle aspect ratio deviates by more than an order of magnitude from that of a sphere.[81] Such nonspherical particles can establish conducting paths easily, leading to a low percolation threshold,[82] and as previously described these composites can exhibit high dielectric constant but high loss (due to near-percolation effects) at low filler fractions. In the presence of an external electric field of 15 kV/mm and 0.5 filler volume fraction, high aspect ratio Pb[Zr$_x$Ti$_{1-x}$]O$_3$ ($0 \leq x \leq 1$) nanowires showed an increase in energy density of 78% over low aspect ratio Pb[Zr$_x$Ti$_{1-x}$]O$_3$ ($0 \leq x \leq 1$) nanorods.[1] Such effects have lead to extensive modeling investigations of the influence of particle shape on dielectric permittivity.[81–84] In one study, the strong mutual interaction between the edges and corners of symmetric angular shapes (such as cubes and square cylinders) has been proposed to result in very low polarization per unit volume.[83] This interaction was not significant between circular cylinders and spheres and hence these tended to have relatively high dipole moments, leading to high dielectric constant.

The dielectric constant is also dependent on the spatial arrangement and any directional alignment of the fillers. Recent computational approaches to analyze these dependences have produced results typified in Figures 17.7 and 17.8[85] where the relationship

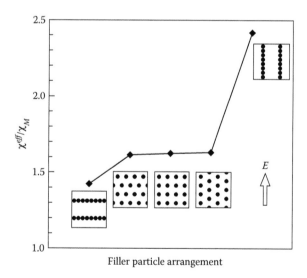

FIGURE 17.7 Effect of filler particle microstructural arrangement on effective composite susceptibility in composites composed of 0.2 volume fraction of circular fillers with dielectric susceptibility $\chi_F/\chi_M = 10$. Insets illustrate the corresponding particle arrangements. Arrow indicates electric field E direction. χ_{eff} is the dielectric susceptibility component in the E direction, χ_F and χ_M are the dielectric susceptibility of filler and matrix, respectively. (Reprinted with permission from Wang, Y.U. and Tan, D.Q., *J. Appl. Phys.*, 109, 104102. Copyright 2011, American Institute of Physics.)

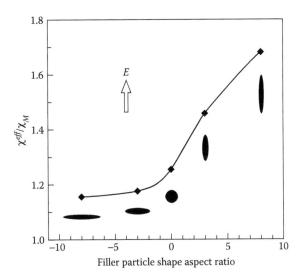

FIGURE 17.8 Effect of filler particle shape and orientation on effective composite susceptibility in composites composed of 0.087 volume fraction of fillers with dielectric susceptibility $\chi_F/\chi_M = 10$. Insets illustrate the corresponding particle shapes and orientations. Shape aspect ratio is defined as $R_{long}/R_{short} -1$, where R_{long} and R_{short} are the radii along long and short axes of the particle, respectively. The sign indicates particle orientation with long axis parallel (+) or perpendicular (–) to electric field E direction as indicated by arrow. χ_{eff} is the susceptibility component in E direction. (Reprinted with permission from Wang, Y.U. and Tan, D.Q., *J. Appl. Phys.*, 109, 104102. Copyright 2011, American Institute of Physics.)

between filler microstructure and effective properties of dielectric composites were studied using a phase field dielectric composite model.[85,86] The results are presented in terms of dielectric susceptibility instead of dielectric constant for convenience, where dielectric constant k is related to dielectric susceptibility χ by $k = 1 + \chi$. From Figures 17.7 and 17.8, composites with closely arranged filler particles and high aspect ratio particles aligned (long axis) in the direction of applied field exhibited the highest dielectric constant. However, predictions from this type of theoretical model and simulation should be compared with experimental data, but this data is often lacking due to the absence of manufacturing processes that can produce such tightly controlled microstructures with low dispersion.

17.4.4 The Interface in Dielectric Composites

The interface between the polymer and the filler might sometimes usefully be considered as a separate phase—an "interphase"—especially where a reaction between filler and matrix, or of the filler, has occurred. Similarly, an interphase might be expected where significant surface modification of the filler has been used as previously described. The idea of this interphase has been developed to try and help understand the dielectric behavior of these types of composite systems. For example, the interphase power law model[87] predicts the effective dielectric constant of a composite by taking into consideration the dielectric constant and volume fraction of the interphase. This model suggested that deviations from standard mixture models in experiments may be due to the interphase, such as enhanced molecular polarizability at the interface due to the formation of interfacial dipoles.[88]

Interfacial polarization, often referred to as Maxwell–Wagner–Sillars polarization[89,90] takes place due to charges[91] that move and pile up at the interfaces in heterogeneous materials. Charge accumulation and separation might also take place at the interfaces due to "trapping" of charges that originate by thermionic emission from metallic fillers/electrode and by quantum mechanical tunneling of charge carriers above an applied threshold electrical field.[92] Defect-rich interfaces have been found to enhance charge trapping around metallic core–oxide shell particles.[93]

These interfacial effects can be expected to become more significant in nanocomposites than composite systems with micron-sized fillers but this is yet to be demonstrated explicitly, and the field is still somewhat in its infancy, mainly due to problems in quantified, nanoscale characterization of the interfaces themselves.

Free charge build up at the interface between the dielectric material and the electrode (current collector) has been shown to increase capacitance and dissipation factor in a capacitor.[94] Enhanced dielectric constant and dissipation factor have also been observed in epoxy–silica composites due to interfacial polarization mechanisms and ionic conductivity by contaminants.[12,95] Water is one such contaminant that fills voids along interfaces and sometimes forms a "water shell" around nanoparticles.[96] When nanoparticles are in close proximity, these water shells overlap and allow charge-carrier movement in between them.[96] Since water is a polar molecule, the presence of water in nanocomposites increases both dielectric constant and dielectric loss.

17.5 PROCESSING TECHNIQUES

17.5.1 Vacuum Deposition

Vacuum deposition is a well-known and widely used technique to produce economically polymer films and monolithic layers of metals. Vacuum deposition can be readily carried out on the laboratory scale, as well as in industrial scale roll-to-roll coaters.[8,97–105] Thermal vacuum deposition involves evaporation of the material (by heating) and its subsequent deposition (condensation) on a cold substrate in vacuum. In the case of preparing PNCs, the polymer and the minority phase (e.g., a metal) are simultaneously heated to form two vapors, as shown at the lab scale in Figure 17.9. Mixing of vapors takes place and the intention is that the film condenses as a composite. Polymers such as PVDF, poly(p-phenylene), poly(methyl methacrylate), Teflon AF 1600, and nylon-6[8,97,98] have been studied previously for their applicability to vacuum deposition. Coevaporation of two or more monomers[106] has also been studied to obtain more complex copolymer films.

Low temperatures in the chamber during evaporation generally facilitate a low deposition rate, and under such conditions, vaporized metal atoms and small metal clusters have greater probability of encountering each other to form larger nanoparticles before deposition.[107] Low temperature is also not favorable for random diffusion (and possible agglomeration) of nanoparticles in the polymer.[108]

While fabricating a composite dielectric film by thermal evaporation, if the rate of evaporation of the majority polymer is increased too much, then there is a tendency to

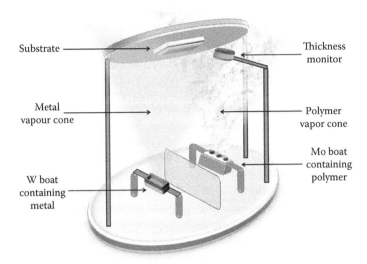

FIGURE 17.9   Schematic diagram of the vacuum codeposition process for PNC films.

"spitting" of polymer, leading to the formation of nonuniform films.[109] During evaporation, the scission of polymer chain can occur randomly or can follow an orderly depolymerization mechanism depending on the nature of the polymer. Increasing the temperature of polymer evaporation has shown to yield a wide array of degradation products (with higher mass).[110]

In electron beam-assisted physical vapor deposition, the target (anode) is bombarded with an accelerated electron beam under high vacuum. The electron beam (produced by a charged W filament) causes evaporation of the metal/polymer from the target, which then deposit on the cool substrate. The mechanism of polymer evaporation by accelerated electrons is essentially thermal in nature, whereas in ion beam evaporation of polymers, the mechanism can be thermal or sputtering.[110]

The main disadvantages of vacuum deposition processes are (1) line of sight deposition; (2) control over the relative deposition rates of two or more materials is difficult; and (3) degradation of polymers can lead to products of unknown chemistry, making characterization and the corelating properties to chemistry of deposited films difficult.

17.5.2 Spraying

In spraying,[76] a high accuracy controlled flow rate syringe pump delivers a suspension (containing polymer and/or monomer) to an atomizing device, such as a hypodermic needle that is placed at the end of an atomizing (spray) nozzle. The suspension is atomized continuously into droplets with diameters of less than 0.1 mm and sprayed onto a heated substrate. Films of required thickness can be produced by programmed repetitive spray cycles. The monomer can be polymerized (e.g., thermally) later to produce a polymer composite film. For example, well-dispersed $BaTiO_3$ nanoparticles embedded in a polydiacrylate matrix with dielectric constant of ~39 at 10 kHz have been produced.[76] Maintaining well-dispersed particles in the suspension is a typical challenge with suspension-based processing, and this is usually addressed by using surfactants and sonication.

17.5.3 Spin Coating

Spin coating involves rotating (spinning) a substrate containing a polymer solution. Due to spinning, the polymer solution flows across the substrate under the centrifugal force and later, the solvent evaporates leaving behind a polymer film.[111] Fillers can be added to the polymer solution to produce composite films. This technique has been employed to produce dielectric composite films[7,20,112] but is generally applied successfully only at a small scale.

17.5.4 Other Techniques

Many other processes to fabricate PNCs such as atomic layer deposition, sol–gel routes, and intercalation have also been investigated. Processes such as extrusion coating and tape casting can be employed to produce high volume and cost-effective composite films, but the thickness of these films is generally always above a few microns, and are not suitable for production of the thin polymer composite films with high capacitances generally preferred for embedded capacitors in printed circuit boards.

17.6 CONCLUSIONS

PNCs provide many attractive attributes for use in the thin film high-energy density capacitors, and there are several routes by which an enhanced capacity over polymer-only materials can be achieved.

Composites containing conductive fillers offer the highest absolute dielectric constants of ~10^7, but controlling the critical filler volume fraction in large-scale production of capacitors is a major challenge and these capacitors are by definition always lossy.

Moderately high-k PNCs have been realized by dispersing ceramics such as $BaTiO_3$ particles in a polymer film to produce PNCs with $k \sim 15$ at 10 kHz and $\upsilon_f \sim 0.25$. Higher k values are difficult to achieve or not always desirable because higher volume fractions of filler are hard to process and the relatively high density of the filler undermines specific energy/power density, as well as breakdown strength. The dielectric properties of such ceramic polymer-based composites generally fall in the regime of EMT, and can be best fitted to a modified rule of mixtures.

Figure 17.10 is a graphical representation of the properties of (1) polymers, (2) ceramics, (3) composites that broadly obey EMT, and (4) composites that do not obey an EMT. The subscripts p and c represent the "parent" phases—polymer and ceramic, respectively, and the size of the circles approximately represents the magnitude of the denoted variable. Although a composite represented in Figure 17.10c inherits the attractive properties of processability into thin films (small d) with large area (high A) and high electric breakdown strength (high E_d) of the polymer, the effective enhancement in the dielectric constant is restricted even when high-k ceramics such as $BaTiO_3$ are used because the ceramic is always a minority phase (especially if breakdown is to be avoided).

Relatively low-loss PNCs with "super normal" k have been fabricated by incorporating surface-modified fillers into a polymer matrix. The high k observed at low frequencies (<1 kHz) has been ascribed to the motion and accumulation of charges in and around the surface-modified fillers, which can include core–shell particles. Although enhancing the polarizability of interfaces in PNCs is a promising approach to obtain the properties envisaged in

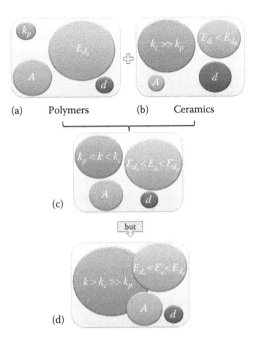

FIGURE 17.10 A graphical representation of the properties of (a) polymers-easily processed into thin films (low d) with large area (high A), high breakdown strength E_{d_p} (~10^5 kV/m), but low k_p (~5); (b) ceramics-hard to process into thin films with large area, low breakdown strength, but some ceramics have k_c ~10^4. (c) composites that obey an effective medium theory (EMT) consisting of ceramic particles embedded in a polymer matrix and (d) composites that do not obey an EMT. The subscripts p and c represent the "parent" phases—polymer and ceramic respectively, and the size of the circles approximately represents the relative magnitude of each variable. Some composites have $k > k_c \gg k_p$.

Figure 17.10d, this approach demands critical control over the structure and chemistry of the interfaces at a nanoscale, stressing the importance of the choice of the processing technique. In some cases, the shell can be described as a third phase, or interphase in the composite, and there is significant research effort on characterizing the features of this interphase to gain the maximum polarizability, while retaining all the advantages of a largely polymeric film.

Vacuum deposition and spraying, in principle, offer good control over the operational parameters and potential for cost-effective scale-up. Most of the other current processing techniques, such as spin coating, face problems related to scalability when large areas of continuous films have to be produced (e.g., continuous composite films for wound electrostatic capacitors).

REFERENCES

1. Tang, H., Lin, Y., Andrews, C., and Sodano, H. A. Nanocomposites with increased energy density through high aspect ratio PZT nanowires. *Nanotechnology* 22(1), 015702 (2011).
2. Waser, R., editor. *Nanoelectronics and Information Technology*, 3rd edition. Wiley, Weinheim, Germany (2012).
3. Wang, N., Lu, F. C., and Li, H. M. Analytical processing of on-line monitored dissipation factor based on morphological filter. *IEEE Transactions on Dielectrics and Electrical Insulation* **11**(5, SI), 840 (2004).

4. Mo, T.-C., Wang, H.-W., Chen, S.-Y., and Yeh, Y.-C. Synthesis and dielectric properties of poly-aniline/titanium dioxide nanocomposites. *Ceramics International* **34**(7), 1767 (2008).

5. Fujisaki, S., Ishiwara, H., and Fujisaki, Y. Low-voltage operation of ferroelectric poly(vinylidene fluoride-trifluoroethylene) copolymer capacitors and metal-ferroelectricinsulator-semiconductor diodes. *Applied Physics Letters* **90**(16), 162902 (2007).

6. Zhou, Y., Wang, H., Xiang, F., Zhang, H., Yu, K., and Chen, L. A poly(vinylidene fluoride) composite with added self-passivated microaluminum and nanoaluminum particles for enhanced thermal conductivity. *Applied Physics Letters* **98**(18) (2011).

7. Xu, J. W. and Wong, C. P. Effect of the polymer matrices on the dielectric behavior of a percolative high-k polymer composite for embedded capacitor applications. *Journal of Electronic Materials* **35**(5), 1087 (2006).

8. Takele, H., Greve, H., Pochstein, C., Zaporojtchenko, V., and Faupel, F. Plasmonic properties of Ag nanoclusters in various polymer matrices. *Nanotechnology* **17**(14), 3499 (2006).

9. Barber, P., Balasubramanian, S., Anguchamy, Y., Gong, S., Wibowo, A., Gao, H., Ploehn, H., and Loye, H. Polymer Composite and Nanocomposite Dielectric Materials for Pulse Power Energy Storage. *Materials* **2,** 1697 (2009).

10. Mahadevegowda, A., Young, N. P., and Grant, P. S. Electron microscopy of multi-layered polymer-nanocomposite based dielectrics. *Journal of Physics: Conference Series* **522,** 012041 (2014).

11. Ma, D. L., Hugener, T. A., Siegel, R. W., Christerson, A., Martensson, E., Onneby, C., and Schadler, L. S. Influence of nanoparticle surface modification on the electrical behaviour of polyethylene nanocomposites. *Nanotechnology* **16**(6), 724 (2005).

12. Sun, Y. Y., Zhang, Z. Q., and Wong, C. P. Influence of interphase and moisture on the dielectric spectroscopy of epoxy/silica composites. *Polymer* **46**(7), 2297 (2005).

13. Xu, J. W., Moon, K. S., Tison, C., and Wong, C. P. A novel aluminum-filled composite dielectric for embedded passive applications. *IEEE Transactions on Advanced Packaging* **29**(2), 295 (2006).

14. Lu, J. and Wong, C. P. Polymer nanocomposites with high dielectric strength and high frequency performance for embedded passive applications. *Proceedings of the IEEE Conference on Polymers and Adhesives in Microelectronics and Photonics and 2008 IEEE Interdisciplinary Conference on Portable Information Devices,* 171. Garmisch Partenkirchen, Germany, August 17–20, 2008.

15. Xu, J. and Wong, C. P. Characterization and properties of an organic-inorganic dielectric nano-composite for embedded decoupling capacitor applications. *Composites Part A-Applied Science and Manufacturing* **38**(1), 13 (2007).

16. Lewis, T. J. Interfaces are the dominant feature of dielectrics at the nanometric level. *IEEE Transactions on Dielectrics and Electrical Insulation* **11**(5, SI), 739 (2004).

17. Montanari, G. C., Fabiani, D., and Palmieri, F. Modification of electrical properties and performance of EVA and PP insulation through nanostructure by organophilic silicates. *IEEE Transactions on Dielectrics and Electrical Insulation* **11**(5, SI), 754 (2004).

18. Li, J., Claude, J., Norena-Franco, L. E., Il Seok, S., and Wang, Q. Electrical energy storage in ferroelectric polymer nanocomposites containing surface-functionalized BaTiO(3) nanoparticles. *Chemistry of Materials* **20**(20), 6304 (2008).

19. Mahadevegowda, A., Young, N. P., and Grant, P. S. Core–shell nanoparticles and enhanced polarization in polymer based nanocomposite dielectrics. *Nanotechnology* **25**(47), 475706 (2014).

20. Kilaru, M. K., Heikenfeld, J., Lin, G., and Mark, J. E. Strong charge trapping and bistable electrowetting on nanocomposite fluoropolymer : BaTiO3 dielectrics. *Applied Physics Letters* **90**(21) (2007).

21. Wu, Y. G., Zhao, X. H., Li, F., and Fan, Z. G. Evaluation of mixing rules for dielectric constants of composite dielectrics by MC-FEM calculation on 3D cubic lattice. *Journal of Electroceramics* **11**(3), 227 (2003).

22. Brosseau, C. Modelling and simulation of dielectric heterostructures: A physical survey from an historical perspective. *Journal of Physics D-Applied Physics* **39**(7), 1277 (2006). *Conference on Modelling Simulation and Design of Dielectrics*. Homerton College, Cambridge, April 2005.
23. Myroshnychenko, V. and Brosseau, C. Finite-element modeling method for the prediction of the complex effective permittivity of two-phase random statistically isotropic heterostructures. *Journal of Applied Physics* **97**(4), 044101 (2005).
24. Rao, Y., Qu, J. M., Marinis, T., and Wong, C. P. A precise numerical prediction of effective dielectric constant for polymer-ceramic composite based on effective-medium theory. *IEEE Transactions on Components and Packaging Technologies* **23**(4), 680 (2000).
25. Ying, K.-L. and Hsieh, T.-E. Sintering behaviors and dielectric properties of nanocrystalline barium titanate. *Materials Science and Engineering B-Solid State Materials for Advanced Technology* **138**(3), 241 (2007).
26. Yoon, D. H., Zhang, J. P., and Lee, B. I. Dielectric constant and mixing model of BaTiO3 composite thick films. *Materials Research Bulletin* **38**(5), 765 (2003).
27. Yamada, T., Ueda, T., and Kitayama, T. Piezoelectricity of a high-content lead zirconate titanate polymer composite. *Journal of Applied Physics* **53**(6), 4328 (1982).
28. Stauffe, D. and Aharony, A. *Introduction to Percolation Theory*. Taylor & Francis Group, London (1994).
29. Pecharroman, C. and Moya, J. S. Experimental evidence of a giant capacitance in insulatorconductor composites at the percolation threshold. *Advanced Materials* **12**(4), 294 (2000).
30. Rao, Y. and Wong, C. P. Material characterization of a high-dielectric-constant polymerceramic composite for embedded capacitor for RF applications. *Journal of Applied Polymer Science* **92**(4), 2228 (2004).
31. Thomas, P., Satapathy, S., Dwarakanath, K., and Varma, K. B. R. Dielectric properties of poly(vinylidene fluoride)/CaCu(3)Ti(4)O(12) nanocrystal composite thick films. *Express Polymer Letters* **4**(10), 632 (2010).
32. Paik, K. W., Cho, S. D., Hyun, J. G., Lee, S., Kim, H., and Kim, J. Epoxy/BaTiO(3) (SrTiO(3)) composite films for high dielectric constant and low tolerance embedded capacitors fabrication in organic substrates. *Proceedings of the 10th International Symposium on Advanced Packaging Materials: Processes, Properties and Interfaces*, 227. Irvine, CA, March 16–18, 2005.
33. George, S., Anjana, P. S., Sebastian, M. T., Krupka, J., Uma, S., and Philip, J. Dielectric, mechanical, and thermal properties of low-permittivity polymer-ceramic composites for microelectronic applications. *International Journal of Applied Ceramic Technology* **7**(4), 461 (2010).
34. Li, J., Seok, S. I., Chu, B., Dogan, F., Zhang, Q., and Wang, Q. Nanocomposites of ferroelectric polymers with TiO(2) nanoparticles exhibiting significantly enhanced electrical energy density. *Advanced Materials* **21**(2), 217 (2009).
35. Ogitani, S., Bidstrup-Allen, S. A., and Kohl, P. A. Factors influencing the permittivity of polymer/ceramic composites for embedded capacitors. *IEEE Transactions on Advanced Packaging* **23**(2), 313 (2000).
36. Cho, S. D. and Paik, K. W. Relationships between suspension formulations and the properties of BaTiO3/epoxy composite films for integral capacitors. *Proceedings of the 51st Electronic Components & Technology Conference*, 1418. Orlando, FL, May 29–June 01, 2001.
37. He, F., Lau, S., Chan, H. L., and Fan, J. High dielectric permittivity and low percolation threshold in nanocomposites based on poly(vinylidene fluoride) and exfoliated graphite nanoplates. *Advanced Materials* **21**(6), 710 (2009).
38. Huang, X., Jiang, P., and Xie, L. Ferroelectric polymer/silver nanocomposites with high dielectric constant and high thermal conductivity. *Applied Physics Letters* **95**(24), 242901 (2009).
39. Dang, Z. M., Lin, Y. H., and Nan, C. W. Novel ferroelectric polymer composites with high dielectric constants. *Advanced Materials* **15**(19), 1625 (2003).
40. Huang, C., Zhang, Q. M., and Su, J. High-dielectric-constant all-polymer percolative composites. *Applied Physics Letters* **82**(20), 3502–3504 (2003).

41. Song, Y., Noh, T. W., Lee, S., and Gaines, J. R. Experimental study of the three-dimensional ac conductivity and dielectric constant of a conductor-insulator composite near the percolation threshold. *Physical Review B* **33**(2), 904 (1986).

42. Grannan, D. M., Garland, J. C., and Tanner, D. B. Critical behaviour of the dielectric constant of a random composite near the percolation threshold. *Physical Review Letters* **46**(5), 375 (1981).

43. Lu, J. X., Moon, K. S., Xu, J. W., and Wong, C. P. Synthesis and dielectric properties of novel high-K polymer composites containing in-situ formed silver nanoparticles for embedded capacitor applications. *Journal of Materials Chemistry* **16**(16), 1543 (2006).

44. Shen, Y., Lin, Y., Li, M., and Nan, C.-W. High dielectric performance of polymer composite films induced by a percolating interparticle barrier layer. *Advanced Materials* **19**(10), 1418 (2007).

45. Zhao, X., Koos, A. A., Chu, B. T. T., Johnston, C., Grobert, N., and Grant, P. S. Spray deposited fluoropolymer/multi-walled carbon nanotube composite films with high dielectric permittivity at low percolation threshold. *Carbon* **47**(3), 561 (2009).

46. Wei, T., Jin, C. Q., Zhong, W., and Liu, J. M. High permittivity polymer embedded with Co/ZnO core/shell nanoparticles modified by organophosphorus acid. *Applied Physics Letters* **91**(22), 222907 (2007).

47. Xu, J. W. and Wong, C. P. Low-loss percolative dielectric composite. *Applied Physics Letters* **87**(8), 082907 (2005).

48. Xu, J. W. and Wong, C. P. Development of a novel aluminum-filled composite for embedded passive applications. *Proceedings of the 53rd Electronic Components & Technology Conference*, 173. New Orleans, LA, May 27–30, 2003.

49. Emtage, P. R. and Tantraporn, W. Schottky emission through thin insulating films. *Physical Review Letters* **8**(7), 267 (1962).

50. Hartman, T. E. and Chivian, J. S. Electron tunneling through thin aluminum oxide films. *Physical Review A: General Physics* **134**(4A), 1094 (1964).

51. Bond, C. D., Guenzer, C. S., and Carosella, C. A. Intermodulation generation by electrontunneling through aluminum-oxide films. *Proceedings of the IEEE* **67**(12), 1643 (1979).

52. Paulus, G. G., Zacher, F., Walther, H., Lohr, A., Becker, W., and Kleber, M. Above-threshold ionization by an elliptically polarized field: Quantum tunneling interferences and classical dodging. *Physical Review Letters* **80**, 484 (1998).

53. Ruseckas, J. Possibility of tunneling time determination. *Physical Review A* **63**, 052107 (2001).

54. Uiberacker, M., Uphues, T., Schultze, M., Verhoef, A. J., Yakovlev, V., Kling, M. F., Rauschenberger, J. et al. Attosecond real-time observation of electron tunnelling in atoms. *Nature* **446**(7136), 627 (2007).

55. MacColl, L. A. Note on the transmission and reflection of wave packets by potential barriers. *Physical Review* **40**, 621 (1932).

56. Winful, H. G. The meaning of group delay in barrier tunnelling: A re-examination of superluminal group velocities. *New Journal of Physics* **8**, 101 (2006).

57. Lewis, T. J. Interfaces and nanodielectrics are synonymous. *Proceedings of the 8th IEEE International Conference on Solid Dielectrics*, 792. IEEE, France, July 5–9, 2004.

58. Nelson, J. K. and Hu, Y. Nanocomposite dielectrics - properties and implications. *Journal of Physics D: Applied Physics* **38**(2), 213 (2005).

59. Kim, D.-W., Lee, D.-H., Kim, B.-K., Je, H.-J., and Park, J.-G. Direct assembly of BaTA poly(methyl methacrylate) nanocomposite films. *Macromolecular Rapid Communications* **27**(21), 1821 (2006).

60. Devaraju, N. G. and Lee, B. I. Dielectric behavior of three phase polyimide percolative nanocomposites. *Journal of Applied Polymer Science* **99**(6), 3018 (2006).

61. Qi, L., Lee, B. I., Samuels, W. D., Exarhos, G. J., and Parler, Jr., S. G. Three-phase percolative silver-BaTiO3-epoxy nanocomposites with high dielectric constants. *Journal of Applied Polymer Science* **102**(2), 967 (2006).

62. Li, L., Takahashi, A., Hao, J. J., Kikuchi, R., Hayakawa, T., Tsurumi, T. A., and Kakimoto, M. A. Novel polymer-ceramic nanocomposite based on new concepts for embedded capacitor application (I). *IEEE Transactions on Components and Packaging Technologies* **28**(4), 754 (2005).

63. Dang, Z. M., Fan, L. Z., Shen, Y., and Nan, C. W. Study on dielectric behavior of a three-phase CF/(PVDF+BaTiO3) composite. *Chemical Physics Letters* **369**(1–2), 95 (2003).

64. Yang, C., Lin, Y., and Nan, C. W. Modified carbon nanotube composites with high dielectric constant, low dielectric loss and large energy density. *Carbon* **47**(4), 1096 (2009).

65. Kim, P., Jones, S. C., Hotchkiss, P. J., Haddock, J. N., Kippelen, B., Marder, S. R., and Perry, J. W. Phosphonic acid-modiried barium titanate polymer nanocomposites with high permittivity and dielectric strength. *Advanced Materials* **19**(7), 1001 (2007).

66. Xie, L., Huang, X., Wu, C., and Jiang, P. Core-shell structured poly(methyl methacrylate)/BaTiO(3) nanocomposites prepared by in situ atom transfer radical polymerization:A route to high dielectric constant materials with the inherent low loss of the base polymer. *Journal of Materials Chemistry* **21**(16), 5897 (2011).

67. Turky, G. M., Ghoneim, A. M., Kyritsis, A., Raftopoulos, K., and Moussa, M. A. Dielectric dynamics of some nylon 6/CaCo3 composites using broadband dielectric spectroscopy. *Journal of Applied Polymer Science* **122**(3), 2039 (2011).

68. Benhadjala, W., Bord-Majek, I., Bechou, L., Suhir, E., Buet, M., Rouge, F., Gaud, V., Plano, B., and Ousten, Y. Improved performances of polymer-based dielectric by using inorganic/organic core-shell nanoparticles. *Applied Physics Letters* **101**(14), 142901 (2012).

69. Callister W. D., Jr. *Materials Science and Engineering—An Introduction.* John Wiley & Sons, New York (2002).

70. Dissado, L. A. and Fothergill, J. C. *Electrical Degradation and Breakdown in Polymers.* Peter Peregrinus, London (1992).

71. Ohki, Y, Fuse, N., and Arai, T. Band gap energies and localized states in several insulating polymers estimated by optical measurements. *Annual Report Conference on Electrical Insulation and Dielectric Phenomena*, 1, West Lafayette, IN, (2010).

72. Sreelatha, K. and Predeep, P. Iodine doped, semi-conducting nylon 6 polymers. *Journal of Plastic Film & Sheeting* **29**(2), 127 (2013).

73. Artbauer, J. Electric strength of polymers. *Journal of Physics D: Applied Physics* **29**(2), 446 (1996).

74. Job, A. E., Alves, N., Zanin, M., and Ueki, M. M. Increasing the dielectric breakdown strength of poly(ethylene terephthalate) films using a coated polyaniline layer. *Journal of Physics D: Applied Physics* **36**(12), 1414 (2003).

75. Zakrevskii, V. A., Sudar, N. T., Zaopo, A., and Dubitsky, Y. A. Mechanism of electrical degradation and breakdown of insulating polymers. *Journal of Applied Physics* **93**(4), 2135 (2003).

76. Zhao, X., Hinchliffe, C., Johnston, C., Dobson, P. J., and Grant, P. S. Spray deposition of polymer nanocomposite films for dielectric applications. *Materials Science and Engineering B: Advanced Functional Solid-State Materials* **151**(2), 140 (2008).

77. Hong, J. I., Winberg, P., Schadler, L. S., and Siegel, R. W. Dielectric properties of zinc oxide/low density polyethylene nanocomposites. *Materials Letters* **59**(4), 473 (2005).

78. Zhao, X. *Spray Deposition of Nanostructured Materials for Energy Storage.* PhD thesis, University of Oxford, Oxford (2009).

79. Hakim, L. F., Portman, J. L., Casper, M. D., and Weimer, A. W. Aggregation behavior of nanoparticles in fluidized beds. *Powder Technology* **160**(3), 149 (2005).

80. Rasul, A., Zhang, J., Gamota, D., Singh, M., and Takoudis, C. Flexible high capacitance nanocomposite gate insulator for printed organic field-effect transistors. *Thin Solid Films* **518**(23), 7024 (2010).

81. Jones, S. B. and Friedman, S. P. Particle shape effects on the effective permittivity of anisotropic or isotropic media consisting of aligned or randomly oriented ellipsoidal particles. *Water Resources Research* **36**(10), 2821 (2000).

82. Xu, P. and Li, Z. Y. Effect of particle shape on the effective dielectric response of nanocomposite close to the percolation threshold. *Physica B: Condensed Matter* **348**(1–4), 101 (2004).

83. Whites, K. W. and Wu, F. Effects of particle shape on the effective permittivity of composite materials with measurements for lattices of cubes. *IEEE Transactions on Microwave Theory and Techniques* **50**(7), 1723 (2002).

84. Sihvola, A. Dielectric polarization and particle shape effects. *Journal of Nanomaterials* **2007**(1), 45090 (2007).

85. Wang, Y. U. and Tan, D. Q. Computational study of filler microstructure and effective property relations in dielectric composites. *Journal of Applied Physics* **109**(10, SI), 104102 (2011).

86. Wang, Y. U. Phase field model of dielectric and magnetic composites. *Applied Physics Letters* **96**(23), 232901 (2010).

87. Todd, M. G. and Shi, F. G. Complex permittivity of composite systems: A comprehensive inter-phase approach. *IEEE Transactions on Dielectrics and Electrical Insulation* **12**(3), 601 (2005).

88. Murugaraj, P., Mainwaring, D., and Mora-Huertas, N. Dielectric enhancement in polymer-nanoparticle composites through interphase polarizability. *Journal of Applied Physics* **98**(5), 054304 (2005).

89. Dang, Z.-M., You, S.-S., Zha, J.-W., Song, H.-T., and Li, S.-T. Effect of shell-layer thickness on dielectric properties in Ag@TiO2 core@shell nanoparticles filled ferroelectric poly(vinylidene fluoride) composites. *Physica Status Solidi A—Applications and Materials Science* **207**(3, SI), 739 (2010).

90. Rahimabady, M., Mirshekarloo, M. S., Yao, K., and Lu, L. Dielectric behaviors and high energy storage density of nanocomposites with core-shell BaTiO3@TiO2 in poly(vinylidene fluoride-hexafluoropropylene). *Physical Chemistry Chemical Physics* **15**(38), 16242 (2013).

91. Taylor, D. M. Space charges and traps in polymer electronics. *IEEE Transactions on Dielectrics and Electrical Insulation* **13**(5), 1063 (2006).

92. Chiguvare, Z. Electric field induced transition from electron-only to hole-only conduction in polymer-fullerene metal-insulator-metal devices. *Journal of Applied Physics* **112**(10), 104508 (2012).

93. Yuan, C. L. and Lee, P. S. Enhanced charge storage capability of Ge/GeO(2) core/shell nano-structure. *Nanotechnology* **19**(35), 355206 (2008).

94. Ray, D. K., Hirnanshu, A. K., and Sinha, T. P. Structural and low frequency dielectric studies of conducting polymer nanocomposites. *Indian Journal of Pure & Applied Physics* **45**(8), 692 (2007).

95. Gonon, P., Sylvestre, A., Teysseyre, J., and Prior, C. Combined effects of humidity and thermal stress on the dielectric properties of epoxy-silica composites. *Materials Science and Engineering B: Solid State Materials for Advanced Technology* **83**(1–3), 158 (2001).

96. Zou, C., Fothergill, J. C., and Rowe, S. W. The effect of water absorption on the dielectric properties of epoxy nanocomposites. *IEEE Transactions on Dielectrics and Electrical Insulation* **15**(1), 106 (2008).

97. Bodhane, S. P. and Shirodkar, V. S. Change in crystallinity of poly vinylidene fluoride) due to thermal evaporation. *Journal of Applied Polymer Science* **64**(2), 225 (1997).

98. Kubono, A. and Okui, N. Polymer thin-films prepared by vapor-deposition. *Progress in Polymer Science* **19**(3), 389 (1994).

99. Affinito, J., Martin, P., Gross, M., Coronado, C., and Greenwell, E. Vacuum deposited polymer metal multilayer films for optical application. *Thin Solid Films* **270**(1–2), 43 (1995).

100. Ludwig, R., Kukla, R., and Josephson, E. Vaccum web coating-state of the art and potential for electronics. *Proceedings of the IEEE* **93**(8), 1483 (2005).

101. Abbas, G., Ding, Z., Mallik, K., Assender, H., and Taylor, D. M. Hysteresis-free vacuumpro-cessed acrylate-pentacene thin-film transistors. *IEEE Electron Device Letters* **34**(2), 268 (2013).

102. Affinito, J. D. Hybridization of the polymer multi-layer (PML) deposition process. *Surface & Coatings Technology* **133**, 528 (2000).

103. Abbas, G., Assender, H., Ibrahim, M., and Taylor, D. M. Hysteresis-free vacuumprocessed acrylate-pentacene thin-film transistors. Organic thin-film transistors with electron-beam cured and flash vacuum deposited polymeric gate dielectric. *Journal of Vacuum Science & Technology B* **29**(5), 052401–1 (2011).

104. Affinito, J. D., Gross, M. E., Coronado, C. A., Graff, G. L., Greenwell, E. N., and Martin, P. M. A new method for fabricating transparent barrier layers. *Thin Solid Films* **290,** 63 (1996).

105. Affinito, J. D., Eufinger, S., Gross, M. E., Graff, G. L., and Martin, P. M. PML/oxide/PML barrier layer performance differences arising from use of UV or electron beam polymerization of the PML layers. *Thin Solid Films* **308,** 19 (1997).

106. Kubono, A., Yuasa, N., Shao, H., Umemoto, S., and Okui, N. In-situ study on alternating vapor deposition polymerization of alkyl polyamide with normal molecular orientation. *Thin Solid Films* **289**(1–2), 107 (1996).

107. Giesfeldt, K. S., Connatser, R. M., De Jesus, M. A., Dutta, P., and Sepaniak, M. J. Goldpolymer nanocomposites: Studies of their optical properties and their potential as SERS substrates. *Journal of Raman Spectroscopy* **36**(12), 1134 (2005).

108. Starke, T. K. H., Johnston, C., and Grant, P. S. Evolution of percolation properties in nanocomposite films during particle clustering. *Scripta Materialia* **56**(5), 425 (2007).

109. Alismail, S. and Hogarth, C. A. Novel sources for metal-loaded polymer-films prepared by vacuum evaporation. *Journal of Physics E: Scientific Instruments* **20**(3), 344 (1987).

110. Gritsenko, K. P. and Krasovsky, A. M. Thin-film deposition of polymers by vacuum degradation. *Chemical Reviews* **103**(9), 3607 (2003).

111. Schubert, D. W. and Dunkel, T. Spin coating from a molecular point of view: Its concentration regimes, influence of molar mass and distribution. *Materials Research Innovations* **7**(5), 314–321 (2003).

112. Lu, J., Moon, K.-S., and Wong, C. P. High-k polymer nanocomposites as gate dielectrics for organic electronics applications. *Proceedings of the 57th Electronic Components & Technology Conference*, 453, Reno, NV, (2007).

V

Bio-Based and Biocompatible Polymer Materials

Synthesis, Characterization, and Adsorption Properties of Gum Polysaccharide-Based Graft Copolymers

Hemant Mittal, Arjun Maity, and Suprakas Sinha Ray

CONTENTS

18.1 INTRODUCTION

Gum polysaccharides consist of different monosaccharide units that are joined together by *O*-glycosidic linkages. Diversity in the structural features of different polysaccharides is because of the difference in the monosaccharide composition, chain shapes, and degree of polymerization and linkage types. Gum polysaccharides are commonly used in food and nonfood industries as stabilizers, thickening and gelling agents, crystallization inhibitors, and encapsulating agents. Therefore, they are also known as hydrocolloids or gums. Gum polysaccharides in nature mainly occur as storage materials, cell wall components, exudates, and extracellular substances from plants or microorganisms. In the past few years, an expanding interest has been devoted to the synthesis of chemically modified polysaccharides through grafting, which combine the advantages of both synthetic and natural polymers. Graft copolymerization of vinyl monomers onto natural polysaccharides is the most promising technique as it functionalizes the biopolymers to their potential by imparting desirable properties onto them (Patel et al., 1999; Pourjavadi et al., 2005). Graft copolymerization can be done by conventional redox grafting methods (Kang et al., 2006), microwave irradiation (Kaith et al., 2012a; Mittal et al., 2015; Sen et al., 2009), γ-irradiation (Kumar et al., 2012a, b), or using electron beams (Vahdat et al., 2007).

Over the last half century, many gum polysaccharides have been investigated by the scientific community for their potential applications in food industries such as emulsifiers, drug delivery devices, stabilizers, and thickeners, and in pharmaceuticals, cosmetics, textiles, and lithography (Krishnaiah and Srinivas, 2008; Prezotti et al., 2014; Singh and Kim, 2007). Natural gums are polysaccharides consisting of multiple sugar units linked together to create large molecules. Gums are frequently produced by higher plants as a result of their protection mechanisms following injury. They are heterogeneous in composition. On hydrolysis, simple sugar units, such as arabinose, galactose, glucose, mannose, xylose, and uronic acids, are obtained (Rana et al., 2011). Gum polysaccharide-based graft copolymers, mainly hydrogel polymers, have been used as adsorbents for the removal of different impurities, such as dyes and heavy metal ions from the aqueous solutions (Deng et al., 2012; Lan et al., 2014; Masoumi and Ghaemy, 2014; Saravanan et al., 2012; Thakur et al., 2014). The main advantage of using gum polysaccharides as adsorbents is their environment-friendly nature, abundant availability, and very low cost (Mittal et al., 2013a, 2014a).

18.2 STRUCTURE AND PROPERTIES OF GUM POLYSACCHARIDES

18.2.1 Xanthan Gum

Xanthan gum is a gram-negative, yellow pigmented bacterium and an anionic polysaccharide obtained from the bacterial coat of *Xanthomonas campestris* and is generally produced by fermentation of glucose, sucrose, or lactose by the *X. campestris* bacterium. The fermented part is precipitated using isopropyl alcohol, dried, converted into fine powder, and added to a liquid medium to form the gum. Xanthan gum has high viscosity even at very low concentrations and is completely soluble in cold as well as hot water.

In the main chain of xanthan gum, pentasaccharide repeating units of D-glucose are joined through the β-1 position of one unit with the fourth position of the next unit similar in the chemical structure of cellulose. In the primary structure of xanthan gum main chain of (1→4)-linked β-D-glucopyranose units is attached with the trisaccharide side chains of alternate sugar residues at the C-3 position, and the side chain is made up of two mannose residues and a glucuronic acid residue. In the terminal portion, β-D-mannopyranose residue is (1→4)-linked β-D-glucuronic acid residue, which in turn is (1→2)-linked to non-terminal α-D-mannopyranose residue (Figure 18.1) (Li and Feke, 2015; Paquet et al., 2014). The 6-OH group of the nonterminal D-mannopyranose residue is present as acetic acid ester, and the pyruvate acetal groups are located on the D-mannopyranosyl end groups of the side chains. Modification of polysaccharide chain confirmations and their characteristics depends on the different glycosidic linkages in the main chain. Pyruvic acid attached to the terminal carbohydrate of the side chains adds another carboxylate group, and the different pyruvate levels of xanthan gum have different rheological properties. The main constitutes of xanthan gum are glucose, mannose, glucoronic acid, acetate, and pyruvate, and they are present in the ratio 37:43.4:19.5:4.5:4.4, respectively (Jindal et al., 2013).

Xanthan gum is mainly used as thickener and stabilizer in the industries because of its high viscosity compared to the other polysaccharides. The viscosity of the xanthan

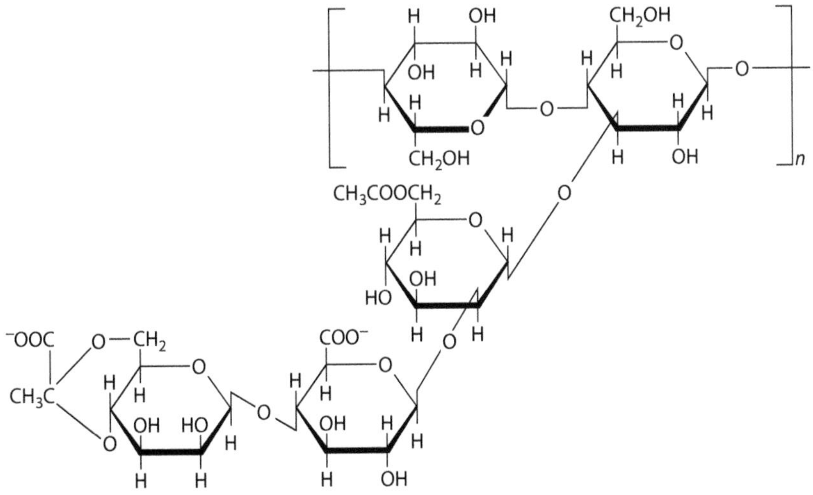

FIGURE 18.1 Structure of xanthan gum. (From Jindal, R. et al., *Polym. Renew. Resour.*, 4, 19–34, 2013. With permission.)

gum solution is independent of the temperature change and is virtually constant at boiling point and freezing point of the pure water. Xanthan gum is completely biodegradable using certain microorganism under controlled conditions, but it is resistant to degradation using different enzymes such as protease, pectinase, and amylase. Due to its extraordinary properties, xanthan gum is used extensively in food and pharmaceutical industries (Katzbauer, 1998; Palaniraj and Jayaraman, 2011).

18.2.2 Gum Ghatti

Gum ghatti, an exudate of the *Anogeissus latifolia* tree belonging to the family Combretaceae, is a complex anionic polysaccharide. Its exact molecular weight is not known. It is also known as Indian gum. It occurs as a gray to reddish-gray powder and granular or light to dark brown lump, and is almost odorless. It has sugars, such as L-arabinose, D-galactose, D-mannose, D-xylose, and D-glucuronic acid in the molar ratio of 48:29:10:5:10 (Aspinall et al., 1955). The structure of gum ghatti has the alternating 4-O-subsituted and 2-O-subsituted α-D-mannopyranose units and chains of (1→6)-linked β-D-galactopyranose units attached to the single L-arabinofuranose residue (Figure 18.2) (Kaith et al., 2010a).

Gum ghatti is exuded naturally in the form of rounded tears of less than 1 cm diameter or in the form of larger vermiform masses. It has a glassy fracture, and the color of the exudates varies from very light to dark brown. It is collected manually in the same way as most of other tree exudate gums (Jefferies et al., 1977; Mittal et al., 2014a).

Gum ghatti powder has been used as a treatment method and an aid for various health conditions. It has been shown to reduce cholesterol levels, to accelerate the healing of wounds, and to build and strengthen the immune system. It is also given to chemotherapy patients to prevent or lessen the side effects of treatment. Gum ghatti powder has been found to reduce and slow down the signs of ageing. It is also being studied for use as a

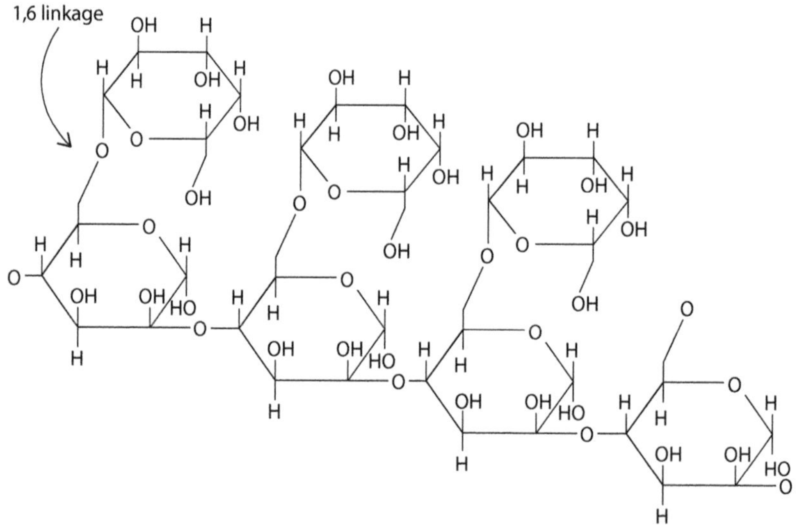

FIGURE 18.2 Structure of gum ghatti. (From Mittal, H. et al., *Carbohydr. Polym.*, 114, 321–329, 2014a. With permission.)

weight loss supplement. This substance also helps to prevent the absorption of sugars in the intestine and has been used to treat diabetes. It is a common ingredient in glyconutrient supplements. Glyconutrients are used to provide the eight essential sugars needed by the body for cell communication. They have many benefits, such as stimulating the immune system and helping to reduce effects of various neurological diseases and autoimmune diseases. There have been no major side effects or drug interactions reported from Gum ghatti powder. However, an excessive use of Gum ghatti powder has been shown to have a laxative effect (Jefferies et al., 1977; Mittal et al., 2014a,b; Mittal and Mishra, 2014).

18.2.3 Gum Arabic

Gum arabic is the dried, gummy exudation obtained from various species of *Acacia* trees of the Leguminosae family. It derived its name from the Arabs who started its trade. About 500 species of *Acacia* are distributed over tropical and subtropical areas of Africa, India, Australia, Central America, and southwest North America, but only a comparatively few are commercially important. Gum arabic is unique among the natural hydrocolloids because of its extremely high solubility in water. Most of the gums cannot be dissolved in water at concentrations higher than about 5%, but gum arabic can yield solutions of upto 50% concentration. Gum arabic is obtained as an exudate from the stems and branches of sub-Saharan (Sahel zone) Acacia senegal (Leguminosae) trees and is produced naturally as large nodules during a process called gummosis to seal wounds in the bark of the tree. It finds a number of applications in confectionery, flavors, brewing, pharmaceuticals, cosmetics, lithography, inks, and textiles (Glicksman and Sand, 1973).

Gum arabic is a highly branched, complex acidic heteropolysaccharide with a main chain of (1→3)-β-D-galactopyranosyl units and side chains containing L-arabinofuranosyl, L-rhamnopyranosyl, D-galactopyranosyl, and D-glucopyranosyl uronic acid units. From a detailed study of its structure, it was found that it consists of mainly three fractions (Fenyo and Vandervelde, 1990; Randall et al., 1988): (1) The major one is a highly branched polysaccharide (microwave [MW] = 3×10^5) consisting of β-(1→3) galactose backbone with linked branches of arabinose and rhamnose, which terminate in glucuronic acid (found in nature as magnesium, potassium, and calcium salt); (2) a smaller fraction (~10 wt.% of the total) is a higher molecular-weight (~1 × 10^6 g/mol) arabinogalactan–protein complex (GAGP-GA glycoprotein) in which arabinogalactan chains are covalently linked to a protein chain through serine and hydroxyproline groups. The attached arabinogalactan in the complex contains ~13% (by mole) glucoronic acid; and (3) the smallest fraction (~1% of the total) having the highest protein content (~50 wt.%) is a glycoprotein, which differs in its amino acid composition from that of the GAGP complex.

Its glycoprotein which is a high-molecular-weight hydroxyproline-rich arabinogalactan containing a repetitive and almost symmetrical 19-residue consensus motif—ser-hyp[a]-hyp[a]-hyp[a]-thr-leu-ser-hyp[b]-ser-hyp[b]-thr-hyp-thr-hyp[a]-hyp[a]-hyp[a]-gly-pro-his- with contiguous hydroxyprolines (hyp[a]) attached to oligo-α-1,3-L-arabinofurans and noncontiguous hydroxyprolines (hyp[b]) attached to galactose residues of oligo-arabinogalactans, combining a β-1,3-D-galactopyran core with rhamnoglucuronoarabinogalactose pentasaccharide side chains joined to the main chain via 1,6-linkages, accounting for its very complex structure

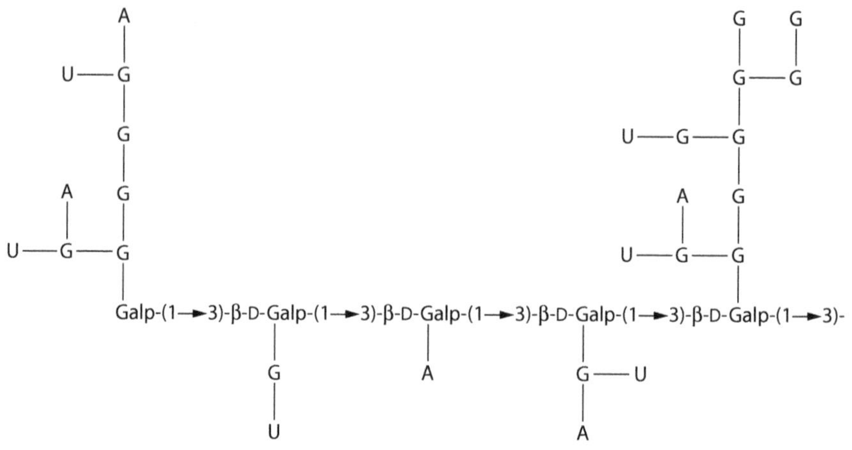

FIGURE 18.3   Structure of guar gum. G = β-D-galactopyranose; A = L-arabinofuranose- or L-arabinopyranose-terminated short chains of (1→3)-linked L-arabinofuranose- or α-D-galactopyranose-(1→3)-L-arabinofuranose; U = α-L-rhamnopyranose-(1→4)-β-D-glucuronic acid or β-D-glucuronic acid (4-OMe). (From Fenyo, J. C. and Vandervelde, M. C., *Gums and Stabilizers for the Food Industry*, Oxford, IRL Press, 1990. With permission.)

(Figure 18.3). As a food additive, it is a useful but expensive hydrocolloid emulsifier, texturizer, and film-former, widely used in the drinks industry to stabilize flavors and essential oils, for example, in soft drink concentrates. The simultaneous presence of hydrophilic carbohydrate and hydrophobic protein enables its emulsification and stabilization properties. Gum arabic is also used in confectionery such as traditional hard (wine) gums and pastilles, and as a foam stabilizer in marshmallows (Phillips et al., 2008).

18.2.4 Guar Gum

Guar gum is extracted from the seed of plant *Cyamopsis tetragonoloba*. It is mainly found in the sandy soils of North India and some parts of Pakistan. It is a nonionic hydrocolloid and is not affected by ionic strength but gets degraded at pH 3.0 and 50°C (Brown and Livesey, 1994).

Guar gum is a galactomannana, and its chemical structure is composed of a (1→4)-linked β-D-mannopyranose backbone with branches of (1→6)-linked α-D-galactopyranose (Figure 18.4). There are between 1.5 and 2 mannose residues for every galactose residue. Guar gum is a water-soluble fiber that acts as a bulk-forming laxative and is claimed to be effective in promoting regular bowel movements and relieve constipation and chronic related functional bowel ailments, such as diverticulosis, Crohn's disease, colitis, and irritable bowel syndrome, among others.

18.2.5 Gum Tragacanth

Gum tragacanth is obtained as a gummy exudate of *Astragalus gummifer* Labillardiere or other asiatic species of *Astragalus* genus. It is a highly branched heterocyclic polysaccharide. It is found in dry dessert and mountainous regions of southwest Asia and mainly in Iran and Turkey (Anderson and Bridgeman, 1985). *Astragalus* genus is a low bushy perennial

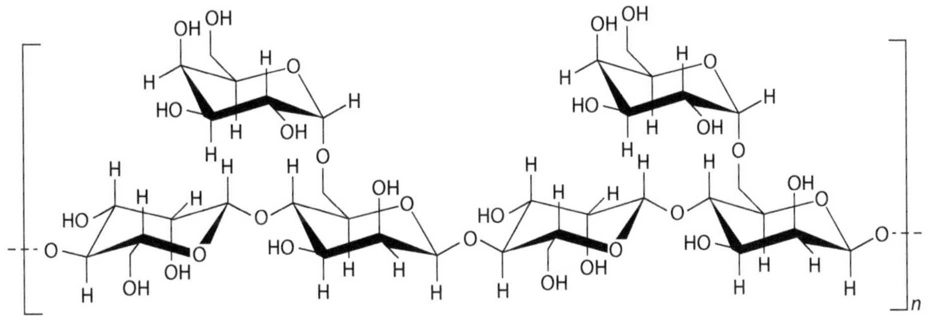

FIGURE 18.4 Structure of guar gum. (From Morrison, N. A. et al., New textures with highly acyl gellan gum, in *Gum and Stabilisers for the Food Industry 11*, Williams, P.A. and Phillips, G.O. (eds.), RSC Special Publication, Cambridge, 2002, pp. 297–305. With permission.)

small shrub and has a large taproot with branches. It consists of a water-soluble component (i.e., tragacanthin) and a water-insoluble component (i.e., tragacanthic acid). When the gum is added to the water, tragacanthin dissolves to form viscous solution, whereas tragacanthic acid swells to give soft and adhesive gel-like state (Anderson and Bridgeman, 1988). Tragacanthin is a highly branched neutral polysaccharide with structure consisting of a main chain of repeated D-galactose residues to which chains of L-arabinose residues are attached and have spheroidal molecular shape, whereas tragacanthic acid consists of a structure with a main chain of repeated unit of (1→4)-linked D-galactopyranosyl with a side chain of substituted xylosyl-linked C-3 position of the glacturonic acid residues (Figure 18.5).

18.2.6 Gum Gellan

Gum gellan is obtained from the microorganism *Aeromonas elodea* as an extracellular bacterial polysaccharide. It is present in both substituted and unsubstituted forms. It is

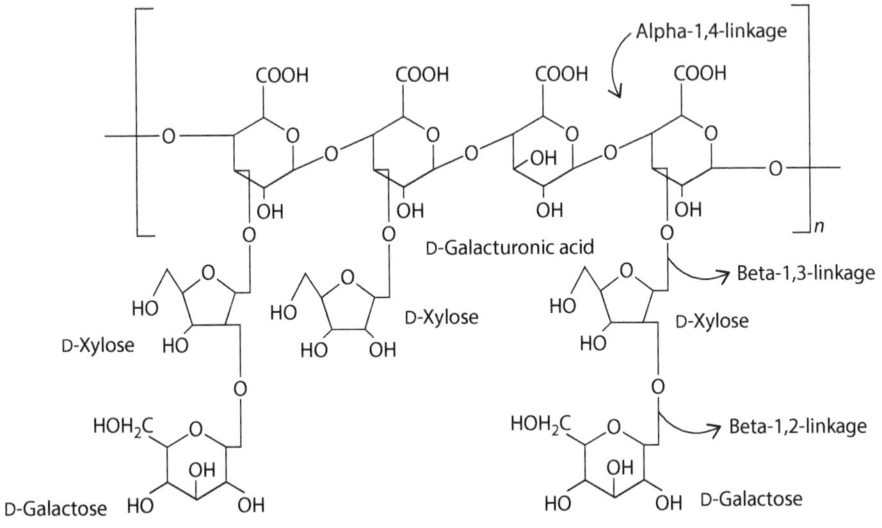

FIGURE 18.5 Structure of tragacanthic acid. (From Sadeghi, S. et al., *Mater. Sci. Eng. C*, 45, 136–145, 2014. With permission.)

produced by inoculating a fermentation medium with *A. elodea*. Fermentation is carried out under controlled conditions, and the product is recovered in a few days. Recovery by alcohol precipitation yields the substituted gum, whereas recovery by alkali produces the unsubstituted gum. When hot solutions of gum gellan are cooled in the presence of gel-promoting cations, they form a gel. The substituted form produces soft, elastic gels, whereas the unsubstituted form produces hard, brittle gels and the gel properties depend on the degree of substitution. The primary structure of gum gellan is composed of a repeating sequence of →3)-β-D-Glcp-(1→4)-β-D-GlcpA-(1→4)-β-D-Glcp-(1→4)-α-L-Rhap-(1→ unit (Figure 18.6). (O'Neill et al., 1983). Gum gellan is used as gelling agents in dessert jellies, dairy products, and sugar confectionery. It is also used to prepare structured lipids.

18.2.7 Locust Bean Gum

Locust bean gum is extracted from the seeds of the carob tree of the *Ceratonia siliqua* family that is found in Mediterranean regions. It is partially soluble in cold water, and its solutions exhibit high viscosity at low concentrations and the gum efficiently binds water. The main chain of locust bean gum consists of (1→4) linked β-D-mannose residues with a side chain of (1→6)-linked α-D-galactose (Figure 18.7). The galactose sugars are not evenly distributed along the chain but tend to be clustered together in blocks. The chains have an irregular structure with alternating "smooth" and substituted zones (Cronin et al., 2002).

Locust bean gum is mainly used in the manufacturing industry as a thickening agent. Common items that contain the powder include salad dressing, syrup, cheese, condiments, and ice cream. In addition to food products, it can also be added to cosmetics, pet food, and even insecticides to prevent the products from becoming too thin or runny during the packaging process (Cronin et al., 2002).

18.2.8 Alginate

Alginate, an anionic natural polysaccharide obtained from the brown seaweed, has been used for many industrial biomedical applications due to its biocompatibility, low toxicity, relatively lower cost, and the mild gelation by the addition of the divalent cations such as Ca^{2+}

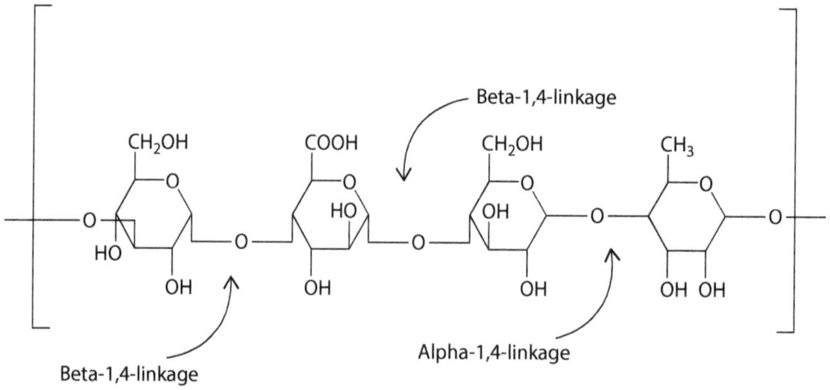

FIGURE 18.6 Structure of gum gellan. (From Sworn, G., Gellan gum, in *Handbook of Hydrocolloids*, Phillips, G. O. and Williams, P. A. (eds.), CRC Press, Boca Raton, FL, 2000, pp. 379–395. With permission.)

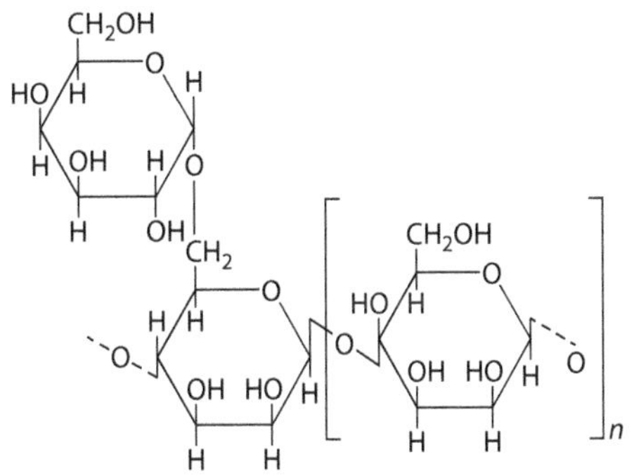

FIGURE 18.7   Structure of locust bean gum. (From Cronin, C. E. et al., Formation of strong gels by enzymic debranching of guar gum in the presence of ordered xanthan, in *Gum and Stabilisers for the Food Industry 11*, Williams, P. A. and Phillips, G. O. (eds.), RSC Special Publication, Cambridge, 2002, pp. 286–296. With permission.)

(Gomboltz et al., 1998). Hydrogels of the alginates can be easily prepared by the cross-linking with some of the cations, and they have potential applications in the wound healing, delivery of drugs and proteins, wastewater treatment, and so on (Agorku et al., 2014). Commercially, alginate is extracted by the alkali treatment of the brown algae, including *Laminaria hyperborea, L. digitata, L. japonica, Ascophyllum nodosum*, and *Macrocystis pyrifera*. After the alkali treatment, the extract filtered with either sodium or calcium chloride is added for the precipitation of alginate salt. Thereafter, the alginate salt is treated with the dilute HCl for its conversion into alginic acid (Rinaudo, 2008). The chemical structure of alginate was extensively studied by Fischer and Dörfel (1955) and was found to have D-mannuronate as the main component. The structure of alginate consists of the block copolymers having the guluronate and mannuronate in different ratios depending on the natural source of the alginate. The alginate consists of the blocks of (1,4)-linked β-D-mannuronate (M) and α-L-guluronate (G) residues. The blocks are having the consecutive G residues (GGGGGG), the consecutive M residues (MMMMMM), and the alternating G and M residues (GMGMGM) (Figure 18.8). Different alginates extracted from the different sources have different M and G contents as well as the length of each block, and more than 200 different alginates exist depending on the length and constitute M and G chains (Fischer and Dörfel, 1955).

18.2.9 Carrageenans

Carrageenan mainly exists in three forms, namely, kappa, iota, and lambda. It is a high-molecular-weight linear polysaccharide having repeating galactose units and sulfated and nonsulfated 3,6-anhydrogalactose (3,6-AG) joined together by the alternating (1,3) and (1,4) glycosidic links. Rees and coworkers (Rees, 1963) distinguished between the different carrageenans on the basis of their chemical structures. Lambda, kappa, and iota, which are considered as the main carrageenans, can be synthesized by the selective extraction techniques by using the mu and nu carrageenans. The alkali treatment of the mu and nu carrageenans forms the kappa and iota carrageenans. The structures of carrageenans are shown in Figure 18.9. Basically, these carrageenans differ in the 3,6-AG and ester sulfate

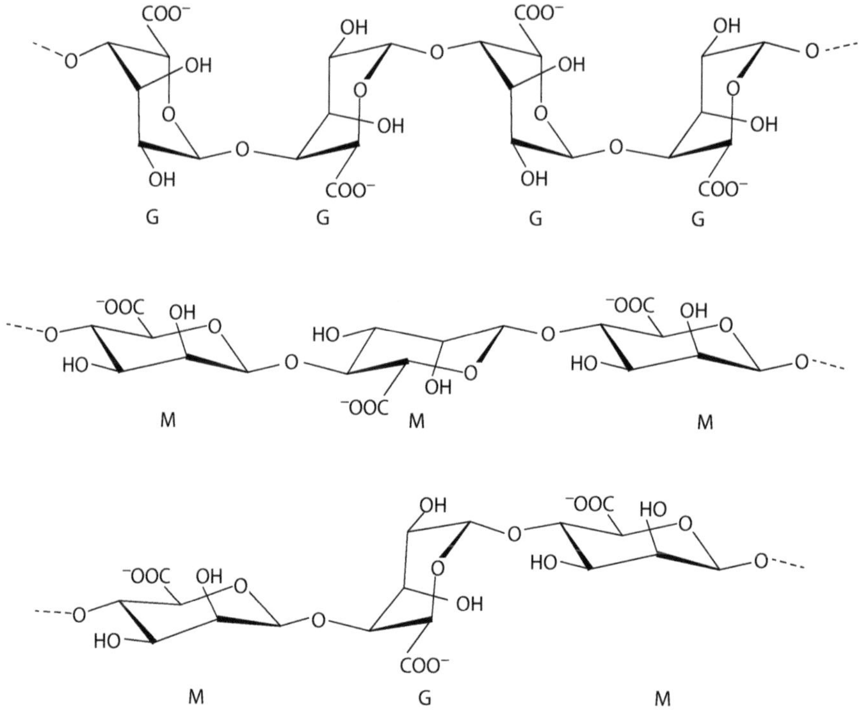

FIGURE 18.8 Chemical structure of G, M, and alternating blocks in alginate.

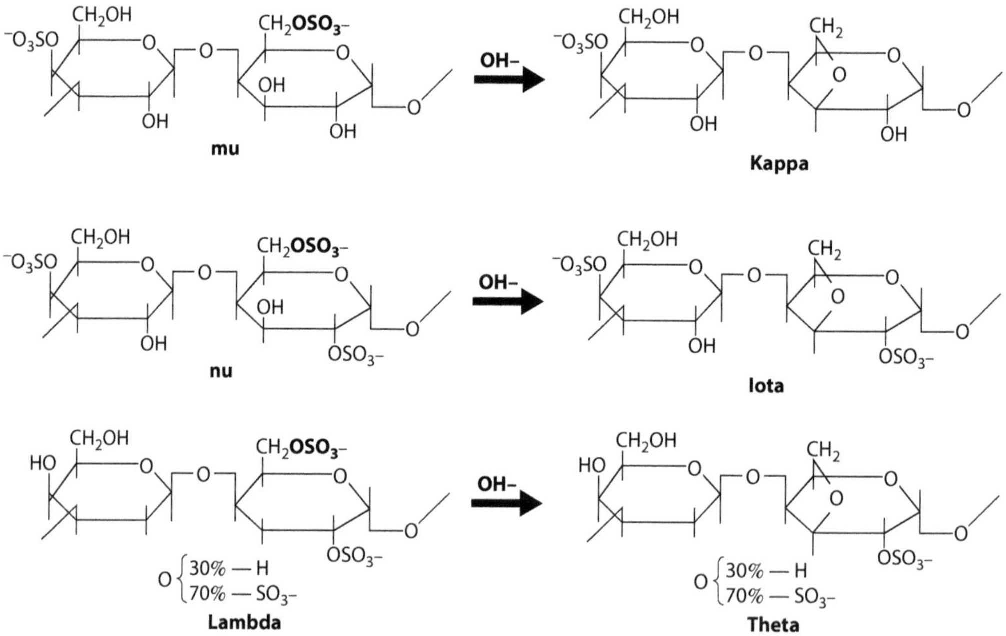

FIGURE 18.9 Structures of different carrageenans.

content, which influences different chemical, physical, and structural properties of these carrageenans. In kappa carrageenans, the ester sulfate and 3,6-AG content is approximately 25% and 34%, whereas in iota carrageenans, it is 32% and 30%. However, in lambda carrageenans, the ester sulfate content is 35% with very little or no content of the 3,6-AG. As the carrageenans are the high-molecular-weight polysaccharides, the commercial kappa carrageenan extracts have the molecular weight between 400 and 560 kDa (Hoffmann et al., 1996). In all carrageenans, the 5% content of the total amount have molecular weight less 100 kDa, which is believed to be inherent in native algal weed. The main applications of the carrageenans are as gelling, thickening, and suspending agents. The kappa carrageenan forms the firm, clear, and brittle gels with poor freeze stability, but these gels can be softened by adding the locust bean gum, whereas the theta carrageenan forms the soft and elastic gel with good freeze–thaw stability. The lambda carrageenan does not form the gel because of the very low content of the 3,6-anhydro group in the (1→4)-linked α-D-galactopyranosyl residues, which is required for the initial formation of double helix.

18.3 GRAFT COPOLYMERIZATION OF GUMS

18.3.1 Graft Copolymerization with Vinyl Monomers

Graft copolymerization of gum polysaccharides with vinyl monomers, which also act as a cross-linking agent, is one of the most studied techniques for their modification. The polymerization reaction is initiated by a chemical initiator and can be carried out in bulk, in solution, or in suspension. The second method involves cross-linking of linear polymer chains by irradiation or by chemical compounds (Peppas and Khare, 1993). The monomers used in the preparation of the ionic polymer network contain an ionizable group, a group that can be ionized, or a group that can undergo a substitution reaction after the polymerization is completed due to which the synthesized hydrogels contain weakly acidic groups such as carboxylic acids or a weakly basic group such as substituted amines or a strong acidic and basic group such as sulfonic acids and quaternary ammonium compounds. A few monomers and cross-linking agents used in the preparation of hydrogels are acrylic acid, methacrylic acid, acrylonitrile, acrylamide derivatives, divinyl benzene, glutaraldehyde, and N,N'-methylene-bis-acrylamide.

18.3.2 Free Radical Graft Copolymerization without a Cross-Linking Agent

This is one of the most widely used and oldest techniques for the synthesis of gum polysaccharide-based graft copolymers. In this technique, the gum polysaccharide is mixed with a monomer, and the reaction is initiated using an initiator such as potassium persulfate, ammonium persulfate, or a mixture of redox initiators such as ascorbic acid–potassium persulfate (Kaith et al., 2012b). One of the best examples of this kind of graft copolymers is the synthesis of Fe^{2+}/BrO^{3-}-initiated graft copolymer of acrylamide with xanthan gum (Behari et al., 2001). The main advantage of synthesis of a noncross-linked polymer network of gum polysaccharides is that it can be dissolved in the water, which can be further utilized for other applications. Graft copolymer of xanthan gum with poly(acrylamide) synthesized using the graft copolymerization technique was utilized as a template for the synthesis of silica nanoparticles (Ghorai et al., 2012).

18.3.3 Free Radical Graft Copolymerization with a Cross-Linking Agent

In this type of preparation method, gum polysaccharides are mixed with the ionic or neutral monomers and multifunctional cross-linking agents. The polymerization reaction is initiated thermally by ultraviolet light or by a redox initiator system. The solvent acts as a heat sink and minimizes temperature control problems. The synthesized graft copolymers are washed with suitable solvent, usually distilled water, to remove the unreacted monomers, cross-linkers, and initiators. In this kind of graft copolymers, the cross-linkers form the links between different polymer chains through covalent bonding. Cross-links act as bridges between polymer chains and form three-dimensional polymeric networks, which are practically insoluble in water (Mittal et al., 2010b). Hydrogels is one of the most important classes of this type of graft copolymers and has diversified applications in many industries. It is basically a kind of superabsorbent, which swells in water up to many hundred times in water without dissolving in them. Gum polysaccharide-based hydrogels are very important for biomedical applications such as soft-tissue engineering, sustained drug delivery, and contact lenses because of their biocompatibility (Kumar et al., 2010; Mi et al., 2002). There are two main categories of the cross-linked gels: (1) physical hydrogels, in which different polymeric chains are joined together by reversible noncovalent bonding and (2) chemical hydrogels, in which different polymeric chains are joined together though nonreversible covalent bonding.

18.3.4 Graft Copolymerization through Irradiation

High-energy radiation such as gamma and electron beams has been used to prepare the hydrogels of unsaturated compounds. The irradiation of aqueous polymer solution results in the formation of radicals on the polymer chains. Moreover, radiolysis of water molecules results in the formation of hydroxyl radicals, which also attack the polymer chains, resulting in the formation of macroradicals. Recombination of the macroradicals on different chains results in the formation of covalent bonds, and finally, a cross-linked structure is formed (Peppas and Mikos, 1986). During radiation polymerization, macroradicals can interact with oxygen, and as a result, radiation-initiated reaction is performed in an inert atmosphere using nitrogen or argon gas. In case of graft copolymerization using microwave radiations, two types of procedures are adopted: microwave assisted, in which the microwave radiation is used for the initiation reactions in addition to the chemical initiators (Singh et al., 2004), and microwave-initiated graft copolymerization, in which only microwave radiations initiate the polymerization reactions (Mittal et al., 2010a). In microwave graft copolymerization, the polysaccharide solution should either be soluble in water or form a suspension. Sometimes, even completely soluble polysaccharide does not form homogeneous solutions, and this loss of homogeneity decreases the overall graft yield.

18.4 CHARACTERIZATION OF GUM-BASED GRAFT COPOLYMERS

18.4.1 Swelling Characteristics

One of the most important properties of gum polysaccharide-based graft copolymers is that they swell in water and biological fluids up to hundred times of their dry volume. This property of swelling in water and biological fluids and biocompatible nature makes them very important in many biomedical applications (Nandkumar et al., 2002; Yokoyama, 2002). The swelling behavior of gum polysaccharide-based graft copolymers shows external environmental responsive behavior, that is, their swelling capacity changes with respect to the changes in the external environmental conditions such as temperature, pH, electrical stimulus, and magnetic field. They are also known as smart, environmental responsive, stimuli-sensitive, and intelligent polymers (Kikuchi and Okano, 2002; Kuhn et al., 1950). Initially, the graft copolymer recognizes the signal and then judges the magnitude of the signal; finally, it changes chain conformation in response to that signal. The external stimuli could be chemical or physical. Chemical stimuli such as pH, chemical agents, or ionic strength of the swelling medium can change the interactions between the polymer chain and solvent molecules, whereas physical stimuli such as temperature and electric or magnetic fields on alteration affect the molecular interactions, thereby causing change in molecular shape.

18.4.2 Spectral Analysis of Graft Copolymers

The graft copolymerization of gum polysaccharides with different vinyl monomer mixtures through the formation of new covalent between the different functionalities of the reactants is basically proved using Fourier transform infrared (FT-IR) spectroscopy. FT-IR is one of the oldest and widely used techniques to show the formation of different covalent bonds between the gum polysaccharides and the vinyl monomers. In case of the FT-IR of the graft copolymers of gum with polyacrylamide, it showed peaks around 1667.57 cm^{-1} (C=O stretching of amide-I band), 1447.5 cm^{-1} (NH in-plane bending of amide-II band), 1239.4 cm^{-1} (CN stretching vibrations of amide-III band), and 769.51 cm^{-1} (OCN deformations of amide-IV band) in addition to the FT-IR peaks of the gum polysaccharide (Kaith et al., 2012b). The graft copolymerization usually occurs through the free hydroxyls of the polysaccharides, and the participation of the free hydroxyl groups in the graft copolymerization can be proved by a shifting of the position of –OH peak as well as the decrease in the intensity of the peak. The graft copolymerization of k-carrageenan with *N*-vinyl-2-pyrrolidone was confirmed by the shifting of the –OH peak at 3428.3 cm^{-1}, which was originally observed at 3504.6 cm^{-1} in the FT-IR spectrum of k-carrageenan. The graft copolymerization was further evidenced by the presence of the characteristic peaks of the *N*-vinyl-2-pyrrolidone at 1673.1 cm^{-1} (due to –CO stretching vibration) and 1459.2 cm^{-1} (due to –CN stretching vibration) in addition to the peaks already observed in the FT-IR of k-carrageenan (Mishra et al., 2010).

18.4.3 Morphology of Graft Copolymers

The changes in the morphology of the gum polysaccharide after grafting and cross-linking with vinyl monomers are usually studied using scanning electron microscopy (SEM). Usually, the surface of the gum polysaccharide is very smooth and homogeneous, but after grafting and cross-linking, the smooth surface of the gum polysaccharide is converted into very rough and heterogeneous due to the formation of different cross-links between different polymeric chains. In a study, Jindal et al. (2013) observed that the morphological characteristics of the xanthan gum almost change after the graft copolymerization with different binary vinyl monomer mixtures such as poly(acrylamide-*co*-acrylic acid), poly(acrylamide-*co*-acrylonitrile), and poly(acrylamide-*co*-methacrylic acid).

18.4.4 Thermal behavior of Graft Copolymers

The changes in the thermal stability of the gum polysaccharides after the graft copolymerization and cross-linking are usually studied using thermogravimetric analysis (TGA)/ differential thermal analysis (DTA)/differential thermogravimetric analysis (DTG). Thermal stability of the gum polysaccharides is expected to increase after the grafting and cross-linking because of the formation of cross-links between different polymeric chains. The enhancement in the thermal stability of the gum polysaccharide is usually determined in terms of the increase in the initial decomposition temperature (IDT) and the final decomposition temperature (FDT). However, sometimes the changes in IDT and FDT also depend on the monomer used. For example, when the polyacrylamide is used as a monomer, the IDT of the graft copolymer is less than that of the gum polysaccharide, whereas the FDT is higher than that of gum polysaccharide. The grafting with poly(acrylamide) lowers the IDT of the graft copolymers and the chains of poly(acrylamide) degraded by the formation of imide groups and the evaluation of ammonia, but after the imide group formation the FDT of the graft copolymer increases (Behari et al., 2001). Moreover, with the graft copolymerization and cross-linking, the rate of weight loss of the polymer decreases with temperature (Mittal et al., 2010b).

18.5 STIMULI-RESPONSIVE GRAFT COPOLYMERS

18.5.1 Temperature-Responsive Graft Copolymers

In case of temperature-responsive graft copolymers, intermolecular interaction between the three-dimensional polymer network and the swelling medium results in the physical cross-links, micelle aggregation, or polymer shrinkage. Intermolecular forces that usually take place between water and graft copolymers are the hydrogen bonding and hydrophobic interactions. Hydrogen bonding by intermolecular association is a random coil-to-helix transition in which different biopolymer chains form a helix conformation and results in the physical cross-links to make a polymer network by lowering the temperature. Also association/dissociation of the hydrogen bonding between different pendant groups can be controlled by change in temperature. Moreover, hydrophilic groups of the hydrogels form hydrogen bonds with water molecules. Such bonds cooperatively form a stable shell of hydration around the hydrophilic groups leading to the greater uptake of water and result in larger swelling. However, the associated interactions among the hydrophilic groups release the entrapped water molecules from the nanocomposite network by change in temperature (Chen and Cheng, 2009; Xu et al., 2006; Zhang et al., 2009).

18.5.2 pH-Responsive Graft Copolymers

pH-responsive nanocomposites are a class of responsive nanocomposites whose swelling properties change in response to small change in pH. They consist of ionizable functional groups that can accept or donate protons upon changing the pH of the swelling medium.

With change in the environmental pH, the pK_a value of the weakly ionizable functional group in the nanocomposite network changes and results in the rapid swelling/deswelling of the respective nanocomposite systems with the alteration of the hydrodynamic volume of the polymer chains. This change in the volume of the nanocomposites is explained on the basis of the osmotic pressure swelling (π_{ion}) theory resulting from difference between the concentration of mobile ions in swollen polymer and the external solution (Mittal et al., 2010b). The polymer backbones with ionizable groups form polyelectrolytes in water. Weak polyacids such as poly(acrylic acid) act as a proton acceptor in the acidic medium, whereas they act as a proton donor in neutral as well as in basic media. However, polybase groups act as a proton donor in the acidic medium and as a proton acceptor in basic as well as in neutral media. Therefore, the selection of polyacids and polybases is very important for the desired applications. In case of a backbone with polyacids, increased swelling was observed in the neutral medium compared to alkaline and acidic media. In dilute electrolyte solutions, the swelling behavior of ionic polymers is due to the osmotic swelling pressure between the nanocomposite phase and the external solution (swelling medium), and the osmotic swelling pressure (π_{ion}) is given as follows (Mittal et al., 2014a, b, 2015):

$$\pi_{ion} = RT \sum \left(C_i^g - C_i^s \right) \tag{18.1}$$

where:

C_i^g and C_i^s respectively, are the molar concentrations of mobile ions in the swollen nanocomposite and the external solution

R is the gas constant

T is the absolute temperature

In the neutral medium, π_{ion} becomes very large because the concentration of mobile ions in the external solution (C_i^s) becomes almost negligible. Also electrostatic repulsion between carboxylate ions promotes the swelling, so larger swelling is observed. However, in the acidic medium, most of the polyacids get protonated; therefore, the concentration of mobile ions within the swollen nanocomposite (C_i^g) is less compared to the concentration of the mobile ions in the solution, resulting in the smaller π_{ion} value. Whereas, in alkaline solutions, polyacid groups completely dissociate, but at the same time the concentration of Na^+ and ^-OH ions is also very high and leads to a decrease in π_{ion} and percentage swelling. Moreover, in the alkaline medium, a screening effect of the counter ions (Na^+) shields the electrostatic repulsion between the carboxylate anions; as a result, a remarkable decrease in percentage swelling (P_s) is observed (Mahdavinia et al., 2004).

18.5.3 Magneto-Sensitive Graft Copolymers

Hydrogels or the graft copolymers of gum polysaccharides, which change their properties in response to variation in the applied magnetic field, are known as magneto-sensitive hydrogels. Magnetically modulated hydrogels are prepared by encapsulation of magnetic particles within the three-dimensional cross-linked polymer matrix of the gum polysaccharide-based graft copolymers. The development of emulsion polymerizations, such as the conventional emulsion polymerization process, mini-emulsion and micro-emulsion polymerization, and soap-free emulsion polymerization, has led to new synthesis methods for the preparation of magneto-sensitive graft copolymers. Magneto-sensitive graft copolymers play a strong role in the health-care and biological applications. The combination of fine particles and magnetism in the field of biology and biomaterial has been found useful in sophisticated biomedical applications such as cell separation, gene and drug delivery, and magnetic intracellular hyperthermia treatment of cancer. Magneto-sensitive graft copolymers have also been widely used for the treatment of wastewater. The basic advantage of using these graft copolymers in wastewater treatment is their easy separation from the water after the adsorption process. These nanocomposites can be easily separated by the application of magnetic field. Magnetic nanocomposites of calcium alginate with iron oxide nanoparticles have been successfully used for the removal of Cu(II) ions from the wastewater (Lim et al., 2009). In another study, magnetic nanocomposites of gum arabic with iron oxide nanoparticles were successfully synthesized and utilized for the adsorption of Cu(II) ions from the wastewater, and the maximum adsorption capacity of 38.5 mg g^{-1} was achieved (Banerjee and Chen, 2007).

18.5.4 Electrosensitive Graft Copolymers

Electrically responsive delivery systems are prepared from polyelectrolytes and are thus pH responsive as well. Under the influence of an electric field, electroresponsive hydrogels generally deswell or bend, depending on the shape and orientation of the hydrogel. The gel bends when it is parallel to the electrodes, whereas deswelling occurs when the hydrogel lies perpendicular to the electrodes. Electrosensitive graft copolymers of gum ghatti with poly(acrylamide) were synthesized, and their swelling–deswelling behavior in response to change in electrical stimulus was successfully investigated. These hydrogels exhibited remarkable deswelling behavior on the application of electrical stimulus (Kaith et al., 2010b).

18.6 BIODEGRADATION OF GUM-BASED GRAFT COPOLYMERS

One of the biggest advantages of the gum polysaccharide-based graft copolymers is their environment-friendly nature. They have the ability to degrade under both aerobic and anaerobic conditions. Generally, the gum polysaccharides were grafted with different synthetic vinyl monomers to improve their mechanical and other properties for their applications in various industries, but the graft copolymerization of the gum polysaccharides should not affect their biodegradable or environment-friendly nature. It was observed that the graft copolymers of different gum polysaccharides with different vinyl monomers made them little bit resistant to the biodegradation, but they were also found biodegradable according to American Society for the Testing of Materials (ASTM) standards of biodegradation (Mittal et al., 2013b, c, 2014a, 2015). Recently, the biodegradable nature of the gum ghatti and different vinyl monomers such as poly(acrylamide)-, poly(acrylic acid)-, poly(methacrylic acid)-, and poly(acrylonitrile)-based hydrogel polymers has been studied, and they are found to be fully biodegradable using the soil composting method.

It was also observed that in the case of the gum polysaccharide-based graft copolymers, initially biodegradation starts from the backbone polymer of the gum polysaccharide itself, whereas in the later stages, the cross-links between polymeric chains break, and finally, the whole three-dimensional polymer network collapses. It was also observed that the rate of degradation was slower in the starting of the biodegradation process, whereas it became much faster in the later stages. Initially, the hydrogel polymer takes time to absorb some water, but after absorbing the water up to its equilibrium capacity, the hydrogel polymer starts degrading due to the action of the microorganisms present in the water of the soil compost. Various processes involved in the degradation of the hydrogel polymer included enzymatic and chemical degradation. The secretion products of the microbes present in the composting media were also found to be involved in the degradation of the gum polysaccharide-based graft copolymers, which caused the cleavage of chemical bonds and bacterial degradation (Mittal et al., 2013b, c, 2014a, 2015). Different characterization techniques such as FT-IR and SEM have been used to check the progress of the biodegradation at different stages. The changes in the intensity as well as the shifting of some of the peaks were observed in the FT-IR of the grafted polymer at different stages of biodegradation; moreover, the SEM images of the samples at different stages showed some cracks.

18.7 APPLICATIONS OF GUM-BASED GRAFT COPOLYMERS AS ADSORBENTS

Different types of toxic pollutants such as heavy metals (Pb^{2+}, Cd^{2+}, Hg^{2+}, Ni^{2+}, Cu^{2+}, Co^{2+}, Cr^{6+}), cationic (methylene blue, rhodamine B, methyl violet, brilliant green, etc.), and anionic dyes (congo red, direct red 81, etc.) exist in large amount in the wastewater of many industries, such as battery manufacturing, printing and pigment, metal plating, tanneries, oil refining, and mining (Schneegurt et al., 2001) mostly in the developing countries. Heavy metals and dyes have adverse effect on the human health. They are not biodegradable and result in a long-term threat to human health (Naiya et al., 2009). Most of the heavy metal ions and synthetic dyes are carcinogenic in nature, and their uptake more than the permissible limit can cause serious health problems such as vomiting, hypotension, melena, gastrointestinal distress, and kidney failure. Dyes from wastewater are stable and more resistant to biological degradation because of their complex structure and very high solubility in water. The dyes also affect the biological activities in waterbodies. Therefore, these toxic pollutants such as heavy metals, dyes from wastewater must be removed before discharge to the environment.

The graft copolymers of different gum polysaccharides such as gum ghatti, xanthan gum, sodium alginate, and carrageenan, with the synthetic vinyl monomers such as acrylic acid, acrylamide, methacrylic acid, and acrylonitrile, have been used as adsorbents for the removal of different impurities such as dyes and heavy metal ions from the aqueous solutions (Deng et al., 2012; Lan et al., 2014; Masoumi and Ghaemy, 2014; Mittal et al., 2014c; Mittal and Mishra, 2014; Saravanan et al., 2012; Thakur et al., 2014). The main advantage of using these materials is their low cost, biodegradable nature, and ability to design the material as per the end user requirement. The graft copolymers of the gum polysaccharides can be tailored according to a particular impurity, which has to be removed from the contaminated water. Moreover, all the gum polysaccharides have abundant anionic functionalities in their structure so they can be used for the highly effective removal of different cationic impurities from wastewater. In spite all the advantages, gum polysaccharides have a lot of disadvantages such as poor mechanical strength so they cannot be used as such without modification for the adsorption of heavy metal ions and dyes. Therefore, they should be modified with some suitable synthetic monomers such as acrylic acid or acrylamide. The modified gum polysaccharides, that is, graft copolymers, have the advantages of both the counterparts—eco-friendly nature of the gum polysaccharides as well as the excellent mechanical and physical properties of vinyl monomers. The graft copolymers of gum polysaccharides have already shown their potential for the removal of different pollutants from the aqueous solutions. The graft copolymers of kappa-carrageenan and N-vinyl-2-pyrrolidone were used for the adsorption of different cations such as Pb^{2+}, Ni^{2+}, and Zn^{2+} from the aqueous solutions (Mishra et al., 2010). In another study, the biodegradable graft copolymers of gum ghatti with poly(acrylic acid) was used for the effective removal of Pb^{2+} (q_m = 84.7 mg g^{-1}) and Cu^{2+} (q_m = 310.55 mg g^{-1}) from aqueous solution (Mittal et al., 2013a).

Recently, a lot of research work has also been reported on the synthesis of gum polysaccharide-based nanocomposites, where in some of the cases the already synthesized metal oxide nanoparticles such as Fe_3O_4 and SiO_2 nanoparticles were incorporated within the hydrogel polymer matrix (Mittal and Mishra, 2014) or in some cases the hydrogel polymer matrix of gum polysaccharides was used for the *in situ* synthesis of the metal oxide nanoparticles (Ghorai et al., 2012). These hydrogel nanocomposites based on gum polysaccharides and the metal oxide nanoparticles have shown their potential in the field of water purification. In a study, the iron oxide magnetic nanoparticles were incorporated within the hydrogel polymer matrix of the graft copolymer of xanthan gum and the copolymer mixture of poly(acrylic acid-*co*-acrylamide), and the synthesized hydrogel nanocomposite was used for the removal of malachite green from the aqueous solution with the maximum adsorption capacity of 497.15 mg g^{-1} (Mittal at el., 2014c). In another study, the hydrogel polymer matrix of the graft copolymer of xanthan gum and poly(acrylamide) was used for the *in situ* synthesis of nanosilica, and the resultant hydrogel nanocomposite was used for the adsorption of Pb^{2+} and congo red from the aqueous solution (Ghorai et al., 2013).

18.8 CONCLUSION

From the aforementioned discussion, it can be concluded that the gum polysaccharide based graft copolymers have a lot of benefits over other materials, and they can be used for a number of applications. As an adsorbent, they have many advantages over other adsorbents such as eco-friendly nature and cost effectiveness. Moreover, they have the ability to be modified as per the requirement. The graft copolymer of gum polysaccharides can be triggered with response to external environment conditions such as pH, temperature, and magnetic field. Gum polysaccharides and metal oxide-based hydrogel nanocomposites have also the ability to be used as an adsorbent. Therefore, it can be concluded that the gum polysaccharide-based graft copolymers have shown their potential as adsorbents for the removal of different adsorbents from the aqueous solution.

ACKNOWLEDGMENTS

The authors are grateful to the National Research Foundation, South Africa, for awarding a postdoctoral research fellowship to Dr. Hemant Mittal. The authors (HM, AM, SSR) also thank the Department of Science and Technology and the Council for Scientific and Industrial Research for financial support.

REFERENCES

Agorku, E. S., Mittal, H., Mamba, B. B., Pandey, A. C., and A. K. Mishra. 2014. Fabrication of biocompatible photocatalyst based on Eu^{3+}-doped ZnS-SiO_2/poly (acrylamide-co-methacrylic acid) core shell nanocomposite. *International Journal of Biological Macromolecules* 70: 143–149.

Anderson, D. M. W. and M. M. E. Bridgeman. 1985. The composition of the proteinaceous polysaccharides exuded by *Astragalus microcephalus, A. gummifer* and *A. kurdicus*—The sources of Turkish gum tragacanth. *Phytochemistry* 24: 2301–2304.

Anderson, D. M. W. and M. M. E. Bridgeman. 1988. The chemical characterisation of the test article used in toxicological studies of gum tragacanth. *Food Hydrocolloids* 2: 51–57.

Aspinall, G. O., Hirst, E. L., and A. Wickstrom. 1955. Gum ghatti (Indian gum). The composition of the gum and the structure of two aldobiouronic acids derived from it. *Journal of the Chemical Society* 1160–1165.

Banerjee, S. S. and D. W. Chen. 2007. Fast removal of copper ions by gum arabic modified magnetic nano-adsorbent. *Journal of Hazardous Materials* 147: 792–799.

Behari, K., Pandey, P. K., Kumar, R., and K. Taunk. 2001. Graft copolymerization of acrylamide onto gum xanthan. *Carbohydrate Polymers* 46: 185–189.

Brown, J. C. and G. Livesey. 1994. Energy balance and expenditure while consuming Guar gum at various fat intakes and ambient temperatures. *American Journal of Clinical Nutricians* 60: 956–964.

Chen, J. P. and T. H. Cheng, 2009. Preparation and evaluation of thermo-reversible copolymer hydrogels containing chitosan and hyaluronic acid as injectable cell carriers. *Polymer* 50: 107–116.

Cronin, C. E., Giannouli, P., McCleary, B. V., Brooks, M., and E. R. Morris. 2002. Formation of strong gels by enzymic debranching of guar gum in the presence of ordered xanthan. In: *Gum and Stabilisers for the Food Industry 11*, eds. P. A. Williams and G. O. Phillips, 286–296. Cambridge: RSC Special Publication.

Deng, C., Liu, J., Zhou, W., Zhang, Y. K., Du, K. F., and Z. M. Zhao. 2012. Fabrication of spherical cellulose/carbon tubes hybrid adsorbent anchored with welan gum polysaccharide and its potential in adsorbing methylene blue. *Chemical Engineering Journal* 200–202: 452–458.

Fenyo, J. C. and M. C. Vandervelde. 1990. *Gums and Stabilizers for the Food Industry*. Oxford: IRL Press.

Fischer, F. G. and H. Dörfel. 1955. Die polyuronsauren der braunalgen- (kohlenhydrate der algen-I). *Z Physiol Chem* 302: 186–203.

Ghorai, S., Sarkar, A. K., Panda, A. B., and S. Pal. 2013. Effective removal of Congo red dye from aqueous solution using modified xanthan gum/silica hybrid nanocomposite as adsorbent. *Bioresource Technology* 144: 485–491.

Ghorai, S., Sinhamahpatra, A., Sarkar, A., Panda, A. B., and S. Pal. 2012. Novel biodegradable nanocomposite based on XG-g-PAM/SiO_2: Application of an efficient adsorbent for Pb^{2+} ions from aqueous solution. *Bioresource Technology* 119: 181–190.

Glicksman, M. and R. E. Sand. 1973. Polysaccharides and their derivatives. In: *Industrial Gums*, eds. R. L. Whistler, and J. N. BeMiller, 119–230. New York: Academic Press.

Gombotz, W. R. and S. F. Wee. 1998. Protein release from alginate matrices. *Advanced Drug Delivery Reviews* 31: 267–285.

Hoffmann, R. A., Russel, A. R., and M. J. Gidley. 1996. Molecular weight distribution of carrageenans. In: *Gums & Stabilisers for the Food Industry 8*, eds. G. O. Phillips, P. J. Williams, and D. J. Wedlock, 137–148. Oxford: IRL Press.

Jefferies, M., Pass, G., and G. O. Phillips. 1977. Viscosity of aqueous of gum ghatti. *Journal of the Science of Food and Agriculture* 28: 173–179.

Jindal, R., Kaith, B. S., Mittal, H., and S. Berry. 2013. In Vacuo synthesis of Gum xanthan based hydrogels with different vinyl monomer mixtures and their swelling behavior in response to external environment conditions. *Polymers from Renewable Resources* 4: 19–34.

Kaith, B. S., Jindal, R., Mittal, H., and K. Kumar. 2010a. Temperature, pH and electric stimulus responsive hydrogels from Gum ghatti and polyacrylamide-synthesis, characterization and swelling studies. *Der Chemica Sinica* 1: 44–54.

Kaith, B. S., Jindal, R., Mittal, H., and K. Kumar. 2012a. Synthesis of crosslinked networks of Gum ghatti with different vinyl monomer mixtures and effect of ionic strength of various cations on its swelling behavior. *International Journal of Polymeric Materials* 61: 99–115.

Kaith, B. S., Jindal, R., Mittal, H., and K. Kumar. 2012b. Synthesis, characterization and swelling behavior evaluation of Gum ghatti and acrylamide based hydrogel for selective absorption of saline from different petroleum fraction–saline emulsions. *Journal of Applied Polymer Science* 124: 2037–2047.

Kaith, B. S., Jindal, R., Mittal, H., Kumar, K., and K. S. Nagla. 2010b. Synthesis and characterization of Gum ghatti-acrylamide based electrical sensitive smart polymer. *Trends in Carbohydrate Research* 2: 35–44.

Kang, H. M., Cai, Y. L., and P. S. Liu. 2006. Synthesis, characterization and thermal sensitivity of chitosan-based graft copolymers. *Carbohydrate Research* 11: 2851–2855.

Katzbauer, B. 1998. Properties and applications of xanthan gum. *Polymer Degradation and Stability* 59: 81–84.

Kikuchi, A. and T. Okano. 2002. Intelligent thermoresponsive polymeric stationary phases for aqueous chromatography of biological compounds. *Progress in Polymer Science* 27: 1165–1193.

Krishnaiah, Y. S. R. and B. P. Srinivas. 2008. Effect of 5-fluorouracil pretreatment on the in vitro drug release from colon-targeted guar gum matrix tablets. *The Open Drug Delivery Journal* 2: 71–76.

Kuhn, W., Hargitay, B., Katchalsky, A., and H. Eisenberg. 1950. Reversible dilation and contraction by the state of ionization of high-polymer acid networks. *Nature* 165: 514–515.

Kumar, K., Kaith, B. S., Jindal, R., and H. Mittal. 2010. Utilization of acrylamide and natural polysaccharide based polymeric networks in pH controlled release of 5-amino salicylic acid. *Journal of Chilean Chemical Society* 55: 522–526.

Kumar, K., Kaith, B. S., Jindal, R., and H. Mittal. 2012a. Gamma-radiation initiated synthesis of *Psyllium* and acrylic acid based polymeric networks for selective absorption of water from different oil-water emulsions. *Journal of Applied Polymer Science* 124: 4969–4977.

Kumar, K., Kaith, B. S., and H. Mittal. 2012b. A study on effect of different reaction conditions on grafting of psyllium and acrylic acid based polymeric networks. *Journal of Applied Polymer Science* 123: 1874–1883.

Lan, S., Leng, Z., Guo, N., Wu, X., and S. Gan. 2014. Sesbania gum-based magnetic carbonaceous nanocomposites: Facile fabrication and adsorption behavior. *Colloids and Surfaces A: Physicochemical and Engineering Aspects* 446: 163–171.

Li, R. and L. D. Feke. 2015. Rheological and kinetic study of the ultrasonic degradation of xanthan gum in aqueous solutions. *Food Chemistry* 172: 808–813.

Lim, S. F., Zheng, Y. M., Zou, S. W., and J. P. Chena. 2009. Removal of copper by calcium alginate encapsulated magnetic sorbent. *Chemical Engineering Journal* 152: 509–513.

Mahdavinia, G. R., Pourjavadi, A., Hosseizadeh, H., and M. J. Zohuriaan. 2004. Modified chitosan 4. Superabsorbent hydrogels from poly(acrylic acid-coacrylamide) grafted chitosan with salt- and pH-responsiveness properties. *European Polymer Journal* 40: 1399–1407.

Masoumi, A. and M. Ghaemy. 2014. Removal of metal ions from water using nanohydrogel tragacanth gum-g-polyamidoxime: Isotherm and kinetic study. *Carbohydrate Polymers* 108: 206–215.

Mi, F. L., Tan, Y. C., Liang, H. F., and H. W. Sung. 2002. In vivo biocompatibility and degradability of a novel injectable-chitosan-based implant. *Biomaterials* 23: 181–191.

Mishra, M. M., Sand, A., Mishra, D. K., Yadav, M., and K. Behari. 2010. Free radical graft copolymerization of N-vinyl-2-pyrrolidone onto k-carrageenan in aqueous media and applications. *Carbohydrate Polymers* 82: 424–431.

Mittal, H., Fosso-Kankeu, E., Mishra, S., and A. Mishra. 2013a. Biosorption potential of Gum ghatti-g-poly(acrylic acid) and susceptibility to biodegradation by *B. subtilis*. *International Journal of Biological Macromolecules* 62: 370–378.

Mittal, H., Mishra, S.B., Mishra, A.K., Kaith, B.S., Jindal, R. and Kalia, S. 2013c. Preparation of poly(acrylamide-co-acrylic acid)-grafted gum and its flocculation and biodegradation studies. *Carbohydrate Polymers* 98:397–404.

Mittal, H., Mishra, S. B., Mishra, A. K., Kaith, B. S., and R. Jindal. 2013b. Flocculation characteristics and biodegradation studies of Gum ghatti based hydrogels. *International Journal of Biological Macromolecules* 58: 37–46.

Mittal, H., Jindal, R., Kaith, B. S., Maity, A., and S. S. Ray. 2014a. Synthesis and evaluation of the properties of gum ghatti and poly(acrylamide-co-acrylonitrile) based biodegradable flocculants. *Carbohydrate Polymers* 114: 321–329.

Mittal, H., Jindal, R., Kaith, B. S., Maity, A., and S. S., Ray. 2015. Flocculation and adsorption properties of biodegradable gum-ghatti-grafted poly(acrylamide-co-methacrylic acid) hydrogels. *Carbohydrate Polymers* 115: 617–628.

Mittal, H., Kaith, B. S., and R. Jindal. 2010a. Microwave radiation induced synthesis of gum ghatti and acrylamide based crosslinked network and evaluation of its thermal and electrical behavior. *Der Chemica Sinica* 1: 59–69.

Mittal, H., Kaith, B. S., and R. Jindal. 2010b. Synthesis, characterization and swelling behavior of poly(acrylamide-co-methacrylic acid) grafted Gum ghatti based superabsorbent hydrogels. *Advances in Applied Science Research* 1: 56–66.

Mittal, H. and S. Mishra. 2014a. Gum ghatti and Fe$_3$O$_4$ magnetic nanoparticles based nanocomposites for the effective adsorption of rhodamine B. *Carbohydrate Polymers* 101: 1255–1264.

Mittal, H., Mishra, S. B., Mishra, A. K., Kaith, B. S., Jindal, R., and S. Kalia. 2014b. Preparation of poly(acrylamide-co-acrylic acid)-grafted Gum and its flocculation and biodegradation studies. *Carbohydrate Polymers* 98: 397–404.

Mittal, H., Parashar V., Mishra, S., and A. Mishra. 2014c. Fe$_3$O$_4$ MNPs and gum xanthan based hydrogels nanocomposites for the efficient capture of malachite green from aqueous solution. *Chemical Engineering Journal* 255: 471–482.

Morrison, N. A., Sworn, G., Clark, R. C., Talashek, T., and Y. L. Chen. 2002. New textures with highly acyl gellan gum. In: *Gum and Stabilisers for the Food Industry 11* eds. P. A. Williams and G. O. Phillips, 297–305. Cambridge: RSC Special Publication.

Naiya, T. K., Bhattacharya, A. K., and S. K. Das. 2009. Clarified sludge (basic oxygen furnace sludge)-an adsorbent for removal of Pb(II) from aqueous solutions—Kinetics, thermodynamics and desorption studies. *Journal of Hazardous Materials* 170: 252–262.

Nandkumar, M. A., Yamato, M., Kushida, A., Konno, C., Hirose, M., Kikuchi, A., and T. Okano. 2002. Two-dimensional cell sheet manipulation of heterotypically co-cultured lung cells utilizing temperature-responsive culture dishes results in long-term maintenance of differentiated epithelial cell functions. *Biomaterials* 23: 1121–1130.

O'Neill, M. A., Selvendran, R. R., and V. J. Morris. 1983. Structure of the acidic extracellular gelling polysaccharide produced by *Pseudomonas elodea*. *Carbohydrate Research* 124: 123–133.

Palaniraj, A. and V. Jayaraman. 2011. Production, recovery and applications of xanthan gum by *Xanthomonas campestris*. *Journal of Food Engineering* 106: 1–12.

Paquet, E., Bédard, A., Lemieux, S., and S. L. Turgeon. 2014. Effects of apple juice-based beverages enriched with dietary fibres and xanthan gum on the glycemic response and appetite sensations in healthy men. *Bioactive Carbohydrates and Dietary Fibre* 4: 39–47.

Patel, G. M., Patel, C. P., and T. C. Trivedi. 1999. Ceric-induced grafting of methyl acrylate onto sodium salt of partially carboxymethylated sodium alginate. *European Polymer Journal* 35: 201–208.

Peppas, N. A. and A. R. Khare. 1993. Preparation, structure and diffusional behavior of hydrogels in controlled release. *Advanced Drug Delivery Reviews* 11: 1.

Peppas, N. A. and A. G. Mikos. 1986. In: *Hydrogels in Medicine and Pharmacy*, ed. N. A. Peppas, 1–25. Boca Raton, FL: CRC Press.

Phillips, G. O. 2008. Mixing hydrocolloids and water: Polymers versus particles. In: *Gums and Stabilisers for the Food Industry 14*, eds. P. A. Williams and G. O. Phillips, 3–28. Cambridge: RSC Special Publication.

Pourjavadi, A., Hosseinzadeh, H., and R. Mazidi. 2005. Modified carrageenan synthesis and swelling behavior of crosslinked kC-*g*-AMPS superabsorbent hydrogel with anti-salt and pH-responsiveness properties. *Journal of Applied Polymer Science* 98: 255–263.

Prezotti, F. G., Cury, B. S. F., and R. C. Evangelista. 2014. Mucoadhesive beads of gellan gum/pectin intended to controlled delivery of drugs. *Carbohydrate Polymers* 113: 286–295.

Rana, V., Rai, P., Tiwary, A. K., Singh, R. S., Kennedy, J. F., and C. J. Knill. 2011. Modified gums: Approaches and applications in drug delivery. *Carbohydrate Polymers* 83: 1031–1047.

Randall, R. C., Phillips, G. O., and P. A. Williams. 1988. The role of the proteinaceous component on the emulsifying properties of gum arabic. *Food Hydrocolloids* 2: 131–140.

Rees, D. A. 1963. The carrageenan system of polysaccharides, Part 1. The relation between kappa and lambda components. *Journal of the Chemical Society* 1821–1832.

Rinaudo, M. 2008. Main properties and current applications of some polysaccharides as biomaterials. *Polymer International* 57: 397–430.

Sadeghi, S., Rad, F. A., and A. Z. Moghaddam. 2014. A highly selective sorbent for removal of Cr(VI) from aqueous solutions based on Fe3O$_4$/poly(methyl methacrylate) grafted Tragacanth gum nanocomposite: Optimization by experimental design. *Materials Science and Engineering: C* 45: 136–145.

Saravanan, P., Vinod, V. T. P., Sreedhar, B., and R. B. Sashidhar. 2012. Gum kondagogu modified magnetic nano-adsorbent: An efficient protocol for removal of various toxic metal ions. *Materials Science and Engineering: C* 32: 581–586.

Schneegurt, M. A., Jain, J. C., Menicucci, J. A., Brown, S. A., Kemner, K. M., Garofalo, D. F., Quallick, M. R., Neal, C. R., and C. F. Kulpa. 2001. Biomass by-products for the remediation of wastewaters contaminated with toxic metals. *Environmental Science & Technology* 35: 3786–3791.

Sen, G., Kumar, R., Ghosh, S., and S. Pal. 2009. A novel polymeric flocculent based on polyacrylamide grafted carboxymethyl starch. *Carbohydrate Polymers* 77: 822–831.

Singh, B. N. and K. H. Kim. 2007. Characterization and relevance of physicochemical interactions among components of a novel multiparticulate formulation for colonic delivery. *International Journal of Pharmaceutics* 341: 68–77.

Singh, V., Tiwari, A., Tripathi, D. N., and R. Sanghi. 2004. Microwave assisted synthesis of guar-g-poly(acrylamide). *Carbohydrate Polymers* 58: 1–6.

Sworn, G., 2000. Gellan gum. In: *Handbook of Hydrocolloids*, eds. G. O. Phillips and P. A. Williams, 379–395. Boca Raton, FL: CRC Press.

Thakur, S., Kumari, S., Dogra, P., and G. S. Chauhan. 2014. A new guar gum-based adsorbent for the removal of Hg(II) from its aqueous solutions. *Carbohydrate Polymers* 106: 276–282.

Vahdat, A., Bahrami, H., Ansari, N., and F. Ziaie. 2007. Radiation grafting of styrene onto polypropylene fibres by a 10MeV electron beam. *Radiation Physics and Chemistry* 76: 787–793.

Xu, F. J., Kang, E. T., and K. G. Neoh. 2006. pH- and temperature-responsive hydrogels from crosslinked triblock copolymers prepared via consecutive atom transfer radical polymerization. *Biomaterials* 27: 2787–2797.

Yokoyama, M., 2002. Gene delivery using temperature-responsive polymeric carriers. *Drug Discovery Today* 7: 426–432.

Zhang, H. F., Zhong, H., Zhang, L. L., Chen, S. B., Zhao, Y. J., and Y. L. Zhu. 2009. Synthesis and characterization of thermosensitive graft copolymer of N-isopropylacrylamide with biodegradable carboxymethylchitosan. *Carbohydrate Polymers* 7: 785–790.

CHAPTER 19

Thermosets from Renewable Resources

Fundamentals, Preparation, and Properties

Arunjunai Raj Mahendran and Nicolai Aust

CONTENTS

19.1 INTRODUCTION AND HISTORY

Currently, the utilization of renewable resources to synthesize bio-based polymers has been increased because of depletion of fossil resources and concerns regarding environmental impacts of crude oil products. The major development of renewable polymers is driven by carbon footprint reduction and a strong desire to replace the crude oil-derived organic chemicals with renewable "green" chemicals. A comprehensive interest in utilizing renewable resources, such as lignins, proteins, polysaccharides, and vegetable oils, as raw materials for synthesizing bio-based polymers and in employing the polymers in automotive, electronic, coating, composite, and synthetic fiber industries is getting broadened now. The commercial utilization of fully renewable polymers in some applications is limited because of high costs and inferior performance compared to synthetic polymers. An alternative choice can be the partial substitution of petrochemical-based raw materials with renewable materials that can reduce costs and help to achieve the desired end use properties.

Synthetic polymers are divided into three major classes on the basis of their physical properties: thermoplastics, thermosets, and elastomers. Around 88% of the polymers produced are thermoplastics, whereas thermosets are the second major class. Thermoset resins such as unsaturated polyesters (UPs), phenol formaldehyde (PF)-, polyurethane-, and epoxy resins are mainly utilized in construction and automotive industries. In this chapter, the significance of replacing petrochemicals with two renewable raw materials, lignin and vegetable oil, in thermosetting polymers such as acrylic-, phenolic-, epoxy-, and polyurethane resins are investigated. The properties of the renewable thermosetting resins for their use in composites, coatings, and other applications are also studied in detail.

Lignin is one of the renewable raw materials that is used to synthesize bio-based thermoset resins. In terms of chemical structure, lignins are complex and irregular polymers present in the cell walls of vascular plants built from three basic monolignols: *p*-coumaryl-, coniferyl-, and sinapyl alcohol. The radical coupling reactions of three primary monolignol precursors generate guaiacyl-, syringl-, and *p*-hydroxyphenyl propane units. Guaiacyl lignin is principally composed of coniferyl alcohol units, whereas guaiacyl–syringyl lignin contains monomeric units from coniferyl- and sinapyl alcohol. Most commonly, guaiacyl lignin is found in softwoods, whereas guiacyl-syringl lignins are present in hardwoods [1]. Lignin with its highly branched three-dimensional phenolic structure is the second most abundant biological material on the earth comprising 15%–30% of the dry weight of woody plants.

Huge amounts of lignin are produced by the paper industries as an industrial waste during different pulping processes. The chemical pulping is the most common process that is done by treating the wood in an alkaline or acidic solution at high temperatures. Some of the most commercial processes are the kraft process (chemicals are Na_2S, NaOH, and water), the sulfite process (chemicals are H_2SO_3 and water), and the soda process (chemicals are NaOH and water). The extracted lignins derived by the pulping processes are further processed to give valuable chemicals or burnt for energy production.

In the paper industry, kraft pulping is used to manufacture high-strength sack kraft papers and absorbed kraft papers. Lignins from the pulping process are called kraft lignins and they have mainly free phenolic groups due to cleavage of ether bonds. Apart from some valuable materials, such as tal oil and terpentine obtained as by-products, the kraft lignins are mainly burnt for energy production because of their high calorific value. Compared to other pulping method, the advantage of kraft process is that it requires shorter cooking times and also efficient in recovering the pulping chemicals.

Lignins from the sulfite pulping process are called lignosulfonates and are mainly used as additives or plasticizers in making concrete and as a raw material for the synthesis of vanillin. The sulfite pulping is the second most widely used method in the paper industries accounting for 5%–6% of the world production. The sulfite pulping is used to manufacture viscose rayon, acetate filament, films, and cellophane. The sulfite pulping follows many of the processing steps as in the kraft pulping. The major difference lies in the chemicals used: H_2SO_3 and HSO_3^-. Therefore, the lignin obtained mainly contains sulfonic-, carboxylic-, and phenolic hydroxyl groups.

Lignins contain three major functional groups: methoxyl-, phenolic hydroxyl-, and terminal aldehyde. The abundance and type of the functional groups in the lignin structure vary depending on the pulping method. Hence, based on the availability and type of functional groups present in the lignin structure, they are used as renewable "green" chemical for synthesizing phenolic-, polyester-, polyurethane-, and epoxy resins [2–6]. Until now, the most examined research work for utilizing lignin as a renewable raw material is for the phenolic industries. Lignins have long been considered as a phenol substituent in various polycondensation resins, that is, PF resins.

In general, the PF resins are prepared by the reaction of phenol or substituted phenol with an aldehyde, especially formaldehyde, in the presence of an acid or basic catalyst. The substitution of phenol in the PF resin with renewable lignin is advantageous in terms of environmental and economic aspects. The toxic phenol in the PF composition is partially substituted by lignin as a cheap raw material from pulping waste. Lignins have been extensively investigated as phenol replacement in both novolac and resol type resins by several authors [3,7–9]. Using kraft lignin, a substitution of up to 60% of phenol for PF adhesives was achieved as reported by several authors, only if the factors affecting the production of such adhesives, such as mixing ratios, are carefully controlled [10,11]. Olivares et al. obtained resins with 20% phenol replacement by kraft lignin for application as a glue in particleboard [12].

The reaction rate of alkali-catalyzed condensation of formaldehyde with phenol, kraft lignin, and steam-exploded hardwood lignin was studied by Gardner and McGinnis [13] through monitoring the consumption of formaldehyde under various conditions. The results show that kraft lignin has a faster reaction rate with formaldehyde than steam-exploded lignin because the rate of reaction between formaldehyde and lignin is dependent on the availability of the reactive groups of the lignin macromolecule. El Mansouri et al. [14] reported that kraft lignin contains more phenolic hydroxyl units than lignosulfonate and is also rich in guaiacyl units. Hence, they are more suitable for phenol replacement in PF resins. The lignosulfonate contains less hydroxyl units than kraft lignin and more syringyl units than guaiacyl units. The formaldehyde addition to the phenolic structure depends on the available reactive groups, which is primarily concerned about the structural units present in the lignin type. The lignosulfonate structure having syringyl-, guiacyl- and p-hydroxyphenyl propane units is shown in Figure 19.1. The points in the figure mark the reactive sites where a possible formaldehyde addition can occur during resin synthesis.

In general, the reactivity of both technical lignins is limited because of the small amount of ortho- and para-reactive sites and their poor accessibility [3]. Few techniques have been developed to enhance the reactivity, either by decreasing the molecular weight, that is, by oxidation followed by isolating the lignins' low molecular weight fragments, or by modifying the structure to increase the amount of reactive functional groups.

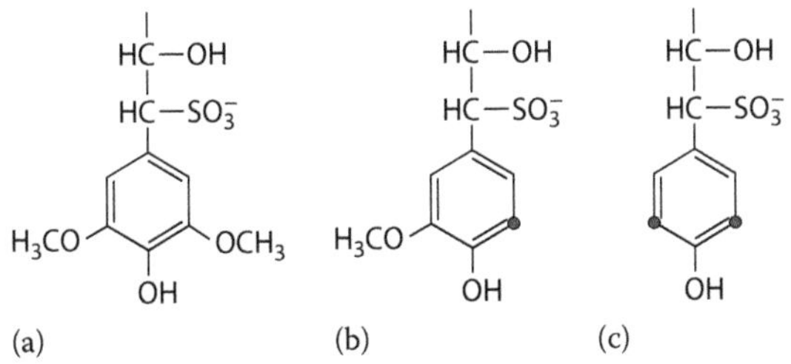

FIGURE 19.1   The structural units of lignosulfonate: syringyl (a), guiacyl (b), and p-hydroxyphenyl propane (c).

The structural modification is the preferred process because of economic reason. There are several investigations done earlier to modify the lignin structure. The structure modification can be carried out by methylolation [15,16], phenolation [17], or demethylation, where the method of methylolation is the preferred one, because it can be carried out under common laboratory conditions.

The goal of lignin methylolation is to introduce hydroxy methyl groups into the lignin structure, which can be achieved in an alkaline medium by reacting lignin with formaldehyde. The methylolated lignins can be used as phenol replacement in the PF resin.

The principal mechanism of formaldehyde addition to the lignosulfonate and the formation of hydroxy methyl groups is shown in Figure 19.2. Zhao et al. [18] studied the hydroxymethylation of pine kraft lignin, and the results showed that about 0.36 mole of $-CH_2OH$ per C_9 units was incorporated into the lignin molecule. About 0.33 mole was introduced into a C_5 unit of guaiacyl moieties via the Lederer–Manasse reaction [19]. Hu et al. [20] and Mu et al. [21] used hydroxymethylated lignin to substitute 40% of phenol in a PF resin, and they obtained a PF resin with low free formaldehyde. Lin et al. [22] found that there are less considerable differences in the properties between the PF resin and hydroxymethylated lignin-substituted PF resin once the level of substitution is higher than 30%. Nevertheless, the use of hydroxymethylated lignin resulted in a higher content of residual formaldehyde.

For lignosulfonate, Alonso et al. [3] found that methylolated lingosulfonate can substitute up to 35% of the phenol in the PF resin when the molar ratios of sodium hydroxide to phenol-methylolated ammonium lignonsulfonate and formaldehyde to phenol-methylolated ammonium lignosulfonate were 0.6 and 2.5, respectively. The synthesized resol resin using ammonium lignosulfonate shows similar properties as those of the the standard PF resin.

From the literature, it is well known that lignin-PF (LPF) resins prepared from methylolated lignins have much better properties than unmodified lignin. In this chapter, the lignosulfonate methylolation in an alkaline medium and their substitution in thermosetting PF resin are studied. The synthesized LPF is used as a resin matrix for nonwoven natural fiber reinforcements, and the properties of the composites are characterized in detail.

Another important versatile renewable raw material, which is currently investigated for synthesizing reactive bio-based thermoset resins, is vegetable oil. Vegetable oil is consumed

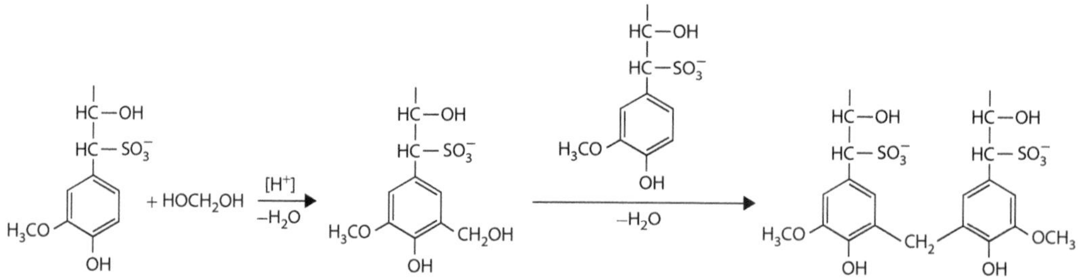

FIGURE 19.2 Methylolation reaction in the lignosulfonate structure. (From Perez J.M. et al., *BioResources*, 2, 270, 2007. With permission)

worldwide primarily in food, animal feed, and industrial sectors. Currently, the global vegetable oil production is around 138 million tons and it is expected to be 186 million tons by the end of the year 2020. The second major consumption of vegetable oil is for industrial applications where it is mainly used to make soaps and skin and other cosmetic products. Some of the oils are also used as drying agents in the paint industry. In the past few years, two new sources of demand have emerged for vegetable oil. One is biodiesel production, which has been rapidly accelerated and now consumes over a tenth of the global vegetable-oil crop, and the other one is the oleochemical industries. The value of vegetable oil in the polymer industry as a "green" raw material is recognized now, so it will be an important component in transitioning the chemical industry from nonrenewable to renewable resources [24]. In order to develop specific industrial applications and to attain some specific properties, the understanding of the chemistry of oil is necessary.

Regarding the chemical structure, vegetable oils are triglycerides formed by esterification of a molecule of glycerol and three fatty acids chains as can be seen in Figure 19.3. The fatty acid composition in vegetable oil varies depending on plant, crop, season, and growing conditions [25]. The most important parameters affecting the physical and chemical properties of the oils are the position and stereochemistry of the double bonds of the fatty acid chains, their degree of unsaturation, and the length of the carbon chain of the fatty acids. The length of the carbon chain varies from 14 to 22 carbons and from 0 to 3 double bonds per fatty acid. The double bonds in the fatty acid units are the important functionality from which interesting properties are expected. Other than double bond functionality, some vegetable oils have hydroxyl- or epoxy groups in the fatty acid chain.

The degree of unsaturation or the available functional groups in the triglyceride can be characterized by the iodine value. Additionally, it is reported that oils with a high iodine value can also be polymerized directly via cationic polymerization [26–28] or via thermal polymerization [29]. The direct polymerization of the double bonds does not provide polymeric materials with good mechanical or thermal properties because of a low degree of cross-linking. To improve the functional performance, the cross-linking density in the cured network needs to be increased, which is done by chemical modification of the available double bonds.

The modification of triglyceride oils and the synthesis of polymers from them are recently reviewed by Güner et al. [29]. A whole chapter in the book by Wool is dedicated to polymers derived from triglyceride oil [30]. There are some more recent reviews that describe the extensive use of modified triglycerides in synthesizing thermoset resins [31–33].

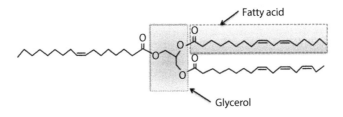

FIGURE 19.3 Chemical structure of a triglyceride.

The cross-linking density can be improved by incorporating suitable reactive groups into the fatty acid unit. The methods reported in the literature for the modification of the double bonds are acrylation, maleination, hydrogenation, halogenation, ozonolysis, dimerization, metathesis, epoxidation, and hydroxylation [34,35]. In this chapter, the following chemical modifications of the double bonds of the linseed oil are described: epoxidation, acrylation, and carbon dioxide insertion. The method of synthesizing epoxides, acrylates, and carbonates from linseed oil and their different characterization methods are reported.

The epoxidation of the double bonds is one of the most important functionalization reactions where a single oxygen atom is added into the double bond. The epoxidized oils are considered as promising intermediates for a broad range of applications such as coatings, casting resins, adhesive, inks, and lubricants. In this chapter, an environment-friendly procedure (chemoenzymatic epoxidation) alternative to the standard epoxidation method is investigated, and the cure kinetics of the epoxidized oil with anhydride curing agent is analyzed.

The epoxidized oil is also a precursor for synthesizing acrylated oil. The ring opening of the epoxy group leads to numerous products where some of them are reported in the literature [36,37]. Ring opening can also be done using acrylic acid, which gives a product with terminal acrylate groups [38]. The acrylated epoxidized oil (AELO) can be UV-cured after the addition of a photoinitiator that guarantees an immediate curing upon exposure. The advantages of UV curing are the solvent-free polymerization with a low energy consumption and fast curing rate. The fast curing of the AELO leads to shorter curing lines and cost reduction. The UV curing process does not need any external thermal energy. Hence, it can be used for coatings of heat-sensitive materials such as wood or wood-based materials. The finished products have unique surface characteristics such as good abrasion, solvent resistance, and high gloss.

The epoxidized oils can be used as precursors for synthesizing five-member cyclic carbonates that are used as reactive components for isocyanate-free polyurethanes. The isocyanate-free urethanes can be prepared through aminolysis reaction either by reacting five- and six-member cyclic bis-carbonates or oligomers terminated with five-member cyclic carbonate groups with polyamine. The resulting polyurethane network contains hydroxy groups in β-position to the urethane group [39,40]. The carbon dioxide insertion into the oxirane group of the epoxidized vegetable oil in the presence of a catalyst provides a five-member cyclic carbonate. The literature mainly focuses on cyclic carbonates derived from vegetable oil where the amines reacting with the cyclic carbonates are still derived from petrochemicals. Hence, for the first time, a bio-based polyamine (phenalkamine) from cashew nut shell liquid was used as an amine reactant. Phenalkamines are a class of Mannich bases obtained by the reaction of a cardanol, an aldehyde compound, and an amine. The phenalkamine contains an aliphatic polyaminic substituent attached to the aromatic ring that is responsible for the aminolysis reaction with cyclic carbonate. The aromatic backbone leads to a high chemical resistance where the hydrophobic side chains support the water resistance of the coating. The aminolysis reaction of carbonated linseed oil (CLSO) with phenalkamine yields to isocyanate-free polyurethanes. In this chapter, different parameters for the effective aminolysis reaction between the phenalkamines and cyclic carbonates are studied, and the coating properties of the isocyanate-free urethanes on an aluminum substrate are evaluated.

19.2 PHENOLIC BINDER FROM LIGNIN

LPF binder resin was synthesized using sodium lignosulfonate methylolated in the presence of formaldehyde. The methylolated lignosulfonate was used as a phenol substituent in the PF resin with an amount of substitution of 30 wt.%. The methylolation reaction was carried out at a pH of 9.4 for the duration of 4 to 5 h at a temperature of 45°C. The detailed experimental procedure of the methylolation reaction and the synthesis of the LPF binder using methylolated lignosulfonate are described in the literature [41].

During the methylolation process, the lignosulfonate reacts with formaldehyde and forms a lignin derivative with methylol groups. In the spectrum obtained by Fourier transform infrared spectroscopy (FTIR), the occurrence of the methylol groups of the modified lignosulfonate was confirmed by the peak in the range of 1460–1470 cm^{-1}. The differential scanning calorimetry (DSC) curing exotherms of both the pure PF resin and the LPF binder resin were compared. The onset of the curing of the LPF resin was observed at 119°C and the maximum peak temperature was found at 150°C, whereas the onset of the unsubstituted PF resin was 107°C and the maximum peak temperature was 144°C. The curing enthalpies of the PF and LPF resins are 95 J/g and 62 J/g, respectively. The LPF binder possesses a broader exothermic peak and a smaller curing enthalpy representing a lower reactivity and a longer curing time. This is caused by the higher average molecular weight of the LPF resin since it contains the high molecular weight lignin. The curing process is slow, still the viscosity of the resin causes an early gelation of the LPF binder. The gelation usually does not inhibit the curing process, but after gelation the transition of the kinetic process from chemical to diffusion-controlled mechanism can progressively decay the reaction rate.

To prepare natural fiber-reinforced composite (NFC), the following three different nonwoven fiber mats, flax (100%), flax/wood (25%/75%), and flax/kenaf (50%/50%) were impregnated with the synthesized LPF binder resin. The impregnated mats were compression molded at a temperature of 160°C, whereas the LPF binder content for the composites was around 30 wt.%. Water absorption (WA) and thickness swelling (TS) measurements were carried out for the composites containing both the PF and the LPF resin matrix. The WA values for the flax fiber composites are much higher than those for the other composites. This might be because of an improper fiber matrix interaction or higher void content. The WA values for the flax/wood and flax/kenaf reinforced composites are in the range of 25%–27%, which is quite acceptable for the NFC specimens having a binder content of only 30 wt.%. The flax/wood composites have higher TS values (26%) than the other composites, which is primarily because of penetration of water molecules into the cell walls of the wood fibers. The flax/kenaf fiber-reinforced composites have the lowest TS value (13%), which is because of a negligible loss in the packing density of the compressed flax/kenaf fibers after WA.

The flexural strengths of the NFC specimens show that the flax fiber-reinforced composite has a lower value than the other two composites, which is primarily because of the poor fiber/matrix interfacial adhesion. Scanning electron microscopy (SEM) also showed a weak interfacial adhesion between the flax fiber and the matrix resin. The reason might be because of improper wetting of the binder on the cell walls of the flax fiber. The flax/kenaf composite shows overall a better fiber/matrix interaction than the two other NFC samples, which was proven by the SEM micrograph [41]. The SEM micrograph of the flax/wood fiber composite shows several scattered domains of the matrix resin and a few protruded fibers. The domains are the region where proper dispersion of the wood-fiber reinforcement in the matrix resins is present.

The thermal stability of the natural fiber-reinforced composites was characterized using thermogravimetric (TG) and differential TG (DTG) analysis. A weight loss of 4%–5% was observed till 150°C for all of the three composites that corresponds to the loss of moisture. The flax and flax/kenaf fiber-reinforced composites have a shoulder between 250°C and 300°C [41], which is because of depolymerization of the hemicellulose and the pectin in the fibers. The incorporation of the LPF binder has increased the thermal stability of the natural fibers, which could be observed in the increase of the decomposition onset temperature. The fiber with a higher content of cellulose (flax) decomposed earlier than the two other fibers, which could be observed in the degradation curve in the temperature range from 150°C to 380°C [41]. The peak observed above 380°C in all DTG curves originates mainly from lignin decomposition and also from decomposition of cellulose and LPF resin residuals.

The renewable lignin can partially replace the toxic phenol in PF resins. The curing rate of the LPF binder resin shows a slower curing rate compared to the PF resin. The mechanical and thermal properties of the NFC samples prepared using LPF binder containing 30 wt.% of lignin are quite comparable with the standard PF resin. The nonwoven natural fiber reinforcements along with the LPF binder resin provide a very good mechanical strength, and also the mixing of two different natural fibers enhances the mechanical strength. The composites with short wood fibers provide a reasonable strength along with flax fibers because of the very good resin/matrix interaction.

19.3 ISOCYANATE-FREE POLYURETHANES FROM VEGETABLE OIL

19.3.1 Synthesis, Characterization, and Coating Properties

The isocyanate-free polyurethanes from vegetable oil were prepared by converting the double bonds of the fatty acid chains into cyclic carbonates and by the subsequent reaction with phenalkamine. This was done by reacting epoxidized linseed oil with carbon dioxide in the presence of tetrabutylammonium bromide as a catalyst. The experimental procedure for the preparation of CLSO is described in the literature [42].

During the carbon dioxide insertion into the oxirane ring, the disappearance of the oxirane ring was observed at 840–820 cm^{-1} and the appearance of the new cyclic carbonate at 1803 cm^{-1} was noted in the FTIR spectrum. After 72 h of reaction time, the viscosity increased from 1.05 to 163 Pa·s and the weight-average molar mass of the CLSO changed from 1100 to 2600 g/mol. The increase of both the viscosity and the molecular weight was primarily because of the conversion of epoxy into five-member cyclic carbonate.

The reaction of the cyclic carbonate with the amine was characterized using real-time FTIR. The synthesized CLSO was mixed with the phenalkamine at a weight ratio of 1:1 and the structural changes during the reaction at 80°C and 100°C were monitored. The C=O group free of hydrogen bonding at the bands of 1725 and 1710 cm^{-1} proved the existence of urethane linkages. The strong C=O stretch at 1710 cm^{-1} and the presence of a band at 1545 cm^{-1} are strongly coupled to the C–N stretching vibration of the urethane linkage. A weak band around 3300 to 3500 cm^{-1} corresponds to the N–H vibration (Figure 19.4). The intensity of the phenolic hydroxyl group at 1364 cm^{-1} has also been reduced to some extent, which might be because of the hydroxy alkylation reaction of phenolic hydroxyl present in the phenalkamine with cyclic carbonate.

The percentage of cyclic carbonate conversion during the reaction time of 129 min at 80°C and 100°C was 83% and 89%, respectively. A fast decay of the cyclic carbonate was observed at the initial stage of curing reaction. After 65% of conversion, the reaction rate decreased and slowly reached the maximum degree of conversion. The rheological studies show that during the cross-linking reaction both the elastic and the viscous moduli increased. The elastic modulus expressed by the storage modulus G' increased faster than the viscous one expressed by the loss modulus G''. This shows the elastic nature of the cross-linked network. The time that is required to reach the crossover of G' and G'' is the gelation time. The crossover or gelation time for the samples at 100°C and 80°C are 8 min and 20 min, respectively (Figure 19.5). The decrease of the curing rate after gelation was because of the formation of a viscoelastic network that was characterized by the change in cure kinetics predicted using Vyazovkin's model-free kinetic method. The activation energy regime shows that during the initial stage a slow increase of the activation energy

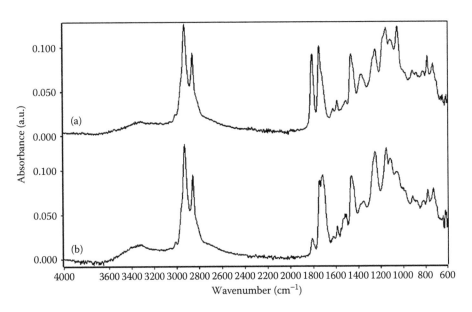

FIGURE 19.4   FTIR spectra of the reaction mixture of CLSO and Phenalkamie at 0 (a) and 120 min (b) of curing time at 80°C. (Reprinted with permission from Mahendran A.R. et al., *J. Poly. Environ.*, 20, 926, 2012.)

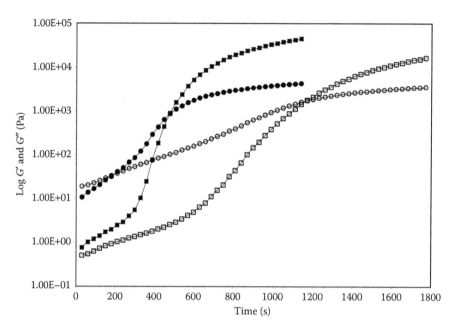

FIGURE 19.5   Elastic G' (open square, filled square) and viscous G'' (open circle, filled circle) modulus of the reaction mixture of CLSO and phenalkamine during curing process at 80°C (open symbols) and 100°C (filled symbols). (Reprinted with permission from Mahendran A. R. et al., *J. Poly. Environ., 20,* 926, 2012.)

was observed until 20% of conversion. After the initiation process, the electron-releasing substituent attached to the cyclic carbonate group decreases the partial positive charge of the carbonyl carbon. This finally results in a decrease of the reactivity of the cyclic carbonate.

Besides the structural changes mentioned earlier, the change in chain mobility during the curing process is also a reason for the kinetic changes. During the initial stage of cure, the viscosity is very low, so high chain mobility is given and the reactive sites are easily available for cross-linking. The aminolysis reaction leads to a rapid increase in the viscosity and finally inhibits the completion of the reaction because of the restricted mobility of the longer chains. The system moves to a glassy stage that leads to a rapid increase in the activation energy at 65% of conversion.

From these investigations, it is concluded that an increase in cure temperature also increases the percentage of cyclic carbonate conversion. The conversion was not directly proportional to the appearance of the new carbonyl peak in the FTIR spectra, which showed that some side reactions occurred during the ring opening reaction of the cyclic carbonate. The rheological and chemical changes during the aminolysis reaction caused a complex activation energy regime that can be well predicted using the model-free kinetic method.

From the previous results, it is understood that amine concentration and temperature of curing influence the properties of the polyurethane network. Hence, porous-free urethane coatings were prepared on an aluminum substrate using synthesized CLSO and polyamine at three different mixing ratios (1:0.5, 1:0.75, and 1:1.25) and at three different cure temperatures (60°C, 80°C, and 100°C). The formation of urethane linkages and the rate of cyclic carbonate conversion during the aminolysis reaction were characterized using FTIR spectroscopy (Figure 19.6).

The thermal behavior of the cured urethane films was characterized using thermogravimetry. The effects of amine concentration and cure temperature on the optical, mechanical,

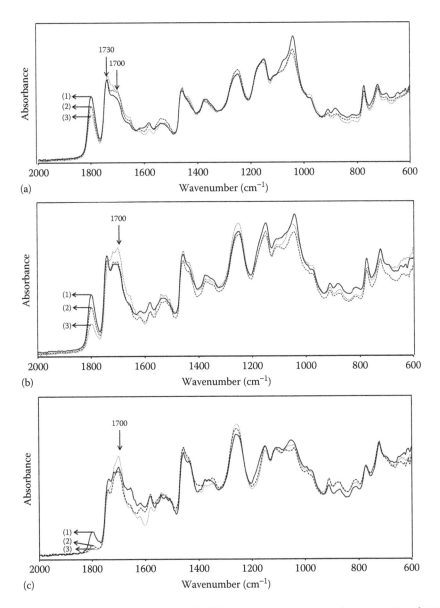

FIGURE 19.6 (a) FTIR spectra of the cured CLSO/phenalkamine samples at a ratio of 1:0.5 at (1) 60°C, (2) 80°C, and (3) 100°C; (b) FTIR spectra of the cured CLSO/phenalkamine samples at a ratio of 1:0.75 at (1) 60°C, (2) 80°C, and (3) 100°C; and (c) FTIR spectra of the cured CLSO/phenalkamine samples at a ratio of 1:1.25 at (1) 60°C, (2) 80°C, and (3) 100°C. (Reprinted with permission from Mahendran A. R. et al., *J. Coat. Technol. Res.* 11, 329, 2014.)

and end-use properties of the coatings were evaluated by measuring specular gloss, solvent resistance, cross-hatch, solvent swelling, and impact strengths [43]. The CLSO/phenalkamine aminolysis reaction rate was highly depending on the cure temperature, which could be observed from real-time FTIR studies. In terms of thermal stability, the increase in the cure temperature shifted the onset decomposition temperature to higher values and improved the thermal stability. On the other hand, the thermal stability decreased with an increase in amine concentration. The coating properties such as solvent resistance and cross-hatch resistance could be improved by increasing the cure temperature. The specular gloss does not show much difference with respect to an increase in the cure temperature, but the swelling resistance could be improved. The impact strength of the coating slightly increased with an increase in amine concentration. It is concluded that the acceleration of the cure process highly depends on the cure temperature. At the same time, the increase in the cure temperature leads to a change in the color of the specimen from light to dark brown. The sample prepared at a CLSO/phenalkamine ratio of 1:0.75 cured at 100°C showed the best overall performance in coating properties with a less color change compared to all other CLSO/amine concentrations. The increase in cure temperature is directly proportional to the rate of the aminolysis reaction and to the coating performances.

19.4 UV-CURABLE ACRYLATES FROM VEGETABLE OIL

Volatile organic compound (VOC)-free photocross-linkable coatings were prepared from linseed oil by converting the epoxy to an acrylate group. To obtain acrylated linseed oil (ALSO), the epoxidized oil was reacted with acrylic acid in the presence of triphenylphosphine. A detailed description for synthesizing acrylated oil can be found in the literature [44]. The acrylated linseed oil was cured under UV light in the presence of three different photoinitiators: benzophenone (photoinitiator A), 2,4-diethyl-9H-thioxanthen-9-one (photoinitiator B), and 2-hydroxy-2-methyl-phenyl-propane-1-one (photoinitiator C), each at two different concentrations (1 = low and 2 = high). The rate of double bond conversion with respect to exposure time was determined using real-time FTIR. The ALSO was coated on wood substrates, and the surface properties were characterized.

The FTIR spectrum of the ALSO shows terminal acrylate groups that were identified by the double bond peak at 1406 cm^{-1}. To compare the efficiency of three photoinitiators, the UV light absorbance for all of the three photoinitiator systems was kept constant, and the photoinitiator concentrations in the ALSO mixtures were adjusted to a uniform absorbance value of 0.1 absorbance units at the wavelength of 365 nm. The extinction coefficient values for the photoinitiators A, B, and C at the wavelength of 365 nm were 46, 2811, and 18 L mol^{-1}cm^{-1}, respectively.

In the real-time FTIR studies, the changes of the double bond peaks at 1406 cm^{-1} were observed and plotted as a function of exposure time [44]. The conversion of the double bonds started very rapidly after an initial induction time of about 20 s for both concentrations A1 and A2 of photoinitiator A. After 600 s of exposure time, the double bond conversion has reached a maximum of 90% and 95% for A1 and A2, respectively [44]. The initial induction time for photoinitiator B was very short compared to photoinitiator A. It was less than 10 s for B1 and less than 20 s for B2. After 600 seconds of exposure time, the double bond conversion has reached a maximum of 78% and 74% for B1 and B2 samples, respectively. The induction time for photoinitiator C was less than 10 s for both concentrations C1 and C2. Photoinitiator C showed a complete disappearance of the acrylate double bonds after 160 s for C1 and 70 s for C2. Among the three photoinitiators, photoinitiator C was found to be the most efficient one for the ALSO because it yielded to a 100% conversion within a very short exposure time.

The surface properties of the cured acrylated linseed oil on wood surfaces were characterized. The samples coated with the acrylated linseed oil show a very high gloss finish except the sample containing the high amount of the phototiniatior A, A2. Samples B1 and A1 show very good scratch resistance, whereas samples C1 and C2 show only a poor scratch resistance. The coatings with photoinitiator A show very good resistance against the solvent methyl ethyl ketone compared to photoinitiators B and C. The adhesion properties were evaluated using a cross-hatch test. The results show that samples cured with photoinitiators A and B had no visible destruction on the cross-hatch patterns, which means none of the squares of the lattice was detached. The samples cured with photoinitiator C show some loss in adhesion with the substrate and also show a class 2 adhesion in the cross-hatch test. The class 2 adhesion denotes that coatings have flaked along the edges and at the intersections of the cuts. A crosscut area significantly greater than 5%, but not significantly greater than 15%, is affected.

From these investigations, it is concluded that ALSO cured under UV light provides very good surface properties. Photoinitiator C shows a complete conversion of the acrylate double bond within a very short exposure time. Hence, it was identified as the most efficient photoinitiator for the ALSO. In terms of surface properties, the wood surfaces coated and cured with photoinitiator A and photoinitiator B show good results in the scratch resistance, solvent resistance, and cross-hatch tests. The coatings with photoinitiator C failed to achieve a good coating performance. Overall, photoinitiator B shows the best technical performance and very good surface properties, which can be achieved even at a low photoinitiator concentration.

19.5 EPOXIDES DERIVED FROM VEGETABLE OIL BY CHEMOENZYMATIC METHODS

Bio-based epoxides from vegetable oil are obtained by converting the double bonds of the fatty acids of the triglycerides to oxirane rings in the presence of a catalyst. The standard industrial procedure is based on an *in situ* epoxidation method [45]. The major drawback of this method is an increased risk of equipment corrosion and potential danger of hydrolysis of the ester groups by an undesired ring opening because of reaction with water. Therefore, the lipase enzyme catalyzed epoxidation was proposed by Rusch gen Klass et al. [46,47]. The advantages are that enzyme catalysts are reusable and regioselective, and the reaction can be carried out under mild conditions. In the first step, the unsaturated fatty acids are converted to peroxy fatty acid as an intermediate, which is followed by an auto oxidation step (Prileshajew epoxidation) that results in the formation of an epoxy and an acid [48]. Some researchers already studied the effect of different reaction parameters on the amount of epoxidation in soybean oil with immobilized *Candida antarctica* lipase B (CALB) enzyme catalyst [49–51]. The results reported from previous studies show that catalyst concentration and temperature are the most important factors influencing the degree of epoxidation. Hence, this was investigated for linseed oil in order to find an optimum catalyst concentration for epoxidation. The experiments were carried out using the CALB catalyst at four different concentrations (2%, 5%, 10%, and 20%) in a solvent for the duration of 24 h. The detailed procedure followed for the epoxidation experiment is described in the literature [52]. The structural changes during epoxidation and the amount of epoxy ring formation were characterized using FTIR and proton nuclear magnetic resonance spectrometer (^{1}H NMR) (Figure 19.7).

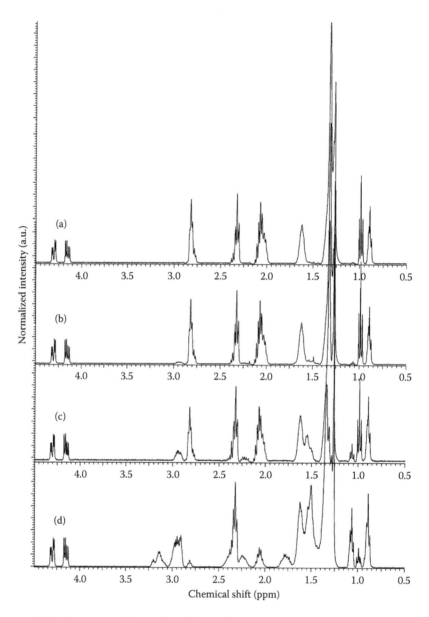

FIGURE 19.7 ^{1}H NMR spectra of epoxidized linseed oil at various catalyst concentrations after 24 h reaction time: (a) 2%, (b) 5%, (c) 10%, and (d) 20%. (Reprinted with permission from Mahendran A.R. et al., *Macromol. Symp.*, 311, 18, 2012.)

The FTIR spectrum showed a decrease in the intensity of the double bonds with an increasing catalyst concentration. Also a new doublet peak at 822 and 833 cm^{-1} was observed because of the occurrence of the oxirane ring. The NMR spectrum did not show a significant change in the spectrum for 2% catalyst concentration, whereas catalyst concentrations of 5%, 10%, and 20% showed a degree of 38%, 76%, and 96% of epoxidation. The percentage of epoxidation was calculated from the amount of double bond conversion. The increase in catalyst concentration also increased the amount of epoxidation without any side reactions [52].

The next step is the curing of the epoxidized linseed oil that can be performed with anhydride, amine, or polyamide as curing agents. Anhydrides are preferred as hardeners for composite matrix resins, where a high heat distortion temperature is needed. Hence, Mahendran et al. [53] investigated the curing kinetics of epoxidized linseed oil with nadic methyl anhydride (NMA) hardener in the presence of 1-methyl imidazole as initiator and catalyst. The advantage of using NMA as hardener is that it is liquid at room temperature and therefore can be mixed homogeneously with the epoxidized linseed oil. Also a high glass transition temperature can be achieved for the cured network. Thermal cure characteristics were investigated using DSC, and the curing kinetics was predicted using the model-free kinetic method. The experiments were carried out by mixing epoxidized linseed oil with NMA and 1-methyl imidazole at a ratio of 1:0.8:0.18 (wt.%). The dynamic DSC measurements were performed for the resin mixtures at three different heating rates, 10°C, 15°C, and 20°C min^{-1}, and the curing kinetics was predicted using the following three model-free kinetic methods: Friedman (FR), Kissinger–Akahira–Sunose (KAS), and Vyazovkin (VA). The theory of model-free kinetics and the methods of predicting apparent activation energy using model-free kinetics are described in the literature [54]. The model-free kinetic method shows the change in the apparent activation energy with respect to conversion during the course of the reaction.

The results show that the predicted activation energy regime using the KAS method had a gradual decrease in activation energy at the initial stage of the cure (until 20% of conversion), and it remained constant until 40% of conversion. After 40% of conversion, it decreases until the final completion of the curing process. The activation energy regime for the FR and VA method showed an increase in activation energy until 20% of conversion and tends to fall until the final completion of the cure (Figure 19.8).

Both the FR and VA methods had a similar trend of the activation energy. The decrease of the activation energy after 20% of conversion was explained by a diffusion-controlled kinetics that can be caused by two phenomena, gelation and vitrification. Gelation does not usually inhibit the process of curing and cannot be detected by the techniques, which are sensitive only to chemical reaction, like DSC. Hence the gelation was characterized using a plate/plate rheometer. During vitrification, the rate of reaction usually undergoes a significant decrease and thereafter the diffusion of reactants is controlled. Hence the occurrence of vitrification during the cure can be ascertained by a significant decrease in complex heat capacity, which was experimentally determined using alternating differential scanning calorimetry (ADSC) [55].

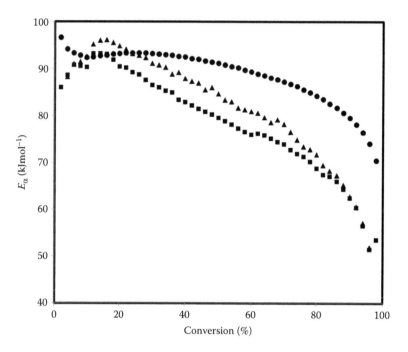

FIGURE 19.8 Dependence of activation energy on conversion for the curing of epoxidized linseed oil using three different isoconversional methods (filled triangle FR, filled circle KAS, and filled square VA). (Reprinted with permission from Mahendran A. R. et al., *J. Therm. Anal. Calorim.* 107, 989, 2012.)

In the ADSC curve, a fall in complex heat capacity was observed for the epoxidized linseed oil, which showed the occurrence of vitrification during cure. The glass transition temperature of the cured epoxidized linseed oil (Tg_{∞}) was determined from the ADSC heat capacity curve, and it was found to be 176°C. The vitrification may occur if it is cured isothermally below this temperature. Hence the kinetic models were evaluated by comparing the measured isothermal values with the predicted isothermal values at the temperature of 180°C (above 176°C has been chosen to avoid vitrification). After comparisons, the FR and VA methods provided more consistent and accurate functions of the activation energy than the KAS method. The VA method resulted in much closer values to the measured value than the other two methods (Figure 19.9).

From these investigations, it is concluded that the FR and VA methods provide consistent apparent activation energy values for the complex epoxidized linseed oil–anhydride cure system. Although the curing mechanism is complex, the model-free kinetic methods provide a complete insight into the curing mechanism of epoxidized linseed oil with anhydride hardener. At the initial stage of conversion, an increase in activation energy was observed, which might be because of the slow initiation mechanism. Gelation occurred at around 18% of conversion and was determined quantitatively from rheological measurements. After gelation, a decrease in activation energy was observed, which is because of the transition from chemical to diffusion-controlled process irrespective of the heating rate. At the lower heating rate, the diffusion-controlled process was predominantly influenced by the

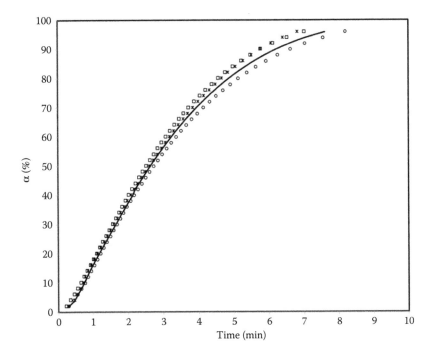

FIGURE 19.9 Comparison of experimental degree of cure with predicted degree of cure by FR, KAS, VA methods for the curing temperature 180°C (open circle KAS, open square VA, open diamond FR, and continuous line isothermal at 180°C). (Reprinted with permission from Mahendran A. R. et al., *J. Therm. Anal. Calorim.* 107, 989, 2012.)

chemo-rheological effect called vitrification, which was observed by the abrupt decrease in the complex heat capacity. During the nonisothermal cure, the vitrification is followed by devitrification, which was also identified using ADSC measurement. Finally, the validation experiments showed that the VA method predicts more precisely the experimental values than the other two methods.

The overall results of this chapter show that renewable resources such as vegetable oils or lignins are capable of substituting the fossil fuel derived organic chemicals in the thermoset industries. In future, advanced characterization and cost-effective industrial production methods could achieve "green" products from renewable resources for their application in composites, construction, and other sectors.

REFERENCES

1. Ragauskas, A. J. Technical review. Lignin Overview, Georgia Institute of Technology, Atlanta, Georgia. http://www.ipst.gatech.edu/faculty/ragauskas_art/technical_reviews/Lignin Overview. pdf.
2. Gruber E., Grundlagen der Zellstofftechnologie, Vorlesungsskriptum zum Lehrgang Papiertechnik, Berufsakademie Karlsruhe. http://www.gruberscript.net/13Sulfitaufschluss.pdf
3. Alonso, M. V., Oliet, M., Rodriguez, F., Astarloa, G., and Echeverria, J. M. 2004. *J. Appl. Polym. Sci.* 94: 643–650.
4. Thring, R. W., Vanderlaan, M. N., and Griffin, S. L. 1997. *Biomass Bioenergy* 13: 125–132.
5. Stewart, D. 2008. *Ind. Crops. Prod.* 27: 202–207.

6. Nonaka, Y., Tomita. B., and Hatano, Y. 1997. *Holzforschung* 51: 183–187.
7. Kou, M., Hse, C. Y., and Huang, D. H. 1991. *Holzforschung* 45: 47–51.
8. Danielson, B. and Simonson, R. J. 1998. *Adhes. Sci. Technol.* 12: 941–946.
9. Ysbrandy, R. E., Sanderson, R. D., and Gerischer, G. F. R. 1992. *Holzforschung* 46: 249–252.
10. Pecina, H., Kuhne, G., Bernaczyk, Z., and Wienhaus, O. 1991. *Holz als Roh- und Werkstoff* 49: 391–397.
11. Pecina, H., Kuhne, G., Bernaczyk, Z., and Wienhaus, O. 1992. *Holz als Roh- und Werkstoff* 50: 407–409.
12. Olivares, M., Aceituno, H., Neimann, G., Rivera, E., and Sellers, T. 1995. *Forest Prod. J.* 45: 63–67.
13. Gardner, D. J. and Mc Ginnis, G. D. 1988. *J. Wood Chem. Technol.* 8: 261–288.
14. El Mansouri, N. E., Yuan, Q., and Huang, F. 2011. *Bioresources* 6: 2647–2662.
15. Dolenko, A. J. and Clarke, M. R. 1978. *Forest Prod. J.* 28: 41–46.
16. Alonso, M. V., Rodriguez, J. J., Oliet, M., Rodriguez, F., Garcia, J., and Gillaranz, M. A. 2001. *J. Appl. Polym. Sci.* 82: 2661–2668.
17. Alonso, M. V., Oliet, M., Rodriguez, F., Garcia, J., Gilarranz, M. A., and Rodriguez, J. J. 2005. *Bioresource Tech.* 96: 1013–1018.
18. Zhao, L. W., Griggs, B. F., Chen, C. L., Gratzl, J. S., and Hse, C. Y. 1994. *J. Wood Chem. Tech.* 14: 127–145.
19. Tyagi R. 2005. Chapter 67, Lederer–Manasse reaction. In: *Organic Reactions (Mechanisms with Problems)*. New Delhi, India: Discovery Publishing House, pp. 289–291.
20. Hu, L., Pan, H., Zhou, Y., and Zhang, M. 2011. *Bioresource Tech.* 6: 3515–3525.
21. Mu, Y. B., Wang, C. P., Zhao, L. W., and Chu, F. X. 2009. *Ind. Forest. Prod.* 29: 38–42.
22. Lin, Z. X., Ouyang, X. P., Yang, D. J., Deng, Y. H., and Qiu, X. Q. 2010. *World Sci-Tech R&D* 32: 348–351.
23. Perez, J. M., Rodriguez, F., Alonso, M. V., Oliet, M., and Echeverria, J. M. 2007. *BioResources* 2: 270–283.
24. Meier, M. A. R., Metzger, J. O., and Schubert, U. S. 2007. *Chem. Soc. Rev.* 36: 1788–1802.
25. Gunstone, F. D., Harwood, J. L., and Dijkstra, A. J. (Eds.). 2007. *The Lipid Handbook*. Boca Raton, FL: Taylor & Francis Group.
26. Li, F. and Larock, R. C. 2001. *J. Appl. Polym. Sci.* 80: 658–670.
27. Li, F. and Larock, R. C. 2002. *Polym. Adv. Technol.* 13: 436–449.
28. Meiorin, C., Aranguren, M. I., and Mosiewicki, A. 2012. *J. Appl. Polym. Sci.* 124: 5071–5078
29. Güner, F. S., Yagci, Y., and Erciyes, A. T. 2006. *Prog. Polym. Sci.* 31: 633–670.
30. Wool, R. P. 2005. Chapter 4. Polymer and composite from plant oils. In: Wool, R. P. and Su, X. S. (Eds.). *Bio-Based polymers and Composites*, 1st edn. Burlington, NJ: Elsevier Academic Press, pp. 56–113.
31. Raquez, J. M., Deléglise, M., Lacrampe, M. F., and Krawczak, P. 2010. *Prog. Polym. Sci.* 35: 487–509.
32. Montero de Espinosa, L. and Meier, M. A. R. 2011. *Eur. Polym. J.* 4: 837–852.
33. Galià, M., de Espinosa, L. M., Ronda, J. C., Lligadas, G., and Cádiz, V. 2010. *Eur. J. Lipid Sci. Technol.* 112: 87–96.
34. Pelletier, H. and Gandini, A. 2006. *Eur. J. Lipid Sci. Technol.* 108: 411–420.
35. Khot, S. N., La Scala, J. J., Can, E., Morye, S. S., Palmese, G. R., Kusefoglu, S. H., and Wool, R. P. 2001. *J. Appl. Polym. Sci.* 82: 703–723.
36. Petrović, Z. S., Guo, A., and Zhang, W. 2000. *J. Polym. Sci. Part A: Polym. Chem.* 38: 4062–4069
37. Behera, D. and Banthia, A. K. 2008. *J. App. Polym. Sci.* 109: 2583–2590.
38. Elliott, W. T. 1993. Alkyd resins. In: Parsons, P. (Ed.), *Surface Coatings: Raw Materials and Their Usage*, 3rd edn. London: Chapman & Hall, pp. 76–109.
39. Rokicki, G. and Łaziński. R. 1989. *Angew. Makromol. Chem.* 170: 211–225.
40. Tomita, H., Sanda, F., and Endo, T. 2001. *J. Polym. Sci. Part A: Polym. Chem.* 39: 860.

41. Mahendran, A. R., Wuzella, G., Aust, N., Müller, U., and Kandelbauer, A. 2013. *Polym. and Polym. Comp.* 21: 199–206.
42. Mahendran, A. R., Aust, N., Wuzella, G., Müller, U., and Kandelbauer, A. 2012. *J. Poly. Environ.* 20: 926–931.
43. Mahendran, A. R., Wuzella, G., Aust, N., and Müller, U. 2014. *J. Coat. Technol. Res.* 11: 329–339.
44. Mahendran, A. R., Wuzella, G., Aust, N., Kandelbauer, A., and Müller, U. 2011. *Prog. Org. Coat.* 74: 697–704.
45. Baumann, H., Bühler, M., Fochem, H., Hirsinger, F., Zoebelein, H., and Falbe, J. 1988. *Angew. Chem. Int. Ed. Engl.* 27: 41–62.
46. Rusch gen Klass, M. and Warwel, S. 1997. *J. Mol. Catal. A Chem.* 117: 311–319.
47. Rusch gen Klass, M. and Warwel, S. 1999. *Ind. Crops. Prod.* 9: 125–132.
48. Prileschajew, N. 1909. *Berichte der deutschen chemischen Gesellschaft* 42: 4811–4815.
49. Rusch gen Klass, M. and Warwel, S. 1995. *J. Mol. Catal. A Chem.* 1: 29–35.
50. Vlcek, T. and Petrovic. 2006. *J. Am. Oil Chem. Soc.* 83: 247–252.
51. Ikhuoria, E. U., Obuleke, R. O., and Okieimen, F. E. 2007. *J. Macromol. Sci. Part A: Pure.* 44: 235–238.
52. Mahendran, A. R., Wuzella, G., Aust, N., and Kandelbauer, A. 2012. *Macromol. Symp.* 311: 18–27.
53. Mahendran, A. R., Wuzella, G., Kandelbauer, A., and Aust, N. 2012. *J. Therm. Anal. Calorim.* 107: 989–998.
54. Kandelbauer, A., Wuzella, G., Mahendran, A. R., Taudes, I., and Widsten, P. 2009. *Chem. Eng. J.* 152: 556–565.
55. Reading, M. and Douglas, J. S. 2006. *Modulated Temperature Differential Scanning Calorimetry*, Vol. 6. Springer, New York.

CHAPTER 20

Nanocellulose-Based Bionanocomposites

Cintil Jose Chirayil, Merin Sara Thomas, B. Deepa,
Laly A. Pothen, and Sabu Thomas

CONTENTS

20.1 INTRODUCTION

Cellulose, the world's most abundant, natural, renewable, biodegradable polymer, is a classic example of reinforcing element, which occurs as whisker-like microfibrils that are biosynthesized and deposited in a continuous fashion. This quite "primitive" polymer can be used to create high-performance nanocomposites presenting outstanding properties. This reinforcing capability results from the intrinsic chemical nature of cellulose and from its hierarchical structure. During the past decade, many works have been devoted to mimic biocomposites by blending cellulose whiskers from different sources with polymer matrices [1]. There has been an expanding search for new materials with high performance at affordable costs in recent years. The potential of nanocomposites in various sectors of research and application is promising and attracting increasing investments. The production of nanoscale cellulose fibers and their application in composite materials have gained increasing attention due to their high strength and stiffness combined with low weight, biodegradability, and renewability. The main reason to utilize cellulose nanofibers in composite materials is due to its high stiffness [2]. Nanocomposites, in general, are two-phase materials in which one of the phases has at least one dimension in the nanometer range (1–100 nm). A number of researchers have, therefore, explored the concept of fully bio-derived nanocomposites as a route for the development of bioplastics or bioresins with better properties and this has been the subject of recent reviews [3–7].

20.1.1 Bionanocomposites

New nanocomposite materials with original properties were obtained by physical incorporation of cellulose whiskers into a polymeric matrix. Similar to multiphase materials, the properties of these cellulosic nanocomposites depend on two constituents, namely, whiskers and matrix [8]. Different types of nanocellulose-based bionanocomposites are reported in the literature.

The first publication reporting the preparation of cellulose nanocrystal (CN)-reinforced polymer nanocomposites was carried out using a latex obtained by the copolymerization of styrene and butyl acrylate (poly(S-co-BuA)) and tunicin whiskers [9] in 1995. The same copolymer has been used in association with wheat straw [10] or sugar beet [11] CNs. Other latexes, such as poly(hydroxyoctanoate) (PHO) [12], polyvinylchloride (PVC) [13], waterborne epoxy [14], natural rubber (NR) [15], and polyvinyl acetate (PVAc) [16], have also been used as matrices. Recently, stable aqueous nanocomposite dispersions containing cellulose whiskers and a poly(styrene-co-hexyl-acrylate) matrix were prepared via miniemulsion polymerization [17]. The addition of a reactive silane was used to stabilize the dispersion. Solid nanocomposite films can be obtained by mixing and casting the two aqueous suspensions followed by water evaporation. Alternative methods consist of freeze-drying and hot-pressing or freeze-drying, extruding, and hot-pressing the mixture. The preparation of cellulosic particle-reinforced starch [18], chitosan [19], silk fibroin [20], poly(oxyethylene) (POE) [21], polyvinyl alcohol (PVA) [22], hydroxypropyl cellulose (HPC) [22], carboxymethyl cellulose (CMC) [23], or soy protein isolate (SPI) [24] has been reported in the literature.

20.2 PROCESSING OF NANOCELLULOSE-BASED BIONANOCOMPOSITES

The first publication reporting the preparation of CN-reinforced polymer nanocomposites described a method using a latex of (poly(S-*co*-BuA)) and tunicin whiskers [25]. The nanocomposite films were obtained by water evaporation and particle coalescence at room temperature, that is, at a temperature higher than glass transition temperature (T_g) of the polymeric matrix (around 0°C). The mechanical properties (shear modulus) of the obtained nanocomposites increased by more than two orders of magnitude in the rubbery state of the polymeric matrix, when the nanocrystal content was 6% (w/w). An alternative way to prepare nonpolar polymer nanocomposites reinforced with CNs consists of their dispersion in an adequate (with regard to the matrix) organic medium. Coating with a surfactant or surface chemical modification of the nanoparticles can be considered. The overall objective is to reduce their surface energy in order to improve their dispersibility and compatibility with the nonpolar media. Long-chain surface chemical modification of cellulosic nanoparticles consisting of grafting agents bearing a reactive end group and a long "compatibilizing" tail has also been reported in the literature [26]. The general objective is, of course, to increase the apolar character of the nanoparticles. Various processing methods of nanocellulose-based composites are discussed in Sections 20.2.1 through 20.2.4

20.2.1 Casting and Evaporation

Casting–evaporation is one of the most common processes used to produce nanocomposite films. In general, CN is dispersed within a given medium, typically water, but various organic media have also been used and then polymer solutions are mixed with the CN dispersion. To achieve good reinforcement, the dispersibility of CN in both the polymeric matrix and processing solvents is critical. Composite films can then be produced from this mixture via casting on a suitable surface followed by evaporation. For example, nanocomposite films based on ethylene oxide/epichlorohydrin and CN were produced by dispersion casting of CN fillers in tetrahydrofuran (THF)/water mixtures [27]. CNs without surface modification have intrinsically strong interactions and have been reported as notoriously difficult to disperse. Moreover, this issue is exacerbated when the CN dispersions are dried before nanocomposite processing, which generally implies that drying and redispersion of CNs without aggregation is challenging. Due to the hydrophilic character of CNs, the simplest polymer systems that incorporate CNs are water-based. Never-dried aqueous dispersions of CNs are simply mixed with aqueous polymer solutions or dispersions. However, these systems suffer from limited utility and are only appropriate for water-soluble or -dispersible polymers such as latexes. The combination of aqueous solutions of polymers with aqueous CN suspensions has been reported [28]. The use of polar solvents, most commonly *N,N*-dimethylformamide (DMF), with CNs with no surface modification has been explored [29]. Solvents nonmiscible with water and with low polarity, such as toluene, have also been widely used. However, a drawback is that the process requires solvent exchange steps.

The properties of nanocomposite films based on ethylene oxide/epichlorohydrin and nanocrystalline cellulose (NCC), which was produced by dispersion-casting of NCC fillers in THF/water mixtures, were reported [30,31]. The polymer grafting is beneficial for dispersing NCC and to formulate nanocomposites in nonpolar solvents. Habibi and Dufresne reported nanocomposite films using unmodified and polycaprolactone (PCL)-grafted NCC nanoparticles as filler and PCL as the matrix, and they found that PCL-grafted nanoparticles were easily dispersed when compared to the unmodified system [32]. They demonstrated that the transformation of NCC nanoparticles into a cocontinuous material through long-chain surface chemical modification represents a new and promising way for the processing of nanocomposite materials.

Owing to the fact that acetone is miscible with water, it usually serves as carrier to transfer CNs from water to organic solvents. Freeze-drying and redispersion of CNs from tunicate in toluene were used to integrate these fillers into atactic polypropylene, but strong aggregation occurred [33]. Freeze-dried CNs were successfully redispersed in dipolar aprotic solvents, such as DMSO and DMF, containing small amounts of water (0.1%), and it has been possible to obtain films of these suspensions by the casting evaporation technique [34]. More recently, Van den Berg et al. investigated the factors limiting the dispersibility of CNs extracted from tunicate via hydrochloric or sulfuric acid hydrolyses [35]. Another approach already considered in this review to change interactions of CNs is via surface modifications. Such approach can break the percolating hydrogen-bonded network and affect the macroscopic mechanical properties of the resulting nanocomposite. Chemical modifications have been explored to improve dispersibility of CNs in a wide range of organic solvents, from medium to low polarity [36]. This approach also allows the manipulation of dried CNs because it facilitates freeze-drying and dispersion.

20.2.2 Electrospinning

Electrostatic fiber spinning or "electrospinning" is a versatile method for preparing fibers with diameters ranging from several microns down to 100 nm through the action of electrostatic forces [37]. Electrospinning is a fast and simple process to produce polymeric filaments, and this approach has been widely studied [38–41]. In general, CN is dispersed within a given medium (typically 0.05–5 wt.% solids) and then polymer solutions are mixed with the CN dispersion. Using this technique, the solvent evaporates as the fiber moves between the source and the collector and the polymer coagulates, forming a composite fiber. The CN-reinforced fibers can be further dried and/or made to go through additional treatments (e.g., heating to cross-link the matrix polymer). Electrospinning shares characteristics of both electrospraying and conventional solution dry spinning of fibers. The process is noninvasive and does not require the use of coagulation chemistry or high temperatures to produce solid threads from the solution. This makes the process particularly suited for the production of fibers using large and complex molecules. Bacterial cellulose whiskers were incorporated into POE nanofibers with a diameter of less than 1 μm by electrospinning to enhance the mechanical properties of the electrospun fibers [38]. The whiskers were found to be globally well embedded and aligned inside the fibers, even though they were partially aggregated. Likewise, electrospun PVA fiber mats loaded with CNs (Figure 20.1), with diameters in the nanoscale range and enhanced mechanical properties, were successfully produced [42]. A nonionic surfactant, sorbitan monostearate, was used to improve the dispersion of the particles in the hydrophobic polystyrene (PS) matrix, while surface grafting of the long chains was used with PCL.

Electrospun PVA and PCL nanofibers reinforced with CN were also obtained by electrospinning [42,43]. The new composites from PCL and nanocellulose (NC) showed a significant increase in the storage modulus at all temperatures and possessed a nonlinear stress–strain deformation behavior. Fiber webs from PCL reinforced with 2.5% unmodified NC showed 1.5-fold increase in Young's modulus and the ultimate strength compared to PCL webs. The reason was explained in terms of differences in the fiber diameter, NC loading, and crystallization processes. A coelectrospinning technique to produce a core-in-shell nanomaterial consisting of a cellulose shell and a core containing the CNs was first reported by Magalhaes et al. [41].

20.2.3 Extrusion and Impregnation

Extrusion and impregnation are two methods used to prepare nanocomposites comprising a polymeric matrix and modified NC filler. The main challenge lies in the poor dispersion and agglomeration of NCC inside the polymeric matrix, which is due to the hydrophilic nature of NC and the formation of interchain hydrogen bonding. These nonoptimized conditions thus limit the mechanical properties of the prepared nanocomposite. One way to address the agglomeration problem is by introducing microcrystalline cellulose (MCC) at the intermediate stage during the extrusion of poly(lactic acid) (PLA), where the suspension of whiskers was pumped into the polymer melt during the extrusion process [3]. *N,N*-dimethylacetamide (DMAc) containing lithium chloride (LiCl) was used to better disperse the MCC; however, it resulted in the degradation of the composites at high temperature. The mechanical properties did not show improvements compared to pure PLA, due to the

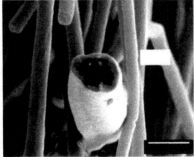

FIGURE 20.1　Cryo-scanning electron micrographs of electrospun PVA loaded with 15% of CNs. (From Peresin, M.S. et al., *Biomacromolecules*, 11, 674, 2010. With permission.)

combined effect of additives and high processing temperature. Wide angle X-ray diffraction showed that the crystal integrity of the cellulose was retained after extrusion. A similar process was performed by employing PVA as the dispersing agent; and the MCC was distributed in the PVA phase [44]. Another method reported is the modification of surface functional groups on NC prior to extrusion [45]. Hydrolysis followed by the grafting of organic aliphatic acid chain of different lengths was used. The mixture of modified NC and low-density polyethylene (LDPE) was extruded and improvement in the dispersion resulting in a more homogeneous mixture was observed, where better dispersion was reported in formulations with longer aliphatic chain. Another possible processing technique to prepare nanocomposites using cellulosic nanoparticles in the dry state consists of the filtration of the aqueous suspension to obtain a film or dried mat of particles followed by immersion in a polymer solution. The impregnation of the dried mat is performed under vacuum. At low pressure, the resin impregnates and fills the cavities within the NCC, which is followed by curing to produce a composite. This technique is mostly used for sample preparation for evaluating the mechanical, thermal, and optical properties of cellulose-filled composites [46–49]. Nakagaito and Yano used NaOH-treated cellulose microfibers impregnated with phenol–formaldehyde resin, and a 20% increase in fracture strain compared to nontreated cellulose microfibers was observed [50].

20.2.4 Other Processing Techniques

A melt-compounding technique is used in the production of hydrophobic polymeric nanocomposites. In general, melt-compounding processes involve the incorporation of CN into thermoplastic polymers by using thermal–mechanical mixing, extrusion of the melt mixture, and optional compression molding into specific test specimen geometries and configurations. In one invention, a concentrated suspension of cellulose nanowhiskers and a plasticizer liquid were pumped at the same time into an extruder, giving a partially molten matrix, to produce nanocomposites [51]. A solid-phase compounding technique was used to mix the nanofibers isolated from a soybean source with polyethylene (PE) or polypropylene (PP) [52,53]. Coated nanofibers were added to the molten polymer (2.5% and 5% by weight) in a laboratory compounder at 170°C. Capadona et al. have recently reported a versatile processing approach consisting of a three-dimensional template through self-assembly of well-individualized CNs and then filling the template with a polymer of choice [54–56]. The first step (Figure 20.2 left, a and b) is the formation of a CN template through a sol–gel process involving the formation of a homogeneous aqueous whisker dispersion that is followed by gelation through solvent exchange with a water-miscible solvent (routinely acetone). In the second step (Figure 20.2 left, d and e), the CN template is filled with a matrix polymer by immersing the gel into a polymer solution (Figure 20.2). It should be noted that the polymer solvent must be miscible with the gel solvent.

The use of twin extrusion as a processing method to prepare CN-based nanocomposites has been attempted by Oksman et al. [58]. The process consists of pumping an aqueous dispersion of CNs coated with PVA into a melt polymer (i.e., PLA) during extrusion [59]. However, such systems have unfortunately shown a lack of compatibility. Starch and CN nanocomposites were processed by extrusion technique [20]. A different method for

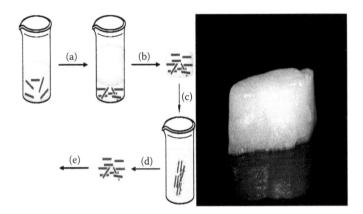

FIGURE 20.2   (left) Schematic representation of the template approach to obtain well-dispersed polymer/CN composites: (a) a nonsolvent is added to the dispersion of CNs in the absence of any polymer, (b) solvent exchange promotes the self-assembly of a gel of CNs, (c) the gelled CNs scaffold is interpenetrated with a polymer by immersion in a polymer solution, before the nanocomposite is (d) dried and (e) compacted (right). (From Capadona, J.R. et al., *Biomacromolecules*, 10, 712, 2009. With permission.)

producing nanocomposites was presented in a patent by using the layer-by-layer (LBL) technique [60]. Glass microscope slides were used as a substrate for LBL assembly. The glass slides were sequentially immersed into an aqueous poly(diallyldimethylammonium chloride) (PDDA) solution, and then into a CN suspension. This cycle was repeated up to 10 times to produce a multilayer nanocomposite.

20.3 PROPERTIES OF NANOCELLULOSE-BASED BIONANOCOMPOSITES

20.3.1 Thermal Properties

The analysis of thermal properties of materials is important to determine their processing temperature range and use. It is possible to determine the main characteristics of polymeric systems such as T_g, melting point (T_m), and thermal stability through differential scanning calorimetry (DSC) experiments.

20.3.1.1 Thermal Stability and Thermal Expansion Coefficient

Nanosized cellulose fibrils have been reported to improve the thermal properties of polymers. Thermal stability of polymers in nanocomposites with cellulose whiskers was reported to be enhanced when compared to those of the corresponding bulk polymers [61]. The thermal expansion of crystalline cellulose (e.g., Iα, Iβ), and nanocellulose composites (e.g., cellulose nanoparticles embedded in a polymer matrix) has been an active area of research. The coefficient of thermal expansion (CTE) of cellulose Iβ crystals has been experimentally determined at temperatures from −173°C to 200°C. In composites, the addition of cellulose nanoparticles to a matrix polymer has been shown to lower the macroscopic CTE [62]. In general, the CTE of composite materials depends on the properties of the matrix material, the polymer, the resulting CN network structure CN–CN bond strength, CN-matrix bond strength, and CN volume fraction [63]. Diaz et al. explored the influence of CN orientation on the in-plane thermal expansion of neat CN films in a contact-free way via polarized light image correlation (PLIC), utilizing the distinct birefringent optical properties of CN films [64]. They found that oriented CN films exhibited a highly anisotropic in-plane thermal expansion as compared with the isotropic response determined in self-organized films, whereas the CTE in the transverse direction approximates to that of polymers.

20.3.1.2 Glass Transition

The effect of NC on T_g of polymers is controversial. Most of the earlier studies did not observe important changes in the T_g of nanocomposites reinforced with cellulose nanofillers. This result was surprising due to the high specific area of such nanofillers [18,65,66]. More recently, Abraham et al. evaluated the thermal properties of NR latex reinforced with cellulose nanofiber, and a negligible change in T_g was detected [67]. Anglès and Dufresne reported a peculiar effect on the T_g of the starch-rich fraction of glycerol-plasticized starch-based nanocomposites reinforced with tunicin whiskers [18]. It was determined that the T_g of nanocomposites depended on moisture conditions. For low loading levels (up to 3.2 wt.%), a classical plasticization effect of water was reported whereas an antiplasticization effect was observed for higher whisker content (6.2 wt.% and above). These observations were due to the possible interactions between hydroxyl groups on the cellulosic surface and starch and the selective partitioning of glycerol and water in the bulk starch matrix or at whisker surface.

The plasticizing effect of water molecules is considered to be responsible for the decrease of T_g values in PVAc and PVA-based nanocomposites reinforced with sisal [16] and cotton whiskers [68]. In spite of the composition, a decrease in T_g was observed as the humidity content increased. On the other hand, in moist atmosphere, the T_g of PVA-based nanocomposites significantly increased when cotton whiskers were added. For glycerol-plasticized starch nanocomposite reinforced with cellulose crystallites, an increase of T_g with filler content was reported and attributed to cellulose/starch interactions [69]. Mathew and Oksman reported that, for tunicin whisker/sorbitol-plasticized starch nanocomposite, T_gs were found to increase slightly up to about 15 wt.% whiskers and to decrease for higher whisker loading [70]. Crystallization of amylopectin chains upon whisker addition and migration of sorbitol molecules to the amorphous domains were proposed to explain the observed modifications. The T_g of electrospun PS fiber with added cellulose whiskers (and surfactant) tends to decrease with the whisker content [71]. The authors attributed the reduction of T_g to the plasticizing effect of the surfactant.

20.3.1.3 Melting/Crystallization

A significant increase in crystallinity of sorbitol-plasticized starch was reported when increasing the CN content [70]. This phenomenon was ascribed to an anchoring effect of the cellulosic filler, which probably acted as a nucleating agent. The crystallization behavior of the semicrystalline PLA with plasticizer and nanofibers were reported [38]. Neat PLA showed glass transition peak, cold crystallization peak, and melting peak in the heating scan with the respective values of $T_g = 62.8°C$, $T_{cc} = 99.0°C$, and $T_m = 174.5°C$, while the plasticized PLAs and PLA nanocomposite only show the T_m peak. The addition of 20 wt.% of glycerol triacetate (GTA) in liquid form to neat PLA resulted in a slight decrease in the T_m, from 174.5°C to 169.1°C. A clear increase in the crystallinity of the neat PLA from 23% to 60% has also been observed. This is because of the increased molecular chain mobility in the presence of plasticizer and can also enhance the crystallization process [72]. It is also reported that reinforcements such as MCC, cellulose fibers, wood flour, and microfibrillated cellulose can act as nucleating agents for the crystallization of PLA [38,70–73].

20.3.2 Mechanical Properties

Cellulose nanoparticles such as whiskers and MFC possess a huge specific area and impressive mechanical properties, rendering an appreciable improvement in mechanical properties of neat matrix. Dynamic mechanical analysis (DMA) is presented as a powerful tool to investigate the linear mechanical behavior of nanocomposites. The mechanical properties of nanocomposites with cellulose fibers have been reported to be strongly associated with the dimensions and consequent aspect ratio of the fibers [12]. Aspect ratios are related to the origin of the cellulose used and whisker preparation conditions [65]. Any factor that affects the formation of the percolating nanonetwork or interferes with it changes the mechanical performances of the composite. Three main parameters have been reported as affecting the mechanical properties of such materials: (1) the morphology and dimensions of the nanoparticles, (2) the processing method, and (3) the microstructure of the matrix and matrix–filler interactions [74].

20.3.2.1 Effect of Morphology and Loading Level of Nanocellulose

Dogan and McHugh observed that an MCC with submicron-sized diameters had a much superior effect on tensile strength of hydroxypropyl methylcellulose (HPMC) than that of a micron-sized MCC counterpart; moreover, the negative impact of a micron-sized MCC on elongation of the films was much more dramatic than that of its submicron-sized counterpart [75]. Chen et al. produced composites of a pea starch matrix added with cellulose whiskers extracted from pea hull fibers with different hydrolysis times, which resulted in different aspect ratios [76]. The composite produced by using whiskers with the highest aspect ratio exhibited the highest transparency and best tensile properties. Jiang et al. demonstrated that the mean aspect ratio cannot be considered without proper assumptions; one must also consider the orientation distribution of the fillers [77]. When the fillers does not follow a symmetric distribution, the overall mechanical properties obtained by the average aspect ratio of the fillers may be greatly different from those obtained when considering the aspect ratio distribution. The orientation of cellulose fibers can greatly improve the tensile properties of a resulting nanocomposite. Kvien and Oksman applied a magnetic field to a nanocomposite of PVA with cellulose whiskers to orient the whiskers, the modulus of the resulting nanocomposite was greatly increased by orientation [78].

20.3.2.2 Effect of Interfacial Interaction between Nanocellulose and Polymeric Matrix

Cellulose nanoreinforcements have been reported to have a great effect in improving the modulus of polymer matrices [79]. For example, Helbert et al. reported that a poly(styrene-*co*-butyl acrylate) latex film containing 30 wt.% of straw cellulose whiskers presented a modulus more than a thousand times higher than that of the bulk matrix [10]. According to the authors, such a great effect is ascribed not only to the geometry and stiffness of the whiskers but also to the formation of a fibril network within the polymer matrix, the cellulose fibers being probably linked through hydrogen bonds. Zimmermann et al. observed that fibril contents of up to 5% resulted in no strength or stiffness improvement of PVA composites, and they suggested that probably minimum fibril content is required to induce intense interactions between fibrils and matrices [22].

20.3.2.3 Effect of Processing Condition

Slow processes such as solvent casting have been reported as producing materials with the highest mechanical performance compared with freeze-drying/molding and freeze-drying/extruding/molding. This effect was ascribed to the probable orientation of the rod-like nanoparticles during film processing resulting from shear stress induced by freeze-drying/molding and freeze-drying/extrusion/molding [80]. During slow water evaporation, because of Brownian motion in the suspension or solution (whose viscosity remains low up to the end of the process when the latex particle or polymer concentration becomes very high), the rearrangement of the nanoparticles is possible. They have enough time to interact and connect to form a continuous network, which is the basis of their reinforcing effect. The resulting structure is completely relaxed and direct contact between the nanocrystals or nanofibrils is created. On the other hand, during the freeze-drying/hot-pressing process, the nanoparticle arrangement in the suspension is first frozen, and then during the hot-pressing stage, because of the polymer melt viscosity, the particle rearrangements are strongly limited. When using a processing route other than casting/evaporation in a water medium, the dispersion of the hydrophilic filler in the polymeric matrix is also involved, and improved filler–matrix interactions generally lead to higher mechanical properties [45]. In nonpercolating systems, for instance, for materials processed from freeze-dried CNs, strong matrix–filler interactions enhanced the reinforcing effect of the filler [72].

20.3.3 Swelling and Barrier Properties

Nowadays, there is an increasing interest in the swelling as well as in the barrier properties due to increased tortuosity provided by nanoparticles. The barrier properties of a material indicate their resistance to sorption and diffusion of moisture and gases across the packaging material. Bionanocomposite films show improved barrier properties because nanoparticles dispersed in the biopolymer matrix provide a tortuous path for water and gas molecules to pass through. This increases the effective path length for diffusion, thereby improving the barrier properties [81]. Barrier properties of interest in food packaging are water vapor permeability (WVP) and oxygen permeability (OP). Figure 20.3 shows a schematic representation of increased diffusion path within the nanocellulose films.

20.3.3.1 Water Sorption and Swelling Properties

The water sorption properties of nanocomposites reinforced with CNs depend on the hydrophilicity of the matrix material. For hydrophilic matrices, both water uptake

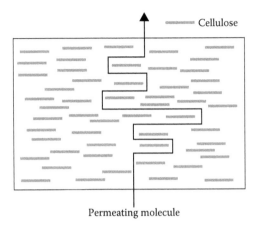

Cellulose

Permeating molecule

FIGURE 20.3   Schematic representation of increased diffusion path within the nanocellulose films.

and diffusion coefficient are found to decrease upon adding CNs. This behavior can be explained on the basis of strong interfacial interaction between the cellulose nanoparticles themselves and the interaction between the polar matrices and nanoparticles. However, in some cases the water uptake is found to increase at higher concentration of nanocrystals due to agglomeration of the nanocrystals on the polymer matrix. For hydrophobic matrices, an opposite trend is observed as expected. A lower water uptake was observed when using nanofibrillated cellulose (NFC) instead of CNs as a reinforcing phase in NR [82]. This observation was explained by the difference in the structure and composition of the two types of nanoparticle and, in particular, by the presence of residual lignin, extractive substances, and fatty acids at the surface of NFC, which limits, comparatively, the hydrophilic character of the filler. In addition, assuming that filler–matrix compatibility was consequently lower for nanocrystal-based nanocomposites, one can imagine that water infiltration is easier at the filler–matrix interface. For NFC-based nanocomposites, despite a higher amorphous cellulose content, the higher hydrophobic character of the filler favors compatibility with NR and restricts, therefore, the interfacial diffusion pathway for water. The swelling or kinetics of solvent absorption can highlight interactions between the filler and the matrix. Generally, the short-time behavior displays a fast absorption phenomenon, whereas at longer times the kinetics of absorption is low and leads to a plateau, corresponding to the solvent uptake at equilibrium.

20.3.3.2 Water Vapor Transfer Rate and Water Vapor Permeability

The water vapor barrier properties of nanocellulose are low. This is mainly due to high affinity between water and the nanocellulose film. NC is a much better water vapor barrier than cellulose fiber. NC has a strong reducing effect on water vapor diffusion due to its size and swelling constraints formed due to rigid network within the films. However, at a high relative humidity, these structural organizations can be disrupted due to high swelling and lose barrier properties for both oxygen and water vapor. Water vapor permeability is the rate of water vapor transmission through the unit area of a flat material of unit thickness induced by unit vapor pressure difference across the material. The incorporation of CNs reduces the water vapor transmission rate (WVTR) of the membranes. However, the results were more interesting when nanocomposites were prepared by mixing PVA matrix with low quantities (10 wt.% or 20 wt.%) of poly(acrylic acid) (PAA) and whiskers [83]. These results can be ascribed to the cross-linking with PAA that reduces the number of hydroxyl groups in the composites and thus the hydrophilicity. The incorporation of CNs into the nanocomposite provides a physical barrier through the creation of a tortuous path for the permeating moisture. Consequently, the addition of 10 wt.% of CNs reduced the WVTR of the matrix more than the addition of PAA. The combination of 10 wt.% whiskers and 10 wt.% PAA gave the best performance. With respect to the chemical vapor transmission rate, nanocomposites containing 10 wt.% cellulose whiskers and 10 wt.% PAA exhibit the same synergetic effects with increasing time lag and decreasing flux. Overall, membrane barrier properties were improved by the addition of cellulosic crystals in combination with PAA.

20.3.3.3 Gas Permeability

Nanocellulose such as cellulose nanofibrils (CNFs) and CNs have opened vast possibilities of utilizing cellulose-based materials. The use of CNFs in films, composites, and coatings has been found to substantially reduce the OP within these materials. The oxygen barrier efficiency of pure CNF films is highly competitive and even comparable with commercial synthetic polymers. The improvement of oxygen barrier properties by CNFs can be attributed to the dense network formed by nanofibrils with smaller and more uniform dimensions. Even though CNs have higher crystallinity than CNFs, mechanically fibrillated CNF films were found to have much lesser OP than CNs. The CNF films have higher entanglements within the film, which increases the diffusion path for gas molecules [84].

The gas molecules should be first dissolved in the membrane or film before diffusing. Even though the surface of films influences the permeating gas molecules, the most dominant factor in molecular migration is bulk flow, that is, rate of molecule diffusion in the membrane [85]. Considering this fact, the pores within the films serve as the major path for permeating oxygen molecules. The dense nanofibrils form more complex and smaller pores compared to pure cellulose fibers, which are in microscale. This complex dense network increases the tortuosity within the film and thereby decreases the permeability within the films [86]. The high crystalline structure within the nanofibrils or whiskers also contributes to the gas barrier properties [87]. Cellulose is composed of both crystalline and disordered regions. High crystallinity ranging from 40% to 90% has been reported for the nanocellulose, with CNs showing higher crystallinity than the CNFs because strong acid hydrolyzes disordered cellulose to result in highly crystalline CNs [88,89]. Even though the CNs have higher crystallinity than CNFs, mechanically fibrillated CNF films were found to have much lesser OP than CNs. Both showed similar solubility, but the oxygen molecules penetrated more slowly through the CNF films. This is mainly due to the structural organization within the films. The CNF films have higher entanglements within the film, which increased the tortuosity factor or increase the diffusion path [90].

CNFs are a strong gas barrier material. Compared to CNs, CNFs consist of crystalline and disordered regions. In most cases, crystallinity ranging from 40% to 75% has been reported for the CNFs obtained from softwoods and hardwoods [88–91]. Films made purely of mechanically fibrillated CNFs have very high air and oxygen barrier property. The oxygen transmission rates of CNF films with thickness of 21 μm were as low as 17 ± 1 ml m^{-2} day^{-1}. These values are competitive with other best synthetic polymers such as ethylene vinyl alcohol and polyvinylidene chloride-coated polyester films [86]. Recently, Osterberg et al. demonstrated a rapid method of making robust CNF films with high oxygen barrier property [92]. The CNF solutions were first filtered followed by hot pressing at high pressure followed by air drying. At a relative humidity below 65%, the OP of these films was below 0.6 cm^3 μmm^{-2} day^{-1} kPa^{-1}. However, OP of CNF films increases with the increase in relative humidity. This is mainly due to the plasticizing and swelling of nanofibrils through the adsorption of water molecules at high relative humidities.

20.3.4 Biodegradation Properties

Biodegradability of bionanocomposites is one of the most interesting issues in the bionanocomposite materials. Polymeric systems based on cellulose with responsive behaviors have shown unique properties such as biocompatibility, biodegradability, and biological functions and have been exploited by many researchers. Biopolymers are widely used as matrix material in the field of packaging industry and drug delivery applications because of their biodegradability. For most biodegradable polymers, some of their physical properties need to be improved in order to fight with petroleum-based materials. Films with antimicrobial activity could help control the growth of pathogenic and spoilage microorganisms [81]. Starch-based materials have been extensively investigated as a choice product to improve biodegradability of a variety of plastics. However, the brittleness of starch requires the use of plasticizers such as polyols, which improves starch flexibility but, on the other hand, decreases its thermomechanical properties. Biodegradable nanocomposites were prepared by casting a mixture of NR and cellulose whiskers isolated from sugarcane bagasse [61]. They found that the presence of cellulose whiskers significantly enhanced biodegradation of rubber in soil.

20.4 APPLICATIONS OF NANOCELLULOSE-BASED BIONANOCOMPOSITES

20.4.1 Biomedical Field

The development of novel biomedical materials from natural polymers for practical and clinical applications is always one of the most concerned topics for biologists and material scientists. Attributed to the properties of biocompatibility and right mechanical properties similar to natural tissue, nanocellulose-based biomaterials can provide a cell-friendly environment to encourage the cells' attachment and proliferation as a special tissue bioscaffold. Promising mechanical properties and good biocompatibility of nanocellulose promote its research and development as a substitute/medical biomaterial, such as the replacement of blood vessels (vascular graft) and soft tissues [93]. Wound dressing systems based on pure oxidized regenerated cellulose (ORC) have been developed such as Tabotamp® and a mixture from collagen and ORC named Promogran®. Tabotamp® is a thin gauze layer from pure ORC that is used in acute wounds like trauma, surgical injuries, and burns [94].

It provides hemostatic and antimicrobial properties. Collagen is recognized as the major component of the ECM and serves as a natural substrate for cell attachment, proliferation, and differentiation. Collagen has been employed as a matrix material for TE and wound dressing [95]. Promogran® is a commercial spongy collagen matrix (55%) containing ORC (45%) that has been recently introduced in the United States and European markets for the treatment of exuding diabetic and ulcer wounds [95,96]. This product is a dressing consisting of a sterile, freeze-dried matrix and combines the properties of its components, such as fluid absorption and hemostatic properties. Recently, nanocellulose has been called as the eyes of biomaterial and has been found to be highly applicable to biomedical industry, which includes skin replacements for burns and wounds, drugs releasing system, blood vessel growth, nerves, gum and dura mater reconstruction, scaffolds for tissue engineering, stent covering, and bone reconstruction [97–101].

The team of Dieter Klemm (Institute of Organic and Macromolecular Chemistry, Friedrich-Schiller-University, Jena, Germany) was the first research organization to investigate and apply artificial vascular substitute obtained with biomaterials from bacterial cellulose. They have discussed the application of BC as blood vessel replacement in some publications [101–103] and especially described a clinical product named BActerial SYnthesized Cellulose (BASYC) with high mechanical strength in wet state, enormous water retention property, and low roughness of inner tube surface. It is reported that BASYC from BC has been successfully used as the artificial blood vessel in rats and pigs for microsurgery [104–110]. Examples of substitutes from nanocellulose are shown in Figure 20.4 [106].

BC in the form of membranes has been applied for guided bone regeneration in bone defects of critical and noncritical size [111], in periodontal lesions [112], and as a resorbable barrier membrane for preventing the invasion of fibroblast cells and fibrous connective tissue into bone defects [113]. Results from the literature indicate that BC membranes promote effective bone formation at the site, besides being a low-cost treatment [114,115]. BC has been widely investigated for wound healing due to its purity and high water retention capacity, and a series of BC-based wound dressings are currently marketed. Their potential as wound dressing has been reviewed by Czaja et al. [116,117]. BC has been used in cartilage [118,119], replacement of blood vessels in rats [101], and wound healing processes. Hydrogels based on bacterial nano cellulose (BNC) mimic basic living processes and are of growing importance as bioactive scaffolds. BC was reported to be developed as biomaterial for the reconstruction of damaged peripheral nerves via cellulosic guidance channels. *In vivo* experiments were conducted on the femoral nerve of Wistar rats for three months. Evaluation of results from histological analysis and postoperative observation of motor recovery showed that BC neurotubes can effectively prevent the formation of neuromas, while allowing the accumulation of neurotrophic factors inside and facilitating the process of nerve regeneration [120]. Barud (2009) has developed a biological membrane with bacterial cellulose and standardized extract of propolis [121]. Propolis has many biological properties including antimicrobial and anti-inflammatory activities. All the aforementioned characteristics present make the membrane a good treatment for burns and chronic wounds.

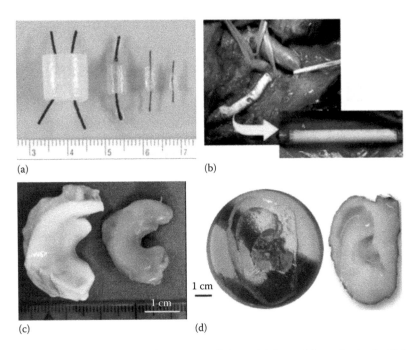

FIGURE 20.4 Examples of substitutes from nanocellulose. (a) Bacterial synthesized cellulose with different dimensions; (b) vascular prostheses made of CNF-polyurethane placed between the brachiocephalic trunk and the right common carotid artery in male; (c) comparison between pig meniscus and BC hydrogel; (d) negative silicone mold used to guide the bacteria during the bacterial culture to reproduce the large-scale features of the outer ear (left), and 3D BC implant prototype (1% effective cellulose content) produced in the shape of the whole outer ear according to the 3T MRI scanning technique (right). (From Ning, L. and Dufresne, A., *European Polymer Journal*, 59, 302, 2014. With permission.)

In another study, this CNF/polypyrrole composite prepared with the same strategy and source of CNF was applied as the hemodialysis membrane to purify blood [122]. It was reported that this biomaterial exhibited an effective removal of small uremic toxins in blood and an improvement in thrombogenic properties with the coating of heparin, which were attributed to superior ion-exchange capability and large surface area of the membrane. It should also be noted that due to the introduction of natural CNF, the hemocompatibility of this composite biomaterial was much better than commercial synthetic membranes (such as polysulfone) [123]. In ophthalmologist area, Huia et al. (2009) explored the potentiality of nanocellulose applied as the scaffold of tissue engineering [124]. They studied the growth of human corneal stromal cells on nanocellulose. The ingrowth of corneal stromal cells into the scaffold was verified by laser scanning confocal microscope. The results suggest the potentiality for this biomaterial as a scaffold for tissue engineering of artificial cornea. Another use of nanocellulose is for nasal reconstruction. The desire for an ideal shape has always been a part of mankind. Nose, centrally located in the face, is better susceptible to traumas, deformities, and thus social disorders. Other than having a major breathing function, it has a great esthetic function, highlighting face's genetics. Amorim et al. evaluated the tissue response to the presence of nanocellulose in the nose bone [125].

20.4.2 Packaging Field

To date, the production and use of nonbiodegradable materials or plastics as food packaging materials have been significantly increased. These types of materials are usually derived from petroleum products and cause problem in waste disposal [126]. To meet the increasing demand for sustainability and environmental safety, a growing number of investigations have been directed toward the development of food packaging materials that can rapidly degrade and completely mineralize in environment [127,128]. Biopolymers have been one of the favorable alternatives to be exploited and developed into eco-friendly food packaging materials due to its biodegradability [129]. Very often packaging materials act as containers and protect food products from water vapor, microorganism, gases, odors, dust, mechanical shock, vibrations, and other unfavorable conditions that cause external and/or internal damages during distribution and storage. In order to suit the complex and busy lifestyle of modern consumers, producers of food products strive hard to develop functional packaging systems having various convenient features and multiple end uses.

Cellulose is a renewable and nontoxic biopolymer having biocompatibility with other substances. Therefore, cellulose nanocomposite materials have a huge potential for a wide range of applications in the food industry. They can be used even for innovative active food packaging having biofunctional and antimicrobial properties [81]. The acceptable structural integrity and barrier properties, along with functional and film-forming properties of cellulose nanocomposite, can be the primary driving force in the development of biopackaging applications.

Nanocomposites are promising to expand the use of edible films and coatings because the addition of nanoparticles can improve their performance [130,131]. In recent years, various organic formulations, such as MCC nanofibers and chitosan nanoparticles, have been used in biodegradable polymer matrices to enhance polymer performance. Several studies have been developed that incorporate MCC nanofibers [75,132–138] or chitosan nanoparticles into nanocomposite edible films [139]. Generally, studies on moisture sorption and water vapor permeability reveal that the addition of CNs reduces the moisture affinity of hydrophilic films, which is very useful for edible packaging applications. Bilbao-Sáinz et al. observed that the water barrier properties of hydrophilic films can be improved by the addition of hydrophobic nanoemulsions into the film matrix [133].

Dehnada et al. prepared chitosan–nanocellulose biocomposites from chitosan having molecular weight of 600–800 kDa, nanocellulose with 20–50 nm diameters, and various levels of 30%, 60%, and 90% (v/wCHT) for glycerol [140]. Agitation and sonication were used to facilitate even dispersion of particles in the polymer matrix. Chitosan–nanocellulose nanocomposites showed high T_g range of 115°C–124°C and were able to keep their solid state until the temperature (T_m) range of 97°C–99°C. X-ray diffraction (XRD) analysis photographs revealed that nanocellulose peak completely disappeared after their addition to chitosan.

20.4.3 Electronic Industry

There has been a surge of interest in the field of ion-conducting solid polymer electrolytes because of their potential application in rechargeable batteries, fuel cells, light-emitting electrochemical cells, electrochromics, and many other electrochemical devices [141,142]. Cellulose whiskers have been used as mechanical reinforcing agents of low-thickness polymer electrolytes for application in lithium batteries application [143,144]. High-performance solid lithium-conducting nanocomposite polymer electrolytes have been prepared from lithium salts such as lithium trifluoromethyl sulfonyl imide (LiTFSI) and polymers such as high-molecular-weight POE and ethylene oxide–epichlorohydrin copolymers (EO–EPI) with addition of high-aspect-ratio CN whiskers. Among various applications studied so far, which have already reached the level of practical use is related to acoustic diaphragms, nanocellulose has been found to bear two essential properties: high sonic velocity and low dynamic loss. In fact, the sonic velocity of pure film was almost equivalent to those of aluminum and titanium [145]. Jonas and Farah stated that SONY® had already been using it in headphones diaphragm [146]. The nanocellulose diaphragms are developed by dehydration and compressed to a thickness of 20 microns in a diaphragm die. The advantage of the ultrathin nanocellulose diaphragm is that it can produce the same sound velocity as an aluminum or titanium diaphragm, along with the warm, delicate sound that a paper diaphragm provides.

The low CTE of nanocellulosics combined with high strength, high modulus, and transparency make them a potential reinforcing material in roll-to-roll technologies (e.g., for fabricating flexible displays, solar cells, electronic paper, and panel sensors) [147]. Organic light-emitting diode (OLED) materials were prepared with wood–cellulose nanocomposites [148–152]. Cellulose has always been the prime medium for displaying information in our society; nowadays efforts have been made to find dynamic display technology, for example, in electronic paper. Nanocellulose is dimensionally stable and has a paper-like appearance that puts it into the leading role for the electronic paper's basic structure [153]. Shah and Brown proved the concept in a device that holds many advantages such as high paper-like reflectivity, flexibility, contrast, and biodegradability [153].

Magnetic nanocomposites based on bacterial cellulose substrates containing large quantities of magnetite particles (Fe_2O_3) have been prepared [154]. In BC membranes, needle-like lepidocrocite (γ-FeOOH) was formed along the cellulose fibrils, using the crystalline surface as a nucleation site. Spherical magnetite particles subsequently formed around the needles. The treated BC composite membranes were superparamagnetic at room temperature.

20.5 CONCLUSIONS

Nanocellulose is the wave of the future and has an exciting potential as reinforcements in nanocomposites. Due to a large number of properties, nanocellulose-based materials are chiefly considered to be in a wide range of applications such as paper and packaging products, construction, automotive, furniture, electronics, pharmacy, and cosmetics, and biomedical applications are also being considered. High strength and stiffness, the surface reactivity (with numerous hydroxyl groups), the specific organization as well as the small dimensions of nanocellulose may well impart useful properties for the reinforcement of composite materials. It was shown that the use of high-aspect-ratio cellulose whiskers induces a mechanical percolation phenomenon leading to outstanding and unusual mechanical properties through the formation of a rigid filler network. In addition to some practical applications, the study of such model systems can help to understand some physical properties such as geometric and mechanical percolation effects.

REFERENCES

1. M.A.S.A. Samir, F. Alloin, A. Dufresne, *Biomacromolecules* 6 (2005): 612–626.
2. S. Kalia, A. Dufresne, B.M. Cherian, B.S. Kaith, L. Avérous, J. Njuguna, *International Journal of Polymer Science* 35 (2011): 37875.
3. K. Oksman, A.P. Mathew, D. Bondeson, I. Kvien, *Composite Science and Technology* 66 (2006): 2776–2784.
4. L. Petersson, I. Kvien, K. Oksman, *Composite Science and Technology* 67 (2007): 2535–2544.
5. D. Plackett, T.L. Andersen, W.B. Pedersen, L. Nielsen, *Composite Science and Technology* 63 (2003): 1287–1296.
6. Z.G.T. Akbari, S. Moghadam, *International Journal of Food Engineering* 3 (2007): 1–24.
7. J.K. Pandey, A.P. Kumar, M. Misra, A.K. Mohanty, L.T. Drzal, R.P. Singh, *Journal of Nanoscience and Nanotechnology* 5 (2005): 497–526.
8. F.P. La Mantia, M. Morreale, *Composites Part A: Applied Science and Manufacturing* 42 (2011): 579–588.
9. V. Favier, G.R.Canova, J.Y. Cavaille, *Polymer for Advanced Technologies* 6 (1995): 351–355.
10. W. Helbert, J.Y. Cavaille, A. Dufresne, *Polymer Composites* 17 (1996): 604–611.
11. M.A.S. Azizi Samir, F. Alloin, M. Paillet, *Macromolecules* 37 (2004): 4313–4316.
12. D. Dubief, E. Samain, A. Dufresne, *Macromolecules* 32 (1999): 5765–5771.
13. A. Dufresne, M.B. Kellerhals, B. Witholt, *Macromolecules* 32 (1999): 7396–7401.
14. M. Matos Ruiz, J.Y. Cavaillé, A. Dufresne, *Macromolecular Symposia* 169 (2001): 211–222.
15. A. Bendahou, Y. Habibi, H. Kaddami, *Journal of Biobased Materials and Bioenergy* 3 (2009): 81–90.
16. N.L. Garcia de Rodriguez, W. Thielemans, A. Dufresne, *Cellulose* 13 (2006): 261–270.
17. A. Ben Elmabrouk, T. Wim, A. Dufresne, *Journal of Applied Polymer Science* 114 (2009): 2946–2955.
18. M.N. Angles, A. Dufresne, *Macromolecules* 33 (2000): 8344–8353.
19. A.P. Mathew, A. Dufresne, *Biomacromolecules* 3 (2002): 609–617.
20. W.J. Orts, J. Shey, S.H. Imam, G.M. Glenn, M.E. Guttman, J.F. Revol, *Journal of Polymers and the Environment* 13 (2005): 301–306.
21. M.A.S. Azizi Samir, F. Alloin, W. Gorecki, *Journal of Physical Chemistry B* 108 (2004): 10845–10852.
22. T. Zimmerman, E. Poehler, T. Geiger, *Advanced Engineering Materials* 6 (2004): 754–761.
23. Y. Choi, J. Simonsen, *Journal of Nanoscience and Nanotechnology* 6 (2006): 633–639.
24. Y. Wang, X. Cao, L. Zhang, *Macromolecular Bioscience* 6 (2006): 524–531.
25. A.K. Mohanty, A. Misra, L.T. Drzal, *Journal of Polymers and the Environment* 10 (2002): 19–26.
26. I. Armentano, M. Dottori, E. Fortunati, S. Mattioli, J.M. Kenny, *Polymer Degradation and Stability* 95 (2010): 2126–2146.
27. M. Schroers, A. Kokil, C. Weder, *Journal of Applied Polymer Science* 93 (2004): 2883–2888.
28. J. Lu, T. Wang, L.T. Drzal, *Composites Part A* 39 (2008): 738–746.
29. M.A.S. Azizi Samir, F. Alloin, J.Y. Sanchez, N. El Kissi, A. Dufresne, *Macromolecules* 37 (2004): 1386.
30. E.S. Medeiros, L.H.C. Mattoso, E. Ito, K.S. Gregorski, G.H. Robertson, *Journal of Biobased Materials and Bioenergy* 2 (2008): 1–12.
31. N. Ljungberg, C. Bonini, F. Bortolussi, C. Boisson, L. Heux, J.Y. Cavaillé, *Biomacromolecules* 6 (2005): 2732–2739.

32. Y. Habibi, A. Dufresne, *Biomacromolecules* 9 (2008): 1974–1980.
33. P. Podsiadlo, S.Y. Choi, B. Shim, J. Lee, M. Cuddihy, N.A. Kotov, *Biomacromolecules* 6 (2005): 2914–2918.
34. D. Viet, B. Candanedo, D.G. Gray, *Cellulose* 14 (2007): 109.
35. O. van den Berg; J.R. Capadona, C.Weder, *Biomacromolecules* 8 (2007): 1353.
36. Y. Habibi, A. Lucia, O.J. Rojas, *Chemical Reviews* 110 (2010): 3479–3500.
37. A. Dufresne, *Molecules* 15 (2010): 4111–4128.
38. O.J. Rojas, G.A. Montero, Y. Habibi, *Journal of Applied Polymer Science* 113 (2009): 927–935.
39. J.O. Zoppe, M.S. Peresin, Y. Habibi, R.A. Venditti, O.J. Rojas, *ACS Applied Materials Interfaces* 1 (2009): 1996–2004.
40. W.-I. Park, M. Kang, H.S. Kim, H.-J. Jin, *Macromolecular Symposia* 249–250 (2007): 289–294.
41. W.L.E. Magalhäes, X. Cao, L.A. Lucia, *Langmuir* 25 (2009): 13250–13257.
42. M.S. Peresin, Y. Habibi, J.O. Zoppe, J.J. Pawlak, O.J. Rojas, *Biomacromolecules* 11 (2010): 674–681.
43. I. Siró, D. Plackett, *Cellulose* 17 (2010): 459–494.
44. D. Bondeson, K. Oksman, *Composites Part A* 38 (2007): 2486–2492.
45. A. de Menezes, G. Siqueira, A.A.S. Curvelo, A. Dufresne, *Polymer* 50 (2009): 4552–4563.
46. A. Henriksson, L.A. Berglund, *Journal of Applied Polymer Science* 106 (2007): 2817–2824.
47. S. Iwamoto, K. Abe, H. Yano, *Biomacromolecules* 9 (2008): 1022–1026.
48. Y. Shimazaki, Y. Miyazaki, Y. Takezawa, M. Nogi, K. Abe, *Biomacromolecules* 8 (2007): 2976–2978.
49. M. Nogi, K. Handa, A.N. Nakagaito, H. Yano, *Applied Physics Letters* 87 (2005): 243110–243112.
50. A.N. Nakagaito, H. Yano, *Cellulose* 15 (2008): 323–331.
51. A. Iwatake, M. Nogi, H. Yano, *Composites Science and Technology* 68 (2008): 2103–2106.
52. B. Wang, M. Sain, *Polymer International* 56 (2007): 538–546.
53. B. Wang, M. Sain, *Composites Science and Technology* 67 (2007): 2521–2527.
54. J.R. Capadona, O. van den Berg, L.A. Capadona, M. Schroeter, S.J. Rohan, D.J Tyler, C. Weder, *Nature Nanotechnology* 2 (2007): 765.
55. C. Weder, J. Capadona, O. van den Berg, (Case Western Reserve University, U., Ed. Application) US 2008/79264 (2008).
56. J.R. Capadona, K. Shanmuganathan, D. Tyler, S.J. Rowan, C. Weder, *Science* 319 (2008): 1370.
57. J.R. Capadona, K. Shanmuganathan, S. Trittschuh, S. Seidel, S.J. Rowan, C. Weder, *Biomacromolecules* 10 (2009): 712.
58. K. Oksman, D. Bondeson, P. Syre, (NTNU Technology Transfer AS, Norway). US Patent Application: US 2006/560190 (2008).
59. A.P. Mathew, A. Chakraborty, K. Oksman, M. Sain, In *Cellulose Nanocomposites: Processing, Characterization, and Properties*; Oksman, K., Sain, M., Eds.; ACS Symposium Series 938; American Chemical Society: Washington, DC, 2006.
60. N. Kotov, B.S. Shim, P. Podsiadlo, Layer-by-layer assemblies having preferential alignment of deposited axially anisotropic species and methods forpreparation and use thereof, US Patent: US 2010/0098902A1 (2010).
61. J. Bras, M.L. Hassan, C. Bruzesse, E.A. Hassan, N.A. El-Wakil, A. Dufresne, *Industrial Crops and Products* 32 (2010): 627–633.
62. A.N. Nakagaito, H. Yano, *Cellulose* 15 (2008): 555–559.
63. R.J. Moon, A. Martini, J. Nairn, J. Simonsen, J. Youngblood, *Chemical Society Reviews* 40 (2011): 3941–3994.
64. J.A. Diaz, X. Wu, A. Martini, J.P. Youngblood, R.J. Moon, *Biomacromolecules* 14 (2013): 2900–2908.
65. A.M.A.S. Samir, F. Alloin, J.Y. Sanchez, A. Dufresne, *Polymer* 45 (2004): 4149–4157.
66. G. Morandi, L. Heath, W. Thielemans, *Langmuir* 25 (2009): 8280–8286.

67. E. Abraham, B. Deepa, L.A. Pothan, M. John, S.S. Narine, S. Thomas, R. Anandjiwala, *Cellulose* 20 (2013): 417–427.
68. M. Roohani, Y. Habibi, N.M. Belgacem, G. Ebrahim, A.N. Karimi, A. Dufresne, *European Polymer Journal* 44 (2008): 2489–2498.
69. Y. Lu, L. Weng, X. Cao, *Macromolecular Bioscience* 5 (2005): 1101–1107.
70. A.P. Mathew, K. Oksman, M. Sain, *Journal of Applied Polymer Science* 101 (2006): 300–310.
71. O.J. Rojas, G.A. Montero, Y. Habibi, *Journal of Applied Polymer Science* 113 (2009): 927–935.
72. N. Ljungberg, B. Wesslén, *Biomacromolecules* 6 (2005): 1789–1796.
73. L. Suryanegara, A.N. Nakagaito, H. Yano, *Composites Science Technology* 69 (2009): 1187–1192.
74. A. Dufresne, *Journal of Nanoscience and Nanotechnology* 6 (2006): 322–330.
75. N. Dogan, T. H. McHugh, *Journal of Food Science* 72 (2007): E16–E22.
76. Y. Chen, C. Liu, P.R. Chang, X. Cao, D.P. Anderson, *Carbohydrate Polymers* 76 (2009): 607–615.
77. B. Jiang, C. Liu, C. Zhang, Z. Wang, *Composites: Part B* 38 (2007): 24–34.
78. I. Kvien, K. Oksman, *Applied Physics A: Materials Science and Processing* 87 (2007): 641–643.
79. A. Bhatnagar, M. Sain, *Journal of Reinforced Plastics and Composites* 24 (2005): 1259–1268.
80. A. Dufresne, *Canadian Journal of Chemistry* 86 (2008): 484–494.
81. J.W. Rhim, P.K.W. Ng, *Critical Reviews in Food Science and Nutrition* 47 (2007): 411–433.
82. A. Bendahou, H. Kaddami, A. Dufresne, *European Polymer Journal* 46 (2010): 609–620.
83. S.A. Paralikar, J. Simonsen, J. Lombardi, *Journal of Membrane Science* 320 (2008): 248–258.
84. S.S. Nair, J.Y. Zhu, Y. Deng, A.J. Ragauskas, *Sustainable Chemical Processes* 2 (2014): 23.
85. J.M. Lagaron, R. Catala, R. Gavara, *Journal of Materials Science and Technology* 20 (2004): 1–7.
86. K. Syverud, P. Stenius, *Cellulose* 16 (2009): 75–85.
87. A. Saxena, T.J. Elder, J. Kenvin, A.J. Ragauskas, *Nano-Micro Letters* 2 (2010): 235–241.
88. S.S. Nair, J.Y. Zhu, Y. Deng, A.J. Ragauskas, *Journal of Nanoparticle Research* 16 (2014): 2349.
89. J. Guo, J.M. Catchmark, *Carbohydrate Polymers* 87 (2012): 1026–1037.
90. S. Belbekhouche, J. Bras, G. Siqueira, C. Chappey, L. Lebrun, B. Khelifi, S. Marais, A. Dufresne, *Carbohydrate Polymers* 83 (2011): 1740–1748.
91. Q.Q. Wang, J.Y. Zhu, R. Gleisner, T.A. Kuster, U. Baxa, S.E. McNeil, *Cellulose* 19 (2012): 1631–1643.
92. M. Osterberg, J. Vartiainen, J. Lucenius, U. Hippi, J. Seppala, R. Serimaa, J. Laine, *ACS Applied Materials and Interfaces* 5 (2013): 4640–4647.
93. N. Lin, A. Dufresne, *Polymer Journal* 59 (2014): 302–325.
94. U. Schönfelder, M. Abel, C. Wiegand, D. Klemm, P. Elsner, U.C. Hipler, *Biomaterials* 26 (2005): 6664–6673.
95. L.S. Nair, C.T. Laurencin. *Progress in Polymer Science* 32 (2007): 762–798.
96. J.L. Lázaro Martínez, E.G. Morales, A. Sánchez, *Cirugia Espanola* 82 (2007): 27–31.
97. J.D. Fontana, A.M. de Souza, C.K. Fontana, I.L. Torriani, J.C. Moreschi, B.J. Gallotti, S.J. de Souza, G.P. Narciso, J.A. Bichara, L.F. Farah, *Applied Biochemistry and Biotechnology* 253 (1990): 24–25.
98. L.R. Mello, Y. Feltrin, R. Selbach, G. Macedo Jr., C. Spautz, L.J. Haas, *Arquivos de Neuro-Psiquiatria* 59 (2001): 372.
99. W.K. Czaja, D.J. Young, M. Kawecki, R.M. Brown Jr., *Biomacromolecules* 8 (2007): 13–17.
100. S.W. Negrão, R.R.L. Bueno, E.E. Guérios, F.T. Ultramari, A.M. Faidiga, P.M.P. Andrade, D.C. Nercolini, J.C. Tarastchuck, L.F. Farah, *Revista Brasileira de Cardiologia Invasiva* 14 (2006): 10–19.
101. D. Klemm, D. Schumann, U. Udhardt, S. Marsch, *Progress in Polymer Science* 26 (2001): 1561–1567.
102. M. Gama, P. Gatenholm, D. Klemm, Multifunctional materials, Boca Raton, FL: CRC Press/Taylor & Francis Group (2013), pp. 263–273.

103. D. Klemm, D. Schumann, F. Kramer, N. Hesler, M. Hornung, H. Schmauder, *Advances in Polymer Science* 205 (2006): 49–96.
104. D.A. Schumann, J. Wippermann, D.O. Klemm, F. Kramer, D. Koth, H. Kosmehl, *Cellulose* 16 (2009): 877–885.
105. J. Wippermann, D. Schumann, D. Klemm, H. Kosmehl, S.S. Gelani, T. Wahlers, *European Journal of Vascular Endovascular Surgery* 37 (2009): 592–596.
106. L. Ning, A. Dufresne, *European Polymer Journal* 59 (2014): 302–325.
107. P. Gatenholm, D. Klemm, *MRS Bulletin* 35 (2010): 208–213.
108. B.M. Cherian, A.L. Leao, S.F. de Souza, L.M.M. Costa, G.M. de Olyveira, M. Kottaisamy, *Carbohydrate Polymers* 86 (2011): 1790–1798.
109. A. Bodi, S. Concaro, M. Brittberg, P. Gatenholm, *Journal of Tissue Engineering and Regenerative Medicine* 1 (2007): 406–408.
110. L. Nimeskern, H.M. Avila, J. Sundberg, P. Gatenholm, R. Muller, K.S. Stok, *Journal of the Mechanical Behavior of Biomedical Materials* 22 (2013): 12–21.
111. S. Saska, H.S. Barud, A.M. Gaspar, R. Marchetto, S.J.L. Ribeiro, Y. Messaddeq, *International Journal of Biomaterials* 1 (2011): 175362–175362.
112. X. Struillou, H. Boutigny, Z. Badran, B.H. Fellah, O. Gauthier, S. Sourice, P. Pilet, T. Rouillon, P. Layrolle, P. Weiss, A. Soueidan, *Journal of Materials Science: Materials in Medicine* 22 (2011): 1707–1717.
113. B. Fang, Y.Z. Wan, T.T. Tang, C. Gao, K.R. Dai, *Tissue Engineering Part A* 15 (2009): 1091–1098.
114. E.L. Batista, A.B. Novaes, J.J. Simonpietri, F.C. Batista, *Journal of Periodontology* 70 (1999): 1000–1007.
115. C. Simonpietri, J.J. Novaes, A.B. Batista, E.L.E.J. Feres, *Journal of Periodontology* 71 (2000): 904–911.
116. W.K. Czaja, D.J. Young, M. Kawecki, R.M. Brown, *Biomacromolecules* 8 (2007): 1–12.
117. W. Czaja, A. Krystynowicz, S. Bielecki, R.M. Brown, *Biomaterials* 27 (2006): 145–151.
118. A. Svensson, E. Nicklasson, T. Harrah, B. Panilaitis, D.L. Kaplan, M. Brittberg, P. Gatenholm, *Biomaterials* 26 (2005): 419–431.
119. L.E. Millon, C.J. Oates, W. Wan, *Journal of Biomedical Materials Research Part B* 90 (2009): 922–929.
120. K.K. Ludwicka, J. Cala, B. Grobelski, D. Sygut, D. Jesionek-Kupnicka, M. Kolodziejczyk et al., *Archives of Medical Science* 9 (2013): 527–534.
121. H. S. Barud. São Paulo Research Foundation—FAPESP, Brazil, 2009.
122. S.H. Oh, C.L. Ward, A. Atala, J.J. Yoo, B.S. Harrison. *Biomaterials* 30 (2009): 757–762.
123. P. Gonzalez, J.P. Borrajo, J. Serra, S. Chiussi, B. Leon, J.M. Fernandez et al. *Journal of Biomedical Materials Research Part A* 88 (2009): 807–813.
124. J. Huia, J. Yuanyuan, W. Jiao, H. Yuan, Z. Yuan, J. Shiru, In *Proceedings of the 2nd International Conference on Biomedical Engineering and Informatics,* Tianjin, China, 12 (2009), pp. 1–5.
125. W.L. Amorim, H.O. Costa, F.C. Souza, M.G. Castro, L. Silva, *Brazilian Journal of Otorhinolaryngology* 75 (2009): 200.
126. M. Avella, J.J.D. Vlieger, M.E. Errico, S. Fischer, P. Vacca, M.G. Volpe, *Food Chemistry* 93 (2005): 467–474.
127. J. Jayaramudu, G.S.M. Reddy, K. Varaprasad, E.R. Sadiku, S.S. Ray, A.V. Rajulu, *Carbohydrate Polymers* 93 (2013): 622–627.
128. K. Majeed, M. Jawaid, A. Hassan, A.A. Bakar, H.P.S.A. Khalil, A.A. Salema, I. Inuwa, *Materials and Design* 46 (2013): 391–410.
129. X.Z. Tang, P. Kumar, S. Alavi, K.P. Sandeep, *Critical Reviews in Food Science and Nutrition* 52 (2012): 426–442.
130. S.D.F. Mihindukulasuriya, L.-T. Lim, *Trends in Food Science & Technology* 40 (2014): 149–167.
131. R.S. Sinha, M. Bousmina, *Progress in Materials Science* 50 (2005): 962–1079.

132. H.M.C. Azeredo, L.H.C. Mattoso, D. Wood, T.G. Williams, R.J. Avena-Bustillos, T.H. McHugh, *Journal of Food Science* 74 (2009): N31–N35.
133. C. Bilbao-Sainz, R.J. Avena-Bustillos, D Wood,. T.G. Williams, T.H. McHugh, *Journal of Agricultural and Food Chemistry* 58 (2010): 3753–3760.
134. M.R. de Moura, R.J. Avena-Bustillos, T.H. McHugh, D.F. Wood, C.G. Otoni, L.H.C. Mattoso, *Journal of Food Engineering* 104 (2011): 154–160.
135. I. Olabarrieta, M. Gallstedt, I. Ispizua, J.R. Sarasua, M.S. Hedenqvist, *Journal of Agricultural and Food Chemistry* 54 (2006): 1283–1288.
136. M. Pereda, G. Amica, I. Racz, N.E. Marcovich, *Journal of Food Engineering* 103 (2011): 76–83.
137. M.D. Sanchez-Garcia, L. Hilliou, J.M. Lagaron, *Journal of Agricultural and Food Chemistry* 58 (2010): 6884–6894.
138. J.G. Siddaramaiah, *Carbohydrate Polymers* 87 (2012): 2031–2037.
139. M.R. de Moura, F.A. Aouada, R.J. Avena-Bustillos, T.H. McHugh, J.M. Krochta, L.H.C. Mattoso, *Journal of Food Engineering* 92 (2009): 448–453.
140. D. Dehnada, H. Mirzaeia, Z.E. Djomehb, S.M. Jafaria, S. Dadashiba, *Carbohydrate Polymers* 109 (2014): 148–154.
141. P.G. Bruce, *Solid State Electrochemistry*. Cambridge University Press, Cambridge, UK, 1995.
142. H. Jain, J. O. Thomas, M. S. Whittingham, *Materials Research Bulletin* 25 (2000): 11–15.
143. F. Alloin, M.A.S.A. Samir, W. Gorecki, J.Y. Sanchez, A. Dufresne, *Journal of Physical Chemistry B* 108 (2004): 10845–10852.
144. M.A.S.A. Samir, F. Alloin, J.Y. Sanchez, and A. Dufresne, *Macromolecules* 37 (2004): 4839–4844.
145. M. Iguchi, S. Yamanaka, A. Budhiono, *Journal of Materials Science* 35 (2000): 261.
146. R.E Jonas, L.F. Farah, *Polymer Degradation and Stability* 59 (1998): 101.
147. S. Ummartyotin, J. Juntaro, M. Sain, H. Manuspiya, *Industrial Crops and Products* 35 (2012): 92–97.
148. M. Nogi, K. Handa, A.N. Nakagaito, H. Yano, *Applied Physics Letters* 87 (2005): 1–3.
149. H. Yano, J. Sugiyama, A. N. Nakagaito, M. Nogi, T. Matsuura, M. Hikita, K. Handa, *Advanced Materials* 17 (2005): 153–155.
150. S.M. Iftekhar, S. Ifuku, M. Nogi, T. Oku, H. Yano, *Applied Physics* A102 (2011): 325–331.
151. Y. Okahisa, A. Yoshida, S. Miyaguchi, H. Yano, *Composites Science and Technology* 69 (2009): 1958–1961.
152. S. Iwamoto, A.N. Nakagaito, H. Yano, *Applied Physics A—Materials Science and Processing* 89 (2007): 461–466.
153. J. Shah, R.M. Brown Jr., *Applied Microbiology and Biotechnology* 66 (2005): 352.
154. E. Sourty, D.H. Ryan, R.H. Marchessault, *Chemistry of Materials* 10 (1998): 1755–1757.

Rheological Properties and Self-Assembly of Cellulose Nanowhiskers in Renewable Polymer Biocomposites

Hesam Taheri and Pieter Samyn

CONTENTS

21.1 INTRODUCTION

Recently, scientists gained high interest in using cellulose nanowhiskers (CNWs) or cellulose nanocrystals (CNCs) as mechanical reinforcement in nanocomposites with a (bio) polymer matrix. This interest is primarily abstracted from the plentiful availability of cellulose and shortage of fossil oil-based polymer resources that elucidate the need of renewable resources. Therefore, CNWs are increasingly used as alternative additives to decrease dependency on petroleum-based products (Goetz et al. 2009). Agricultural residues, water plants, grasses, or marine organisms are sources from which cellulose can be obtained for the production of nanofibers. Different definitions for nanocellulose are often referred to (1) CNWs or CNCs are short crystalline rodlike nanoparticles, while (2) microfibrillated cellulose (MFC), nanofibrillated cellulose (NFC), or cellulose nanofiber (CNF) are long flexible fibers having crystalline and amorphous zones with micro- to nanometer fiber diameters forming a dense fiber network. The beneficial use of CNW additives in combination with biopolymers has evolved from their multifunctional properties and ability to control the self-assembly of CNW into well-defined architectures. Moreover, the high crystallinity of CNW and induced variations in crystallization kinetics within a polymer nanocomposite may influence the morphological and thermomechanical properties. However, main challenges in adding CNW to polymer matrices include a good control on the development of a percolation network and good interface compatibility. A drawback of CNW is poor dispersibility within hydrophobic or nonpolar polymer composites due to the high density of hydroxyl groups on its surfaces, which leads to aggregation and agglomeration. While the compatibility of hydrophilic CNW within hydrophobic polymer matrices is initially controlled through chemical surface modification, the suitable processing of polymer blends with CNW additives requires good understanding of the physical compounding and mixing properties.

In a more fundamental approach, detailed knowledge about the role of CNW on structural and mechanical properties of polymer nanocomposites should be obtained by focusing on the rheological properties. Rheological data may help to better control mixing, aggregation, flocculation, and percolation problems of CNW and it may also provide a guidance to define specific processing conditions under which the flow of CNW can lead to production of functional biocomposites with specific alignment and distribution of CNW into highly ordered structures. As such, the application range of nanocomposite materials in demanding engineering designs may be broadened. It has been demonstrated that the alignment of CNW in thin films or coatings by using a conductive-shear method could increase the surface strength (and consequently wear resistance or lifetime) as a result of the cellulose orientation (Hoeger et al. 2011). The creation of oriented CNW films with shear-oriented orientation allows to vary the coefficient of thermal expansion with low values parallel and high values perpendicular to the CNW alignment as a primary result of single crystal expansion (Diaz et al. 2013). Otherwise, the controlled self-assembly of CNW can potentially be used as a supramolecular templating method to formulate functional materials (Giese et al. 2014). The creation of lateral surface structures in micro- and nanoscale ranges has become important for technological applications such as microsystems or biological devices, as it allows to control adhesion, friction, wettability, biocompatibility, and optical characteristics among others. Therefore, CNWs are projected as promising materials for advanced applications (Eichhorn 2011).

This chapter will review the rheological effects of CNWs in both aqueous suspension and in polymer matrices in order to better understand the factors that are helpful for processing and alignment of CNW fillers by melt extrusion, injection molding, or electrospinning. In this chapter, the preparation and characterization, rheological features, self-assembly/orientation, and processing of polymer/CNW or cellulose/CNW nanocomposites are highlighted. Some main parameters such as size, aspect ratio, polydispersity, concentration, and surface charge affecting the rheological behavior of CNWs are discussed. The better insight on influences of surface properties and interface modification is a key to advance processing of functional biocomposites.

21.1.1 Preparation of CNWs

CNWs are traditionally obtained by acid hydrolysis (Bondeson et al. 2006) or alternatively by ionic liquid treatment (Mao et al. 2013) of native cellulose fibers. During acid hydrolysis, the amorphous zones are chemically removed while the crystalline zones have higher resistance to acid attack and remain intact (Anglès and Dufresne 2001). The actual acid cleavage is attributed to differences in the kinetics of hydrolysis for the amorphous and crystalline domains. This process generally results in a rapid decrease in degree of polymerization (DP) of the cellulose molecules toward a constant level-of DP (LODP) with values depending on the cellulose origin (Habibi et al. 2010). To remove any free acid molecules, the dispersion is subsequently diluted with distilled water, washed, centrifuged, and finally filtrated (Bai et al. 2009). Different acids such as sulfuric, hydrochloric, phosphoric, and hydrobromic acids or their combinations can be used: the nature of the acid and the ratio of acid-to-cellulosic fibers importantly affect the processing and final properties of CNWs, including differences in process yield, mechanical properties, surface properties, and rheological features.

The CNW can be produced from different media, including MFC, NFC, or MCC (Beck-Candanedo et al. 2005; De Rodriguez et al. 2006; Bai et al. 2009) originating from various sources such as wood (Revol et al. 1992; Araki et al. 1998), sisal (De Rodriguez et al. 2006), ramie (Whistler and BeMiller 1997; Habibi et al. 2008), cotton stalks (El-Sakhawy and Hassan 2007), wheat straw (Helbert et al. 1996), bacterial cellulose (Stromme et al. 2002; Bondeson et al. 2006), sugar beet (Dufresne et al. 1997; Azizi Samir et al. 2004a), chitin (Gopalan and Dufresne 2003; Gopalan et al. 2003), potato pulp (Dufresne and Vignon 1998; Dufresne et al. 2000), and tunicin (Favier et al. 1997; Heux et al. 2000). Among the many sources, cotton is most favorable to extract CNWs (Savadekar et al. 2014), due to the high cellulose content avoiding intensive purification and leading to high yield (Eichhorn et al. 2010). The structure of CNW after sulfuric acid hydrolysis is also affected by preparation conditions such as time, temperature, and ultrasound treatment (Dong et al. 1998). A similar hydrolysis procedure under mild and diluted acid conditions can be used to first provide microcrystalline cellulose (MCC) as microscale powder particles

(Iijima and Takeo 2000; Kamel 2007), from which the CNW are subsequently isolated (Lee et al. 2013). By comparing CNW produced from MCC or bio-residues, the latter have lower surface charges and lower crystallinity, while having a higher degradation onset temperature (218°C compared with 155°C) and maximum degradation temperature (Herrera et al. 2012). The morphology and geometrical characteristics (see Table 21.1) of CNWs from different cellulose sources are illustrated in Figure 21.1. The main features of the CNWs are their high aspect ratio and very high surface area: this might be a reason to form a stable interface and control the surface properties by which mechanical properties of the matrix are modified (Chazeau et al. 1999). A high aspect ratio of CNWs is desirable as a critical length is required to transfer applied stress from the matrix to the reinforcing rods (Eichhorn et al. 2010).

The stability of CNW suspensions depends on the introduced number of negative surface charges after sulfuric acid hydrolysis, through which the cellulose hydroxyl groups are esterified to sulfate groups. After a process of controlled freeze-drying, the CNW with sulfate-esterified groups can be fully re-dispersed without agglomeration while maintaining the crystalline structure and high surface area (Lu and Hsieh 2010). After freezing the dispersed state of CNW in water, the spray freeze-drying technique was most successful in obtaining a nonagglomerated porous powder, as capillary forces and ice crystal growth during normal spray drying and freeze drying play key roles in CNW aggregation (Khoshkava and Kamal 2014). The freeze-dried CNW can also be homogenized by sonication and re-dispersed in organic medium suspensions such as DMF (Azizi Samir et al. 2004c), or DMSO and formamide (Viet et al. 2007). The dispersion of CNW in low-polarity solvents (e.g., ethanol) is required in order to incorporate them in nonwater-soluble polymers. The occurrence of flow birefringence in different solvents is a proof for homogeneous dispersion at the nanoscale (Petersson et al. 2009). The sulfate content and surface charges are fundamentally studied by conductometry and influence the viscosity behavior of the CNW suspension. The degree of sulfonation can be tuned by the ratio of acid to cellulose, reaction times, and temperatures: it was found that the total sulfur content and negative surface charge of the nanowhiskers gradually increase when increasing the acid hydrolysis reaction time from 10 to 240 min at 45°C (Dong et al. 1998). A further increase in number of sulfate groups was observed after post-sulfonation at 40°C for 2 h (Araki et al. 1999). As a result, the whiskers with lower content of sulfate groups showed a slight viscosity increase with time at high solid contents, though it was not as significant for whiskers prepared by HCl. Compared to the hydrochloric acid procedure, sulfuric acid hydrolysis needs lower temperature, acid to cellulose ratio, and less time to produce

TABLE 21.1 Geometrical Characteristics of CNWs Obtained by Transmission Electron Microscope

Source	Average Width of CNW (nm)	Average Length of CNW (nm)	Reference
Cotton	7	120	Araki et al. (2001)
Ramie	7	200	Habibi et al. (2008)
Sisal	5	210	De Rodriguez et al. (2006)
Soft wood	4	150	Araki et al. (1998)
Sugar beet	30	210	Azizi Samir et al. (2004a)

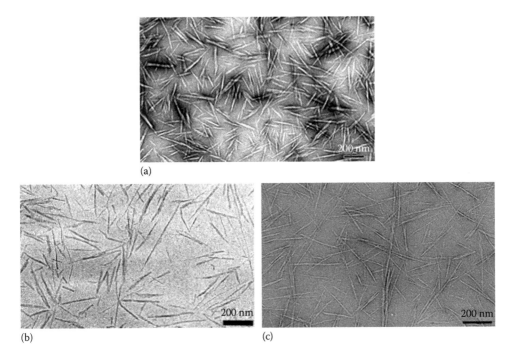

FIGURE 21.1 Transmission electron microscope (TEM) images of cellulose nanowhiskers obtained from (a) ramie (Reprinted with permission from Habibi, Y. et al., *J. Mater. Chem.*, 18, 5002, 2008.); (b) sugar beet (Reprinted with permission from Azizi Samir, M.A.S. et al., *Macromolecules,* 37, 4313, 2004a.); and (c) sisal (Reprinted with permission from De Rodriguez, N.L.G. et al., *Cellulose,* 13, 261, 2006.)

whiskers of similar dimensions. The pH value of HCl-whiskers suspension is about 6, while that of H_2SO_4 whiskers is at about 2 to 3 due to the presence of sulfate esters (Araki et al. 1998). The overall mass yield of HCl-whiskers (10%–20%) is lower than for H_2SO_4-whiskers (70%–75%) (Araki et al. 2000). The CNW obtained by HCl hydrolysis generally present lower surface charges and reduced dispersibility, as the latter is related to the inherent electrostatic stabilization. Therefore, the posttreatment of CNW with sulfuric acid has been used to introduce a controlled number of surface charges. Consequently, the electrostatic repulsion between negatively charged sulfate esters results in a more stable suspension instead of aggregated whiskers. Otherwise, the thermostability of CNW after introduction of sulfate groups is reduced. Therefore, the combination of H_2SO_4 and HCl hydrolysis steps followed by sonication treatment are able to provide spherical CNW structures with better thermal stability (Wang et al. 2007).

The dispersibility and interface compatibility of CNW in a polymer phase can be controlled by chemical surface modifications. By introducing stable negative or positive electrostatic charges, the surface energy properties can be tuned especially when further used in combination with more hydrophobic polymer matrixes. Traditional sizing agents known from papermaking industry such as alkenyl succinic anhydride (ASA) can be added to CNW followed by drying, heating, and freeze-drying, which directly improves the dispersibility in organic solvents without changing the crystalline CNW structure (Yuan et al. 2006). Most chemical surface modifications rely on a direct chemical modification of the hydroxyl groups through covalent binding of functional groups, including, for example, esterification, etherification, acetylation, silylation, and maleination. While partial silylation may improve the dispersibility of CNW in low-polar solvents, the long reaction times may destroy its morphology and crystallinity: therefore, strict control on the reaction conditions is necessary (Goussé et al. 2002). Other surface modifications by polymer grafting on the surface of CNWs have been mainly followed, including two different strategies explained as "grafting from" and "grafting onto" (Dufresne 2010). In the first method, the polymer chains are prepared with the use of an *in situ* surface-initiated polymerization from the immobilized initiators on the substrate (Habibi et al. 2008). In the second strategy, a pre-synthesized polymer chain is attached to the nanowhisker surface (Azzam et al. 2010). As such, specific polymers such as amines, isocyanates, oxazonline, polycaprolactone, polyaniline, vinyl, and acrylates can be linked to the CNW surface. Otherwise, TEMPO-mediated oxidation is frequently used to convert the surface hydroxyl groups into carboxylic moieties. Through the high selectivity of this reaction, only the primary hydroxyl groups are modified while the secondary hydroxyl groups remain unaffected. The higher amount of carboxyl groups relates with higher dispersion stability, in parallel with higher shear stress and viscosity (Hirota et al. 2010). New mechanisms of noncovalent surface modifications include the physical adsorption of surfactants and polymer coatings onto the CNW surface. As such, the dispersibility of CNW in toluene or cyclohexane improves after coating with an alkaline surfactant (Elazzouzi-Hafraoui et al. 2009). Alternatively, positive charges can be introduced on the CNW surface by ammonium treatment (Hasani et al. 2008). The latter cationic CNWs provide stable aqueous suspensions with unexpected rheological properties of thixotropy and gelling: the cationic CNW suspensions show shear birefringence, but no liquid crystalline chiral nematic phase separation was detected due to high viscosity of the suspension in contrast with negatively charged surfaces.

21.1.2 Application of CNWs

For biomedical applications, CNWs have beneficial properties such as hydrophilicity, biocompatibility, and biodegradability. The cellulosic materials with nanofibrillar structures have been developed for tissue engineering and scaffolds (Dugan et al. 2013b, Jia et al. 2013). The concept of using biocompatible and biodegradable materials as molecular carriers especially provides a platform that can be widely adapted for the controlled delivery of amine-containing drugs, or enzymes: for example, amine-containing nanoscale drug molecules were functionalized with CNW for targeted delivery (Dash and Ragauskas 2012). The nanocrystalline cellulose can be used in cooperation with other nanomaterials such as peptides and proteins for biosensor applications. The biosensors can be used for electrical, optical, and mechanical applications with a wide range of detector technologies such as colorimetric, fluorescent, bioluminescent, electrochemical, piezoelectric, quartz microbalance, acoustic, and conductometric signals (Tamayo et al. 2013). In clinical research, the potential for cellular uptake has been studied by using fluorescently labeled CNWs for primary human brain microvascular endothelial cells (Roman et al. 2009) and for cytotoxicity and uptake of both fluorescein and rhodamine B conjugated CNWs in human and insect cells (Mahmoud et al. 2010). Otherwise, clinically relevant proteases were prepared with conjugated CNWs that have enzyme affinity (Edwards et al. 2013).

For nanocomposite applications, the CNW can be introduced during the electro-spinning process in combination with a polymer matrix to increase the conductivity and mechanical properties of electrospun fibers. In addition, new functional groups can be added to electrospun nanofibers after chemical modification of the CNW (Liu et al. 2012). Conductive fiber mats of PMMA/CNW composites with different nanowhisker contents could be produced with fiber diameters in the range of several hundred nanometers (Dong et al. 2012), where the conductivity consequently increased with higher CNW content and good dispersion in a PMMA/DMF solution. Depending on the processing conditions, the degree of CNW alignment in electrospun cellulose acetate fibers can be varied between random orientation and aligned CNW for concentrations up to 5 wt.% (Herrera et al. 2011). During the electrospinning process, the nanowhiskers can be aligned along the fiber axis, providing orientations with preferential mechanical properties. The alignment of CNW fibers greatly improves the mechanical properties of electrospun protein fibers both in tangential and normal directions by not only altering the piling up pattern, but also by promoting phase separation and interface interactions (Wang and Chen 2014). The nanofiber webs of electrospun polyvinylalcohol (PVA) with different degrees of CNW orientations show higher modulus and tensile strength along the fiber direction compared with isotropic PVA electrospun fiber webs, in parallel with different load transfer mechanisms for aligned fibers. Irrespective of the orientation, however, the inclusion of CNW in electrospun fiber webs show improvement in mechanical properties (Lee and Deng 2012). The fabrication of polyethylene oxide (PEO)/CNW nanofibrous mats showed a transition in homogeneous toward heterogeneous microstructures at higher concentrations of the electrospinning solution, whereas the CNW were highly aligned along the fiber axis and the heterogeneous composites were composed of rigid-flexible bimodal nanofibers (Zhou et al. 2011a). Based on PEO/CNW derived from chitosan, high porosity nonwoven nano-composite fiber mats have been produced for wound dressings, where the addition of CNW shows a general increment in mechanical properties (Naseri et al. 2014). The CNW have also a role as nucleation agent in the electrospinning process. The PEO/CNW electrospun fibers developed a rare shish-kebab crystallite structure, where the CNW facilitated the growth of the PEO lamellar crystals (Xu et al. 2014). In combination with poly(lactic acid) (PLA), the nucleation properties have been observed by significant shifting of the spectral absorption peaks of amorphous phase to crystalline absorption peaks (Liu et al. 2012). In parallel, the nucleation of PLA occurs with decrement of the cold crystallization onset temperature. Other researchers have reported the slow crystallization of poly(L-lactide) or PLLA with addition of CNW as bio-based nucleating agent. After subsequent function-alization of CNW by partial silylation to improve the dispersion in PLLA, the degree of crystallinity and tensile modulus or strength increases abruptly (Pei et al. 2010).

For pulp, paper, and wood applications, nanoscale cellulose additives are used as rheology modifiers and pigments to improve the attractiveness of surfaces. Especially, a homogeneous distribution of CNW in polymer coatings is beneficial for improving the barrier properties and reducing moisture sensitivity of food packaging (Martinez-Sanz et al. 2013). A study on the novel carrageenan nanobiocomposites showed better water resistance after incorporation of CNWs (Sanchez-Garcia et al. 2010). Otherwise, the presence of CNW promoted the aggregation in the carrageenan gel structure while MFC did not show significant differences on rheological properties (Gomez-Martinez et al. 2012). The nanocrystalline cellulose is also used in clear wood coatings, where the distribution and alignment of the CNW rodlike particles on the surface of the wood can improve their optical effect. Optical microscopy with polarized light on coated surfaces showed the changes in chiral nematic and confirmed by the CD spectra (Vlad-Cristea et al. 2013). The iridescent effect of coatings depends on the nature and concentration of the pigments and additives that are selected to improve CNW orientation.

In many of the above applications, the hierarchical bottom-up organization of CNW can provide an efficient way in increasing the performance and functionality of structural nanocomposites and devices. Therefore, good knowledge and strict control on the suspension behavior of CNW is required to induce spontaneous organization of the rodlike nanoparticles.

21.2 ORGANIZATION OF CNWs OR CNCs

21.2.1 Self-Assembly and Chiral Nematic Nature of CNW Suspensions

The nanoscale rodlike materials form a lyotropic liquid crystalline phase under specific concentrations. In general, the lyotropic liquid crystal is formed as a dispersion of anisotropic macromolecules in a solvent. The phase transitions in such a system are governed by variations in the so-called anisotropic mesogen concentration, as a mesogen represents the fundamental unit of a liquid crystal that induces local structural order. Depending on the range of organization of rodlike nanoparticles, the liquid crystalline phase shows different characteristics with (1) nanoparticle orientations along the director with long-range orientational order and short-range positional order (i.e., *nematic phase*); or (2) twisting of the nematic microstructure along an axis perpendicular to the director (i.e., *chiral nematic or cholesteric phase*); or (3) both long-range and short-range orientational order (i.e., *smectic phase*). The self-organization of CNW in suspending media results in the formation of a chiral nematic liquid crystalline phases. However, phase transitions in ordered lyotropic suspensions happen depending on the nature, aspect ratio, concentration, and surface charges of the rodlike particles.

The use of sulfuric acid as hydrolyzing agent induces negative charges on the surface of CNWs, which provide perfectly uniform dispersion of the whiskers in water via electrostatic repulsions (Revol et al. 1992). The formation of a nematic liquid crystalline alignment in colloidal suspensions of cellulose crystallites obtained by acid hydrolysis has been discovered in 1959 (Marchessault et al. 1959). By removing the water phase, self-organization into liquid crystalline structures is introduced in concentrated suspensions by the change into new nanocrystal configurations (Habibi et al. 2010). The main driving force for reorganization of CNW upon drying is a minimization of existing electrostatic forces. However, the evaporation of water has been identified as a more complex two-stage process (Mu and

Gray 2014), where the variation in chiral nematic pitch mainly occurs in the first phase as the concentration of the CNW in the suspension gradually increases during evaporation, while finally a concentration is reached where the formation of ordered gels and glasses prevents further major changes in pitch. The formation of a chiral nematic ordering of the suspensions has been demonstrated by the appearance of "fingerprint" patterns, as observed by polarized microscopy (Revol et al. 1992). This chiral nematic structure originally formed in suspension can be preserved after complete drying, creating iridescent organized CNW layers with potential applications as coating materials for decorative materials or security papers. Also after freeze-drying of nanocrystalline colloids, a microporous foam with ordered layers having a period of 1 to 2 μm is formed (Liu et al. 2011): this clarifies that the self-alignment of the CNWs along a vector director in a packed nematic planar and the period of the cholesteric liquid crystalline phase does not change with its drying state. In suspension, different types of order develop depending on the concentration of the CNW (Figure 21.2a). Under very dilute conditions, the CNWs are randomly oriented in an isotropic medium as they move freely under Brownian motion: they appear as spheroids or ovaloids and the initial ordered domains are similar to tactoids (Habibi et al. 2010).

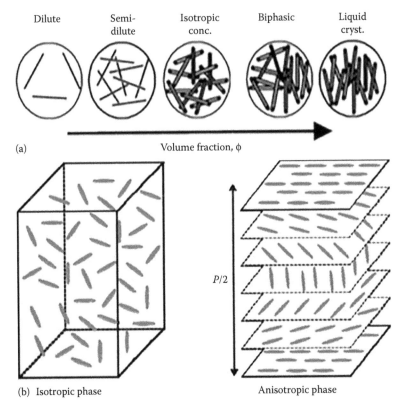

FIGURE 21.2   (a) Schematic representation of phase behavior of fluid dispersed rods (Reprinted with permission from Davis, V.A. et al., *Macromolecules*, 37, 154, 2004.) and (b) schematic representation of CNWs orientation in both the isotropic and anisotropic (chiral nematic) phases (Reprinted with permission from Habibi, Y. et al., *Chem. Rev.*, 110, 3479, 2010.)

In semidiluted systems, the rotational freedom of the nanorods gets inhibited and they are confined into smaller volumes near the isotropic concentration. When the concentration gradually increases, a nematic liquid crystalline alignment is adopted through coalescence of the previously formed tactoids and unidirectional self-orientation within an anisotropic phase. Above a critical concentration, the CNW suspension starts to form a chiral nematic order with features resembling a cholesteric liquid crystal (Hirai et al. 2009). The latter biphase system develops as an equilibrium state between particles in the anisotropic (chiral nematic) and isotropic phase, but it spontaneously separates into an upper isotropic and a lower anisotropic phase upon standing (Figure 21.2b), causing characteristic shear birefringence properties. Finally, the system becomes a fully homogeneous liquid crystalline phase above a second critical concentration resulting in complete anisotropy. The two critical concentrations depend on the CNW aspect ratio (Wu et al. 2014): the rods with higher aspect ratio transition into a biphase system and form a hydrogel at lower concentrations than those with small aspect ratio. Exceptionally, deviations in the behavior of aqueous CNW suspensions from a lyotropic characteristic were observed where the suspensions remained isotropic over a range of concentrations (Lu et al. 2014), attributed to the strong electrostatic repulsions between and the weak tendency or driving force of anisotropy formation as a result of the small aspect ratio of CNW particles.

The alignment of CNWs into a nematic order is caused by entropically driven self-orientation above a critical concentration, including translational and orientational contributions of the nanorods (Onsager 1949). The coupling of these phenomena makes a parallel configuration of the rods, which is favored since the effective excluded volume becomes zero. At low concentrations, the translational component is predominant as the contribution from the gain in free volume is low. At high concentrations, the entropy of parallel alignment is lower and thermodynamically favored. The phase transitions in liquid crystalline dispersion of rodlike colloidal particles have been fundamentally explained as a function of their surface interactions (Van Bruggen et al. 1996). The isotropic-to-anisotropic phase transformation is sensitive to the electrolytes and does not often occur in free electrolyte suspensions as shown in Figure 21.3a and b. The specific chemical nature of the electrolyte affects the stability of the ordered phases, the temperature dependence of the phase separation, and the chiral nematic pitch size. Typically, the ordered nematic phase in suspensions without electrolyte develops at critical concentrations of 1 to 10 wt.% (Habibi et al. 2010). Otherwise, the biphase system develops at concentrations up to 13 wt.%, where the system turns into complete anisotropy. The CNW obtained from bacterial cellulose that was hydrolyzed by sulfuric acid, spontaneously separates in a nematic phase. This phase separation can persist for up to seven days, while adding NaCl electrolyte decreases the separation time to only two days as the anisotropic phase turns into a chiral nematic order (Hirai et al. 2009). The formation of a chiral nematic phase is strongly affected by the nature and density of charges on the surface of CNWs: for example, CNWs hydrolyzed by post-sulfated HCl show a birefringent glassy phase with a crosshatch pattern while directly sulfated CNWs show chiral nematic phases as shown in Figure 21.3c (Araki et al. 2000).

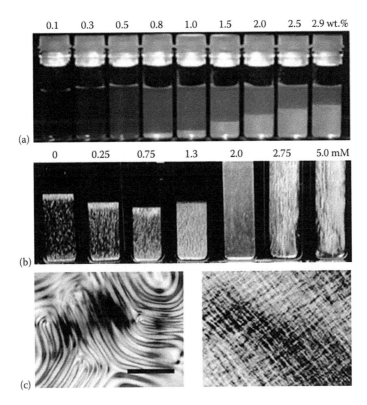

FIGURE 21.3 (a) Phase behavior of bacterial cellulose nanowhiskers dispersed in deionized water as a function of the total concentration after 25 days of standing. (b) Effect of added NaCl on the phase behavior of suspensions of CNWs from bacterial cellulose for a fixed total cellulose concentration of 3 wt.% after 25 days of standing (Reprinted with permission from Hirai, A. et al., *Langmuir*, 25, 497, 2009.); (c) polarized-light micrographs of cellulose suspensions. (Left) Fingerprint pattern in the chiral nematic phase of the directly H_2SO_4-hydrolyzed suspension (initial solid content 5.4%). Scale bar: 200 μm. (Right) Crosshatch pattern of postsulfated suspension (solid content 7.1%) (Reprinted with permission from Araki, J. et al., *Langmuir*, 16, 2413, 2001.)

The chiral nematic phase of CNW suspensions is characterized by the spontaneous formation of a helical arrangement. The helicoidal structures were initially assumed to form an asymmetry or twist in the cellulose nanorods (Revol and Marchessault 1993). On the other hand, negative charges were thought to be responsible for chiral ordering since the uncharged CNW generated from hydrochloric acid did not tend to give this ordering (Revol et al. 1992). However, the suspensions of CNW with a surfactant or polymer surface coating also show tendency for chiral nematic order (Heux et al. 2000). Therefore, the intrinsic morphology of single CNW as screw-like rods has been confirmed by small angle neutron scattering (SANS) in aqueous suspension (Orts et al. 1998). The exact path for self-assembly has been recently revised by taking into account the complete phase diagram of CNW suspensions, including the competition between gelation or glass formation and liquid crystalline ordering (Lagerwall et al. 2014). Certainly, at high CNW concentrations, the transition into a nonequilibrium glassy state needs to be handled. As such, the organization into short- and high-specified pitch of the helix and photonic bandgap can be enhanced by controlled drying under variable atmospheric humidity and alternative solvent removal procedures with adding nonvolatile cosolvents. The uniformly layered helical structures can also be preserved as internal structures of dried CNW films by a novel vacuum-assisted self-assembly technique, with variable organization depending on the sonication time, suspension volume, and degree of vacuum (Chen et al. 2014).

21.2.2 Mechanical and Interfacial Properties of CNWs and Nanocomposites

The determination of the mechanical properties for single and ultrafine CNWs is very difficult. After having removed the amorphous states of native cellulose fibers by acid hydrolysis, the remaining nanowhiskers should theoretically have a modulus close to a cellulose crystal (Eichhorn 2011). According to experimental AFM, bending stiffness measurements on a single CNW from tunicate with cross-sectional diameters of 8 to 20 nm, the elastic moduli of single microfibrils prepared by TEMPO-oxidation and acid hydrolysis were 145.2 ± 31.3 and 150.7 ± 28.8 GPa, respectively. The result showed that the experimentally determined modulus was in agreement with the elastic modulus of native cellulose crystals (Iwamoto et al. 2009; Lahiji et al. 2010). Experimental measurements of the CNW modulus with use of sound velocities by inelastic X-ray scattering yielded a value of 15 to 220 GPa (Diddens et al. 2008), which is much higher than theoretical estimates in the range of 100 to 160 GPa (Šturcová et al. 2005), due to influences of molecular dynamics/mechanics methods, ordered/disordered states, and anisotropy. Based on these values, the CNW may provide significant reinforcement in nanocomposite materials, comparable to alternative reinforcing materials such as Kevlar fiber, carbon fiber, glass fiber, or carbon nanotubes (CNTs).

The outstanding mechanical properties of CNW nanocomposites rely on the nature of the interactions of CNWs within a polymeric matrix. Based on micromechanical models for prediction of the elastic modulus in a reinforced polymer matrix, CNW rods behave similarly as CNTs (Loos and Manas-Zloczower 2012). The mechanical properties of CNW within nanocomposite materials depend on the interaction and affinity of cellulosic nanomaterial and ability to induce a rigid mechanical percolation network structure (Chauhan and Chakrabarti 2012), mainly explaining the high moduli of CNW nanocomposites (Azizi Samir et al. 2004b). Similarly, templating of CNW in conductive polymers can reduce the electrical percolation threshold over several magnitudes (Tkalya et al. 2013). The interactions between whiskers and polymer matrix are sensitive to the whisker concentration and environment: (1) the absence of interactions during exposure in water can be explained by a mechanism involving competitive hydrogen bonding of water and whiskers, which reduces the whisker/whisker and whisker/matrix interactions (Capadona et al. 2008); while (2) the superior reinforcing effect of CNW additives beyond a critical percolation threshold results from the formation of a stiff and continuous three-dimensional network at critical filler volumes, held by hydrogen bonding (Rojas et al. 2009). As such, extremely low filled contents of 0.2 wt.% can provide significant strength and rigidity in different microstructures such as scaffolds (Pooyan et al. 2012). Depending on the geometry of CNW, the higher aspect ratios ensure better percolation with resulting mechanical improvements and thermal stability at lower fiber loads (De Rodriguez et al. 2006): for example, the high aspect ratio value of CNW from sisal thus provides superior properties. A microstructure-based finite element analysis model was developed to predict the effective elastic property of all-cellulose composites reinforced with CNW: with the increase of aspect ratio, the effective elastic modulus increases in isotropic microstructure. The elastic property anisotropy increases with the aspect ratio and anisotropy of nanowhisker orientation (Li et al. 2011). The local mechanical performance of electrospun PVA fibers with 20 wt.% CNW was investigated by nanoindentation, which indicates an increase in modulus from 2.1 GPa for a pure PVA electrospun fiber toward 7.6 GPa for reinforced fibers. The modules of the CNW are 60%–80% higher than the isotropic model predictions but lower than longitudinal model prediction, suggesting that the nanowhiskers are partially aligned to the electrospun fiber direction (Lee and Deng 2013). In the presence of CNW, the thermomechanical stability of the nanocomposite is simultaneously enlarged as the modulus improves at temperatures above the glass transition temperature of the matrix (Helbert et al. 1996).

The reinforcement of a nanocomposite by CNW is caused by interfacial stress-transfer processes within a continuous network of nanorods and strong interactions at the interface between the CNW and polymer matrix by hydrogen bonding (Favier et al. 1995): as such, even few percentages of CNWs homogeneously dispersed as reinforcement into polymer matrix remarkably improve the mechanical properties. The different mechanisms for interfacial energy dissipation were investigated by cyclic tensile and compressive deformation of CNW/epoxy matrices and a theoretical model is used to determine the dissipation of energy at the interfaces between whiskers and at the whisker–matrix interface (Rusli and Eichhorn 2012). The local molecular deformation and interfacial interactions of CNW can also be illustrated with Raman spectroscopy, which shows clear distinction between fibers under uniform stress or strain conditions (Young and Eichhorn 2007). It has been demonstrated that the extent of stress transfer and modulus of nanocomposite is mainly influenced by level of dispersion (Ljungberg et al. 2006) and local orientation of the CNWs (Rusli et al. 2010), which is governed by shear conditions and affected by the environmental conditions such as temperature or moisture.

21.3 RHEOLOGICAL BEHAVIOR OF CNW SUSPENSIONS

In general, the rheological properties of well-dispersed CNW in aqueous media can be described in parallel with the behavior of rodlike particles in colloidal suspensions. Colloidal suspensions are defined as a dispersed phase of a two-component system, where the dispersed particles do not simply sediment within a suspending liquid medium. Each component in the system is too small to be observed separately and the stability is attributed to the balance between hydrodynamic and thermodynamic interactions as well as Brownian forces, electrical, thermal, and magnetic fields. Moreover, other parameters of the dispersed particles such as rotation, orientation, and particle shape play an important role in the rheological features of the medium (Mewis and Wagner 2012). Early work on the rheology of nonspherical suspensions in a shear flow can be found in the works of Zirnsak et al. (1994) and Wierenga and Philipse (1998). The rheology and particle motion of nonspherical suspension is greatly influenced by particle orientation under given flow conditions. When compared to the flow properties of spherical particle suspensions, the orientation vector of rodlike CNW particles relative to the flow direction should be considered. The orientation and alignment of the nanoparticles into the flow is governed by the balance of hydrodynamic forces, which macroscopically leads to shear thinning effects. The shear thinning behavior of CNW suspensions under permanent shear flow may be further described with the Herschel–Bulkley model and scales with the rotational Péclet number (Willenbacher and Georgieva 2013). For instance, the Brownian motion may result in deviations of the rotational Péclet number (Pe), with a random orientation at low Pe and more orientation at higher Pe. Such alignment reduces the viscosity, but can also lead to highly nonlinear rheological properties (Mewis and Wagner 2012). As mentioned before, the CNW colloidal suspensions form a chiral nematic structure above a critical concentration (Revol et al. 1994; Beck-Candanedo et al. 2005), which depends on size, polydispersity, physical dimension, surface charge, and the ionic strength of the medium (Dong et al. 1998). These orientations can eventually lead to the formation of a gel-like material at higher concentrations (Liu et al 2011; Urena-Benavides et al. 2011). The gelling properties of semidilute CNW suspensions were characterized by a critical gel time resulting from the growth and percolation of self-similar CNW clusters at 0.2 to 0.3 vol.% (Le Goff et al. 2014). The viscoelastic properties of the suspension were further confirmed by dynamic rheology experiments (Gong et al. 2011): the elastic modulus (G') and viscous modulus (G'') were frequency independent in the low-frequency region and G' was almost tenfold higher than G'', showing a typical elastic gel behavior.

Previous rheological studies on the shear alignment of microcrystal suspensions have been done for chitin whiskers (200 nm of length and aspect ratio of about 139 nm) (Li et al. 1996) and black spruce cellulose microcrystals (180–280 nm length and aspect ratio of about 30) with use of SANS (Orts et al. 1998). The X-ray scattering results of microcrystal alignment under shear rate confirmed the alignment model based on rheological data: prior to shearing, the microcrystals appeared randomly oriented and went through a two-step alignment process with increasing shear rate (Ebeling et al. 1999): (1) at low shear rates, the X-ray scattering patterns indicated little or no preferential alignment in the shear direction; however, in the plane perpendicular to the shear direction, the microcrystals were preferentially oriented in the vertical direction; and (2) at high shear rates, the microcrystals were aligned horizontally along the shear direction. These phenomena were completely reversible. In general, the suspension concentration, applied shear rate, and temperature affect the flow properties and orientation of CNW in its medium and ultimately the cholesteric pitch, which defines the optical properties of ordered CNW structures.

21.3.1 Effect of Different Parameters on Rheological Behavior

The relationship between the microstructure and rheological features of CNW suspensions are extensively influenced by parametrical processing factors and need to be tightly controlled within given values relative to the suspension properties. For rapid characterization of diluted aqueous CNW suspensions in industrial applications, a method of viscosity measurements by a rolling-ball viscometer has been developed (Gonzalez-Labrada and Gray 2012): the suspensions are characterized by an intrinsic viscosity that directly relates to the hydrodynamic dimensions of the CNW. More fundamental characterization requires comprehensive information about the type and selection of suitable rheometers, including rotational and oscillatory measurements (Barnes et al. 1989; Macosko 1994; Larson 1999; Faith 2001). There are also many reports on specific rheological data and experimental methods for characterization of nanofiber suspensions (Iotti et al. 2011; Saarikoski et al. 2012; Saarinen et al. 2014). In contrast, some analytical models have been adapted to describe the viscosity and rheological functions of CNW aqueous suspensions, based on dimensionless numbers and correction parameters to match the theoretical viscosity under alignment conditions with experimental values (Noroozi et al. 2014).

21.3.1.1 Effect of Brownian Dynamic on Rheological Behavior

The Brownian dynamic behavior of nanorods and entanglement effects have been fundamentally discussed in the Doi–Edwards theory (Doi and Edwards 1978a), depending on concentration and aspect ratio of the particles in suspension. According to this theory, the CNWs follow a superposition curve in reduction of viscosity or rotational diffusivity versus particle concentration (Doi and Edwards 1978b). This theory can be applied to any nanorod or nanofiber suspensions where Brownian motions exist in the suspending liquid, including CNTs, CNWs, and polymer fibers. In brief, two different concentration regimes are defined depending on the number of rods per unit volume (C), length of rods (L), and diameter of rods (d), where the rod concentration in suspension is proportional to $1/L^3$. The first regime is characterized as a very dilute system when $C \ll 1/L^3$ and the Brownian motion of the nanorods is independent and determined by the viscosity of the solvent. Therefore, the nanorods are able to rotate freely without any interaction with other nanorods. For the second regime with higher concentration (semidilute system) when $1/L^3 \ll C \ll 1/dL^2$, the motion of each nanorod is severely restricted and the dynamics of the system relates to the nanorod entanglements. The condition $C \gg 1/L^3$ means that the rods are completely entangled, while $C \ll 1/dL^2$ presents a system that resembles a thermodynamically ideal solution. In order to define the transition from a dilute to a semidilute regime, a dimensionless parameter $\beta \approx CL^3$ has been defined that was experimentally close to $\beta \approx 30$ in the transition region (Doi and Edwards 1986). The rotary diffusivity of nanorods can be described in the dilute regime (D_{r0}) and the semidilute regime (D_r) according to Equations 21.1 and 21.2, respectively (Bitsanis et al. 1990):

$$D_{r0} = \frac{3k_B T (\ln(L/d) - \gamma)}{\pi L^3 \eta_s} \tag{21.1}$$

$$D_r = AD_{r0}(CL^3)^{-2} \tag{21.2}$$

where:
T is the temperature of suspension
k_B is the Boltzmann constant
γ is a constant coefficient ($\gamma \approx 0.8$)
η_s is the viscosity of the solvent (or matrix)

L/d is the aspect ratio

A is a dimensionless constant with large values that can be calculated via theoretical predictions and simulations methods

The zero shear viscosity (η_0) of semidilute suspensions is given by Equation 21.3:

$$\eta_0 = \frac{Ck_BT}{10D_r} \tag{21.3}$$

where:

D_r is rotary diffusivity in a semidilute regime

T is the temperature of suspension

k_B is the Boltzmann constant

C is concentration of suspension

Ultimately, the zero shear viscosity (η_0) can be formulated as a combination of previous Equations 21.1 through 21.3 resulting in Equation 21.4:

$$\eta_0 = \frac{\eta_s \pi}{30\,A(\ln(L/d) - 0.8)}(CL^3)^3 \tag{21.4}$$

According to a study shown in Table 21.2, the rotary diffusivity (D_r) and the constant A of a semidilute system with CNWs have been calculated from the zero shear viscosity (Cassagnau et al. 2013) in combination with experimental work (Bercea and Navard. 2000). The rigidity or flexibility of the nanorods has a strong effect on the constant A, acceleration of nanorods, and the Brownian dynamics: it has been observed that with decreasing of A and increment of nanorod concentration, the power law model leading to a single value for the rotary diffusion D_r cannot be applied precisely over the whole range of CL^3 values (Ma et al. 2008). Therefore, different values for the rotary diffusion D_r have been proposed,

TABLE 21.2   Zero Shear Viscosity η_0, Rotary Diffusivity D_r, and Constant A for Aqueous Cellulose Nanowhisker Suspensions with Different Volume Fraction ϕ

ϕ (%)	η_0 (Pa.s)	CL^3	D_r (s^{-1})	A
0.02	1.5×10^{-3}	5	–	–
0.04	1.8×10^{-3}	10	–	–
0.06	2.1×10^{-3}	15	–	–
0.08	3.0×10^{-3}	20	3.0×10^{-1}	90
0.10	4.0×10^{-3}	25	2.8×10^{-1}	100
0.20	1.2×10^{-2}	50	1.8×10^{-1}	260
0.40	4.0×10^{-2}	100	1.1×10^{-1}	620
0.60	1.0×10^{-1}	150	6.7×10^{-2}	840

Sources: Cassagnau, P. et al., *Rheol. Acta*, 52, 815, 2013; Bercea, M. and Navard, P., *Macromolecules*, 33, 6011, 2000.

The calculations are based on $d = 15$ nm, $L = 2$–100 nm, $L/d = 140$, $T = 25°C$, and $D_{r0} = 1.75$ s^{-1}.

indicating that the rigid rods with high aspect ratio have very low rotary diffusion coefficient and shear is able to align the rods in flow direction.

The dynamic rheological behavior of stiff polymers and nanowhisker suspensions shows that the Doi–Edwards theory is valid for different rigid rodlike structures at $30 < CL^3 < 230$ (Cassagnau et al. 2013). The variation with CL^3 obeys a master curve at temperatures below T_g for the reduction in zero shear viscosity (Figure 21.4a) and rotary diffusivity (Figure 21.4b), irrespective of the nature of materials. These results elucidate that CNWs can be considered as straight rigid rods with Brownian motion. On the opposite, stiff polymers and cross-linked polymer fibers at temperatures above T_g do not obey the Doi–Edwards theory and show two distinct regimes of dynamic behavior, resulting in a dynamical behavior close to the physics

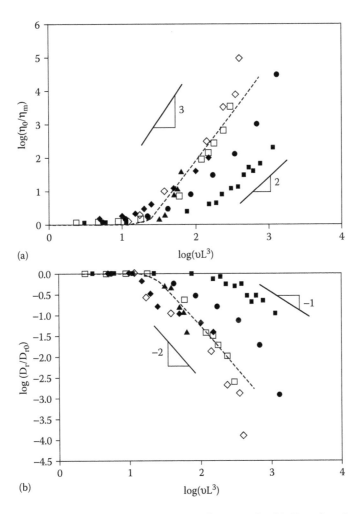

FIGURE 21.4 Variation versus the concentration of nanorods. (a) For the viscosity reduction according to Equation 21.4 and (b) for the rotary diffusivity according to Equation 21.2. White square: Cross linked polymer nanofiber ($T = 25°C$). Triangle: CNTs in PDMS. Circle: PBLG in m-cresol. Black diamond: Cellulose nanowhisker. White diamond: Polymer nanofiber ($T = 25°C$). Black square: Cross linked polymer nanofiber ($T = 130°C$). (Reprinted with permission from Cassagnau, P. et al., *Rheol. Acta*, 52, 815, 2013.)

of polymer solutions in terms of power laws (Larson 1999; Cobb and Butler 2006). The rheo-optical studies of CNT suspensions under shear flow indeed confirmed that the Doi–Edwards theory can be successfully used at low Péclet number, Pe < 200 (Fry et al. 2006).

21.3.1.2 Effect of Acid Hydrolysis and Surface Charges on Rheological Behavior

The hydrolysis strength affects the size and specific surface properties of CNWs with concentration of anionic sulfur groups (Dong et al. 1998), which led to electrostatically stabilized CNW suspensions (Revol et al. 1992). Consequently, differences in rheological behavior of CNW suspensions obtained by sulfuric acid and hydrochloric acid hydrolysis have been observed. The H_2SO_4/CNW suspensions have time-independent viscosity over a broad range of concentrations (Revol et al. 1992). The HCl/CNW suspensions are thixotropic at concentrations > 0.5% (w/v) and anti-thixotropic at concentrations < 0.3% (w/v) (Araki et al. 1998): the thixotropy was attributed to interparticle agglomeration under highly concentrated static conditions and break-up under flow. Similarly, the anti-thixotropy results from reduced particle interactions under low concentrations and individual alignment under flow. In conclusion, CNW with similar size and shapes can be prepared irrespective of the preparation method, but the introduction of surface charges drastically reduces the viscosity of the whisker suspension and removes its time dependency.

The rheological behavior of CNW suspensions hydrolyzed toward different degrees of sulfation, is related to the physicochemical properties of these suspensions. The effect of the sulfation degree on the isotropic-to-liquid crystal and liquid crystal-to-gel microstructure transitions of CNW suspensions has been demonstrated by rheometry and polarized optical microscopy (Shafiei-Sabet et al. 2013): two suspensions with 0.69% and 0.85% sulfur, respectively, are isotropic at low concentration and experience two different transitions as concentration increases. The formation of a chiral nematic liquid crystal structure for both suspensions has been evidenced above a first critical concentration in parallel with the formation of a gel phase at higher concentrations. According to steady-state shear tests on sets of CNW suspensions with different degrees of sulfation at various concentrations (1 to 7 wt.%), the isotropic 1 wt.% CNW suspensions show a Newtonian plateau at low shear rates. At intermediate shear rates, shear thinning occurs due to alignment of the CNWs in the shear direction resulting in a progressive decrease of viscosity. At high shear rates where almost all CNWs have been aligned, another plateau with steady viscosity has been observed. As shown in Figure 21.5, the hydrolysis strength affects the profile of the three-region viscosity of CNW liquid crystal suspensions and critical concentrations: according to oscillatory tests, the initiation of gel-like behavior has been observed at concentrations of 12 and 10 wt.% for high and low degrees of sulfation, respectively (Shafiei-Sabet et al. 2013). In conclusion, the degree of sulfation of CNWs significantly affects the critical concentrations at which transitions from isotropic to liquid crystal and liquid crystal to gel occur.

21.3.1.3 Effect of Ionic Strength on Rheological Behavior

The ionic strength of the medium influences the stability of CNW suspensions and high electrolyte concentrations may induce instabilities due to interparticle interactions (Dong et al. 1996). The increasing salt concentrations directly relates to a decrease in absolute

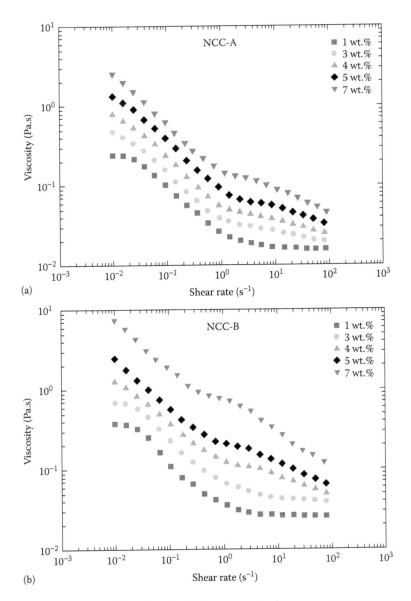

FIGURE 21.5 The rheological behavior of cellulose nanowhisker suspensions at concentrations of 1–7 wt.% (sonicated for 1,000 J/g) with different degree of sulfation: (a) CNC-A suspensions with strong hydrolysis and (b) CNC-B suspensions with weak hydrolysis. (Reprinted with permission from Shafiei-Sabet, S. et al., *Rheol. Acta*, 52, 741, 2013.)

value of the zetapotential of the suspended particles. At low CNW concentrations, the salt concentration had no influence on the flow properties of the system. At higher CNW concentrations, however, the yield stress exponentially decreased with lower zetapotential values (Tatsumi et al. 1999). These facts indicate that the shear stress consists of two different contributions, caused by (1) friction of the effective volume of the particles, and (2) overlapping in electric double layer of the particles. The isotropic-to-chiral nematic phase equilibrium is sensitive to the nature of the counterions (Dong and Gray 1997).

Suspensions with H^+ counterions form an ordered phase at the lowest CNW concentrations. For inorganic counterions, the critical concentration for ordered phase formation increases in the order $H^+ < Na^+ < K^+ < Cs^+$. For the organic counterions, the critical concentration in general increases with counterion size, suggesting that the equilibrium is governed by balance between hydrophobic attraction and steric repulsion forces. The chiral nematic pitch of the anisotropic phase decreases with increasing electrolyte concentration and a decrease in double layer thickness increases the chiral interactions between the crystallites. The specific effects of electrolyte on the intrinsic bulk viscosity of CNW suspensions and transitions from isotropic into chiral nematic anisotropic phase have been demonstrated (Boluk et al. 2011): the formation of an electrical double layer contributes to the viscosity of the CNW suspension as the viscosity of dilute and charge-free nanowhisker suspensions is much lower than charged rods, which is caused by electroviscous effects. As such, it was demonstrated that the electrolyte concentration of suspension has a strong effect on the relative viscosity.

The addition of NaCl electrolyte to an aqueous CNW suspension changes the microstructures, the critical concentration of liquid crystalline phase formation and ultimately the rheological behavior of suspensions (Araki and Kuga 2001; Hirai et al. 2009). The addition of NaCl reduces the electroviscous effects and destabilizes the suspension because of coagulation at concentrations higher than 10 mM. In parallel, the viscosity measurements and electroviscous effects could be used as a method to calculate the shape factor of a CNW suspension. Other effects of ionic strength on rheological behavior of CNW suspensions in different regimes of isotropic, anisotropic chiral nematic and gelling phase have also been studied (Shafiei-Sabet et al. 2014): an increase of the ionic strength up to 5 mM of NaCl within the isotropic suspension leads to a decrement of electroviscous effects and viscosity. For biphase samples with chiral nematic liquid crystal domains, the higher ionic strength decreases the size of the chiral nematic domains and leads to higher viscosity at low shear rates. Finally, all ordered domains are broken and addition of NaCl causes a decrement of viscosity at high shear rates. In conclusion, the increase of ionic strength up to 5 mM of NaCl for gel regime results in reduced viscosity and formation of a weaker gel structure.

21.3.1.4 Effect of Sonication on Rheological Behavior

Sonication has an effect on the physicochemical properties of CNW suspensions without affecting the dimensions of the individual rods, as they do not break with different intensities of ultrasound energy. In parallel, the physicochemical properties of the CNW suspensions result in higher electrophoretic mobility and conductivity when higher ultrasound energy is applied, which suggests that sonication has an effect on surface charge and zeta potential (Beck et al. 2011). According to their research, the applied ultrasound energy less than 5000 J/g is not sufficient to break the covalent sulfate ester–cellulose bonds at ambient temperature. In contrast, the use of ultrasonication treatments in combination with acid hydrolysis for synthesis of CNW is nonselective, which means it can remove amorphous cellulose and crystalline cellulose (Li et al. 2012). After mild sonication of CNW suspensions, a significant drop in viscosity was observed with three typical regions for lyotropic liquid crystal polymer structures (Onogi and Asada 1980): (1) the first region at low shear rates shows shear thinning behavior due to the alignment of the chiral nematic liquid crystal, (2) the second region at intermediate shear rates shows a plateau due to orientation of crystals along the shear direction, and (3) the third region at high shear rates presents a second shear thinning regime, where the liquid crystal domains are destroyed by the shear stress and crystal rods are aligned along the shear flow direction. In general, the critical concentration is much higher for sonicated compared to unsonicated suspensions (Shafiei-Sabet et al. 2012). The high amounts of ultrasound energy affect the chiral nematic liquid crystalline domain structure, which is responsible for the rheological behavior within the first region. It has been also reported that sonication levels above 1000 J/g significantly decrease the viscosity of the CNW suspension (Dong et al. 1998). They have also reported that sonication time longer than 5 min did not decrease the size of CNWs but it affects the liquid crystal suspension behavior. By increasing the sonication energy, the size of chiral nematic and pitch becomes larger with weaker interactions, which results in easy shearing and lower viscosity. The systematic increase in pitch size with ultrasound treatment can be explained by the liberation of excess ions that are trapped within the bound layer of CNWs, leading to expansion of electrical double layer and weaker chiral interaction (Beck et al. 2011).

21.3.1.5 Effect of Concentration on Rheological Behavior

The concentration of CNW in suspension plays an important role due to the phase separation of liquid crystalline and isotropic domains. The transition from an isotropic into a biphase system occurs at the critical concentration, where the smallest particles remain in the isotropic phase and the larger ones form a liquid crystalline phase (Hirai et al. 2009). In this regard, the size distribution of rodlike structures influences the phase behavior and rheological behavior (Beck-Candanedo et al. 2005). In addition, both storage modulus and loss modulus of cellulose nanofiber suspensions rapidly increased with increasing concentration because of the gradual formation of a stronger network structure. As such, the sol-gel transformation and the viscoelastic transition depend on the hydroxyl bonding and the cross-linking extent of cellulose nanofibers in various concentrations (Chen et al. 2013).

The effect of concentration on shear response of CNW suspensions has been investigated (Urena-Benavides et al. 2011): the viscoelastic properties were evaluated in addition to the steady shear response by means of the Cox–Merz rule, which is helpful to understand more about rheology and phase behavior at various concentrations. According to this rule, the values of dynamic complex viscosity $|\eta^*(\omega)|$ and steady-state viscosity $\eta(\dot{\gamma})$ overlap with each other at equal absolute values of rotational frequency and shear rate (Cox and Merz 1958). In general, lyotropic macromolecular liquid crystals do not obey the Cox–Merz rule. The complex viscosity of CNC suspensions is generally higher than their steady viscosities and the Cox–Merz rule is not valid, which may be attributed to the existence of a liquid crystal domain (Wu et al. 2014). At lower concentrations for CNW suspensions or isotropic systems, little deviation from the Cox–Merz rule (at low shear rate/frequency) and greater deviation (at high shear rate/frequency) has been observed after a critical concentration (Urena-Benavides et al. 2011), which is expected for suspensions containing oriented domains (Davis et al. 2004).

One main feature of lyotropic suspensions with a biphase region at low shear viscosity is the steep increment in viscosity versus concentration because the resistance to flow increases with higher concentration, while this maximum viscosity then decreases with increasing of shear rate. However, in the biphase concentration interval, the increasing volume fraction of liquid crystalline phase results in a lower viscosity, storage modulus, and loss modulus at higher concentrations (Larson 1999). As shown in Figure 21.6, the steady shear viscosity and complex viscosity versus concentration do not have a maximum value in the biphase region boundaries in contrast to many lyotropic suspensions: the absence

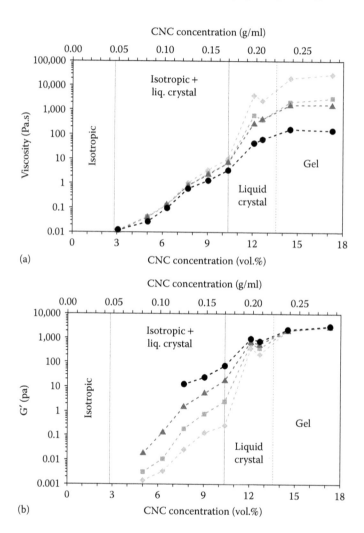

FIGURE 21.6 Rheological properties versus concentration for cellulose nanowhisker suspensions: (a) The steady-state shear and complex viscosities at 0.1 s⁻¹ (triangles), 1 s⁻¹ (circles), 0.1 rad/s (diamonds), and 1.0 rad/s (squares) and (b) storage modulus at 0.1 rad/s (diamonds), 1.0 rad/s (squares), 10 rad/s (triangles), and 100 rad/s (circles). The solid vertical lines indicate a phase transition; the discontinuous vertical line indicates that the liquid crystal to gel transition is approximate. (Reprinted with permission from Urena-Benavides, E. E. et al., *Macromolecules*, 44, 8990, 2011.)

of an explicit maximum is due to very weak ordering of CNW that can be disrupted at a shear rate of 0.1 s^{-1}. Only a small change in slope is observed at concentrations of 7.7 vol.%. On the other hand, complex viscosity, storage, and loss modulus show regular maxima at concentrations that correspond with a fully liquid crystalline phase (Urena-Benavides et al. 2011): the complex viscosity and storage modulus have smooth maxima at a concentration of 12.1 vol.% where a purely liquid crystalline phase has been confirmed by optical microscopy. A plateau for both complex viscosity and storage modulus with concentration has also been observed in the gel region. These observations suggest that the interface in biphasic samples affects the elastic relaxation but not the viscous response. At higher concentrations, the fingerprint texture of the liquid crystal phase was absent, and the dispersions behaved as rheological gels. Similar behavior has been reported for CNW suspensions with micrometer length rods, which show only a very small decrement of viscosity at the critical concentration and the sample becomes biphasic (Bercea and Navard. 2000). The maximum packing concentration of CNW that corresponds to the experimentally measured isotropic-to-anisotropic transition can be estimated from the plateau values in viscosity versus concentration: at low shear rates, the CNW suspensions are in the semidilute regime at above 0.02 wt.%, while high shear rate plateau viscosity data show that the CNW suspension is still in the dilute state above 0.6 wt.%. The critical concentration for a dilute–semidilute transition strongly depends on the state of order, which suggests that care has to be taken when measuring parameters extracted from flowing suspensions (e.g., with Ubbelohde viscometry).

21.3.1.6 Effect of Temperature on Rheological Behavior

The alteration in microstructural arrangements of CNW with temperature has direct effects on rheological behavior of medium. At temperatures below 35°C, little influences on rheology and phase behavior have been reported. A significant change in both the fraction of isotropic phase and the rheological properties was noticed between 35°C and 40°C (Urena-Benavides et al. 2011). The viscosity of CNW suspensions generally decreases within specific temperature regimes at various applied shear intervals (Shafiei-Sabet et al. 2012), but the temperature dependency strongly depends on the state of order and concentration of the CNW suspension. At concentrations where the amount of ordered phase increase and suspensions become anisotropic, a three-region behavior was apparent at all temperatures (Shafiei-Sabet et al. 2012): the decrement of viscosity with increasing temperature mainly occurs in the second and third concentration regions, which were aforementioned as transition regions. However, in the first region, an abrupt increase in viscosity was observed at 40°C due to microstructural changes. They have also reported, for higher concentrated suspensions that the viscosity is independent of temperature between 10°C and 40°C, suggesting that the temperature has no effect on the mesogen rearrangements of the CNW. However, the occurrence of increasing viscosities at higher temperature in more concentrated suspensions may be due to the formation of a more densely packed chiral nematic state with lower pitch size (Pan et al. 2010). According to dynamic temperature sweep tests, the complex viscosity decreases continuously for lower concentration while temperature increases. For higher concentrations, the complex viscosity decreases with increasing temperature up to a critical temperature beyond which the viscosity starts increasing again. These critical temperatures occur at about 27°C, 36°C, and 42°C for 5, 7, and 10 wt.% CNW suspensions, respectively (Shafiei-Sabet et al. 2012). This temperature dependency can be attributed to variations in local mobility and microstructural changes: in particular, an abrupt increment in the number of isotropic regions at a critical temperature leads to higher viscosity. A change in viscosity due to alteration of isotropic and liquid crystalline regions at different temperatures has been mentioned (Papkov et al. 1974).

21.3.2 Individual Shear-Orientation Measurements

Polarized rheo-optical devices are mostly used to study the individual alignment of CNW in suspension. As referred to before, polarized rheo-optical observations reveal a "fingerprint" of the chiral nematic phase that are deformed and gradually disappear with increasing shear rate.

The alignment of CNW in colloidal aqueous suspension under shear rate could be demonstrated by small angle synchrotron radiation scattering (Ebeling et al. 1999). Following this technique, a nearly transparent Couette cell is placed in the neutron beam to visualize the effects of shear rate on the organization of rods in the CNW suspension. Two distinct beam paths are scattered through the samples parallel and perpendicular to the shear flow direction. The results showed ordering of microcrystals within the vertical direction and the flow direction. Furthermore, data reveals that high shear rates can disrupt the chiral nematic phase while it reforms into a nematic ordering phase. In the tangential position, the X-ray scattered parallel to the flow direction or the shear velocity and showed ordering perpendicular to the flow direction. The alignment of microcrystals with applied shear rate followed a two-step process: (1) first, the cellulose crystals form anisotropic domains at rest while they are in a uniplanar organization within these domain; (2) second, the disruption of the anisotropic domains accompanied by alignment microcrystals occurred. At low shear rates, the X-ray scattering patterns of this study indicate little alignment of microcrystals in the shear direction and preferentially vertical orientation perpendicular to the shear direction. At higher shear rates, the microcrystals show horizontal orientation or alignment along the shear direction (Ebeling et al. 1999). Based on neutron scattering technique, high contrast and resolution data can be obtained as samples are suspended D_2O. Moreover, neutron scattering is rather nondestructive compared to X-ray scattering and samples do not change or degrade during longer testing times and continuous monitoring.

An ordering parameter (orientational distribution function) can be defined as the alignment of fibrils into the same direction along the flow direction (Orts et al. 1998), or their tendency to be parallel in the nematic suspension (Oldenbourg et al. 1988). The ordering parameter S could be calculated from SANS, which is a key experimental technique to determine the degree of orientation under applied shear rate. The ordering parameter is correlated to the sharpening of SANS peaks by measuring the relative width of azimuthal peak traces. Generally, the azimuthal peak width intensity $G(\psi)$ along an arc can be defined by the angle from the equator (ψ), which is related to the orientational probability distribution $f(\beta)$ of finding a rod tilted at an angle (β) with respect to the director. Thus, the azimuthal peak width intensity can be described as Equation 21.5:

$$G(\psi) = \int I_s(\omega) f(\beta) \sin \omega \, d\omega \qquad (21.5)$$

where $I_s(\omega)$ is the single rod intensity function of a rod tilted at an angle (ω) with respect to the beam direction (Oldenbourg et al. 1988). For a Gaussian probability distribution $f(\beta)$ with a very small peak width, the angular spread of the rod axis around the director is small. If the neutron beam direction is perpendicular to the director or shear flow direction, the relationship between the angle from the equator (ψ), rod tilted at an angle respect to the beam direction (ω), and rod tilted at an angle with respect to the director (β) can be simplified as Equation 21.6:

$$\beta = \cos^{-1}(\cos \psi \sin \omega) \qquad (21.6)$$

According to this simplification, the azimuthal peak width intensity can be described as Equation 21.7, where A and (α) are adjustable fitting parameters:

$$G(\psi) = A \exp\left(-\sin^2 \frac{\psi}{2\alpha^2}\right)\left[\frac{1}{\cos \psi} + \frac{\alpha^2}{2\cos^3 \psi} + \ldots\right] \qquad (21.7)$$

The parameters A and α are used to describe the Gaussian orientational distribution $f(\beta)$, which is used to compute the ordering parameter S, given by Equation 21.8 (Oldenbourg et al. 1988):

$$S = f(\beta)\frac{1}{2}(3\cos^2 \beta - 1)d\beta \qquad (21.8)$$

The SANS patterns show anisotropic interference peaks for CNW from black spruce microfibrils while sheared at 7000 s^{-1}. These peaks sharpen with increment of shear rate and an ordering parameter $S = 0.9$ has been reported for shear rates above 1000 s^{-1} (Orts et al. 1998). In parallel, the spacing measurements between single oriented rods illustrates that the packing behavior is tighter along the cholesteric axis compared to the perpendicular axis. This can be related to the hypothesis that the cellulose crystals are twisted by strain in their crystalline microstructure (Revol and Marchessault 1993).

The better understanding of the rheological behavior of CNW suspensions reveals from a combination of multiple online measurements, providing simultaneous information on microscopic structure from dynamics point of view and macroscopic mechanical properties from rheological standpoint. For instance, dynamic light scattering (DLS) and oscillatory shear testing have been combined for compressed emulsions (Hébraud et al. 1997), foams (Höhler et al. 1997), hard sphere glasses (Petekidis et al. 2002) and colloidal gels (Smith et al. 2007). In more recent work, the rheological behavior of CNW suspensions has been studied by light scattering (LS)-echo and polarized optical microscopy coupled with rotational rheometry (Derakhshandeh et al. 2013): the irreversible microscopic orientation of CNWs has been illustrated with onset of shear, which coincides with the strain at storage and loss modulus cross-over in the nonlinear viscoelastic regime. The intensity correlation function $g^{(2)}(t)-1$ has been measured via LS-echo at various strains in the linear and nonlinear viscoelastic regimes at frequencies of 1 Hz (6.28 rad/s) and 10 Hz (62.8 rad/s) to demonstrate time-dependent aging and yielding transitions. As an example, Figures 21.7a and b show the intensity of the correlation function of the first echo over time (i.e., one period of oscillation) corresponding to the short-term response of the samples: this data can be used to evaluate the reversibility of CNW motions. The value of the intensity correlation function is around 1 at strain amplitudes smaller than 15% for low frequencies and at strain amplitudes smaller than 25% for high frequencies: these values indicate elastic deformation and reversible particle rearrangements. Otherwise, the intensity correlation

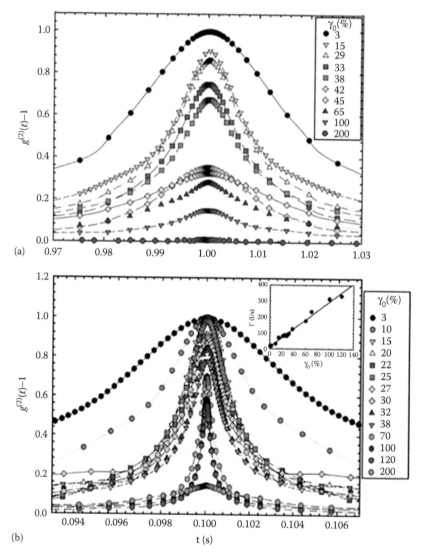

FIGURE 21.7 The evolution of the first echo for various strains using 10 wt.% cellulose nanowhisker suspension under oscillatory shear measured from 0 to 300 s: (a) at frequency 1 Hz and (b) at frequency 10 Hz. The inset in (b) shows the initial decay rate with strain at 10 Hz. (Reprinted with permission from Derakhshandeh, B. et al., *J. Rheol.*, 57, 131, 2013.)

function has values around 0.82 at higher strain amplitude of 18% for low frequency and at higher strain amplitude of 27% for high frequency: the values represent the onset of irreversible particle rearrangements. Thus, the particle motions are reversible at strain amplitudes lower than the yield strain and they are irreversible at strain amplitudes beyond the yield strain. The initial decay rate (Γ_{ini}) related to the oscillatory motion of CNW increased with the strain amplitude at constant frequency, as shown in the inset of Figure 21.7b. The linear increase of initial decay versus shear rate provided clear evidence that the bulk of the sample is uniformly sheared without wall slip phenomena (Derakhshandeh et al. 2013).

21.4 CNWs IN POLYMER NANOCOMPOSITES

21.4.1 Processing and Rheological Properties

The influences of CNW on the structural and rheological properties of biopolymer matrices play an important role where polymer processing is involved. The rheological changes due to the addition of CNW have a significant effect on the formation of a matrix network structure and ultimately on the mechanical properties of the nanocomposite.

The changes in rheological properties of hydrogels have been studied in the presence of CNW. The hydrogel networks retain large quantities of water due to the formation of permanent (chemical hydrogel) and transient (physical hydrogel) junctions, while these network structures may be further stabilized with CNW fillers. The gelation process is mainly affected by temperature (Tako and Nakamura 1988), but different concentrations of CNW additives may change the hydrogel properties and rheological features. Based on viscoelasticity measurements indicating higher linear elastic moduli than loss moduli, the gel formation of CNW particles and presence of internal structures dominate the behavior of polymer solutions. The formation of a weak gel structure was due to nonadsorbing macromolecules that caused a depletion-induced interaction among the particles (Boluk et al. 2012). The addition of CNW to polyacrylamide nanocomposite hydrogels accelerated the onset of gelation and acted as a multifunctional cross-linking agent during gelation (Zhou et al. 2011b): the CNW with lower aspect ratio promoted the onset time and sol–gel transition as they facilitated the network formation and also toughened the matrix. Furthermore, hemicellulose hydrogels reinforced with CNW show excellent improvements in mechanical properties and recovery behavior (Karaaslan et al. 2011). The pure agarose hydrogels with relatively low volume fractions of CNW revealed significant reinforcement effects according to small-amplitude oscillatory shear tests: there was an increase of more than one decade of the storage modulus by adding 0.13 vol.% CNW to 0.2 wt.% agarose hydrogel (Le Goff et al. 2015). As these concentrations are far below the critical percolation threshold value mentioned before, the high mechanical stability are mainly caused by the negative surface charges of CNW and their influences on the gelling process and matrix network structure. In other work, cross-linked CNWs were prepared with poly(methyl vinyl ether-*co*-maleic acid) (PMVEMA) and poly(ethylene glycol) (PEG) by an esterification reaction between the hydroxyl groups on the cellulose (Goetz et al. 2009). The reaction resulted in the formation of a homogeneous film-like material with unique water-absorbing properties, which shows that the nanocomposite can swell in water and reach stable gel properties.

The water-soluble polymers are a most conventional choice as matrix materials to prepare CNW nanocomposites (Anglès and Dufresne 2000). Aqueous and solvent solution casting is a common method to prepare cellulose nanocomposites (Dufresne et al. 1999; Grunert and Winter 2002; Mathew and Dufresne 2002; Petersson et al. 2007; Pu et al. 2007): after having mixed both components from aqueous suspensions, a solid nanocomposite film can be cast by solvent evaporation while forming a percolating filler network. For instance, a reinforced polymer matrix of PVA with CNW develops strong interactions between filler and matrix with increment of glass temperature and mechanical strength (Roohani et al. 2008), where the interactions between CNW and the polymer matrix are due to the polar nature of both components. In contrast, the CNW can be also used as a reinforcement into water-insoluble biopolymer matrices such as PLA (Pandey et al. 2009) and poly(L-lactide) (Pei et al. 2010). However, the melt-processing approach for water-insoluble polymers has not been much exploited due to aggregation and lack of compatibility between the hydrophilic CNW and the hydrophobic matrix. Twin-screw melt extrusion of cellulose nanocomposites with PLA and cellulose acetate butyrate has been achieved properly (Oksman et al. 2006; Bondeson et al. 2007).

The rheological properties of poly(3-hydroxybutyrate-co-3-hydroxyvalerate) (PHBV) with CNW have been extensively studied (Ten et al. 2012), and typical melt behavior in the terminal region has been observed for pure PHBV. For pure PHB, both storage and loss moduli increase at higher frequency while the loss modulus remains higher than the storage modulus. Due to the large loss modulus, the viscous behavior of molten PHBV dominates over the elastic features. For PHBV/CNW nanocomposites, the storage modulus within the terminal region is higher and shear frequency independency of both storage and loss modulus has been observed. This behavior corresponds to a typical transition from liquid-like to solid-like rheological features, which occurs with addition of 0.5 to 2 wt.% CNW. The dissipation factor (tan δ) reduces at 1.2 wt.% CNW compared to the lower contents, as this concentration represents a transition beyond which the material becomes more elastic and dissipates less energy (Ten et al. 2012).

In a recent study, the preparation of fully bio-based nanocomposites with covalent linkage between CNW and chitosan has been reported (De Mesquita et al. 2012). The covalent bonds were created by functionalization of the CNW surface with methyl adipoyl chloride to create reactive end groups that directly react with the amino groups of the chitosan, resulting in significantly higher mechanical strength. The rheological behavior of extruded poly(ε-caprolactone) nanocomposites reinforced by surface-grafted CNW has been studied (Goffin et al. 2011): it was reported that the ungraft CNWs did not affect the viscoelastic properties of the polymer matrix due to poor dispersive mixing and the lack of interactions between matrix and the CNW. In contrast, grafted CNWs improved the mechanical and rheological performances when only few percentages were added, with an excellent dispersion of the nanofillers in the hydrophobic matrix and good interfacial compatibility. They have also reported a solid-like behavior of the composite through the formation of a physical polymer network structure. The entanglement of the polymer chains on the grafted CNW with the matrix polymer contributed to a decrease in chain mobility and depressed relaxation phenomena. In general, the homogeneous dispersion of

CNW in polymer blends can be improved by combination of solvent casting and extrusion, as proven for PLA/CNW/natural rubber blends (Bitinis et al. 2013).

21.4.2 Processing and Mechanical Properties

The main parameters affecting mechanical properties are interfacial compatibility of CNWs and polymer matrix, the molecular structure of the matrix, aspect ratio of CNW, and the nanocomposite preparation method. The molecular structure of the matrix affects the interaction between polymer matrix and CNW (Miao and Hamad 2013), while a higher aspect ratio gives more reinforcement potential and better mechanical properties. In neat CNW films prepared by shear-based film casting, the CNW alignment was efficiently controlled through the casting shear rates and pH: these parameters could increase several mechanical features including elastic modulus, but had no effects on ultimate tensile strength, elongation at failure, and work of fracture (Reising et al. 2012). The variation in mechanical properties of CNW/latex composites under different processing conditions are given in Table 21.3 (Hajji et al. 1996). The solution casting provides superior mechanical properties of CNW composites allover, while intermediate freeze-drying followed by hot pressing and/or in combination with extrusion provides lower reinforcement, likely due to dispersing problems. The unfilled latex shows a nonlinear elastic behavior, which is not affected by processing conditions. On the other hand, nanocomposites with 6 wt.% CNWs and processed by solution casting shows a pseudoplastic behavior with a clear yield point in the stress–strain curve. For the same processing method with 1 wt.% CNW, the nonlinear elastic (rubberlike) behavior of the matrix dominates, but higher stress levels are observed.

21.4.3 Controlling Alignment Properties in Nanocomposites

The future significance of CNW additives in functional bio-based nanocomposites strongly depends on the control over their alignment during processing. In thin film and bulk polymer composites alignment can be achieved by application of load at elevated temperature. Otherwise, thin acrylic copolymer films with highly packed oriented CNW were obtained by a shear-convective assembly method (Khelifa et al. 2012). The simple formation of films with orientated CNW surfaces may be obtained by spin coating, allowing the modulation of CNW adsorption and relative orientation (Dugan et al. 2013a). More advanced techniques to induce orientation of CNW in nanocomposites include the combination with external stimuli, for example, under modulated electric or magnetic fields (Kvien and

TABLE 21.3 Mechanical Properties of Latex/Cellulose CNWs Composites Processed by Different Methods

Cellulose Nanowhiskers Content (wt.%)	Solution Casting		Freeze-Drying Followed by Hot Compression		Freeze Drying, Extrusion, and Hot Compression	
	Modulus (MPa)	Tensile Strength (MPa)	Modulus (MPa)	Tensile Strength (MPa)	Modulus (MPa)	Tensile Strength (MPa)
0	0.2	0.15	0.2	0.12	0.2	0.12
1	0.6	0.49	0.5	0.31	0.4	0.23
6	32.3	5.00	5.2	1.63	1.5	1.06

Source: Hajji, P. et al., *Polym. Compos.*, 17, 612, 1996.

Oksman 2007, Pullawan et al. 2012). However, the spontaneous organization can most easily be controlled by physicochemical aspects. Therefore, an alternative method has recently been developed, where the self-assembly of the CNW can be tuned by the surface tension torque close to the drying boundary line during evaporation of the liquid solution phase (Mashkour et al. 2014): by careful advancement of the drying contact line, special linear and curved patterns of both CNWs and the molecular chains of the PVA polymer matrix can be obtained.

The wide-angle X-ray diffraction (WAXD) is an efficient technique to measure the concentration and orientation of crystalline states into nanocomposites. The intensity distribution in the WAXD pattern has been used to quantify the alignment of the nanowhiskers into the matrix, as sharp and broad peak distributions indicated the higher and lower degree of alignment, respectively. As such, the alignment of cellulose nanorods into alginate nanocomposite fibers has been followed during processing (Urena-Benavides and Kitchens 2011): depending on the fiber stretching conditions during spinning, the degree of orientation of the CNW decreased at higher loads, as the interparticle interactions induced twisting around the longitudinal axis away from the drawing direction (schematic in Figure 21.8). At high CNW concentration and sufficiently low jet stretches, the CNW assembled within the alginate matrix in a spiral orientation around the longitudinal fiber axis. Similar spiral assembly patterns have been observed for microfibrils within native cellulose fibers. The increased fiber stretching during spinning retarded the appearance of a spiral assembly and increased the CNW alignment. Similarly, the extrusion of PVA/CNW suspensions in

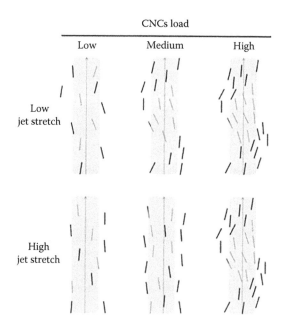

FIGURE 21.8 Schematic representation of the assembly of CNW in a calcium alginate fiber. (Reprinted with permission from Urena-Benavides, E.E. and Kitchens, C.L., *Macromolecules* 44, 3478, 2011.)

cold methanol forming gel fibers was followed by a hot drawing process to induce CNW orientation (Uddin et al. 2011): the as-spun fiber with small amounts of CNW showed highest drawability, leading to an extremely high orientation. The orientation under strain has been adopted to many materials including agarose hydrogels (Osorio-Madrazo et al. 2012). The controlled drying of the agarose hydrogel is essential to establish good interfacial interactions with the CNW, which allow efficient stress transfer during stretching thereby promoting the alignment.

While most reports on the fabrication of cellulose nanocomposites involve melt processes or solvent casting processes, the potential of gas phase processes for assembling CNW in combination with a polymerization process has been recently explored (Samyn et al. 2011, 2012): the *in situ* pulsed plasma polymerization of CNW with maleic anhydride precursors yields nanocomposite films with organized microstructures. The formation of different microstructural patterns can be directly controlled by the concentration ratio or deposition parameters. The organized domains can further be manipulated by changing environmental humidity conditions, suggesting local reorganization of the constituents. Therefore, most of the organized structures obtained under plasma processing have been considered as metastable phases.

21.5 CONCLUSION

This chapter provides an overview of the preparation, application, characterization, rheological features, self-assembly, and orientation of CNWs in combination with a polymer matrix. The influences of processing conditions of polymer/CNW or cellulose/CNW nanocomposites are discussed. Controlling the rheological features of CNW suspensions next to the type of polymer matrices, surface modification, surfactant coating, or grafting is one of the key parameters to improve the processing characteristics of the rodlike fillers into biocomposite materials. In this regard, the significant parameters that affect the rheological behavior and orientation of CNW under shear flow have been reviewed, in order to control the bottom-up organization and provide functional materials. The alignment of CNW above a critical concentration in aqueous suspension is entropically driven toward a thermodynamic equilibrium. Depending on the CNW concentration, a nematic liquid crystalline ordering is adopted that changes into the formation of a chiral nematic phase. In general, the rheology of CNWs can be categorized in parallel with particle motions of nonspherical suspension, which is greatly influenced by particle orientation under specific flow conditions. The three-region viscosity profile of CNW suspensions is characterized by shear thinning, shear thickening, and a plateau regime while transitions occur from isotropic to liquid crystal and liquid crystal to gel phases. These flow regimes are strongly affected by parametrical and processing factors, such as Brownian dynamic, hydrolysis strength, ionic strength, sonication, concentration, and temperature. The orientation of CNW particles is governed by balance of hydrodynamic forces, which tends to align the particles in flow direction. The alignment leads to shear thinning that scales with the rotational Péclet number.

The mechanical properties of a composite reinforced by CNW depend on different mechanisms, including traditional percolation theories, interfacial phenomena, and indirect influences on the internal network structure of the matrix. The control over good embedding and organization of the CNW in specific structures requires good understanding of the rheological properties in combination with polymer melts. Moreover, the influences of CNW on the structural and rheological properties of biopolymer matrices play an important role where the polymer processing is involved. Orientation of CNW into the polymeric matrix is taken to account while shear rates are applied during processing. The postprocessing conditions such as fiber drawing may further imply reorientations of the CNW fillers due to induced twisting.

The importance of rheological behavior of CNWs in getting more fundamental knowledge about the behavior in suspension or in combination with polymer processing techniques has been highlighted. The full understanding of different phase behaviors may help in defining optimum processing conditions of nanocomposite materials and controlling filler orientation in combination with the nanocomposite microstructure. However, more detailed knowledge on the influences of hydrophobic surface modifications of CNW and their effect on rheology and mixing behavior with different polymer matrixes is still mainly experimentally approached and needs to be developed more generally in combination with environmentally friendly chemical processes. Furthermore, the modeling of rheological behavior of CNW suspensions and microfibrillar suspensions can lead to better extrapolation rules and more general insights in the processing of such materials, which remains an open field of research.

ACKNOWLEDGMENTS

This work is supported by the Robert Bosch Foundation in the framework of Sustainable Use of Natural Materials ('Foresnab'-project, 2011–2016) and Juniorprofessorenprogramm Baden-Württemberg ('NaCoPa'-project, 2012–2015).

REFERENCES

Anglès, M. N. and A. Dufresne. 2000. Plasticized starch/tunicin whiskers nanocomposite materials. 1. Structural analysis. *Macromolecules* 33: 8344–8353.

Anglès, M. N. and A. Dufresne. 2001. Plasticized starch/tunicin whiskers nanocomposite materials. 2. Mechanical behavior. *Macromolecules* 34: 2921–2931.

Araki, J. and S. Kuga. 2001. Effect of trace electrolyte on liquid crystal type of cellulose microcrystals. *Langmuir* 17: 4493–4496.

Araki, J., M. Wada, and S. Kuga. 2001. Steric stabilization of a cellulose microcrystal suspension by poly(ethylene glycol) grafting. *Langmuir* 17: 21–27.

Araki, J., M. Wada, S. Kuga, and T. Okano. 1998. Flow properties of microcrystalline cellulose suspension prepared by acid treatment of native cellulose. *Colloids Surf. A* 142: 75–82.

Araki, J., M. Wada, S. Kuga, and T. Okano. 1999. Influence of surface charge on viscosity behaviour of cellulose microcrystal suspension. *J. Wood Sci.* 45: 258–261.

Araki, J., M. Wada, S. Kuga, and T. Okano. 2000. Birefringent glassy phase of a cellulose microcrystal suspension. *Langmuir* 16: 2413–2415.

Azizi Samir, M. A. S., F. Alloin, M. Paillet, and A. Dufresne. 2004a. Tangling effect in fibrillated cellulose reinforced nanocomposites. *Macromolecules* 37: 4313–4316.

Azizi Samir, M. A. S., F. Alloin, J. Y. Sanchez, and A. Dufresne. 2004b. Cellulose nanocrystals reinforced poly(oxyethylene). *Polymer* 45: 4149–4157.

Azizi Samir, M. A. S., F. Alloin, J. Y. Sanchez, N. El Kissi, and A. Dufresne. 2004c. Preparation of cellulose whiskers reinforced nanocomposites from an organic medium suspension. *Macromolecules* 37: 1386–1393.

Azzam, F., L. Heux, J. L. Putaux, and B. Jean. 2010. Preparation by grafting onto, characterization, and properties of thermally responsive polymer-decorated cellulose nanocrystals. *Biomacromolecules* 11: 3652–3659.

Bai, W., J. Holbery, and K. C. Li. 2009. A technique for production of nanocrystalline cellulose with a narrow size distribution. *Cellulose* 16: 455–465.

Barnes, H. A., J. F. Hutton, and K. Walters. 1989. *An Introduction to Rheology*. Amsterdam, the Netherlands: Elsevier Science Publishers.

Beck, S., J. Bouchard, and R. Berry. 2011. Controlling the reflection wavelength of iridescent solid films of nanocrystalline cellulose. *Biomacromolecules* 12: 167–172.

Beck-Candanedo, S., M. Roman, and D. G. Gray. 2005. Effect of reaction conditions on the properties and behavior of wood cellulose nanocrystal suspensions. *Biomacromolecules* 6: 1048–1054.

Bercea, M. and P. Navard. 2000. Shear dynamics of aqueous suspensions of cellulose whiskers. *Macromolecules* 33: 6011–6016.

Bitinis, N., R. Derdejo, J. Bras, E. Fortunati, J. M. Kenny, L. Torre, and M. A. Lopez. 2013. Poly(lactic acid)/natural rubber/cellulose nanocrystal bionanocomposites. *Carbohydr. Polym.* 96: 611–620.

Bitsanis, I., H. T. Davis, and M. Tirrell. 1990. Brownian dynamics of nondilute solutions of rodlike polymers. 2. High concentrations. *Macromolecules* 23: 1157–1165.

Boluk, Y., R. Lahiji, L. Zhao, and M. T. McDermott. 2011. Suspension viscosities and shape parameter of cellulose nanocrystals (CNC). *Colloids Surf. A* 377: 297–303.

Boluk, Y., L. Zhao, and V. Incani. 2012. Dispersions of nanocrystalline cellulose in aqueous polymer solutions: Structure formation of colloidal rods. *Langmuir* 28: 6114–6123.

Bondeson, D., A. Mathew, and K. Oksman. 2006. Optimization of the isolation of nanocrystals from microcrystalline cellulose by acid hydrolysis. *Cellulose* 13: 171–180.

Bondeson, D., P. Syre, and K. Oksman. 2007. All cellulose nanocomposites produced by extrusion. *J. Biomater. Bioenergy* 1: 367–371.

Capadona, J. R., K. Shanmuganathan, D. J. Tyler, S. J. Rowan, and C. Weder. 2008. Stimuli-responsive polymer nanocomposites inspired by the sea cucumber dermis. *Science* 319: 1370–1374.

Cassagnau, P., W. Zhang, and B. Charleux. 2013. Viscosity and dynamics of nanorod (carbon nanotubes, cellulose whiskers, stiff polymers and polymer fibers) suspensions. *Rheol. Acta* 52: 815–822.

Chauhan, S. V. and S. K. Chakrabarti. 2012. Use of nanotechnology for high performance cellulosic and papermaking products. *Cell. Chem. Technol.* 46: 389–400.

Chazeau, L., M. Paillet, and J. Y. Cavaillé. 1999. Plasticized PVC reinforced with cellulose whiskers. I. Linear viscoelastic behavior analyzed through the quasi-point defect theory. *J. Polym. Sci., Part B: Polym. Phys.* 37: 2151–2164.

Chen, P., H. P. Yu, Y. X. Lu, W. S. Chen, X. Q. Wang, and M. L. Ouyang. 2013. Concentration effects on the isolation and dynamic rheological behavior of cellulose nanofibers via ultrasonic processing. *Cellulose* 20: 149–157.

Chen, Q., P. Liu, F. Nan, L. Zhou, and J. Zhang. 2014. Tuning the iridescence of chiral nematic cellulose nanocrystal films with a vacuum-assisted self-assembly technique. *Biomacromolecules* 15: 4343–4350.

Cobb, P. D. and J. E. Butler. 2006. Simulations of concentrated suspensions of semirigid fibers: Effect of bending on the rotational diffusivity. *Macromolecules* 39: 886–892.

Cox, W. P. and E. H. Merz. 1958. Correlation of dynamic and steady flow viscosities. *J. Polym. Sci.* 28: 619–622.

Dash, R. and A. J. Ragauskas. 2012. Synthesis of a novel cellulose nanowhisker-based drug delivery system. *RSC Adv.* 2: 3403–3409.

Davis, V. A., L. M. Ericson, A. N. G. Parra-Vasquez et al. 2004. Phase behavior and rheology of SWNTs in superacids. *Macromolecules* 37: 154–160.

De Mesquita, J. P., C. L. Donnici, I. F. Teixeira, and F. V. Pereira. 2012. Bio-based nanocomposites obtained through covalent linkage between chitosan and cellulose nanocrystals. *Carbohydr. Polym.* 90: 210–217.

De Rodriguez, N. L. G., W. Thielemans, and A. Dufresne. 2006. Sisal cellulose whiskers reinforced polyvinyl acetate nanocomposites. *Cellulose* 13: 261–270.

Derakhshandeh, B., G. Petekidis, S. Shafiei Sabet, W. Y. Hamad, and S. G. Hatzikiriakos. 2013. Ageing, yielding, and rheology of nanocrystalline cellulose suspensions. *J. Rheol.* 57: 131–148.

Diaz, J. A., X. Wu, A. Martini, J. P. Youngblood, and R. J. Moon. 2013. Thermal expansion of self-organized and shear-oriented cellulose nanocrystal films. *Biomacromolecules* 14: 2900–2908.

Diddens, I., B. Murphy, M. Krisch, and M. Müller. 2008. Anisotropic elastic properties of cellulose measured using inelastic X-ray scattering. *Macromolecules* 41: 9755–9759.

Doi, M. and S. F. Edwards. 1978a. Dynamics of rod-like macromolecules in concentrated solution. Part 1. *J. Chem. Soc. Faraday Trans. II* 74: 560–570.

Doi, M. and S. F. Edwards. 1978b. Dynamics of rod-like macromolecules in concentrated solution. Part 2. *J. Chem. Soc. Faraday Trans. II* 74: 918–932.

Doi, M. and S. F. Edwards. 1986. *The Theory of Polymer Dynamics.* Oxford University Press: London.

Dong, H., K. E. Strawhecker, J. F. Snyder, J. A. Orlicki, R. S. Reiner, and A. W. Rudie. 2012. Cellulose nanocrystals as a reinforcing material for electrospun poly(methyl methacrylate) fibers: Formation, properties and nanomechanical characterization. *Carbohydr. Polym.* 87: 2488–2495.

Dong, X. M. and D. G. Gray. 1997. Effect of counterions on ordered phase formation in suspensions of charged rodlike cellulose crystallites. *Langmuir* 13: 2404–2409.

Dong, X. M., T. Kimura, J. F. Revol, and D. G. Gray. 1996. Effects of ionic strength on the iso-tropic-chiral nematic phase transition of suspensions of cellulose crystallites. *Langmuir* 12: 2076–2082.

Dong, X. M., J. F. Revol, and D. G. Gray. 1998. Effect of microcrystallite preparation conditions on the formation of colloid crystals of cellulose. *Cellulose* 5: 19–32.

Dufresne, A. 2010. Processing of polymer nanocomposites reinforced with polysaccharide nano-crystal. *Molecules* 15: 4111–4128.

Dufresne, A., J. Y. Cavaillé, and M. R. Vignon. 1997. Mechanical behavior of sheet prepared from sugar beet cellulose microfibrils. *J. Appl. Polym. Sci.* 64: 1185–1194.

Dufresne, A., D. Dupeyre, and M. R. Vignon. 2000. Cellulose microfibrils from potato tuber cells: Processing and characterization of starch-cellulose microfibril composites. *J. Appl. Polym. Sci.* 76: 2080–2092.

Dufresne, A., M. B. Kellerhals, and B. Witholt. 1999. Transcrystallization in mcl-PHAs/cellulose whiskers composites. *Macromolecules* 32: 7396–7401.

Dufresne, A. and M. R. Vignon. 1998. Improvement of starch film performances using cellulose microfibrils. *Macromolecules* 31: 2693–2696.

Dugan, J. M., R. F. Collins, J. E. Gough, and S. J. Eichhorn. 2013a. Oriented surfaces of adsorbed cellulose nanowhiskers promote skeletal muscle myogenesis. *Acta Biomater.* 9: 4707–4715.

Dugan, J. M., J. E. Gough, and S. J. Eichhorn. 2013b. Bacterial cellulose scaffolds and cellulose nanowhiskers for tissue engineering. *Nanomedicine* 8: 287–298.

Ebeling, T., M. Paillet, and R. Borsali. 1999. Shear-induced orientation phenomena in suspensions of cellulose microcrystals, revealed by small angle X-ray scattering. *Langmuir* 15: 6123–6126.

Edwards, J. V., N. Prevost, A. French, M. Concha, A. DeLucca, and Q. Wu. 2013. Nanocellulose-based biosensors: Design, preparation, and activity of peptide-linked cotton cellulose nanocrystals having fluorimetric and colorimetric, elastase detection sensitivity. *Engineering* 5: 20–28.

Eichhorn, S. J. 2011. Cellulose nanowhiskers: Promising materials for advanced applications. *Soft Matter* 7: 303–315.

Eichhorn, S. J., A. Dufresne, and M. Aranguren. 2010. Review: Current international research into cellulose nanofibres and nanocomposites. *J. Mater. Sci.* 45: 1–33.

Elazzouzi-Hafraoui, S., J. L. Putaux, and L. Heux. 2009. Self-assembling and chiral nematic properties of organophilic cellulose nanocrystals. *J. Phys. Chem. B* 113: 11069–11075.

El-Sakhawy, M. and M. L. Hassan. 2007. Physical and mechanical properties of microcrystalline cellulose prepared from local agricultural residues. *Carbohydr. Polym.* 67: 1–10.

Faith, A. M. 2001. *Understanding Rheology*. New York: Oxford University Press.

Favier, V., H. Chanzy, and J. Y. Cavaille. 1995. Polymer nanocomposites reinforced by cellulose whiskers. *Macromolecules* 28: 6365–6367.

Favier, V., R. Dendievel, G. Canova, J. Y. Cavaillé, and P. Gilormini. 1997. Simulation and modeling of three- dimensional percolating structures: Case of a latex matrix reinforced by a network of cellulose fibers. *Acta Materiala* 45: 1557–1565.

Fry, D., B. Langhorst, and H. Wang 2006. Rheo-optical studies of carbon nanotube suspensions. *J. Chem. Phys.* 124: 054703.

Giese, M., L. K. Blush, M. K. Khan, and M. J. MacLachlan. 2014. Functional materials from cellulose-derived liquid-crystal templates. *Angew. Chem. Int.*, 54(10):2888–2910.

Goetz, L., A. Mathew, K. Oksman, P. Gatenholm, and A. J. Ragauskas. 2009. A novel nanocomposite film prepared from crosslinked cellulosic whiskers. *Carbohydr. Polym.* 75: 85–89.

Goffin, A. L., J. M. Raquez, E. Duquesne, G. Siqueira, Y. Habibi, A. Dufresne, and P. Dubois. 2011. Poly(ε-caprolactone) based nanocomposites reinforced by surface-grafted cellulose nanowhiskers via extrusion processing: Morphology, rheology, and thermo-mechanical properties. *Polymer* 52: 1532–1538.

Gomez-Martinez, D., M. Stading, and A. M. Hermansson. 2012. Correlation between viscoelasticity and microstructure of a hierarchical soft composite based on nanocellulose and carrageenan. *Trans. Nord. Rheol. Soc.* 20: 117–121.

Gong, G., A. P. Mathew, and K. Oksman. 2011. Strong aqueous gels of cellulose nanofibers and nanowhiskers isolated from softwood flour. *TAPPI J.* 10: 7–14.

Gonzalez-Labrada, E. and D. G. Gray. 2012. Viscosity measurements of dilute aqueous suspensions of cellulose nanocrystals using a rolling ball viscometer. *Cellulose* 19: 1557–1565.

Gopalan, N. K. and A. Dufresne. 2003. Crab shell chitin whisker reinforced natural rubber nanocomposites. 1. Processing and swelling behavior. *Biomacromolecules* 4: 657–665.

Gopalan, N. K., A. Dufresne, A. Gandini, and M. N. Belgacem. 2003. Crab shell chitin whisker reinforced natural rubber nanocomposites. 3. Effect of chemical modification of chitin whiskers. *Biomacromolecules* 4: 1835–1842.

Goussé, C., H. Chanzy, G. Excoffier, L. Soubeyrand, and E. Fleury. 2002. Stable suspensions of partially silylated cellulose whiskers dispersed in organic solvents. *Polymer* 43: 2645–2651.

Grunert, M. and W. T. Winter. 2002. Nanocomposites of cellulose acetate butyrate reinforced with cellulose nanocrystals. *J. Polym. Environ.* 10: 27–30.

Habibi, Y., A. L. Goffin, N. Schiltz, E. Duquesne, P. Dubois, and A. Dufresne. 2008. Bionanocomposites based on poly(ε-caprolactone)-grafted cellulose nanocrystals by ring-opening polymerization. *J. Mater. Chem.* 18: 5002–5010.

Habibi, Y., L. A. Lucia, and O. J. Rojas. 2010. Cellulose nanocrystals: Chemistry, self-assembly, and applications. *Chem. Rev.* 110: 3479–3500.

Hajji, P., J. Y. Cavaille, V. Favier, C. Gauthier, and G. Vigier. 1996. Tensile behavior of nanocomposites from latex and cellulose whiskers. *Polym. Compos.* 17: 612–619.

Hasani, M., E. D. Cranston, G. Westman, and D. G. Gray. 2008. Cationic surface functionalization of cellulose nanocrystals. *Soft Matter* 4: 2238–2244.

Hébraud, P., F. Lequeux, J. P. Munch, and D. J. Pine. 1997. Yielding and rearrangements in concentrated emulsions. *Phys. Rev. Lett.* 78: 4657–4660.

Helbert, W., J. Y. Cavaillé, and A. Dufresne. 1996. Thermoplastic nanocomposites filled with wheat straw cellulose whiskers. Part I: Processing and mechanical behavior. *Polym. Compos.* 17: 604–611.

Herrera, M. A., A. P. Mathew, and K. Oksman. 2012. Comparison of cellulose nanowhiskers extracted from industrial bio-residue and commercial microcrystalline cellulose. *Mater. Lett.* 71: 28–31.

Herrera, N. V., A. P. Mathew, L. Y. Wang, and K. Oksman. 2011. Randomly oriented and aligned cellulose fibres reinforced with cellulose nanowhiskers prepared by electrospinning. *Plast. Rub. Compos.* 2: 57–64.

Heux, L., G. Chauve, and C. Bonini. 2000. Nonocculating and chiral-nematic self-ordering of cellulose microcrystals suspensions in nonpolar solvents. *Langmuir* 16: 8210–8212.

Hirai, A., O. Inui, F. Horii, and M. Tsuji. 2009. Phase separation behavior in aqueous suspensions of bacterial cellulose nanocrystals prepared by sulfuric acid treatment. *Langmuir* 25: 497–502.

Hirota, M., N. Tamura, T. Saito, and A. Isogai. 2010. Water dispersion of cellulose II nanocrystals prepared by TEMPO-mediated oxidation of mercerized cellulose at pH 4.8. *Cellulose* 17: 279–288.

Hoeger, I., O. Rojas, K. Efimenko, O. D. Velev, and S. S. Kelley. 2011. Ultrathin film coatings of aligned cellulose nanocrystals from a convective-shear assemble system and their surface mechanical properties. *Soft Matter* 7: 1957–1967.

Höhler, R., S. Cohen-Addad, and H. Hoballah. 1997. Periodic nonlinear bubble motion in aqueous foam under oscillating shear strain. *Phys. Rev. Lett.* 79: 1154–1157.

Iijima, H. and K. Takeo. 2000. Microcrystalline cellulose: An overview. In: G. O. Phillips and P. A. Williams (eds.). *Handbook of Hydrocolloids*, pp. 331–346. Cambridge, UK: Woodhead Publishing Limited.

Iotti, M., Ø. Gregersen, S. Moe, and M. Lenes. 2011. Rheological studies of microfibrillar cellulose water dispersions. *J. Polym. Environ.* 19: 137–145.

Iwamoto, S., W. H. Kai, A. Isogai and T. Iwata. 2009. Elastic modulus of single cellulose microfibrils from tunicate measured by atomic force microscopy. *Biomacromolecules* 10: 2571–2576.

Jia, B., Y. Li, B. Yang et al. 2013. Effect of microcrystal cellulose and cellulose whisker on biocompatibility of cellulose-based electrospun scaffolds. *Cellulose* 20: 1911–1923.

Kamel, S. 2007. Nanotechnology and its applications in lignocellulosic composites, a mini review. *eXPRESS Polym. Lett.* 1: 546–575.

Karaaslan, M. A., M. A. Tshabalala, D. J. Yelle, and G. Buschle-Diller. 2011. Nanoreinforced biocompatible hydrogels from wood hemicelluloses and cellulose whiskers. *Carbohydr. Polym.* 86: 192–201.

Khelifa, F., Y. Habibi, P. Leclère, and P. Dubois. 2012. Convection-assisted assembly of cellulose nanowhiskers embedded in an acrylic copolymer. *Nanoscale* 5: 1082–1090.

Khoshkava, V. and M. R. Kamal. 2014. Effect of drying conditions on cellulose nanocrystal (CNC) agglomerate porosity and dispersibility in polymer nanocomposites. *Powder Technol.* 261: 288–298.

Kvien, I. and K. Oksman. 2007. Orientation of cellulose nanowhiskers in polyvinyl alcohol. *Appl. Phys. A* 87: 641–643.

Lagerwall, J. P. F., C. Schütz, M. Salajkova, J. Noh, J. H. Park, G. Scalia, and L. Bergström. 2014. Cellulose nanocrystal-based materials: From liquid crystal self-assembly and glass formation to multifunctional thin films. *NPG Asia Mater.* 6, e80.

Lahiji, R. R., X. Xu, R. Reifenberger, A. Raman, A. Rudie, and R. J. Moon. 2010. Atomic force microscopy characterization of cellulose nanocrystals. *Langmuir* 26: 4480–4488.

Larson, R. G. 1999. *The Structure and Rheology of Complex Fluids*. New York: Oxford University Press.

Le Goff, K. J., C. Gaillard, W. Helbert, C. Garnier, and T. Aubry. 2015. Rheological study of reinforcement of agarose hydrogels by cellulose nanowhiskers. *Carbohydr. Polym.* 116: 117–123.

Le Goff, K. J., D. Jouanneau, C. Garnier, and T. Aubry. 2014. Gelling of cellulose nanowhiskers in aqueous suspension. *J. Appl. Polym. Sci.* 131: 40676.

Lee, J. and Y. Deng. 2012. Increased mechanical properties of aligned and isotropic electrospun PVA nanofiber webs by cellulose nanowhisker reinforcement. *Macromolec. Res.* 20: 76–83.

Lee, J. and Y. Deng. 2013. Nanoindentation study of individual cellulose nanowhisker-reinforced PVA electrospun fiber. *Polymer Bull.* 71: 1205–1219.

Lee, J. H., S. H. Park, and S. H. Kim. 2013. Preparation of cellulose nanowhiskers and their reinforcing effect in polylactide. *Macromolec. Res.* 21: 1218–1225.

Li, D., X. Sun, and M. A. Khaleel. 2011. Materials design of all-cellulose composite using microstructure based finite element analysis. *J. Eng. Mater. Technol.* 134: 010911 (9p).

Li, J., J. F. Revol, and R. H. Marchessault. 1996. Rheological properties of aqueous suspensions of chitin crystallites. *J. Colloid Interface Sci.* 183: 365–373.

Li, W., J. Yue, and S. Liu. 2012. Preparation of nanocrystalline cellulose via ultrasound and its reinforcement capability for poly(vinyl alcohol) composites. *Ultrason. Sonochem.* 19: 479–485.

Liu, D., X. Chen, Y. Yue, M. Chen, and Q. Wu. 2011. Structure and rheology of nanocrystalline cellulose. *Carbohydr. Polym.* 84: 316–322.

Liu, D., Y. X. Yuan, and D. Bhattacharyya. 2012. The effects of cellulose nanowhiskers on electrospun poly (lactic acid) nanofibres. *J. Mater. Sci.* 47: 3159–3165.

Ljungberg, N., J. Y. Cavaillé, and L. Heux. 2006. Nanocomposites of isotactic polypropylene reinforced with rod-like cellulose whiskers. *Polymer* 47: 6285–6292.

Loos, M. R. and I. Manas-Zloczower. 2012. Micromechanical models for carbon nanotube and cellulose nanowhisker reinforced composites. *Polym. Eng. Sci.* 53: 882–887.

Lu, A., U. Hemraz, Z. Khalili, and Y. Boluk. 2014. Unique viscoelastic behaviors of colloidal nanocrystalline cellulose aqueous suspensions. *Cellulose* 21: 1239–1250.

Lu, P. and Y. L. Hsieh. 2010. Preparation and properties of cellulose nanocrystals: Rods, spheres and network. *Carbohydr. Polym.* 82: 329–336.

Ma, A. W. K., F. Chinesta, A. Ammar, and M. R. Mackley. 2008. The rheology and modeling of chemically treated carbon nanotubes suspensions. *J. Rheol.* 52: 1311–1330.

Macosko, C. W. 1994. *Rheology: Principles, Measurements, and Applications.* New York: Wiley-VCH.

Mahmoud, K. A., J. A. Mena, K. B. Male, S. Hrapovic, A. Kamen, and J. H. T. Luong. 2010. Effect of surface charge on the cellular uptake and cytotoxicity of fluorescent labeled cellulose nanocrystals. *ACS Appl. Mater. Interfaces* 2: 2924–2932.

Mao, J., A. Osorio-Madrazo, and M. P. Laborie. 2013. Preparation of cellulose I nanowhiskers with a mildly acidic aqueous ionic liquid: Reaction efficiency and whiskers attributes. *Cellulose* 20: 1829–1840.

Marchessault, R. H., F. F. Morehead, and N. M. Walter. 1959. Liquid crystal systems from fibrillar polysaccharides. *Nature* 184: 632–633.

Martinez-Sanz, M., A. Lopez-Rubio, and J. M. Lagaron. 2013. High-barrier coated bacterial cellulose nanowhiskers films with reduced moisture sensitivity. *Carbohydr. Polym.* 15: 1072–1082.

Mashkour, M., T. Kimura, F. Kimura, M. Mashkour, and M. Tajvidi. 2014. Tunable self-assembly of cellulose nanowhiskers and polyvinyl alcohol chains induced by surface tension torque. *Biomacromolecules* 15: 60–65.

Mathew, A. P. and A. Dufresne. 2002. Morphological investigations of nanocomposites from sorbitol plasticised starch and tunicin whiskers. *Biomacromolecules* 3: 609–617.

Mewis, J. and N. J. Wagner. 2012. *Colloidal Suspension Rheology.* New York: Cambridge University Press.

Miao, C. and W. Y. Hamad. 2013. Cellulose reinforced polymer composites and nanocomposites: A critical review. *Cellulose* 20: 2221–2262.

Mu, X. Y. and D. G. Gray. 2014. Formation of chiral nematic films from cellulose nanocrystal suspensions is a two-stage process. *Langmuir* 30: 9256–9260.

Naseri, N., A. P. Mathew, L. Girandon, M. Fröhlich, and K. Oksman. 2014. Porous electrospun nanocomposite mats based on chitosan–cellulose nanocrystals for wound dressing: Effect of surface characteristics of nanocrystals. *Cellulose*, DOI 10.1007/s10570-014-0493-y.

Noroozi, N., D. Grecov, and S. Shafiei-Sabet. 2014. Estimation of viscosity coefficients and rheological functions of nanocrystalline cellulose aqueous suspensions. *Liquid Cryst.* 41: 56–66.

Oksman, K., A. P. Mathew, D. Bondeson, and I. Kvien. 2006. Manufacturing process of cellulose whiskers/polylactic acid nanocomposites. *Compos. Sci. Technol.* 66: 2766–2784.

Oldenbourg, R., X. Wen, R. B. Meyer, and D. L. D. Caspar. 1988. Orientational distribution function in nematic tobacco-mosaic-virus liquid crystals measured by X-ray diffraction. *Phys. Rev. Lett.* 16: 1851–1854.

Onogi, S. and T. Asada. 1980. Rheology and rheo-optics of polymer liquid crystals. In: G. Astarita, G. Marrucci, and L. Nicolais (eds.). *Proceedings of the 8th International Congress on Rheology*, pp. 127–147. New York: Plenum Press.

Onsager, L. 1949. The effects of shape on the interaction of colloidal particles. *Ann. N.Y. Acad. Sci.* 51: 627–659.

Orts, W. J., L. Godbout, R. H. Marchessault, and J. F. Revol. 1998. Enhanced ordering of liquid crystalline suspensions of cellulose microfibrils: A small angle neutron scattering study. *Macromolecules* 31: 5717–5725.

Osorio-Madrazo, A., M. Eder, M. Rueggeberg et al. 2012. Reorientation of cellulose nanowhiskers in agarose hydrogels under tensile loading. *Biomacromolecules* 13: 850–856.

Pan, J., W. Y. Hamad, and S. K. Straus. 2010. Parameters affecting the chiral nematic phase of nano-crystalline cellulose films. *Macromolecules* 43: 3851–3858.

Pandey, J. K., W. S. Chu, S. C. Kim, C. S. Lee, and S. H. Ahn. 2009. Bio-nano reinforcement of environmentally degradable polymer matrix by cellulose whiskers from grass. *Composites, Part B* 40: 676–680.

Papkov, S. P., V. G. Kulichikhin, V. D. Kalmykova, and A. Y. Malkin. 1974. Rheological properties of anisotropic poly(para-benzamide) solutions. *J. Polym. Sci. Polym. Phys.* 12: 1753–1770.

Pei, A., Q. Zhou, and L. A. Berglund. 2010. Functionalized cellulose nanocrystals as biobased nucleation agents in poly(L-lactide) (PLLA)—Crystallization and mechanical property effects. *Compos. Sci. Technol.* 70: 815–821.

Petekidis, G., A. Moussaid, and P. N. Pusey. 2002. Rearrangements in hard-sphere glasses under oscillatory shear strain. *Phys. Rev. E* 66: 051402.

Petersson, L., I. Kvien, and K. Oksman. 2007. Structure and thermal properties of polylactic acid/cellulose whiskers nanocomposite materials. *Compos. Sci. Technol.* 67: 2535–2544.

Petersson, L., A. P. Mathew, and K. Oksman. 2009. Dispersion and properties of cellulose nanowhiskers and layered silicates in cellulose acetate butyrate nanocomposites. *J. Appl. Polym. Sci.* 112: 2001–2009.

Pooyan, P., R. Tannenbaum, and H. Garmestani. 2012. Mechanicl behavior of a cellulose-reinforced scaffold in vascular tissue engineering. *J. Mech. Behavior Biomed. Mater.* 7: 50–59.

Pu, Y., J. Zhang, T. Elder, Y. Deng, P. Gatenholm, and A. J. Ragauskas. 2007. Investigation into nanocellulosics versus acacia reinforced acrylic films. *Composites B* 38: 360–366.

Pullawan, T., A. N. Wilkinson, and S. J. Eichhorn. 2012. Influence of magnetic field alignment of cellulose whiskers on the mechanics of all-cellulose nanocomposites. *Biomacromolecules* 13: 2528–2536.

Reising, A. B., R. J. Moon, and J. P. Youngblood. 2012. Effect of particle alignment on mechanical properties of neat cellulose nanocrystalline films. *J. Sci. Technol. Forest Prod. Process.* 2: 32–41.

Revol, J. F., H. Bradford, J. Giasson, R. H. Marchessault, and D. G. Gray. 1992. Helicoidal self ordering of cellulose microfibrils in aqueous solution. *Int. J. Biol. Macromol.* 14: 170–172.

Revol, J. F., L. Godbout, X. M. Dong, D. G. Gray, H. Chanzy, and G. Maret. 1994. Chiral nematic suspensions of cellulose crystallites; phase separation and magnetic field orientation. *Liq. Cryst.* 16: 127–134.

Revol, J. F. and R. H. Marchessault. 1993. In vitro chiral nematic ordering of chitin crystallites. *Int. J. Biol. Macromol.* 15: 329–335.

Rojas, O., G. A. Montero, and Y. Habibi. 2009. Electrospun nanocomposites from polystyrene loaded with cellulose nanowhiskers. *J. Appl. Polym. Sci.* 113: 927–935.

Roman, M., S. P. Dong, A. Hirani, and Y. W. Lee. 2009. Cellulose nanocrystals for drug delivery. In: K. J. Edgar, T. Heinze, and C. M. Buchanan (eds.). *Polysaccharide Materials: Performance by Design*, pp. 81–91. Washington, DC: ACS Publications.

Roohani, M., Y. Habibi, N. M. Belgacem, G. Ebrahim, A. N. Karimi, and A. Dufresne. 2008. Cellulose whiskers reinforced polyvinyl alcohol copolymers nanocomposites. *Eur. Polym. J.* 44: 2489–2498.

Rusli, R. and S. J. Eichhorn. 2012. Interfacial energy dissipation in a cellulose nanowhisker composite. *Nanotechnology* 22: 325706.

Rusli, R., K. Shanmuganathan, S. J. Rowan, C. Weder, and S. J. Eichhorn. 2010. Stress-transfer in anisotropic and environmentally adaptive cellulose whisker nanocomposites. *Biomacromolecules* 11: 762–768.

Saarikoski, E., T. Saarinen, J. Salmela, and J. Seppälä. 2012. Flocculated flow of microfibrillated cellulose water suspensions: An imaging approach for characterisation of rheological behavior. *Cellulose* 19: 647–659.

Saarinen, T., S. Haavisto, A. Sorvari, J. Salmela, and J. Seppälä. 2014. The effect of wall depletion on the rheology of microfibrillated cellulose water suspensions by optical coherence tomography. *Cellulose* 21: 1261–1275.

Samyn, P., A. Airoudj, M. P. Laborie, A. P. Mathew, and V. Roucoules. 2011. Plasma deposition of polymer composite films incorporating nanocellulose whiskers. *Eur. Phys. J. Appl. Phys.* 56: 24015 (5p).

Samyn, P., A. Airoudj, A. P. Mathew, V. Roucoules, H. Haidare, and M. P. Laborie. 2012. Metastable patterning of plasma nanocomposite films by incorporating cellulose nanowhiskers. *Langmuir* 28: 1427–1438.

Sanchez-Garcia, M. D., L. Hilliou, and J. M. Lagaron. 2010. Morphology and water barrier properties of nanobiocomposites of κ/ι-hybrid carrageenan and cellulose nanowhiskers. *J. Agric. Food Chem.* 58: 12847–12857.

Savadekar, N. R., V. S. Karande, N. Vigneshwaran, P. G. Kadam, and S. T. Mhaske. 2014. Preparation of cotton linter nanowhiskers by high-pressure homogeneization and its application in thermoplastic starch. *Appl. Nanosci.*, 5(3):281–290.

Shafiei-Sabet, S., W. Y. Hamad, and S. G. Hatzikiriakos. 2012. Rheology of nanocrystalline cellulose aqueous suspensions. *Langmuir* 28: 17124–17133.

Shafiei-Sabet, S., W. Y. Hamad, and S. G. Hatzikiriakos. 2013. Influence of degree of sulfation on the rheology of cellulose nanocrystal suspensions. *Rheol. Acta* 52: 741–751.

Shafiei-Sabet, S., W. Y. Hamad, and S. G. Hatzikiriakos. 2014. Ionic strength effects on the microstructure and shear rheology of cellulose nanocrystal suspensions. *Cellulose* 21: 3347–3359.

Smith, P., G. Petekidis, S. U. Egelhaaf, and W. C. K. Poon. 2007. Yielding and crystallisation of colloidal gels under oscillatory shear. *Phys. Rev. E* 76: 041402.

Stromme, M., A. Mihranyan, and R. Ek. 2002. What to do with all these algae? *Mater. Lett.* 57: 569–572.

Šturcová, A., G. R. Davies, and S. J. Eichhorn. 2005. Elastic modulus and stress-transfer properties of tunicate cellulose whiskers. *Biomacromolecules* 6: 1055–1061.

Tako, M. and S. Nakamura. 1988. Gelation mechanism of agarose. *Carbohydr. Res.* 180: 277–284.

Tamayo, J., P. M. Kosaka, J. J. Ruz, A. S. Paulo, and M. Calleja. 2013. Biosensors based on nanomechanical systems. *Chem. Soc. Rev* 42: 1287–1311.

Tatsumi, D., S. Ishioka, and T. Matsumoto. 1999. Effect of particle and salt concentrations on the rheological properties of cellulose fibrous suspensions. *Nihon Reoroji Gakkaishi* 27: 243–248.

Ten, E., D. F. Bahr, B. Li, L. Jiang, and M. P. Wolcott. 2012. Effects of cellulose nanowhiskers on mechanical, dielectric, and rheological properties of poly(3-hydroxybutyrate-co-3-hydroxyvalerate)/cellulose nanowhisker composites. *Ind. Eng. Chem. Res.* 51: 2941–2951.

Tkalya, E., M. Ghislandi, W. Thielemans, P. van der Schoot, G. de With, and C. Koning. 2013. Cellulose nanowhiskers templating in conductive polymer nanocomposites reduces electrical percolation threshold 5-fold. *ACS Macro Lett.* 2: 157–163.

Uddin, J. A., J. Araki, and Y. Gotoh. 2011. Toward "strong" green nanocomposites: Polyvinyl alcohol reinforced with extremely oriented cellulose whiskers. *Biomacromolecules* 12: 617–624.

Urena-Benavides, E. E., G. Ao, V. A. Davis, and C. L. Kitchens. 2011. Rheology and phase behavior of lyotropic cellulose nanocrystal suspensions. *Macromolecules* 44: 8990–8998.

Urena-Benavides, E. E. and C. L. Kitchens. 2011. Wide-angle X-ray diffraction of cellulose nanocrystal-alginate nanocomposite fibers. *Macromolecules* 44: 3478–3484.

Van Bruggen, M. P. B., F. M. van der Kooij, and H. N. W. Lekkerkerker. 1996. Liquid crystal phase transitions in dispersions of rod-like colloidal particles. *J. Phys. Condens. Matter* 8: 9451–9456.

Viet, D., S. Beck-Candanedo, and D. G. Gray. 2007. Dispersion of cellulose nanocrystals in polar organic solvents. *Cellulose* 14: 109–113.

Vlad-Cristea, M. S., V. Landry, P. Blanchet, and C. Ouellet-Plamondon. 2013. Nanocrystalline cellulose as effect pigment in clear coatings for wood. *ISRN Nanomaterials* 1: 12–23.

Wang, N., E. Ding, and R. Cheng. 2007. Thermal degradation behaviors of spherical cellulose nanocrystals with sulfate groups. *Polymer* 48: 3486–3493.

Wang, Y. and L. Chen. 2014. Cellulose nanowhiskers and fiber alignment greatly improve mechanical properties of electrospun prolamin protein fibers. *ACS Appl. Mater. Interfaces* 6: 1709–1718.

Whistler, R. L. and J. M. BeMiller. 1997. *Carbohydrate Chemistry for Food Scientists.* St. Paul, MN: American Association of Cereal Chemists.

Wierenga, A. M. and A. P. Philipse. 1998. Low-shear viscosity of isotropic dispersions of (Brownian) rods and fibres: A review of theory and experiments. *Colloid Surf. A* 137: 355–372.

Willenbacher, N. and K. Georgieva. 2013. Rheology of disperse systems. In: U. Bröckel, W. Meier, and G. Wagner (eds.). *Product Design and Engineering: Formulation of Gels and Pastes,* Weinheim, Germany: Wiley-VCH Verlag GmbH & Co. KGaA.

Wu, Q., Y. Meng, S. Wang, Y. Li, S. Fu, L. Ma, and D. Harper. 2014. Rheological behavior of cellulose nanocrystal suspension: Influence of concentration and aspect ratio. *J. Appl. Polym. Sci.* 131: 40525.

Xu, X., H. Wang, L. Jiang, X. Wang, S. A. Payne, J. Y. Zhu, and R. Li. 2014. Comparison between cellulose nanocrystal and cellulose nanofibril reinforced poly(ethylene oxide) nanofibers and their novel shich-kebab crystalline structure. *Macromolecules* 47: 3409–3416.

Young, R. J. and S. J. Eichhorn. 2007. Deformation mechanisms in polymer fibres and nanocomposites. *Polymer* 48: 2–18.

Yuan, H., Y. Nishiyama, M. Wada, and S. Kuga. 2006. Surface acetylation of cellulose whiskers by drying aqueous emulsions. *Biomacromolecules* 7: 696–700.

Zhou C., R. Chu, R. Wu, and Q. Wu. 2011a. Electrospun polyethylene oxide/cellulose nanocrystal composite nanofibrous mats with homogeneous and heterogeneous microstructures. *Biomacromolecules* 12: 2617–2625.

Zhou, C., Q. Wu, and Q. Zhang. 2011b. Dynamic rheology studies of in situ polymerization process of polyacrylamide-cellulose nanocrystal composite hydrogels. *Colloid Polym. Sci.* 289: 247–255.

Zirnsak, M. A., D. U. Hur, and D. V. Boger. 1994. Normal stresses in fibre suspensions. *J. Non-Newtonian Fluid Mech.* 54: 153–193.

CHAPTER 22

Injectable Biomaterials for Endovascular Applications

Devarshi Kashyap and S. Kanagaraj

CONTENTS

22.1 INTRODUCTION

Biomaterial is defined as A synthetic material used to replace part of a living system or to function in intimate contact with living tissue and does not trigger any adverse biological reactions. The ultimate goal of a biomaterial is to improve the quality of life by restoring the function of natural living tissues and organs in the human body. The distinctive characteristic of a biomaterial is the biocompatibility, which is defined as "the ability of a biomaterial to perform its desired function with respect to a medical therapy, without eliciting any undesirable local or systemic effects in the recipient or beneficiary of that therapy, but generating the most appropriate beneficial cellular or tissue response in that specific situation, and optimising the clinically relevant performance of that therapy" (Williams 2008, p. 2952). Though the biomaterial field seems to be relatively young, archaeologists have found evidence of dental implants in humans as early as 200 A.D. Significant advancement in the biomaterials began at the end of World War II, and biomedical devices such as intraocular lenses, hip and knee prostheses, dental implants, artificial hearts, artificial kidneys, and vascular grafts saw great development during the past 60 years. Materials such as silicones, polyurethanes, hydroxyapatite, teflon, nylon, methacrylate, titanium, and stainless steel have been used successfully under *in vivo* conditions (Ratner et al. 2004).

Biomaterials science is an interdisciplinary field involving chemists, biologists, engineers, and physicians. The rapid development in all the spheres of traditional science and materials field has aided in the continuous progress in the field of biomaterials. The advancement in the field of biomaterials and deeper understanding on the principles of biocompatibility helped in the treatment of diseases and medical conditions such as aneurysm, osteoporotic vertebral compression, and arteriovenous malformation (AVM), which were earlier deemed to be untreatable. A wide range of medical devices, diagnostic products, and new techniques has been introduced to help in the clinical practices.

Polymers with their tailorable properties, low cost, easier manufacturing, and versatility are rapidly replacing other conventional biomaterials such as metals and ceramics. The following sections discuss the polymeric biomaterials that are being used in interventional radiology techniques for endovascular embolization in special regard to aneurysm and other vascular conditions.

22.1.1 Endovascular Embolization

It is an alternate to open surgery where small tubes called *catheters* are inserted into the blood vessels, preferably the femoral artery, and it is guided to the required area with the help of medical imaging techniques. Injectable embolic agents such as PVA micro particles, metallic coils, and adhesives travel through the catheter and occlude the damaged vessel. It can be used to treat aneurysm, AVMs, certain fistulas, and tumors. Endovascular embolization techniques have become quite popular because of their many advantages over open surgery. These procedures have lesser risk, pain, and recovery time compared to open surgery.

22.2 REQUIREMENTS OF INJECTABLE EMBOLIC BIOMATERIALS

The embolic agents help in blocking the flow of blood to the desired area and are a vital part of the procedure. The characteristics of an ideal embolic agent are definitive occlusion, able to occlude the desired vessel perfectly, and prevent flow of blood to the desired area. It should be painless and safe and should not cause any problem under *in vivo* conditions. It should be inexpensive and easily available for the people. Moreover, the embolic material should be nontoxic and should not have any undesirable effect on the neighboring tissues. An ideal embolic agent should not fail and promote unintended recanalization of the blocked vessel. The embolic agent should be easy to inject and should not have any complications while being injected through the catheter.

The clinically relevant properties of an injectable embolization agent have been described by Jordan et al. (2005). The delivery of the embolic agent is assisted through small needles and microcatheters. The injection pressure required to push the embolic agent through the narrow tubing should not cause failure of the microcatheter. Viscosity of the embolic agent is an important criterion for both delivery of the agent and the penetration in the vascular spaces. Low viscosity not only facilitates easier delivery but also higher penetration in the vascular spaces. However, higher viscosity is preferred if distal progression is not intended. The setting mechanism of the injectable embolic agent also plays an important part in the occlusion. Polymerizing liquid like cyanoacrylate sets in a predefined time range and is triggered off once the implant encounters the reaction initiator. Precipitating embolics such as ethylene vinyl alcohol (EVAL) form a thin skin, and it cannot harden the deepest part of the agent at once. Shape memory polymers (SMPs) initiate the shape memory effect once it gets the required stimuli. The setting time of the agents influences the choice of embolic agents. Fast setting materials are preferred where undue penetration of the agent is not required. The injectable agent should be biocompatible and should not have any adverse effect on the surrounding tissues. However, in some cases, a controlled or mild tissue reaction can promote scar tissue, resulting in better embolization. Biodegradable characteristic of an embolic agent is another factor of relevance. An embolic agent degrades or disappears and is replaced by biological tissue. However, the embolic agents should be bioactive in contrast to the conventional bio-inert materials to support tissue repair and accelerate the wound healing. The mechanical properties of embolic agents should be comparable to the target tissue and should be able to blend under *in vivo* environment without any failure. Radiopacity of the embolic agents is critical for interventional radiology technique and its safe delivery in the body.

A number of embolization agents has been developed, which are being used clinically. The choice of the embolization agent for treating a certain condition depends on the clinical scenario and the goal of embolization procedure. The factors influencing the choice of embolization agent are the period of time for occlusion, size of the vessel, and functionality of the tissue (Lubarsky et al. 2009). These factors influenced the selection of the embolic agent. In the following discussions, vascular conditions in which embolization techniques are being used have been described.

22.3 VASCULAR MALFORMATIONS

22.3.1 Aneurysm

An aneurysm is a blood-filled dilation of a blood vessel caused due to reduced strength of the vessel's wall. The weak section of the artery wall bulges out because of the pressure of the blood flow and results in an abnormal widening, ballooning, or bleb. Although arteries have thick walls to withstand normal blood pressure, the general wear and tear, certain medical problems, genetic conditions, and trauma can damage or injure them, leading to significant decrease in the strength of the blood vessels at certain places. The sections that are subjected to higher blood pressure are more susceptible to aneurysm development, such as the branching points of arteries, called *bifurcations*. The force of blood, hemodynamic force, pushing against the weakened or injured walls can cause an aneurysm. In the same way, a balloon expands, the aneurysm sac balloons out from the vessel wall slowly and it becomes weaker progressively. The biggest danger is rupturing of the aneurysm with time. The blood flowing through the blood vessels redirects into the aneurysm sac at high pressure, causing an eddy-like effect, leading to its rupture in due course of time.

There are several procedures for treating deformation such as surgical clipping, grafting and occlusion, and bypass. These are all open surgical techniques where the surgeon cut opens the body near the aneurysm and performs the required procedure. These procedures cannot be performed if the aneurysm is inaccessible and the risk factor is also high. Serious complications related to open surgery such as heart-related problems, swelling or infections at the site, and respiratory or urinary infections are always probable in open surgery. Minimally invasive technique using radiology tools called *interventional radiology* is another procedure for treating these deformations. Interventional radiologist occludes the aneurysm with the help of catheterization devices, and this procedure is called *endovascular embolization*.

22.3.2 Arteriovenous Malformations

AVMs are abnormal connection between veins and arteries. In general, the oxygenated blood is pumped from the heart through the arteries to the whole body. The arteries branch off into fine network of tiny vessels called *capillaries*, where the blood nourishes the tissue by exchanging oxygen and nutrients and taking waste from it. The deoxygenated blood flows back into the heart through the veins. The AVM is formed in the absence of capillaries and joins the arteries to veins without exchanging oxygen and nutrients. The connection replacing the capillaries between the arteries and veins is called the *shunt* and the area of the tissues is called the *nidus*. When the high-pressure blood in the arteries directly enters the thin-walled veins through AVM without the damping effect of the capillaries, it might cause rupture of the vein resulting in bleeding. AVM is considered to be hereditary and occurs with equal frequency in both man and woman.

22.3.3 Tumors

The embolization techniques are helpful to treat vascular tumors of the head and neck such as juvenile nasopharyngeal angiofibromas, paragangliomas, and hemangiomas, as well as to treat tumors in the liver, kidney, and uterus (Gemmete et al. 2009, Mavrogenis et al. 2012). The embolization is performed prior to a planned surgical procedure. Tumor embolization has a number of advantages; for example, it can be used on surgically inaccessible vessels. It decreases surgical morbidity by reducing blood loss and also shortens the operative procedure time. The damage to adjacent normal tissue is reduced and also helps in relieving intractable pain. Tumor recurrence decreases and also allows for better visualization of the surgery with decreased overall surgical complication.

Chemoembolization and radioembolization are the techniques where tumors are directly treated with interventional radiology techniques. Small beads containing chemotherapy drugs or radioactive isotope are directly injected near the tumor cells. They give chemotherapy/radiation at the site reducing damage to the surrounding tissue. The radiation travels a short distance and hence the effects are limited to the tumor cells.

The injectable embolic agents play an important role for the effective treatment of vascular imperfections. Different embolic agents that are being used in the medical industry are discussed below.

22.4 PRESENT EMBOLIC AGENTS

22.4.1 Endovascular Coils

One of the most common embolic agents is the metallic coils, which are the permanent occlusion devices, and they were first used as an embolization agent by Gianturco et al. (1975). The main objective of providing the coils is to reduce the blood flow to the malformations and provide a clotting surface for thrombosis. They are generally made of steel, platinum, tungsten, and their alloys. A platinum (92%)/tungsten (8%) alloy has become the basic material for most of the currently used coil designs (White et al. 2008). These coils are attached with synthetic fibers of polymeric materials such as dacron, nylon, polyester, wool, silk, polyvinyl alcohol (PVA), and polyglycolic-polylactic acid (PGLA) to promote quicker and increased thrombogenicity.

Wool fibers were also used with the coils for the better occlusion, but it was followed by an intense chronic inflammatory reaction (Barth et al. 1978). The addition of PGLA, a bioabsorbable polymer, on the bare coils was used to increase the occlusion rate by promoting cellular reaction, which helped in the formation of stable intra-aneurysmal scar tissue formation (Fiorella et al. 2006). There are three different bioactive coils available in the market: (1) matrix detachable coil (Boston Scientific), (2) cerecyte coil (DuPuy Synthes, CA), and (3) nexus coil (MicroTherapeutics Inc., Irvine, CA). A number of clinical interventions has been done to study the efficacy of recanalization as compared to bare platinum coils in both short and long term, but none of the results have been able to confirm the decrease in recanalization rate (Kang et al. 2005, Van Rooij et al. 2008, Piotin et al. 2012, Bose et al. 2012, and Gory and Turjman 2014).

HydroCoils, the embolic coils made by MicroVention, Tustin, California, are platinum coils coated with expendable hydrogels. When the coils come in contact with blood, it causes disentanglement of polymer chains of the hydrogel and then expands. The expansion of the coils potentially helps in increasing the packing density and volume filling for improvement of occlusion efficacy.

The issues with endovascular coils are the complications associated with nontarget embolization, coil migration resulting in pulmonary embolism, and stroke and myocardial infarction (Leyon et al. 2014). The coil migrations have been reported in 3% of the case (Dutton et al. 1995, Prasad et al. 2004). Furthermore, depending on the coil delivery technique, there is about 5% chances of the rupture of aneurysms during endovascular coil embolization (Brilstra et al. 1999, Tummala et al. 2001, Cloft and Kallmes 2002, Murayama et al. 2003, Henkes et al. 2004, Brisman et al. 2005, 2006). In addition, Horowitz et al. (1997) and Kallmes et al. (1998) claimed occlusion of less than 50% of the aneurysm volume while using Guglielmi detachable coil.

22.4.2 Tissue Adhesives (Cyanoacrylates)

Cyanoacrylates were accidentally discovered by Harry Coover in 1942 and later were marketed as "Super Glue" by Kodak in 1958. Cyanoacrylates polymerizes when they come in contact with water, and the monomers react with hydroxyl ions to form long polymer chains. Medical grade cyanoacrylates were developed and used in the Vietnam War to close up wounds of the soldiers. *N*-butyl cyanoacrylates, octyl cyanoacrylates, and isobutyl cyanoacrylates are common tissue adhesives used for clinical practice. Tissue adhesives are being used as an alternate to standard wound closure such as sutures, staplers, adhesive tapes for closing laceration, and surgical incisions.

N-butyl-2 cyanoacrylate (NBCA) such as TruFill (Cordis, Miami Lakes, FL) and Glubran 2 (Gem, Lucca, Italy) are being used successfully for endovascular emboliza-tion. TruFill is a cyanoacrylate-based adhesive that was approved by the Food and Drug Administration (FDA) in 2000 for cerebral AVM embolization. Glubran 2 is a tissue adhesive approved for use in Europe, and has been used successfully for embolization of brain and spinal cord tumors, cerebral AVM, and brain and spinal cord dural arte-riovenous fistula. Glubran 2 contains methacryloxy sulfolane that increases stability of the mixture and decreases the rate of polymerization. Cyanoacrylates are being used as embolization agents as they rapidly solidify on contact with ionic fluids. Cyanoacrylate polymerize in the presence of ionic fluids to form a covalently cross-linked polymer, where lipidol or tungsten is mixed in order to have the radiopaque characteristics. The mixture with the radiopaque agents also helps in altering the level of embolization and improves the handling (Stoesslein et al. 1982). A high number of clinical interven-tions has been successfully performed (Casasco et al. 1994, Paulsen et al. 1999, Kerber and Wong 2000), and Pollak and White 2001). 2-hexyl cyanoacrylate (Neuracryl M) is a new cyanoacrylate-based embolizing agent being tested for embolization (Komotar et al. 2006).

However, the precise amount of delivery of embolization material becomes difficult due to its rapid polymerization and the need to add radiopaque agents that interfere with polymerization. Furthermore, its adhesive properties led to adhesion with the catheter and vascular walls (Gruber et al. 1996, Laurent 1998, and Casasco et al. 1999). Acute and chronic inflammatory reactions were also experienced by the patient due to the toxic reaction products and the exothermic polymerization (Pollak and White 2001).

22.4.3 Onyx (Ethylene Vinyl Alcohol)

Ethylene vinyl alcohol (EVOH) is a transparent, hydrophilic and nonbiodegradable copo-lymer of ethylene and vinyl alcohol. It has excellent oxygen barrier characteristics and is used in the fabrication of contact lenses.

Taki et al. (1990) and Terada et al. (1991) used an EVOH solution for embolization of AVM and aneurysm. The EVOH solution precipitated the copolymer when it came in con-tact with physiological fluids. Micro Therapeutics Inc. (now Ev3) (Irvine, CA) developed an EVOH solution dissolved in dimethyl sulfoxide (DMSO) and mixed with micron-sized tantalum powder (35% weight) to provide opacity under fluoroscopy and commercialized the same in the name of Onyx. Onyx is delivered through a microcatheter to the target location. The DMSO solvent diffuses in the blood, when it comes in contact with the inter-stitial fluid and the EVAL copolymer precipitate along with the suspended tantalum pow-der into a spongy, coherent solid. Onyx, unlike cyanoacrylates is nonadhesive, resulting in lesser chance of the catheter to adhere to the tissues. In addition, Velat et al. (2008) suggested that handling of Onyx was better, as it can be delivered in a more precise and controlled manner. Onyx has been successfully used in the treatment of large aortic aneurysm, controlling the bleeding in the pelvis, kidney, and mesenteric region (Nürnberg et al. 2010, Henrikson et al. 2011, and Müller-Wille et al. 2012). Onyx was approved by the FDA in 2005 but in Europe, it has been used for clinical application since 1999.

Onyx is available in three different viscosities: Onyx 18 (18 centipoise with 6% of EVOH), Onyx 34 (34 centipoise, with 8% of EVOH), and Onyx 500HD (500 centipoise). The viscosity of Onyx 34 is almost twice as viscous as that of Onyx 18. Onyx 34 is expected to offer better control while it is injected whereas the lower viscosity Onyx allows for deeper penetration. Onyx is being successfully used for a number of endovascular embolization applications from treating intracranial aneurysm, AVMs to tumours.

Sedimentation of the tantalum and insufficient radiopacity are the inherent limitations of the above technique (Murayama et al. 2000, Velat et al. 2008). In addition to that Onyx is an expensive agent. Furthermore, concerns have been raised about the toxicity of DMSO. A recanalization rate of 36% in giant aneurysm has also been reported by Cekirge et al. (2006). The Onyx was injected very slowly because large volume of DMSO can cause tissue necrosis (Murayama et al. 1988).

22.4.4 Polyvinyl Alcohol Particles

PVA is a synthetic polymer and is produced commercially from polyvinyl acetate, where the acetate groups are hydrolyzed fully or partially by ester interchange with methanol in the presence of anhydrous sodium methylate or aqueous sodium hydroxide. The amount of hydroxylation determines the physical, chemical, and mechanical properties of the PVA. PVA is used in medical devices because of its biocompatible, nontoxic, noncarcinogenic, swelling properties, and bioadhesive characteristics (Hassan and Peppas 2000). Contact lenses, artificial pancreases, haemodialysis, and implantable material to replace cartilage are some of the areas where PVA has been developed for the medical industry (Baker et al. 2012).

PVA has been used as an embolic agent during 1970s when Tadavarthy et al. (1975) used IVALON® (trade name) foam for emobolization of hepatic artery, AVMs of spinal artery of his patients. It was concluded that the Ivalon foam was an ideal agent for permanent haemostatic and promoting thrombosis. The first PVA microparticles were irregularly shaped, which promoted occlusion by mechanical obstruction and foreign body inflammatory reaction. The PVA particles caused inflammatory reaction on the surrounding tissue, helping in the embolization process. They were relatively inexpensive and easy to use. Furthermore, regular spherical-shaped PVA particles ranging from 45 to 1800 μm were produced, which were compressible, hydrophilic, and microporous and helped in more consistent penetration. The PVA particles further evolved to a smoother contour embolic agent, Bead Block, which is a PVA hydrogel-based microsphere containing acrylamido polyvinyl. The microspheres are colored blue for visualization within the syringe and are available in sizes ranging from 100 to 1,200 μm. DC

Bead (Biocompatibles UK Ltd) is another spherical PVA-based bead that can be loaded with chemotherapeutic agents (doxorubicin/epirubicin/irinotecan) to treat primary and metastatic liver cancer. LC Bead (BTG International, West Conshohocken, PA), and EmboGoldsphere, Embosphere, QuadraSphere microspheres (Merit Medical Systems, Inc., South Jordan, UT), and Embozene Tandem microspheres (CeloNova BioSciences, Inc., San Antonia, TX) are other polymeric microspheres that are being successfully used. LC Bead is hydrogel microspheres whereas EmboGoldsphere, Embosphere, and QuadraSphere microspheres are composed of trisacryl gelatin. Embozene Tandem microspheres have a hydrogel core coated with polyzene-F, which is proprietary formulation of poly[bis(trifluoroethoxy) phosphazene] and it has thrombo-resistant characteristics, and it promotes rapid endothelialization.

Currently, micro-sized embolic agents are commonly used for embolization of uterine arteries, bronchial arteries, external carotid artery branches, renal arteries, and in the pre-operative embolization of primary bone tumors. Additionally, they can also be used for the management of epistaxis, benign or malignant liver neoplasms, AVMs, and preoperative portal vein embolization (Guimaraes et al. 2013).

PVA is inherently not having radiopaque characteristics. Thus, it is prepared by mixing a dilute suspension of contrast agents for catheter injection. The contrast agents not only provide acceptable opacity but also help in decreasing the overall viscosity of the solution, resulting in decreasing the risk of microcatheter occlusion. Lu et al. (2013) produced a radiopaque embolic agent by encapsulating lipidol in PVA. The lipiodol-containing PVA microcapsules were investigated in a living mice as shown in Figure 22.1.

PVA particle embolization has issues with recanalization, nontarget embolization, duration of occlusion, and tissue ischemia (Vernon and Riley 2011, Guimaraes et al. 2013). As the particle size decreases, the risk of complications is increased because of the greater penetration in the vascular bed. Selecting microparticles of appropriate sizes is important to prevent any unwanted complications.

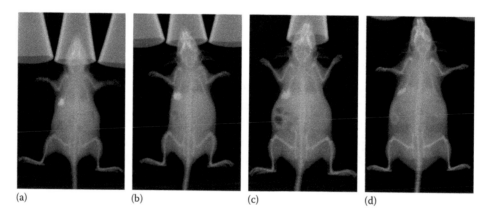

(a) (b) (c) (d)

FIGURE 22.1   X-ray images of the same mouse at (a) first day, (b) first week, (c) fourth week, and (d) third month after subcutaneous injection of lipiodol-containing polyvinyl alcohol microcapsules. (From Lu, X.J. et al., *Int. J. Pharm.*, 452[1–2], 211–219, 2013. With permission.)

22.5 PROMISING EMBOLIC AGENTS

22.5.1 Calcium Alginate

Alginates are naturally occurring copolymer typically extracted from brown seaweed. They are chain-forming heteropolysaccharides with mannuronic and guluronic acid blocks. Calcium alginate has shown great potential in wound healing, drug delivery, *in vitro* cell culture, and tissue engineering. The properties that make it attractive are biocompatibility, mild gelation conditions, and ability to manipulate the new types of alginate derivatives. It is being safely used for clinical applications such as wound healing dressing material and pharmaceutical component (Lee and Mooney 2012). Calcium ions exchange for sodium ions with calcium alginate when they come in contact with the blood. The extra calcium ions help in blood clotting and encourage healing.

Becker et al. (2001) investigated the potential of calcium alginate as an embolic agent. It was reported that the calcium alginate has the characteristics to become a mechanically stable and biocompatible endovascular embolic agent. Becker et al. (2002) tested the stability of calcium alginate for endovascular application in an acute swine AVM model. Furthermore, Becker et al. (2005, 2007) continued on the swine model and reported the complete occlusion and long-term biocompatibility and stability of the calcium alginate even after 6 months. The appearance of minor fibrous tissue around the alginate after 6 months helped in the occlusion of the vessel. A double lumen catheter was used to deliver sodium alginate and calcium chloride initiator to the lesion site. The Ca^{2+} replaces the Na^+ ion on the alginate resulting in cross-linking as each Ca^{2+} ion can be attached with two polymer strands. Forster et al. (2010) reported on the effectiveness of calcium alginate beads as an embolic agent on sheep uterine arterial embolization model. They reported that high molecular weight and the high purity alginate beads were biodegradable in nature and they can be considered as semipermanent embolization beads. Xuan et al. (2015) produced thrombin loaded alginate calcium microspheres and reported good *in vivo* and *in vitro* biocompatibility.

Barnett and Gailloud (2011) evaluated EmboGel, an alginate-based biomaterial for various embolic applications and Emboclear, a solution of alginate lyase and ethylenediaminetetraacetic acid (EDTA), which can selectively dissolve EmboGel. EDTA dissolves polymerized alginate by chelating the calcium alginate and the alginate lyase enzyme hydrolyses with alginate creating polymannuronic acid. The degradation of the alginate creates smaller molecules which can be readily absorbed by the blood stream and eliminated. The effectiveness of the embolic agent was confirmed by both *in vitro* and *in vivo* experiments. The selective dissolving of EmboGel provides an advantage in removing the nontarget embolization and also can be used for temporary occlusion of vasculature.

Alginate-based materials in clinical treatment and medicine are likely to evolve considerably. A continuous challenge is to match the physical properties of alginate-based gel to a particular application. Further understanding on the properties of alginate and the ability to synthesize controlled physical and chemical properties of the same may revolutionize its use in biomedical science and engineering.

22.5.2 Shape Memory Polymer

SMPs have been proposed as an alternative embolization material for the treatment of aneurysms. SMP has become a popular choice because of its numerous advantages, such as light weight, large shape recovery of up to 400% plastic strain, nontoxicity, nonmutagenicity, ease of processing, and inexpensive, which make them preferable compared to shape memory alloys. Furthermore, the ease in tailoring the characteristics of SMP, in addition to its biocompatibility and biodegradability, make it an exciting material. Shape memory materials are capable of changing the shape on the application of certain external stimuli such as light, temperature, moisture, magnetic, or electric field. It is able to recover from a distorted shape to a predefined original shape when an external stimulus is applied. It has been found in a number of alloys, polymers, and ceramics.

The shape memory effect is not an inherent property; that is, polymers do not display the said effect naturally. Shape memory results from the combination of polymer structure and processing, which can be obtained via polymer synthesis route. The SMP is processed into its permanent shape during extrusion or injection molding process. Subsequently, the sample is deformed and fixed into the temporary shape. The initial permanent shape of the polymer can be recovered on the application of external stimuli. SMP can be considered as phase-segregated linear block copolymers consisting of a hard segment and a soft segment. The hard segment or netpoint acts as the permanent shape of the polymer network and the soft segment or molecular switch acts as a switching segment. On application of external stimuli such as temperature or pH, the soft segment changes phase and the shape memory effect is triggered.

Metcalfe et al. (2003) used cold hibernated elastic memory (CHEM) polyurethane based foams for embolization on experimental canine aneurysms. It was found that CHEM polyurethane has nontoxic, nonmutagenic, and poor thrombogenic characteristics that could be used as a material for endovascular application. Furthermore, the porous foam permits cellular invasion and secondary neointima formation. Small et al. (2007) demonstrated a new concept for endovascular embolization of nonnecked fusiform aneurysms in the cerebral vasculature using a laser-actuated SMP device. The SMP was mixed with a light absorbing dye (Epolight 4121, Epolin, Inc., Newark, New Jersey) to increase its light absorption. The prototype SMP stent-foam device is deployed in an *in vitro* model, which indicated its ability to embolize the aneurysm and maintain an open lumen in the parent artery. However, the use of powerful laser was found to be unsuitable due to its impact on the blood vessels and further research on an optimized device to increase its light absorption to reduce laser power was advised. Singhal et al. (2012) synthesized ultra low density and highly cross-linked biocompatible SMPU foams. The foam showed high glassy storage modulus of 200–300 kPa even at such low density. These foams recorded excellent shape recovery of 97%–98% and up to a 70 times volume expansion was observed which is a significant improvement over other known SMP foams.

A critical factor in the feasibility of SMPU foams as an embolic agent is the pressure exerted by the SMP foam on the aneurysm wall during expansion. As the aneurysm wall strength has been reported to be in the range of 700–5,000 kPa by MacDonald et al. (2000)

and Humphrey and Na (2002), the rupture of blood vessels was not expected to occur because of foam expansion (Ortega et al. 2007). As a result, SMP foams have potential to improve the treatment of aneurysms. Hwang et al. (2012) studied with the help of finite element model under different conditions such as the stresses in a human aneurysm wall with a reasonable minimum thickness, 1.5 and 2 times oversized embolization, and linear elastic material properties for the aneurysm wall. They reported that the maximum circumferential stress was 350 kPa at a circumferential Green strain of 0.21. The results indicated that SMP foam samples having 1.5–2 times the size of an aneurysm can be safely implanted in the aneurysm without a risk of rupture, thereby providing a high packing volume in the treatment. Muschenborn et al. (2013) reported that SMP foams with an average cell size of 0.7 and 1.1 mm provide a larger flow resistance than mock embolic coils arranged with volumetric packing density of 28% due to increased viscous and inertial losses.

Researchers have proposed new shape memory composites to be used in interventional radiology techniques by adding radiopaque filler to improve radiopacity and the properties of the polymer. Hampikian et al. (2006) investigated a commercially available polyurethane SMP, Calomer™, with 3 wt.% Ta metal filler for radiopacity. The modulus of the composites was observed to be increased below the glass transition temperature (T_g); however, it was increased above the T_g in the case of pure polymer. The reduction of maximum recovery stress was observed to be 33%. The thermomechanical data showed that the maximum recovery of composite, and unfilled SMP material occurred in a range of 3°C–5°C above T_g. It was concluded that the highest recovery force at body temperature can be achieved with a SMP having the T_g in the temperature range of 32°C and 35°C. It was also observed that addition of metal filler did not affect the shape recovery behavior of SMP, but it reduced the T_g and maximum recovery stress. The examination of coils having 3% metal composition with clinical fluoroscopic X-rays demonstrated that the composite material was opaque with conventional medical imaging.

Romero-Ibarra et al. (2012) prepared 1 wt.% polyurethane nanocomposite with nano-sized barium sulfate spherical particles and fibers via melt extrusion process. It was observed that the required level of radiopaque nature and homogeneous dispersion of nanoparticles were achieved when the size of the particles or fibers was reduced/controlled. It was also noted that the nanoparticles of $BaSO_4$ compared to micron-sized particles did not diminish the transparency of the polyurethane matrix.

Rodriguez et al. (2012) reported on increasing the radiopacity of SMP foam by adding the high atomic number element tungsten to attenuate X-rays, as shown in Figure 22.2. It was observed that dispersing 4 vol.% tungsten powder in SMP foams made them visible at a thickness greater than 8 mm when superimposed with the skull and tissues. However, the 4 vol.% tungsten dispersed 6 mm SMP foam was visible in the crimped state, and the sample may not be visible when expanded under *in vivo* condition. The addition of 4 vol.% tungsten to the SMP foam increased its tensile strength, breaking strain, and Young's modulus by 67%, 60%, and 43%, respectively. Complete healing of the aneurysm site after 90 days and the lack of inflammation were indicative of the biocompatibility of the tungsten-dispersed SMP foam.

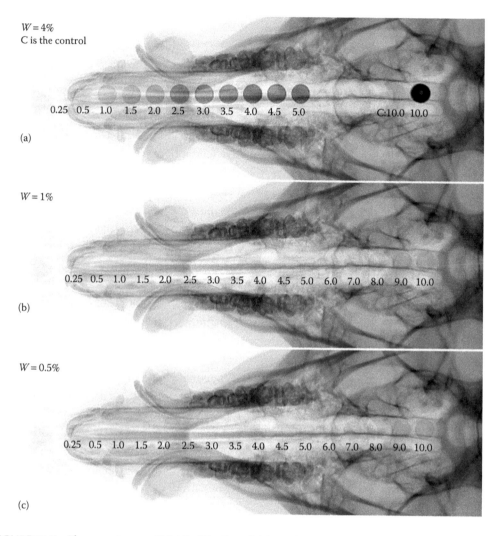

FIGURE 22.2   Fluroscopic test of (a) 4%, (b) 1%, and (c) 0.5% volume concentration of tungsten for different thickness. (From Rodriguez, J.N. et al., *Ann. Biomed. Eng.*, 40[4], 883–897, 2012. With permission.)

SMP can be an important group of materials for endovascular applications such as stents, vascular plugs, and embolic agents. However, very few shape memory products have been approved for *in situ* applications in human beings. Understanding and overcoming the problems associated with shape memory loss during storage and sterilization as well as improving upon the kinetics of shape memory effect may further revolutionize its use in health sector.

22.6 CONCLUSION

The present embolic agents such as endovascular coil, tissue adhesive and PVA particles, and Onyx have been successfully used to treat a number of patients, but each one of them has its own drawback, which affect their *in vivo* behavior. Clinicians have to identify the best embolic agent in order to improve the drawbacks of the present materials. SMPs and calcium alginate are newer potential embolic materials that are being developed to improve the drawbacks of the present materials. It is evident that there is a vast scope to improve the present agents, techniques, and also to find newer materials, which have the potential to be safer and efficient embolic agents.

REFERENCES

Baker MI, Walsh SP, Schwartz Z, Boyan BD; A review of polyvinyl alcohol and its uses in cartilage and orthopedic applications; *Journal of Biomedical Materials Research B: Applied Biomaterials*; 2012; 100(5):1451–1457.

Barnett BP, Gailloud P; Assessment of EmboGel—A selectively dissolvable radiopaque hydrogel for embolic applications; *Journal of Vascular and Interventional Radiology*; 2011; 22:203–211.

Barth KH, Strandberg JD, Kaufman SL, White RI Jr; Chronic vascular reactions to steel coil occlusion devices; *American Journal of Roentgenology*; 1978; 131(3):455–458.

Becker TA, Kipke DR, Brandon T; Calcium alginate gel: A biocompatible and mechanically stable polymer for endovascular embolization; *Journal of Biomedical Material Research*; 2001; 54(1):76–86.

Becker TA, Kipke DR, Preul MC, Bichard WD, McDougall CG; In vivo assessment of calcium alginate gel for endovascular embolization of a cerebral arteriovenous malformation model using the Swine rete mirabile; *Neurosurgery*; 2002; 51(2):453–458; discussion 458–459.

Becker TA, Preul MC, Bichard WD, Kipke DR, McDougall CG; Calcium alginate gel as a biocompatible material for endovascular arteriovenous malformation embolization: Six-month results in an animal model; *Neurosurgery*; 2005; 56(4):793–801; discussion 793–801.

Becker TA, Preul MC, Bichard WD, Kipke DR, McDougall CG; Preliminary investigation of calcium alginate gel as a biocompatible material for endovascular aneurysm embolization in vivo; *Neurosurgery*; 2007; 60(6):1119–1127; discussion 1127–1128.

Black J; Biological performance of materials: Fundamentals of biocompatibility, 2nd ed; Marcel Dekker, New York, 1992.

Bose RS, Dowling RJ, Yan B, Mitchell PJ; A single centre study of coil embolization of intracranial aneurysms comparing bare platinum and PGLA-coated coils; *Journal of Clinical Neuroscience*; 2012; 19(2):271–276.

Brilstra EH, Rinkel GJ, van der Graaf Y, Van Rooij WJ, Algra A; Treatment of intracranial aneurysms by embolization with coils: A systematic review; *Stroke*; 1999; 30(2):470–476.

Brisman JL, Niimi Y, Song JK, Berenstein A; Aneurysmal rupture during coiling: Low incidence and good outcomes at a single large volume center; *Neurosurgery*; 2005; 57(6):1103–1109.

Brisman JL, Song JK, Newell DW; Cerebral aneurysms; *New England Journal of Medicine*; 2006; 355(9):928–939.

Casasco A, Herbreteau D, Houdart E, George B, Tran Ba Huy P, Deffresne D, Merland JJ; Devascularisation of craniofacial tumours by percutaneous tumour puncture; *American Journal of Neuroradiology*; 1994; 15:1233–1239.

Casasco A, Houdart E, Biondi A, Jhaveri HS, Herbreteau D, AymardA, Merland JJ; Major complications of percutaneous embolization of skull-base tumors; *American Journal of Neuroradiology*; 1999; 20:179–181.

Cekirge SH, Saatci I, Ozturk HM, Cil B, Arat A, Mawad M, Ergungor F. et al.; Late angiographic and clinical follow-up results of 100 consecutive aneurysms treated with Onyx reconstruction: largest single-center experience; *Neuroradiology*; 2006; 48:113–126.

Cloft HJ, Kallmes DF; Cerebral aneurysm perforations complicating therapy with Guglielmi detachable coils: A meta-analysis; *American Journal of Neuroradiology*; 2002; 23(10):1706–1709.

Dutton JA, Jackson JE, Hughes JM; Pulmonary AVM results of treatment with coil embolization in 53 patients; *American Journal of Roentgenology*; 1995; 165:1119–1125.

Fiorella D, Albuquerque FC, McDougall CG. Durability of aneurysm embolization with matrix detachable coils; *Neurosurgery*; 2006; 58:51–59.

Forster REJ, Thurmer F, Wallrapp C, Lloyd AW, Macfarlane W, Phillips GJ, Boutrand JP, Lewis AL; Characterisation of physico-mechanical properties and degradation potential of calcium alginate beads for use in embolization; *Journal of Material Science: Materials in Medicine*; 2010; 21:2243–2251.

Gemmete JJ, Ansari SA, McHugh J, Gandhi D; Embolization of vascular tumors of the head and neck; *Neuroimaging Clinics of North America*; 2009; 19(2):181–198.

Gianturco C, Anderson JH, Wallace S; Mechanical devices for arterial occlusion; *American Journal of Roentgenology Radium Therapy and Nuclear Medicine*; 1975; 124:428–435.

Gory B, Turjman F; Endovascular treatment of 404 intracranial aneurysms treated with nexus detachable coils: Short-term and mid-term results from a prospective, consecutive, European multicenter study; *Acta Neurochirurgica (Wien)*; 2014; 156(5):831–837.

Gruber A, Mazal PR, Bavinzski G, Killer M, Budka H, Richling B; Repermeation of partially embolized cerebral arteriovenous malformations: A clinical, radiologic, and histologic study; *American Journal of Neuroradiology*; 1996; 17:1323–1331.

Guimaraes M, Arrington D, MacFall T, Yamada R, Schönholz C; Particulate embolics the background, available options, and ideal applications of particulate embolics; *Endovascular Today*; 2013; 12(4):70–74.

Hampikian JM, Heaton BC, Tong FC, Zhang Z, Wong CP; Mechanical and radiographic properties of a shape memory polymer composite for intracranial aneurysm coils; *Materials Science and Engineering C*; 2006; 26:1373–1379.

Hassan CM, Peppas NA. Structure and applications of poly(vinyl alcohol) hydrogels produced by conventional crosslinking or by freezing/thawing methods; *Advance in Polymer Science*; 2000; 153:35–65.

Henkes H, Fischer S, Weber W, Miloslavski E, Felber S, Brew S, Kuehne D; Endovascular coil occlusion of 1811 intracranial aneurysms: Early angiographic and clinical results; *Neurosurgery*; 2004; 54(2):268–280.

Henrikson O, Roos H, Falkenberg M; Ethylene vinyl alcohol copolymer (Onyx) to seal type 1 endoleak. A new technique; *Vascular*; 2011; 19:77–81.

Horowitz M, Samson D, Purdy P; Does electrothrombosis occur immediately after embolization of an aneurysm with Guglielmi detachable coils?; *American Journal of Neuroradiology*; 1997; 18(3):510–513.

Humphrey JD, Na S; Elastodynamics and arterial wall stress; *Annals of Biomedical Engineering*; 2002; 30(4):509–523.

Hwang W, Volk BL, Akberali F, Singhal P, Criscione JC, Maitland DJ; Estimation of aneurysm wall stresses created by treatment with a shape memory polymer foam device; *Biomechanics and Modelling in Mechanobiology*; 2012; 11:715–729.

Jordan O, Doelker E, Rufenacht DA; Biomaterials used in injectable implants (Liquid Embolics) for percutaneous filling of vascular spaces; *Cardiovascular Interventional Radiology*; 2005; 28(5):561–569.

Kallmes DF, Williams D, Cloft HJ, Lopes MBS, Hankins GR, Helm G; Platinum coil mediated implantation of growth factor-secreting endovascular tissue grafts: An in vivo study; *Radiology*; 1998; 207(2):519–523.

Kang HS, Han MH, Kwon BJ, Kwon OK, Kim SH, Choi SH, Chang KH; Short-term outcome of intracranial aneurysms treated with polyglycolic acid/lactide copolymer-coated coils compared to historical controls treated with bare platinum coils: A single-center experience; *American Journal of Neuroradiology*; 2005; 26(8):1921–1928.

Kerber CW, Wong W; Liquid acrylic adhesive agents in interventional neuroradiology; *Neurosurgery Clinical North America*; 2000; 11:85–99.

Komotar RJ, Ransom ER, Wilson DA, Connolly ES Jr, Lavine SD, Meyers PM; 2-Hexyl cyanoacrylate (neuracryl M) embolization of cerebral arteriovenous malformations; *Neurosurgery*; 2006; 59:464.

Laurent A; Materials and biomaterials for interventional radiology; *Biomedical Pharmacotherapy*; 1998; 52:76–88.

Lee KY, Mooney DJ; Alginate: Properties and biomedical applications; *Progress in Polymer Science*; 2012; 37(1):106–126.

Leyon JJ, Tracey L, Balaji R, Edward T, Ganeshan A; Endovascular Embolization: Review of currently available Embolization Agents; *Current Problems in Diagnostic Radiology*; 2014; 43(1):35–53.

Lu XJ, Zhang Y, Cui DC, Meng WJ, Du LR, Guan HT, Zheng ZZ et al.; Research of novel biocompatible radiopaque microcapsules for arterial embolization; *International Journal of Pharmaceutics*; 2013; 452(1–2):211–219.

Lubarsky M, Ray CE, Funaki B; Embolization Agents—Which one should be used when? Part 1: Large-vessel embolization; *Seminar Interventional Radiology*; 2009; 26(4):352–357.

MacDonald DJ, Finlay HM, Canham PB; Directional wall strength in saccular brain aneurysms from polarized light microscopy; *Annals of Biomedical Engineering*; 2000; 28(5):533–542.

Mavrogenis AF, Rossi G, Calabrò T, Altimari G, Rimondi E, Ruggieri P; The role of embolization for hemangiomas; *Musculoskeletal Surgery*; 2012; 96(2):125–35.

Metcalfe A, Desfaits AC, Salazkin I, Yahia L, Sokolowskic WM, Raymond J; Cold hibernated elastic memory foams for endovascular interventions; *Biomaterials*; 2003; 24:491–497.

Müller-Wille R, Heiss P, Herold T; Endovascular treatment of acute arterial hemorrhage in trauma patients using ethylene vinyl alcohol copolymer (Onyx); *Cardiovascular Interventional Radiology*; 2012; 35:65–75.

Murayama Y, Nien YL, Duckwiler G, Gobin YP, Jahan R, Frazee J, Martin N, Vinuela F; Guglielmi detachable coil embolization of cerebral aneurysms: 11years' experience; *Journal Neurosurgery*; 2003; 98(5):959–966.

Murayama Y, Vinuela F, Tateshima S, Akiba Y; Endovascular treatment of experimental aneurysms by use of a combination of liquid embolic agents and protective devices; *American Journal of Neuroradiology*; 2000; 21:1726–1735.

Murayama Y, Viñuela F, Ulhoa A, Akiba Y, Duckwiler GR, Gobin YP, Vinters HV, Greff RJ; Nonadhesive liquid embolic agent for cerebral arteriovenous malformations: Preliminary histopathological studies in swine rete mirabile; *Neurosurgery*; 1998; 43(5):1164–1175.

Muschenborn AD, Ortega JM, Szafron JM, Szafron DJ, Maitland DJ; Porous media properties of reticulated shape memory polymer foams and mock embolic coils for aneurysm treatment; *BioMedical Engineering Online*; 2013; 12:103.

Nürnberg RA, Erlangen MU, Kleinschmidt T; Embolization of acute abdominal and thoracic hemorrhages with ethylene vinyl alcohol copolymer (Onyx): Initial experiences with arteries of the body trunk; *RöFo*; 2010; 182:900–904.

Ortega J, Maitland D, Wilson T, Tsai W, Savas O, Saloner D; Vascular dynamics of a shape memory polymer foam aneurysm treatment technique; *Annals of Biomedical Engineering*; 2007; 35(11):1870–1884.

Paulsen RD, Steinberg GK, Norbash AM, Marcellus ML, Marks MP; Embolization of basal ganglia and thalamic arteriovenous malformations; *Neurosurgery*; 1999; 44:991–996.

Piotin M, Pistocchi S, Bartolini B, Blanc R; Intracranial aneurysm coiling with PGLA-coated coils versus bare platinum coils: Long-term anatomic follow-up; *Neuroradiology*; 2012; 54(4):345–348.

Pollak JS, White RI Jr; The use of cyanoacrylate adhesives in peripheral embolization; *Journal of Vascular Interventional Radiology*; 2001; 12:907–913.

Prasad V, Chan RP, Faughnan ME; Embolotherapy of pulmonary arteriovenous malformations: Efficacy of platinum versus stainless steel coils; *Journal of Vascular Interventional Radiology*; 2004; 15:153–160.

Ratner BD, Hoffman AS, Schoen FJ, Lemons JE; *Biomaterials Science: An Introduction to Materials in Medicine*, 2nd Edition; Elsevier Academic Press, Amsterdam, the Netherlands, 2004.

Rodriguez JN, Ya J, Miller MW, Wilson T, Hartman J, Clubb FJ, Gentry B, Maitland DJ; Opacification of shape memory polymer foam designed for treatment of intracranial aneurysms; *Annals of Biomedical Engineering*; 2012; 40(4):883–897.

Romero-Ibarra IC, Bonilla-Blancas E, Sánchez-Solís A, Manero O; Influence of the morphology of barium sulphate nanofibers and nanospheres on the physical properties of polyurethane nanocomposites; *European Polymer Journal*; 2012; 48:670–676.

Singhal P, Rodriguez JN, Small W, Eagleston S, Water JV, Maitland DJ, Wilson TS; Ultra low density and highly crosslinked biocompatible shape memory polyurethane foams; *Journal of Polymer Science Part B: Polymer Physics*; 2012; 50:724–737.

Small W, IV, Buckley PR, Wilson TS, Benett WJ, Hartman J, Saloner D, Maitland DJ; Shape memory polymer stent with expandable foam: A new concept for endovascular embolization of fusiform aneurysms; *IEEE Transactions on Biomedical Engineering*; 2007; 54(6):1157–1160.

Stoesslein F, Ditscherlein G, Romaniuk PA; Experimental studies on new liquid embolization mixtures (Histoacryl-Lipiodol, Histoacryl-Panthopaque); *Cardiovascular Interventional Radiology*; 1982; 5:264–267.

Tadavarthy SM, Moller JH, Amplatz K; Polyvinyl alcohol (Ivalon)—A new embolic material; *American Journal of Roentgenology, Radium Therapy and Nuclear Medicine*; 1975; 125:609–616.

Taki W, Yonekawa Y, Iwata H, Uno A, Yamashita K, Amemiya H; A new liquid material for embolization of arteriovenous malformations; *American Journal of Neuroradiology*; 1990; 11:163–168.

Terada T, Nakamura Y, Nakai K, Tsuura M, Nishiguchi T, Hayashi S, Kido T, Taki W, Iwata H, Komai N; Embolization of arteriovenous malformations with peripheral aneurysms using ethylene vinylalcohol copolymer; *Journal of Neurosurgery*; 1991; 75:655–660.

Tummala RP, Chu RM, Madison MT, Myers M, Tubman D, Nussbaum ES; Outcomes after aneurysm rupture during endovascular coil embolization; *Neurosurgery*; 2001; 49(5):1059–1066.

Van Rooij WJ, de Gast AN, Sluzewski M; Results of 101 aneurysms treated with polyglycolic/polylactic acid microfilament nexus coils compared with historical controls treated with standard coils; *American Journal of Neuroradiology*; 2008, 29(5):991–996.

Velat GJ, Reavey-Cantwell, JF, Fautheree, GL, Whiting J, Lewis SB, Hoh BL, Sistrom C, Smullen D, Firment, CS, Mericle RA; Comparison of N-butyl cyanoacrylate and onyx for the embolization of Intracranial Arteriovenous Malformations: Analysis of fluoroscopy and procedure times; *Neurosurgery*; 2008; 63(1):73–80.

Vernon BL, Riley C; Vascular applications of injectable biomaterials; In *Injectable Biomaterials Science and Applications*; Ed., Brent Vernon; Woodhead Publishing, Oxford, 2011.

White JB, Ken CGM, Cloft HJ, Kallmes DF; Coils in a nutshell: A review of coil physical properties; *American Journal of Neuroradiology*; 2008; 29:1242–1246.

Williams DF; On the mechanisms of biocompatibility; *Biomaterials*; 2008; 29:2941–2953.

Xuan F, Zhao L, Rong J, Liang M, Sun J, Han Y; *In vitro* and *in vivo* biocompatibility evaluation of new vascular embolization agent; *Heart*; 2015; 101:A36.

VI

New Polymer Applications

CHAPTER 23

Graphene-Based Polymer Composites for Biomedical Applications

R. Rajesh, Y. Dominic Ravichandran,
A.M. Shanmugharaj, and A. Hariharasubramanian

CONTENTS

23.1 INTRODUCTION

The discovery of novel polymer nanocomposites often termed as *bionanocomposites* forms a fascinating interdisciplinary area that brings together biology, materials science, and nanotechnology (Hule and Pochan, 2007; Paul and Robeson, 2008). These polymer nanocomposites have established themselves as a promising class of hybrid materials derived from natural and synthetic biodegradable polymers and organic/inorganic fillers (Kuilla et al., 2010). Generally, polymer nanocomposites are the result of the combination of polymers and inorganic/organic fillers at the nanometer scale. Existing bionanocomposites include combination of biopolymers namely polysaccharides, aliphatic polyesters, polypeptides and proteins, and polynucleic acids and inorganic nanofillers such as clays, hydroxyapatite, and metal nanoparticles (Hule and Pochan, 2007). Alternatively, carbon-based nanofillers such as carbon black, expanded graphite (EG), carbon nanotube (CNT), and carbon nanofiber have been proven as the effective nanofiller in comparison to the inorganic/organic fillers because of their improved physicochemical, mechanical, electrical, and thermal properties (Armentano et al., 2010; Kuilla et al., 2010). Recently, two-dimensional (2D) carbon nanostructures, such as graphene (GNS), have been extensively explored for the preparation of polymer nanocomposites with tremendous application potential (Ryu and Shanmugharaj, 2014a, b; Ryu et al., 2014; Alam et al., 2015).

GNS, a single-atom-thick layer of sp^2 hybridized carbon atoms arranged in a honeycomb lattice, is the basic building block of other carbon allotropes including 0D fullerenes, 1D CNTs, and 3D graphite (Zhou et al., 2012). New GNS-based materials have profound impact in energy technology, sensors, catalysis, and bioscience/biotechnologies due to their unique physicochemical properties namely high surface area (2630 m²/g), excellent electrical conductivity (1738 S/m), strong mechanical strength (about 1100 GPa), unparalleled thermal conductivity (5000 W/m/K), and ease of functionalization (Yang et al., 2013a). Though, GNS was originally developed for nanoelectronics applications, it has received wider attention in constructing biosensors and loading drugs due to its excellent electrochemical and optical properties, as well as the capability to adsorb a variety of aromatic biomolecules through a π–π stacking interaction and/or electrostatic interaction (Kim et al., 2011; Labroo and Cui, 2013; Neelgund et al., 2013; Sattarahmady et al., 2013). Moreover, presence of the abundant oxygen-containing groups in the honeycomb structure of graphene oxide (GO) reacts with targeting ligands facilitating biofunctionalization, biocompatibility, and nontoxicity toward several cell lines, thus making it as suitable candidate for biomedical applications (Chen et al., 2012a), including bioassays (Cai et al., 2011), targeted bio-imaging (Deepachitra et al., 2013), drug delivery (Sun et al., 2008), and tissue engineering (Nayak et al., 2011; Goenka et al., 2014) (Figure 23.1).

Owing to its large surface area, GNS has been widely used to selectively enrich and detect aromatic molecules and single-stranded DNA (ssDNA) through π–π stacking interactions (Liu et al., 2008; Tang et al., 2010; Goenka et al., 2014). GO has also been used as a photothermal agent for cancer treatment with encouraging therapeutic outcomes due to its high, intrinsic near-infrared absorbance (Robinson et al., 2011; Li et al., 2012; Goenka et al., 2014).

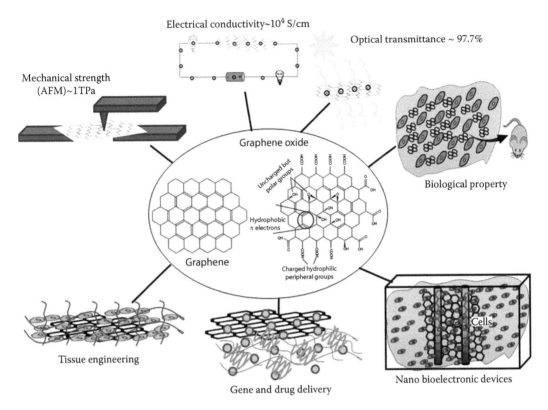

FIGURE 23.1   Schematic overview of various applications of graphene. Graphene-based nanomaterials have been explored for various nonmedical and biomedical applications due to their excellent mechanical, electrical, and optical properties. (Reproduced with permission from Goenka, S. et al., *J. Control. Release*, 173, 75–88, 2014.)

23.2 SYNTHESIS OF GNS AND GO

23.2.1 Mechanical Cleavage

The excellent properties of GNS depend on the exfoliation of graphite into single-layer GNS sheet. The major hurdle in the bulk synthesis of GNS sheets is the aggregations. If the GNS is not well separated from each other, it tends to form irreversible aggregation or even restack to form graphite through van der Walls interactions (Hakimi and Alimard, 2010). Although sporadic attempts to study GNS can be traced back to 1859 (Brodie, 1859), active and focused investigation of this material started only a few years ago, after a simple and effective way to produce relatively large isolated GNS samples using *Scotch tape method*, which is often termed as *micromechanical cleavage* technique (Figure 23.2).

The top layer of the high-quality graphite crystal is removed by a piece of adhesive tape, which—with its graphitic crystallites—is then pressed against the substrate of choice. If the adhesion of the bottom GNS layer to the substrate is stronger than that between the layers of GNS, then a layer of GNS can be transferred onto the surface of the substrate, producing extremely high-quality GNS crystallites via an amazingly simple procedure (Novoselov

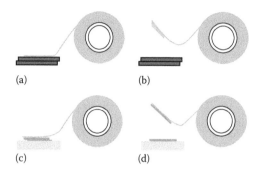

FIGURE 23.2 The micromechanical cleavage technique (*Scotch tape* method) for producing graphene. Top row (a, b): Adhesive tape is used to cleave the top few layers of graphite from a bulk crystal of the material. Bottom left (c): The tape with graphitic flakes is then pressed against the substrate of choice. Bottom right (d): Some flakes stay on the substrate, even on removal of the tape. (Reproduced with permission from Novoselov, K.S., *Angew. Chemie. Int. Ed.*, 50, 6986–7002, 2011.)

et al., 2004; Novoselov, 2011). In principle, this technique works with practically any surface that has reasonable adhesion to GNS (Figure 23.3).

GNS layers with few to 100 nm thickness on silicon substrates (Si/SiO$_2$ with a 300 nm SiO$_2$ layer, for instance) can produce an optical contrast of up to 15% for some wavelengths of incoming light (Figure 23.3). Alternatively, new technique of mechanical cleavage leading to the few layers of GNS from highly ordered pyrolytic graphite sample could be produced using ultrasharp single crystal diamond wedge assisted by ultrasonic oscillations (Jayasena and Subbiah, 2011). Chemical treatment of micromechanically cleaved graphite using sulfuric acid resulted in the hydrogenation reaction simulating surfactant effect on the process cleavage of graphite leading to the individual layers of

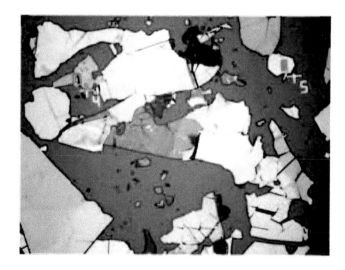

FIGURE 23.3 Thin graphitic flakes on a surface of Si/SiO$_2$ wafer. (Reproduced with permission from Novoslev, K.S., *Angew. Chemie. Int. Ed.*, 50, 6986–7002, 2011.)

GNS (Torres et al., 2014). Although the micromechanical cleavage technique is more efficient in producing pure GNS sheets, it is not suitable for large-scale synthesis (Rao et al., 2009; Kuila et al., 2012).

23.2.2 Chemical Vapor Deposition and Epitaxial Growth

Large-area, single or few layers, high-quality GNS has been produced recently using chemical vapor deposition (CVD), which involves a high-temperature decomposition step of a carbon source (feedstock) in the presence of a transition-metal catalyst such as ruthenium, platinum, iridium, nickel, or copper (Park et al., 2010; Avouris and Dimitrakopoulos, 2012; Petrone et al., 2012; Ambrosi et al., 2014). Although GNS grows epitaxially on most metals, on polycrystalline metal substrates, it has a polycrystalline structure in 2D, that is, within the same GNS layer there are single crystal domains of GNS azimuthally rotated relative to neighboring domains and stitched together with defective domain boundaries, such as alternating pentagon–heptagon structures. CVD GNS grown on copper (Cu) has the advantage of an almost exact 1 mL growth, but even when it is grown on single crystal Cu, it still has a multi-domain structure, because of the rotational disorder between domains and the many grain boundaries. Under the appropriate conditions, large single crystals of GNS can be grown as shown in Figure 23.4.

Alternatively, large-area GNS with domains having an area of tens of square micrometers can be synthesized by two-step CVD process on Cu foils (Li et al., 2010). Furthermore, it has been shown that high-temperature and low methane flow rate and partial pressure are preferred to generate a low density of GNS nuclei, whereas high methane flow rate or partial pressure are preferred for continuous large-area GNS films (Li et al., 2010). Oxygen moiety free continuous GNS monolayer films on copper foils can be synthesized using various alcohol

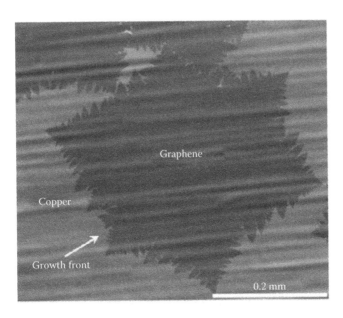

FIGURE 23.4 Single crystal of CVD graphene on Cu. (Reproduced with permission from Li et al., *J. Am. Chem. Soc.*, 133, 2861–2819, 2011b.)

precursors by employing a vacuum-assisted CVD technique (Guermoune et al., 2011). A two-step CVD method involving toluene as the carbon precursor leads to a continuous large-area monolayer GNS film on a very flat, electropolished Cu foil surface at 600°C (Zhang et al., 2012). Ago et al. demonstrated that the use of a crystalline Co film obtained by high-temperature sputtering followed by the H_2 annealing on a sapphire c-plane substrate resulted in a defect-free single-layer GNS over the Co film (Ago et al., 2010). Reduction of silicon carbide at high-temperature results in small islands of the graphitized epitaxial GNS, as silicon desorbs around 1000°C in ultrahigh vacuum (Rollings et al., 2006; Kedzierski et al., 2008). They have also studied the influence of interfacial effects in epitaxial growth of GNS, which are heavily dependent on both the silicon carbide substrate and several growth parameters (Rollings et al., 2006; Kedzierski et al., 2008).

23.2.3 Chemical Exfoliation

Chemical exfoliation is one of the main strategies, which involves oxidation of graphite or thermal expansion of graphite oxide and presumed to be an economic way to produce GNS from bulk graphite in a large quantity. During the process, oxygen-containing functional groups such as hydroxyl and epoxides have been introduced, which in turn reduces the interlayer interaction, leading to an increase in the d-spacing of GO and thereby resulting in the complete exfoliation of single GO layers. This method involves the chemical modification of graphite into water dispersible graphite oxide by Hummer's method (Hummer and Offeman, 1958; Allen et al., 2010) (Figure 23.5).

Chemical exfoliation of GO via ultrasonic treatment and followed by chemical reduction with hydrazine hydrate resulted in single-layer GNS (Figure 23.6) (Stankovich et al., 2006).

Schniepp et al. demonstrated the process of exfoliation by oxidizing and subsequent thermal expansion of graphite (Schniepp et al., 2006). Preparation of GNS nanoribbons by thermal exfoliation of graphite and followed by chemomechanical breaking using ultrasonication resulted in GNS sheets (Li et al., 2008). The lateral size and crystallinity of the starting graphite play major role in chemical exfoliation process for the number of GNS layers obtained. For example, artificial graphite, flake graphite powder, Kish graphite, and natural flake graphite as starting materials produced 80% of the final products as single-layer, single- and double-layer, double- and triple-layer, and few-layer (4–10 layers) GNS, respectively. Alternatively, highly oriented pyrolytic graphite produces few layers (4–10 layers) and thick GNS (>10 layers), which conclude that smaller the size and lower the crystallinity of the starting graphite produce fewer layers of GNS (Wu et al., 2009).

Chemical reduction of GNS introduces numerous defects and other functional groups, which naturally disrupts the planarity of them and makes GO an insulator. In order to overcome this, other methods such as direct sonication of graphite utilizing different solvent have been investigated (Economopoulos et al., 2010). Hsu et al. used 0.05% of benzylamine in isopropanol/water system to exfoliate the graphite in order to functionalize with Pt nanoparticles (Hsu et al., 2011). Covalent functionalization of GP sheets has been demonstrated by Feng et al. in which the graphite was reduced by alkali metal in the presence

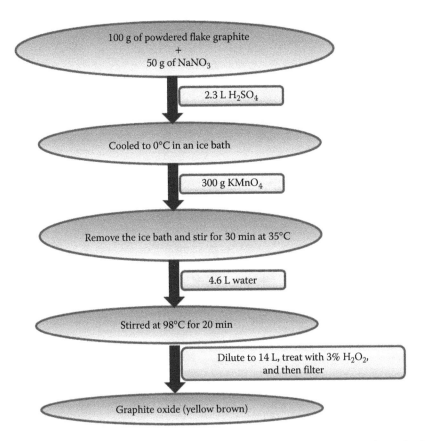

FIGURE 23.5 Schematic representations for the preparation of GO by Hummer's method.

of naphthalene in tetrahydrofuran. Exfoliation of GNS sheets occurs as the alkylation reaction slowly propagates from the graphite edges, resulting exfoliated and alkylated GNS sheets (Feng et al., 2013).

23.2.4 Exfoliation

Solvent exfoliation method is one of the techniques which is investigated enormously in recent days. GO obtained from exfoliation of graphite oxide can form stable colloid aqueous or nonaqueous suspension with the help of sonication. Defect-free monolayer GNS can be achieved by exfoliation of graphite using certain solvents including N-methylpyrrolidone and water. These solvents have high boiling points and are difficult to remove in the next step (Liang and Hersam, 2010). Lotyfa et al. demonstrated defect-free and nonoxidative method to disperse and exfoliate graphite to give GNS using water–surfactant system with the help of ultrasonication (Lotyfa et al., 2009). The adsorbed surfactant controls the reaggregation of graphitic flakes and remains stable for six weeks. Solvent-exfoliated GNS dispersions were reported by Umar Khan et al. where graphite flakes were dispersed in N,N-dimethyl formamide (DMF) in a sonic bath for 150 h and the concentration of dispersed GNS is determined by Raman spectroscopy (Khan et al., 2010).

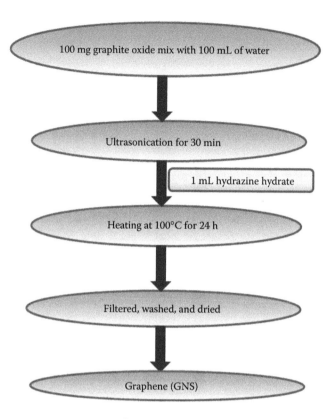

FIGURE 23.6   Schematic representations for the preparation of GNS from GO.

Li et al. demonstrated the preparation of single or few layers of GNS obtained by solvent exfoliation method using DMF with assistance of supercritical carbon dioxide, and then the GNS was noncovalently functionalized with pyrene derivatives with the assistance of supercritical carbon dioxide. They also used density functional theory to investigate the interaction energy between pyrene derivatives and exfoliated GNS. Further studies revealed that the pyrene derivatives act as a *molecular wedge* in GNS exfoliation process (Li et al., 2013c). O'Neil et al. demonstrated the GNS exfoliation using ethanol as solvent and a stabilizing polymer, ethyl cellulose (O'Neil et al., 2011). Recently, exfoliation of graphite using low boiling point volatile solvents such as acetone, chloroform, and isopropanol has also been investigated. A typical concentration of dispersed GNS is as high as 0.5 mg/mL, which is reasonably stable even after 100 h (Liang and Hersam, 2010).

23.3 PREPARATION OF POLYMER–GNS NANOCOMPOSITES

23.3.1 *In Situ* Polymerization

In situ polymerization method is considered to be one of the important techniques to improve the dispersion of the graphitic fillers in the polymeric matrix. During the process the monomer (and/or oligomer) is polymerized in the presence of filler (Sengupta et al., 2011). Das et al. demonstrated the coating of the surface of dispersed pristine GNS sheets with nylon 6,10 and nylon 6,6 using *in situ* polymerization technique. The prepared nanocomposite showed better dispersion quality and very low reaggregation in aqueous solution (Das et al., 2011). Wei et al. reported the preparation of GNS nanosheet/polymer (vinyl chloride/vinyl acetate copolymer) by *in situ* reduction–extractive dispersion method in which the GNS was obtained by reduction of GO (Wei et al., 2009). Bai et al. recently prepared GO/polypyrrole (PPy), GO/poly (3, 4-ethylenedioxythiophene) (PEDOT), and GO/polyaniline hydrogels by *in situ* chemical polymerization in aqueous dispersions of GO sheets (Bai et al., 2011). Wang et al. prepared poly (butylene succinate) (PBS)/GO nanocomposites on the basis of carboxyl and epoxy groups of GO via *in situ* polymerization (Wang et al., 2012c). In addition to that, GO/poly (vinyl alcohol) (PVA) nanocomposites by direct esterification among the hydroxyl group in PVA and the carboxylic group in GO were reported by Salavagione et al. (2009). Thin, lightweight, and flexible polyaniline (PANI)-GO and PANI-GNS hybrid papers were synthesized by Yan et al. via polymerization of aniline on the surfaces of GO and the prepared composites also exhibited excellent electrochemical properties and biocompatibility (Yan et al., 2010). Also, poly (methyl methacrylate) (PMMA)/GO nanocomposite was prepared through *in situ* polymerization method by Potts et al. and their thermomechanical properties were investigated (Potts et al., 2011b). Thermally reduced GO/ poly(L-lactide) composites were prepared by Yang et al. via *in situ* ring-opening polymerization of lactide in which the residual oxygen-containing functional groups on the GNS sheets act as an initiator for the ring-opening polymerization reaction (Yang et al., 2012).

23.3.2 Solution Mixing

Solution mixing method is majorly used for preparation of polymer composites because of simple method without any need of special equipment and allows for large-scale production. Solubility or dispersibility of the GNS and GO in polymers solution is an important hurdle in this technique but can be overcome by ultrasonication. Solution mixing method involves the mixing of GO platelets or GNS-based materials with the desired polymer in two ways which are either themselves already in solution or by dissolving in the polymer in the suspension of GO platelets, by simple stirring or shear mixing. The resulting suspension can be precipitated using nonsolvent for the polymer, filtered and dried or directly cast into a mold and the solvent removed, but it leads to aggregation of the filler in the composites. There are several new methods including lyophilization methods, phase transfer techniques, and surfactants that have been used to facilitate solution mixing of GNS-based composites (Moniruzzaman and Winey, 2006; Wei et al., 2009; Cao et al., 2010; Huang et al., 2011; Potts et al., 2011a).

Liang et al. using a simple water solution processing method prepared a new PVA nanocomposites with GO, which showed 76% increase in tensile strength and a 62% improvement of Young's modulus with the addition of only 0.7 wt.% of GO (Liang et al., 2009). Highly conductive and thermally stable GNS/polyethylene nanocomposite was reported by Kuila et al. which involves the preparation of GNS from oxidation and reduction of graphite powder and then dispersed in the polymer matrix at various concentrations by sonication (Kuila et al., 2011a). Fan et al. reported the preparation of GNS/chitosan composite using solution processing method that involves the dispersion of GNS sheets in acetic acid solution followed by the dissolution of chitosan in the GNS dispersion. The elastic modulus of chitosan increased over about 200% by the addition of 0.1%–0.3% GNS in chitosan. (Fan et al., 2010). The modulus, strength, elongation, and toughness of the chitosan/poly (vinyl pyrrolidone) (PVP) polymer blend are improved when GO was dispersed in the matrix. The chitosan/PVP/GO nanocomposite was prepared through solution processing method by Achaby et al., in which the chemically oxidized GO has been utilized. (Achaby et al., 2014). Yang et al. studied the interaction exists between chitosan and GO which are prepared by mixing them in aqueous media. Furthermore, they have shown that the tensile strength and Young's modulus of the GNS/chitosan composite materials are significantly improved by about 122% and 64%, respectively, with incorporation of 1 wt.% GO when compared with pure chitosan (Yang et al., 2010b). Using solution casting method, biodegradable poly (3-hydroxybutyrate-*co*-4-hydroxybutyrate) [PHBV]/GNS nanocomposites were prepared and their thermal and morphology were reported by Sridhar et al. (2013).

23.3.3 Melt Blending

Melt blending methods involve the mixing of polymer and filler material using high shear forces at elevated temperatures, which is considered to be more economical because of better compatibility and the process does not require any solvent. It has the drawback of lower dispersion of the fillers in polymer when compared to *in situ* polymerization and solution mixing (Kuilla et al., 2010; Huang et al., 2011; Potts et al., 2011a). A biocompatible ultrahigh molecular weight polyethylene (UHMWPE)/GO nanocomposite has been synthesized by Chen et al., and the mechanical property of the nanocomposite also has been evaluated. The preparation involves 30 min ultrasonication of GO in alcohol solution in order to get well-dispersed GO followed by sonication of the latter with UHMWPE for 1 h (Chen et al., 2012b). Sun and He, recently, demonstrated the ring opening reaction between oxygen-containing functional groups and commercial D-lactide (PDLA) monomers which furnish GO-D-lactide (GO-PDLA) nanocomposite. Later, the GO-PDLA was blended with commercial poly(L-lactide) (PLLA) in chloroform in order to get the stereocomplex nanocomposite (Sun and He, 2012). Murariu et al. reported the synthesis of polylactide/GO nanocomposite through melt-blending method using polylactide (PLA) with commercially available EG. The new composite has shown high rigidity and the thermomechanical properties of the composite also been evaluated (Murariu et al., 2010). GP was prepared from successive oxidation and reduction of graphite by Bao et al., and the resulted GP was dispersed at various concentrations into polylactic acid. The nanocomposite has shown excellent mechanical properties, electrical conductivity, and fire resistance which are dependent on the dispersion and loading content of GP (Bao et al., 2012).

23.3.4 Surface Grafting

Grafting of GP and GO with polymers create strong bonding between them, which involves two methods either *grafting to* or *grafting from*. It has the drawback of producing additional defects and hence weakens the structure of GP. The *grafting to* approach involves a synthesis of a polymer with a specific molecular weight terminated with reactive groups or radical precursor, whereas the *grafting from* approach involves polymerization of monomer from surface-derived initiators on GP (Hsiao et al., 2010; Spitalsky et al., 2010). A new GP-pyrene-terminated poly 2-*N,N'*-(dimethyl amino ethyl acrylate) (PDMAEA) polymer nanocomposite has been prepared by Liu et al. The composites were found to possess phase transfer behavior between aqueous and organic media at different pH values (Liu et al., 2010). Kumar et al. recently reported the preparation of PANI-*g*-GO (polyaniline-grafted GO) nanocomposite using well-known "graft-from" method. In this method, the oxygen-containing functional groups were acylated using thionyl chloride in order to get acylated GO. Furthermore, the acyl groups were reacted with amine-protected 4-aminophenol. Then, PANI-G-reduced GO was prepared by *in situ* oxidative polymerization of aniline in the presence of an oxidant and reduced GO as an initiator, after deprotecting the amino group (Kumar et al., 2012). Cheng et al. prepared PVA-grafted GO/PVA (PVA-G-GO/PVA) polymer nanocomposite. First, PVA was surface grafted on GO with the aid of oxygen-containing functional groups in order to get PVA-G-GO. This surface-grafted GO was dispersed in the PVA matrix to form PVA-G-GO/PVA by solution processing method. The new nanocomposite reported to be stronger and tougher than PVA (Chen et al., 2012c). Goncalves et al. used chemically exfoliated GO to synthesize poly (methyl methacrylate) (PMMA) grafted from the surface of GO via atom transfer radical polymerization method. The PMMA was grafted on the surface of the GO through the oxygen-containing functional groups and the PMMA-grafted-GO nanocomposite is soluble in chloroform (Goncalves et al., 2010). Sun et al. reported the synthesis of polystyrene-grafted GO (PS-G-GO) using click chemistry. Polystyrene was grafted on the surface of the GO via Cu(I)-catalyzed 1,3-dipolar azide–alkyne cycloaddition reaction and the resulting nanocomposite is shown to be soluble in tetrahydrofuran (Sun et al., 2010).

23.4 CYTOTOXIC PROPERTIES OF GNS AND ITS POLYMER NANOCOMPOSITES

The biocompatibility and response to the biologically targeted tissue/cells of the nanomaterials make them as emerging biomaterials (Li et al., 2011). GNS has created a novel group of nanomaterials with superior electrical, mechanical, and biological properties for cell and tissue engineering applications. The attractivity of GNS toward cellular stimulation is due to the stable conductivity in the aqueous physiological environment. Moreover, the strong and flexible nature of GNS makes it attractive toward medical device and implant applications. Furthermore, the GNS can boost neuronal cell activity by providing a shortcut for electrical coupling between somatic and dendritic neuronal compartments. The unique electrical properties and chemical stability of GNS facilitate the integration with neural tissues (Figure 23.7) (Li et al., 2011; Sebaa et al., 2013).

Cytotoxicity is the crucial parameter of the materials to be used in biomedical applications. GNS and GO show nontoxicity toward several cell lines for both *in vitro* and *in vivo* studies. Yuan et al. compare the single-walled CNTs (SWCNTs) and GNS for the protein profile change on human hepatoma HepG2 cells. SWCNT causes the induced oxidative stress, thereby activating p53-mediated DNA damage checkpoint signals and

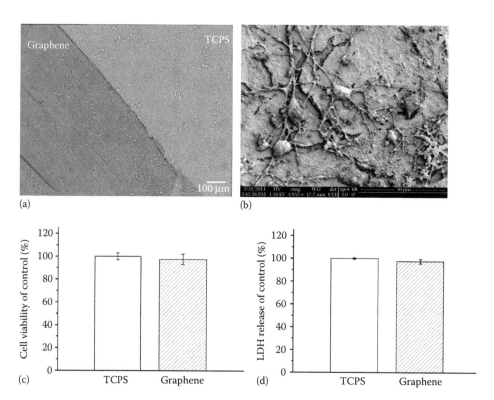

FIGURE 23.7 Neurons cultured on different substrates. (a) An optical image of neurons cultured on the border of graphene (left) and tissue culture polystyrene (TCPS) (right), (b) scanning electron microscopy image of neurons on graphene, (c) MTT-measured viability of neurons cultured on TCPS and graphene after 7 days, (d) LDH activity of neurons after 7 days incubation on TCPS and graphene. Data are expressed as mean ± SEM ($n = 9$, $p > .05$). (Reproduced with permission from Li, N. et al., *Biomaterials*, 32, 9374–9382, 2011.)

leading to apoptosis. However, the GNS shows moderate variation in the protein levels in the treated cells (Figure 23.8).

This proves GNS was less toxic than the SWCNTs (Yuan et al., 2011). The *in vitro* studies of GNS on mouse hippocampal cells by Li et al. show an excellent biocompatibility and cell viability. Meanwhile, neurite numbers and average neurite length on GNS were significantly enhanced during 2–7 days after cell seeding compared with tissue culture polystyrene (TCPS) substrate. The advanced neurite sprouting and outgrowth were confirmed from the western blot analysis of GNS with enhanced growth-associated protein-43 (GAP-43) than the TCPS (Li et al., 2011). The ultra section of A549 cells treated with GO was observed under TEM (Figure 23.9), which shows similar structure to the control cell, and none of the GO sheets were present inside cells (Chang et al., 2011).

The *in vitro* and *in vivo* studies of GO and dextran (DEX)-functionalized GO prove their biocompatibility and cell proliferation. It was found that GO did not induce obvious cell death even at high concentration up to 200 mg/L (Figure 23.10).

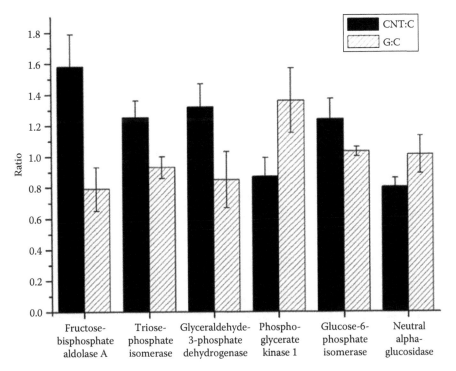

FIGURE 23.8 Differentially expressed metabolic enzymes in HepG2 cells incubated with graphene and SWCNTs. CNT:C is the ratio of protein expression level in the cells treated with SWCNTs relative to control cells; G:C is the ratio of protein expression level in the cells treated with graphene relative to the control cells. (Reproduced with permission from Yuan, J. et al., *Toxicol. Lett.*, 207, 213–221, 2011.)

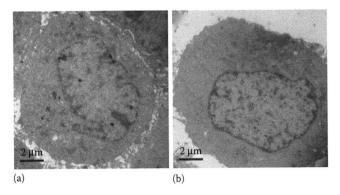

FIGURE 23.9 TEM images of the GO treated A549 cells (a) and the control cells (b). (Reproduced with permission from Chang, Y. et al., *Toxicol. Lett.*, 200, 201–210, 2011.)

The *in vivo* studies prove the excretion of GO-DEX through urine and feces (Figure 23.11) (Zhang et al., 2011b). Furthermore, covalently modified GNS with polyacrylic acid and fluorescein o-methaacrylate (Gollavelli and Ling, 2012) and *in vivo* studies of GO-functionalized polyethylene glycol (Yang et al., 2013b) confirmed the nontoxicity of GNS, GO, and its composites. Hence, a GNS-based composite material might be promising for biomedical applications.

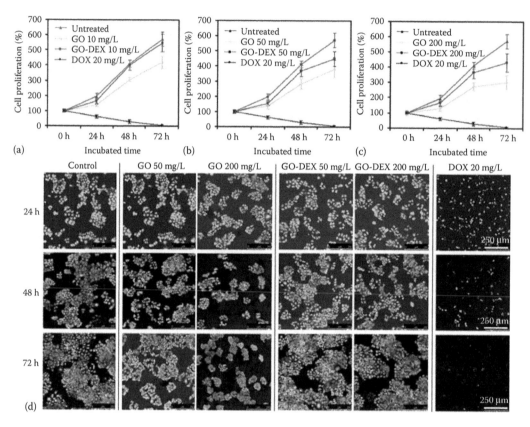

FIGURE 23.10 *In vitro* cell proliferation and toxicity assays. (a–c) The proliferation of HeLa cells in the presence of 10 mg/L (a), 50 mg/L (b), and 200 mg/L (c) of GO or GO–Dextran (DEX). Doxorubicin (DOX) at 20 mg/L was used as the positive control. The relative cell proliferation was determined by cell counting. Standard deviations were based on five parallel samples. (d) Confocal fluorescence images of calcein AM/PI stained HeLa cells 24, 48, and 72 h after various treatments indicated. Live and dead cells were stained by calcein AM and PI, respectively. (Reproduced with permission from Zhang, R. et al., *Carbon*, 49, 1126–1132, 2011a.)

23.5 BIOMEDICAL APPLICATIONS OF GNS AND ITS POLYMER NANOCOMPOSITES

23.5.1 Antibacterial Activity

Recently, there has been a report regarding the antibacterial activity of GNS-based materials (Hu et al., 2010; Liu et al., 2011a; Bykkam et al., 2013). The antibacterial activity of GNS-based composites was attributed by the membrane stress induced by sharp edges of GNS sheets, which may result in physical damage to cell membranes, leading to the loss of bacterial membrane integrity and the leakage of RNA. On the other hand, GNS may induce oxidative stress on neural phaeochromocytoma-derived PC12 cells. Furthermore, the interaction of GNS-based materials with bacterial cells such as solubility, dispersion, and size influences the antibacterial activity. There are several reports to support that the carbon-based nanomaterials are cytotoxic to the bacteria (Liu et al., 2011a; Gurunathan et al., 2012). In 2011, Liu et al. studied the antibacterial activity of GNS-based materials toward an

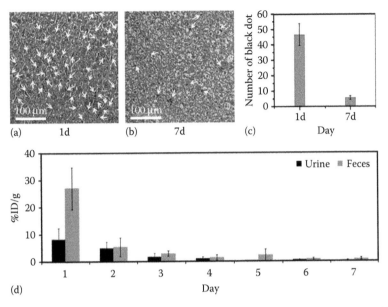

FIGURE 23.11 Excretion of GO–Dextran (DEX) from mice. (a and b) H&E stained liver images from mice injected with GO–DEX at 1 day p.i. (a) and 7 days p.i. (b). Black spots, which were associated with GO aggregates, were pointed by white arrows. (c) Statistics of average numbers of black dots per image field in the mouse liver 1 day and 7 days after GO–DEX injection. Error bars were based 10 slices per each sample. A remarkable decrease in the density of black spots was noticed 7 days after injection, indicating the clearance of GO–DEX from the mouse liver. (d) Distribution of 125I–GO–DEX in urine and feces of Balb/c mice collected by metabolism cages. The egesta was wet weighed and measured by a gamma counter. (Reproduced with permission from Zhang, R. et al., *Carbon*, 49, 1126–1132, 2011a.)

Escherichia coli. The scanning electron microscope (SEM) images show that GO aggregates have the smallest average size and exhibit higher antibacterial activity (Figure 23.12).

The GP-based materials can oxidize glutathione, which serves as redox state mediator in bacteria. Hence, Liu et al. proposed that the antibacterial activity is because of the initial cell deposition on GNS-based materials, membrane stress caused by direct contact with sharp nanosheets, and the ensuing superoxide anion-independent oxidation (Liu et al., 2011a). They studied the antibacterial activity of GO toward a bacteria *E. coli* and confirmed that the antibacterial activity was depended on the lateral dimension of GO sheets (Figure 23.13) (Liu et al., 2012a).

Large GO sheets more easily cover the cells and the cells cannot proliferate resulting in the bacterial cell apoptosis (Liu et al., 2012b). GO–silver nanoparticle composite shows an improved antibacterial activity against *E. coli*, *Listonella anguillarum*, *Bacillus cereus*, and *Staphylococcus aureus* (Nguyen et al., 2012; Kumar et al., 2013).

23.5.2 Biosensors

Field-effect transistors (FETs) have gained great interest in the area of biosensing as these can provide full electronic detection that is fully integrated into the electronic chips

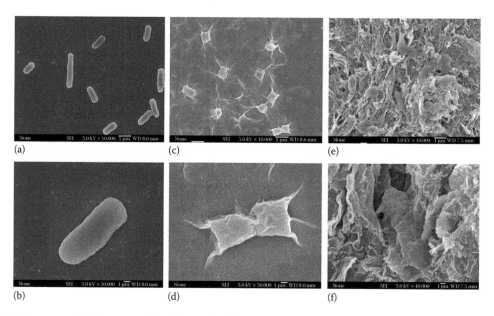

FIGURE 23.12 SEM images of (a, b) *E. coli* after incubation with saline solution for 2 h without graphene-based materials, (c, d) *E. coli* cells after incubation with GO dispersion (40 μg/mL) for 2 h, (e, f) *E. coli* cells after incubation with rGO dispersion (40 μg/mL) for 2 h. (Reproduced with permission from Liu, S. et al., *ACS Nano*, 5, 6971–6980, 2011a.)

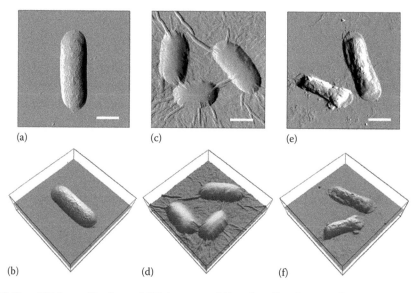

FIGURE 23.13 AFM amplitude and 3D images of *E. coli* cells after incubation with GO sheets. (a, b) *E. coli* incubation with deionized water for 2 h, (c, d) *E. coli* incubation with the 40 μg/mL GO-0 suspension for 2 h, and (e, f) *E. coli* after incubation with the 40 μg/mL GO-240 suspension for 2 h. The scale bars are 1 μm. (Reproduced with permission from Liu, S. et al., *Langmuir*, 28, 12364–12372, 2012b.)

produced by semiconductor companies. GNS is an ideal material for the construction of FET biosensors because of its zero-gap semiconductor property. The band gap can be tuned by the surface modification of GNS with polymers or metal nanoparticle. The carbon-based atomic structure of GNS and π stacking interaction between its hexagonal cells make it absorb the biomolecules containing of carbon-based ring structures (Pumera, 2011; Maharana and Jha, 2012). Moreover, the extremely high surface-to-volume ratio and excellent electrical conductivity of GNS allow it to act as an electron wire between the redox centers of an enzyme or protein and an electrode surface, which causes the accurate and selective detection of biomolecules. It provides a sensitive platform for interfacing with biological cells to detect intra- and extracellular phenomena and the mechanism of GNS/cell detection systems is based on carrier doping via the cell wall's electronegativity or dipole moment (Figure 23.14).

The GNS is used in electronic and sensor applications because of electroactivity, transparency and flexibility (Kuila et al., 2011b; Nguyen and Berry, 2012; Kim et al., 2013). Hence, GNS has been used for a wide variety of biosensing schemes, including transducer in the bio-field-effect transistors, electrochemical biosensors, impedance biosensors, electrochemiluminescence, electrochemical immunosensor (Figure 23.15), and fluorescence biosensors, as well as biomolecular labels (Liu et al., 2011b; Maharana and Jha, 2012). The GNS-based composites used in biosensors applications are listed in Table 23.1.

23.5.3 Drug Delivery

The ideal drug delivery system combines a strong affinity for target cells with controlled release such that the drug is delivered and released in a selective and discriminatory fashion. Recently, several nanoscale drug delivery systems have been developed (Zhang et al., 2009). The π–π stacking interaction of GNS and GO makes them as a simplified noncovalent functionalization with drugs and enables high drug loading with subsequent controlled release of drugs (Depan et al., 2011b; An et al., 2013). The gelatin–GNS nanocomposite was prepared by An et al. with good dispersibility and stability in water, as well as

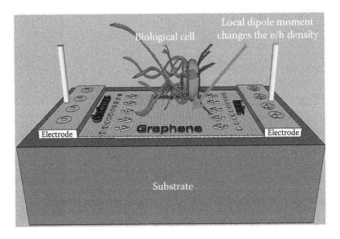

FIGURE 23.14 A generalized schematic for a graphene device interfaced with a biological cell. (Reproduced with permission from Nguyen, P. and Berry, V., *J. Phys. Chem. Lett.*, 3, 1024–1029, 2012.)

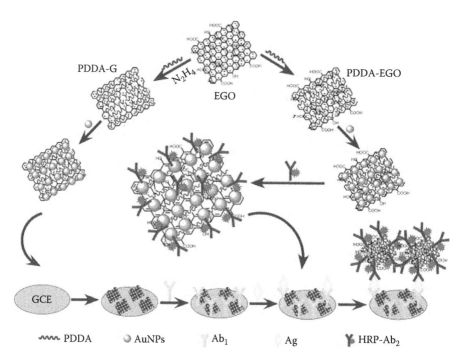

FIGURE 23.15 Schematic illustration of the preparation of graphene-based hybrids and the construction of the sandwich-type electrochemical immunosensor. PDDA-Poly (diallyldimethylammonium chloride), PDDA-*g*-PDDA functionalized graphene nanosheets, AuNPs–Gold nanoparticles, Ab$_1$–Rabbit anti-human IgG and HRP-Ab$_2$–HRP-labeled monoclonal goat antihuman IgG. (Reproduced with permission from Liu, K. et al., *Biosens. Bioelectr.*, 26, 3627–3632, 2011b.)

TABLE 23.1   List of GNS-Based Composites as Biosensors

S. No.	Sensing Composites	Detected Element	References
1	Screen printed GNS electrode	Ascorbic acid, dopamine, and uric acid	Ping et al. (2012)
2	GNS-coated silver electrode	Neural sensors	Chiu et al. (2013)
3	Silver nanoparticle-GO-GCE	Tryptophan	Li et al. (2013a)
4	CoPt-reduced GO	Thrombin	Wang et al. (2011)
5	GNS-MnO$_2$-carbon ionic liquid electrode	Rutin	Sun et al. (2013b)
6	CuO-GNS	Glucose	Hsu et al. (2012)
7	GNS-TiO$_2$ Composite	Glucose	Jang et al. (2012)
8	GNS-AuNPs-glucose oxidase-chitosan	Glucose	Chan et al. (2010)
9	Chitosan-ferrocene/GO/glucose oxidase	Glucose	Qiu et al. (2011)
10	Chitosan-Prussian blue-GNS	Glucose	Zhong et al. (2012)
11	Glucose oxidase-GNS-chitosan	Glucose	Kang et al. (2012)
12	Hemin-GNS	Glucose and H$_2$O$_2$	Guo et al. (2011)
13	Cuprous oxide-reduced GO	Non enzymatic H$_2$O$_2$	Xu et al. (2013)
14	Fe$_3$O$_4$-reduced GO-hemoglobin	H$_2$O$_2$	Zhu et al. (2013)
15	Hemoglobin/chitosan-ionic liquid-ferrocene/ GNS/ GCE	H$_2$O$_2$	Huang et al. (2012)

(Continued)

TABLE 23.1 (*Continued*)   List of GNS-Based Composites as Biosensors

S. No.	Sensing Composites	Detected Element	References
16	GNS-AuNPs-chitosan	H_2O_2 and O_2	Chan et al. (2010)
17	GNS-polyaniline	Dopamine	Liu et al. (2012d)
18	GO-AuNPs-chitosan/silica gel-GCE	Dopamine and uric acid	Liu et al. (2012c)
19	Poly (acridine)-GNS- GCE	Uric acid	Wang et al. (2012b)
20	Thionine-GO	Uric acid	Sun et al. (2013a)
21	GNS-chitosan- GCE	Acetaminophen	Zheng et al. (2013)
22	GNS-1-butyl-3-methylimidazolium hexafluorophosphate	Metronidazole	Peng et al. (2012)
23	GNS-thionine- horseradish peroxidase-antiprostate-specific antigen antibody	Cancer	Yang et al. (2010a)
24	GNS-ionic liquid-chitosan	Adenine and guanine	Niu et al. (2012)
25	GNS-Pt nanoparticle	Cholesterol	Dey and Raj (2010)
26	GO-polyaniline-GCE	DNA	Bao et al. (2011)

GNS, graphene; GO, graphene oxide; AuNPs, gold nanoparticle; GCE, glassy carbon electrode.

various physiological solutions and was studied for methotrexate drug delivery application. Gelatin–GNS composite shows excellent biocompatibility and stability. The pH-dependent release behavior of gelatin–GNS-loaded methotrexate showed the higher release amount of methotrexate under acidic condition than the neutral condition (An et al., 2013). GO has been utilized for the drug release of rhodamine B by Zhang et al., and 0.5 mg/g rhodamine B can be loaded on GO. The drug releases were studied in water, pH 4.5 phosphate buffer solution (PBS), and pH 7.4 PBS. The pH-sensitive drug release was observed; the higher pH values lead to weaker hydrophobic force and hydrogen bonds, which causes higher drug release rate (Zhang et al., 2011a). In another study, Wang et al. have synthesized carboxylic polystyrene latex/Fe_3O_4/GO composite for rhodamine B drug releases. Carboxyl-functionalized polystyrene latex was used as a template to deposit of higher loading layers of Fe_3O_4 and then GO on spheres (Wang et al., 2012a). Ghosh et al. showed GO-p-amino benzoic acid nanosheet as a drug delivery system to treat drug-resistant bacteria using tetracycline. The percentage of drug loading and drug releasing efficiency were found to be 64.05 ± 2.74 and 38.35 ± 0.07, respectively, and the minimal inhibitory concentration against *E. coli* XL-1 was found to be 100 µg/mL (Ghosh et al., 2010).

Nowadays, major attention has been given to life-threatening disease of cancer which continues to increase with increasing age of the population and urbanization. Medical science and biomedical engineering have gained advanced state for the therapeutic development of anticancer strategies but often lack by administrative problem of the drugs. A GP-based composite of polymers includes cyclodextrin, polylactide, and chitosan which enhance the biocompatibility of the matrix for controlled release of therapeutic molecules, and folic acid is a ligand for targeting cell membrane used as an effective delivery of anticancer drugs (Depan et al., 2011b). Transferin (Tf)-conjugated ployethylene glycol (PEG)-based–nanoscaled GO has been developed to deliver anticancer doxorubicin (Dox) drug. *In vitro* and *in vivo* studies of Tf-PEG-GO-Dox exhibited significant improvement in therapeutic efficiency to deliver Dox to brain tumors (Figure 23.16) (Liu et al., 2013).

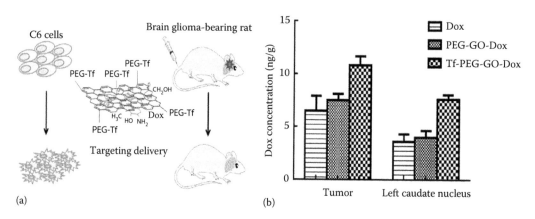

(a)   (b)

FIGURE 23.16 Schematic representation of Dox delivery to brain tumor (a). Distribution of Dox in the tumor and left caudate nucleus in brain, after i.v. of Dox, PEG-GO-Dox, and Tf-PEG-GO-Dox ($n = 3$) (b). (Reproduced with permission from Liu, G. et al., *ACS Appl. Mater. Interfaces*, 5, 6909–6914, 2013.)

Zhao et al. have developed 3D biocompatible, reduction-responsive Dox-loaded cytamine (Cy)-modified PEGylated alginate (ALG) with GO nanoparticle (GON-Cy-ALG-PEG). The *in vitro* release showed that the platform could not only prevent the leakage of the loaded DOX under physiological conditions but also detach the Cy-modified PEGylated alginate (Cy-ALG-PEG) moieties, as response to glutathione (GSH) (Figure 23.17) (Zhao et al., 2014).

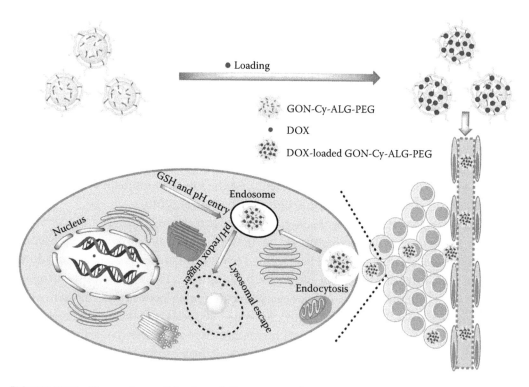

FIGURE 23.17 Drug release of the drug delivery system by GSH and pH trigger in the HepG2 Cell. (Reproduced with permission from Zhao, X. et al., *Langmuir*, 30, 10419–10429, 2014.)

Furthermore, Depan et al. synthesized folic acid-conjugated chitosan-GO composite, and Qin et al. synthesized folic acid-GO-poly (vinyl pyrrolidone) composite for the delivery of an anticancer drug doxorubicin. The *in vitro* drug delivery of both the composites show pH dependence and imply hydrogen bonding interaction between GO and doxorubicin. At low pH about 5.3 to 5.5, the composite shows higher drug delivery because of the reduced interaction between doxorubicin and composites (Depan et al., 2011b; Qin et al., 2013). There are several biocompatible GNS-based composites that have been developed for anticancer drug doxorubicin delivery for cancer treatment, which include GO-Fe_3O_4 (Yang et al., 2009), chitosan-GO (Wang et al., 2013), PEG-grafted GO (Miao et al., 2013), and GP-thiol grafted pluronic (Al-Nahain et al., 2013).

23.5.4 Tissue Engineering

The natural extracellular matrix (ECM) gives structural support for cells with facilitation of the surrounding tissue for tissue growth and recovering the damaged tissue. ECM is composed of collagen, proteoglycans, glycosaminoglycans, adhesion protein, and signaling molecule; the carbohydrate plays a major role in tissue engineering (Lee and Mooney, 2001; Eslaminejad and Bagheri, 2009; Rajesh et al., 2012). Hence, several biodegradable and biocompatible natural as well as synthetic polymers include chitosan, collagen, gelatin, hyaluronate, fibrin, poly lactic acid, poly(DL-lactide), and poly(lactide-*co*-glycolide) have been used to mimic the ECM in tissue engineering (Lee and Mooney, 2001; Nair and Laurencin, 2007; Armentano et al., 2010; Dash et al., 2011). However, the poor mechanical strength of polymers makes it unfit for major load-bearing applications. Hence, the higher mechanical strength of GNS and GO can be used as an excellent reinforcement material. Liao et al. studied the cytotoxicity of GNS and GO on human skin fibroblast by measuring mitochondrial activity. The GNS sheets are more damaging to mammalian fibroblast cells than the less densely packed GO. The toxicity of GNS and GO depends on the exposure environment and mode of interaction with cells (Liao et al., 2011). Even though, there is a report of toxicity of GNS and GO to human cells, Ku et al. proved the effective behavior of GO and reduced GO on mouse myoblast C2C12 cells including cell adhesion, proliferation, differentiation, and biocompatibility by the obtained results of enhanced myogenic differentiation on GO from serum protein adsorption and nanotopographical cues (Figure 23.18).

The ability of GO to stimulate myogenic differentiation suggesting the potential application of GO in tissue engineering (Ku and Park, 2013). Moreover, GNS and GO sheets are soft membranes with high in-plane stiffness and can potentially serve as a biocompatible, transferable, and implantable platform for stem cell culture. The strong noncovalent binding abilities of GNS allow it to act as a preconcentration platform for osteogenic inducers, which accelerate mesenchymal stem cells growing on it toward the osteogenic lineage (Lee et al., 2011). Hence, GNS-based nanomaterials can be effectively used for tissue engineering.

Chitosan, a natural polysaccharide, can be used in tissue engineering and has an advantage of forming highly porous and interconnected pore structure, osteoconductivity, biocompatibility, and the ability to enhance bone formation. Moreover, chitosan has similar structure of glycosaminoglycans, which is majorly present in the ECM (Depan et al., 2011a).

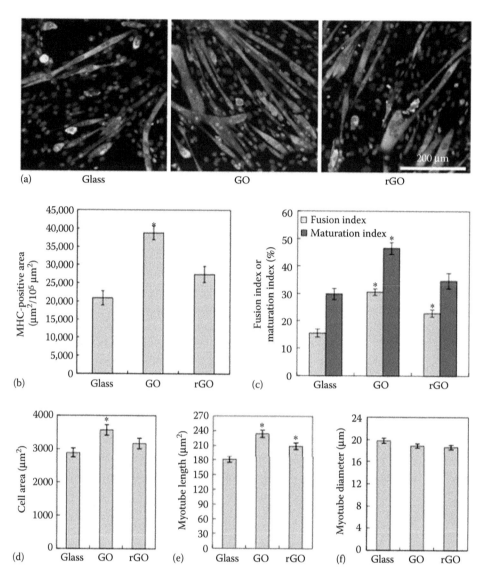

FIGURE 23.18 Myotube formation on unmodified, GO-, and reduced GO (rGO)-modified glass substrates. (a) Immunofluorescence staining for myosin heavy chain (MHC). (b) Quantitative analysis of total MHC-positive area. (c) Quantification of fusion index and maturation index. Quantification of (d) cell area, (e) length, and (f) diameter of multinucleate myotubes. C2C12 cells were grown in GM for 1 day and then incubated in DM for 5 days. * denotes a significant difference compared with unmodified glass ($p < .05$). (Reproduced with permission from Ku, S.H. and Park, C.B., *Biomaterials*, 34, 2017–2023, 2013.)

Hence, chitosan-based GNS composite has been prepared for tissue engineering application with better biocompatibility. Depan et al. (2011a) prepared chitosan-GO scaffold and studied *in vitro* biocompatibility and cell proliferation on mouse pre-osteoblast cell line. The scaffold material shows excellent cell adhesion, cell proliferation, mineralization, and *in vitro* biodegradation. The negatively charged and polar carboxylic groups on the GO are believed to be an important factor enhancing osteoblast–biomaterial interactions. Fan et al.

(2010) prepared GNS-chitosan composite and studied the *in vitro* biocompatibility on L-929 cells, and Li et al. (2013b) prepared GO-hydroxyapatite (HA) and GO-chitosan-HA composite and studied the *in vitro* cytotoxicity on murine fibroblast L-929 cells and human osteoblast-like MG-63 cell line as well as alkaline phosphatase activity (ALP) on MG-63 cell line (Figure 23.19).

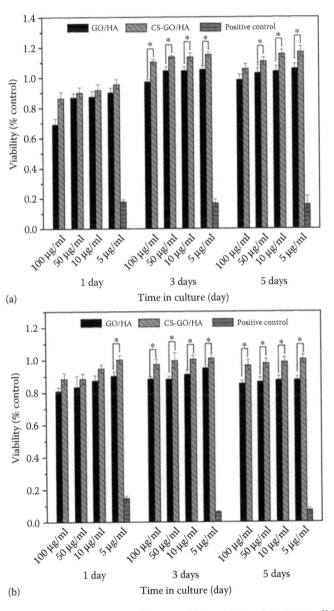

FIGURE 23.19 The cell proliferation rates of L-929 cell lines (a) and MG-63 cell lines (b) cultured in media for 1 day, 3 days, and 5 days, with different concentrations (100 μg/mL, 50 μg/mL, 10 μg/mL, and 5 μg/mL) of GO–HA and CS-GO–HA. Asterisks indicate significance level obtained by *t*-test ($p < .05$). (Reproduced with permission from Li, M. et al., *J. Mater. Chem.* B, 2013, 475–484, 2013b.)   *(Continued)*

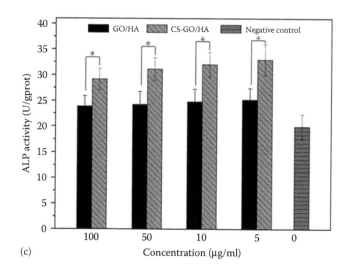

FIGURE 23.19 (Continued)   The ALP activity of MG-63 cells (c) after cultivating with different concentrations of GO–HA and CS-GO–HA for 7 days. Asterisks indicate significance level obtained by t-test ($p < .05$). (Reproduced with permission from Li, M. et al., *J. Mater. Chem.* B, 2013, 475–484, 2013b.)

Both the composites show higher cell proliferation and biocompatibility. ALP is an important feature of osteoblast cells expressed in their differentiation phase and a significant quantitative marker of osteogenesis. The higher ALP expression of GO-chitosan-HA indicated that the functionalization of the GO sheets with chitosan could enhance the mineralization and cell activation of the MG-63 cells (Li et al., 2013b).

Freestanding polyaniline conducting polymer-based GNS hybrid paper (PANI-GNS) showed excellent biocompatibility toward mouse fibroblast cell line L-929, which is commonly used to assess cytotoxicity of potential for cell growth (Yan et al., 2010) (Figure 23.20a). Yen et al. have showed that PANI coatings on GO and GNS papers can enhance the cell proliferation of the corresponding parent papers (Figure 23.20b). In addition, they have also studied the cell growth onto the PANI-GNS hybrid paper and the cell morphology using fluorescence microscopy and SEM, respectively. The fluorescent staining shown in Figure 23.8c revealed that L-929 cells are well adhered and proliferated on the PANI-GNS hybrid paper with the increase of the culture time. Other biocompatible GNS-based polymer composites including GO-polyethylene (Chen et al., 2012b), fibrin-decorated GO, and GO-poly (propylene-carbonate) (Yang et al., 2013c) composites are also prepared by various research groups for tissue engineering applications.

23.6 CONCLUSION

As a result, fascinating properties of GNS, including unique mechanical strength, electrical properties, large surface area, π–π stacking interaction with biomolecules, and biocompatibility make it to offer some advantages to use in biomedical applications such as antibacterial agents, drug delivery, biosensors, bioelectronics, and tissue engineering.

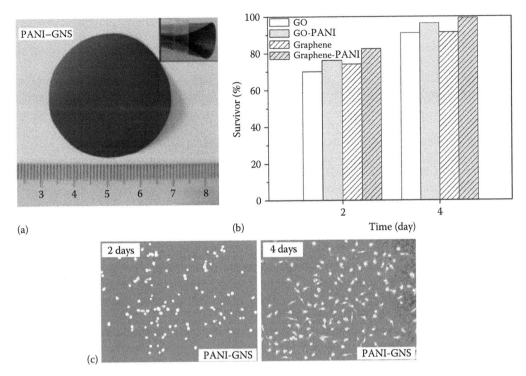

FIGURE 23.20 (a) Digital camera image of PANI-graphene paper (PANI-GNS) (The insert image indicate the flexibility of the sample) (b) Survivor ratio of L-929 cells on GO, PANI-GO, GNS, PANI-GNS papers after 2 and 4 days respectively (c) fluorescence microscopy images of L-929 cells grown on PANI-GNS paper for 2 and 4 days (Magnification-100 X). (Reproduced with permission from Yan, X. et al., *ACS Appl. Mater. Interfaces*, 9, 2521–2529, 2014.)

REFERENCES

Achaby, M. E., Essamlali, Y., Miri, N. E. et al. 2014. Graphene oxide reinforced chitosan/polyvinyl-pyrrolidone polymer bio-nanocomposites. *J. Appl. Polym. Sci.*, 133:41042.

Ago, H., Ito, Y., Mizuta, N. et al. 2010. Epitaxial chemical vapor deposition growth of single-layer graphene over cobalt film crystallized on sapphire. *ACS Nano* 4:7407–7414.

Alam, J., Ryu, S. H., Shanmugharaj, A. M., 2015. Influence of hexamethylene diamine functionalized graphene oxide on structural characteristics and properties of epoxy nanocomposites. *Sci. Adv. Mater.*, 7:993–1001.

Allen, M. J., Tung, V. C., Kaner, R. B., 2010. Honeycomb carbon: A review of graphene. *Chem. Rev.*, 110:132–145.

Al-Nahain, A., Lee, S. Y., In, I., Lee, K. D., Park, S. Y., 2013. Triggered pH/redox responsive release of doxorubicin from prepared highly stable graphene with thiol grafted Pluronic. *Inter. J. Pharm.*, 450:208–217.

Ambrosi, A., Chua, C.K., Bonanni, A., Pumera, M., 2014. Electrochemistry of graphene and related materials. *Chem. Rev.*, 114:7150–7188.

An, J., Gou, Y., Yang, C., Hu, F., Wang. C., 2013. Synthesis of a biocompatible gelatin functionalized graphene nanosheets and its application for drug delivery. *Mater. Sci. Engg. C*, 33:2827–2837.

Armentano, I., Dottori, M., Fortunati, E., Mattioli, S., Kenny, J. M., 2010. Biodegradable polymer matrix nanocomposites for tissue engineering: A review. *Polym. Degrad. Stab.*, 95:2126–2146.

Avouris, P., Dimitrakopoulos, C., 2012. Graphene: Synthesis and applications. *Mater. Today,* 15:86–97.

Bai, H., Sheng, K., Zhang, P., Li, C., Shi. G., 2011. Graphene oxide/conducting polymer composite hydrogels. *J. Mater. Chem.,* 21:18653–18658.

Bao, C., Song, L., Xing, W. et al. 2012. Preparation of graphene by pressurized oxidation and multiplex reduction and its polymer nanocomposites by masterbatch-based melt blending. *J. Mater. Chem.,* 22:6088–6096.

Bao, Y., Yuan, H., Hu, Y., Yao, T., Huang, H., 2011. A novel electrochemical DNA biosensor based on graphene and polyaniline nanowires. *Electrochim. Acta* 56:2676–2681.

Brodie, B. C., 1859. On the atomic weight of graphite. *Philos. Trans. Royal Soc. London* 149:12

Bykkam, S., Rao, K. V., Chakra, C. H. S., Thunugunta, T., 2013. Synthesis and characterization of graphene oxide and its antimicrobial activity against klebseilla and staphylococcus. *Inter. J. Adv. Biotech. Res.,* 4:142–146.

Cai, Y., Li, H., Du, B. et al. 2011. Ultrasensitive electrochemical immunoassay for BRCA1 using BMIM.BF4-coated SBA-15 as labels and functionalized graphene as enhancer. *Biomaterials* 32:2117–2123.

Cao, Y., Feng, J., Wu, P., 2010. Preparation of organically dispersible graphene nanosheet powders through a lyophilization method and their poly(lactic acid) composites. *Carbon* 48:3834–3839.

Chan, C., Yang, H., Han, D., Zhang, Q., Ivaska, A., Niu, L., 2010. Graphene/AuNPs/chitosan nanocomposites film for glucose biosensing. *Biosens. Bioelectr.,* 25:1070–1074.

Chang, Y., Yang, S. -T., Liu, J.-H., Dong, E., Wang, Y., Cao, A., Liu, Y., Wang, H. 2011. *In-vitro* toxicity evaluation of graphene oxide on A549 cells. *Toxicol. Lett.,* 200:201–210.

Chen, G. -Y., Pang, D. W. -P., Hwang, S. -M., Tuan, H. -Y., Hu, Y. -C., 2012a. A graphene-based platform for induced pluripotent stem cells culture and differentiation. *Biomaterials,* 33:418–427.

Chen, Y., Qi, Y., Tai, Z., Yan, X., Zhu, F., Xue, Q., 2012b. Preparation, mechanical properties and biocompatibility of graphene oxide/ultrahigh molecular weight polyethylene composites. *Euro. Polym. J.,* 48:1026–1033.

Chen, K. H. F., Sahoo, N. G., Tan, Y. P. et al. 2012c. Poly (vinyl alcohol) nanocomposites filled with poly (vinyl alcohol) grafted graphene oxide. *ACS Appl. Mater. Interfaces,* 4:2387–2394.

Chiu, C. W., He, X. L., Liang, H. 2013. Surface modification of a neural sensor using graphene. *Electrochim. Acta,* 94:42–48.

Das, S. Wajid, A. S., Shelburne, J. L., Liao, Y. -C., Green, M. J., 2011. Localized in-situ polymerization on graphene surfaces for stabilized graphene dispersions. *ACS Appl. Mater. Interfaces,* 3:1844–1851.

Dash, M., Chiellini, F., Ottenbrite, R. M., Chiellini, E., 2011. Chitosan-A versatile semi-synthetic polymer in biomedical applications. *Prog. Polym. Sci.,* 36:981–1014.

Deepachitra, R., Chamundeeswari, M., Kumar, B. S. et al. 2013. Osteo mineralization of fibrin-decorated graphene oxide. *Carbon,* 56:64–76.

Depan, D., Girase, B., Shah, J. S., Misra, R. D. K., 2011a. Structure-process-property relationship of the polar graphene oxide-mediated cellular response and stimulated growth of osteoblasts on hybrid chitosan network structure nanocomposite scaffolds. *Acta Biomaterialia,* 7:3432–3445.

Depan, D., Shah, J., Mishra, R. D. K., 2011b. Controlled release of drug from folate-decorated and graphene mediated drug delivery system: Synthesis, loading efficiency, and drug release response. *Mater. Sci. Engg C,* 31:1305–1312.

Dey, R. S., Raj, C. R., 2010. Development of an amperometric cholesterol biosensor based on graphene-Pt nanoparticle hybrid material. *J. Phys. Chem. C,* 114:21427–21433.

Economopoulos, S. P., Rotas, G., Miyata, Y., Shinohara, H., Tagmatarchis, N., 2010. Exfoliation and chemical modification using microwave irradiation affording highly functionalized graphene. *ACS Nano,* 12:7499–7507.

Eslaminejad, M. B., Bagheri, F. 2009. Tissue engineering approach for reconstructing bone defects using mesenchymal stem cells. *Yakhte Medicinal J.,* 11:263–272.

Fan, H., Wang, L., Zhao, K. et al. 2010. Fabrication, Mechanical properties, and biocompatibility of graphene-reinforced chitosan composites. *Biomacromolecules,* 11:2345–2351.

Feng, L., Liu, Y. -W., Tang, X. -Y. et al. 2013. Propagative exfoliation of high quality graphene. *Chem. Mater.,* 25:4487–4496.

Ghosh, D., Chandra, S., Chakraborty, A., Ghosh, S. K., Pramanik, P. 2010. A novel graphene oxide-para amino benzoic acid nanosheet as effective drug delivery system to treat drug resistant bacteria. *Inter. J. Pharma. Sci. Drug Res.,* 2:127–133.

Goenka, S., Sant, V., Sant, S. 2014. Graphene based nanomaterials for drug delivery and tissue engineering. *J. Control. Release,* 173:75–88.

Gollavelli, G., Ling. Y. -C., 2012. Multi-functional graphene as an *in-vitro* and *in-vivo* imaging probe. *Biomaterials,* 33:2532–2545.

Goncalves, G., Marques, P.A.A.P., Barros-Timmons, A. et al., 2010. Graphene oxide modified with PMMA via ATRP as a reinforcement filler. *J. Mater. Chem.,* 20:9927–9934.

Guermoune, A., Chari, T., Popescu, F. et al. 2011. Chemical vapor deposition synthesis of graphene on copper with methanol, ethanol, and propanol precursors. *Carbon* 49:4204–4210.

Guo, Y., Li, J., Dong, S. 2011. Hemin functionalized graphene nanosheets-based dual biosensor platforms for hydrogen peroxide and glucose. *Sens. Actuat. B* 160:295–300.

Gurunathan, S., Han, J. W., Dayem, A. A., Eppakayala, V., Kim, J. -H., 2012. Oxidative stress-mediated antibacterial activity of graphene oxide and reduced graphene oxide in Pseudomonas aeruginosa. *Inter. J. Nanomed.,* 7:5901–5914.

Hakimi, M., Alimard, P. 2012. Graphene: Synthesis and applications in biotechnology -A review. *World Appl. Program.,* 2:377–388.

Hsiao, M. -C., Lio, S. -H., Yen, M. -Y. et al. 2010. Preparation of covalently functionalized graphene using residual oxygen-containing functional groups. *ACS Appl. Mater. Interfaces,* 2:3092–3099.

Hsu, Y. -W., Hsu, T. -K., Sun, C. -L., Nien, Y. -T., Pu, N. -W., Ger, M. -D., 2012. Synthesis of CuO/graphene nanocomposites for non-enzymatic electrochemical glucose biosensor applications. *Electrochim. Acta,* 82:152–157.

Hsu, C. -H., Liao, H. -Y., Wu, Y. -F., Kuo, P. -L., 2011. Benzylamine-assisted non-covalent exfoliation of graphite-protecting pt nanoparticles applied as catalyst for methanol oxidation. *ACS Appl. Mater. Interfaces,* 3:2169–2172.

Hu, W., Peng, C., Luo, W. et al., 2010. Graphene-based antibacterial paper. *ACS Nano,* 4:4317–4323.

Huang, K.-J., Miao, Y. -X., Wang, L., Gan, T., Yu, M., Wang, L. -L., 2012. Direct electrochemistry of hemoglobin based on chitosan-ionic liquid-ferrocene/graphene composite film. *Process Biochem.,* 47:1171–1177.

Huang, X., Yin, Z., Wu, S. et al., 2011. Graphene-based materials: Synthesis, characterization, properties, and applications. *Small,* 7:1876–1902.

Hule, R. A., Pochan, D. J., 2007. Polymer nanocomposites for biomedical applications. *MRS Bull.,* 32:354–358.

Hummer, W. S., Offeman, R. E., 1958. Preparation of graphitic oxide. *J. Am. Chem. Soc.,* 80:1339.

Jang, H. D., Kim, S. K., Chang, H., Roh, K. -M., Choi, J. -W., Huang, J. 2012. A glucose biosensor based on TiO_2–Graphene composite. *Biosensors and Bioelectronics* 38:184–188.

Jayasena, B., Subbiah, S., 2011. A novel mechanical cleavage method for synthesizing few-layer graphenes. *Nanoscale Res. Lett.* 6:95.

Kang, X., Wang, J., Wu, H., Aksay, I. A., Liu, J., Lin, Y., 2009. Glucose oxidase-graphene-chitosan modified electrode for direct electrochemistry and glucose sensing. *Biosens. Bioelect.* 25:901–905.

Kedzierski, J., Hsu, P. L., Healeye, P. et al. 2008. Epitaxial graphene transistors on SIC substrates. *IEEE Trans. Electron Dev.,* 55:2078–2085.

Khan, U., May, P., O'Neill, A., Coleman, J. N., 2010. Development of stiff, strong, yet tough composites by the addition of solvent exfoliated graphene to polyurethane. *Carbon,* 48:4035–4041.

Kim, J. H., Hwang, T., Dugasani, S. R. et al. 2013. Graphene based fiber optic surface plasmon resonance for biochemical sensor applications. *Sens. Actuat. B* 187:426–433.

Kim, S., Ku, S. H., Lim, S. Y., Kim, J. H., Park, C. B., 2011. Graphene-biomineral hybrid materials. *Adv. Mater.,* 23:2009–2014.

Ku, S. H., Park, C. B., 2013. Myoblast differentiation on graphene oxide. *Biomaterials,* 34:2017–2023.

Kuilla, T., Bhadra, S., Yao, D., Kim, N. H., Bose, S., Lee, J. H. 2010. Recent advances in graphene based polymer composites. *Prog. Polym. Sci.,* 35:1350–1375.

Kuila, T., Bose, S., Hong, C. E. et al. 2011a. Preparation of functionalized graphene linear low density polyethylene composites by a solution mixing method. *Carbon,* 49:1033–1037.

Kuila, T., Bose, S., Khanra, P., Mishra, A. K., Kim, N. H., Lee, J. H. 2011b. Recent advances in graphene-based biosensors. *Biosens. Bioelectr.,* 26:4637–4648.

Kuila, T., Bose, S., Mishra, A.K., Khanra, P., Kim, N. H., Lee, J. H. 2012. Chemical functionalization of graphene and its applications. *Prog. Mater. Sci.,* 57:1061–1105.

Kumar, N. A., Choi, H. -J., Shin, Y. R., Chang, D. W., Dai, L., Baek, J. -B. 2012. Polyaniline-grafted reduced graphene oxide for efficient electrochemical supercapacitors. *ACS Nano,* 2:1715–1723.

Kumar, S. V., Huang, N. M., Lim, H. N., Marlinda, A. R., Harrison, I., Chia, C. H. 2013. One-step size-controlled synthesis of functional graphene oxide/silver nanocomposites at room temperature. *Chem. Engg. J.,* 219:217–224.

Labroo, P., Cui, Y. 2013. Flexible graphene bio-nanosensor for lactate. *Biosens. Bioelectr.,* 41:852–856.

Lee, W. C., Lim, C. H. Y. X., Shi, H. et al. 2011. Origin of enhanced stem cell growth and differentiation on graphene and graphene oxide. *ACS Nano,* 5:7334–7341.

Lee, K. Y., Mooney, D. J. 2001. Hydrogels for tissue engineering. *Chem. Rev.,* 110:1869–1879.

Li, J., Kuang, D., Feng, Y. et al. 2013a. Green synthesis of silver nanoparticles–graphene oxide nanocomposite and its application in electrochemical sensing of tryptophan. *Biosens. Bioelectr.* 42:198–206.

Li, X., Magnuson, C. W., Venugopal, A. et al. 2010. Graphene films with large domain size by a two-step chemical vapor deposition process. *Nano Lett.,* 10:4328–4334.

Li, M., Wang, Y., Liu, Q. et al. 2013b. In situ synthesis and biocompatibility of nano hydroxyapatite on pristine and chitosan functionalized graphene oxide. *J. Mater. Chem. B* 2013:475–484.

Li, X. L., Wang, X. R., Zhang, L., Lee, S. W., Dai, H. J. 2008. Chemically derived, ultra-smooth graphene nanoribbon semiconductors. *Science,* 319:12291232.

Li, M., Yang, X., Ren, J., Qu, K., Qu, X. 2012. Using graphene oxide high near-infrared absorbance for photothermal treatment of Alzheimer's disease. *Adv. Mater.,* 24:1722–1728.

Li, N., Zhang, X., Song, Q. et al. 2011a. The promotion of neurite sprouting and outgrowth of mouse hippocampal cells in culture by graphene substrates. *Biomaterials,* 32:9374–9382.

Li, X., Mangnuson, C. W., Venugopal, A. et al. 2011b. Large-area graphene single crystals grown by low-pressure chemical vapor deposition of methane on copper. *J. Am. Chem. Soc.,* 133:2861–2819.

Li, L., Zheng, X., Wang. J., Sun, Q., Xu, Q. 2013c. Solvent-exfoliated and functionalized graphene with assistance of supercritical carbon dioxide. *ACS Sustain. Chem. Eng.,* 1:144–151.

Liang, Y. T., Hersam, M. C. 2010. Highly Concentrated graphene solutions via polymer enhanced solvent exfoliation and iterative solvent exchange. *J. Am. Chem. Soc.,* 132:17661–17663.

Liang, J., Huang, Y., Zhang, L. et al. 2009. Molecular-level dispersion of graphene into poly(vinyl alcohol) and effective reinforcement of their nanocomposites. *Adv. Funct. Mater.,* 19:2297–2302.

Liao, K. -H., Lin, Y. -S., Macosko, C. W., Haynes, C. L. 2011. Cytotoxicity of graphene oxide and graphene in human erythrocytes and skin fibroblasts. *ACS Appl. Mater. Interfaces,* 3:2607–2615.

Liu, J., Guo, S., Han, L., Ren, W., Liu, Y., Wang, E., 2012a. Multiple pH-responsive graphene composites by non-covalent modification with chitosan. *Talanta,* 101:151–156.

Liu, S., Hu, M., Zeng, T. H. et al. 2012b. Lateral dimension-dependent antibacterial activity of graphene oxide sheets. *Langmuir,* 28:12364–12372.

Liu, Z., Robinson, J. T., Sun, X., Dai, H. 2008. PEGylated nano graphene oxide for delivery of water-insoluble cancer drugs. *J. Am. Chem. Soc.,* 130:10876–10877.

Liu, G., Shen, H., Mao, J. et al. 2013. Transferrin modified graphene oxide for glioma-targeted drug delivery: *In-vitro* and *in-vivo* evaluations. *ACS Appl. Mater. Interfaces,* 5:6909–6914.

Liu, J., Tao, L., Yang, W. et al. 2010. Synthesis, characterization, and multilayer assembly of pH sensitive graphene-polymer nanocomposites. *Langumir,* 26:10068–10075.

Liu, X., Xie, L., Li, H., 2012c. Electrochemical biosensor based on reduced graphene oxide and Au nanoparticles entrapped in chitosan/silica sol–gel hybrid membranes for determination of dopamine and uric acid. *J. Electroanalyt. Chem.,* 682:158–163.

Liu, S., Xing, X., Yu, J., 2012d. A novel label-free electrochemical aptasensor based on graphene-polyaniline composite film for dopamine determination. *Biosens. Bioelectr.,* 36:186–191.

Liu, S., Zeng, T. H., Hofmann, M. et al. 2011a. Antibacterial activity of graphite, graphite oxide, graphene oxide, and reduced graphene oxide: Membrane and oxidative stress. *ACS Nano,* 5:6971–6980.

Liu, K., Zhang, J. -J., Wang, C., Zhu. J. -J. 2011b. Graphene-assisted dual amplification strategy for the fabrication of sensitive amperometric immunosensor. *Biosens. Bioelectr.* 26:3627–3632.

Lotyfa, M., Hernandez, Y., King, P. J. et al. 2009. Liquid phase production of graphene by exfoliation of graphite in surfactant/water solutions. *J. Am. Chem. Soc.,* 131:3611–3620.

Maharana, P. K., Jha, R., 2012. Chalcogenide prism and graphene multilayer based surface plasmon resonance affinity biosensor for high performance. *Sensors and Actuat. B,* 169:161–166.

Miao, W., Shim, G., Lee, S., Lee, S., Choe, Y. S., Oh, Y. -K., 2013. Safety and tumor tissue accumulation of PEGylated graphene oxide nanosheets for co-delivery of anticancer drug and photosensitizer. *Biomaterials,* 34:3402–3410.

Moniruzzaman, M., Winey, K. I., 2006. Polymer nanocomposites containing carbon nanotubes. *Macromolecules,* 39:5194–5205.

Murariu, M., Dechief, A. L., Bonnaud, L. et al. 2010. The production and properties of polylactide composites filled with expanded graphite. *Polym. Degrad. Stab.,* 95:889–900.

Nair, L. S., Laurencin, C. T., 2007. Biodegradable polymers as biomaterials. *Prog. Polym. Sci.,* 2007:762–798.

Nayak, T. R., Andersen, H., Makam, V. S. et al. 2011. Graphene for controlled and accelerated osteogenic differentiation of human mesenchymal stem cells. *ACS Nano,* 5:4670–4678.

Neelgund, G. M., Oki, A., Luo, Z. 2013. In situ deposition of hydroxyapatite on graphene nanosheets. *Mater. Res. Bull.,* 48:175–179.

Nguyen, P., Berry, V. 2012. Graphene interfaced with biological cells: Opportunities and challenges. *J. Phys. Chem. Lett.,* 3:1024–1029.

Nguyen, V. H., Kim, B. -K., Jo, Y. -L., Shim, J. -J., 2012. Preparation and antibacterial activity of silver nanoparticles-decorated graphene composites. *J. Supercritical Fluids* 72:28–35.

Niu, X., Yang, W., Renet, J. et. al., 2012. Electrochemical behaviors and simultaneous determination of guanine and adenine based on graphene-ionic liquid-chitosan composite film modified glassy carbon electrode. *Electrochim. Acta,* 80:346–353.

Novoselov, K. S., 2011. Graphene: Materials in the flatland (Nobel lecture). *Angew. Chemie. Inter. Ed.,* 50:6986–7002.

Novoselov, K. S., Geim, A. K., Morozov, S. V. et al. 2004. Electric field effect in atomically thin carbon films. *Science,* 306:666–669.

O'Neil, A., Khan, U., Nirmalraj, P. N., Boland, J., Coleman, J. N., 2011. Graphene dispersion and exfoliation in low boiling point solvents. *J. Phys. Chem. C,* 115:5422–5428.

Park, H., Meyer, J., Roth, S., Skakalova, V. 2010. Growth and properties of few-layer graphene prepared by chemical vapor deposition. *Carbon,* 48:1088–1094.

Paul, D. R., Robeson, L. M., 2008. Polymer nanotechnology: Nanocomposites. *Polymer,* 49:3187–3204.

Peng, J., Huo, C., Hu, X., 2012. Determination of metronidazole in pharmaceutical dosage forms based on reduction at graphene and ionic liquid composite film modified electrode. *Sens. Actuat. B* 169:81–87.

Petrone, N., Dean, C. R., Meric, I. et al. 2012. Chemical vapor deposition-derived graphene with electrical performance of exfoliated graphene. *Nano Lett.*, 12:2751–2756.

Ping, J., Wu, J., Wang, Y., Ying, Y., 2012. Simultaneous determination of ascorbic acid, dopamine and uric acid using high-performance screen-printed graphene electrode. *Biosens. Bioelectr.*, 34:70–76.

Potts, J. R., Dreyer, D. R., Bielawski, C.W., Ruoff, R. S., 2011a. Graphene-based polymer nanocomposites. *Polymer*, 52:5–25.

Potts, J. R., Lee, S. H., Alam, T. M. et al. 2011b. Thermomechanical properties of chemically modified graphene/poly (methyl methacrylate) composites made by in situ polymerization. *Carbon*, 49:2615–2623.

Pumera, M. 2011. Graphene in biosensing. *Mater. Today*, 14:309–315.

Qin, X. C., Guo, Z. Y., Liu, Z. M., Zhang, W., Wan, M. M., Yang, B. W., 2013. Folic acid-conjugated graphene oxide for cancer targeted chemo-photothermal therapy. *J. Photochem. Photobiol. B Biol.*, 120:156–162.

Qiu, J. -D., Huang, J., Liang, R. -P., 2011. Nanocomposite film based on graphene oxide for high performance flexible glucose biosensor. *Sens. Actuat. B* 160:287–294.

Rajesh, R., Hariharasubramanian, A., Ravichandran, Y. D., 2012. Chicken bone as bioresource for the bioceramic (Hydroxyapatite). *Phosphor. Sulfur Silicon Related Elements*, 187:914–925.

Rao, C. N. R., Sood, A. K., Subrahmanyam, K. S., Govindaraj, A., 2009. Graphene: The new two-dimensional nanomaterial. *Angew. Chemie*, 48:7752–7777.

Robinson, J. T., Tabakman, S. M., Liang, Y., Wang, H., Casalongue, H. S., Vinh, H. et al., 2011. Ultrasmall reduced graphene oxide with high near-infrared absorbance for photothermal therapy. *J. Am. Chem. Soc.*, 133:6825–6831.

Rollings, E., Gweon, G. -H., Zhou, S. Y. et al. 2006. Synthesis and characterization of atomically thin graphite films on a silicon carbide substrate. *J. Phys. Chem. Solids* 67:2172–2177.

Ryu, S. H., Shanmugharaj, A. M., 2014a. Influence of long-chain alkylamine-modified graphene oxide on the crystallization, mechanical and electrical properties of isotactic polypropylene nanocomposites. *Chem. Engg. J.*, 244:552–560.

Ryu, S. H., Shanmugharaj, A. M., 2014b. Influence of hexamethylene diamine functionalized graphene oxide on the melt crystallization and properties of polypropylene nanocomposites. *Mater. Chem. Phys.*, 146:478–486.

Ryu, S. H., Sin, J. H., A. M. Shanmugharaj, S. H., 2014. Study on the effect of hexamethylene diamine functionalized graphene oxide on the curing kinetics of epoxy nanocomposites. *Euro. Polym. J.*, 52:88–97.

Salavagione, H. J., Gomez, M. A., Martinez, G., 2009. Polymeric modification of graphene through esterification of graphite oxide and poly (vinyl alcohol). *Macromolecules*, 42:6331–6334.

Sattarahmady, N., Heli, H., Moradi, S. E., 2013. Cobalt hexacyanoferrate/graphene nanocomposite-Application for the electrocatalytic oxidation and amperometric determination of captopril. *Sens. Actuat. B* 177:1098–1106.

Schniepp, H. C., Li, J. L., McAllister, M. J. et al. 2006. Functionalized single graphene sheets derived from splitting graphite oxide. *J. Phy. Chem. B*, 110:8535–8539.

Sebaa, M., Nguyen, T. Y., Paul, R. K., Mulchandani, A., Liu, H. 2013. Graphene and carbon nanotube-graphene hybrid nanomaterials for human embryonic stem cell culture. *Mater. Lett.*, 92:122–125.

Sengupta, R., Bhattacharya, M., Bandyopadhyay, S., Bhowmick, A. K. 2011. A review on the mechanical and electrical properties of graphite and modified graphite reinforced polymer composites. *Prog. Polym. Sci.*, 36:638–670.

Spitalsky, Z., Tasia, D., Papagelis, K., Galiotis, C., 2010. Carbon nanotube–polymer composites: Chemistry, processing, mechanical and electrical properties. *Prog. Polym. Sci.*, 35:357–401.

Sridhar, V., Lee, I., Chun, H. H., Park, H., 2013. Graphene reinforced biodegradable poly (3-hydroxy-butyrate-co-4-hydroxybutyrate) nano-composites. *Express Polym. Lett.,* 7:320–328.

Stankovich, S., Dikin, D. A., Dommett, G. H. B. et al. 2006. Graphene-based composite materials. *Nat. Lett.,* 442:282–286.

Sun, S., Cao, Y., Feng, J., Wu, P., 2010. Click chemistry as a route for the immobilization of well-defined polystyrene onto graphene sheets. *J. Mater. Chem.,* 20:5605–5607.

Sun, Z., Fu, H., Deng, L., Wang, J., 2013a. Redox-active thionine-graphene oxide hybrid nanosheet: One-pot, rapid synthesis, and application as a sensing platform for uric acid. *Anal. Chim. Acta,* 761:84–91.

Sun, Y., He, C., 2012. Synthesis and stereocomplex crystallization of poly (lactide)/graphene oxide nanocomposites. *ACS Macro Lett.,* 1:709–713.

Sun, X., Liu, Z., Welsher, K. et al. 2008. Nano-graphene oxide for cellular imaging and drug delivery. *Nano Res.,* 1:203–212.

Sun, W., Wang, X., Zhu, H. et al. 2013b. Graphene MnO_2 nanocomposite modified carbon ionic liquid electrode for the sensitive electrochemical detection of rutin. *Sens. Actuat. B,* 178:443–449.

Tang, Z., Wu, H., Cort, J. R., Buchko, G. W., Zhang, Y., Shao, Y. et al., 2010. Constraint of DNA on functionalized graphene improves its biostability and specificity. *Small,* 6:1205–1209.

Torres, T., Armas, L. G., Seabra, A. C., 2014. Optimization of micromechanical cleavage technique of natural graphite by chemical treatment. *Graphene,* 3:1–15.

Wang, C., Mallela, J., Garapati, U. S., 2013. A chitosan-modified graphene nanogel for noninvasive controlled drug release. *Nanomed. Nanotechnol. Biol. Med.,* 9:903–911.

Wang, J., Tang, B., Tsuzuki, T., Liu, Q., Hou, X., Sun, L., 2012a. Synthesis, characterization and adsorption properties of superparamagnetic polystyrene/Fe_3O_4/graphene oxide. *Chem. Eng. J.,* 204–206:258–263.

Wang, Z., Xia, J., Zhu, L. et al. 2012b. The fabrication of poly (acridine orange)/graphene modified electrode with electrolysis micelle disruption method for selective determination of uric acid. *Sens. Actuat. B,* 161:131–136.

Wang, Y., Yuan, R., Chai, Y., Yuan, Y., Bai, L., Liao, Y., 2011. A multi-amplification aptasensor for highly sensitive detection of thrombin based on high-quality hollow CoPt nanoparticles decorated graphene. *Biosens. Bioelectr.* 30:61–66.

Wang, X. W. Zhang, C. -A., Wang, P. L. et al. 2012c. Enhanced performance of biodegradable poly (butylene succinate)/graphene oxide nanocomposites via in situ polymerization. *Langmuir,* 28:7091–7095.

Wei, T., Luo, G., Fan, Z. et al. 2009. Preparation of graphene nanosheet/polymer composites using in situ reduction-extractive dispersion. *Carbon,* 47:2290–2299.

Wu, Z. -S., Ren, W., Gao, L., Liu, B., Jiang, C., Cheng. H. -M., 2009. Synthesis of high-quality graphene with a pre-determined number of layers. *Carbon,* 47:493–499.

Xu, F., Deng, M., Li, G., Chen, S., Wang, L., 2013. Electrochemical behavior of cuprous oxide-reduced graphene oxide nanocomposites and their application in non-enzymatic hydrogen peroxide sensing. *Electrochim. Acta,* 88:59–65.

Yan, X., Chen, J., Yang, J., Xiu, Q., Miele, P., 2010. Fabrication of free-standing, electrochemically active, and biocompatible graphene oxide-polyaniline and graphene-polyaniline hybrid papers. *ACS Appl. Mater. Interfaces,* 9:2521–2529.

Yang, G., Su, J., Gao, J., Hu, X., Geng, C., Fu, Q., 2013c. Fabrication of well-controlled porous forms of graphene oxide modified poly (propylene-carbonate) using supercritical carbon dioxide and its potential tissue engineering applications. *J. Supercrit. Fluids,* 73:1–9.

Yuan, J., Gao, H., Ching, C. B., 2011. Comparative protein profile of human hepatoma HepG2 cells treated with graphene and single-walled carbon nanotubes: An iTRAQ-coupled 2D LC-MS/MS proteome analysis. *Toxicol. Lett.,* 207:213–221.

Yang, J. -H. Lin, S. -H., Lee, Y. -D., 2012. Preparation and characterization of poly (L-lactide)-graphene composites using the in situ ring-opening polymerization of PLLA with graphene as the initiator. *J. Mater. Chem.,* 22:10805–10815.

Yang, K., Gong, H., Shi, X., Wan, J., Zhang, Y., Liu, Z., 2013b. *In-vivo* biodistribution and toxicology of functionalized nano-graphene oxide in mice after oral and intraperitoneal administration. *Biomaterials,* 34:2787–2795.

Yang, M., Javadi, A., Li, H., Gong, S., 2010a. Ultrasensitive immunosensor for the detection of cancer biomarker based on graphene sheet. *Biosens. Bioelectr.,* 26:560–565.

Yang, X., Tu, Y., Li, L., Shang, S., Tao. X. -M., 2010b. Well-dispersed chitosan/graphene oxide nanocomposites. *ACS Appl. Mater. Interfaces,* 2:1707–1713.

Yang, X., Zhang, X., Ma, Y., Huang, Y., Wang, Y., Chen, Y., 2009. Super paramagnetic graphene oxide-Fe3O4 nanoparticles hybrid for controlled targeted drug carriers. *J. Mater. Chem.,* 19:2710–2714.

Yang, Y., Asin, A. M., Tang, Z., Du, D., Lin, Y., 2013a. Graphene based materials for biomedical applications. *Mater. Today,* 16:365–373.

Zhang, B., Lee, W. H., Piner, R. et al. 2012. Low-temperature chemical vapor deposition growth of graphene from toluene on electropolished copper foils. *ACS Nano,* 6:2471–2476.

Zhang, R., Hummelgard, M., Lv, G., Olin, H., 2011a. Real time monitoring of the drug release of rhodamine B on graphene oxide. *Carbon,* 49:1126–1132.

Zhang, S., Yang, K., Feng, L., Liu, Z., 2011b. *In-vitro* and *in-vivo* behaviors of dextran functionalized graphene. *Carbon,* 49:4040–4049.

Zhang, X., Meng, L., Liu, Q., Fei, Z., Dyson, P. J., 2009. Targeted delivery and controlled release of doxorubicin to cancer cells using modified single wall carbon nanotubes. *Biomaterials,* 30:6041–6047.

Zhao, X., Liu, L., Li, X., Zeng, J., Jia, X., Liu, P. 2014. Biocompatible graphene oxide nanoparticle-based drug delivery platform for tumor microenvironment-responsive triggered release of doxorubicin. *Langmuir,* 30:10419–10429.

Zheng, M., Gao, F., Wang, Q., Cai, X., Jiang, S., Huang, L., 2013. Electrocatalytical oxidation and sensitive determination of acetaminophen on glassy carbon electrode modified with graphene–chitosan composite. *Mater. Sci. Eng. C,* 33:1514–1520.

Zhong, X., Yuan, R., Chai. Y. -Q., 2012. Synthesis of chitosan-Prussian blue-graphene composite nanosheets for electrochemical detection of glucose based on pseudobienzyme channeling. *Sens. Actuat. B,* 162:334–340.

Zhou, H., Zhao, K., Li, W. et al. 2012. The interactions between pristine graphene and macrophages and the production of cytokines/chemokines via TLR- and NFkB-related signaling pathways. *Biomaterials,* 33:6933–6942.

Zhu, S., Guo, J., Dong, J. et al. 2013. Sonochemical fabrication of Fe_3O_4 nanoparticles on reduced graphene oxide for biosensors. *Ultrasonics Sonochem.,* 20:872–880.

CHAPTER 24

Application of Polymer/Nanocomposites for Sensitive Electronic Device Packaging

Issues and Challenges Involved in Achieving Ultralow Barrier Property

Satyajit Gupta, S. Sindhu, S. Saravanan, Praveen
C. Ramamurthy, and Giridhar Madras

CONTENTS

24.1 INTRODUCTION

Barrier materials are an important class of materials that have versatile applications. One of the applications is in the field of organic devices. The flexible organic electronics have been an important and interesting research domain for the past few decades.[1-4] Due to the flexibility, lightweight, and the ability to work under diffuse sunlight, organic devices have an exceptional scope for the future. One of the major problems in the introduction of flexible electronics and in particular organic photovoltaic devices commercially in the market is their limited lifetime. This is due to the degradation of the organic conducting polymers (such as polyphenylvinylene) and other small organic molecules leading to device failure (a schematic of an organic photovoltaic device is shown in Figure 24.1), under atmospheric conditions, which is mainly caused by the intrusion of moisture and oxygen into the active device layers. Hence, barrier materials are essential to reduce the exposure of these devices to moisture and oxygen in the atmosphere, which will enhance the lifetime and reduce the degradation of these materials. There are other factors (apart from degradation) contributing to this decrease in efficiency as follows:

1. Photogenerated excitons that are strongly bound, resisting dissociation into separate charges

2. Low charge carrier mobility ($\mu \sim 10^{-3}$–10^{-5} cm^2/V s)

3. Short exciton diffusion length (LD $\sim$ 5–10 nm)

4. The deposition of a metal cathode (usually Ag or Al) that can cause defects at the acceptor/cathode interface

5. The interface stability and physical changes in morphology due to thermodynamic instability

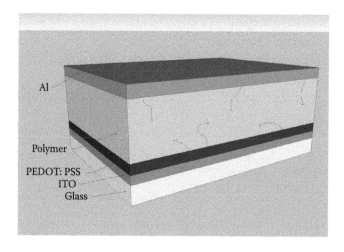

FIGURE 24.1 Schematic of degradation in organic devices. PEDOT:PSS, poly(3,4-ethylenedioxythiophene) polystyrene sulfonate; ITO, indium tin oxide.

The efficiency of organic photovoltaics has reached 10.6% (Konarka Technologies, Inc., Massachusetts) as opposed to 45% efficiency observed in inorganic photovoltaics. Thus, there is a scope for the improvement of organic photovoltaics because of their simple processability, flexibility, and lightweightness. Commercialization is basically based on achieving better efficiencies suitable for optimal device performance following some strategies listed in Table 24.1 and also increasing the stability of the device throughout its lifetime. Apart from increasing the efficiency of the device, increasing the device lifetime can be looked upon as an intriguing challenge as it offers larger scope to maintain the device performance for longer periods and to mitigate the problem of device instability/failure, caused by external factors such as ultraviolet (UV) light, environmental conditions, and agents such as moisture, oxygen, and other reactive gases.

There are many factors and mechanisms that have been proposed for the degraded performance of organic devices,[5] finally leading to device failure, which is caused due to the oxidation of low work function cathode materials such as Al and Ca, and the delamination of the active layer. In addition, the diffusion of the emitting layer material, electrical shorts, electrochemical reactions at the electrode interfaces, oxygen-activated photochemical damage,[6–8] and oxidation of polymers by oxygen entering through the grain boundaries in aluminum electrode and oxygen originating from indium tin oxide (ITO) also lead to device failure. Hence, proper encapsulation of active materials in organic electronics is required to assure the prevention of contact of the active materials with deterrents such as moisture, oxygen, carbon dioxide, and fluorocarbon vapors. It has been reported that the diffusion of moisture into the active organic layer increases the conductivity of sample,[9,10] and the adsorption of water alters the morphology of the organic film.[11] Low work function electrode materials such as aluminum and calcium also degrade in the presence of moisture to form layer of oxides (Al_2O_3 and CaO). In the absence of an efficient encapsulation, the active material degrades, reducing efficiency, and thus limits the lifetime of the device. The extent of allowance given to the gases depends on the deterring nature of the gas and the active materials used. Therefore, there is a broad range of barrier permeation requirements for different active materials used in various applications.

In this chapter, we will discuss the advantages of single-layered polymer nanocomposites as barrier packaging materials over other present encapsulation strategies. We have also highlighted various nanomaterial-reinforced polymer composites as potential encapsulants for organic device packaging. This chapter also discusses methods available for measuring permeation.

TABLE 24.1   The Controlling Parameters for Organic Photovoltaics

Parameter	Strategies
Absorption and V_{OC} (open circuit voltage)	Band gap engineering utilizing a donor–acceptor–donor-type organic molecular architecture
J_{SC} (short circuit current)	Enhancement in mobility by film and device treatments, and device architecture (inverted structure)
FF (fill factor)	Tuning of the contacts and morphology for decreasing series resistance

24.2 ENCAPSULATION MATERIALS: ISSUES AND CHALLENGES

The primary and the most elementary understanding in barrier technology has originated from food/beverage/pharmaceutical packaging. Polymers, such as polyethylene (PE), polypropylene (PP), poly(vinyl alcohol) (PVA), poly(ethylene terephthalate) (PET), and poly(ethylene naphthalate) (PEN), are used for packaging applications in food and pharmaceutical industries as barrier materials. These flexible polymer matrices are also cost effective compared to metal foils. In the case of organic electronics, the most difficulty for the commercialization of flexible electronics into the market is the restricted lifetime, primarily due to the degradation. Hence, proper barrier materials are required to encapsulate the organic devices in order to protect from atmospheric moisture and oxygen. Encapsulation of the active materials in order to protect them against these deterrents has a wide range of possible attempted solutions during the last decade. The proposed mechanisms/techniques differ with the nature of the detrimental gas as well as with the extent to which the encapsulant is sought to be impermeable to these gases. In order to achieve an active device lifetime of >10,000 h, a water vapor transmission rate (WVTR) of ~10^{-6} g/m^2/day and an oxygen transmission rate (OTR) of ~10^{-3} cm^3/m^2/day is an unofficial standard for the organic light-emitting diode (OLED) industry, which is calculated from the amount of oxygen and water required to degrade the reactive cathode.

24.2.1 Major Requirements of Encapsulants

The major requirements of encapsulants are as follows:

1. *Hermetic encapsulation*: Encapsulation that needs to be free of pinholes and should adhere well to the surface of the active device component, in order to isolate it from the odd environment

2. *Conformal coating*: An uniform and homogeneous coating that protects the active component and keeps the shape/architecture intact after encapsulation

3. Good chemical, electrical, environmental, and thermal stability

4. Must accommodate for coefficient of thermal expansion mismatch between the encapsulant and the active component throughout the device lifetime

5. Compatibility with mass production techniques

6. Room-temperature deposition condition, in order to avoid the degradation of the active component

7. Minimum shrinkage of the encapsulant upon curing

8. Should not affect the active device characteristics such as electron–hole mobility and luminescence of OLEDs

Based on these requirements, various different materials have been used as enclosure materiale such as glass,[12] epoxy resins,[13,14] polyimides,[15] silicones,[16] ethylene vinyl acetate

(EVA),[17] and polyurethanes.[18] Obviously, none of these materials suffice the need of all the abovementioned requirements. Other polymers that can also be included in the list are nylon 6, PET, PEN, polyethersulfone (PES), parylene, polyacrylate, polyvinyl fluoride, polyvinylidene fluoride, hydroxyl polyester polyurethanes under the trade name Desmocoll® (Bayer AG, Germany), bismaleimide, surlyn, and cyanoester-containing epoxy.

Although glass is an ultrahigh barrier material, due to rigidity, it cannot fulfill the applications demanding the flexibility of the devices. Bare polymer films, even though flexible and transparent, have very high permeability values (WVTR of 10^{-1}~40 $g/m^2/day$ and OTR of 10^{-2}–10^2 $cm^3/m^2/day$),[19,20] and thus, they cannot be used as such in device packaging. In addition to the issue of designing encapsulant materials with ultralow permeability, another problem that exists is to measure accurately such lower magnitudes of permeability. For this reason, development of a better instrumental technique is required to measure the same at such a low level.

Organic devices can be enclosed by sealing the device under an inert condition using a metal and can be sealed by epoxy resin by UV-curing technique, and finally incorporate moisture scavengers into the system to remove water[21] that permeates through the matrix. However, these methods of encapsulation are complex. Glass ceramic films are currently used by the OLED industries for encapsulation application. Commercial barrier film Barix™ uses alternating layers of polymer film and few-nanometer-thick metal oxides as a multilayer protective film solution. An arrangement of inorganic layers with polymer foils has a great potential to reduce the permeability, but these sequences of different layers are too expensive. Other methods of encapsulation include the use of barrier-coated polymer films, having a transparent or opaque organic/inorganic multilayer structure. However, these encapsulation strategies are highly complex and time consuming when large area deposition is required.

Meyer et al.[21] studied the atomic layer deposition (ALD) method of depositing Al_2O_3/ZrO_2 nanolaminates for OLED encapsulation; they evaluated the use of the ALD method to deposit Al_2O_3 and Al_2O_3/ZrO_2 individually and achieved WVTRs up to 5×10^{-7} $g/m^2/$day. Their evaluation of barrier material by ALD deposition was dependent on the coating thickness and the temperature at which these inorganic layers were grown on the plastic substrates. Wang et al. described the use of multilayer hybrid encapsulation of PI/TiN/SS foil compared to glass encapsulation and correlated the measured time-dependent reflectivity of encapsulated calcium with oxygen and WVTRs. Their results show that the moisture and OTR of hybrid encapsulation are similar to those of glass-capped substrates.[22] Lee et al. made a multilayer barrier stack of plasma-enhanced chemical vapor-deposited CF_x layer, followed by magnetron cosputtering of Al_2O_3/SiO_2 for OLED encapsulation by optimizing the co-oxide composition and considering the surface roughness for defect-free CF_x deposition. The OLED device lifetime of 18,000 h that is comparable to conventional glass-encapsulated device[23] could be achieved by this method, but this process is highly energy extensive. Baik et al. used optically transparent SnO_2 films by ion beam-assisted deposition to encapsulate pentacene organic thin-film transistors (TFT)[24] and also achieved high device lifetimes.

Compared to multilayered encapsulation (Vitex technology) and ALD/molecular layer deposition-based technique for encapsulation, polymer/composite-based encapsulation is simpler, less expensive, and less energy intensive. However, polymer-based composite

TABLE 24.2   Barrier Properties of Various Materials

Material	WVTR (g/m²/day)	OTR (cm³/m²/day)
PET	3.9–17	1.7–7.7
PEN	7.3	3.0
PE	1.2–5.9	70–550
PP	1.2–5.9	93–300
PES	14	0.04
PI	0.4–21	0.04–17
Polystyrene	7.9–40	200–540
15 nm Al/PET	0.18	0.2–2.9

materials can be processed by solution casting, *in situ* curing, extrusion molding, and blow molding, and can easily be commercialized and integrated into the devices. Inorganic oxide-based thin films (Al_2O_3, SiO_x) also exhibit good barrier property, but generation of pinholes and cleavages prevents them to be used as a barrier material. An approach to use ultrabarrier thin films uses nanoparticle reinforcements in plastic substrates and in barrier oxide materials to reduce the multilayer architecture of encapsulation in devices. The presence of nanoparticles gives torturous pathways to moisture and oxygen, which increases the penetration time of these deterrent gases.

It is critical that the barrier property can be a function of the permeate species. Permeation through the polymers occurs by various mechanisms for different gas molecules. Oxygen, which is nonpolar, can interact with the polymer matrix differently than water, which is polar and condensable.[25,26] It is also evident that the permeation mechanisms through barrier films can be different for oxygen and moisture.[21] The WVTR and OTR of some of the polymers are listed in Table 24.2.

24.3 ULTRALOW PERMEABILITY MEASUREMENT UNIT: TECHNIQUES AND CHALLENGES

An obstacle is the development of these economic, flexible, ultralow permeable barrier materials, which will have applications in versatile fields. In addition, the required transmission rates for organic devices are very small and also several orders of magnitude lower than what can be measured using commercially available equipment designed for this purpose. The Mocon measurement technique, which is an industry standard, does not provide accurate measurements at the lower levels of permeability that are required for organic devices.[27] Other than Mocon, permeation rates can also be measured by Brugger,[28] but this method also cannot measure in the limit of ultralow permeability. Spectroscopic techniques have been recently used for the moisture analysis in semiconductor industries. Instruments that are currently used for the measurement of trace amount of moisture are Fourier transform infrared spectroscopy (FTIR), tunable diode laser absorption spectroscopy (TDLS), intracavity laser spectroscopy, electron impact ionization mass spectrometry, and atmospheric pressure ionization mass spectrometry. These techniques can be adapted to measure the moisture that permeates through the barrier layer. Most of these methods need to be developed and integrated into this application (Table 24.3).

TABLE 24.3 WVTRs and OTRs of Different Techniques

Technique	WVTR (g/m²/day)	OTR (cm³/m²/day)
Mocon limit	0.0005	0.005
FTIR	200	–
TDLS	200	–
OPVD requirement	10^{-6}	10^{-5}–10^{-3}

The evaluation of the encapsulant is carried out either by evaluating the device under test (DUT) or by evaluating the permeation response of the material. Much of the research is done by evaluating the performance of the DUT after encapsulation. The performance of DUT can change with time and may not be only dependent on the performance of the encapsulant but can also depend on other parameters. At present, the direct evaluation of the encapsulant is carried out by the following techniques:

- Differential pressure method (Mocon)
  - Equal pressure method employing infrared sensor (ASTM F1249)
  - Equal pressure method employing coulometric sensor
- Calcium (degradation) test
- Gravimetric method
- Spectroscopic method

These techniques used for the measurement of permeability are briefly discussed in Sections 24.3.1 through 24.3.4.

24.3.1 Differential Pressure Method

In this method, a high-pressure chamber and a low-pressure chamber are separated by the test sample for the measurement of permeability. Pressure gauges are used in each chamber to determine the pressure. The high-pressure chamber side is filled with test gas at 0.1 MPa, and the low-pressure chamber side, having a known volume, is maintained under vacuum. After sealing with the sample, the pressure increase in the low-pressure side is measured with the pressure gauge. Thus, the gas permeability through the sample as a function of time can be determined.

24.3.1.1 Equal Pressure Method Employing Coulometric Sensor

Coulometry is a methodology for the measurement of unknown concentration of any analyte by converting it from one oxidation state to another. This is an absolute measurement technique like gravimetry or titrimetry. By using the same principle as in ASTM 1249, using an absolute coulometric sensor instead of the concentration-based pulse modulated infrared sensor, the sensitivity in measuring the WVTR can be increased to 5×10^{-4} g/m²/day.

24.3.2 Calcium Degradation Test

This technique is based on the degradation of calcium metal reacting with atmospheric gases. In this technique, a thin calcium layer is deposited on a cleaned glass substrate under inert atmosphere (inside glove box), and then the aluminum is deposited over that as electrode (Figure 24.2). The barrier material is then used to seal the calcium, acting as a moisture sensor (Figure 24.3). Then the encapsulated device is exposed to a humidity chamber. The water vapor passing through the barrier film oxidizes the opaque calcium metal to form translucent CaO and Ca(OH)$_2$, respectively. Then either the optical transmission or the resistance[29] is measured as a function of time, and the WVTR is estimated. Langereis et al. studied the plasma-assisted ALD of Al$_2$O$_3$ for moisture barriers. In this study, they used the calcium degradation test, and they observed that the barrier properties improved with the decrease in the substrate temperature and measured the WVTR of 5×10^{-3} g/m^2/day.[30] OLED industries use the calcium test in order to verify the barrier properties of encapsulants.[31]

24.3.3 Gravimetric Method

Gravimetric method was an initiative by Paul Luchinger in his patent work on the instrument for gravimetric moisture determination.[32] This instrument consists of a test cell with

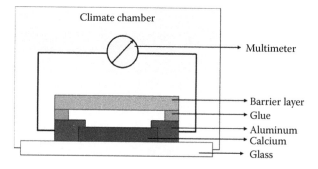

FIGURE 24.2   Schematic setup of calcium test for measurement of permeability of a barrier material.

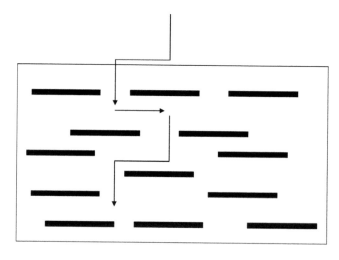

FIGURE 24.3   Schematic of retardation of a permeant species due to the torturous pathway.

a halogen lamp as radiator and a sample receiver, which is connected to a gravimetric weighing cell. To accurately predict the moisture content, the sample is initially dried by heating and the weight of the released moisture in the test compartment/cell is determined by the weighing cell.

Zhang et al. studied WVTR for OLEDs (Alq_3) encapsulated by SiO_X/Si_XN_Y thin films by observing the changes in the photoluminescent intensity of organic materials caused due to the reactivity of penetrated water vapor with the device organics. The resulting changes are observed by recording the photoluminescent spectra before and after the test condition (85% RH and 38°C) because of the sensitivity of organic materials toward moisture.[33]

These existing testing methods are inefficient in obtaining leakage-free measurements. Thus, they produce unreliable results in the ultralow range of permeate concentration. Lowest reliable quantity that can be measured is in the range of 5×10^{-4} g/m²/day, but a WVTR of 1×10^{-6} g/m²/day is the desirable order of sensitivity in permeability testing equipments used for organic electronics.

24.3.4 Spectroscopic Method

The other methods of measurements include the spectroscopic method, which provide faster response times and high sensitivities.[34] Quantitative detection by this method is possible by following Beer–Lambert's law from which the concentration of an analyte can be obtained using the following equation: $A = \varepsilon \times C \times L$, where A is the absorbance, ε is a constant (specific absorptivity constant), C is the concentration of the analyte, and L is the path length. A precalibration is required to determine the concentration of an unknown species.

24.4 POLYMER NANOCOMPOSITES AS BARRIER MATERIALS: ADVANTAGES AND LITERATURE REVIEW

The barrier properties of polymer-based composite materials have been examined by various groups. Addition of the high aspect ratio to uniformly distributed and exfoliated/intercalated layered nanosilicates considerably reduces the gas permeation in the polymer matrix by giving a torturous pathway to permeating/penetrating moisture and oxygen.[35–37] Hence, the barrier properties of the polymers can be modified by incorporating an appropriate inorganic phase, as it reduces the diffusion of the incoming gas molecules, by providing an enhanced and extended pathway (Figure 24.3).[38,39]

Macosko et al. showed that thermoplastic polyurethane reinforced with exfoliated graphene exhibits a reduction in N_2 permeability.[40] Triantafyllidis et al.[41] synthesized epoxy–clay composite film, having a high oxygen barrier property and observed a reduction in O_2 permeability by 2–3 orders compared to the neat epoxy polymer. The dispersion of nanofillers not only improves the barrier property but also improves the thermomechanical properties of the composite matrix.

Generally, there are two possible mechanisms that can explain the gas barrier property for polymer/clay nanocomposites. First, inorganic nanoclay layers (impermeable to gases), having high degree of dispersion in the polymer matrix, provide a tortuous pathway (an elongated diffusive pathway) that retards the movement of gas molecules through the composite. Second, the exfoliated and intercalated clay layers constrain the polymer chain motion, which reduces the coefficient of diffusion of the impermeant gas molecules. Experimental and theoretical studies[42] have proposed that the specific aspect ratio of fillers is the key factor in determining the gas permeation mechanism through the nanocomposites. The filler having the higher aspect ratio will have the higher gas barrier property. Many polymers have been used in preparing the nanoclay composites such as nylon 66, polyacrylates, polyurethanes, and epoxy. A reduction in permeability was observed by Barrer et al.,[43] when various weight percentages of micron-sized ZnO particles were incorporated into natural rubber.

Kim et al. examined the moisture barrier properties of epoxy-based nanocomposites with various compatibilized organoclays.[44] They observed that the type of organoclay determines the rate of moisture absorption. This is because the varying interlayer distances in the composite determines the diffusion path of molecules. The diffusivity of moisture decreased with increasing clay content, which was in accordance with the tortuous path model. In an optically clear PVA/sodium montmorillonite (MMT) nanocomposite, a 5 wt.% of MMT reduced water permeability by 60%.[45] Chang et al. observed that 2 wt.% of dodecyl amine (C_{12})- and hexadecyl amine (C_{16})-modified MMT in polyimide showed about 94%–95% reduction in O_2 permeability compared to that in pure polyimide along with enhancement in the tensile property.[46] Water vapor permeability had also shown a reduction in the permeability behavior. The modified organoclay addition (6 wt.% of dodecyl amine [C_{12}]-modified MMT) reduced the rate of relative permeability to half of that of pristine polyimide and hexadecyl amine (C_{16})-modified MMT reduced to less than one-third of that of the pristine polyimide.[46]

Polyimide/clay hybrid composite[47] material has been prepared with various different sizes of clay materials (Hectrite [460 A°], saponite [1650 A°], MMT [2180 A°], and synthetic mica [12300 A°]). They observed that as the length of the inorganic clay increased, the properties of the hybrid also improved. In the case of polyimide/synthetic mica, a loading of 2 wt.% of mica reduced the coefficient of permeability to more than 90% compared to pure polyimide. In addition to the reduction in permeability coefficient, they also observed a reduction in thermal expansion coefficient. Storage elastic modulus was also observed to increase with the increase in the length of clay material. Organically modified mica-type silicate/poly(ε-caprolactone) nanocomposite also showed a significant reduction in water vapor permeability, which was due to the dispersed silicate layers of high aspect ratios.[48] Reduction of water vapor permeability by an order of magnitude was observed at 4.8% silicate by volume. A 2 wt.% addition of MMT to a polyimide matrix reduced the permeability coefficients of various gases to values that are lower than a half of that in polyimide, as observed by Yano et al.[49] They observed from transmission electron microscopy (TEM) that MMT is oriented in parallel to the film surface, which increases the diffusion pathway of gases. The thermal expansion coefficient was found to decrease with the increase in the MMT content in the matrix. This is a critical property in the context of electronic device packaging, as the coefficient of thermal expansion of the encapsulant should match with that of the device. Recently, it has been shown that synthetic HCE (Li-hectorites) is a good filler for gas barrier applications.[50]

Silicone/organo-odified clay loading of 8.0 wt.% reduced only 25% of oxygen permeability of that of the pure elastomer, whereas the tensile properties improved.[51] This little reduction in the barrier property was attributed to the random distribution of clay layers in the polymer matrix, which had been observed from TEM imaging. The effect of interactions between sodium MMT (MMT-Na$^+$) and modified PVA was observed by Grunlan et al.,[52] in which they found that a 10 wt.% concentration of clay content was sufficient to achieve oxygen permeability as low as 0.001 cc mil/m^2/day (mil = 25.4 μm) at 55% relative humidity.

Dai et al. studied epoxy-organoclay (Na$^+$-MMT) composite for their barrier properties at different compositions.[53] They drop casted the epoxy-organoclay composite in N,N-dimethylacetamide onto the substrate and cured it for 6 h at 140°C. Their study showed that the nanocomposite with 7 wt.% Na$^+$-MMT-modified dodecyltriphenyl-phosphonium bromide in epoxy exhibited better barrier properties.

Epoxy resin (siloxane-modified) with layers of MMT had been prepared by a thermal ring-opening polymerization using 1,3-bis(3-aminopropyl)-1,1,3,3-tetramethyldisiloxane as a curing agent.[54] The resultant solution was drop casted onto the substrate and cured at 120°C for 5 h, and the cast film was used for barrier property characterization. These two groups worked on the epoxy-clay nanocomposites, but the curing method is not suitable to encapsulate a photovoltaic cell. The ideal material for photovoltaic encapsulants is something that can be cured faster at room temperature.

Decker et al. photopolymerized multifunctional acrylate or epoxy monomers containing ~3 wt.% clay, which provided cross-linking at room temperature under UV light, which forms chemical and heat-resistant composite.[55] Ceccia et al. showed *in situ* intercalative polymerization along with the UV-curing technique to synthesize epoxy-organoclay composite.[56] Ethylene glycol to epoxy-organophilic MMT was added to increase photopolymerization curing kinetics and improve organoclay intercalation. The dispersion was coated onto a substrate that was exposed to UV light of 20 mW/cm^2 for 2 min for curing. However, this UV-curing methodology for *in situ* device encapsulation is not possible, as UV causes degradation of the organic active material. Canto et al. showed that epoxy functionalized by siloxane formed hybrid films with good homogeneity and coating properties.[57] Further, the barrier properties of these materials were not determined.

Polyesteramide containing octadecylammonium-treated MMT clays (5 and 13 wt.%) was prepared by injection molding and studied by Hedenqvist et al.,[58] and found that the oxygen barrier property improved along with an increase in in-plane stiffness and strength caused due to higher degree of clay exfoliation and their parallel orientation to the film. A 80% reduction in permeability with respect to the pure polymer matrix was observed with 13 wt.% of clay. Zielecka et al. studied silicone-containing polymer matrices as coating materials,[59] using organic functional modifiers to improve their coating properties along with water resistance.

In PVA (PVOH), low O$_2$ permeability is observed to have strong inter- and intramolecular cohesive energy arising out of –OH functionalities.[60] However, a fluorinated polymer such as polytetraflouroethylene repels water due to its hydrophobicity. Computational calculations for the transport properties of a small molecule in polymers are very important for the designing of appropriate materials. Hence, molecular modeling simulation tools can be used to determine the diffusion properties of the small molecules such as oxygen, water vapor in the polymers, and nanocomposites. Meunier et al. used molecular dynamics (MD) simulations on 3D periodic plots to determine the self-diffusion coefficient of small gas molecules in amorphous polymer over a temperature range theoretically.[61] By arriving at optimum simulation results, from rigorous calculations, which would match with experimental results for diffusivities, MD can be used for selecting polymer nanocomposite economically.

24.5 POLYMER NANOCOMPOSITES AS ENCAPSULATION MATERIALS: A NONMETALLIZATION APPROACH

The diffusion of gas molecules into the polymer is dependent on factors such as polarity, hydrophobicity, molecular weight, and architecture of a polymer matrix. The choice of the polymer in this context for the encapsulation is key issue. Variations in polymer synthesis, molecular weight, branching, film thickness, tacticity, and quantity of nanofillers lead to different barrier properties. The choice of polymer and nanocomposites will not only influence the ultimate barrier properties but also affect the encapsulated device performance (Figure 24.4).

Polyisobutene and EVA have been studied for organic device encapsulation.[62,63] Yan et al. used cross-linked PVA as an encapsulant for copper-phthalocyanine-based TFT devices, which were found resistant to air for 1 year based on their unchanged TFT characteristics.[64] Nguyen et al. used cellulose acetate and polyvinylidene chloride (PVDC) films as barrier materials to protect OLEDs of poly(p-phenylene vinylene) (PPV) and poly[2-methoxy-5-(2-ethylhexyloxy)-1,4-phenylenevinylene] (MEH-PPV) films. They evaluated changes observed in the optical properties of the PPV films after 28 days of exposure to air and observed that there were no significant improvements, except that the protective layer showed delay in the degradation pathway of oxygen through the matrix.[65] Kim et al. used photopolymerizable polyacrylate

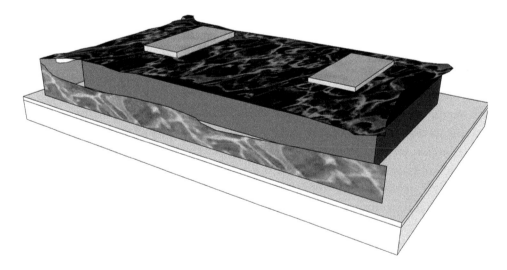

FIGURE 24.4   Schematic of monolithic encapsulation.

films to encapsulate OLEDs using spin coating and noted that by changing initiator concentration and film thickness, moisture permeation as observed from device characterization is significantly influenced only when thickness of the films was increased.[66] In another study, they used multilayer structures of polyacrylate adhesive and aluminum for simple adhesive-based lamination encapsulation process and observed increased lifetime of the device by maintaining the electroluminescence characteristics of the device.[67] Cytop™ is a transparent, hydrophobic perfluorinated polymer, which had been shown to be a promising candidate for the encapsulation of organic devices.[68] This polymer is solution processable and hence easy to deposit. Epoxy and siloxane[69] are a class of polymers, having several advantages, such as easy processability as well as chemical modifiability, wide range of curing temperatures, low shrinkage, high thermal stability, low volatility, UV resistant, and hydrophobic, which have been used in various commercial applications. Polymer/nanocomposite materials based on various silicone polymers, using organo-functionalized alumina and silica nanoparticles, have also been developed, which can be explored for the encapsulation of organic devices.[70-73] In these studies, amine and allyl functionalized alumina nanoparticles and allyl functionalized mesoporous silica particles were used as filler materials for the silicone matrix (epoxy and hydride-terminated silicone polymer was used as a polymer matrix). Even though the oxygen barrier property is not desirable for silicone polymer, the intended inorganic phase was supposed to provide a torturous pathway for the incoming oxygen molecules. The permeability properties of those composites were not measured, but a good thermal stability and a good mechanical flexibility were observed, which are critical for an encapsulant. The loading of the filler was optimized, in terms of thermal and mechanical properties, for the encapsulation application.

It has also been observed that inclusion of inorganic oxide nanoparticles in a hybrid organic nanocomposite matrix reduces the gas permeability[74] because the inclusion of alumina retards the gas penetration. This also increases the adhesive strength as well as transparency and enhances the hardness but decreases the refractive index value. Uniform dispersion of surfactant-coated TiO_2 (40 nm diameter) in the epoxy matrix having a refractive index ($n = 1.67$ at $\lambda = 500$ nm) can be used for encapsulation of OLEDs.[75] It was also observed that the surfactant-coated TiO_2 showed lower agglomeration inside the epoxy matrix than that of pristine TiO_2. Saran (a copolymer of vinylidene chloride and acrylonitrile) with boron nitride (BN) nanotube composites, having an excellent transparency in the visible region and thermally stable, were prepared by a solution casting method (methyl ethyl ketone as a solvent) and have been shown that the composites can be good encapsulating materials[76] for OPVDs. In this study, the barrier performance was evaluated by encapsulating poly(3-hexylthiophene), which is an active material for organic device. Silica nanoparticle-reinforced transparent organic–inorganic hybrid nanocomposites for OLED encapsulant application have been prepared, having a WVTR of 0.24 g/m²/day,[38] as estimated from Ca degradation test. Chen et al. reported a transparent and low-moisture permeable, photo-curable *co*-polyacrylate/silica nanocomposite matrix by following an *in situ* sol–gel process for encapsulation of OLEDs.[77] Moisture and oxygen reactive inorganic semiconductor materials such as zinc oxide, dispersed in polymer matrix, have also been used as encapsulants for organic devices.[38]

Biodegradable polylactic acid/Cloisite 20A composite was prepared by extrusion molding, for organic solar cell encapsulation,[78] and was anticipated to reduce the permeation of oxygen and moisture, but no improvement in performance or stability of the device was observed. This indicates that this could be due to the improper dispersion of clay or that the selected clay may not be compatible for the process. Therefore, dispersion of the nanomaterials in the polymer matrix is a key variable that can influence many properties. Flexible and photostable PVA films have good transparency, oxygen barrier property[79] and can be used for solar cell encapsulation.[80] Therias et al. showed that photooxidative degradation in PVA films can occur up to a surface level of <5 µm.[63] PVA/clay nanocomposites have been synthesized using a solution casting method (using water as a solvent) and used for ultrahigh barrier multilayer encapsulation of organic solar cells.[81] In this method, the clay content (MMT-Na$^+$) up to 10 w.t% in a PVA matrix was loaded to obtain a transparent and gas barrier material. It was also observed that 5 wt.% of MMT-Na$^+$ in PVA reduces the helium, oxygen, and moisture permeability, and an encapsulated organic solar cell with this composition showed an improved performance.

24.6 CONCLUSIONS

A significant issue preventing the commercialization of conducting polymer-based devices is encapsulation. A possible solution may be the use of suitable polymer nanocomposite-based barrier layers. The choice of polymer as well as inorganic oxide (filler) phase is important, and the method of the fabrication determines the encapsulation property. Various groups have reported different combinations of polymers as well as filler matrices and used different procedures of fabrication. This chapter summarizes the various studies in this field. In addition, the field needs an appropriate measurement technique for the evaluation of the permeability values of the barrier materials at an ultralow level. The development of the field of the encapsulation cannot progress independently, until a development of a proper measurement technique is envisaged. This chapter also discusses the current methods available for the measurement of ultralow permeability values of materials.

ACKNOWLEDGMENTS

SG acknowledges the Council of Scientific & Industrial Research, New Delhi, India, for financial support and fellowship. The authors also gratefully acknowledge the financial support from department of science and technology (DST) No. SR/S3/ME/022/2010-(G). This work was also partially supported by the Ministry of Communication and Information Technology under a grant for the Centre of Excellence in Nanoelectronics, Phase II.

REFERENCES

1. Brabec, C. J.; Sariciftci, N. S.; Hummelen, J. C. Plastic solar cells. *Adv. Funct. Mater.*, **2001**, *11*, 15–26.
2. Spanggaard, H.; Krebs, F. C. A brief history of the development of organic and polymeric photovoltaics. *Sol. Energy Mater. Sol. Cells*, **2004**, *83*, 125–146.
3. Ramamurthy, P. C.; Malshe, A. M.; Harrell, W. R.; Gregory, R. V.; McGuire, K.; Rao, A. M. Polyaniline/single-walled carbon nanotube composite electronic device. *Solid State Electronics*, **2004**, *48*, 2019–2024.

4. Ramamurthy, P. C.; Harrell, W. R.; Gregory, R. V. The influence of solvent content in polyaniline on conductivity and electronic device performance. *Electrochem. Solid State Lett.*, **2003**, *6(9)*, G113–G116.
5. Jorgensen, M.; Norrman, K.; Krebs, F. C. Stability/degradation of polymer solar cells. *Sol. Energy Mater. Sol. Cells.*, **2008**, *92*, 686–714.
6. Cumpston, B. H.; Parker, I. D.; Jensen, K. F. In situ characterization of the oxidative degradation of a polymeric light emitting device. *J Appl. Phys.*, **1997**, *81*, 3716–3720.
7. Bliznyuk, V. N.; Carter, S. A.; Scott, J. C.; Klarner, G.; Miller, R. D.; Miller, D. C. Electrical and photoinduced degradation of polyfluorene based films and light-emitting devices. *Macromolecules*, **1999**, *32*, 361–369.
8. Scurlock, R. D.; Wang, B.; Ogilby, P. R.; Sheats, J. R.; Clough, R. L. Singlet oxygen as a reactive intermediate in the photodegradation of an electroluminescent polymer. *J. Am. Chem. Soc.*, **1995**, *117*, 10194–10202.
9. Ye, R. B.; BaBa, M.; Suzuki, K.; Ohishi, Y.; Mori, K. Effects of O_2 and H_2O on electrical characteristics of pentacene thin film transistors. *Thin Solid Films*, **2004**, *464–465*, 437–440.
10. Qiu, Y.; Hu, Y. C.; Dong, G. F.; Wang, L. D.; Xie, J. F.; Ma, Y. L. H_2O effect on the stability of organic thin-film field-effect transistors. *Appl. Phys. Lett.*, **2003**, *83*, 1644–1646.
11. Zhu, Z. T.; Mason, J. T.; Dieckmann, R.; Malliaras, G. G. Humidity sensors based on pentacene thin-film transistors. *Appl. Phys. Lett.*, **2002**, *81*, 4643–4645.
12. Decroux, H. M.; Van Den Vlekkert, H. H.; De Rooij, N. F. Glass encapsulation of chemfets: A simultaneous solution for chemfet packaging and ion selective membrane fixation. *Proceedings of the Second International Meeting on Chemical Sensors*, Bordeaux, France, **1986**, pp. 403–406.
13. Sibbald, A.; Whalley, P. D.; Covington, A. K. A miniature flow-through cell with a four-function chemfet integrated circuit for simultaneous measurements of potassium, hydrogen, calcium and sodium ions. *Anal. Chim. Acta.*, **1984**, *159*, 47–62.
14. Dumschat, C.; Müller, H.; Rautschek, H.; Timpe, H. J.; Hoffmann, W.; Pham, M. T.; Hüller, J. Encapsulation of chemically sensitive field-effect transistors with photocurable epoxy resins. *Sens. Actuators B: Chem.*, **1990**, *2*, 271–276.
15. Ho, N. J.; Kratochvil, J. Encapsulation of polymeric membrane-based ion-selective field effect transistors. *Sens. Actuators*, **1983**, *4*, 413–421.
16. Shimada, K.; Yano, M.; Shibatani, K.; Komoto, Y.; Esashi, M.; Matsuo, T. Application of catheter-tip ISFET for continuous "in vivo" measurement." *Med. Biol. Eng. Comput.*, **1980**, *18*, 741–745.
17. Sadeghi, M.; Khanbabaeia, G.; Saeedi Dehaghani, A. H.; Sadeghi, M.; Aravand, M. A.; Akbarzade, M.; Khatti, S. Gas permeation properties of ethylene vinyl acetate–silica nanocomposite membranes. *J. Membrane Sci.*, **2008**, *322*, 423–428.
18. Corcione, C. E.; Prinari, P.; Cannoletta, D.; Mensitieri, G.; Maffezzoli, A. Synthesis and characterization of clay-nanocomposite solvent-based polyurethane adhesives. *Int. J. Adhes. Adhes.*, **2008**, *28*, 91–100.
19. Lewis, J. S.; Weaver, M. S. Thin-film permeation-barrier technology for flexible organic light-emitting devices. *IEEE J. Select. Top. Quant. Electron.*, **2004**, *10*, 45–57.
20. Leterrier, Y. Durability of nanosized oxygen-barrier coatings on polymers. *Prog. Mater. Sci.*, **2003**, *48*, 1–55.
21. Riedl, T.; Winkler, T.; Schmidt, H.; Meyer, J.; Schneidenbach, D.; Johannes, H. H.; Kowalsky, W.; Weimann, T.; Hinze, P. Reliability aspects of organic light emitting diodes. *IEEE*, **2010**, 3F 1.1–3F 1.7.
22. Wu, Z.; Wang, L.; Chang, C.; Qiu, Y. A hybrid encapsulation of organic light-emitting devices. *J. Phys. D: Appl. Phys.*, **2005**, *38*, 981–984.
23. Wong, F. L.; Fung, M. K.; Ng, C. Y.; Ng, A.; Bello, I.; Lee, S. T.; Lee, C. S. Co-sputtered oxide thin film encapsulated organic electronic devices with prolonged lifetime. *Thin Solid Films*, **2011**, *520*, 1131–1135.

24. Kim, W. J.; Koo, W. H.; Jo, S. J.; Kim, C. S.; Baik, H. K.; Lee, J.; Im, S. Enhancement of long-term stability of pentacene thin-film transistors encapsulated with transparent SnO_2. *Appl. Surf. Sci.*, **2005**, *252*, 1332–1338.

25. Erlat, A. G.; Spontak, R. J.; Clarke, R. P.; Robinson, T. C.; Haaland, P. D.; Tropsha, Y.; Harvey, N. G.; Vogler, E. A. SiOx gas barrier coatings on polymer substrates: Morphology and gas transport considerations. *J. Phys. Chem. B*, **1999**, *103*, 6047–6055.

26. Ouyang, H.; Wu, M. T.; Ouyang, W. The study of mass transport of acetone in polycarbonate. *J. Appl. Phys.*, **2004**, *96*, 7066–7070.

27. Park, J. S.; Chae, H.; Chung, H. K.; Lee, S. I. Thin film encapsulation for flexible AM-OLED: A review. *Semicond. Sci. Technol.*, **2011**, *26*, 034001 (8 pp).

28. Charton, C.; Schiller, N.; Fahland, M.; Hollander, A.; Wedel, A.; Noller, K. Development of high barrier films on flexible polymer substrates. *Thin Solid Films.*, **2006**, *502*, 99–103.

29. Paetzolda, R.; Winnacker, A.; Henseler, D.; Cesari, V.; Heuser, K. Permeation rate measurements by electrical analysis of calcium corrosion. *Rev. Sci. Instrum.*, **2003**, *74*, 5147–5150.

30. Langereis, E.; Creatore, M.; Heil, S. B. S.; Van de Sanden, M. C. M.; Kessels, W. M. M. Plasma-assisted atomic layer deposition of Al_2O_3 moisture permeation barriers on polymers. *Appl. Phys. Lett.*, **2006**, *89*, 081915

31. Nisato, G.; Bouten, P. C. P; Slikkerveer, P. J.; Bennett, W. D; Graff, G. L; Rutherford, N; Wiese, L. Evaluating high performance diffusion barriers: The calcium test *Proc. Int. Display workshop/Asia Display*, **2001**, 1435–1438.

32. Paul, L. Measuring instrument for gravimetric moisture determination. US 7591169 B2, 2009.

33. Zhang, H.; Zhang, M.; Sang, R.; Li, L.; Wee, B.; Feng, T.; Zhang. J. New permeation rate measurement for thin film encapsulation of organic light emitting diodes. *12th Int. Conf. Electron. Packaging Technol. High Density Packaging*, **2011**, *12*, 1179–1181.

34. Funke, H. H.; Grissom, B. L.; McGrew, C. E.; Raynor, M. W. Techniques for the measurement of trace moisture in high-purity electronic specialty gases. *Rev. Sci. Instrum.*, **2003**, *74*, 3909–3933.

35. Picard, E.; Gerard, J. F.; Espuche, E. Water transport properties of polyamide 6 based nanocomposites prepared by melt blending: On the importance of the clay dispersion state on the water transport properties at high water activity. *J. Membrane Sci.*, **2008**, *313*, 284–295.

36. Xu, B.; Zheng, Q.; Song, Y.; Shangguan, Y. Calculating barrier properties of polymer/clay nanocomposites: Effects of clay layers. *Polymer*, **2006**, *47*, 2904–2910.

37. Choudalakis, G.; Gotsis, A. D. Permeability of polymer/clay nanocomposites: A review. *Eur. Polymer J.*, **2009**, *45*, 967–984.

38. Jin, J.; Lee, J. J.; Bae, B. S.; Park, S. J.; Yoo, S.; Jung, K. H. Silica nanoparticle-embedded sol–gel organic/inorganic hybrid nanocomposite for transparent OLED encapsulation. *Org. Electron.*, **2012**, *13*, 53–57.

39. Gupta, S.; Seethamraju, S.; Kesavan, A.; Ramamurthy, P. C.; Madras, G. Hybrid nanocomposite films of polyvinyl alcohol and ZnO as interactive gas barrier layers for electronics device passivation. *RSC Adv.*, **2012**. doi:10.1039/C2RA21714G.

40. Kim, H.; Miura, Y.; Macosko, C. W. Graphene/polyurethane nanocomposites for improved gas barrier and electrical conductivity. *Chem. Mater.*, **2010**, *22*, 3441–3450.

41. Triantafyllidis, K. S.; LeBaron, P. C.; Park, I.; Pinnavaia, T. J. Epoxy-clay fabric film composites with unprecedented oxygen-barrier properties. *Chem. Mater.*, **2006**, *18*, 4393–4398.

42. Solovyov, S. E. Reactivity of gas barrier membranes filled with reactive particulates. *J. Phys. Chem. B*, **2006**, *110*, 17977–17986.

43. Barrer, R. M.; Barrie, J. A.; Rogers, M. G. Heterogeneous membranes: Diffusion in filled rubber. *J. Polym. Sci., A: Polym. Chem.*, **1963**, *1*, 2565–2586.

44. Kim, J. K.; Hu, C.; Woo, R. S. C.; Sham, M. L. Moisture barrier characteristics of organoclay-epoxy nanocomposites. *Compos. Sci. Technol.*, **2005**, *65*, 805–813.

45. Strawhecker, K. E.; Manias, E. Structure and properties of poly (vinyl alcohol)/Na⁺ montmo-rillonite nanocomposites. *Chem. Mater.*, **2000**, *12*, 2943–2949.
46. Chang, J. H.; Park, K. M.; Cho, A. D. Preparation and characterization of polyimide nano-composites with different organo-montmorillonites. *Polym. Eng. Sci.*, **2001**, *41*, 1514–1520.
47. Yano, K.; Usuki, A.; Okada, A. Synthesis and properties of polyimide-clay hybrid films. *J. Polym. Sci A: Polym. Chem.*, **1997**, *35*, 2289–2294.
48. Messersmith, P. B.; Giannelis, E. P. Synthesis and barrier properties of poly(ε-caprolactone)-layered silicate nanocomposites. *J. Polym. Sci. A: Polym. Chem.*, **1995**, *33*, 1047–1057.
49. Yano, K.; Usuki, A.; Okada, A.; Kurauchi, T.; Kamlgalto, O. Synthesis and properties of poly-imide-clay hybrid. *J. Polym. Sci. Part A: Polym. Chem.*, **1993**, *31*, 2493–2498.
50. Möller, M. W.; Kunz, D. A.; Lunkenbein, T.; Sommer, S.; Nennemann, A.; Breu, J. UV-cured, flexible, and transparent nanocomposite coating with remarkable oxygen barrier. *Adv. Mater.*, **2012**, *24*, 2142–2147.
51. Lebaron, P. C.; Pinnavaia, T. J. Clay nanolayer reinforcement of a silicone elastomer. *Chem. Mater.*, **2001**, *13*, 3760–3765.
52. Grunlan, J. C.; Grigorian, A.; Hamilton, C. B.; Mehrabi, A. R. Effect of clay concentration on the oxygen permeability and optical properties of a modified poly(vinyl alcohol). *J. Appl. Polym. Sci.*, **2004**, *93*, 1102–1109.
53. Dai, C. F.; Li, P. R.; Yeh, J. M. Comparative studies for the effect of intercalating agent on the physical properties of epoxy resin-clay based nanocomposite materials, *Eur. Polym. J.*, **2008**, *44*, 2439–2447.
54. Yeha, J. U, Huanga, H. Y., Chena, C. L., Sua, W. F., Yu, Y. H. Siloxane-modified epoxy resin-clay nanocomposite coatings with advanced anticorrosive properties prepared by a solution dispersion approach. *Surf. Coatings Technol.*, **2006**, *200*, 2753–2763.
55. Decker, C.; Keller, L.; Zahouily, K.; Benfarhi, S. Synthesis of nanocomposite polymers by UV-radiation curing. *Polymer*, **2005**, *46*, 6640–6648.
56. Ceccia, S., Turcato, E. A., Maffettone, P. L., Bongiovanni, R. Nanocomposite UV-cured coat-ings: Organoclay intercalation by an epoxy resin. *Prog. Org. Coatings*, **2008**, *63*, 110–115.
57. Canto, C. F.; Prado, L. A. S.; Radovanovic, E.; Yoshida, I. V. P. Organic–inorganic hybrid materials derived from epoxy resin and polysiloxanes: Synthesis and characterization. *Polym. Eng. Sci.*, **2008**, *48*, 141–148.
58. Krook, M.; Morgan, G.; Hedenqvist, M. S. Barrier and mechanical properties of injection molded montmorillonite/polyesteramide nanocomposites. *Polym. Eng. Sci.*, **2005**, *45*, 135–141.
59. Zielecka, M.; Bujnowska, E. Silicone-containing polymer matrices as protective coatings properties and applications. *Prog. Org. Coatings*, **2006**, *55*, 160–167.
60. Kollen, W.; Gray, D. Controlled humidity for oxygen barrier determinations. *J. Plast. Film Sheet*, **1991**, *7*, 103–117.
61. Meunier, M. Diffusion coefficients of small gas molecules in amorphous cis-1,4 polybutadiene estimated by molecular dynamics simulations. *J. Chem. Phys.*, **2005**, *123*, 134906.
62. Toniolo, R.; Hummelgen, I. A. Simple and fast organic device encapsulation using polyisobu-tene, *Macromol. Mater. Eng.*, **2004**, *289*, 311–314.
63. Agroui, K; Collins, G. Characterisation of EVA encapsulant material by thermally stimulated current technique. *Sol. Energy Mater. Sol. Cells*, **2003**, *80*, 33.
64. Yan, X.; Wang, H.; Yan, D. An investigation on air stability of copper phthalocyanine-based organic thin-film transistors and device encapsulation, *Thin Solid Films*, **2006**, *515*, 2655–2658.
65. Rendu, P. L.; Nguyen, T. P.; Carrois, L. Cellulose acetate and PVDC used as protective layers for organic diodes. *Synthetic Met.*, **2003**, *138*, 285–288.
66. Kim, G. H.; Oh, J.; Yang, Y. S.; Do, L. M.; Suh, K. S. Encapsulation of organic light-emitting devices by means of photopolymerized polyacrylate films. *Polymer*, **2004**, *45*, 1879–1883.
67. Kim, G. H.; Oh, J.; Yang, Y. S.; Do, L. M.; Suh, K. S. Lamination process encapsulation for longevity of plastic-based organic light-emitting devices. *Thin Solid Films*, **2004**, *467*, 1–3.

68. Granstrom, J.; Swensen, J. S.; Moon, J. S.; Rowell, G.; Yuen, J.; Heeger, A. J. Encapsulation of organic light-emitting devices using a perfluorinated polymer, *Appl. Phys. Lett.*, **2008**, *93*, 193304.

69. Pocius, A. V. *Adhesion and Adhesive Technology: An Introduction*; 2nd edn.; Munich/Hauser Publishers, New York, **1997**, pp. 201–217 (Chapter 8).

70. Gupta, S.; Ramamurthy, P. C.; Madras, G. Synthesis and characterization of flexible epoxy nanocomposites reinforced with amine functionalized alumina nanoparticles: A potential encapsulant for organic devices. *Polym. Chem.*, **2011**, *2*, 221–228.

71. Gupta, S.; Ramamurthy, P. C.; Madras, G. Covalent grafting of polydimethylsiloxane over surface-modified alumina nanoparticles. *Ind. Eng. Chem. Res.*, **2011**, *50*, 6585–6593.

72. Gupta, S.; Ramamurthy, P. C.; Madras, G. Synthesis and characterization of silicone polymer/functionalized mesostructured silica composites. *Polym. Chem.*, **2011**, *2*, 2643–2650.

73. Gupta, S.; Ramamurthy, P. C.; Madras, G. Mechanistic overview of the curing behavior of hydride terminated polydimethylsiloxane with allyl functionalized alumina by calorimetry and rheometry. *Thermochimica Acta*, **2011**, *524*, 74–79.

74. Lin, J. S.; Chung, M. H.; Chen, C. M.; Juang, F. S.; Liu, L. C. Microwave-assisted synthesis of organic/inorganic hybrid nanocomposites and their encapsulating applications for photo-electric devices. *J. Phys. Org. Chem.*, **2011**, *24*, 193–202.

75. Mont, F. W.; Kim, J. K.; Schubert, M. F.; Schubert, E. F.; Siegel, R. W. High-refractive-index TiO$_2$-nanoparticle-loaded encapsulants for light-emitting diodes. *J. Appl. Phys.*, **2008**, *103*, 083120 (5 pp).

76. Ravichandran, J.; Manoj, A. G.; Liu, J.; Manna, I.; Carroll, D. L. A novel polymer nanotube composite for photovoltaic packaging applications. *Nanotechnology*, **2008**, *19*, 085712 (5 pp).

77. Wang, Y. Y.; Hsieh, T. E.; Chen, I. C.; Chen, C. H. Direct encapsulation of organic light-emitting devices (OLEDs) using photo-curable co-polyacrylate/silica nanocomposite resin. *IEEE Trans. Adv. Packaging*, **2007**, *30*, 421–427.

78. Strange, M.; Plackett, D.; Kaasgaard, M.; Krebs, F. C. Biodegradable polymer solar cells. *Solar Energy Mater. Solar Cells*, **2008**, *92*, 805–813.

79. Yeun, J. H.; Bang, G. S.; Park, B. J.; Ham, S. K.; Chang, J. H. Poly (vinyl alcohol) nanocomposite films: Thermooptical properties, morphology, and gas permeability. *J. Appl. Polym. Sci.*, **2006**, *101*, 591–596.

80. Gaume, J.; Wong-Wah-Chung, P.; Rivaton, A.; Therias, S.; Gardette, J. L. Photochemical behavior of PVA as an oxygen-barrier polymer for solar cell encapsulation. *RSC Adv.*, **2011**, *1*, 1471–1481.

81. Gaume, J.; Taviot-Gueho, C.; Cros, S.; Rivaton, A.; Therias, S.; Gardette, J. L. Optimization of PVA clay nanocomposite for ultra-barrier multilayer encapsulation of organic solar cells. *Solar Energy Mater. Solar Cells*, **2012**, *99*, 240–249.

CHAPTER 25

Polymer Quenchants for Industrial Heat Treatment

K. Narayan Prabhu, Vignesh Nayak, and Pranesh Rao

CONTENTS

25.1 INTRODUCTION

Quench heat treatment is a technologically important process in the manufacturing of metal components. The process involves heating of a metal alloy to an elevated temperature at which microstructurally a single phase is formed and holding at this temperature to obtain a uniform phase throughout the component followed by controlled cooling. The cooling rate or behavior during the quench must be controlled in order to obtain superior mechanical properties with minimal distortion or other ill effects that are inevitable during the quench process. Generally, a liquid quench medium is commonly used to facilitate/control the heat transfer during quenching. The most commonly used quenchants in industries worldwide is water followed by oils. On immersion quenching in these liquids, the metal experiences three typical stages of quenching: vapor stage, nucleate boiling stage, and the convective cooling stage. However, the severity of cooling and the duration of the stages in both classes are very much different. Water, because of its severe nature of cooling, provides ample, or exceeds, mechanical properties over that which is required. Moreover, in many circumstances, such as cooling of complex objects or high-alloyed steels, it causes undue distortion, residual stresses, and, in the most severe case, crack formation. Oils, by their very viscous nature, deliver lower cooling performance and reduces is the propensity towards distortion and cracking. The low quench severity of oils results in reduced properties. These limitations compelled the quenchant suppliers to develop polymer quench media that offer the benefit of both water and oils.

A number of aqueous polymer-based quench media have been developed. In this chapter, an attempt has been made to review the cooling mechanisms, types, advantages, and limitations and stability aspects of aqueous polymers and their applications with regard to quenching steels and aluminum alloys.

25.2 COOLING MECHANISMS

Figure 25.1 shows the sequence of cooling for quenching of an austenized steel probe in an aqueous polymer quench medium. Unlike water quench, the cooling mechanism of quenching in aqueous polymer quenchant comprises a vapor phase encapsulated by a polymer layer or a polymer-enriched viscous layer (in cases of polyalkylene glycol [PAG]) or a

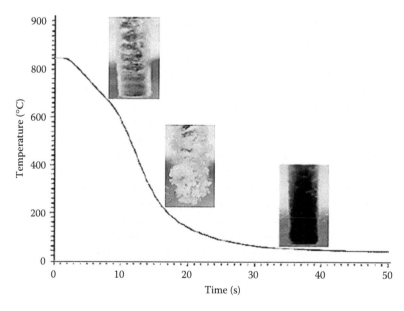

FIGURE 25.1 Cooling mechanism in a polymer quenchant. (From Totten, G.E. et al., Importance of quench bath maintenance, http://getottenassociates.com/pdf_files/Importance of Quench Bath Maintenance.pdf. With permission.)

film composed of a gel (polyvinyl alcohol [PVA], polyvinyl pyrrolidone [PVP], or alkali polyacrylate [ACR]) surrounding the hot component, depending on the concentration of the polymer in the quenchant (Liscic et al. 2010). Further, based on the type of polymer used, the severity of quenching varies (Figure 25.2). The severity of quenching is defined as $H = h/2k$, where H, h, and k are the quench severity, heat transfer coefficient, and thermal conductivity of the test probe, respectively.

The result of this encapsulation is a more uniform heat transfer than that provided in the vapor stage during quenching with oils and water. The heat flux at the metal–quenchant interface during this stage is typically of the order 10^5 W/m². As the metal cools further, wetting of the liquid on the metal surface occurs by the film rupture process. Bubble nucleation, its growth, and departure from the metal surface are the characteristic features of this stage, and the heat flow at the interface reaches its maximum value with the order of the flux being 10^6 W/m². With further loss of heat from the metal surface, bubble dynamics ceases, marking the end of nucleate boiling and the interfacial heat flow drops to 10^5 W/m², signaling the final stage of cooling by convection mechanism. The other factors that affect the cooling performance are agitation rate and the quenchant temperature (Tottten et al. 1993). Obtaining desired properties and achieving low distortion are usually balancing acts. Often, optimal properties are obtained at the expense of high residual stresses or high distortion likewise, low distortions or low residual stresses are usually obtained at the cost of mechanical properties. Therefore, the optimum quench rate is the one at which the best possible properties are obtained with minimum distortion. The cooling characteristics of a quenchant can be measured using probes instrumented with

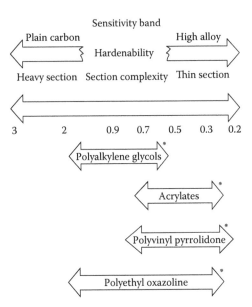

FIGURE 25.2 Severity band for selection of quenchants. *indicated compounds are influenced by polymer concentration, agitation, and bath temperature. (From Steel quenching technology, in: *Heat Treater's Guide: Practices and Procedures for Irons and Steels*, 2nd edition, Chandler, H. ed., ASM International, Materials Park, OH, 1995, pp. 81–82. With permission.)

thermocouples. Various techniques have been used, including both cylindrical and spherical probes manufactured from a variety of metals, including stainless steel, silver, and nickel alloys. One of the most widely used and accepted methods is based upon the use of a 12.5 mm diameter cylindrical probe manufactured from INCONEL® 600 material (Mackenzie 2003).

Some of the polymer quenchants exhibit inverse solubility. As the temperature of a polymer solution rises, a point is reached when the thermal energy of the system becomes greater than the energy of the hydrogen bonding interaction between water and the polymer chain molecules. Thus, the hydrogen bonding is broken, and the polymer chain separates as a polymer-rich phase in a water-rich phase. The two phases have some concentration of the other component. The temperature of a quenchant above which water and the polymer separate out is called the cloud point or separation temperature. The polymer redissolves in the solution when the temperature of the solution reduces below the cloud point (Tottten et al. 1993).

25.3 TYPES OF POLYMER QUENCH MEDIA

The first polymer-based quench medium to be used was PVA in 1950s in the United States and France. Table 25.1 shows the classification of commercially available polymer quenchants in the United States in 1978 (Liscic et al. 2010).

Synthetic polymers typically used as a quenchant in conjunction with water include PVA, PAG, polyethyloxazoline (PEOX), ACR, polysodium acrylate (PSA), and PVP. PVA or PVA + PAG are less popular because they leave a rubber-like film on the quenched part that is difficult to remove (Mackenzie 2003).

PAG quenchants are the most widely used polymer-based quenchants, and they exhibit the inverse solubility phenomenon. The inverse solubility property refers to the

TABLE 25.1 Polymer Quenchants Available in the US Market (1978)

Polymer Type	Polyvinyl Alcohol	Polyvinyl Alcohol + Poly-Oxyalkylene Glycols	Poly-Oxyalkylene Glycols	Polyvinyl Pyrrolidone	Alkali Polyacrylate
Abbreviation	PVA	PVA and GLY(PAG)	GLY(PAG) (range of some different products)	PVP	ACR
Polymer concentration (wt.%)	16	12	40–60	10	10
Solubility in water	Normal	Normal	Inverse	Normal	Normal
Dry residue	Brittle solids	Elastic solids	Viscous liquid	Brittle solids	Brittle solids
Viscosity of concentrate (cst) at 37.8°C	10,000	890	300–500	10.6	565
Viscosity of aqueous solutions (cst) at 37.8°C	30% 25.5 20% 8.4 10% 2.4 5% 1.3	5.2 2.7 1.4 1.0	8.7–11.5 4.4–5.3 1.9–2.2 1.1–1.3	26 1.7 1.1 0.9	51 31 16 9

Source: W. Luty, Cooling media and their properties, in: *Quenching Theory and Technology*, 2nd edition, Liscic, B. et al., (eds.), CRC Press, Boca Raton, FL, 2010, pp. 377–388.

precipitation or separation of PAG dissolved in water at temperatures exceeding the critical temperature (typically 63°C–85°C). PAG has a lower molecular weight compared to ACR, PVP, and POEX. During film boiling, a discontinuous film is formed, and they also do not undergo film rupture.

Since polymer quenchants exhibit inverse solubility phenomenon, they dissolve completely in water when the temperature at the metal quenchant interface decreases below the cloud point. The cloud point of PAG is dependent on the ratio of ethylene oxide to propylene oxide in the polymer. It was observed that as the amount of propylene oxide increases in the solution, the cloud point temperature decreases (Mackenzie 2003). PAG quenchants such as the 10 wt.% concentration exhibit a pseudo-Newtonian cooling process, implying that the same cooling mechanism occurs all over the surface at a given instant of time. Under such a scenario, the polymer film surrounding the part will rupture explosively, and rapid cooling results because of nucleate boiling, and the wetting exhibited is called Newtonian wetting (Sahay et al. 2009).

ACR-based polymer quenchants have oil-like quenching characteristics that enable the quenching of a wide range of alloy steels and higher hardenability steels. They are ionic, so they do not show inverse solubility and are thermally stable, corrosion resistant, and provide better ease of waste disposal (Mackenzie 2003). The water-soluble polymer typically coats the hot metal, forming an insulating film around the work piece, which reduces heat transfer rates and increases the duration of film boiling, favoring the formation of pearlite (Sahay et al. 2009). One of the disadvantages of ACRs is that they form polyelectrolyte complexes with polyvalent cations such as ferric or calcium ions in the solution. Polyelectrolyte complexes either precipitate or form insoluble oil (Tottten et al. 1993).

PVP-based media exhibit faster cooling rates compared to PAG in the higher temperature range, whereas their cooling rates are slower in the lower temperature range (Table 25.2). Their thermokinetic properties are similar to fast oils (Liscic et al. 2010). These are nonionic polymer quenchants and do not exhibit inverse solubility. A thick and strong polymer-rich film is formed near the surface, thus reducing the heat transfer rate (Mackenzie 2003).

POEX is a nonionic polymer and exhibits the inverse solubility phenomenon. A continuous polymer film is formed around the surface of the quenched specimen. During a film rupture process, the polymer dissolves in the solution. As in PAG, when the temperature

TABLE 25.2 Quench Data of Fast Oil, PVP, and PAG

Type and Concentration of Quenchant	Bath Temperature (°C)	Velocity (m/s)	Cooling Rate at 700°C (K/s)	Cooling Rate at 200°C (K/s)
Fast oil	60	0	46.1	5.5
16% PVP	52	0.5	32.8	5.5
		0.75	57.3	12.5
20% PVP	55	0.75	47.3	12.5
16% PAG	52	0	16.1	10.6
	52	1.08	34.8	16.1

Source: W. Luty, Cooling media and their properties, in: *Quenching Theory and Technology*, 2nd edition, Liscic, B. et al., (eds.), CRC Press, Boca Raton, FL, 2010, pp. 377–388.

of the interface decreases below the cloud point, the polymer dissolves in the solution (Tottten et al. 1993). A small concentration of PEOX is enough to obtain desirable quenchant properties (Mackenzie 2003).

PVA, when used as a quench medium, produces acetic acid due to thermal decomposition, causing an uncontrolled effect on cooling power, and it forms a rubber-like film on the surface of the part, fixtures, and quench tank (Mackenzie 2003).

Apart from the above-mentioned quenchants, cellulosic derivatives, polyethylene oxide, latex polymers, polyurethane, and so on are other important polymer quenchants currently used (Tottten et al. 1993).

25.4 EFFECT OF ADDITIVES OF POLYMERIC SOLUTION

Tensi et al. (1993) reported that addition of alkyl phosphate enhances the wetting properties of the polymer solution, thus increasing heat transfer into the quench media. A significant reduction in heat transfer was observed when ionic salts were added to the polymer solution.

Gestwa and Przylecka (2010) determined the cooling rate for quenching of a standard INCONEL probe into water, 10% ACR, and 10% ACR + 1% alumina (50 nm) (Figure 25.3). They also investigated the wetting angle by the pendent drop technique. The result of the center cooling data shows that ACR + alumina solution provides the slowest cooling with an increased duration of the vapor stage compared to water and ACR quenchant. The reduced cooling in the beginning of the martensitic start temperature was found to be the least for quenching with addition of alumina. Further, the cooling rate between 500°C and 600°C with nano addition (90°C/s) was greater than the ACR quenchant (50°C/s). This implies faster cooling rates at the lowest austenitic temperature resulting in a less amount of retained austenite in the case of most structural steels. Water, ACR, and ACR + Al_2O_3 had wetting angles of 72°, 77°, and 65°. The least wetting angle of ACR + Al_2O_3 (65°) implies less heat transfer, and the results obtained were in agreement with the cooling study.

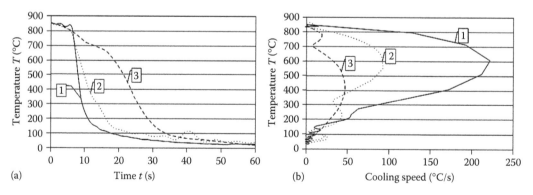

FIGURE 25.3 (a) Cooling curve and (b) cooling rate for distill water (1), 10% polymer water solution with (3) and without (2) 1% Al_2O_3 nanoparticle system. (From Gestwa, W. and Przylecka, M., *Adv. Mater. Sci. Eng.*, 1, 2010. With permission.)

25.5 ADVANTAGES OF POLYMER QUENCHANTS

Water quench is most often used for quenching simple shaped, low alloyed metal objects that require a severe quench in the range of 0.9–2.0 (Totten 2005). Its use is limited because of simultaneous occurrence of all the three cooling mechanisms on the part surface. The inset in Figure 25.4 shows a pictorial diagram of the cylindrical probe instrumented with thermocouples at various locations. The difference in the cooling behavior of the probe during quenching with water and 5 wt.% PAG are also shown in this figure. The dotted profiles of cooling measured at various locations shown in the Figure 25.4 show that PAG quench provides a more uniform cooling than water quench (solid lines) (Sahay et al. 2009).

Quenching of crack sensitive parts and parts that require slow cooling rates can be done either with quench oils or with polymer-based aqueous quenchants. For such metal alloys, oil quench proves satisfactory in terms of mechanical properties and reduced probability of distortion compared to water quench. However, oil quench comes at the expense of inherit demerits such as fire hazard, environment pollution, nonrenewability, disposal problem, and smoke emission during quenching (Ikkene et al. 2013). Moreover, postheat treatment, the oil drenched parts need to be cleaned with alkali solutions leading to waste management issues (Liscic et al. 2010). These drawbacks compelled the formulation of polymer quenchants that have oil-like properties without the disadvantages.

Reduced cracking and distortion of parts quenched in oils are attributed to low convective heat transfer in the martensitic transformation range for steels. Heat transfer rates decrease exponentially with increasing viscosity; thus, the low convective heat transfer is due to high viscosity of oils. The same principle is used in the development of aqueous polymer quenchants to replace water (Tottten et al. 1993).

The increase in the temperature of a polymer-based aqueous quenchant is about one-half of that of oil. It can be used with both sealed quench furnace and continuous atmosphere furnaces (Totten et al. 1997). Moreover, unlike oils, polymer quenchants such as

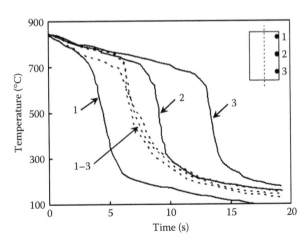

FIGURE 25.4   Cooling curves at different positions within an instrumented cylindrical probe: the solid line is for water and the dashed line is for a 5% aqueous solution of a PAG polymer quenchant. (From Sahay, S.S. et al., *J. ASTM Int.*, 6, 1, 2009. With permission.)

PVA (Totten 2005) and PAG can be used in spray quenching, and they have more recently been employed satisfactorily in intensive-quenching and time-quenching technologies (Totten et al. 1998). Steel parts that have been quenched in polymer quenchants have shown inverse hardening.

In contrast to water quench, small variations in the temperature of the water–polymer solution do not affect the quench performance significantly (Croucher 1982).

Another benefit of using aqueous polymer-based quenchants is that their properties can be tailored by controlling the concentration of the polymer in water to provide cooling rates obtained between water and oils, and in some cases cooling rates slower than oils are possible (Liscic et al. 2010). Moreover uniform heat removal during quenching results in elimination of soft spots, reduced thermal gradients, and reduced distortion (Totten 2005).

On quenching a hot metal into PAG, a liquid polymer coating is formed around the cooling metal (Canter 2010). As the temperature is reduced, the coating is ruptured and heat transfer occurs by convection, and below the inverse solubility temperature, the polymer redissolves in water. Apart from providing fire safety, distortion control, desirable mechanical properties, cooling rate flexibility (Gestwa and Przylecka 2010). Quenching in PAG, is advantageous as the drag out losses are minimum and the parts can be cleaned easily, and, to maintain bath concentration, a top-up of fresh PAG can be added.

25.6 DISADVANTAGES OF POLYMER QUENCHANTS

Disadvantages of implementing polymer-based water quenchants are corrosion of the quench tank and bacterial degradation of the polymer. These issues have been addressed by using pH boosters and other additives. Barberi et al. (2012) have demonstrated the deleterious effect bacterial and fungicidal action can have on the quench properties of PAG quenchant (Aqua-Quench 140 and 145). They compared the cooling curves (ASTM D 6481) for quenching in two similar molecular weight PAG quenchants. In their study, two control test samples were prepared with each of the PAG quenchant, one to assess the effect of bacteria and the other to measure the effect of fungus. The polymer concentration used in their work was 5 wt.% (typically used in induction of hardening process). The test for bacterial contamination was carried out in accordance with the test standard ASTM D3946; the preparation of the test fluid was made by filling a 1 L French square bottle with PAG up to its brim to which 10 gm of metal chips and 100 ml of inoculum were added. The inoculum was prepared by mixing equal proportions of soya bean Casein Digest broth with a contaminated metal working fluid and allowed to stand for 2 days to facilitate bacterial growth. Aeration to the liquid in the bottle was provided using an aquarium pump via a cotton-plugged sterile pipette that was inserted through a hole on the bottle cover. The test solution thus prepared was incubated for about 48 h at approximately 35°C. For fungus testing, the inoculum was prepared and a concentrated spore suspension of *Aspergillus niger*, a resistant mold (fungus) species, was added directly to 100 ml of Sabouraud's dextrose broth, a medium designed to propagate yeasts and molds. This medium contained 2 g/L chloramphenicol to retard the bacteria growth. The incubation time in this case was more than 3 days. Figure 25.5 shows the result of quenching with the two PAG quenchants in accordance with ASTM D 6481. Three sets of tests were conducted with each PAG.

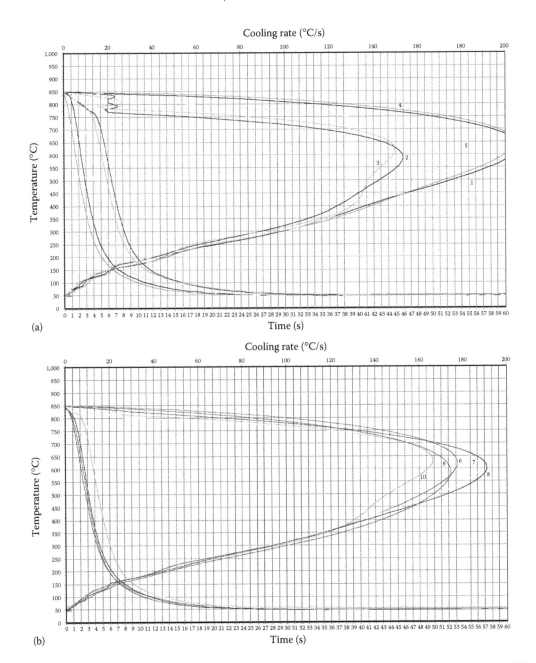

FIGURE 25.5 Comparison cooling curves of Aqua-Quench 140 and Aqua-Quench 145 at 10%. Aqua-Quench 140 (a): 1—fresh; 2—prior aeration; 3—final bacteria; 4—initial fungus; 5—final fungus. Aqua-Quench 145 (b): 6—fresh; 7—prior aeration; 8—final bacteria; 9—initial fungus; 10—Final fungus. (From Barberi, J. et al., Extending the life of polymer quenchants: Cause and effect of microbiological issues, in: *Proceedings of the 6th International Quenching and Control of Distortion Conference Including the 4th International Distortion Engineering Conference*, MacKenzie, D.S. ed., ASM International, Materials Park, OH, 2012, pp. 539–546. With permission.)

Cooling curves were also measured with a fresh quenchant, and with an inoculated-metal chip contaminated quenchant before aeration and a final 13th day test (5 days of aeration followed by 3 days of un-aeration and finally 5 days with aeration). Similar tests were carried with the fungus-contaminated quenchant. The results show that, for Aqua-Quench 140, there was an initial large drop in the maximum cooling rate due to the inoculation of bacteria. Subsequently, there was very little change from the initial drop in the maximum cooling rate due to the bacteria. There was little change upon inoculation of fungus; however, a slight decrease in the overall cooling rate occurred due to the growth of the fungus. Aqua-Quench 145 showed very little change in the maximum cooling rate from the control control (curve 6, fresh and curve 9, initial fungus). A slight increase was observed after inoculation with bacteria, and after the test was complete. However, these changes are not statistically significant. This result was repeated for the fungal inoculation, with the Aqua-Quench 145 showing little change due to fungal contamination. These outcomes indicate that Aqua-Quench 145 is more resilient to the attack of bacteria and fungicides and is therefore better suited for an induction-hardening process.

25.7 IMMERSION QUENCHING

PAGs are used in immersion quenching, induction hardening, and in spray quenching of steel parts. Applications of other polymers include forging, open tank quenching of high hardenability steels used in integral quenching furnaces, patenting of high-carbon steel wire and rod, and in quenching railroad rails (Chandler 1995).

In the case of immersion quenching, cooling rates in quenchants can be adjusted to requirements by changing the concentration of polymer in the solution, quenchant bath temperature, and the degree of agitation of the bath.

Concentration increases the polymer film thickness. With increasing concentrations, the maximum rate of cooling and the cooling rate in the convection phase drops. Wettability of work piece surfaces is improved with 5% solutions of PAG, which is beneficial to quench uniformity. At this concentration, problems with soft spotting associated with water quenching are avoided. Concentrations in the 10% to 20% range accelerate cooling rates to the level of fast quenching oils. These concentrations are suitable for quenching low hardenability steels that require maximum mechanical properties. Concentrations of 20% to 30% boost cooling rates suitable for a wide range of through hardening and case-hardening steels (Chandler 1995).

Figure 25.6 shows a comparison of the maximum cooling rate at the geometric center of 12.5 mm diameter and 60 mm high cylindrical INCONEL probe with varying concentrations of polymer quenchants. PAG and PVP quenchants provide higher cooling rates in all concentrations compared to ACR media. The higher cooling rates at a higher temperature of PAG and PVP quenchants are advantageous to avoid pearlitic transformation, but higher cooling rates at a lower temperature (300°C) may result in distortion and cracking. As shown in Figure 25.7, PVP quenchants provide the advantage of low cooling rates at a lower temperature compared to PAG quenchants (Liscic et al. 2010).

Bath temperature has an influence on the quenching speed of solutions. The effects of three different temperatures on a 25% PAG concentration with vigorous agitation

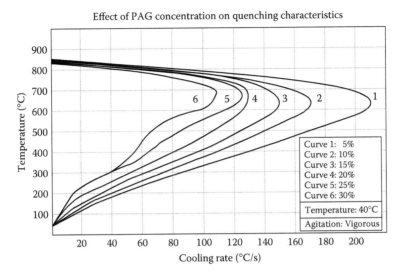

FIGURE 25.6 Effect of PAG concentration on the cooling performance of a polymer quenchant. (From Steel quenching technology, in: *Heat Treater's Guide: Practices and Procedures for Irons and Steels*, 2nd edition, Chandler, H. ed., ASM International, Materials Park, OH, 1995, pp. 81–82.)

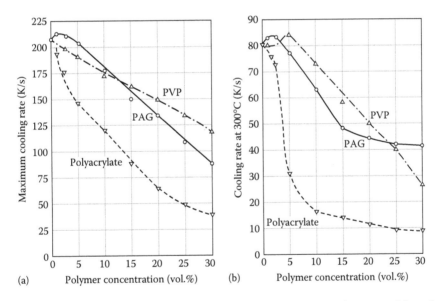

FIGURE 25.7 Effect of polymer concentration on the maximum cooling rate (a) and cooling rate at 300°C (b). (From W. Luty, Cooling media and their properties, in: *Quenching Theory and Technology*, 2nd edition, Liscic, B. et al., eds., CRC Press, Boca Raton, FL, 2010, pp. 377–388. With permission.)

are shown in Figure 25.8. The maximum cooling rate decreases with increasing temperature. PAG solutions must be cooled to prevent them from reaching the inversion temperature. A maximum operating temperature of approximately 55°C is normally recommended. Ikkene et al. (2013) have measured the cooling performance of a PEG-3000 quenchant for quenching of an INCONEL 600 probe from 850°C into quenchants

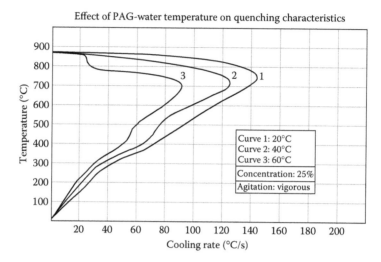

FIGURE 25.8 Effect of bath temperature on the cooling performance of a polymer quenchant. (From *Steel quenching technology*, in: *Heat Treater's Guide: Practices and Procedures for Irons and Steels*, 2nd edition, Chandler, H. ed., ASM International, Materials Park, OH, 1995, pp. 81–82. With permission.)

maintained at 20°C, 30°C, and 40°C and inferred that the maximum cooling rate would decrease with increasing temperatures of the quenchant (Chandler 1995).

At higher temperatures, the effect of bath temperature on the cooling rate in a polymer is similar to that of the effect of concentration. PAG shows (Figure 25.9) a linear relation between cooling rate at 300°C and bath temperature, whereas no such relation exists for quenching with a PVP or ACR aqueous solution (Liscic et al. 2010).

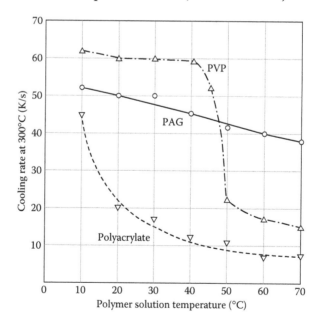

FIGURE 25.9 CR300 as a function of bath temperature. (From W. Luty, Cooling media and their properties, in: *Quenching Theory and Technology*, 2nd edition, Liscic, B. et al., eds., CRC Press, Boca Raton, FL, 2010, pp. 377–388. With permission.)

25.8 EFFECT OF AGITATION

A quenchant can be agitated with the aid of a pump or a stirrer. Typically, propeller-type stirrers are preferred with a directional draft tube meant for guiding the flow. The two important parameters that affect the agitation are agitation rate and quenchant flow rate. The agitation rate provided must be such as to avoid eddy formation and maximize the linear flow rate not exceeding 20–25 ft/min and about 150 ft/min for ferrous and nonferrous materials, respectively. Along with the agitation rate, the flow rate provided must be such that adequate quenchant comes in contact with the quenched part at any given instant. Totten and researchers recommend that the flow rate should be one volume turn-over/min. For example, a 100 L tank must have an agitation system capable of providing at least 100 L/min flow rate (Totten et al. 1991). Agitation ensures a uniform temperature distribution within the bath. The peak cooling rate increases, and an increase in agitation rate results. The duration of the vapor phase (film phase) is shortened and eventually made zero. During the convection phase, agitation has comparatively little effect (Chandler 1995) (Figure 25.10). The effect of agitation not only depends on the velocity of the quenchant but also on the direction of impingement relative to the surface of the quenched part and the degree of turbulence in the quenchant near the quenched part (Tottten et al. 1993).

Hilder (1988) has determined the empirical relation that describes maximum cooling rate (MR) and cooling rate (CR_{300}) at 300°C as a function of polymer concentration (C), quenchant bath temperature, and quenchant velocity (V).

For PAG,

$$MR = 244.6 - 4.3C - 1.7T + 47.2V$$

$$CR_{300} = 83 - 1.6 - .36T + 14.1V$$

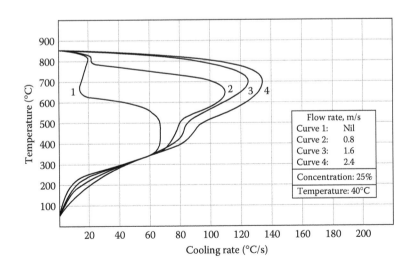

FIGURE 25.10   Effect of agitation on the cooling performance of a polymer quenchant. (From Steel quenching technology, in: *Heat Treater's Guide: Practices and Procedures for Irons and Steels*, 2nd edition, Chandler, H. ed., ASM International, Materials Park, OH, 1995, pp. 81–82. With permission.)

For PVP,

$$MR = 223.7 - 2.94C - 1.9T + 71.9V$$

$$CR_{300} = 96.2 - 1.8 - 1.02T + 37.68V$$

For ACR,

$$MR = 160.59 - 5.63C - 1.1T + 114.8V$$

$$CR_{300} = 58.6 - 2.5C - 0.57T + 43.4V$$

As it is clear from the above expressions, lowering of bath temperature and concentration results in an increase in cooling rates, whereas a decrease in agitation leads to a decrease in cooling rates (Liscic et al. 2010).

25.8.1 Mechanodegradation

Polymer chains coil around each other providing a relatively long-range order that increases viscosity. The degree of polymer-chain entanglement increases with its increasing molecular weight. When the shear force is applied by the fan agitator/pump, shear-induced thinning of the quenchant occurs. This occurs when a sufficient energy due to a mechanical interaction of the agitation system with polymer quenchants breaks the C–C covalent bond. The breakage of carbon covalent bonds leads to a reduction in the average molecular weight of the polymer. This reduction in the molecular weight in turn decreases the viscosity of the polymer quenchant. The decrease in the viscosity thus affects the cooling performance of the quenchant.

It is difficult to develop a procedure to exactly describe the mechanodegradation of a polymer in a quenching system. The worst case scenarios can be evaluated by ASTM D 3519 test. The results of ASTM D 3519 warring blender test conducted on three polymer tests are shown in Table 25.3. In this test, a polymer quenchant is stirred using a blender operated at 9000 rpm for a specified time. The kinematic viscosity measured at time intervals of 1, 2, and 3 min are tabulated. It can be concluded from the table that PAG copolymer is more shear stable compared to the other two quenchants (Tottten et al. 1993).

TABLE 25.3   Effect of Shear Time on Viscosity of Polymer Solution

		Viscosity at 40°C in cSt		
		Shear time, minutes		
Polymer	**Polymer Concentration %**	**1**	**2**	**3**
Polyacrylamide 20% hydrolyzed	2.5	6.3	6.3	6
Polyacrylamide 90% hydrolyzed	0.25	6.3	6.3	5.6
PAG copolymer	7.6	6	6	6.1

Source: Polymer quenchants, in: *Handbook of Quenchants and Quenching Theory*, Tottten, G.E. et al., (eds.), ASM International, Materials Park, OH, 1993, pp. 161–189.

25.8.2 Effect of Aging and Contamination on the Performance of Polymer Quenchants

After repetitive thermal cycling of a polymer quenchant and aeration, the original macromolecule will change to a highly oxidized oligomeric or monomeric derivative. Such breakage affects the cooling behavior of the quenchants. Figure 25.11 shows the effect of a degraded polymeric quenchant that had been used for more than 11 years. One can observe from the figure that the cooling curve of the used quenchant is similar to water, whereas that of the fresh quenchant has a distinct nature of cooling. Further the impact of degradation is more severe in the case of longer polymer (higher molecular weight) quenchants as compared to the shorter ones. This is due to the higher impact of the breakage on viscosity in the case of the polymers that have a higher molecular weight M_{II} relative to M_I, as shown in Figure 25.12. Totten et al. (1997) performed experiments using a 15.9 mm by 200 mm steel rod to assess the probability of degradation due to repetitive quenching. The rod was electrically heated to a test temperature of 845°C and quenched into a 2500 gallons of polymeric quenchant under agitated condition. The dip time was restricted to 2.5 min to ensure resolution of the precipitated polymer into water. The test was conducted totally for 3 days, and the degradation was assessed at an interval of 24 h using the viscosity data. The quenchant temperature during the experiments was maintained at 43°C ± 5°C in the quench tank. The result of their study is shown in Figure 25.13. It can be observed that degradation rate is more severe and higher in the case of a heavier polymer, and the compositions used had relatively a lower effect on degradation as compared to the molecular weights.

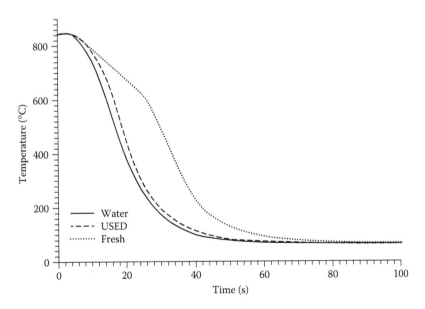

FIGURE 25.11 Cooling curve of water, severely degraded polymer quenchant and fresh polymer quench media. (From Totten, G.E. et al., Thermal/oxidative stability and polymer drag out behavior of polymer quenchants, in: *17th ASM Heat Treating Society Conference Proceedings Including the 1st International Heat Treating Symposium*, Milam, D. et al., eds., ASM International, Materials Park, OH, 1997, pp. 443–448. With permission.)

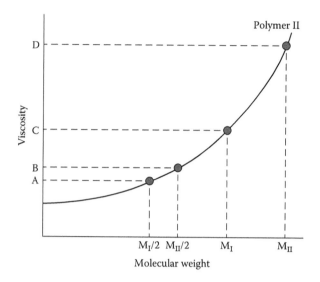

FIGURE 25.12   Viscosity change in polymer quench media as a function of molecular weight. (From Totten, G.E. et al., Thermal/oxidative stability and polymer drag out behavior of polymer quenchants, in: *17th ASM Heat Treating Society Conference Proceedings Including the 1st International Heat Treating Symposium*, Milam, D. et al., eds., ASM International, Materials Park, OH, 1997, pp. 443–448. With permission.)

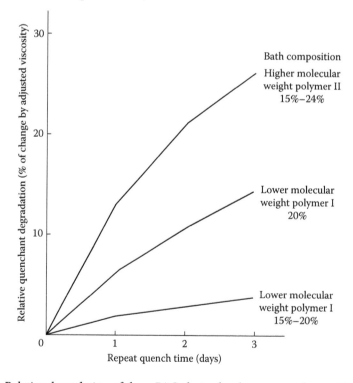

FIGURE 25.13   Relative degradation of three PAG-derived polymer quenchants. (From Totten, G.E. et al., Thermal/oxidative stability and polymer drag out behavior of polymer quenchants, in: *17th ASM Heat Treating Society Conference Proceedings Including the 1st International Heat Treating Symposium*, Milam, D. et al., eds., ASM International, Materials Park, OH, 1997, pp. 443–448. With permission.)

Troell and Kristoffersen (2010) investigated the effect of contamination, thermal degradation, and comparison between fresh and in-use PAG quenchants using cooling curve analyzes. For comparing the quench properties between fresh and quenchants that are in use, three polymer quenchants, Aquaquench 365, Aquatensid, and Feroquench 2000, were analyzed. Figure 25.14 shows their cooling performance measured at 30°C and at an agitation rate of 0.3 m/s.

It is to be noted that with the exception of plot (g), all other in-use quenchants showed an increased cooling rate relative to their respective new quenchant below 400°C. A plausible explanation for the increased cooling rate below 400°C shown by most in-use quenchants is that, with repeated usages, the original polymer chain breaks down and results in a lower viscosity of the polymer-based quenchant. The reduced cooling rate below 400°C for the four-month-old quenchant shown relative to the new quenchant in (g) could be due to the contamination by oil or by the formation of an emulsion. The existence of the vapor phase tendency was observed for the in-use sample.

For studying the effect of aging, tests were conducted in an induction heat treatment test rig. A 75 mm long and 25 mm diameter INCONEL rod was heated and quenched into a 3 L quench tank in which a 10 wt.% Aquatensid BW/RB quenchant was used at 20°C temperature. The bath temperature and agitation rate were kept constant for 400 cycles. The quenchant cooling curve was measured as shown in Figure 25.15. From the plot of cooling curve or its rate, we can observe that, with aging, the cooling rates of the quenchant increased at all temperatures relative to the un-aged sample.

Cooling curves (Figure 25.16) were measured for quenching in an Aquatensid 15 wt.% quenchant administered with 0.5, 2, and 5 wt.% detergent and cutting fluid emulsion to demonstrate the effect of contamination on the cooling behavior of the polymer quenchant.

A concentration of contaminants up to 2% showed no influence on the cooling curve. The maximum cooling rate was lowered with the addition of 5% of contaminants. For detergent added quenchant, no influence could be seen at lower temperatures. At the cutting fluid contamination level of 5%, the cooling rate below 400°C was higher. A higher cooling rate at lower temperatures was also observed at a contamination level of 2%. These studies show that careful monitoring on a periodic basis of polymer quench media is necessary to ascertain their validity for an intended application.

25.9 POLYMER DRAG OUT

When a part is quenched in a polymer quenchant, some amount of polymer is removed from the quenchant solution. This process is called polymer drag out. Polymer drag out may result in a decrease in the concentration of polymer in the solution resulting in increased cooling rates of the quenched part. The degree of redissolution of a polymer in a quenchant after quenching, shape of the work piece, concentration of the polymer in the solution quenchant, degree of wetting of the polymer on the work piece surface, agitation, and bath temperature are some of the important variables that control polymer drag out (Tottten et al. 1993).

The minimum or ideal polymer drag out can be said to be the amount of polymer that leaves the system with the quenchant that adheres to the surface of the work piece due to wetting. The polymer drag out associated with this ideal condition can be calculated as

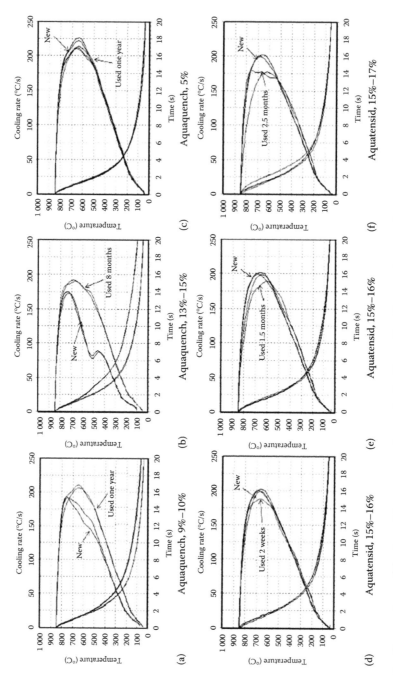

FIGURE 25.14 Comparison of fresh and in-use quenchants (a–f). (From Troell, E. and Kristoffersen, H., *BHM*, 155, 114, 2010. With permission.)

(Continued)

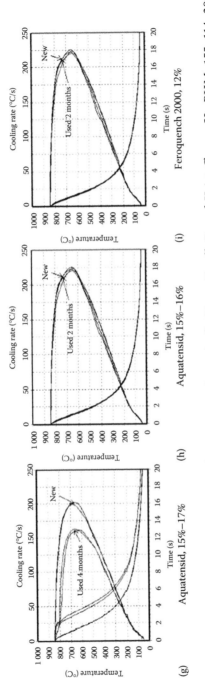

FIGURE 25.14 (Continued) Comparison of fresh and in-use quenchants (a–f). (From Troell, E. and Kristoffersen, H., BHM, 155, 114, 2010. With permission.)

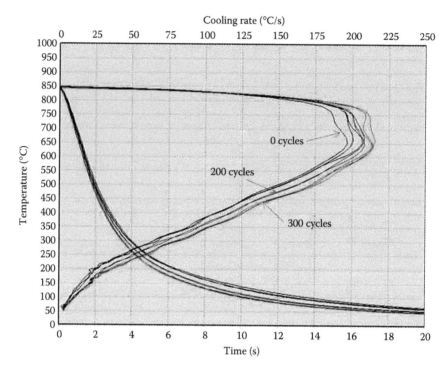

FIGURE 25.15 Effect of thermal cycling on the quench performance of 10 wt.% Aquatensid BW/RB quenchant. (From Troell, E. and Kristoffersen, H., *BHM*, 155, 114, 2010. With permission.)

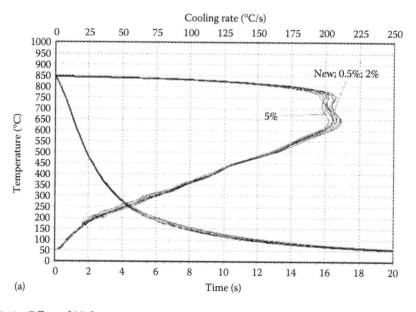

(a)

FIGURE 25.16 Effect of (a) detergent. *(Continued)*

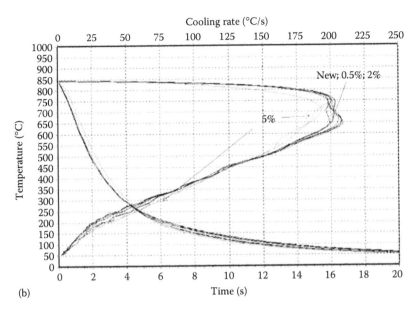

(b)

FIGURE 25.16 (Continued)   Effect of (b) cutting fluid on cooling performance of Aquatensid quenchant. (From Troell, E. and Kristoffersen, H., *BHM*, 155, 114, 2010. With permission.)

the volume of quenchant removed multiplied by the polymer concentration in the solution. It is observed that the polymer drag out is always higher than calculated in the ideal condition. This is due to the fact the extent to which a polymer re-dissolves in the quenchant is dependent on agitation and temperature of the quenchant. Nevertheless, the variation in the structure of polymer near the surface of the quenched part is the root cause for this increased polymer drag out (Tottten et al. 1993).

Figure 25.17 shows the results of experiments conducted by Hilder (1988). An intricate-shaped work piece was heated and quenched, and the amount of polymer drag out associated with static and agitated quenchants was reported. As shown in the figure, as the concentration of polymer increases, the polymer drag out increases. Also for a given concentration, PVP quenchants have the lowest drag out, whereas ACR have highest. Hilder (1988) also reported a linear relation between the kinematic viscosity of the polymer quenchant and the polymer drag out associated with it (Hilder 1988).

25.10 MAINTENANCE OF POLYMER SOLUTION IN QUENCH TANK

Disadvantages of oil include cost, quenching process inflexibility, and disposal and fire hazard. Polymer quenchants are most favorable alternatives to water and oil. PAG quenchants are most commonly used, as they can be removed easily from the quenched parts by dunking in a water tank or by using compressed air spray or water spray. They offer better quench uniformity relative to water and, unlike oils, they are environmental friendly. Polymer quenchants can provide a wide variety of cooling severity between water ($H = 1.0$) and quench oil ($H = 0.4$) by a proper selection of concentration, agitation, and temperature of the polymer. The effect of concentration, agitation, and bath temperature variations for quenching of AISI 1045 quenched in PAG are shown in Figure 25.18.

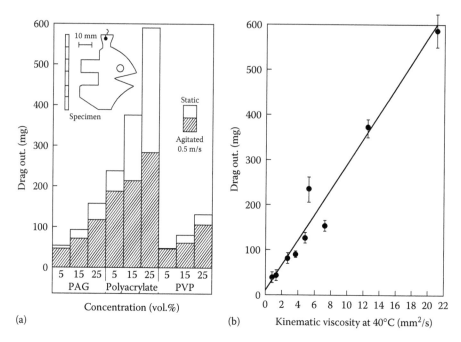

FIGURE 25.17 Effect of polymer concentration (a) and kinematic viscosity (b) of the quenchant on polymer drag out. (From Hilder, N.A, The behavior of polymer quenchants, PhD Thesis, University of Birmingham [Aston], UK, 1988. With permission.)

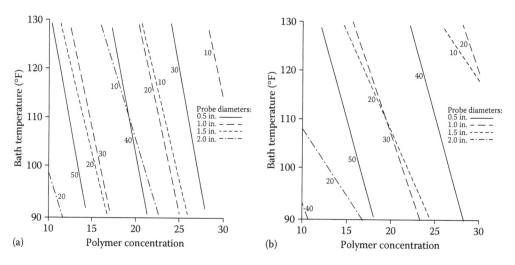

FIGURE 25.18 Effect of concentration and bath temperature variations on Rockwell C hardness for quenching of 1045 steel with (a) 0.1m/s and (b) 0.25m/s agitations. (From Totten, G.E. et al., *Ind. Heat.*, 37, 1991. With permission.)

Steels with a high hardenability (4140) and/or small section thickness require a lower polymer concentration than low hardenability and/or thick-sectioned steels. Also, increasing the rate of agitation and decreasing quenchant temperature increase the prospect of achieving higher material properties. The plots such as those shown in this chapter, when developed for a specific part-quench facility system, play a vital role in the selection of a proper quenchant.

As most PAG quenchants show the inverse solubility phenomenon, it is recommended to restrict the quenchant temperature below 60°C to avoid polymer precipitation out of water. Further the rise in the quenchant temperature during postquench must be limited to 5°C–8°C, so as not to forbid sequential batch quench operation. Also periodic agitation of the bath is recommended to minimize the effect of oxidation and entrapped air in the quench tank.

The control of the concentration of a polymer in a solution, which is one of the important variables that affects the heat extraction rate, is to be maintained with a precision of ±1%. The measurement of the concentration of a polymer in a solution can be realized by measuring kinematic viscosity, density, or refractive index. Of these the measurement of refractive index is the simplest to use. In this, droplet of quenchant is placed on the prism of a small handheld optical device called refractometer. When the device is directed against light, the refractive index can be determined. A standard table, which relates refractive index with polymer concentration, is used. Refractive index is used widely in the industry, as it requires very less time to measure polymer concentration.

Figure 25.19 shows one such standard plot showing a linear relation between concentration and refractive index. However, the disadvantage of this method is that it cannot detect the change in refractive index due to the presence of contaminants such as sludge or oil.

The viscosity of quenchant, which is the most precise measure of the concentration of a polymer, is also used to measure the concentration of the polymer in the quenchant. Figure 25.20 shows the standard curve displaying the relation between viscosities of three polymer quench media as a function of their concentration.

In the case of quenchant contamination with salts, the inverse solubility phenomenon can be used to separate the polymer from them. This is done by heating the solution above the cloud point to separate polymer and water phases. Thus, the dissolved salt in water can be separated from the polymer, and the polymer-rich phase that settles at the bottom can be reused to prepare a fresh quenchant. pH of the polymer solution has an important effect

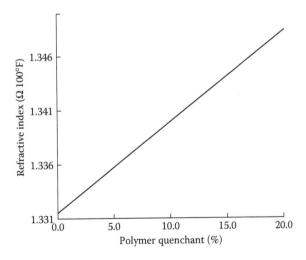

FIGURE 25.19 Refractive index versus PAG quenchants concentration (%). (From Totten, G.E. et al., *Ind. Heat.*, 37, 1991. With permission.)

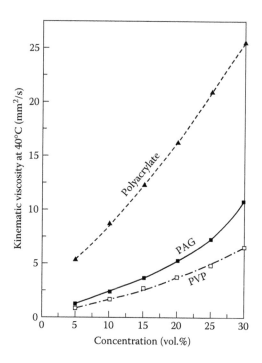

FIGURE 25.20   Variation of concentration as a function of kinematic viscosity of polymer quenchants. (From W. Luty, Cooling media and their properties, in: *Quenching Theory and Technology*, 2nd edition, Liscic, B. et al., eds., CRC Press, Boca Raton, FL, 2010, pp. 377–388. With permission.)

on its quench severity. Mikita et al (1987) have shown that the cloud point of a polymer quenchant is affected by the pH of the aqueous solution. A variation in the cloud point results in a change in the temperature-dependent viscosity of the solution, thus affecting the cooling performance of the quenchant (Figure 25.21). Thus it is important to monitor and maintain the pH of the solution at the desired level, which is achieved by the addition of required inhibitors.

25.11   INVERSE MODELING OF HEAT TRANSFER FOR ESTIMATING INTERFACIAL HEAT FLUX DURING QUENCHING

We know that the heat transfer characteristics of the quenchant play a dominant role in the evolution of mechanical and metallurgical properties of the quenched metal. This heat flow is a surface phenomenon and is difficult to measure directly due to the limitation involved in placing a thermocouple at the surface of the metal. One of the techniques to estimate the heat flow at the metal/quenchant interface is the inverse heat conduction method. It stands out from the direct heat conduction method, in that boundary conditions are estimated by the knowledge of the thermal history in the interior of the metal. The difference between these two methods is illustrated in Figure 25.22. Further, inverse methodology is advantageous compared to the lumped heat capacitance method as the temperature distribution in the metal with complex geometries and parts with thick or variable sections can be estimated. However, the inverse heat conduction analysis is an

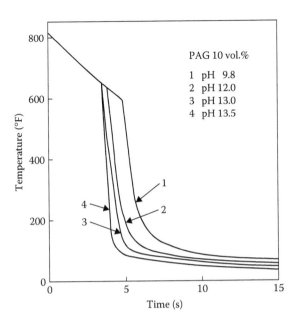

FIGURE 25.21 Cooling curves in polymer quenchants of various pH solutions. (From Polymer quenchants, in: *Handbook of Quenchants and Quenching Theory*, Tottten, G.E. et al., eds., ASM International, Materials Park, OH, 1993, pp. 161–189. With permission.)

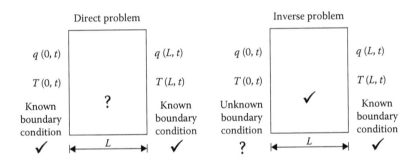

FIGURE 25.22 Pictorial depiction between direct and inverse problem. (From Prabhu, K.N. and Ashish, A.A. *J. Mater. Manuf. Process.*, 17, 469, 2002. With permission.)

ill-posed problem; it does not satisfy the general requirement of existence, uniqueness, and stability under small changes to the input data and is very sensitive to measurement errors. For obtaining a reasonable estimate of the interfacial heat flux and to account for the lag in the temperature drop between the surface and the interior of the metal, thermocouples are required to be positioned as close as possible to the surface (Prabhu and Ashish 2002).

Sarmiento et al. (2008) used the inverse heat transfer conduction method to determine the heat transfer coefficient of Type I PAG quenchant of concentrations varying from 0% to 60%. In Figure 25.23, one can observe that the addition of a mere 12% of PAG to water reduced the peak heat transfer coefficient from 22500 $W/m^2 °C$ to below 7500 $W/m^2 °C$.

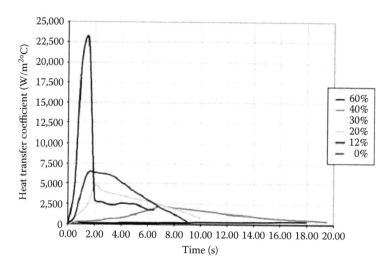

FIGURE 25.23   Heat Transfer coefficient versus time obtained from the inverse heat conduction technique for quenching of an Al alloy in PAG of various concentrations. (From Sarmiento, G.S. et al., *J. ASTM Int.*, 6, 1, 2008. With permission.)

25.12 APPLICATION OF POLYMERIC QUENCHANTS IN HEAT TREATMENT OF STEELS

Kakhki et al. used 1:8, PAG-water quenchant to quench Iranian grade steel alloys of low (1.1191), medium (1.7227), and high hardenability (1.6582 and 1.7765). These steels had section diameters of 60, 63, 65, and 38, respectively. The chemical compositions of the steels used are shown in Table 25.4, and the heat treatment conditions are reported in Table 25.5. The length of all the test specimens was about 200 mm. The specimens were quenched in a 1:8 PAG-water solution, water, and oil. The as-quenched and tempered hardness were measured at the center of the midsection of the block. The results of their study are shown in Figure 25.24.

TABLE 25.4   Chemical Composition of Steels

Grade	%C	%Mn	%Si	%F	%S	%Cr	%Ni	%Mo	%Al
1.7227	0.43	0.74	0.26	0.011	0.034	1.11	0.077	0.17	0.024
1.6582	0.353	0.62	0.274	0.0076	0.0078	1.49	1.51	0.1644	0.0335
1.1191	0.451	0.613	0.227	0.023	0.0199	0.21	0.224	0.0515	0.0168
1.7765	0.325	0.563	0.313	0.0083	0.0085	2.98	0.094	0.965	0.052

Source: Eshraghi-Kakhki, M. et al., *Int. J. ISSI*, 6(1), 34, 2009.

TABLE 25.5   Size and Heat Treating Parameters Applied to the Specimens

Steel Grade	1.7227	1.6582	1.1191	1.7765
Diameter (mm)	63	65	60	38
Austenitizing temperature (°C)	850	845	845	910
Austenitizing time (h)	1	1	1	1
Tempering temperature (°C)	620	610	570	680
Tempering time (h)	3.5	2.5	2.5	2.5

Source: Eshraghi-Kakhki, M. et al., *Int. J. ISSI*, 6(1), 34, 2009.

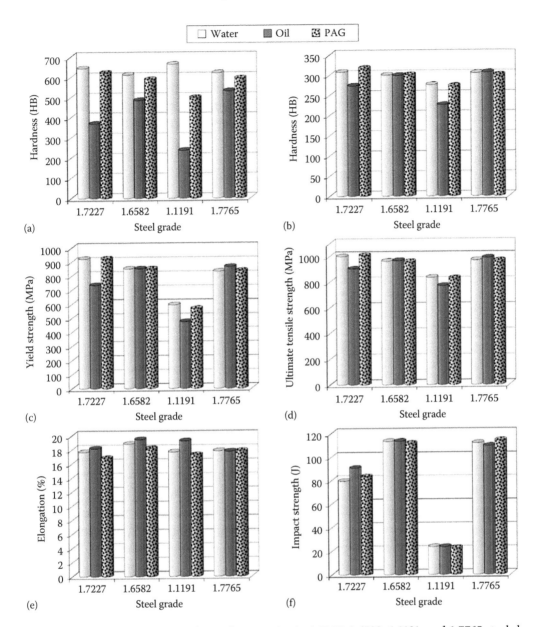

FIGURE 25.24 The resulting mechanical properties in 1.7227, 1.6582, 1.1191, and 1.7765 steels by quenching in water, oil, and PAG quenchants; (a) as-quenched hardness, (b) as-tempered hardness, (c) yield strength, (d) ultimate tensile strength, (e) elongation, and (f) impact strength. (From Eshraghi-Kakhki, M. et al., *Int. J. ISSI*, 6, 34, 2009. With permission.)

For the low alloy steel grade, the as-quenched hardness for quenching in water was the highest followed by PAG and water. The hardness of all the specimens for quenching in all the quenchants dropped on tempering. The as-tempered hardness plot shows comparable values for quenching in water and PAG, while oil quench resulted in lower hardness. However, the slow cooling effect of oils resulted in increased ductility of the oil-quenched samples. The difference in the impact strength of all the specimens was insignificant.

For the medium hardenability steel grade, as-quenched hardness for quenching in water and PAG is similar, while that in oil it is significantly low. The tempered hardness also showed similar behavior with a reduced magnitude for quenching in all types of quench media. But cracks were reported for water quenching. The impact and elongation for oil quench are comparable to that of PAG quench. For steels having a high hardenability, the tempered hardness and mechanical properties were similar for quenching in all the three quenchants.

It is well known that high hardenability steels such as 5XHM that are used for making dyes are quenched in a hot oil medium as they provide slow cooling and reduce the propensity towards crack formation. As the hardness and other mechanical properties of a material are linked to the cooling performance of a quenchant, quenching in hot oils resulted in low hardness. Moreover, the durability of the dyes was found to be insufficient. To overcome this issue, Kobasko (1989) made an attempt to replace hot oil with aqueous polymer quenchants. His study showed that a high concentration of polymer (20%–35%) can be used successfully in conjunction with an intensive quench set-up to provide localized intensive agitation to replace hot oil and also increase the dye durability by about five to six times.

25.13 APPLICATION OF POLYMERIC QUENCHANTS IN HEAT TREATMENT OF ALUMINUM ALLOYS

The most prolific use of polymer quench media has been in the reduction of distortion and warpage that accompany quench treatment on solutionizing of aluminum sheet metal parts. Croucher (2008) describes distortion as a manifestation of the "lack of dimensional stability" and warpage as "to bend or twist out of shape, especially from a straight or flat form." Distortion can occur immediately upon heat treatment, sometime later in a subsequent processing operation, or possibly on a shelf or in service. For example, parts that may be machined, cast, or formed to accurate dimensions prior to heat treatment often emerge from the heat-treating operation with significant changes in their dimensional configuration, or posttreatment process. Bowing or twisting of a sheet metal, canning of machined pocket areas, and other forms of part movement that necessitate extensive "check and straighten" procedures after quenching are examples of warpage that occurs due to the lack of uniformity during quenching.

The two most critical factors for heat treating a delicate, thin gage aluminum sheet metal, forged, and cast parts are as follows:

- Quench rates that are controlled by quenchant selection

- The integration of racking procedures with immersion rate

The most common source of heat-treating warpage occurs during the quench when parts are too thin to resist the movement, which is triggered by "unequal" expansion or contraction of the metal. If one area of the part is contracting as its temperature is lowered, and another area of the part has not yet started to achieve the lower temperature, or has already completely cooled, the part will bend toward the area that is starting to shrink as it is cooled, causing it to warp

Previously hot water was used for solution heat-treating of aluminum sheet metal parts. However, the implication of using water, for thin parts, causes undue distortion, residual stress, and warpage. These undesirable side effects require expensive check and straightening operations, causing a loss of time and money for the heat treater. Efforts directed to minimize these ill effects by quenching with water at elevated temperatures proved to be a disappointment as the cooling performance of water alters significantly by raising its temperature. Figure 25.25 shows the outcome of increased water temperature during quenching of half an inch of 7075 plate material. An increased vapor stage duration can be observed with an increased temperature (75°F [24°C] to 160°F [71°C]). Also, quenching at elevated temperatures caused slow cooling during subsequent stages. Moreover, water heated beyond 160°F (71°C) caused sustainable of vapor stage (200°F [93°C]), and completely altered its cooling performance. The effect of these reduced cooling performances on mechanical properties was reflected on the tensile strength of the parts, shown in Figure 25.26. Although water up to a temperature of 71°C can be used satisfactorily, a precise control of quenchant conditions for batch processing would be a tedious job.

These drawbacks can be satisfactorily answered by using polymer quenchants. The cooling performance can be controlled by adjusting the concentration of polymer addition. For example, in a quench tank, if the concentration is beyond the requirement, the polymer concentration in the solution can be reduced by heating the tank and removing the excess polymer content. Similarly to heat-treated thin sectioned or slender parts, an additional polymer can be used to increase its concentration to reduce the cooling rate.

This advantage eliminates the requirement to heat or cool the quenchant for consecutive load treatment. Figure 25.27 shows the variation of cooling curves obtained for quenching

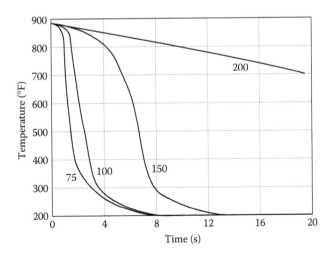

FIGURE 25.25   Cooling curves for water at different temperatures. (From Croucher, T., *J. ASTM Int.*, 5, 1, 2008. With permission.)

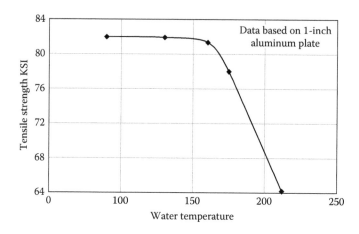

FIGURE 25.26   Effect of quench water temperature on the tensile strength of 1 in. 7075.

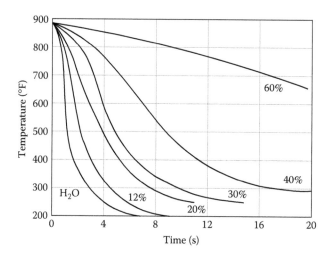

FIGURE 25.27   Cooling curves for polymers at various concentrations. (From Croucher, T., *J. ASTM Int.*, 5, 1, 2008. With permission.)

of an aluminum alloy in a polymer quenchant of various concentrations. These curves can be used by the heat treater with the knowledge of quench sensitivity of an alloy to select the proper concentration. Also unlike water at elevated temperatures, the cooling performance of aqueous polymer quench media is affected to a very limited extent by small temperature variations.

25.14 SUMMARY

Quench hardening is an important industrial process that enables us to achieve increased mechanical properties in metal alloys. The side effects of quenching must be controlled to a minimum so as not to adversely affect the quenched parts. The mechanism of cooling in polymer-based quenchants follows a polymer encapsulation stage, pseudo-nucleate boiling, and finally convective cooling. Water and oil quenchants that are most widely used cause distortion, warping, crack formation, and reduced properties, respectively, because of uneven heat extraction. Various polymer types that are soluble in an aqueous solution have been used with varying degrees of success. PAG is the most popular polymer, and it exhibits inverse solubility with ease of removal from the part postquenching. The cooling behavior of a polymer quenchant can be changed by concentration, bath temperature, agitation rate and nanoparticle additions, and other additives. The performance of polymer quenchants is affected by mechanodegradation, thermal cycling, oxidation, contamination, and drag out. Refractive index and viscosity measurements are needed to be done periodically to ensure the quenchant stability. The inverse solution of the transient heat conduction equation for estimating interfacial surface heat flux allows a comparison of polymer media and quenchants of various concentrations in selecting the suitable quench medium. PAG-based aqueous solution can be used for heat-treating steels of all grades. They can also be successfully employed for heat-treating of aluminum alloys.

REFERENCES

J. Barberi, C. Faulkner, and D. S. MacKenzie. 2012. Extending the life of polymer quenchants: Cause and effect of microbiological issues. In: D. S. MacKenzie (Ed.). *Proceedings of the 6th International Quenching and Control of Distortion Conference Including the 4th International Distortion Engineering Conference*, ASM International, Materials Park, OH, pp. 539–546.

N. Canter. 2010. MWF heat treatment: Options for a critical process. *TLT*, 66(3), 26–28.

T. Croucher. 1982. Polymer quenchants: Their advantages for aluminum alloys. *Heat Treating*, 14(11), 18–19.

T. Croucher. 2008. Using polyalkyleneglycol quenchants to effectively control distortion and residual stresses in heat treated aluminum alloys. *J. ASTM Int.*, 5(10), 1–16.

M. Eshraghi-Kakhki, M. A. Soltani, K. Amini, H. R. Mirjalili, R. Rezaei, T. Haghir, and M. R. Zamani. 2009. Application of polymeric quenchant in heat treatment of steels. *Int. J. ISSI*, 6(1), 34–38.

W. Gestwa and M. Przylecka. 2010. The modification of sodium polyacrylate water solution cooling properties by Al_2O_3. *Adv. Mater. Sci. Eng.*, 2010, 1–5.

N. A. Hilder. 1988. The behavior of polymer quenchants. PhD. Thesis, University of Birmingham (Aston), Birmingham, UK.

R. Ikkene, Z. Koudil, and M. Mouzali. 2013. Cooling power of polyethylene glycol (peg) aqueous solutions used as quenching media. Evaluation of heat transfer coefficient. In: N. M. Johari, M. J. M. M. Noor, T. I. M. Ghazi, R. M. Yusuff, T. A. Mohammed, A. A. Aziz, and R. Wagiran (Eds.). *3rd International Conference Engineering Education Proceedings*, Federation of Engineering Institutions of Islamic Countries, Madinah, Kingdom of Saudi Arabia, pp. 1–8.

N. I. Kobasko. 1989. Quenching media. *Metalloved Term. Obrabotka*, 23, 127–166.

W. Luty. 2010. Cooling media and their properties. In: B. Liscic, H. M. Tensi, L. C. F. Canale, and G. E. Totten (Eds.). *Quenching Theory and Technology*, 2nd edition, CRC Press, Boca Raton, FL, pp. 377–388.

D. S. Mackenzie. 2003. Advances in quenching-a discussion of present and future technologies. In: *Proceedings of the 22nd Heat Treating Society Conference and the 2nd International Surface Engineering Congress*, ASM International, Materials Park, OH, pp. 228–239.

Y. Mikita, I. Nakabayashi, N. Ohga, and K. Ohsaka. 1987. Study of quench cracking of a high carbon chromium steel bar. *Trans. of the Japan Soc. of Mech. Eng. Series A.* 53(496), 2211–2215.

Polymer quenchants. In: G. E. Tottten, C. E. Bates, and N. A. Clinton (Eds.). *Handbook of Quenchants and Quenching Theory*, ASM International, Materials Park, OH, 1993, pp. 161–189.

K. N. Prabhu and A. A. Ashish. 2002. Inverse modeling of heat transfer with application to solidification and quenching. *J. Mater. Manuf. Process.*, 17(4), 469–481.

S. S. Sahay, G. Mohapatra, and G. E. Totten. 2009. Overview of pearliticrail steel: Accelerated cooling, quenching, microstructure, and mechanical properties. *J. ASTM Int.*, 6(7), 1–26.

G. S. Sarmiento, C. Bronzini, A. C. Canale, L. C. F. Canale, and G. E. Totten. 2008. Water and polymer quenching of aluminum alloys: Are view of the effect of surface condition, water temperature, and polymer quenchant concentration on the yield strength of 7075-T6 aluminum plate. *J. ASTM Int.*, 6(1), 1–18.

Steel quenching technology. In: H. Chandler (Ed.). *Heat Treater's Guide: Practices and Procedures for Irons and Steels*, 2nd edition, ASM International, Materials Park, OH, 1995, pp. 81–82.

G. E. Totten. 2005. Polymer quenchants. In: K. H. J. Buschow, R. W. Cahn, M. C. Flemings, B. Ilschner, E. Kramer, S. Mahajan, P. Veyssiere (Eds.). *Encyclopedia of Materials: Science and Technology*, 2nd edition, Pergamon Press, Oxford, UK, pp. 1–11.

G. E. Totten, B. Liscic, N. I. Kobasko, S. W. Han, and Y. H. Sun. 1998. Advances in polymer quenching technology. In: R. Colas, K. Funatani, and C. A. Stickels (Eds.). *The 1st International Automotive Heat Treating Conference*, ASM International, Materials Park, OH, pp. 37–44.

G. E. Totten, K. B. Oraszak, L. M. Jarvis, and R. R. Blackwood. 1991. How to effectively use polymer quenchants. *Ind. Heat.*, 1991, 37–41.

G. E. Totten, G. M. Webster, R. R. Blackwood, and L. M. Jarvis. 2015. Importance of quench bath maintenance. http://getottenassociates.com/pdf_files/Importance of Quench Bath Maintenance.pdf.

G. E. Totten, G. M. Webster, L. M. Jarvis, S. H. Kang, and S. W. Han. 1997. Thermal/oxidative stability and polymer drag out behavior of polymer quenchants. In: D. Milam, D. Poteet, G. Plaffmann, W. Albert, A. Muhlbauer and V. Rudnev (Eds.). *17th ASM Heat Treating Society Conference Proceedings Including the 1st International Heat Treating Symposium*, ASM International, Materials Park, OH, pp. 443–448.

E. Troell and H. Kristoffersen. 2010. Influence of ageing and contamination of polymer quenchants on cooling characteristics. *BHM*, 155(3), 114–118.

CHAPTER 26

Polymer Nanocomposites for Food Packaging Applications

Gibin George, B. Sachin Kumar, and Anandhan Srinivasan

CONTENTS

POLYMERS ARE WIDELY USED in packaging of food, chemicals, medicines, industrial components, agriculture, and household items as listed in Table 26.1. Polymer consumption has an average growth rate of 5% and it will touch a figure of 227 million metric tons in 2015. Criteria for selection of a particular polymer for packaging is its biodegradability, migration of hazardous compounds, barrier properties, processability, strength,

TABLE 26.1 Common Polymers Used in Packaging

Application		Polymer
Food and beverage packaging	Ready meals	HIPS, PVC, PS
	Flour, spices, sugar	PP/HDPE woven bags, PET/PE, PET/PET/PE, PET/PA/PE, PET/AL/PE layered bags
	Meats and seafood	PP/EVOH blend, HDPE
	Snacks and confectioneries	PET/PE, PET/MET-PET/PE, PET/MET-BOPP, BOPP/BOPP, PET/PA/AL/PE, PA/PE, MET-PET/PE layered bags
	Dairy and cheese	PET, HDPE
	Ghee, oils, pickles	PET/PE, PET/PA/PE, HDPE
	Beverage	Packages: PET/AL/PE, PET/PA/AL/PE, PA/AL/PE, PA/PE Bottles: PET, PEN
	Frozen desserts	PP, HDPE
	Vegetables, fruits	PP
Cosmetics and personal care	Creams, lotions	PVC, HDPE, PMMA
Retail and consumer goods	Blisters or clamshells	PVC, PET
	Skin or stretch packaging	LDPE
Medical and pharmaceutical	Tablets and capsules	PVC, PET
	Syrup and solutions	PP, HDPE
Agriculture	Seeds	PP, BOPP, VMPET, PE
	Fertilizers	PP, HDPE/PP
	Pesticides	PET/VMPET/CPP
Automotive	Lubricants	PET/PE, PET/MET-PET/PE layered, HDPE
	Gasoline	HDPE, EVOH/HDPE layered
	Metal parts	PP
Construction materials	Cement, sand	PP/HDPE woven bags, PP
Household	Water bottles	PET
	Paints	HDPE, PP
	Cleaning liquids, shampoo	HDPE

HIPS, high-impact polystyrene; PVC, poly(vinyl chloride); PS, polystyrene; HDPE, high-density polyethylene; PET, polyethylene terephthalate; PE, polyethylene; PA, polyamide; AL, aluminum; PP, polypropylene; EVOH, ethylene vinyl alcohol copolymer; MET, methyl–ethylene terephthalate; LDPE, low-density polyethylene; PMMA, poly(methyl methacrylate); BOPP, biaxially oriented polypropylene; VMPET, vacuum metallized PET.

and inertness. Polymers are made to order in a variety of fashions using additives, functionalization, crosslinking, and so on for a particular application. A successful packaging material should be able to maintain the quality of the packed commodities to meet the customers' expectation as it moves from manufacturer to customer.

26.1 INTRODUCTION

One of the major applications of polymers in packaging is for food. There are several polymers used in packaging of food, but all of them are used as either blends or multilayer films, since no polymer in its pure form exhibits adequate mechanical strength, and barrier properties required for food packaging. For instance, an application where high-level oxygen barrier and mechanical strength is required, a single pure polymer film cannot serve the purpose; instead, a combination of polymers, as a multilayer film, is used. The production of such a multilayer film is complex and expensive, because of the additional requirements of adhesives and additives, also it has to meet the considerations of the federal law. Polymers are used as packaging films as well as sealants and adhesives. Figure 26.1 shows the structure of a multilayered polymer film.

Polymer nanocomposites are capable of solving the bottlenecks in packaging applications, because they have a polymer matrix reinforced with nanosized materials. Nanomaterials are classified into three major classes according to their quantum confinement, namely, one dimensional, two dimensional, and three dimensional. Spherical nanoparticles (NPs) such as metal oxides are placed under the group of three-dimensional nanomaterials, which have all the dimensions within the nanometer range. Whiskers and nanotubes represent two-dimensional nanomaterials, which have two dimensions within the nanometer range. The third classification is flakes and sheets, which have one dimension in the nanometer range. Graphene sheets and layered silicates are examples of this class. Nanomaterials are preferred over their micro counterparts because of their high surface area. Imagine a cube of length 1 μm is divided into cubes of 1 nm; then, the surface area is increased by 1000-fold as depicted in Figure 26.2. Thus, the amount of filler required to enhance properties of a matrix is much less than that is required by the microsized fillers. The incorporation of nanosized fillers does not alter the esthetics of a polymer film in terms of transparency and flexibility, and the amount of fillers required is the minimum to reach the required quality, whereas in the case of microsized fillers, the entire properties of the polymers are changed and also it is required in abundance to achieve the preferred properties.

Nanosized reinforcements will improve mechanical and barrier properties of polymers. Besides the aforementioned properties, polymer nanocomposites exhibit many

FIGURE 26.1   Construction of a multilayer film used in food packaging.

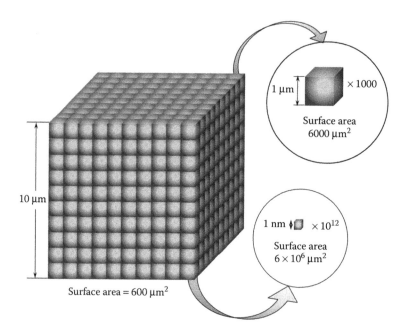

FIGURE 26.2   Increase in surface area with reduction in size.

other unique properties, such as enhanced electrical conductivity, degradation stability, antimicrobial activities, sensing, flame retardancy, and so on. For instance, titanium dioxide (TiO_2) filled nanocomposites exhibit a unique photodegradation property. For a brittle polymer, such as PVC, a rubber-based toughener is used to improve its toughness, which reduces the heat resistance, modulus, and processability of PVC. To overcome these difficulties, heat stabilizer, plasticizer, and lubricants are added to it. Thus, the chain of additives prolongs till the desired properties of the pristine polymer is achieved. Many nanofillers can act as a plasticizer to improve processability, while improving toughness, Young's modulus, and barrier properties of polymers. Thus, the amount of additives required to meet the requirements for packaging can be considerably reduced by using nanofillers. The application of a particular polymer nanocomposite is not limited to a specific application such as food packaging; instead, it can be used in a wide range of applications.

26.2 FABRICATION OF POLYMER NANOCOMPOSITES

26.2.1 *In Situ* Polymerization

During this process, the polymerization of monomers takes place in the presence of nanofillers. Initially the particles of nanofillers are dispersed well in the monomer or precursor solution, the mixture is then allowed to polymerize by adding suitable initiator/catalysts and the necessary conditions for polymerization. Physical properties of the nanocomposite can be tailored by controlling the polymerization process. Uniform dispersion and good polymer–filler interaction could be achieved by the *in situ* polymerization.

26.2.2 Emulsion Polymerization

This is analogous to *in situ* polymerization except for the medium of dispersion of nanofillers. In the former technique, the reinforcements are dispersed in the monomer or the precursor itself, whereas in emulsion polymerization the nanofillers are dispersed in a medium in which the monomer and the nanofiller disperse well.

26.2.3 Solution Casting

In solution casting, the desired polymer matrix is dissolved in a suitable solvent followed by dispersing the nanofillers in the polymer solution. Then the nanofiller dispersed solution is cast as films. The loading of the fillers is limited in this case because the chance of agglomeration is high. The solution casting process is done at ambient temperature, thus, the degradation of the filler surface modifications during the processing can be avoided.

26.2.4 Melt Mixing/Melt Blending

This is a widely used technique for the fabrication of polymer nanocomposites and it is similar to the existing industrial processing techniques and it doesn't require any solvent. Melt mixing is suitable for large-scale production and is a versatile one. The chances of agglomeration of nanofillers are more in melt mixing, even though a higher shear rate enables a greater extent of filler loading as compared with the other fabrication techniques. A good dispersion of fillers is achieved by careful control over the processing conditions such as temperature, time, shear rate, and sometimes redesigning the processing equipment itself. Good compatibilizers are always preferred to improve the adhesion and the interaction between the matrix and fillers.

26.2.5 Masterbatch Processing

Masterbatch process is similar to melt mixing, in which the composites are prepared in two stages of melt mixing. Initially, a masterbatch with a known high filler loading is melt processed, and then the masterbatch is diluted to the required filler content by melt mixing of the virgin polymer. This process is helpful in the dispersion of layered nanofillers in polymers with low melt viscosities, where the exfoliated dispersion of platelets is difficult to achieve.

26.3 COMMON PACKAGING POLYMERS, POLYMER BLENDS, AND THEIR NANOCOMPOSITES

There is a large variety of polymers and their nanocomposites used in packaging applications. At many instances, polymer nanocomposite themselves can be used as packaging materials, but its application can be extended as heat sealable gums (Manias et al. 2009), the coating on other packaging materials, for example, paper (Youssef et al. 2013b), and as sensors to pathogens. The most common polymers and their nanocomposites, used in packaging applications, on their chronological order are discussed in the following paragraphs. The classification of polymer composites for packaging applications are made on the basis of their biodegradability.

26.3.1 Nonbiodegradable Polymers

The majority of the nonbiodegradable polymers are derived from petroleum-based products. The advantages of these polymers are their stability at rough environmental conditions; esthetic appearance; barrier properties; resistance to oils, chemicals, microbial, and fungi interference. In spite of their advantages, strewing of nonbiodegradable polymers has become a serious threat to the living organisms. The major source of nondegradable plastics waste is from packaging applications, especially those related to food. The reuse or recyclability of such polymers for food packaging is limited because of the health concerns. The incorporation of nanofillers has the potential to reduce the use of such polymers by enhancing the barrier properties, strength, and so on; thus, reducing the thickness of the conventional films to meet the requirements.

26.3.1.1 Polyethylene

Polyethylene (PE) is one of the most common thermoplastics used in packaging applications. The consumption of PE around the world is more than 35 million tons, that is one-third of the world's plastics production. PE grades are mainly classified according to their density as low-density PE (LDPE), high-density PE (HDPE), medium density PE, linear low-density PE (LLDPE), very-low-density PE, ultra-low density PE, and ultra-high molecular weight PE. All the grades of PE have excellent resistance to strong acids or strong bases and also to oxidizing and reducing agents, making them the ideal material for packaging of foods. Therefore, PE is always used as the inner layer in multilayer films, which has direct contact with the packed food material, and the barrier properties of PE alone is not good enough to prevent the contact of surrounding oxygen and water vapor.

The addition of nanoclay in PE matrix significantly improves its barrier properties. The nonpolar nature of PE necessitates a suitable compatibilizer for the fine dispersion of clay layers. LLDPE/org-clay (org-montmorillonite, MMT) nanocomposite with oxidized PEs (OxPEs) as the compatibilizer has a remarkably enhanced O_2 barrier and mechanical properties than the pure LLDPE (Durmuş et al. 2007, Golebiewski et al. 2008), also the presence of 3% clay brings about a reduction of 30%–40% in the peak heat release rate (Zhang and Wilkie 2003). HDPE loaded with org-MMT with maleic anhydride-grafted HDPE (MAH-g-HDPE) as a compatibilizer has a gas permeability coefficient higher than that of neat HDPE (Picard et al. 2007). LDPE/Mg–Al layered double hydroxide (LDH) nanocomposites with MAH-g-PE as compatibilizer exhibit improved fracture toughness and failure properties (Costa et al. 2006) and also a reduction in the heat release rate. Itaconic acid (IA) and 2-[2-(dimethylamino)-ethoxy] ethanol (Sánchez-Valdes et al. 2012) were also used as compatibilizers in PE/clay nanocomposite.

An *in situ* polymerized PE nanocomposite containing silver NPs (Ag-NPs) exhibited a fantastic antimicrobial property at 5% silver content against *Escherichia coli* bacteria (Zapata et al. 2011). PE/ethylene vinyl acetate (EVA)/clay nanocomposite is used in hermetic seals with a general peelable/easy-open character across the broadest possible sealing temperature range (Manias et al. 2009, Zhang et al. 2009); such a material is useful in flexible packaging of fresh-cut vegetables, processed foods, and biomedical devices. Incorporation of EVA and

MMT in the PE-based nanocomposite sealant forms regions with weak interfaces, where the cohesive failure of the nanocomposite is promoted as shown in Figure 26.3.

HDPE loaded with silane-modified carbon nanofibers (CNFs) exhibit a good dispersion and adhesion between the matrix and the filler; thereby, offers good wear resistance and low coefficient of friction (Liu et al. 2011), making them suitable for heavy-duty packaging. The mechanical properties of graphite-loaded PE nanocomposite exhibited a significant improvement in tensile strength by a maximum of 53% and modulus by 30% (Sarikanat et al. 2011).

There are many PE nanocomposites having qualities applicable in packaging applications, but not exclusively studied for packaging. HDPE loaded with expanded graphite particles exhibit a remarkable improvement in tensile properties and storage modulus (Sever et al. 2013). Ag-NPs in addition to the layered fillers in PE matrix improve its antibacterial properties (Becaro et al. 2015). It is also proven that the migration of these silver as ions from the PE nanocomposites to the food stimulants is less than the cytotoxicity level after 30 days (Jokar and Abdul Rahman 2014). In the hybrid nanocomposite of nanostructured TiO_2–clay–LDPE, the exfoliation of the clay layers was improved in the presence of TiO_2 NPs (Moghaddam et al. 2014), thereby an improvement in the barrier and tensile properties were observed in this composite as compared with neat PE. HDPE/Cu-nanofiber nanocomposites exhibited enhanced antibacterial properties, as well as oxygen barrier properties (Bikiaris and Triantafyllidis 2013). The addition of nanosized $CaCO_3$ improved the mechanical properties of the matrix (Zaman and Beg 2014).

It is interesting to note that, in LLDPE/TiO_2 nanocomposites, TiO_2 acts as a photocatalyst for the accelerated degradation of the matrix (Zapata et al. 2014) and it will be

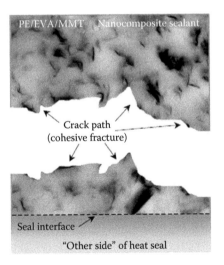

FIGURE 26.3 A schematic representation of the proposed mechanism for the general peelable character of the PE/EVA/MMT nanocomposite sealants. (From Manias, E. et al.: Polyethylene nanocomposite heat-sealants with a versatile peelable character. *Macromol. Rapid Commun.* 2009. 30. 17–23. Copyright Wiley-VCH Verlag GmbH & Co. KGaA. Reprinted with permission.)

an added advantage, especially in the packaging application, because majority of the waste plastics originate from packaging.

26.3.1.2 Polypropylene

Polypropylene (PP) is also a widely used polymer in packaging applications (films, bottles, screw caps, hinged joints, etc.) because of its optimal mechanical properties, processability, and low cost, but the low resistance to crack propagation results in the easy breakdown of the final product in the presence of inherent crack or mechanical failure. The introduction of nano $CaCO_3$ with maleic anhydride-grafted PP as a compatibilizer to PP matrix resulted in better impact strength, modulus, and tensile strength (Fuad et al. 2010). Polyether-treated clay-loaded PP nanocomposites in the presence of maleic anhydride-grafted PP as a coupling agent had a remarkable decrease in the oxygen permeability as compared with the pristine PP. The oxygen permeability decreased with the subsequent increase in the clay loading (Pannirselvam et al. 2008). The mechanical properties, such as flexural and impact strength of PP increased considerably by the addition of organo-MMT (OMMT), which is inferior in pristine PP and the thermal stability of the composite improved distinctly over pure PP (Zhu et al. 2011). Halloysite nanotubes can act as a β-nucleating agent for isotactic PP and at 20 phr of halloysite, content of β-iPP phase was found to be the maximum. The nucleation behavior of these composites was the highest at a crystallization temperature of 135°C and it was also a function of cooling rate (Liu et al. 2009). There was an increase in the gas barrier properties of fumed nanosilica/PP composites using maleated PP as a compatibilizer, with the filler content. The addition of silica not only improved the storage modulus of the composite, but also their crystallinity. The change in storage modulus as well as crystallinity was a function of the amount of the compatibilizer and the nanofiller (Vladimirov et al. 2006). The improvement in tensile strength by 70% and crystallinity by 50% of PP was achieved by the incorporation of cellulose nanowhiskers. The cellulose content in PP matrix also enhanced the degradation temperature, hydrophilicity, and thermal conductivity of the composite over control PP samples. The addition of nanocellulose whiskers to PP matrix improved the mechanical and thermal properties of PP (Bahar et al. 2012)

The presence of nisin in PP/MMT clay nanocomposite exhibited antibacterial properties as compared with conventional packages (Meira et al. 2014). PP coated with nanocomposites of other polymers is capable of improving the gas barrier properties of PP. PP layers coated with corn zein/clay composite (Ozcalik and Tihminlioglu 2013) is the best example for the same. PP/TiO_2 nanocomposite coatings prepared by one-step dipping exhibited a permanent superhydrophobicity (Contreras et al. 2014), which could be helpful in protecting the food material from several harmful environmental simulants. The addition of silica–silver core–shell NPs to PP matrix increases the mechanical properties of PP, along with its antibacterial properties (Suktha et al. 2013).

26.3.1.3 Polyvinyl Chloride

Polyvinyl chloride (PVC) is a commercial polymer widely used in industries and household applications. PVC is used in blister or clamshell packaging of medicines and other commodities. PVC is brittle in nature and its processability, barrier properties, and thermal properties are inferior to other thermoplastics. Therefore, PVC is always used in conjunction with tougheners, plasticizers, heat stabilizers, lubricants, and fillers. Incorporation of nanofillers in PVC is a cheap and efficient method to improve its deficits. *In situ* polymerized nanocomposite of PVC/$CaCO_3$ exhibited an improvement in stiffness and toughness over the pristine PVC (Xie et al. 2004). The improved mechanical properties, such as tensile yield strength, elongation at break, and Young's modulus were observed at 5 wt.% of $CaCO_3$. Also, the shear viscosity was reduced simultaneously due to the ball bearing action of the spherical NPs, which enhances the processability of the composite. PVC reinforced with silica NPs exhibited good thermal properties (Mohagheghian et al. 2014), and a reduction in diffusivity even though the permeability of the gases through the film was increased. Silica NPs grafted with PMMA (poly[methyl methacrylate]) and PSBA (poly[styrene butyl acrylate]) (Zhu et al. 2008a) possessed good interfacial properties with the PVC matrix, hence, an improved thermal and mechanical properties were obtained.

TiO$_2$ nanowire-grafted Ag-NP-filled PVC had antibacterial properties against *E. coli* (Liu et al. 2012), which is a major cause of food poisoning in humans. *Rhodamine B*, a carcinogenic dye, is degraded in the presence of UV light because of the photocatalytic action of the TiO$_2$ nanowires. PVC/PMMA-grafted nanocomposites had good thermal stability (Liu et al. 2011) and also improved toughness, strength, and modulus. Polystyrene (PS)-grafted ZnO nanoparticulate-filled PVC composites exhibited a good absorption of UV light (Tang et al. 2012). Nanoclay-loaded PVC composite exhibited a remarkable smoke reduction during combustion and the thermal stability was improved over the pristine polymer (Awad et al. 2009). Melt-intercalated PVC/compatibilizer/organophilic-MMT composite had dramatically enhanced impact strength over pure PVC (Ren et al. 2005). PVC/OMMT nanocomposites showed better thermal stability and mechanical properties than neat PVC and also a significant decrease in softening temperature. Nontoxic nano zinc borate in PVC matrix significantly improved the flame retardancy of the PVC, without affecting its mechanical strength significantly (Madaleno et al. 2010). PVC matrices loaded with multiwalled carbon nanotubes (MWCNTs) exhibited excellent electrical conductivity, thermal and mechanical properties (Broza et al. 2007, Mamunya et al. 2010), and an improvement in T_g (Sterzyński et al. 2010). Single-walled carbon nanotube (SWCNT)-loaded PVC composites had improved thermal conductivity as compared with the pristine PVC.

26.3.1.4 Polystyrene

PS is used in the protective packaging system of fragile commodities, disposable cups, and meat trays as an expanded foam. Expanded PS has good shock absorbance; nevertheless, low heat resistivity of expanded PS along with its low barrier properties is a major concern in the packaging application. The addition of NPs to PS can reduce the shrinkage and warpage in preforms during the reheating process to improve the blow molding properties of PS (Sun et al. 2011).

PS/clay nanocomposites were prepared by several methods, such as melt intercalation (Wu et al. 2002), solution intercalation (Su et al. 2004), injection molding (Sorrentino et al. 2006), and *in situ* polymerization (Uthirakumar et al. 2005). PS/clay nanocomposites with compatibilizers showed an improvement in T_g by 8°C–9°C, dynamic storage modulus (Ogunsona et al. 2011), tensile strength, and impact strength (Yilmazer and Ozden 2006). Bulk polymerized PS nanocomposite loaded with polybutadiene-modified Na-MMT had superior properties over the other nanocomposite forming techniques (Su et al. 2004). Tetracalcium aluminate, an LDH, filled PS-PMMA exhibited improved T_g values at low filler contents (Matusinović et al. 2009). PS/Mg-Al LDH composite had an enhanced thermal stability and an increase in decomposition temperature with reduced flammability (Tai et al. 2011). The tensile properties of PS were increased by the addition of TiO_2 NPs and the photocatalytic properties of TiO_2 caused the yellowing of PS (Chandra et al. 2007) and this property can be exploited to increase the degradation rate of the polymer. PS loaded with SWCNT grafted with styrene showed significant enhancement in tensile and flexural strength as well as electrical conductivity over pristine PS (Nayak et al. 2007). The addition of 1% of CNT to PS increased the mechanical properties of PS significantly (Qian et al. 2000). Silica NP-filled high-impact PS (HIPS) had better thermal stability and reduced flammability (Katančić et al. 2011). TiO_2 nanofiber-reinforced PS is suitable for packaging because of its ability to remove organochlorine pesticide contaminants (Youssef et al. 2013c). Zn–Al LDH can improve the structural stability of PS and also the gas barrier properties (Youssef et al. 2013a). The mechanical and thermal properties of PS reinforced with modified cellulose nanocrystals (CNCs) are impressive (Lin and Dufresne 2013). Organo-modified vermiculite/PS nanocomposites obtained via *in situ* polymerization showed a significant enhancement on thermal stability and dynamic mechanical properties (Wang et al. 2013). PS loaded with Ag-NPs exhibited antibacterial properties against gram positive and gram negative bacteria, fungus, microbes, and so on and good mechanical strength (Youssef and Aziz 2013).

26.3.1.5 Polyamides (Nylon 6 and Nylon 12)

Nylon 6 and nylon 12 are the most common polymers used in the packaging application from this class. Nylon 6 is the most common among them; because of its wide melting range, it is heat sealable and coextrudable. Nylon 6 is a clear film with pretty good gas and aroma barrier but has poor moisture barrier properties. It also has superior strength and outstanding tear and puncture resistance at low temperature. Biaxial orientation is used to improve the mechanical strength of nylon. When nylon 6 is filled with SiO_2 NPs, the storage and loss modulus decreases but the melting temperature is slightly increased (García et al. 2004). The hardness, elastic modulus, and creep behavior of nylon 6 were improved by the increase in clay loading (Shen et al. 2004). Polyamide 6/OMMT nanocomposite has three to four times less solvent permeability than pristine nylon and this can be exploited in the field of barrier containers for food packaging (Jiang et al. 2005). Halloysite-loaded nylon 6 has flame inhibition properties (Marney et al. 2008). Exfoliated graphite nanosheets act as a nucleating agent in nylon 6. It was also reported that the presence of graphite sheets could hinder the movement of the polymer chains, thereby a decrease in the crystallization growth rate can also occur (Weng et al. 2003). The water vapor permeability of the nylon 12 loaded with MMT nanoclay decreased as the filler content was increased (Alexandre et al. 2009). Polyamide 6 loaded with cloisite 30B (a modified MMT) had reduced oxygen permeability by a thousand times (Pereira et al. 2009). The addition of clay in polyamide can optimize the barrier properties to the required level. The food packed in films with good barrier properties will have an extended shelf life and this enables the long distance transport of the packed food.

Electrospun composite nanofibrous membrane of MMT–nylon 6 on PP films improves the intrinsic oxygen carrier properties of PP slightly (Agarwal et al. 2014). Nanoclay-loaded nylon films exhibit improved barrier properties on oxygen, carbon dioxide, and water vapor. The nanoclay simultaneously acts as a reinforcement and a nucleating agent, since both crystallization temperature and storage modulus increased with clay loading (Allafi and Pascall 2013).

26.3.1.6 PE Terephthalate

PET is one of the most commonly used polymers in packaging, especially for food, because of its inertness and good mechanical properties. PET in biaxially oriented form has excellent strength, toughness, and better oxygen and CO_2 barrier properties. Exfoliated PET/MMT nanocomposite and PET/SiO_2 nanocomposite obtained through *in situ* polymerization had improved mechanical properties and thermal stability than the virgin PET. Both these fillers affect the crystallinity and the molecular weight of PET in the composites (Vassiliou et al. 2010). *In situ* polymerization can be used to synthesize PET/LDH intercalated with terephthalate or dodecyl sulfate nanocomposite. The microwave irradiation improves the delamination of LDH and such nanocomposites showed good thermal stabilities at low temperatures (Herrero et al. 2011). High-energy ball milling is also an alternate way to improve the dispersion of LDH in PET matrix to improve the barrier properties (Tammaro et al. 2014). The exfoliation of clay platelets are improved by equibiaxial stretching of the PET/nanoclay composites, also, the preferential orientation of the clay platelets can be achieved by stretching as shown in the TEM images (Figure 26.4). The improved mechanical and barrier

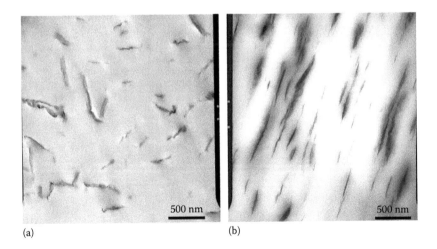

(a)   (b)

FIGURE 26.4 TEM images of the unstretched (a) and equibiaxially stretched and (b) PET nanocomposites. Stretch ratio is 3. Samples were collected from the cross-section of the sheets. (Reprinted from *Eur. Polym. J.*, 45, Rajeev, R.S., Harkin-Jones, E., Soon, K., McNally, T., Menary, G., Armstrong, C.G., and Martin, P.J., Studies on the effect of equi-biaxial stretching on the exfoliation of nanoclays in polyethylene terephthalate, 332–340, Copyright 2009, with permission from Elsevier.)

properties of the composite during stretching are helpful when the polymer composite is used in the packaging applications (Rajeev et al. 2009). The incorporation of graphite nano-platelets improves the barrier properties of PET to a large extent. The induced crystallization in PET matrix by the graphite nanoplatelets also has an important role in improving the barrier properties further (Al-Jabareen et al. 2013).

In situ prepared TiO$_2$ nanoflower on PET fabric grafted with nanocrystalline cellulose hybrid nanocomposite exhibited good self-cleaning performance against methyl orange in the presence of solar light (Peng et al. 2012). Graphene sheets dispersed in PET matrix can significantly improve the oxygen barrier properties of PET (Al-Jabareen et al. 2013). PET/halloysite nanotubes loaded with sodium benzoate, a model antibacterial drug, is a promising material for active packaging application, where the antibacterial drug can be released at a controlled rate, the addition of halloysite improves the mechanical and thermal properties of PET (Gorrasi et al. 2014).

26.3.1.7 Ethylene Vinyl Acetate

EVA/clay nanocomposites, containing different percentage of organo-clay (OC), which are prepared via solution mixing, and films prepared via thermal phase inversion method showed reduced permeability to oxygen. EVA chains enter into interlayer space of the clays to form intercalated or semiexfoliated structures even in the absence of any compatibilizer. Thus, small fractions of OC can significantly reduce the permeability of oxygen through it (Figure 26.5). It is also confirmed that the polarity of the EVA and the basal spacing of clay in nanocomposites can significantly affect the gas barrier properties of EVA/clay nanocomposites. EVA/MMT nanocomposites showed a reduced peak heat release rate, a measure of the flame retardancy (Shafiee and Ramazani 2008), as compared with neat EVA.

EVA-based hybrid material containing silicon and phosphorus has been developed in order to improve the fire retardancy of EVA. Phosphorus promotes the formation of a

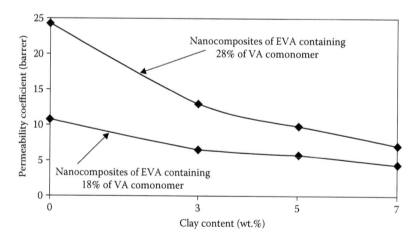

FIGURE 26.5 Oxygen permeability through EVA/organo-clay nanocomposite. (From Shafiee, M. and Ramazani, A.: Investigation of barrier properties of poly(ethylene vinyl acetate)/organoclay nanocomposite films prepared by phase inversion method. *Macromol. Symp.* 2008. 274. 1–5. Copyright Wiley-VCH Verlag GmbH & Co. KGaA. Reproduced with permission.)

charred layer acting as a thermal and diffusion barrier; meanwhile silicon atoms present in the cross-linked network participate in the formation and strengthening of the char (silicophosphorated bonds), leading to a better fire retardance of EVA with very low Si and P contents (less than 3 wt.%) (Tang et al. 2002).

26.3.1.8 Ethylene-Vinyl Alcohol Copolymer

Ethylene vinyl alcohol (EVOH) is known for its exceptional oxygen barrier properties. EVOH is used as barrier films in the packages of dry foods. The antimicrobial films developed for active packaging applications containing the natural antimicrobial compound LAE (lauramide arginine ethyl ester) in EVOH copolymers with different mol % ethylene contents were more effective in inhibiting the growth of *Listeria monocytogenes* than *Salmonella enterica*, this inhibition being more acute at the end of the storage time (Muriel-Galet et al. 2012). Water contact angle measurements demonstrated that the hydrophilicity of EVOH/PVP membranes was improved compared with that of the neat EVOH membrane. The result of *bovine serum albumin* protein adsorption experiments also showed improved antifouling property of EVOH/PVP membranes (Lv et al. 2006). β-Cyclodextrin was successfully incorporated in EVOH with a 44% molar percentage of ethylene by regular extrusion with the aid of glycerol to improve the dispersion of the oligosaccharide. The barrier properties of gases and water decreased slightly due to the increase of free volume in the matrix. Also, the CO_2/O_2 permselectivity increased, an attribute with interesting uses in active packaging (López-de-Dicastillo et al. 2010). The EVOH films containing 20% and 30% of β-cyclodextrins successfully retained cholesterol from pasteurized and UHT milk at 4°C and 23°C, respectively, and showed applicability to scavenging oxidative by-products, when these films were tested with fried peanuts (López-de-Dicastillo et al. 2011). The EVOH nanocomposite obtained with 2% of bentonite nanoclay, containing 5% of carvacrol as an antimicrobial compound, was characterized in terms of antimicrobial solubility and release, water vapor solubility and permeability, oxygen and carbon dioxide permeability, diffusivity and solubility, thermal properties, and microstructural morphology and exhibited a potential application for food packaging (Cerisuelo et al. 2012). EVOH/organically modified MMT nanoclay composite exhibited good mechanical, thermal, and improved oxygen and water vapor barrier properties. The moisture-bound deterioration in barrier properties of EVOH is suppressed in the presence of clay (Kim and Cha 2014).

26.3.2 Biodegradable Polymers

Even though one of the major inventions of this era, the benefits of plastics eclipse their non-biodegradable nature and their inappropriate waste management. The invention of natural polymers, biopolymers, and synthetic polymers from renewable resources is an attempt to overcome the ill effects of the nondegradable plastics on the environment. Biodegradable polymers can be divided into two classes: (1) those, which are produced artificially from petroleum products or other chemicals, such as poly(vinyl alcohol) (PVA), polyurethane (PU), and poly(ε-caprolactone) (PCL) and (2) those, which are produced from natural resources, such as starch, poly(lactic acid) (PLA), cellulose, and chitin. The strength, barrier properties, and the processability of biodegradable polymers are the major concern in the packaging application. By adding nanosized reinforcements to biodegradable plastics, plausible properties of the existing polymers in packaging can be attained. The major biodegradable plastics familiar to the packaging industry are PVA, PU, starch, cellulose, PLA, and so on because of their low cost, renewability, processability, and abundance.

26.3.2.1 Poly(Vinyl Alcohol)

PVA is a water-soluble and biodegradable synthetic polymer derived from petroleum-based products. It has good resistance to oils and solvents, and also good barrier properties against oxygen and aroma. The limiting factor in the application of PVA is atmospheric humidity. The absorbed water in the polymer can act as a plasticizer and degrade all the properties of PVA mentioned earlier. PVA has been used as binding and coating agents in moisture barrier films for food packaging applications, where moisture can damage the contents. It has been used as CO_2 barrier films in PET bottles. PVA is nontoxic and it is used as food contact materials in the packaging of vegetables and fruits to prevent their decay during the transport. PVA/sodium MMT (Na-MMT) nanocomposite exhibited an improvement in tensile properties by 300%, without compromising any other properties, such as stress at break and transparency. The water vapor transmission rates were also remarkably reduced by the addition of Na-MMT to the polymer matrix (Strawhecker and Manias 2000). CNCs can increase the degree of crystallinity in PVA matrix. A good level of compatibility exists between cellulose crystals and polymer matrix (Fortunati et al. 2013). The excellent barrier properties of graphene oxide (GO) nanosheet/PVA nanocomposite against oxygen and water vapor is favorable for food packaging. The high barrier properties, as shown in the Figure 26.6, of the composite films are attributed to the full exfoliation, uniform dispersion, and good alignment of GO nanosheets in the PVA matrix and the strong interfacial adhesion between GO nanosheets and PVA matrix (Huang et al. 2012).

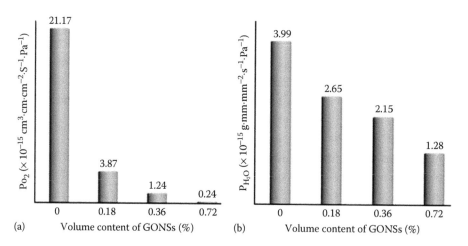

FIGURE 26.6 (a) Permeability coefficient of O2 (P_{O_2}) and (b) coefficient of moisture permeability (P_{H_2O}) for GONS/PVA nanocomposite films as a function of GONS loadings. (Reprinted from *J. Membr. Sci.*, 409–410, Huang, H.-D., P.-G. Ren, J. Chen, Zhanga, W.-Q., Jia, X., and Li, Z.-M., High barrier graphene oxide nanosheet/poly(vinyl alcohol) nanocomposite films, 156–163, Copyright 2012, with permission from Elsevier.)

The composite materials of PVA and bacterial cellulose (BC) were prepared in the form of films and were studied for the situation of irradiation using ionizing radiation of pre-packaged foods. The irradiation had no influence on the transparency of those films, which were transparent enough for food packaging. The swelling ratio of the film (PVA:dry BC = 1:0.05 w/w) was observed to be more stable even after the irradiation enhancing PVA properties (Stoica-Guzun et al. 2013)

The PVA–sodium zirconate (Na_2ZrO_3) nanocomposite films prepared by solution inter-calation technique were reported to have a good physical interaction between PVA and Na_2ZrO_3, and the dielectric constant (ε') and dielectric loss (ε'') were enhanced in the presence of NPs of Na_2ZrO_3 (Chandrakala et al. 2013). CNCs were extracted from Okra bamia (*Abelmoschus esculentus*) bast fibers, which are incorporated in PVA matrix. The 5 wt.% CNC content promoted a direct mechanical interaction between the PVA and cel-lulose structures. IR spectroscopy confirmed that during exposure to UV light, the pres-ence of CNC does not affect the stability of the neat PVA matrix to the photodegradation (Fortunati et al. 2013).

Invasion of larvae into cookies was prevented by the insect-repellent films containing cinnamon oil microencapsulated with PVA. The efficiency of this film was significant at repelling moth larvae and *Plodia interpunctella*, and such microencapsulation did not have any influence on tensile properties of the pristine film. This demonstrated a potential application for insect-repellent food packaging (Jo et al. 2013, Kim et al. 2013). The films of carboxymethyl cellulose (CMC)–PVA–clove oil showed negligible oxygen transmission rate. The meat samples packed in these films had lower total viable counts and displayed a shelf life of 12 days, whereas the pristine film-packed meat sample was spoiled within 4 days of refrigerated storage. The efficacy of these films was also demonstrated by packed inoculum studies against *Staphylococcus aureus* and *Bacillus cereus* in ground chicken meat and this study demonstrated the immense potential of such films to be used as active packaging material for meat preservation (Muppalla et al. 2014).

Antibacterial properties of PVA films were improved by the addition of ZnO NPs. These properties were enhanced by increasing ZnO content up to 5 wt.% at the expense of transparency. The production of hydrogen peroxide at the surface of these films resulted in an outstanding antibacterial activity. Thus, the biocompatible ZnO NPs are promising candidates for substitution of silver-based NPs in food packaging applications (Ahangar et al. 2014).

Oxygen consumption of an active film, PVA/cobalt(II) complex with the ligand L-thre-onine was found to be equal to the complex alone, $Co(II)(L\text{-}Thr)_2(OH_2)_2$ (2.5 mg of O_2 per gram of complex), after water activation (90.5% relative humidity). This study exhibited that, this kind of oxygen scavengers can prevent oxidative damage to flavor as well as color in a wide range of food packaging (Damaj et al. 2014). A composite of sugar beet pulp (SBP) and PVA exhibited an enhanced elongation at break (12.45%), lower water vapor permeability (1.55×10^{-10} gs^{-1} $m^{-1}Pa^{-1}$), and similar tensile strength as that of a pristine polymer (59.68 MPa). The study also showed that 50% of PVA blended with SBP is most suitable for food packaging applications (Shen et al. 2015). The bio-inspired structure of borate ions are used to cross-link PVA and GO to produce flexible, transparent high-barrier composite films. PVA/GO films with only 0.1 wt.% GO and 1 wt.% cross-linker exhibited an O_2 transmission rate <0.005 cc m^{-2} day^{-1} and O_2 permeability rate <5.0 × 10^{-20} cm^3 c m cm^{-2} Pa^{-1} s^{-1}. Further, the transmittance at 550 nm was >85% suggesting them useful materials in food packaging and flexible electronics (Lai et al. 2015).

26.3.2.2 Polyurethane

PUs are available as foams, coatings, adhesives, fibers, rubbers, or thermoplastic elastomers for many industrial as well as packaging applications. PU adhesives are used as laminates for food packaging to hold polymer films together in multilayered packages. PU has poor barrier properties and these have to be improved without increasing the thickness of the multilayer or the number of layers in a food packing laminate. The permeability of PU against water vapor and oxygen has been improved by reinforcing OMMT into the polymer (Osman et al. 2003). Melt compounded thermoplastic PU/clay has been reported to exhibit good tensile properties than the pristine polymer (Dan et al. 2006). Fumed nanosilica/PU exhibited good mechanical, rheological, and adhesion properties over virgin polymer. The enhancement of these properties was due to the formation of hydrogen bonds between silanol groups of nanosilica and ester carbonyl groups of the polymer (Vega-Baudrit et al. 2006). PU/precipitated calcium carbonate nanocomposites exhibited a moderate increase in the rheological and viscoelastic properties due to weak interaction between filler and matrix. PU was used to adhere PVC joints in this study and adhesion between PVC joints increased in the case of PU/precipitated calcium carbonate nanocomposite compared to a pristine polymer (Robles and Martín-Martínez 2011).

The metal particles incorporated PU thin films exhibited excellent antimicrobial property toward *S. aureus, E. coli, Salmonella typhimurium*, and *Klebsiella pneumoniae* bacterial strains. PU films containing Ag, Cu, Mg, and Zn were able to significantly reduce the number of pathogenic cells adherent per polymer unit surface with respect to the pristine PU (Figure 26.7). This study showed the potentiality of the prepared PU films with metal particles as antimicrobial food packaging material (Nirmala et al. 2013).

PU composite film reinforced with functionalized low-defect graphene nanoribbons (HDG) exhibited a decrease in the nitrogen gas diffusivity by three orders with the least filler content of 0.5 wt.%. The phase separation as shown in Figure 26.8 has improved the mechanical and barrier properties of the thermoplastic and has a potential application in food packaging and lightweight mobile gas storage containers (Xiang et al. 2013).

Nonfluorinated PUs, one with a plurality of silicon polyol units and other with a plurality of organic polyol units, wherein, at least one of the organic polyol units consisting an ionizable group were synthesized. These polymers have been used to impart liquid repellency, stain resistance, and bleach resistance to a cardboard storage or a food packaging device or material (Bartley et al. 2014). Waterborne PU/modified nano-ZnO composites membranes were synthesized from modified nano-ZnO, poly(butyleneadipate) glycol (PBA), PCL, isophorone diisocyanate (IPDI), dimethylol propionic acid (DMPA), and ethylenediamine (EDA) by self-emulsified method. The phase property, mechanical, antibacterial, and anti-ultraviolet properties of the composites were enhanced, thus promising a greater potential for food packaging application (Li et al. 2013).

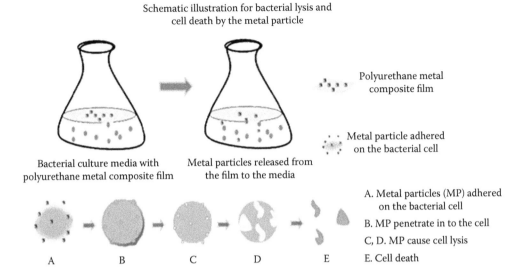

Schematic illustration for bacterial lysis and
cell death by the metal particle

Polyurethane metal
composite film

Metal particle adhered
on the bacterial cell

Bacterial culture media with
polyurethane metal composite film

Metal particles released from
the film to the media

A. Metal particles (MP) adhered
on the bacterial cell

B. MP penetrate in to the cell

C, D. MP cause cell lysis

E. Cell death

A   B   C   D   E

FIGURE 26.7   The antibacterial properties of PU films loaded with metal nanoparticles. (With kind permission from Springer Science+Business Media: *Macromol. Res.*, Multipurpose polyurethane antimicrobial metal composite films via wet cast technology, 21, 2013, 843–851, Nirmala, R., H.Y. Kim, D. Kalpana, Navamathavan, R., and Lee, Y.S.)

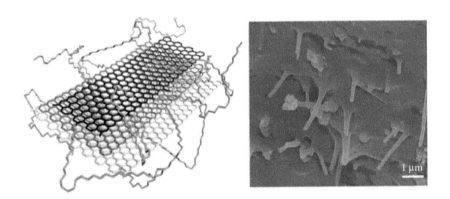

FIGURE 26.8   The PU films loaded with graphene nanoribbons. (Reprinted with permission from Xiang, C. et al., 2013, *ACS Nano* 7, 10380–10386. Copyright 2013 American Chemical Society.)

26.3.2.3 Poly(ε-Caprolactone)

PCL is a biodegradable polyester and semicrystalline in nature. PCL is resistant to water, oil, solvent, and chlorine. Solvent cast films of PCL/nanoclay composite show high thermal properties over pristine PCL (Ahmed et al. 2012). The reinforcement of nanofiller such as MMT into PCL improved the mechanical properties of *in situ* polymerized nanocomposite. Further, as the clay content was increased, the permeability of water and dichloromethane were reduced remarkably. Such a nanocomposite material is widely recommended in biodegradable food packaging (Gorrasi et al. 2002). PCL reinforced with *attapulgite*, a natural fibrillar silicate clay mineral, has a high potential to be used as food packing material (Liu et al. 2007). The reinforcement of cellulose (Siqueira et al. 2011) and TiO_2 (Wei et al. 2012) NPs in PCL drastically improves the crystallization behavior in PCL and it can improve the mechanical properties of the PCL composite films over the neat PCL films. Thermal stabilization of the chitin whisker-reinforced PCL is observed above the melting point of PCL (Morin and Dufresne 2002) explores the melt processability of the PCL. In PCL/$CaCO_3$ nanocomposites, $CaCO_3$ acts as a nucleating agent for crystallization. Thus, mechanical properties of the composite over neat PCL matrix were enhanced (Liu et al. 2010).

26.3.2.4 Starch

Starch is a biodegradable thermoplastic polymer, which can be processed easily. The price of the starch is comparable with that of PET, PS, and LDPE. But the poor barrier and mechanical properties of starch encumber them behind the synthetic petroleum-based polymers in packaging applications. Melt intercalation of clay into starch improved the mechanical properties of starch. Also, incorporation of plasticizers (glycerol, hectorite) could overcome the crack formation noticed in pristine starch or clay/starch nanocomposite during long-term storage. Melt processed starch/clay nanocomposite exhibited good mechanical properties and it was reported that the migration of ions from the composite film to food simulants was within the limit (Avella et al. 2005). Also, the water vapor barrier properties of the composites were enhanced over neat starch. Starch/OMMT nanocomposite exhibited good tensile strength and modulus at an expense of elongation at break. The amount of OMMT controls crystallization in starch, which is favorable for long-term storage (Ren et al. 2009), and also the water vapor barrier properties of the film (Maksimov et al. 2009). Further, it was reported that the reinforcement of CNCs into starch matrix not only increased tensile strength and Young's modulus, but also resistance to water, a measure of water uptake by the composite films was enhanced (Cao et al. 2008). The incorporation of ZnO in *sago* starch matrix improved the overall properties of the matrix such as water vapor permeability of the film. Further, it also induced UV protectability and antibacterial activity to the virgin sago starch film (Nafchi et al. 2012).

A crystalline bio-nanocomposite of starch/SiC exhibited a significant improvement in thermal stability and oxygen barrier property in the presence of SiC NPs as a filler. These bio-nanocomposites were resistant to mineral acid and alkali with little sacrifice in biodegradability and enabled the materials suitable for thermal resistant packaging and adhesive applications (Dash and Swain 2013). The challenge of making starch-based material foams for biodegradable food packaging was investigated using isothermal treatments with supercritical CO_2 with PCL as a blend. PCL (highly foamable polymer) blended with starch increased the foamability of starch. This study showed the valuability of this process to foam starch and starch-based materials that were compostable but not CO_2 miscible. The thermal and mechanical behaviors were enhanced in such foam materials (Ogunsona and D'Souza 2014).

26.3.2.5 Cellulose

Cellulosic materials are widely used in the packaging applications, because of their credibility in biocompatibility, barrier properties, aesthetics, biodegradability and innocuous nature. Hydroxypropyl methylcellulose (HPMC) is a derivation of cellulose, which has good film forming properties approved by US Food and Drug Administration and European Food Safety Authority to use them in food packaging. HPMC reinforced with Ag-NPs exhibited good mechanical and barrier properties, which in turn improved the shelf life of the food products. The antibacterial properties of the film promise its potential for antimicrobial internal coatings in food packaging (de Moura et al. 2012). Edible HPMC blended with chitosan/tripolyphosphate NPs exhibited enhanced mechanical and barrier properties. Also, the thermal degradation temperatures of these composites were significantly improved (de Moura et al. 2009). Cellulose/F-substituted hydroxyapatite (HA) and cellulose/calcium silicate nanocomposites are fabricated by microwave assistance in the presence of an ionic liquid. This is a simple, rapid, and an environment-friendly way to synthesize cellulose-based nanocomposites (Jia et al. 2011, 2012). Ag-NPs reinforced on cellulose filter paper grafted with acrylamide had a fair biocidal action against *E. coli*. This study suggested that it could be used as a food packaging material that prevents the bacterial infection (Tankhiwale and Bajpai 2009). BC/laponite clay nanocomposites exhibited an excellent interaction between organic and inorganic phases. Mechanical properties of the composite films were improved with the increase in clay content (Gorrasi et al. 2002).

Microfibrillated cellulose (MFC) were also used to prepare films containing an active molecule, *lysozyme*, which is a natural antimicrobial agent. The results indicated that due to electrostatic, hydrogen, and ion–dipole interactions, the MFC retained *lysozyme*, presumably in water as well as in ethanol medium. And the largest release of *lysozyme* was just 14% of its initial amount. Thus, MFC-based films could be considered as suitable candidates for controlled-release packaging systems (Cozzolino et al. 2013). Solution-casted edible films of cellulose sulfate (CS) and glycerol with ordered microstructure were transparent, homogeneous, flexible, water-soluble, and also resistant to oils and fats. By the addition of glycerol and increasing CS molecular weight, water vapor permeability was tuned, which revealed the application of CS-based degradable coatings in food packaging (Chen et al. 2014). Agar cellulose-based bio-nanocomposite incorporated with savory essential oil (SEO) reduced the UV transmission of the film and had influenced the antibacterial property by reducing its inhibition on to the film. Such films have great potential to be used as active food packaging materials for improving the safety and shelf life of food products (Atef et al. 2015).

26.3.2.6 Polylactic Acid

Polylactide or poly(lactic acid) (PLA) are emerging and well-liked bioplastics in the market because of their abundant availability, ease in production, and better mechanical properties as compared with other bioplastics. PLA is produced from the monomer lactic acid and it is produced by the fermentation of sugars from carbohydrate. The mechanical properties of PLA are similar to that of PS and PET. The successful dispersion of cellulose nanofibers in PLA can improve the properties of the composite (Bondeson and Oksman 2007) and thereby the potential application in food packaging. PLA/cellulose whisker nanocomposite along with PVA are used to improve dispersion and does not show any improvement in the composite properties since the majority of the cellulose particles are located in the PVA phase alone (Petersson et al. 2007). The surface modification of cellulose fibers with *tert*-butanol or surfactants improves the thermal and mechanical properties of the composite (Jonoobi et al. 2010). Melt-blended PLA nanocomposites based on MMT does not change the optical transparency of the films and the barrier properties are improved remarkably in the presence of MMT (Fukushima et al. 2012). The presence of MMT fillers in PLA also resulted in the sorption of ethanol, methanol, water vapor, and aromatic and aliphatic hydrocarbons (Friess et al. 2006). Ag-NP-loaded PLA films display strong antibacterial activity with the increase in the percentage of the Ag-NPs (Shameli et al. 2010). The nanocomposite of PLA filled with chitosan-modified MMT show improved heat stability, storage, and loss modulus over pristine PLA films. The degradation time required for the composite is higher than that of virgin PLA (Wu and Wu 2006). Melt compounding of PLA with zinc aluminum LDH is an efficient way of improving the flame retardancy of PLA (Wang et al. 2010). A PLA nanocomposite with GO is electrically conductive (Shen et al. 2012) and it can be used in sensors, conducting tags, and so on. PLA/MWCNT conductive biopolymer has the potential application as sensors for the detection of organic vapor molecules and glucose (Kumar et al. 2012, Oliveira et al. 2012). The degradation of harmful chemicals such as methyl orange were achieved using PLA/TiO_2 nanocomposite films through photocatalysis. Despite photocatalysis, the presence of TiO_2 improved the tensile strength and barrier properties of the film (Zhu et al. 2011).

Toughening of the PLA is always required to improve its properties such as processability, film forming characteristics, deformation at break, and so on. Palm oil plasticized PLA/clay (fatty nitrogen compound modified) nanocomposite showed significant improvements in thermal and mechanical properties (Al-mulla 2011). PLA toughened with LLDPE and reinforced with organophilic-modified MMT enhanced thermal stability of the composite. Further, mechanical and thermal properties deteriorated as the clay content was increased (Balakrishnan et al. 2010). Natural rubber toughened PLA/MMT nanocomposites was synthesized by melt processing which had better stiffness and improved permeability for O_2 and CO_2, used in packaging of fresh vegetables and fruits. PLA/nano-calcium carbonate blended with maleated styrene–ethylene/butylene–styrene (SEBS-*g*-MAH) was prepared by melt compounding. This composite exhibited a remarkable enhancement in elongation at break, impact strength, and thermal stability (Chow et al. 2012).

PLA plasticized with tri-(butanediol-monbutyrate) citrate (TBBC) up to 20 wt.% exhibited a good storage stability and had retained the original transparency and mechanical behavior. The flexibility of PLA had been influenced by the content of TBBC and the study revealed that there was a good miscibility with PLA promising its application in PLA-based packaging materials (Wan et al. 2015). As a compatibilizing agent, PCL induced a better dispersion of the modified nanoclay, which indeed improved the surface free energy compared to polyolefins and the obtained materials thus do not require surface activation before printing. Further, improved dispersion reduced the transmission rate of oxygen and water vapor due to elongation of diffusion tortuosity path (Olewnik and Richert 2015).

PLA films were also fabricated by melt processing with the addition of plasticizing agent hexadecyl lactate (HL). The results showed an increase in the water vapor permeability of PLA/HL blend films with HL content. These films could effectively extend the shelf life of fresh-cut pears as compared to commercial LDPE films. The properties indicated that these films could serve as an alternative for food packaging materials to reduce environmental problems associated with synthetic packaging films (Qin et al. 2014b).

26.3.2.7 Chitosan

Chitosan is produced commercially by the acetylation of chitin, which is a structural element in the exoskeleton of crustaceans. Chitosan is known for its inherent resistance against the fungi and pathogens. Solution cast chitosan/MMT-K10 (MMTK-10) clay nanocomposite had a significant improvement in its tensile strength over pristine chitosan. Hardness, elastic modulus, and thermal stability of the composite were improved. The water permeability of the composite was reduced whereas the oxygen permeability was increased (Cabedo et al. 2006). Chitosan/OMMT nanocomposites reported to have a good absorption to Congo red (Wang and Wang 2007), a toxic dye, exploring the possibility of the absorption of hazardous chemicals by the polymer nanocomposite food packages. Solution cast chitosan–silver oxide nanocomposite exhibited good antibacterial properties (Tripathi et al. 2011). These composite films were best suited for wrapping the food items, which are liable to microbial growth or could be used as a coating over the conventional packaging. Mechanical and water vapor barrier properties of chitosan nanocomposite films reinforced with unmodified MMT and organically modified MMT, nano-silver, and silver zeolite were improved over the control chitosan films (Rhim et al. 2006). Chitosan/organic rectorite (a layered hydroxide) nanocomposite films were reported to be used as controlled drug release carriers in antimicrobial food packaging because of its antibacterial properties and water permeation (Wang et al. 2007). Mechanical and vapor barrier properties of the chitosan films were improved by the addition of nanosized cellulose fibers as fillers, but the properties were degraded in the presence of a plasticizer (Azeredo et al. 2010). Solution cast starch nanocrystals/carboxymethyl chitosan nanocomposite films had a tendency to increase tensile strength as the starch content is increased. The dispersion of starch crystals in the matrix is uniform till 30 wt.%. Water absorption and water vapor permeability were improved significantly by the addition of starch nanocrystals (Duan et al. 2011).

Caffeic acid-incorporated films of water-soluble chitosan (N, O-carboxymethyl chitosan, NOCC)/cellulose (methylcellulose, MC) blend showed 20-fold increase in antioxidant activity and 6-fold increase in antibacterial activity as compared to caffeic acid-free composite films. The NOCC/MC composite films were cross-linked using calcium ions. The releasable caffeic acid from the film showed a significant inhibitory effect on lipid oxidation of menhaden oil-in-water emulsion. Caffeic acid-based active packaging was reported to be a potential method to reduce lipid oxidation in fish oils (Yu et al. 2013).

Using carbodiimide-mediated coupling reaction, ferulic acid-coupled chitosan (FA-CTS) was fabricated which had an improvement in thermal stability of ferulic acid. Water vapor permeability of these biodegradable films increased by 11.7%–29.8%, while oxygen permeability decreased by 0.1%–20.6% with FA-CTS loading content due to its hydrophilicity. Further, the films exhibited ~3-fold superior radical scavenging activity compared with control film containing naked ferulic acid. These observations predicted that the FA-CTS could potentially be used as an antioxidant for active packaging film (Woranuch et al. 2015). The nanosized chitin whiskers were reinforced in chitosan film reinforced by the casting–vaporation method and cross-linked using tannic acid. Tannic acid and chitin whisker fillers have been reported to greatly reduce the moisture content by 294% and water solubility by 13% (Rubentheren et al. 2015).

There are several biodegradable polymer or polymer blend nanocomposites that have been synthesized to date. But the possibilities of using such polymers in packaging are not exposed subsequently.

26.3.3 Nanocomposite Blends

Antimicrobial film prepared by blending chitosan (CS) and PVA with glutaraldehyde as the cross-linker have a good molecular miscibility between PVA and chitosan, it was suggested that this film would be an alternative in shelf life extension of minimally processed tomato combined with other types of controls. For instance, quality raw material, hygienic processing conditions, and storage temperatures; especially against food pathogenic bacteria, namely, *E. coli*, *S. aureus*, and *Bacillus subtilis* (Tripathi et al. 2009). Chitosan/PVA/pectin ternary film was prepared successfully by solution casting method. IR spectra and SEM analyses of ternary film indicated PEC (polyelectrolyte complex) formation between pectin and chitosan. The XRD study revealed that chitosan–PVA–pectin ternary films are crystalline. The thermal properties were also quite reasonable for its food-packaging applications, which depict the weight losses at 200°C–300°C. The microbiological screening had demonstrated the positive antimicrobial activity of film against pathogenic bacteria (Tripathi et al. 2010). The films of LDPE/OC were analyzed for permeability measurements exhibited that addition of OC even at low level had a significant effect on barrier properties of the nanocomposites (Dadbin et al. 2008). Blending amorphous PLA with PCL lead to an improvement in mechanical properties and thermal stability without a significant decrease in barrier properties (Cabedo et al. 2006). Improvement in oxygen permeability was significantly noticed in PLA/poly(butylene succinate)/clay nanocomposites (Bhatia et al. 2012). LDPE/EVA/ OC nanocomposite films were prepared using two-step solution method in which permeability coefficient measurements exhibited that adding OC improved barrier properties with a saturation limit (Shafiee and Ramazani 2008). In case of PLA/OMMT/poly(ethylene glycol) (PEG), the addition of OMMT and PEG into PLA eventually increased the tortuous paths, and enhanced water-barrier properties of PLA nanocomposites (Leu and Chow 2011). The use of chitosan/clay nanocomposite coatings improved water vapor barrier properties of PLA films (Park et al. 2012). Also, it was reported that PLA/poly(vinyl acetate-*co*-vinyl alcohol) significantly improved the barrier and mechanical properties of PLA for possible use as a packaging material (Razavi et al. 2012). Further, interfacial compatibilized PLA and PLA/ PCL/OC were observed to show enhanced barrier properties and enzymatic degradation (Sabet and Katbab 2009). In the case of PPMA-grafted silica/PVC, nanosilica were treated through graft polymerization, oxygen, and water permeability rates through thin films of PVC nanocomposites, which were decreased at low filler content of 3 wt.%. Also, mechanical properties were improved effectively (Zhu et al. 2008b). The gasoline permeability of the nylon 6-nanoclay/PVA blends decreased with increase in PVA content. Interestingly, PVA content exhibited the macroscopic barrier property similar to that of microscopic free volume. The relation between free volume fraction and permeability coefficient can be described exactly as an exponent function and this implied that microscopic free volume played an important role in determining macroscopic barrier properties (Cui et al. 2009). Styrene/maleic anhydride (SMA) copolymers could easily be processed in conventional PS equipment. These thermoformed parts give satisfactory mechanical properties that required in food packaging. In addition, these type of polymers impart superior thermal properties and have been used in microwave reheating of food items (Roberts and Kwok 2007). It was reported that water barrier novel blends of chitosan with EVOH copolymers were prepared by solution casting from water/isopropanol solution of acetic acid. These blends exhibited antimicrobial performance by the release of protonated glucosamine fractions and in some cases by a synergistic mild release of entrapped acetic acid (Fernandez-Saiz et al. 2010). The composite of silane-modified nanoclay and starch/PE blend exhibited good mechanical properties and barrier properties as well as rapid biodegradability after an initial lag (Manjunath and Sailaja 2014).

It has been reported that PVA/rice starch film obtained was transparent with good mechanical properties and low water solubility. The reinforcement of silk fibroin to PVA/Rice starch matrix increased the permeability of oxygen as well as degradation of films (Kuchaiyaphum et al. 2013). A blend of poly(L-lactide) (PLLA) with star-shaped poly(ε-caprolactone-co-L-lactide) (s-PCLA) exhibited a better elongation at break than neat PLLA. Further, water vapor permeability of this blend increased with the increase in content of s-PCLA (Qin et al. 2014a). The effects of hydrophilic nanoclay, *Nanomer PGV*, on mechanical properties of PLA/PCL blends were investigated and compared with hydrophobic clay, *Montmorillonite K10*. SEM micrographs in Figure 26.9 revealed that the presence of *Nanomer PGV* in polymer blend influenced the miscibility of polymers. In this case, PLA/PCL blends became more homogeneous and exhibited single-phase morphology (Eng et al. 2013).

The physicochemical and microbial quality of button mushrooms (*Agaricus bisporus*) were investigated which was stored at $4°C \pm 1°C$ for 16 days. CO_2 level inside the PLA/

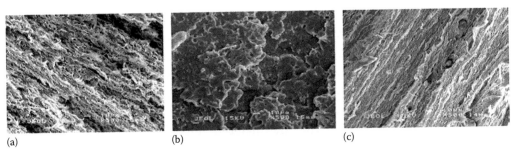

(a)   (b)   (c)

FIGURE 26.9   SEM micrograph of (a) PLA/PCL, (b) PLA/PCL/1 wt.% Montmorillonite K10, and (c) PLA/PCL/1 wt.% Nanomer PGV. (Reprinted under Creative Commons Attribution License© 2013, Eng, C.C. et al., *Ind. J. Mater. Sci.*, 2013, e816503.)

PCL film with cinnamaldehyde (0, 3, and 9 wt.%) was lower than that compared to control (LDPE) and PLA/PCL films, whereas O_2 level was similar in all packages. PLA/PCL/cinnamaldehyde film containing 9 wt.% of cinnamaldehyde was more effective in reducing microbial counts and preserving the color of mushrooms than other films. The higher water vapor permeability of such films could be used to maintain the quality of fresh button mushroom and extend its post-harvest life (Qin et al. 2015).

26.4 ACTIVE PACKAGING APPLICATIONS OF POLYMER NANOCOMPOSITES

Packaging with additional features, such as barrier properties, antimicrobial activity, sensing, and controlled release are called *active packaging*. Active packaging can play a great role in averting food-borne diseases by sensing and antimicrobial actions. Active packaging helps to maintain the quality of food and also, it increases the shelf life of the packed food. Thus, better overall benefits are achieved in food packaging.

26.4.1 Barrier Properties

The major role of packaging is to retain the qualities of commodities from the manufacturer to the end users, especially in food and chemicals, where the atmospheric gases or moisture can harm the quality of the packed items. Also, barrier properties against aroma and flavors are always a challenge in food packaging applications. No polymer in its virgin form exhibit the qualities required for a good food packaging film as polymers with excellent mechanical strength may not have good barrier properties and vice versa. The multilayer concept is exploited in such case, as discussed earlier where a compound layer of polymer with good mechanical strength on a polymer with good barrier properties are laminated to each other. Another technique which is used widely to improve the barrier property is blending, where a polymer with low barrier property but good processability and mechanical strength is blended with polymers with high barrier properties. The production expense and quantity of polymer required are high for fabricating such multilayer films and blends to achieve high-quality packaging.

The idea of polymer nanocomposites for manufacturing packaging film is a cost-effective technique than the existing ones which are employed in the production of packaging films. The amorphous phase in any polymer favors the transport of gases through it, thus lessening the overall performance of the polymer in packaging. The introduction of nanosized fillers to the polymer matrix favors to improve the barrier properties in two ways, one is through the induced crystallization in a polymer matrix by the nanofillers and the second is due to the tortuous path created by nanofillers against permeating gases as shown in the Figures 26.10 and 26.11. In the former case, the overall crystallinity of composite increases and the amount of amorphous phase favoring permeation is subsequently decreased. Thus, the barrier properties are enhanced. In the latter case, barrier properties depend on the nature of nanofillers used in composite. A nanofiller with high aspect ratio can perform well in inducing barrier properties to the polymer film rather than the fillers with small aspect ratio.

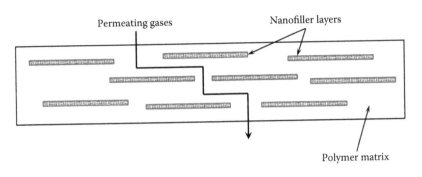

FIGURE 26.10 Tortuous path created by layered nanomaterials to permeating gases.

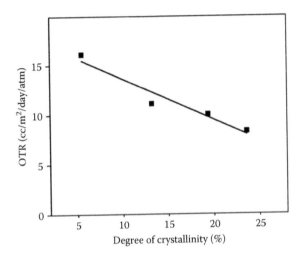

FIGURE 26.11 Oxygen transmission rate as a function of crystallinity in PET film. (From Al-Jabareen, A., Al-Bustami, H., Harel, H., and Marom, G.: Improving the oxygen barrier properties of polyethylene terephthalate by graphite nanoplatelets. *J. Appl. Polym. Sci.* 2013. 128. 1534–1539. Copyright Wiley-VCH Verlag GmbH & Co. KGaA. Reproduced with permission.)

26.4.2 Anti-Microbial Nature

Food is always subjected to degradation by microorganisms; also, it can act as a carrier of pathogens. Food-borne diseases are common all around the world. Antimicrobial packaging conquers or slackens the growth of microorganisms when it comes in contact with the package. The usefulness of metals as sterilizing agents is known from the ancient times. NPs of metals such as silver, gold, and copper exhibits good antibacterial property against the microorganism. A large number of studies on the antibacterial action of silver nanoparticulate polymer composites are reported till date. The Ag-NPs were reported to resist both gram positive and gram negative bacteria (Shameli et al. 2010, Tripathi et al. 2011, Nafchi et al. 2012). The copper NP dispersed polymer nanocomposites were observed to effectively oppose *Pseudomonas* spp. (Longano et al. 2012). ZnO NP-reinforced polymer matrix exhibited good resistance against *E. coli* bacteria (Li et al.

2009). Copper NPs also inhibit the action of gram positive and gram negative bacteria, and the zone of inhibition diameters were found to be 15 mm for *E. coli* (gram positive bacteria) and 5 mm for *Bacillus megaterium* (gram negative bacteria) (Theivasanthi and Alagar 2011). The addition of TiO$_2$ to packaging polymers improved photocatalytic antibacterial properties of packaging films (Bodaghi et al. 2013). The addition of Ag-NPs to BC and BC/polyacrylamide blend, improved the antibacterial properties of both nanocomposites and zone of inhibition as shown in Figure 26.12 (Yang et al. 2015).

26.4.3 Sensors

Films of conducting nanocomposites can be used in the detection of water vapor, oxygen, and carbon dioxide, which are not preferred in many packed food commodities. The presence of food degradation products, toxic chemicals, and pathogens are also determined by such sensors which can be incorporated into food packages. The conducting polymer nanocomposite of PLA/MWCNT could be a good candidate for sensing vapor and volatile organic compounds (Kumar et al. 2012). SWCNT-COOH nanocomposites of PVC, cumene-terminated poly(styrene-*co*-maleic anhydride) (cumene-PSMA), poly(styrene-*co*-maleic acid) partial isobutyl/methyl mixed ester (PSE), and polyvinylpyrrolidon (PVP) exhibited a change in its resistance in the presence of volatile amines. Volatile amines are produced during the degradation of food (Hong et al. 2010). Copper/polyaniline nanocomposite had a potential to detect the chlorinated hydrocarbons, a toxic chemical that could be present in food packages (Sharma et al. 2002). Detection of food-borne pathogens could be identified by quantifying odors, produced by the microorganism during its action. The conducting polymer nanocomposite polyaniline/carbon black gave different responses depending on the type of microorganisms present in food (Arshak et al. 2007). Laponite clay/chitosan entrapped with polyphenol oxidase (PPO) amperometric sensors were used

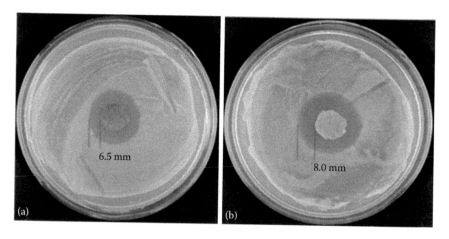

FIGURE 26.12   Zone of inhibition of composite against *Staphylococcus aureus* bacteria (a) bacterial cellulose/Ag nanoparticle and (b) bacterial cellulose/polyacrylamide/Ag nanoparticle. (Reprinted from *Carbohydr. Polym.*, 115, Yang, G., Wang, C., Hong, F., Yang, X., and Cao. Z., Preparation and characterization of BC/PAM-AgNPs nanocomposites for antibacterial applications, 636–642, Copyright 2015, with permission from Elsevier.)

to detect the phenolic vapor, a toxic chemical present in the food processing environment (Fan et al. 2007). Porous clay heterostructures (PCHs)-reinforced PP detected the presence of ethylene in freshly packed vegetables and fruits by varying its resistance (Srithammaraj et al. 2012). Gold–PVA core–shell nanocomposites had a potential to monitor humidity and thus, were integrated with the food packaging, where the controlled atmosphere is important (Yao et al. 2010). There are several polymer nanocomposite sensors which are reported beyond the scope of this chapter, such as polypyrrole/$ZnSnO_3$ nanocomposites for sensing ammonia (Song et al. 2011), $NiFe_2O_4$/CuO/FeO-chitosan nanocomposite for sensing cholesterol (Singh et al. 2012), TiO_2/polyaniline for oxygen sensing (Huyen et al. 2011), and poly(methyl methacrylate)/functionalized CNT composite for humidity sensing (Pandis et al. 2011).

26.5 RISK ASSESSMENT OF POLYMER NANOCOMPOSITES IN PACKAGING

Adding nanomaterials to the packaging polymers are beneficial in terms of mechanical properties, gas barrier properties, and antibacterial properties. But the presence of these nanomaterials can lead to migration of several ions to food items, which in turn depend on the duration of storage and temperature. Many researches prove that migration of ions from the packaging material to food is below cytotoxicity level. Silver as ions from PE nanocomposites to food simulants was reported to be less than the cytotoxicity level after 30 days (Jokar and Abdul Rahman 2014). Similarly, the migration of Ti took place from nano-TiO_2–PE composite packaging into food simulants. Further, as temperature increased the rate of this migration also increased (Lin et al. 2014). A similar study of PVC also revealed the presence of silver ions in food, when PVC/Ag-NP nanocomposite was used for the packaging (Cushen et al. 2013). The migration of aluminum and silicon from PET/clay nanocomposites to acidic food simulant were dependent on storage time and temperature (Farhoodi et al. 2014). Even though, the results were well below the current human exposure limits, but a deep study is necessary to quantify the human risk involved when nanocomposites are used for packaging.

Recyclability of the nanocomposites used for packaging applications is another field, where attention is needed. Studies show that due to the presence of NPs in common packaging polymers such as PE, PP, and PET; mechanical properties are lost after recycling as compared with the recycled pristine polymer. Yellowness index of these polymers are higher than that of the recycled pristine polymer, and also degradation fumes and pinholes (Sánchez et al. 2014). Therefore, new technologies should be adapted to separate nanomaterials from nanocomposites before they are recycled.

Photocatalytic activity of many NPs for the degradation of organic matters and pathogens, which are reinforced in the polymer matrices for improved properties, could lead to degradation of the polymer matrices themselves. Degradation products from the polymers are usually harmful, especially when they come in contact with food. For instance, the photocatalytic degradation of PS (Chandra et al. 2007) and PE (Zapata et al. 2014) in the presence of TiO_2 nanomaterials brings an adverse effect on the properties of a polymer matrix, even though, it is considered as an environmentally benign approach to the degradation of waste plastic. Therefore, it is mandatory to consider catalytic properties of nanomaterials towards the degradation of polymer, within the shelf life of products, before any nanocomposite is prepared for packaging applications.

26.6 SUMMARY

Polymer packaging plays an important role in reducing food waste. It has been reported that around billion tons of food are wasted every year during processing, transport, storage, and consuming. Innovations in packaging will protect food and extend its shelf life from farmers to consumers along the supply chain. A proactive approach to plastics engineering has given immense opportunities to reduce food wastes with active and intelligent polymer nanocomposites for food packaging applications as discussed above. These solutions of food packaging help maintain the quality and expectation of supply chain, reduce food waste, and indeed deliver significant environmental, social, and economic benefits.

REFERENCES

Agarwal, A., A. Raheja, T.S. Natarajan et al. 2014. Effect of Electrospun Montmorillonite-Nylon 6 Nanofibrous Membrane Coated Packaging on Potato Chips and Bread. *Innov. Food Sci. Emerg. Technol.* 26:424–430.

Ahangar, E.G., M.H. Abbaspour-Fard, N. Shahtahmassebi et al. 2014. Preparation and Characterization of PVA/ZnO Nanocomposite. *J. Food Process. Preserv.* 60:144–149.

Ahmed, J., R. Auras, T. Kijchavengkul et al. 2012. Rheological, Thermal and Structural Behavior of Poly(ε-caprolactone) and Nanoclay Blended Films. *J. Food Eng.* 111:580–589.

Alexandre, B., D. Langevin, P. Médéric et al. 2009. Water Barrier Properties of Polyamide 12/montmorillonite Nanocomposite Membranes: Structure and Volume Fraction Effects. *J. Membr. Sci.* 328:186–204.

Al-Jabareen, A., H. Al-Bustami, H. Harel et al. 2013. Improving the Oxygen Barrier Properties of Poly(Ethylene Terephthalate) by Graphite Nanoplatelets. *J. Appl. Polym. Sci.* 128:1534–1539.

Allafi, A.R. and M.A. Pascall. 2013. The Effect of Different Percent Loadings of Nanoparticles on the Barrier and Thermal Properties of Nylon 6 Films. *Innov. Food Sci. Emerg. Technol.* 20:276–280.

Al-Mulla, E.A.J. 2011. Polylactic Acid/epoxidized Palm Oil/Fatty Nitrogen Compounds Modified Clay Nanocomposites: Preparation and Characterization. *Korean J. Chem. Eng.* 28:620–626.

Arshak, K., C. Adley, E. Moore et al. 2007. Characterisation of Polymer Nanocomposite Sensors for Quantification of Bacterial Cultures. *Sens. Actuators, B* 126:226–231.

Atef, M., M. Rezaei, and R. Behrooz. 2015. Characterization of Physical, Mechanical, and Antibacterial Properties of Agar-Cellulose Bionanocomposite Films Incorporated with Savory Essential Oil. *Food Hydrocolloid* 45:150–157.

Avella, M., J.J. De Vlieger, M.E. Errico et al. 2005. Biodegradable Starch/Clay Nanocomposite Films for Food Packaging Applications. *Food Chem.* 93:467–474.

Awad, W.H., G. Beyer, D. Benderly et al. 2009. Material Properties of Nanoclay PVC Composites. *Polymer* 50:1857–1867.

Azeredo, H.M.C., L.H.C. Mattoso, R.J. Avena-Bustillos et al. 2010. Nanocellulose Reinforced Chitosan Composite Films as Affected by Nanofiller Loading and Plasticizer Content. *J. Food Sci.* 75:N1–N7.

Bahar, E., N. Ucar, A. Onen et al. 2012. Thermal and Mechanical Properties of Polypropylene Nanocomposite Materials Reinforced with Cellulose Nano Whiskers. *J. Appl. Polym. Sci.* 125:2882–2889.

Balakrishnan, H., A. Hassan, M.U. Wahit et al. 2010. Novel Toughened Poly(Lactic Acid) Nanocomposite: Mechanical, Thermal and Morphological Properties. *Mater. Des.* 31:3289–3298.

Bartley, J.R., V.W. Brown, J.K. Calhoun, Jr. et al. 2014. Nonfluorinated Polyurethanes and Methods of Making and Using Thereof. http://www.google.com/patents/US8716392.

Becaro, A.A., F.C. Puti, D.S. Correa et al. 2015. Polyethylene Films Containing Silver Nanoparticles for Applications in Food Packaging: Characterization of Physico-Chemical and Anti-Microbial Properties. *J. Nanosci. Nanotechnol.* 15:2148–2156.

Bhatia, A., R.K. Gupta, S.N. Bhattacharya et al. 2012. Analysis of Gas Permeability Characteristics of Poly(Lactic Acid)/Poly(Butylene Succinate) Nanocomposites. *J. Nanomater.* 2012:e249094.

Bikiaris, D.N. and K.S. Triantafyllidis. 2013. HDPE/Cu-Nanofiber Nanocomposites with Enhanced Antibacterial and Oxygen Barrier Properties Appropriate for Food Packaging Applications. *Mater. Lett.* 93:1–4.

Bodaghi, H., Y. Mostofi, A. Oromiehie et al. 2013. Evaluation of the Photocatalytic Antimicrobial Effects of a TiO_2 Nanocomposite Food Packaging Film by In Vitro and In Vivo Tests. *LWT—Food Sci. Technol.* 50:702–706.

Bondeson, D. and K. Oksman. 2007. Poly(Lactic Acid)/Cellulose Whisker Nanocomposites Modified by Poly(Vinyl Alcohol). *Compos. Part A* 38:2486–2492.

Broza, G., K. Piszczek, K. Schulte et al. 2007. Nanocomposites of Poly(Vinyl Chloride) with Carbon Nanotubes (CNT). *Compos. Sci. Technol.* 67:890–894 (Carbon Nanotube (CNT)—Polymer Composites).

Cabedo, L., J.L. Feijoo, M.P. Villanueva et al. 2006. Optimization of Biodegradable Nanocomposites Based on aPLA/PCL Blends for Food Packaging Applications. *Macromol. Symp.* 233:191–197.

Cao, X., Y. Chen, P. Chang et al. 2008. Starch-Based Nanocomposites Reinforced with Flax Cellulose Nanocrystals. *Exp. Polym. Lett.* 2:502–510.

Cerisuelo, J.P., J. Alonso, S. Aucejo et al. 2012. Modifications Induced by the Addition of a Nanoclay in the Functional and Active Properties of an EVOH Film Containing Carvacrol for Food Packaging. *J. Membr. Sci.* 423–424:247–256.

Chandra, A., L.-S. Turng, S. Gong et al. 2007. Study of Polystyrene/Titanium Dioxide Nanocomposites via Melt Compounding for Optical Applications. *Polym. Compos.* 28:241–250.

Chandrakala, H.N., B. Ramaraj, Shivakumaraiah et al. 2013. Investigation on the Influence of Sodium Zirconate Nanoparticles on the Structural Characteristics and Electrical Properties of Poly(Vinyl Alcohol) Nanocomposite Films. *J. Alloys Compd.* 551:531–538.

Chen, G., B. Zhang, J. Zhao et al. 2014. Development and Characterization of Food Packaging Film from Cellulose Sulfate. *Food Hydrocolloid* 35:476–483.

Chow, W.S., Y.Y. Leu, and Z.A.M. Ishak. 2012. Effects of SEBS-g-MAH on the Properties of Injection Moulded Poly(lactic acid)/Nano-calcium Carbonate Composites. *eXPRESS Polym. Lett.* 6:503–510.

Contreras, C.B., G. Chagas, M.C. Strumia et al. 2014. Permanent Superhydrophobic Polypropylene Nanocomposite Coatings by a Simple One-Step Dipping Process. *Appl. Surf. Sci.* 307:234–240.

Costa, F.R., B.K. Satapathy, U. Wagenknecht et al. 2006. Morphology and Fracture Behaviour of Polyethylene/Mg–Al Layered Double Hydroxide (LDH) Nanocomposites. *Eur. Polym. J.* 42:2140–2152.

Cozzolino, C.A., F. Nilsson, M. Iotti et al. 2013. Exploiting the Nano-Sized Features of Microfibrillated Cellulose (MFC) for the Development of Controlled-Release Packaging. *Colloids Surf. B* 110:208–216.

Cui, L., J.-T. Yeh, K. Wang et al. 2009. Relation of Free Volume and Barrier Properties in the Miscible Blends of Poly(Vinyl Alcohol) and Nylon 6-Clay Nanocomposites Film. *J. Membr. Sci.* 327:226–233.

Cushen, M., J. Kerry, M. Morris et al. 2013. Migration and Exposure Assessment of Silver from a PVC Nanocomposite. *Food Chem.* 139:389–397.

Dadbin, S., M. Noferesti, and M. Frounchi. 2008. Oxygen Barrier LDPE/LLDPE/Organoclay Nano-Composite Films for Food Packaging. *Macromol. Symp.* 274:22–27.

Damaj, Z., C. Joly, and E. Guillon. 2014. Toward New Polymeric Oxygen Scavenging Systems: Formation of Poly(Vinyl Alcohol) Oxygen Scavenger Film. *Packag. Technol. Sci.* http://onlinelibrary.wiley.com/doi/10.1002/pts.2112/pdf.

Dan, C.H., M.H. Lee, Y.D. Kim et al. 2006. Effect of Clay Modifiers on the Morphology and Physical Properties of Thermoplastic Polyurethane/Clay Nanocomposites. *Polymer* 47:6718–6730.

Dash, S. and S.K. Swain. 2013. Synthesis of Thermal and Chemical Resistant Oxygen Barrier Starch with Reinforcement of Nano Silicon Carbide. *Carbohydr. Polym.* 97:758–763.

de Moura, M.R., F.A. Aouada, R.J. Avena-Bustillos et al. 2009. Improved Barrier and Mechanical Properties of Novel Hydroxypropyl Methylcellulose Edible Films with Chitosan/Tripolyphosphate Nanoparticles. *J. Food Eng.* 92:448–453.

de Moura, M.R., L.H.C. Mattoso, and V. Zucolotto. 2012. Development of Cellulose-Based Bactericidal Nanocomposites Containing Silver Nanoparticles and Their Use as Active Food Packaging. *J. Food Eng.* 109:520–524.

Duan, B., P. Sun, X. Wang et al. 2011. Preparation and Properties of Starch Nanocrystals/Carboxymethyl Chitosan Nanocomposite Films. *Starch* 63:528–535.

Durmuş, A., M. Woo, A. Kaşgöz et al. 2007. Intercalated Linear Low Density Polyethylene (LLDPE)/Clay Nanocomposites Prepared with Oxidized Polyethylene as a New Type Compatibilizer: Structural, Mechanical and Barrier Properties. *Eur. Polym. J.* 43:3737–3749.

Eng, C.C., N.A. Ibrahim, N. Zainuddin, H. Ariffin, W.M.Z.W. Yunus, Y.Y. Then, and C.C. Teh. 2013. Enhancement of Mechanical and Thermal Properties of Polylactic Acid/Polycaprolactone Blends by Hydrophilic. Nanoclay. *Ind. J. Mater. Sci.* 2013:e816503.

Fan, Q., D. Shan, H. Xue et al. 2007. Amperometric Phenol Biosensor Based on Laponite Clay-chitosan Nanocomposite Matrix. *Biosens. Bioelectron.* 22:816–821.

Farhoodi, M., S.M. Mousavi, R. Sotudeh-Gharebagh et al. 2014. Migration of Aluminum and Silicon from PET/Clay Nanocomposite Bottles into Acidic Food Simulant. *Packag. Technol. Sci.* 27:161–168.

Fernandez-Saiz, P.F., M.J. Ocio, and J.M. Lagaron. 2010. Antibacterial Chitosan-Based Blends with Ethylene–Vinyl Alcohol Copolymer. *Carbohydr. Polym.* 80:874–884.

Fortunati, E., D. Puglia, M. Monti et al. 2013. Cellulose Nanocrystals Extracted from Okra Fibers in PVA Nanocomposites. *J. Appl. Polym. Sci.* 128:3220–3230.

Friess, K., M. Šípek, V. Hynek et al. 2006. Sorption of VOCs and Water Vapors in Polylactate Membrane with Nanocomposite Fillers. *Desalination* 200:265–266.

Fuad, M.Y.A., H. Hanim, R. Zarina et al. 2010. Polypropylene/Calcium Carbonate Nanocomposites—Effects of Processing Techniques and Maleated Polypropylene Compatibiliser. *Exp. Polym. Lett.* 4:611–620.

Fukushima, K., A. Fina, F. Geobaldo et al. 2012. Properties of Poly (Lactic Acid) Nanocomposites Based on Montmorillonite, Sepiolite and Zirconium Phosphonate. *Exp. Polym. Lett.* 6:914–926.

García, M., G. van Vliet, M.G.J. ten Cate et al. 2004. Large-Scale Extrusion Processing and Characterization of Hybrid Nylon-6/SiO$_2$ Nanocomposites. *Polym. Adv. Technol.* 15:164–172.

Golebiewski, J., A. Rozanski, J. Dzwonkowski et al. 2008. Low Density Polyethylene–Montmorillonite Nanocomposites for Film Blowing. *Eur. Polym. J.* 44:270–286.

Gorrasi, G., V. Senatore, G. Vigliotta et al. 2014. PET–Halloysite Nanotubes Composites for Packaging Application: Preparation, Characterization and Analysis of Physical Properties. *Eur. Polym. J.* 61:145–156.

Gorrasi, G., M. Tortora, V. Vittoria et al. 2002. Transport and Mechanical Properties of Blends of Poly(ε-Caprolactone) and a Modified Montmorillonite-Poly(ε-caprolactone) Nanocomposite. *J. Polym. Sci. B Polym. Phys.* 40:1118–1124.

Herrero, M., S. Martínez-Gallegos, F.M. Labajos et al. 2011. Layered Double Hydroxide/ Poly(Ethylene Terephthalate) Nanocomposites: Influence of the Intercalated LDH Anion and the Type of Polymerization Heating Method. *J. Solid State Chem.* 184:2862–2869.

Hong, L., Y. Li, and M. Yang. 2010. Fabrication and Ammonia Gas Sensing of Palladium/Polypyrrole Nanocomposite. *Sens. Actuator B* 145:25–31.

Huang, H.-D., P.-G. Ren, J. Chen et al. 2012. High Barrier Graphene Oxide Nanosheet/Poly(Vinyl Alcohol) Nanocomposite Films. *J. Membr. Sci.* 409–410:156–163.

Huyen, D.N., N.T. Tung, N.D. Thien et al. 2011. Effect of TiO_2 on the Gas Sensing Features of TiO2/ PANi Nanocomposites. *Sensors (Basel)* 11:1924–1931.

Jia, N., S.-M. Li, M.-G. Ma, and R.C. Sun. 2011. Microwave-Assisted Ionic Liquid Preparation and Characterization of Cellulose/Calcium Silicate Nanocomposites in Ethylene Glycol. *Mater. Lett.* 65:918–921.

Jia, N., S.-M. Li, M.-G. Ma, and R.C. Sun. 2012. Rapid Microwave-Assisted Fabrication of Cellulose/ F-Substituted Hydroxyapatite Nanocomposites Using Green Ionic Liquids as Additive. *Mater. Lett.* 68:44–46.

Jiang, T., Y. Wang, J. Yeh et al. 2005. Study on Solvent Permeation Resistance Properties of Nylon6/ Clay Nanocomposite. *Eur. Polym. J.* 41:459–466.

Jo, H.-J., K.-M. Park, S.C. Min et al. 2013. Development of an Anti-Insect Sachet Using a Poly(Vinyl Alcohol)–Cinnamon Oil Polymer Strip against *Plodia interpunctella. J. Food Sci.* 78:E1713–E1720.

Jokar, M. and R. Abdul Rahman. 2014. Study of Silver Ion Migration from Melt-Blended and Layered-Deposited Silver Polyethylene Nanocomposite into Food Simulants and Apple Juice. *Food Addit. Contam.* 31:734–742.

Jonoobi, M., J. Harun, A.P. Mathew et al. 2010. Mechanical Properties of Cellulose Nanofiber (CNF) Reinforced Polylactic Acid (PLA) Prepared by Twin Screw Extrusion. *Compos. Sci. Technol.* 70:1742–1747.

Katančić, Z., J. Travaš-Sejdić, and Z. Hrnjak-Murgić. 2011. Study of Flammability and Thermal Properties of High-Impact Polystyrene Nanocomposites. *Polym. Degrad. Stab.* 96:2104–2111.

Kim, I.-H., J. Han, J.H. Na et al. 2013. Insect-Resistant Food Packaging Film Development Using Cinnamon Oil and Microencapsulation Technologies. *J. Food Sci.* 78:E229–E237.

Kim, S.W. and S.-H. Cha. 2014. Thermal, Mechanical, and Gas Barrier Properties of Ethylene–Vinyl Alcohol Copolymer-Based Nanocomposites for Food Packaging Films: Effects of Nanoclay Loading. *J. Appl. Polym. Sci.* 131. http://onlinelibrary.wiley.com/doi/10.1002/app.40289/pdf.

Kuchaiyaphum, P., W. Punyodom, S. Watanesk et al. 2013. Composition Optimization of Polyvinyl Alcohol/rice starch/silk Fibroin-blended Films for Improving its Eco-friendly Packaging Properties. *J. Appl. Polym. Sci.* 129:2614–2620.

Kumar, B., M. Castro, and J.F. Feller. 2012. Poly(Lactic Acid)–Multi-Wall Carbon Nanotube Conductive Biopolymer Nanocomposite Vapour Sensors. *Sens. Actuator B* 161:621–628.

Lai, C.-L., J.-T. Chen, Y.-J. Fu et al. 2015. Bio-Inspired Cross-Linking with Borate for Enhancing Gas-Barrier Properties of Poly(Vinyl Alcohol)/Graphene Oxide Composite Films. *Carbon* 82:513–522.

Leu, Y.Y., and W.S. Chow. 2011. Kinetics of Water Absorption and Thermal Properties of Poly(lactic acid)/organomontmorillonite/poly(ethylene glycol) Nanocomposites. *J. Vinyl Add. Tech.* 17:40–47.

Li, X., S.C. Li, and L.Q. Huang. 2013. Synthesis and Property of Water Borne Polyurethane/Modified Nano-ZnO Composites Packaging Membranes. *Appl. Mech. Mater* 469:167–170.

Li, X., Y. Xing, Y. Jiang et al. 2009. Antimicrobial Activities of ZnO Powder-Coated PVC Film to Inactivate Food Pathogens. *Int. J. Food Sci. Technol.* 44:2161–2168.

Lin, N. and A. Dufresne. 2013. Physical and/or Chemical Compatibilization of Extruded Cellulose Nanocrystal Reinforced Polystyrene Nanocomposites. *Macromolecules* 46:5570–5583.

Lin, Q.-B., H. Li, H.-N. Zhong et al. 2014. Migration of Ti from Nano-TiO₂-Polyethylene Composite Packaging into Food Simulants. *Food Addit. Contam. A* 31:1284–1290.

Liu, C., Y. Luo, Z. Jia et al. 2011. Enhancement of Mechanical Properties of Poly (Vinyl Chloride) with Polymethyl Methacrylate-Grafted Halloysite Nanotube. *Exp. Polym. Lett.* 5:591–603.

Liu, F., H. Liu, X. Li et al. 2012. Nano-TiO₂@Ag/PVC Film with Enhanced Antibacterial Activities and Photocatalytic Properties. *Appl. Surf. Sci.* 258:4667–4671.

Liu, H., C. Han, and L. Dong. 2010. Preparation and Characterization of Poly(ε-Caprolactone)/ Calcium Carbonate Nanocomposites and Nanocomposite Foams. *Polym. Compos.* 31:1653–1661.

Liu, M., B. Guo, M. Du et al. 2009. Halloysite Nanotubes as a Novel β-Nucleating Agent for Isotactic Polypropylene. *Polymer* 50:3022–3030.

Liu, Q., Z. Peng, and D. Chen. 2007. Nonisothermal Crystallization Behavior of Poly(ε-Caprolactone)/Attapulgite Nanocomposites by DSC Analysis. *Polym. Eng. Sci.* 47:460–466.

Liu, T., W. Wood, and W.-H. Zhong. 2011. Sensitivity of Dielectric Properties to Wear Process on Carbon Nanofiber/High-Density Polyethylene Composites. *Nanoscale Res. Lett.* 6:1–9.

Longano, D., N. Ditaranto, N. Cioffi et al. 2012. Analytical Characterization of Laser-Generated Copper Nanoparticles for Antibacterial Composite Food Packaging. *Anal. Bioanal. Chem.* 403:1179–1186.

López-de-Dicastillo, C., M. Gallur, R. Catalá et al. 2010. Immobilization of β-cyclodextrin in Ethylene-Vinyl Alcohol Copolymer for Active Food Packaging Applications. *J. Membr. Sci.* 353:184–191.

López-de-Dicastillo, C., R. Catalá, R. Gavara et al. 2011. Food Applications of Active Packaging EVOH Films Containing Cyclodextrins for the Preferential Scavenging of Undesirable Compounds. *J. Food Eng.* 104:380–386.

Lv, R., J. Zhou, Q. Du et al. 2006. Preparation and Characterization of EVOH/PVP Membranes via Thermally Induced Phase Separation. *J. Membr. Sci.* 281:700–706.

Madaleno, L., J. Schjødt-Thomsen, and J.C. Pinto. 2010. Morphology, Thermal and Mechanical Properties of PVC/MMT Nanocomposites Prepared by Solution Blending and Solution Blending + Melt Compounding. *Compos. Sci. Technol.* 70:804–814.

Maksimov, R.D., A. Lagzdins, N. Lilichenko et al. 2009. Mechanical Properties and Water Vapor Permeability of Starch/Montmorillonite Nanocomposites. *Polym. Eng. Sci.* 49:2421–2429.

Mamunya, Y.P., V.V. Levchenko, A. Rybak et al. 2010. Electrical and Thermomechanical Properties of Segregated Nanocomposites Based on PVC and Multiwalled Carbon Nanotubes. *J. Non-Cryst. Solids* 356:635–641.

Manias, E., J. Zhang, J.Y. Huh et al. 2009. Polyethylene Nanocomposite Heat-Sealants with a Versatile Peelable Character. *Macromol. Rapid Commun.* 30:17–23.

Manjunath, L. and R.R.N. Sailaja. 2014. Starch/Polyethylene Nanocomposites: Mechanical, Thermal, and Biodegradability Characteristics. *Polym. Compos.* http://onlinelibrary.wiley. com/doi/10.1002/pc.23307/pdf.

Marney, D.C.O., L.J. Russell, D.Y. Wu et al. 2008. The Suitability of Halloysite Nanotubes as a Fire Retardant for Nylon 6. *Polym. Degrad. Stab.* 93:1971–1978.

Matusinović, Z., M. Rogošić, and J. Šipušić. 2009. Synthesis and Characterization of Poly(Styrene-co-Methyl Methacrylate)/Layered Double Hydroxide Nanocomposites via In Situ Polymerization. *Polym. Degrad. Stab.* 94:95–101.

Meira, S.M.M., G. Zehetmeyer, A.I. Jardim et al. 2014. Polypropylene/Montmorillonite Nanocomposites Containing Nisin as Antimicrobial Food Packaging. *Food Bioprocess Technol.* 7:3349–3357.

Moghaddam, H.M., M.H. Khoshtaghaza, A. Salimi et al. 2014. The TiO₂–Clay-LDPE Nanocomposite Packaging Films: Investigation on the Structure and Physicomechanical Properties. *Polym. Plast Technol. Eng.* 53:1759–1767.

Mohagheghian, M., M. Sadeghi, M.P. Chenar et al. 2014. Gas Separation Properties of Polyvinyl-chloride (PVC)-Silica Nanocomposite Membrane. *Korean J. Chem. Eng.* 31:2041–2050.

Morin, A. and A. Dufresne. 2002. Nanocomposites of Chitin Whiskers from Riftia Tubes and Poly(Caprolactone). *Macromolecules* 35:2190–2199.

Muppalla, S.R., S.R. Kanatt, S.P. Chawla et al. 2014. Carboxymethyl Cellulose–Polyvinyl Alcohol Films with Clove Oil for Active Packaging of Ground Chicken Meat. *Food Packag. Shelf Life* 2:51–58.

Muriel-Galet, V., G. López-Carballo, R. Gavara et al. 2012. Antimicrobial Food Packaging Film Based on the Release of LAE from EVOH. *Int. J. Food Microbiol.* 157:239–244.

Nafchi, A.M., A.K. Alias, S. Mahmud et al. 2012. Antimicrobial, Rheological, and Physicochemical Properties of Sago Starch Films Filled with Nanorod-Rich Zinc Oxide. *J. Food Eng.* 113:511–519.

Nayak, R.R., K.Y. Lee, A.M. Shanmugharaj et al. 2007. Synthesis and Characterization of Styrene Grafted Carbon Nanotube and Its Polystyrene Nanocomposite. *Eur. Polym. J.* 43:4916–4923.

Nirmala, R., H.Y. Kim, D. Kalpana et al. 2013. Multipurpose Polyurethane Antimicrobial Metal Composite Films via Wet Cast Technology. *Macromol. Res.* 21:843–851.

Ogunsona, E. and N.A. D'Souza. 2014. Characterization and Mechanical Properties of Foamed Poly(ε-Caprolactone) and Mater-Bi Blends Using CO_2 as Blowing Agent. *J. Cell. Plast.* doi:10.11770021955X14537658.

Ogunsona, E., S. Ogbomo, M. Nar et al. 2011. Thermal and Mechanical Effects in Polystyrene-Montmorillonite Nanocomposite Foams. *Cell. Polym.* 30:79–93.

Olewnik, E. and J. Richert. 2015. Influence of the Compatibilizing Agent on Permeability and Contact Angle of Composites Based on Polylactide. *Polym. Compos.* 36:17–25.

Oliveira, J.E., L.H.C. Mattoso, E.S. Medeiros et al. 2012. Poly(Lactic Acid)/Carbon Nanotube Fibers as Novel Platforms for Glucose. *Biosensors* 2:70–82.

Osman, M.A., V. Mittal, M. Morbidelli et al. 2003. Polyurethane Adhesive Nanocomposites as Gas Permeation Barrier. *Macromolecules* 36:9851–9858.

Ozcalik, O. and F. Tihminlioglu. 2013. Barrier Properties of Corn Zein Nanocomposite Coated Polypropylene Films for Food Packaging Applications. *J. Food Eng.* 114:505–513.

Pandis, C., V. Peoglos, A. Kyritsis et al. 2011. Gas Sensing Properties of Conductive Polymer Nanocomposites. *Procedia Eng.* 25: 243–246.

Pannirselvam, M., A. Genovese, M. Jollands et al. 2008. Oxygen Barrier Property of Polypropylene-Polyether Treated Clay Nanocomposite. *Exp. Polym. Lett.* 2:429–439.

Park, S.-H., H.S. Lee, J.H. Choi et al. 2012. Improvements in Barrier Properties of Poly(Lactic Acid) Films Coated with Chitosan or Chitosan/Clay Nanocomposite. *J. Appl. Polym. Sci.* 125:E675–E680.

Peng, X., E. Ding, and F. Xue. 2012. In Situ Synthesis of TiO_2/polyethylene Terephthalate Hybrid Nanocomposites at Low Temperature. *Appl. Surf. Sci.* 258:6564–6570.

Pereira, D., P.P. Losada, I. Angulo et al. 2009. Development of a Polyamide Nanocomposite for Food Industry: Morphological Structure, Processing, and Properties. *Polym. Compos.* 30:436–444.

Petersson, L., I. Kvien, and K. Oksman. 2007. Structure and Thermal Properties of Poly(Lactic Acid)/Cellulose Whiskers Nanocomposite Materials. *Compos. Sci. Technol.* 67:2535–2544.

Picard, E., H. Gauthier, J.-F. Gérard et al. 2007. Influence of the Intercalated Cations on the Surface Energy of Montmorillonites: Consequences for the Morphology and Gas Barrier Properties of Polyethylene/Montmorillonites Nanocomposites. *J. Colloid Interf. Sci.* 307:364–376.

Qian, D., E.C. Dickey, R. Andrews et al. 2000. Load Transfer and Deformation Mechanisms in Carbon Nanotube-Polystyrene Composites. *Appl. Phys. Lett.* 76:2868–2870.

Qin, Y., D. Liu, Y. Wu et al. 2015. Effect of PLA/PCL/cinnamaldehyde Antimicrobial Packaging on Physicochemical and Microbial Quality of Button Mushroom (*Agaricus bisporus*). *Postharvest Biol. Technol.* 99:73–79.

Qin, Y., S. Liu, Y. Zhang et al. 2014a. Effect of Poly(ε-Caprolactone-*co*-L-Lactide) on Thermal and Functional Properties of Poly(L-Lactide). *Int. J. Biol. Macromol.* 70:327–333.

Qin, Y., Y. Wang, Y. Wu et al. 2014b. Effect of Hexadecyl Lactate as Plasticizer on the Properties of Poly(L-Lactide) Films for Food Packaging Applications. *J. Polym. Environ.* 23:1–9.

Rajeev, R.S., E. Harkin-Jones, K. Soon et al. 2009. Studies on the Effect of Equi-Biaxial Stretching on the Exfoliation of Nanoclays in Polyethylene Terephthalate. *Eur. Polym. J.* 45:332–340.

Razavi, S.M., S. Dadbin, and M. Frounchi. 2012. Oxygen-Barrier Properties of Poly(Lactic Acid)/Poly(Vinyl Acetate-Co-Vinyl Alcohol) Blends as Biodegradable Films. *J. Appl. Polym. Sci.* 125:E20–E26.

Ren, J., Y. Huang, Y. Liu et al. 2005. Preparation, Characterization and Properties of Poly(Vinyl Chloride)/Compatibilizer/Organophilic-Montmorillonite Nanocomposites by Melt Intercalation. *Polym. Test.* 24:316–323.

Ren, P., T. Shen, F. Wang et al. 2009. Study on Biodegradable Starch/OMMT Nanocomposites for Packaging Applications. *J. Polym. Environ.* 17:203–207.

Rhim, J.-W., S.-I. Hong, H.-M. Park et al. 2006. Preparation and Characterization of Chitosan-Based Nanocomposite Films with Antimicrobial Activity. *J. Agric. Food Chem.* 54:5814–5822.

Roberts, R.D. and J.C. Kwok. 2007. Styrene–Maleic Anhydride Copolymer Foam for Heat Resistant Packaging. *J. Cell. Plast.* 43:135–143.

Robles, J.D. and J.M. Martín-Martínez. 2011. Addition of Precipitated Calcium Carbonate Filler to Thermoplastic Polyurethane Adhesives. *Int. J. Adhes. Adhes.* 31:795–804.

Rubentheren, V., T.A. Ward, C.Y. Chee et al. 2015. Processing and Analysis of Chitosan Nanocomposites Reinforced with Chitin Whiskers and Tannic Acid as a Crosslinker. *Carbohydr. Polym.* 115:379–387.

Sabet, S.S. and A.A. Katbab. 2009. Interfacially Compatibilized Poly(Lactic Acid) and Poly(Lactic Acid)/polycaprolactone/organoclay Nanocomposites with Improved Biodegradability and Barrier Properties: Effects of the Compatibilizer Structural Parameters and Feeding Route. *J. Appl. Polym. Sci.* 111:1954–1963.

Sánchez, C., M. Hortal, C. Aliaga et al. 2014. Recyclability Assessment of Nano-Reinforced Plastic Packaging. *Waste Manage.* 34:2647–2655.

Sánchez-Valdes, S., E. Ramírez-Vargas, M.C. Ibarra-Alonso et al. 2012. Itaconic Acid and Amino Alcohol Functionalized Polyethylene as Compatibilizers for Polyethylene Nanocomposites. *Compos. Part B* 43:497–502.

Sarikanat, M., K. Sever, E. Erbay et al. 2011. Preparation and Mechanical Properties of Graphite Filled HDPE Nanocomposites. *Arch. Mater. Sci. Eng.* 50:120–124.

Sever, K., İ.H. Tavman, Y. Seki et al. 2013. Electrical and Mechanical Properties of Expanded Graphite/High Density Polyethylene Nanocomposites. *Composites Part B* 53:226–233.

Shafiee, M., and A. Ramazani. 2008. Investigation of Barrier Properties of Poly(Ethylene Vinyl acetate)/Organoclay Nanocomposite Films Prepared by Phase Inversion Method. *Macromol. Symp.* 274:1–5.

Shameli, K., M.B. Ahmad, W.M.Z.W. Yunus et al. 2010. Silver/Poly(Lactic Acid) Nanocomposites: Preparation, Characterization, and Antibacterial Activity. *Int. J. Nanomed.* 5:573–579.

Sharma, S., C. Nirkhe, S. Pethkar et al. 2002. Chloroform Vapour Sensor Based on Copper/Polyaniline Nanocomposite. *Sens. Actuators B* 85:131–136.

Shen, L., I.Y. Phang, L. Chen et al. 2004. Nanoindentation and Morphological Studies on Nylon 66 Nanocomposites. I. Effect of Clay Loading. *Polymer* 45:3341–3349.

Shen, Y., T. Jing, W. Ren et al. 2012. Chemical and Thermal Reduction of Graphene Oxide and Its Electrically Conductive Polylactic Acid Nanocomposites. *Compos. Sci. Technol.* 72:1430–1435.

Shen, Z., M. Ghasemlou, and D.P. Kamdem. 2015. Development and Compatibility Assessment of New Composite Film Based on Sugar Beet Pulp and Polyvinyl Alcohol Intended for Packaging Applications. *J. Appl. Polym. Sci.* 132. http://onlinelibrary.wiley.com/doi/10.1002/app.41354/pdf.

Singh, J., M. Srivastava, P. Kalita et al. 2012. A Novel Ternary NiFe$_2$O$_4$/CuO/FeO-Chitosan Nanocomposite as a Cholesterol Biosensor. *Process Biochem.* 47:2189–2198.

Siqueira, G., C. Fraschini, J. Bras et al. 2011. Impact of the Nature and Shape of Cellulosic Nanoparticles on the Isothermal Crystallization Kinetics of Poly(ε-Caprolactone). *Eur. Polym. J.* 47:2216–2227.

Song, P., Q. Wang, and Z. Yang. 2011. Ammonia Gas Sensor Based on PPy/ZnSnO$_3$ Nanocomposites. *Mater. Lett.* 65:430–432.

Sorrentino, A., R. Pantani, and V. Brucato. 2006. Injection Molding of Syndiotactic Polystyrene/ Clay Nanocomposites. *Polym. Eng. Sci.* 46:1768–1777.

Srithammaraj, K., R. Magaraphan, and H. Manuspiya. 2012. Modified Porous Clay Heterostructures by Organic–Inorganic Hybrids for Nanocomposite Ethylene Scavenging/Sensor Packaging Film. *Packag. Technol. Sci.* 25:63–72.

Sterzyński, T., J. Tomaszewska, K. Piszczek et al. 2010. The Influence of Carbon Nanotubes on the PVC Glass Transition Temperature. *Compos. Sci. Technol.* 70:966–969.

Stoica-Guzun, A., M. Stroescu, I. Jipa et al. 2013. Effect of Γ Irradiation on Poly (Vinyl Alcohol) and Bacterial Cellulose Composites Used as Packaging Materials. *Radiat. Phys. Chem.* 84:200–204.

Strawhecker, K.E. and E. Manias. 2000. Structure and Properties of Poly(Vinyl Alcohol)/Na$^+$ Montmorillonite Nanocomposites. *Chem. Mater.* 12:2943–2949.

Su, S., D.D. Jiang, and C.A. Wilkie. 2004. Polybutadiene-Modified Clay and Its Polystyrene Nanocomposites. *J. Vinyl. Addit. Technol.* 10:44–51.

Suktha, P., K. Lekpet, P. Siwayaprahm et al. 2013. Enhanced Mechanical Properties and Bactericidal Activity of Polypropylene Nanocomposite with Dual-Function Silica–Silver Core-Shell Nanoparticles. *J. Appl. Polym. Sci.* 128:4339–4345.

Sun, L., J. Sosa, J. Aguirre et al. 2011. Polystyrene Nanocomposites for Blow Molding Applications. http://www.google.com/patents/US20110020571.

Tai, Q., L. Chen, L. Song et al. 2011. Effects of a Phosphorus Compound on the Morphology, Thermal Properties, and Flammability of Polystyrene/MgAl-Layered Double Hydroxide Nanocomposites. *Polym. Compos.* 32:168–176.

Tammaro, L., V. Vittoria, and V. Bugatti. 2014. Dispersion of Modified Layered Double Hydroxides in Poly(Ethylene Terephthalate) by High Energy Ball Milling for Food Packaging Applications. *Eur. Polym. J.* 52:172–180.

Tang, E., C. Fu, S. Wang et al. 2012. Graft Polymerization of Styrene Monomer Initiated by Azobis(4-Cyanovaleric Acid) Anchored on the Surface of ZnO Nanoparticles and Its PVC Composite Film. *Powder Technol.* 218:5–10.

Tang, Y., Y. Hu, S. Wang et al. 2002. Preparation and Flammability of Ethylene-Vinyl Acetate Copolymer/montmorillonite Nanocomposites. *Polym. Degrad. Stab.* 78:555–559.

Tankhiwale, R. and S.K. Bajpai. 2009. Graft Copolymerization onto Cellulose-Based Filter Paper and Its Further Development as Silver Nanoparticles Loaded Antibacterial Food-Packaging Material. *Colloids Surf. B* 69:164–168.

Theivasanthi, T. and M. Alagar. 2011. Studies of Copper Nanoparticles Effects on Micro-Organisms. http://arxiv.org/abs/1110.1372.

Tripathi, S., G.K. Mehrotra, and P.K. Dutta. 2009. Physicochemical and Bioactivity of Cross-Linked Chitosan–PVA Film for Food Packaging Applications. *Int. J. Biol. Macromol.* 45:372–376.

Tripathi, S., G.K. Mehrotra, and P.K. Dutta. 2010. Preparation and Physicochemical Evaluation of Chitosan/Poly(Vinyl Alcohol)/Pectin Ternary Film for Food-Packaging Applications. *Carbohydr. Polym.* 79:711–716.

Tripathi, S., G.K. Mehrotra, and P.K. Dutta. 2011. Chitosan–silver Oxide Nanocomposite Film: Preparation and Antimicrobial Activity. *Bull. Mater. Sci.* 34:29–35.

Uthirakumar, P., M.-K. Song, C. Nah et al. 2005. Preparation and Characterization of Exfoliated Polystyrene/clay Nanocomposites Using a Cationic Radical Initiator-MMT Hybrid. *Eur. Polym. J.* 41:211–217.

Vassiliou, A.A., K. Chrissafis, and D.N. Bikiaris. 2010. In Situ Prepared PET Nanocomposites: Effect of Organically Modified Montmorillonite and Fumed Silica Nanoparticles on PET Physical Properties and Thermal Degradation Kinetics. *Thermochim. Acta* 500:21–29.

Vega-Baudrit, J., V. Navarro-Bañón, P. Vázquez et al. 2006. Addition of Nanosilicas with Different Silanol Content to Thermoplastic Polyurethane Adhesives. *Int. J. Adhes. Adhes.* 26:378–387.

Vladimirov, V., C. Betchev, A. Vassiliou et al. 2006. Dynamic Mechanical and Morphological Studies of Isotactic Polypropylene/fumed Silica Nanocomposites with Enhanced Gas Barrier Properties. *Compos. Sci. Technol.* 66:2935–2944.

Wan, T., G. Yang, T. Du et al. 2015. Tri-(Butanediol-Monobutyrate) Citrate Plasticizing Poly(Lactic Acid): Synthesis, Crystallization, Thermal, and Mechanical Properties. *Polym. Eng. Sci.* 55:205–213.

Wang, D.-Y., A. Leuteritz, Y.-Z. Wang et al. 2010. Preparation and Burning Behaviors of Flame Retarding Biodegradable Poly(Lactic Acid) Nanocomposite Based on Zinc Aluminum Layered Double Hydroxide. *Polym. Degrad. Stab.* 95:2474–2480.

Wang, L., X. Wang, Z. Chen et al. 2013. Effect of Doubly Organo-Modified Vermiculite on the Properties of Vermiculite/Polystyrene Nanocomposites. *Appl. Clay Sci.* 75–76:74–81.

Wang, L. and A.-Q. Wang. 2007. Removal of Congo Red from Aqueous Solution using a Chitosan/ Organo- montmorillonite Nanocomposite. *J. Chem. Technol. Biotechnol.* 82:711–720.

Wang, X., Y. Du, J. Luo et al. 2007. Chitosan/Organic Rectorite Nanocomposite Films: Structure, Characteristic and Drug Delivery Behaviour. *Carbohydr. Polym.* 69:41–49.

Wei, Z., G. Wang, P. Wang et al. 2012. Crystallization Behavior of Poly(ε-caprolactone)/ TiO_2 Nanocomposites Obtained by in Situ Polymerization. *Polym. Eng. Sci.* 52:1047–1057.

Weng, W., G. Chen, and D. Wu. 2003. Crystallization Kinetics and Melting Behaviors of Nylon 6/ Foliated Graphite Nanocomposites. *Polymer* 44:8119–8132.

Woranuch, S., R. Yoksan, and M. Akashi. 2015. Ferulic Acid-Coupled Chitosan: Thermal Stability and Utilization as an Antioxidant for Biodegradable Active Packaging Film. *Carbohydr. Polym.* 115:744–751.

Wu, T.-M., S.-F. Hsu, and J.-Y. Wu. 2002. Crystalline Forms in Melt-Crystallized Syndiotactic Polystyrene/Clay Nanocomposites. *Polym. Eng. Sci.* 42:2295–2305.

Wu, T.-M. and C.-Y. Wu. 2006. Biodegradable Poly(Lactic Acid)/Chitosan-Modified Montmorillonite Nanocomposites: Preparation and Characterization. *Polym. Degrad. Stab.* 91:2198–2204.

Xiang, C., P.J. Cox, A. Kukovecz et al. 2013. Functionalized Low Defect Graphene Nanoribbons and Polyurethane Composite Film for Improved Gas Barrier and Mechanical Performances. *ACS Nano* 7:10380–10386.

Xie, X.-L., Q.-X. Liu, R.K.-Y. Li et al. 2004. Rheological and Mechanical Properties of PVC/ $CaCO_3$ Nanocomposites Prepared by In Situ Polymerization. *Polymer* 45:6665–6673.

Yang, G., C. Wang, F. Hong et al. 2015. Preparation and Characterization of BC/PAM-AgNPs Nanocomposites for Antibacterial Applications. *Carbohydr. Polym.* 115:636–642.

Yao, W., X. Chen, and J. Zhang. 2010. A Capacitive Humidity Sensor Based on Gold–PVA Core– Shell Nanocomposites. *Sens. Actuators B* 145:327–333.

Yilmazer, U. and G. Ozden. 2006. Polystyrene–Organoclay Nanocomposites Prepared by Melt Intercalation, In Situ, and Masterbatch Methods. *Polym. Compos.* 27:249–255.

Youssef, A.M. and M.S. Abdel-Aziz. 2013. Preparation of Polystyrene Nanocomposites Based on Silver Nanoparticles Using Marine Bacterium for Packaging. *Polym. Plast Technol. Eng.* 52:607–613.

Youssef, A.M., T. Bujdosó, V. Hornok et al. 2013a. Structural and Thermal Properties of Polystyrene Nanocomposites Containing Hydrophilic and Hydrophobic Layered Double Hydroxides. *Appl. Clay Sci.* 77–78:46–51.

Youssef, A.M., S. Kamel, and M.A. El-Samahy. 2013b. Morphological and Antibacterial Properties of Modified Paper by PS Nanocomposites for Packaging Applications. *Carbohydr. Polym.* 98:1166–1172.

Youssef, A.M., F.M. Malhat, and A.F.A. Abd El-Hakim. 2013c. Preparation and Utilization of Polystyrene Nanocomposites Based on TiO_2 Nanowires. *Polym. Plast. Technol. Eng.* 52:228–235.

Yu, S.-H., H.-Y. Hsieh, J.-C. Pang et al. 2013. Active Films from Water-Soluble Chitosan/Cellulose Composites Incorporating Releasable Caffeic Acid for Inhibition of Lipid Oxidation in Fish Oil Emulsions. *Food Hydrocolloid* 32:9–19.

Zaman, H.U. and M.D.H. Beg. 2014. Effect of CaCO₃ Contents on the Properties of Polyethylene Nanocomposites Sheets. *Fibers Polym.* 15:839–846.

Zapata, P.A., F.M. Rabagliati, I. Lieberwirth et al. 2014. Study of the Photodegradation of Nanocomposites Containing TiO₂ Nanoparticles Dispersed in Polyethylene and in Poly(Ethylene-Co-Octadecene). *Polym. Degrad. Stab.* 109:106–114.

Zhu, S., J. Chen, Y. Zuo et al. 2011. Montmorillonite/Polypropylene Nanocomposites: Mechanical Properties, Crystallization and Rheological Behaviors. *Appl. Clay Sci.* 52:171–178.

Zapata, P.A., L. Tamayo, M. Páez et al. 2011. Nanocomposites Based on Polyethylene and Nanosilver Particles Produced by Metallocenic *In Situ* Polymerization: Synthesis, Characterization, and Antimicrobial Behavior. *Eur. Polym. J.* 47:1541–1549.

Zhang, J., E. Manias, G. Polizos et al. 2009. Tailored Polyethylene Nanocomposite Sealants: Broad-Range Peelable Heat-Seals Through Designed Filler/Polymer Interfaces. *J. Adhes. Sci. Technol.* 23:709–737.

Zhang, J. and C.A. Wilkie. 2003. Preparation and Flammability Properties of Polyethylene–Clay Nanocomposites. *Polym. Degrad. Stab.* 80:163–169.

Zhu, A., A. Cai, J. Zhang et al. 2008a. PMMA-Grafted-silica/PVC Nanocomposites: Mechanical Performance and Barrier Properties. *J. Appl. Polym. Sci.* 108:2189–2196.

Zhu, A., A. Cai, W. Zhou et al. 2008b. Effect of Flexibility of Grafted Polymer on the Morphology and Property of Nanosilica/PVC Composites. *Appl. Surf. Sci.* 254:3745–3752.

Zhu, Y., G.G. Buonocore, M. Lavorgna et al. 2011. Poly(Lactic Acid)/Titanium Dioxide Nanocomposite Films: Influence of Processing Procedure on Dispersion of Titanium Dioxide and Photocatalytic Activity. *Polym. Compos.* 32:519–528.

Index

Note: Page numbers followed by f and t refer to figures and tables, respectively.